SIXTH EDITION

CARTOGRAPHY

Thematic Map Design

BORDEN D. DENT
Georgia State University

JEFFREY S. TORGUSON
St. Cloud State University

THOMAS W. HODLER
University of Georgia

Boston Burr Ridge, IL Dubuque, IA New York San Francisco St. Louis
Bangkok Bogotá Caracas Kuala Lumpur Lisbon London Madrid Mexico City
Milan Montreal New Delhi Santiago Seoul Singapore Sydney Taipei Toronto

The McGraw·Hill Companies

CARTOGRAPHY: THEMATIC MAP DESIGN, SIXTH EDITION

Published by McGraw-Hill, a business unit of The McGraw-Hill Companies, Inc., 1221 Avenue of the Americas, New York, NY 10020.

This book is printed on recycled, acid-free paper containing 10% postconsumer waste.

1 2 3 4 5 6 7 8 9 0 QPD/QPD 0 9 8

ISBN 978-0-07-294382-5
MHID 0-07-294382-3

Publisher: *Thomas Timp*
Executive Editor: *Margaret J. Kemp*
Director of Development: *Kristine Tibbetts*
Senior Developmental Editor: *Joan M. Weber*
Senior Marketing Manager: *Lisa Nicks*
Project Coordinator: *Mary Jane Lampe*
Lead Production Supervisor: *Sandy Ludovissy*
Senior Designer: *David W. Hash*
Cover/Interior Designer: *Ellen Pettergell*
Cover Illustration: Map courtesy Reynolds et al., *Exploring Geology,* 2008. Data from Hargrove, W. and F. Hoffman, 2004, "The potential of multivariate quantitative methods for delineation and visualization of ecoregions," *Environmental Management* 34(5): S39–S60.
Lead Photo Research Coordinator: *Carrie K. Burger*
Compositor: *Aptara®, Inc.*
Typeface: *10/12 Times*
Printer: *Quebecor World Dubuque, IA*

Library of Congress Cataloging-in-Publication Data

Dent, Borden D.
Cartography : thematic map design/Borden D. Dent, Jeffrey S. Torguson, Thomas W. Hodler.—6th ed.
p. cm.
Includes index.
ISBN 978-0-07-294382-5—ISBN 0-07-294382-3 (hard copy : alk. paper) 1. Cartography. 2. Thematic maps. I. Torguson, Jeffrey. II. Hodler, T. W. III. Title.
GA105.3.D45 2009
526—dc22 2008025397

www.mhhe.com

In memory of

Borden D. Dent

1938–2000

teacher and leader in the world of cartography.

CONTENTS

ABOUT THE AUTHORS

BORDEN D. DENT, above all, was three things—a geographer, a cartographer, and an educator. His geography training spawned an interest in cartography, which he conceptualized as a manner to richly visualize the world we live in.

Born in Arkansas, Dr. Dent attended elementary and high school in Maryland. He earned his B.A. and M.A. in Geography from Towson State University and the University of California at Berkeley, respectively. He received his Ph.D. from Clark University in Worcester, Massachusetts, where he specialized in thematic cartography.

Dr. Dent published articles on cartography and geography in professional journals, including the *Annals* of the Association of American Geographers, *The American Cartographer, The Cartographic Journal,* and *The Journal of Geography*. He presented papers at professional meetings, the National Geographic Society, the U.S. Census Bureau, and at several universities. He designed maps for *The Linguistic Atlas of the Gulf States*, a project directed at Emory University and funded by the National Endowment for the Humanities.

Dr. Dent authored five editions of *Cartography: Thematic Map Design* and served as chair of the Department of Anthropology and Geography at Georgia State University. His work inspired his students. Following a long and successful career, Dr. Dent passed away in August of 2000.

JEFFREY S. TORGUSON received his B.S. in Geography at St. Cloud State University and his M.A. and Ph.D. in Geography from the University of Georgia. He is currently Professor of Geography at St. Cloud State University, where he teaches courses in thematic cartography, map design and presentation, and GIS. He also previously taught GIScience courses at the University of Wisconsin—Oshkosh. His research and publication activity centers around maps and map design, particularly thematic, virtual, and Web cartography. He has served on several advisory boards concerned with cartography and GIS within higher education systems, and has held office with several related specialty groups within the Association of the American Geographers. He also works with local governments, high schools, and private firms on mapping and GIS issues. Dr. Torguson currently resides in St. Cloud, Minnesota, with his family.

THOMAS W. HODLER received his B.S. in Secondary Education and his M.A.T. in Geography from Indiana University and a Ph.D. in Geography from Oregon State University. He first taught cartography at Bemidji State University following the awarding of his master's degree and later at Western Michigan University after completing the doctorate. He has been teaching thematic cartography for twenty-five years at the University of Georgia. His research focus has been on thematic map design and production, specifically concentrating on state atlas design.

Dr. Hodler was co-author of *The Atlas of Georgia,* the first official atlas of the state. Later he launched the *Interactive Atlas of Georgia,* which included a database that could be queried to produce new maps and reports. He teaches classes in thematic map design, cartographic visualization, map animation, physical geography, and Georgia geography. Dr. Hodler has been recognized for his quality teaching and was awarded the University of Georgia's Bronze Medallion for distinguished service. He served as the Associate Dean of the Graduate School at the University of Georgia and currently serves as cartographic editor for three journals—the *Annals* of the Association of American Geographers, *The Professional Geographer,* and *Cartography and Geographic Information Science*.

PREFACE

WHAT SETS THIS BOOK APART?

The sixth edition of *Cartography: Thematic Map Design* is a completely revised version of previous editions. While the focus remains on the principles of thematic map design, all chapters are updated, now providing a more integrated, practical link between cartographic theory and practice for users of GIS, computer mapping, and graphic design software. The goal of this new edition is to provide an in-depth discussion of the process of designing maps for displaying spatial information.

THE APPROACH OF THIS BOOK

The intended audience of this text is rather broad based. It includes instructors and students of higher education classes in cartography focusing on the design and creation of thematic maps. Both instructors and students will appreciate the straightforward presentation of material, including over 120 newly designed graphics and map examples, as well as the authors' suggestions for the *do's* and *don'ts* of map design. Other users include a range from cartographic professionals to novice users of software that produce maps. For those professionals who wish to have a modern reference book to which they can turn for review of thematic map design principles, this text will serve them well. The book is written in user-friendly language that even the most novice generator of maps can easily follow. It is especially beneficial to those individuals who wish an understanding of map design that goes beyond the use of software default settings. This text will provide each user with an overview of cartographic design of thematic maps based upon the three authors' combined professional and academic experience of over fifty years.

The text's 17 chapters are presented in five parts, each of which has a focus that bonds the chapters together. Users of the previous edition of the text will notice that the chapter on Geographic Information Systems has been removed since the text is now GIS-friendly throughout. Chapter 16, Introduction to Virtual and Web Mapping, is entirely new to the sixth edition, and replaces the fifth edition's Chapter 17. In addition, a number of topics have been consolidated differently which permits the pedagogical approach familiar to many courses in thematic map design. For example, the topic of data classification has been moved into Chapter 5, Descriptive Statistics and Data Classification, since classification can be applied to other mapping techniques as well. While some of the major points and changes are highlighted in this preface, readers for whom those changes from the fifth edition are important (such as instructors) are encouraged to examine the outline for each chapter in the text's table of contents for more specific details.

WHAT'S NEW IN THE SIXTH EDITION?

This sixth edition has retained the focus on excellence in the principles of thematic map *design.* At the same time, this edition has undergone important revisions providing

- a more integrated, practical link between cartographic theory and practice for users of GIS, computer mapping, and graphic design software;
- an improved internal organization within the chapters;
- a NEW chapter on virtual and web mapping (Chapter 16);
- many NEW and updated maps and graphics; and
- an EXPANDED Color Plate section with new map examples illustrating practical cartographic design principles.

DETAILED CONTENT COVERAGE

The five chapters in Part I provide a foundation for thematic mapping. **Chapter 1** presents the underpinnings of thematic cartography, including discussion of the various kinds of maps including qualitative and quantitative thematic maps, map communication and visualization, and other topics. New additions include a brief taxonomy of thematic map types, a preview of the basic map elements, and a discussion on the use of GIS, mapping, and artistic drawing software. The chapter concludes with a presentation on cartographic abstraction and generalization and a discussion on ethics in cartography.

Chapters 2 and 3 are still the "projection chapters" but have been reorganized and rewritten with digital map compilation in mind. **Chapter 2** explores basic geodesy, coordinate systems, and scale. Included is a discussion of reference ellipsoids, datum concepts, and a presentation on the positional shifts of boundaries with the conversion from North American Datum 1927 (NAD27) to NAD83. The coordinate system section includes an expanded discussion of decimal degrees. The chapter concludes by revisiting scale and its impact on line generalization in digital boundaries. **Chapter 3** examines the map projection process, with particular attention paid to projections and projection parameter terminology included in most software, and addresses the often confusing concept of GIS "projection on the fly." A substantial new component of this chapter includes the addition of the UTM coordinate system and an updated discussion of the State Plane coordinate system, with a number of new figures illustrating each. The chapter concludes with a new discussion on the impact of modifying the standard parallels and central meridians, which can produce greater accuracy and an improved aesthetic look to the map.

Chapters 4 and 5 examine the nature and classification of data. In **Chapter 4,** spreadsheet examples are used to visualize the spatial characteristics and attributes of location. Data are characterized by their location (point, line, and area) and by their form. The chapter discusses data based on the characteristic view (qualitative versus quantitative), spatial view (discrete versus continuous), and attribute view (totals versus derived). The cartographer maps data and frequently manipulates and transforms data from one form or measurement scaling to another. The relationship between data, map symbology, and thematic map type is presented in a concise table that enables the user to translate between data and the visual variables of map symbolization. Chapter 4 concludes with an examination of data sources, GIS clearinghouses, and the United States Census Bureau and Geological Survey. The reader is presented with a discussion of cartographic errors that can occur when matching attribute data to location using FIPS codes. **Chapter 5** examines the descriptive statistics of associated data without presenting information common to textbooks in statistics. What is reviewed is an overview of a data set and how the characteristics of the data distribution can be used in data classification. Nine data classification schemes are examined prior to linking them with a particular map form. A summary and comparison of the major classification techniques is presented using a common data set. Tables of the data legends and a discussion of advantages and disadvantages of the schemes provide an overview of the techniques. The user is cautioned about potential problems encountered when data sets contain voids, no data designations, zeros, and the impact of outlier extremes.

Part II examines in six chapters the techniques of quantitative thematic mapping. Although the internal revisions to each chapter are substantial, the ordering of this section will be familiar to users of previous editions. The techniques discussed include: choropleth mapping (**Chapter 6**), dot density mapping (**Chapter 7**), proportional symbol mapping (**Chapter 8**), isarithmic and surface mapping (**Chapter 9**), value-by-area or cartogram mapping (**Chapter 10**), and flow mapping (**Chapter 11**). The most dramatic structural change within this section is the moving of the classification techniques from the choropleth mapping chapter to Chapter 5, as noted previously. Since classification is so integral to the choropleth technique, there is a new discussion of comparison of classification methods as applied to choropleth mapping within Chapter 6. Two other important changes can be inferred in the modification of chapter titles from the fifth edition. In Chapter 7 (dot density mapping), we place a greater emphasis on mapping total data within enumeration units. In Chapter 9 (isarithmic and surface mapping), the discussion has been expanded to include various surfaces. In all of the chapters in Part II we stressed the importance of appropriate data associated with each technique. In the section's conclusion at the end of Chapter 11, the importance of selecting the mapping technique based on the type of data that will be used in the map is emphasized.

Part III looks at the map from the standpoint of its overall design, the type used to communicate information, and the use of color for data visualization and aesthetic appeal. **Chapter 12** provides ideas for total map organization and figure-ground relationships. Of the visual design ideas discussed in this book, an approach to a design problem by way of having a visual hierarchy plan is the most fundamental and necessary. The design approach taken in this part specifically targets the page size map in terms of organization of the visual elements, contrast, and visual acuity. The selection and use of type on the map is the focus of **Chapter 13.** A series of rules to follow in deciding type placement and positioning is provided with a series of graphic examples of how to and how not to label the map. **Chapter 14** is completely redesigned and examines the principles of color for thematic mapping. The additive and subtractive color theories are applied to both the printed map and also the map viewed on the computer monitor. Six color models are examined and their applications are explored, specifically in terms of colors selected for mapping on the Web. The chapter concludes with an examination of color in design especially for developing figure-ground associations.

Immediately following the chapters on design, type, and applications of color, Part IV examines the production of maps in both the hardcopy, **Chapter 15,** and virtual environments, **Chapter 16.** Color plays an important role in both of these procedures. Desktop printers are examined as the standard mode of cartographic production for a small number of copies. For larger quantities, the printing press plays both its traditional role for sheet-fed and web presses that use printing plates and offset lithography, as well as the use of digital printing for direct to press technology. The map production process is laid out in a six-step process that follows the map from design and layout to post-press operations.

Chapter 16, a totally new chapter in the sixth edition, explores virtual and web mapping techniques. Both raster and vector graphic formats and structures are examined including the popular file formats, such as TIFF, JPG, GIF, SVG, EPS, PDF, and others. Key Internet concepts are presented, as well as the design constraints imposed by screen resolutions. Map animation, map interactivity, and cybercartography are also explored as possible solutions to the virtual presentation of maps.

Part V contains a single chapter devoted to the development and design of effective graphs. Graphs are also a product of GIS, mapping, and artistic software. Just as with maps, there are guidelines that should be followed to generate a well-designed graph that communicates the data. Graphs that are both simple and complex, that have two or three axes or no axis at all, are explored in **Chapter 17.**

KEY FEATURES

Study Aids

Key terms set in bold where they first appear, and defined in the Glossary at the end of each chapter
End-of-chapter selected references
End-of-chapter glossary

Color Plate Section

16-page Color Plate section showing 4-color images

Appendices

Appendix A: Worked Problems
Appendix B: Georgia Data

TEACHING AND LEARNING SUPPLEMENTS

For Instructors

Website at http://www.mhhe.com/dent6e

The **Cartography: Thematic Map Design** website offers a wealth of teaching and learning aids for instructors. Instructors will appreciate:

- an online Laboratory Manual containing practical, hands-on ArcGIS exercises
- sample completed maps and answers to accompany the online lab manual exercises
- access to the online **Presentation Center** including illustrations, photographs, and tables from the text in convenient jpg and PowerPoint format

Electronic Textbook

CourseSmart is a new way for faculty to find and review eTextbooks. It's also a great option for students who are interested in accessing their course materials digitally and saving money. CourseSmart offers thousands of the most commonly adopted textbooks across hundreds of courses from a wide variety of higher education publishers. It is the only place for faculty to review and compare the full text of a textbook online, providing immediate access without the environmental impact of requesting a print exam copy. At CourseSmart, students can save up to 50% off the cost of a print book, reduce their impact on the environment, and gain access to powerful web tools for learning including full text search, notes and highlighting, and email tools for sharing notes between classmates. Visit www.CourseSmart.com

For Students

Website at http://www.mhhe.com/dent6e

The **Cartography: Thematic Map Design** website offers a wealth of study aids for students, including:

- chapter outlines
- an online lab manual containing practical, hands-on ArcGIS exercises
- career opportunities
- links to professional organizations
- links to cartography-related topics

ACKNOWLEDGMENTS

This edition bears much of the cartographic philosophy of the late Borden D. Dent. We greatly appreciate his contributions to this book.

All graphics in this book were created using one or more of the following software: ArcGIS™ Desktop 9.2 including ESRI Data and Maps, and ArcView™ 3.3, both registered trademarks of the ESRI Corporation; Golden Software's MapViewer™, Surfer™ and Grapher™ by Golden Software, Inc.; Adobe Illustrator™ and Photoshop™ by the Adobe Corporation; and Microsoft's Excel™ and Word™ software. Many of the software have been mentioned in the text without the trademark notation but that registration is noted here.

The authors wish to acknowledge the support of the Department of Geography at both St. Cloud State University and the University of Georgia for their assistance in the preparation of the manuscript and graphics. Many individuals contributed directly and indirectly to development of the design of maps and graphs that appear in this text, or provided suggestions and corrections to the text. We wish to thank the following: Sherri Anderson, Dwight Lanier, Karl Larsen, Lane Pille, Nathan Schutz, Kunwar Singh, Richard Smith, Chris Sutton, and Tracy Tisbo. We also thank all of our students in our thematic cartography classes over the years—you have been an inspiration to us in writing this text.

The authors wish to express special thanks to McGraw-Hill for editorial support through Marge Kemp and Joan Weber;

the marketing expertise of Lisa Nicks; and the production team led by Mary Jane Lampe, David Hash, Carrie Burger, Sandy Ludovissy, and Judi David.

We would like to extend our thanks to the following reviewers who provided recent feedback on the text and illustrations. Their help has been invaluable in shaping the sixth edition of *Cartography: Thematic Map Design*. We are grateful to these reviewers for their careful evaluation and useful suggestions for improvement.

Ola Ahlqvist, *Ohio State University*

Nicholas Chrisman, *University of Washington*

Mathew A. Dooley, *University of Wisconsin—River Falls*

Roy R. Doyon, *Ball State University*

Chuck Geiger, *Millersville University*

Francis Harvey, *University of Minnesota—Minneapolis*

Gail L. Hobbs, *Pierce College*

Robert M. Hordon, *Rutgers University*

Darin Jensen, *University of California—Berkeley*

Ryan R. Jensen, *Indiana State University*

Nicholas P. Kohler, *University of Oregon*

Olaf Kuhlke, *University of Minnesota—Duluth*

Robert J. Legg, *Northern Michigan University*

Dave Lemberg, *Western Michigan University*

Shun Lin Liang, *University of Maryland*

Nancy Hoalst Pullen, *Kennesaw State University*

Martin Mitchell, *Minnesota State University—Mankato*

Thomas W. Paradis, *Northern Arizona University*

Margaret W. Pearce, *Ohio University*

Paporn Thebpanya, *Towson University*

Jill Freund Thomas, *Illinois State University*

Lin Wu, *California State Polytechnic University—Pomona*

Angela Yao, *University of Georgia*

We also wish to thank our cartographic peers who have participated in many discussions pertaining to map design and production over the years. We salute you. Lastly, we wish to acknowledge our families Jennifer and Andy, Daniel, Andrew, Bethany, Belinda, and especially our wives, Sue and Mary Kay, for their unselfish support while we toiled at a labor of love.

Jeffrey S. Torguson

Thomas W. Hodler

PART I
THEMATIC MAPPING ESSENTIALS

THE FIRST CHAPTER in this part contains an introduction to the world of the thematic map and presents a backdrop of major concepts involved in this map form. Various map examples highlight the rich variety. A discussion is included about map design and the role the cartographer plays in the design activity. After the introductory chapter, this part provides material necessary to begin the thematic mapping task. Three chief components of thematic maps include the base map, the thematic overlay, and a set of ancillary map elements, such as a title, legend, and other elements. The next two chapters provide background and techniques to organize and develop the base-map portion of the complete thematic map, and include a discussion of the variety of map projections useful for the thematic designer. Knowledge of map projections and characteristics is required for successful map design.

Before the thematic cartographer can begin mapping, a fundamental awareness of the nature of geographical inquiry and the ways geographic data present themselves must be grasped. Ability to recognize data forms assists the map designer in the important activity of symbol design logic. For most cartographic design tasks the cartographer needs to develop and transform original data into forms that can be mapped. To do this, knowledge of basic descriptive statistics is essential, and these are discussed in Chapter 5.

1 INTRODUCTION TO THEMATIC MAPPING

CHAPTER PREVIEW Maps are graphic representations of the cultural and physical environment. Two subclasses of maps exist: general purpose (reference) maps and thematic maps. This text concerns the design of the thematic map, which shows the spatial distribution of a particular geographic phenomenon. Map scale, or the amount of reduction of the real world, is critical to the cartographer, because it determines selection and generalization. Mapmaking may be viewed in a larger context of cartographic thinking and cartographic communication. The designer plays an important role in this context as one who deals with the appropriate and creative symbolization of the image selected for communication after visualization and experimentation have taken place. Generalization takes place in the process, which includes selection, classification, simplification, and symbolization. Thematic map design is the aggregate of all the mental processes that lead to solutions in the abstraction phase of cartographic communication. Ethics in cartography is a subject of concern and plays a role in the way one views maps. ■

The purpose of this book is to introduce several principles of thematic map design. The thematic map, only one of many map forms, will be more precisely defined later in this chapter. For now, some general comments about all maps will help set the stage. Maps seem to be everywhere we look: in daily newspapers and weekly newsmagazines, in books, on the World Wide Web, on television, on trains, in kiosks, and even on table place mats. There is also great variety in types of maps. Some are greatly detailed and look like engineering drawings; others appear as freehand sketches or simple wayfinding diagrams. Some show the whole world, others an area no larger than your backyard.

As our society has become more complex, the needs and uses for maps of all kinds have increased. Local governments and planning agencies use them for plotting environmental and resource data. Soil, geology, and water resource professionals use them daily in their work and planning. Public utility and engineering firms consult technical maps in order to complete their tasks. Land-use maps are utilized by planners; detailed cadastral maps are indispensable to city tax recorders (see Figure 1.1).

All of cartography, including thematic mapping, is undergoing rapid changes today. Advances in computer hardware and software technology, increased Internet access, along with the prevalence of geographic information systems (GIS) software that can create many types of maps, have dramatically changed the way maps are created and used. In 2001 it was estimated that over 20 million maps were downloaded each day from the Internet (Peterson 2003). In addition local, state, and federal government agencies are making available online maps and map data in ever increasing amounts. As a result these agencies have become more responsive to the needs of map users, both public and private, across the country. Indeed, maps empower people through their very use (Wood 1992).

Since the 1960s we have also seen a proliferation of state and national atlases. These have particular relevance to our discussion because they consist mainly of thematic maps,

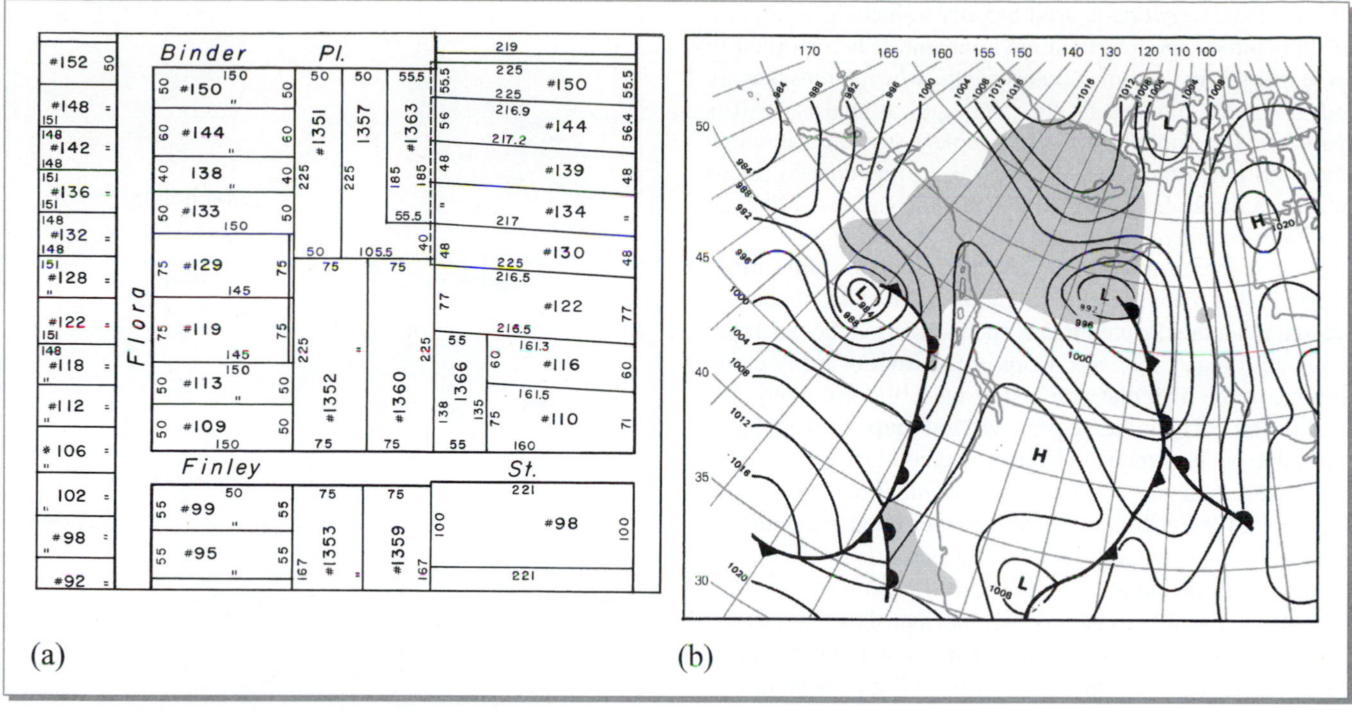

FIGURE 1.1 MAPS SATISFY A VARIETY OF NEEDS.

A record of city land lots is kept on cadastral maps (a), and the National Weather Service produces a daily weather map, as in (b). These are but two of the hundreds of uses of maps.

collected to assist educators and other professionals in displaying historical and geographic information about the region. This trend continues with many web-based map collections that range from a few select county themes to complete online atlases that contain interactive thematic maps.

Mapmaking is an interesting subject to study and an activity that has enjoyed a long history that is closely tied to the history of civilization itself. Maps date as far back as the fifth or sixth century B.C. (Bagrow 1966). Mapmaking has come to be respected as a disciplined field of study in its own right. In the United States, mapmaking has been closely associated with geography curricula since the early decades of the twentieth century. Although still linked in many ways to the study of geography, and rightfully so, cartography is recognized as both a distinct academic field and an integral part of GIS. With the growth of mapmaking has come a partitioning of the field into several component parts, each with its own scope, educational requirements, technology, and philosophical underpinnings. The practice of mapping today is far different from that used by our ancestors, although there are some common bonds.

In this chapter we will look at the foundations of the **thematic map.** Thematic maps came late in the development of cartography; they were not widely introduced until the early nineteenth century. Today such maps are produced quickly and cheaply, not only because the difficult process of base mapping has already been done by others but also because of the benefits of computer technology and the ever increasing availability of data used in their creation. The last 35 years have been referred to as the "era of thematic mapping," and this trend is expected to continue in the future. The ultimate impact on the development and production of thematic maps by the easy access to both data and mapping software on the World Wide Web has probably not yet reached its full potential, although the impact thus far has been dramatic. Thematic maps make it easier for professional geographers, planners, and other scientists and academicians to view the spatial distribution of phenomena.

THE REALM OF MAPS

What is astounding in written language is that new, often provocative, concepts and meanings can be generated by the combination of just a few simple words ("the universe began with a big bang"). The same can be said for a map with just a few graphic elements, and this is the essence of the cartographic instruction presented here. Professional cartographers think of maps as vehicles for the transmission of knowledge and for analysis or discovery. This idea of transmitting or communicating information (a central theme to this book), and examining the visualization of spatial data will serve as our point of departure for introducing thematic cartography.

The Map Defined

Although there are many kinds of maps, one description can be adopted that defines all maps: "A **map** is a graphic representation of the milieu" (Robinson and Petchenik 1976, 16).

In this context, *milieu* is used broadly to include all aspects of the cultural and physical environment. It is important to note that this definition includes *mental abstractions* that are not physically present on the geographical landscape. It is possible, for example, to map people's attitudes, although these do not occupy physical space. In this context, maps also have been described as "models of reality"—although one's reality may be different than another's.

One of the most basic map distinctions in modern cartography is that maps are either tangible (also called "real") or virtual in nature. Printed maps, such as maps in books, hardcopy atlas maps, the large maps at the front of many classrooms, or the map that you just printed, all have a physical reality and are tangible maps. A **virtual map** is a map that is viewable but is without a physical or tangible reality, such as a map that is displayed on a computer monitor display or as a projected graphic in a lecturer's Power Point presentation. Virtual maps originally had several definitions, depending on the map's permanence and how the map information was stored, and could even include original map data (after Moellering 1984). However, in a practical, contemporary sense, the virtual map has come to mean most impermanent, nonprint maps that appear on a display. The term includes static map images, such as scanned historical maps in an online map collection, as well as interactive maps, such as those maps that can be zoomed in or out from, or can be clicked on and queried, such as those maps found in the *National Atlas of the United States* (2007) or *Encarta* (Microsoft 2007). The term also includes elaborate animated maps and visual terrain flythroughs, such as the weather maps on television. Virtual maps are perhaps most commonly found on the Internet but are also found on other sources, such as CD and DVD-ROMs and television broadcasts. Most static maps can be printed, of course, but until they are actually in hardcopy form, they are considered virtual maps.

There are also such things as **mental maps,** generally described as mental images that have spatial attributes. Mental maps are developed in our minds over time by the accumulation of many sensory inputs, including tangible or virtual maps. Figure 1.2, for example, is an interesting composite mental map. The definition used throughout this book, however, assumes that the map is a tangible or virtual *object* that can be viewed by many people.

FIGURE 1.2 MENTAL MAP.
In this case, a composite or shared view of the neighborhoods of Manhattan is shown, and varies from the actual map space of the area. Mental maps of this sort develop stereotypes and provide additional meaning to people not familiar with the mapped areas.
Source: Redrawn from Rooney, *et al.* 1982, 99.

What Is Cartography?

The newcomer will probably be confused by the array of terms, names, and descriptions associated with mapmaking. It is unlikely that all those involved will ever reach complete agreement on terminology. For present purposes, we will adopt the following definitions. First, **mapmaking,** or mapping, refers to all of the processes of producing a map, whether the person is collecting data, performing the design of the map, or preparing the map for distribution in hardcopy or for the Web. Mapping, then, is the process of "designing, compiling, and producing maps" (Monmonier 1977, 9). The mapmaker may also be called a *cartographer* (Robinson and Petchenik 1976).

A second problem is the proper definition of **cartography.** As the discipline has matured and become broader in scope, many professional cartographers have come to make a distinction between mapmaking and cartography. In general, cartography is viewed as broader than mapmaking, for it requires the study of the philosophical and theoretical bases of the rules for mapmaking. It is often thought to be the study of the artistic and scientific foundations of mapmaking. The International Cartographic Association defines cartography as follows:

> *The art, science, and technology of making maps, together with their study as scientific documents and works of art. In this context may be regarded as including all types of maps, plans, charts, and sections, three-dimensional models and globes representing the Earth or any celestial body at any scale* (Meyen 1973, 1).

This definition is broad enough in scope to be acceptable to most practitioners. It certainly will serve us adequately in this introduction.

> Maps provide us with a structure for storing geographic knowledge and experience. Without them, we would find it difficult, if not impossible, to orient ourselves in larger environments. We would be dependent upon the close, familiar world of personal experience and would be hesitant—since many of us lack the explorer's intrepid sense of adventure—to strike out into unknown, uncharted terrain. Moreover, maps give us a means not only for storing information but for analyzing it, comparing it, generalizing or abstracting from it. From thousands of separate experiences of places, we create larger spatial clusters that become neighborhoods, districts, routes, regions, countries, all in relation to one another (Southworth and Southworth 1982, 11).

Geographic Cartography

Geographic cartography, although certainly a part of all cartography, should be defined a bit further. *Geographic cartography* is distinct from other branches of cartography in that it alone is the tool and product of the geographer. The geographic cartographer understands the spatial perspective of the physical environment and has the skills to abstract and symbolize this environment. The cartographer specializing in this branch of cartography is skillful in map projection selection and the mapping and understanding of areal relationships, and has a thorough knowledge of the importance of scale to the final presentation of spatial data (Robinson 1954). Furthermore, geographic cartography, involving an intimacy with the abstraction of geographical reality and its symbolization to the final map product, is capable of "unraveling" or revising the process, that is, geographic cartographers are very adept at map reading (Muehrcke 1981).

Geographers have for many years embraced the study of maps and often suggested that maps are at the heart of their discipline. For example, the famous Berkeley geographer Carl O. Sauer writes:

> *Show me a geographer who does not need them [maps] constantly and want them about him, and I shall have my doubts as to whether he has made the right choice in life. The map speaks across the barriers of language* (Sauer 1956, 289).

British geographer Peter Haggett writes, while exploring new ways of mapping accessibility in the Pacific Basin:

> *Although the new map is unfamiliar, it shows the locational forces at work in the Pacific in a dramatic way. In the physical world the earth's crust is reshaped by the massive slow forces of plate tectonics. So also technological changes of great speed are grinding and tearing the world map into new shapes. Capturing those spatial shifts [on maps] is at the heart of modern geography* (Hagget 1990, 54).

And American geographer John Borchert comments on the importance of maps to geography:

> *In short, maps and other graphics comprise one of three major modes of communication, together with words and numbers. Because of the distinctive subject matter of geography, the language of maps is the distinctive language of geography. Hence sophistication in map reading and composition, and ability to translate between the languages of maps, words, and numbers are fundamental to the study and practice of geography* (Borchert 1987, 388).

Hodler (1994) points out that with the current availability of point and click map production, it may be too easy to ignore fundamental cartographic and geographic principles and knowledge that have been at the traditional core of our discipline. The technological advances that we are experiencing provide the potential for new and better map designs, but without proper cartographic education not only will map quality ultimately suffer but the geographic understanding that is obtained in practicing geographic cartography may be lost as well.

Atlas Mapping

One prominent opportunity for thematic mapping is found in atlas production, as evidenced by the proliferation of state and national atlases. *The Atlas of Georgia* (Hodler and Schretter 1986) and the *Atlas of Pennsylvania* (Cuff *et al.* 1990), for example, contain hundreds of maps, most of them thematic, and are considered classic examples of high quality printed atlases. Electronic atlases, which are distributed online or on CD-ROMs, can contain not only static maps but may also have interactive or animated maps, as well as other multimedia features such as video and sound. The CD-ROM version of the *Atlas of Oregon* (Loy *et al.* 2001) is an excellent example of an electronic atlas that contains numerous

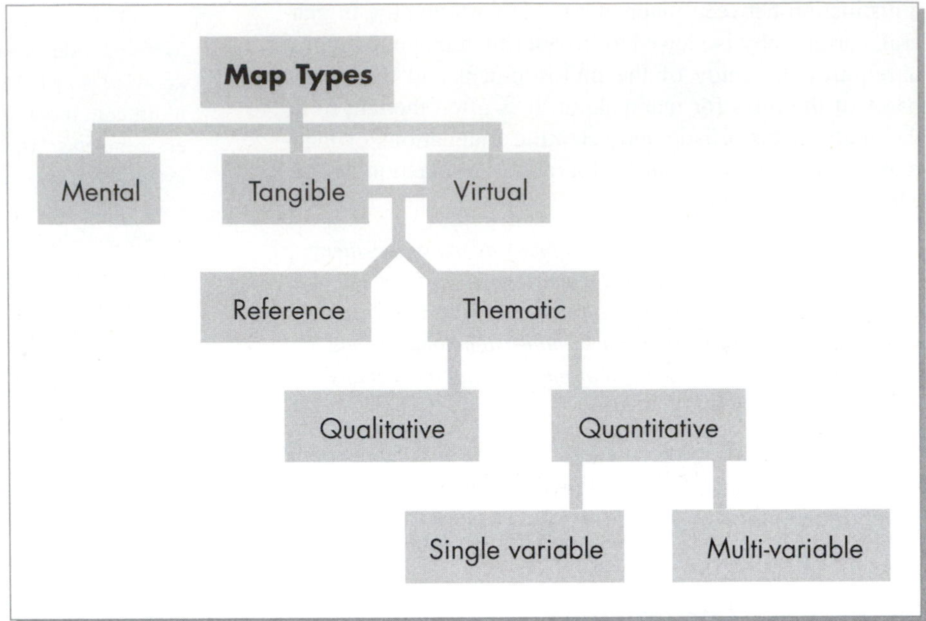

FIGURE 1.3 A CLASSIFICATION OF KINDS OF MAPS.
This organization is helpful in understanding the different types of maps.

interactive and animated thematic maps. The *National Atlas of the United States* (2007) is a prominent online atlas that allows its user to access maps (some of them animated) and data. The atlas also allows its users to decide what features to combine, as well as some of the maps' display options. Those map users making cartographic decisions notwithstanding, the thematic cartographer will continue to need sound design principles, regardless of the medium.

Kinds of Maps

A review of the variety of maps leads to a better understanding of where quantitative thematic mapping fits into the larger realm of maps and mapping. At one time, thematic mapping was estimated to be only 10 percent of the field (Bickmore 1975, 330). In general, most maps may be classified as either general purpose or thematic types. The two categories of maps can overlap at times, making some maps appear as if they can fall into either category. Air photos and other remotely sensed imagery, which are often available on websites featuring maps (such as the popular *Google Earth*), are not generally considered maps as such but are often used in the mapmaking process. However, it is possible to develop the classification further to include a greater variety, and one such scheme is depicted here (see Figure 1.3). These map types are discussed in the sections that follow.

General Purpose Maps

Another name commonly applied to the **general purpose map** is *reference map* (Robinson 1975). Such maps customarily display objects (both natural and man-made) from the geographical environment. The emphasis is on location, and the purpose is to show a variety of features of the world or a portion of it (Robinson and Petchenik 1976). Examples of such maps are topographic maps, maps of countries and states contained in online map collections such as the Perry–Castañeda Library Map Collection (2006), and many atlas maps (see Figure 1.4). The kinds of features found on these maps include coastlines, water features, political boundaries, roads, cities, and other similar objects that are appropriate to a map's scale.

Historically, the general purpose or reference map was the prevalent form until the middle of the eighteenth

FIGURE 1.4 A GENERAL PURPOSE OR REFERENCE MAP.
Source: Allen 2008, xiii, Map 2.

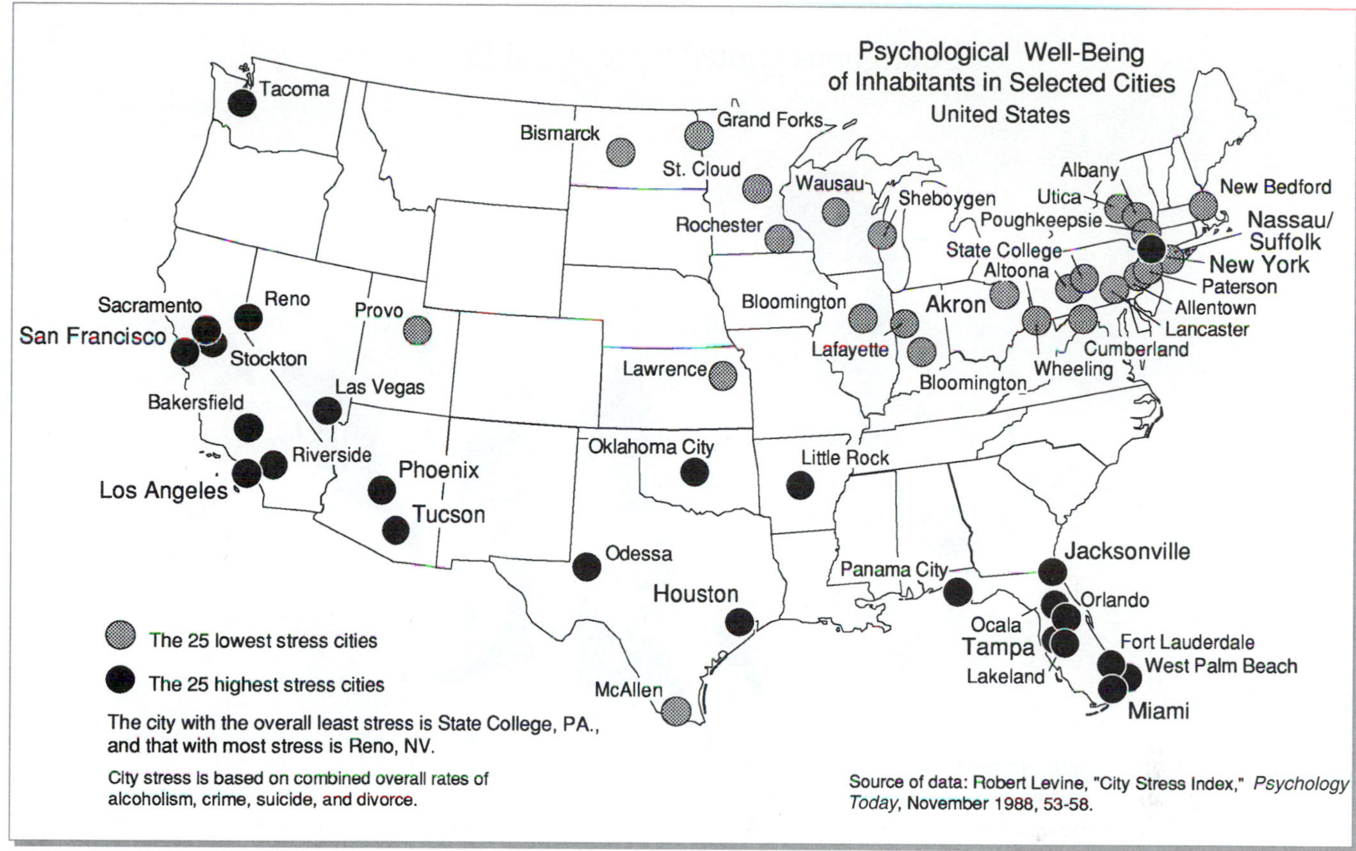

FIGURE 1.5 THEMATIC MAPS PORTRAY A VARIETY OF TOPICS.

Thematic maps often show topical subjects to elicit questions and develop working hypotheses. In this case one might ask why so many so-called Sunbelt cities are in the high-stress category. Sociologist Levine suggests that new arrivals and retirees to these predominately southern cities brought their problems with them.

century. Geographers, explorers, and cartographers were preoccupied with "filling in" the world map. Because knowledge about the world was still accumulating, emphasis was placed on this form. It was not until later, when scientists began to seize the opportunity to express the spatial attributes of social and scientific data that thematic maps began to appear. Such subjects as climate, vegetation, geology, population, and trade, to mention a few, were mapped.

Thematic Maps

The other major class of map is the thematic map, also called a *special-purpose, single-topic,* or *statistical* map. The International Cartographic Association defines the thematic map this way: "A map designed to demonstrate particular features or concepts. In conventional use this term excludes topographic maps" (Meynen 1973, 291).

The purpose of all thematic maps is to illustrate the "structural characteristics of some particular geographical distribution" (Robinson 1975, 11). This involves the mapping of physical and cultural phenomena or abstract ideas about them. Structural features include distance and directional relationships, patterns of location, or spatial attributes of magnitude change.

A thematic map, as its name implies, presents a *graphic theme* about a subject. In GIS terminology, it is a graphic display of *attribute data*, such as population density, average income, or daily temperature range. These attributes are also called variables. It must be remembered that a single theme is chosen for such a map; this is what distinguishes it from a reference map (see Figure 1.5).

Thematic maps may be subdivided into two groups, **qualitative** and **quantitative.** The principal purpose of a qualitative thematic map is to show the spatial distribution or location of single theme of *nominal data*. These types of thematic maps do not show any quantities at all but rather purely qualitative information, and are usually rather generalized in its record. Maps of ecoregions, geology, soil types, and land use/land cover maps are all common types of qualitative maps (see Figure 1.6). With qualitative maps, the reader cannot determine quantity, except as shown by relative areal extent.

Quantitative thematic maps, on the other hand, display the spatial aspects of numerical data. In most instances a single variable, such as corn, people, or income, is chosen,

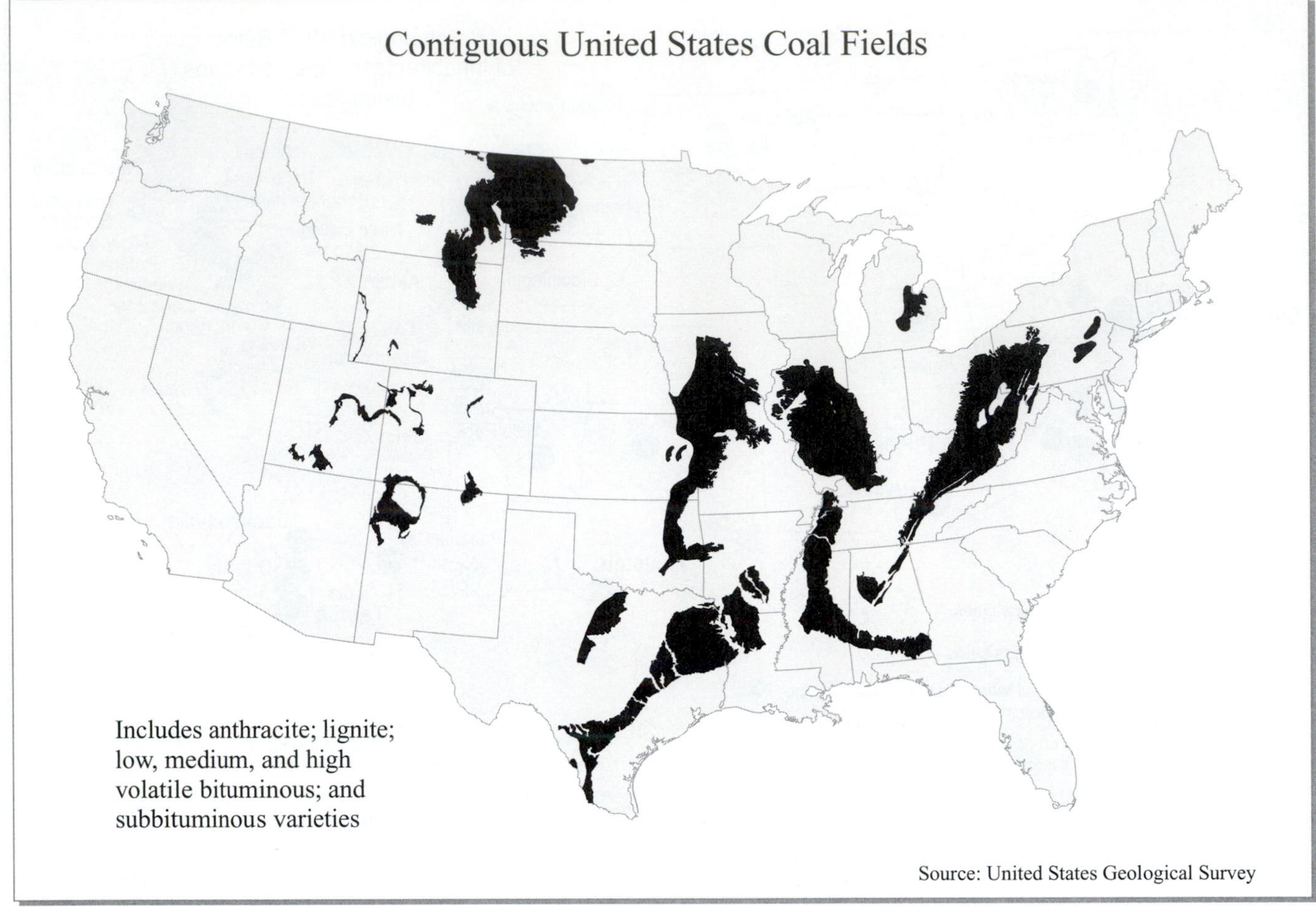

FIGURE 1.6 A QUALITATIVE THEMATIC MAP.

and the map focuses on the variation of the feature from place to place. These maps may illustrate numerical data on the ordinal (less than/greater than) scale or the interval/ratio (how much different) scale (see Figure 1.7). These measurement scales will be treated in depth in a later chapter.

Quantitative mapping, as already pointed out, functions to show *how much* or *to what degree* something is present in the mapped area. The principal operation in quantitative thematic mapping is in the transformation of tabular data (an aspatial format) into the spatial format of the map (Jenks 1976). The qualities that the map format provides (distance, direction, shape, and location) are not easily obtainable from the aspatial tabular listing. In fact, this is the quantitative map's primary reason for existing. If the transformation does not add any spatial understanding, the map should not be considered an alternative form for the reader; the table will suffice.

Furthermore, if the reader requires exact amounts, a quantitative thematic map is not the answer. The results of the transformations in mapping are *generalized* pictures of the original data. "A statistical map is a symbolized generalization of the information contained in a table" (Jenks 1976, 12). Yet the map is the only graphic means we have of showing the spatial attributes of quantitative geographic phenomena. The special process of abstracting, generalizing, and mapping data (even at the expense of losing detail) is therefore justified when the spatial dimension is to be communicated.

Common Quantitative Thematic Map Forms. Quantitative thematic maps often take one of several common forms. The **choropleth map** is a common map type for mapping data collected in **enumeration units.** Each unit, such as a county, state, country, province, is shaded according to a variable or attribute, such as population density (see Figure 1.8). **Dot maps** attempt to show variations in spatial density. These maps have a relatively simple rationale: that one dot represents so many units of some commodity, such as wheat (see Figure 1.9). **Proportional symbol maps** have symbols (usually circles) that are scaled to values at points. The point may be an actual point feature such as a city, or the point may also be derived from an area unit, such as the center of a state (see Figure 1.7). **Isarithmic maps** attempt to map 3-D continuous volumes, such as elevation, temperature, or precipitation. *Isarithms* (also *isolines*)

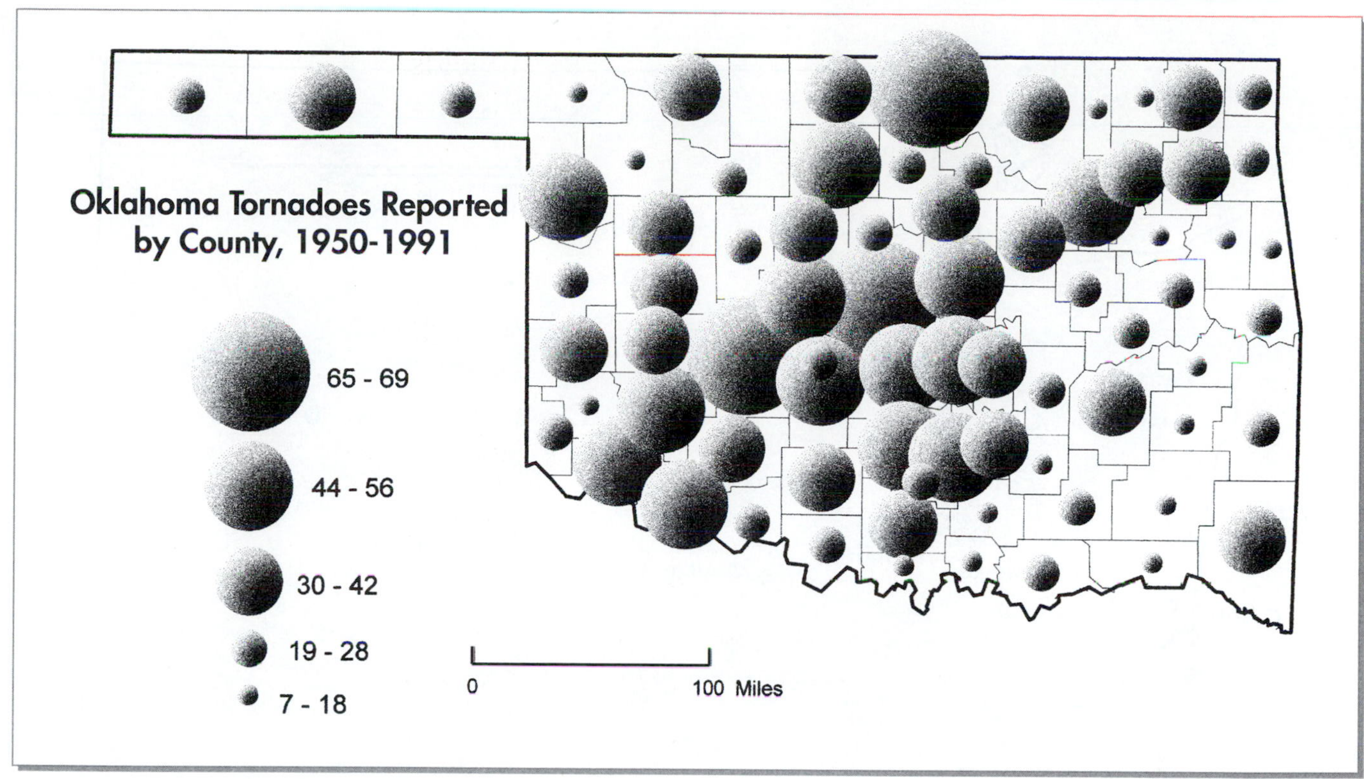

FIGURE 1.7 QUANTITATIVE THEMATIC MAP.

The clustering of tornadoes in the central part of Oklahoma is apparent on this map. This map provides an experimental approach to symbolization within the proportional symbol map form. The reader is not being asked to respond to the volume of the spheres but to the areas beneath them. Symbolizing by spheres is dealt with later in this book. Source: After Johnson and Duchon 1995, Figure 4.13.

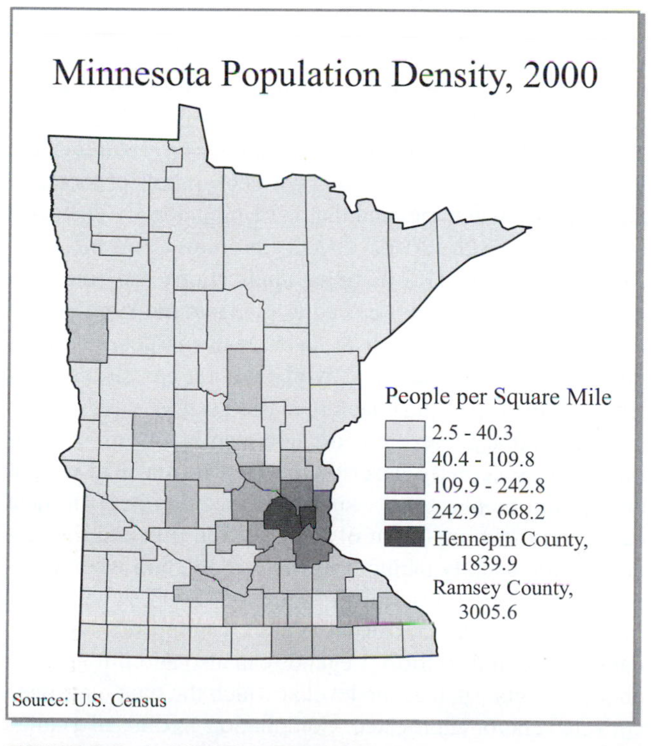

FIGURE 1.8 A CHOROPLETH MAP.

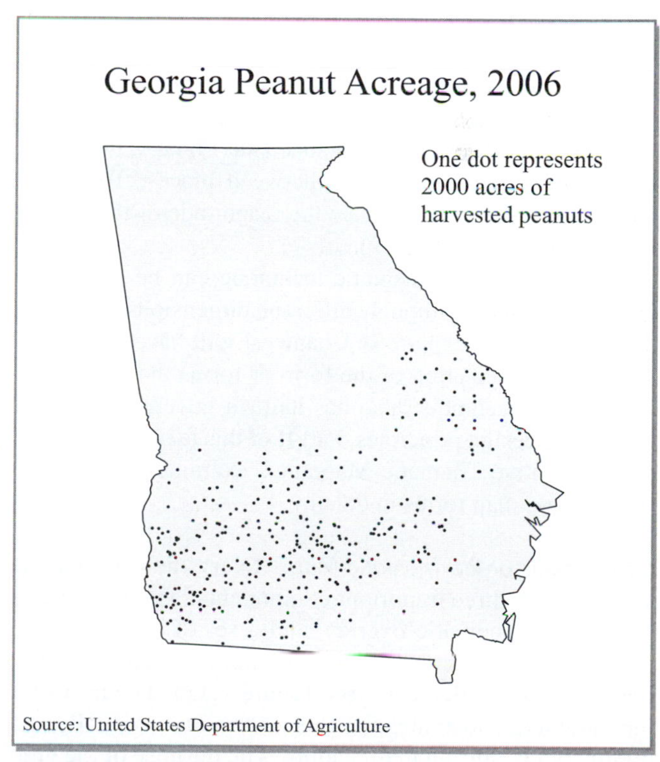

FIGURE 1.9 A DOT MAP.

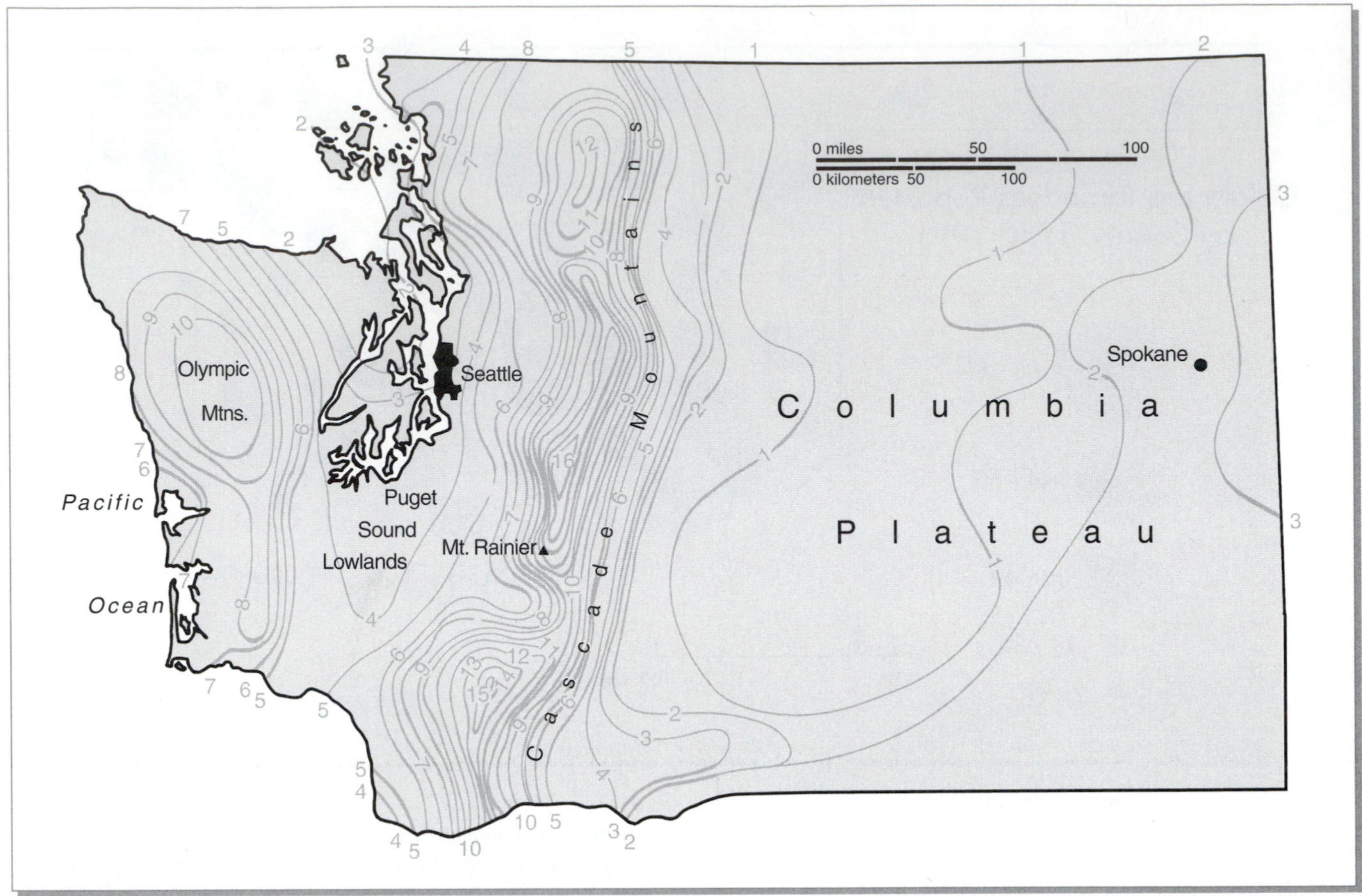

FIGURE 1.10 AN ISARITHMIC MAP. Source: Getis *et al.* 2008, 100, Figure 4.22.

connect points or places of equal value (see Figure 1.10). In the **Value-by Area Cartogram,** the enumeration units' area values are replaced by the variable being represented, often creating a very striking appearance (see Figure 1.11). **Flow maps** show linear movement between places. The lines' thickness and/or color indicate the magnitude of the flow or movement (see Figure 1.12).

Each quantitative thematic technique can be used for a variety of topics, although different dimensions of spatial data that we will explore in Chapter 4 will have a dramatic impact in the selection of the form or forms that are appropriate. Each technique also has its own governing body of theory and design principles. Part II of this text, "Techniques of Quantitative Thematic Mapping," examines each of the six thematic map forms in detail.

Components of the Thematic Map. Every thematic map is composed of three important components: a geographic or base map, a thematic overlay, and a set of ancillary map elements, such as titles, legends, compilation credits, neatlines, and other elements (see Figure 1.13). The user of a thematic map must integrate this information, visually and intellectually, during map reading. The purpose of the geographic base map is to provide locational information to which the thematic overlay can be related. The base map must be well-designed *and include only the amount of information thought necessary to convey the map's message.* Also central to the base map is employment of a correct map projection.

Simplicity and clarity are important design features of the thematic overlay, as is choosing correct symbols or thematic map type to match the data that is being mapped. The symbolization should also be visually prominent on the map. Note that in many GIS mapping applications, the base map and thematic data that are used to generate the symbols are tied together in a single data layer. Major issues with base maps and thematic data and symbolization are discussed at length in the following chapters in this section.

The ancillary map elements in a map layout must accurately and succinctly describe the map to the map reader, and should be placed to visually balance the map. In most cases, the most important of these are the title and the legend. A title usually includes a what, where, and when component, such as Iowa Corn Production, 2008. A legend is important to help the reader correctly interpret symbols, data ranges, and so forth. Legends can also add information about the data, such as the level at which the data have been collected and/or aggregated. Compilation credits, also called a source statement, are also an extremely important map element, although they occupy a small space in the layout, and

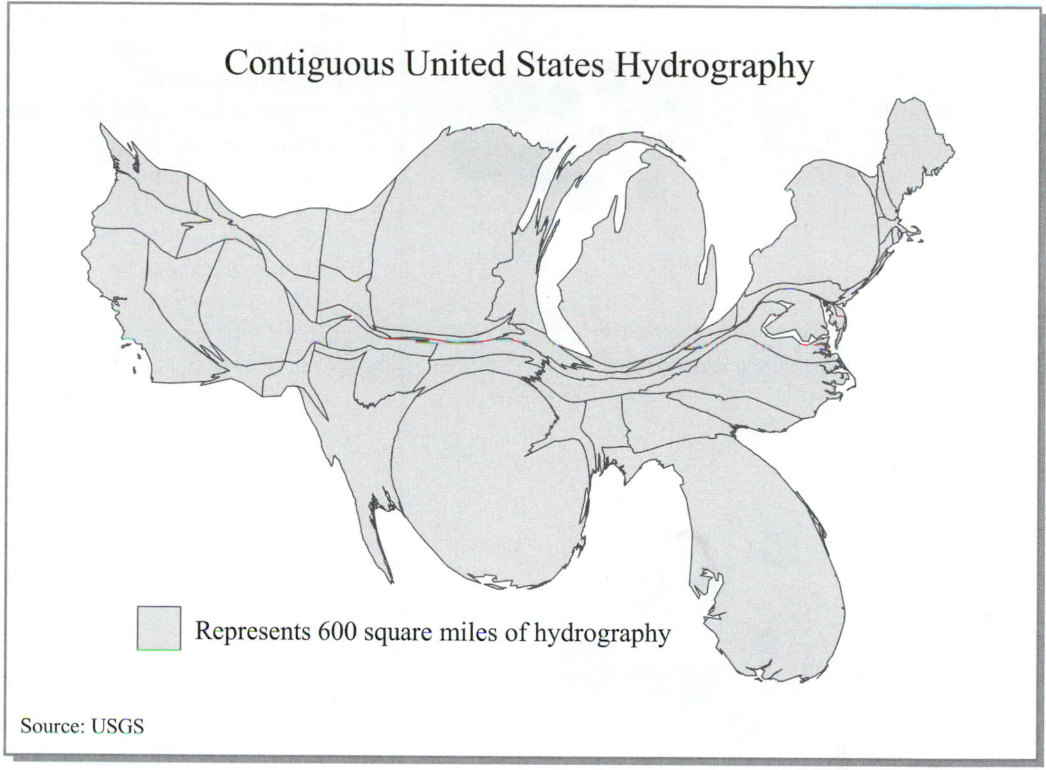

FIGURE 1.11 A VALUE-BY-AREA CARTOGRAM.

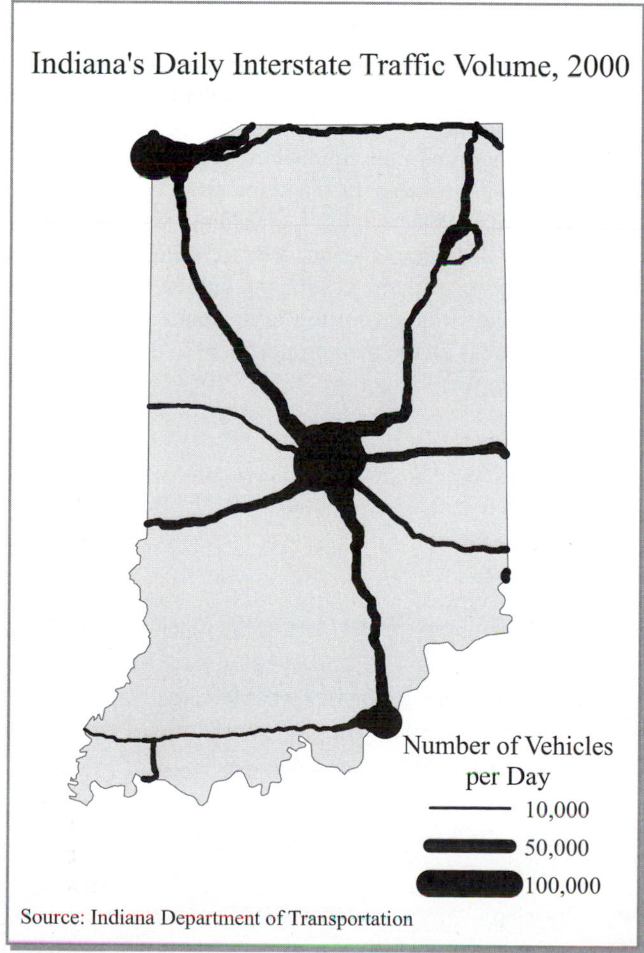

FIGURE 1.12 A FLOW MAP.

are placed in an unobtrusive portion of the map (such as smaller text in the lower left or right corner of the layout). The source statement simply acknowledges the source(s) for any data used in the map. The credits may also include a brief statement about the map authorship. The neatline is usually a thin, unobtrusive line that surrounds the mapped area. All other map objects are balanced within the neatline. An expression of map scale, such as a scale bar or representative fraction is useful if the cartographer wishes to impart a sense of distance, especially if the mapped area is not intuitive to the map reader. A few other common map elements include inset maps, ancillary explanatory text, and a north arrow or other indication of direction. Selection of these or other map elements (for example, pictures or graphs in larger format poster) will depend on the map's purpose and scale, and the medium used by the cartographer. A more exhaustive list of map elements and their appropriate use is found in Chapter 12.

Map Scale

When cartographers decide on graphic representation of the environment or a portion of it, an early choice to be made is that of **map scale.** Scale is the amount of reduction that takes place in going from real-world dimensions to the new mapped area on the map plane. Technically, map scale is defined as a ratio of map distance to earth distance, with each distance expressed in the same units of measurement and customarily reduced so that unity appears in the numerator (for example, 1:25,000). A more detailed discussion of scale appears in the next chapter. For now, our

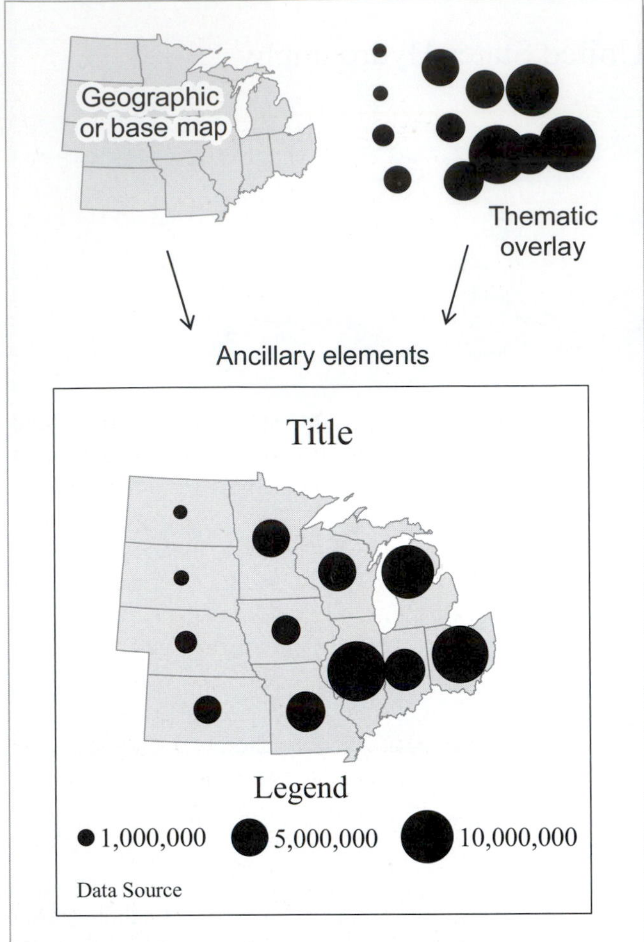

FIGURE 1.13 THE MAJOR COMPONENTS OF THE THEMATIC MAP.

The composite or completed thematic map is made up of three important components; the geographic or base map, the thematic overlay (often tied to the base map in GIS; see text for discussion), and a set of ancillary map elements, such as titles, legends, compilation credits, neatlines, and other elements.

discussion is focused simply on the idea of scale and its relationship to such design considerations as generalization and symbolization.

Scale selection has important consequences for the map's appearance and its potential as a communication device. Scale operates along a continuum from large scale to small scale (see Figure 1.14 and Color Plate 1.1). Large-scale maps show small portions of the Earth's surface; *detailed information* may therefore be shown. Small-scale maps show large areas, so only *limited detail* or generalized situations can be carried on the map.

It is important to note that the definition of scale as discussed here and in other textbooks is specific to cartography and maps. The cartographer's approach to scale is somewhat different from that used by people in defining the scope of what they do or events that occur. For example, the newscaster may report that the disaster was contained to a small scale (meaning held to a limited area) or that the cure for polio was implemented on a large scale (meaning great breadth). Since a cartographer's view is the "opposite" of these popular connotations of scale, it may be helpful to remember the often-used pneumonic for many cartographers: large scale, large in detail, but small in area—small scale, small in detail, but large in area.

Which final scale is selected for a given map design problem will depend on the map's purpose and physical size. The amount of geographical detail necessary to satisfy the purpose of the map will also act as a constraint in scale selection. Generally, the scale used will be a compromise between these two controlling factors.

Another important consequence of scale selection is its impact on *symbolization.* In changing from large scale to small scale, map objects must increasingly be represented with symbols that are no longer true to scale and thus are more *generalized.* At large scales, the outline and area of a city may be shown in proportion to its actual size—that is, may occupy areas on the map proportional to the city's area.

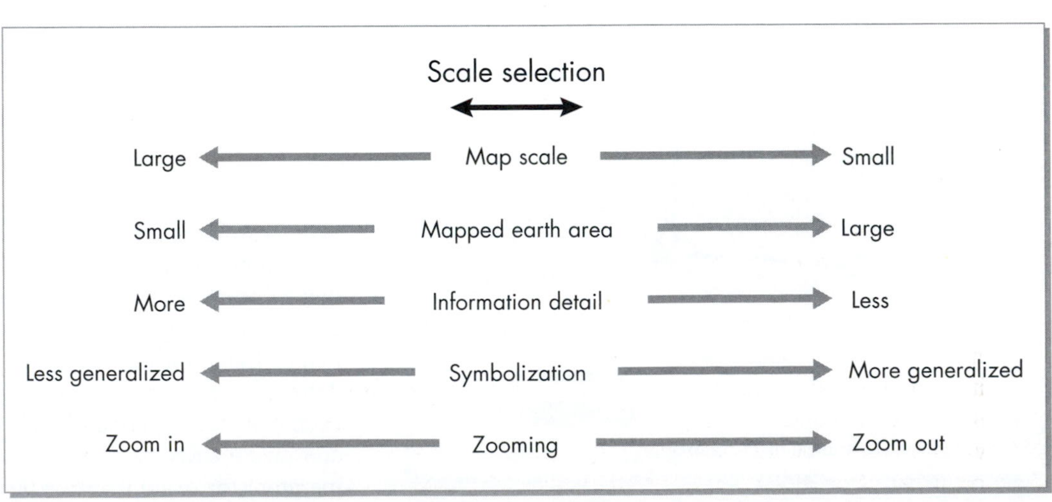

FIGURE 1.14 MAP SCALE AND ITS EFFECT ON MAPPED EARTH AREA, MAP INFORMATION, AND SYMBOLIZATION.

The selection of a map scale has definite consequences for map design. For example, small-scale maps contain large earth areas and less specific detail and must use symbols that are more generalized. Selection of map scale is a very important design consideration because it will affect other map elements.

> Making a road relatively wider than it is on the Earth makes it visible; distorting distances on a projection enables the map user to see the whole Earth at once; separating features by greater than Earth distances allows representation of relative positions. Distortion is necessary in order that the map reader be permitted to comprehend the meaning of the map (Monmonier 1977, 7).

At smaller scales, whole cities may be represented by a single dot having no size relation to the city's real size.

Scale varies over the map depending on the projection used, as will be explained in Chapters 2 and 3. Scale, symbolization, and map projection are thus interdependent, and the selection of each will have considerable effect on the final map. *The selection of scale is perhaps the most important decision a cartographer makes about any map.*

In general, there is an inverse relationship between reference maps and thematic maps regarding scale. In other words, most thematic maps are made at small scales and reference maps tend to be made at larger scales. As thematic cartographers generally work at small map scales, they must be especially attentive to the operations of **cartographic generalization.** This is especially true for geographic cartographers.

Modern Views of Map Communication

Views of how maps may function today in the contexts of visualization and communication are changing. The relatively simple *linear* interpretation of map communication as data field to map to map reader has given way to new concepts. With reference to new models, several ideas important to design are presented here.

Map Communication and Visualization

Considerable discussion, debate, and even disagreement exists today about the roles of *communication* and *visualization* within the context of modern cartography. Scientific visualization, in general, can be thought of as a method that incorporates computers that can transform data into visual models that could not have been seen ordinarily. Scientific visualization (SVIS) in cartography and geography, in general, has led to other terms, such as cartographic visualization (CVIS) and geographic visualization (GVIS). Within the discipline over the past 20 years, there generally has been a sort of demise of *cartographic communication* as the appropriate model for cartography (especially thematic). Many cartographers have not completely given up this view, and some have accommodated it within their new constructs or definition of cartography, GIS, and cartographic visualization.

David DiBiase's view of visualization suggests that visualization takes place along a continuum, with exploration

> The map provides an information source which can offer up the makings of many unexpected patterns from which insights may emerge. As even the most complex maps contain limited input data, when compared to the real world and do not have simply defined messages, observed patterns must be the result of an interactive cognitive dialogue between the map and the viewer's mind. The perfect visual working environment for this process is data-rich and well designed (Hearnshaw and Unwin 1994, 15).

and confirmation in the private realm, and synthesis and presentation in the public realm. For him the private realm constitutes "visual thinking" and the public realm is "visual communication" (DiBiase 1990). According to cartographic researcher Alan MacEachren this view is using the term visualization in a new way to describe cartography as a research tool.

MacEachren places cartographic communication within a "cartography cube" anchored at one vertex, with visualization at the other (MacEachren 1994a; see Figure 1.15). An interesting aspect of the discussion of cartography and visualization is that cartographic communication is not dead but is incorporated into more complex descriptions of cartography, indeed, as an important component. As MacEachren says, "All authors, however, seem to agree that visualization includes both an analysis/visual thinking component and a communication/presentation component and suggest (or at least imply) that communication is a subcomponent of visualization (MacEachren 1994, 5).

Where does map communication fit? Do we expect the map user (or map reader) to interact with the map in a cognitive way? One contribution that cartographers have made in

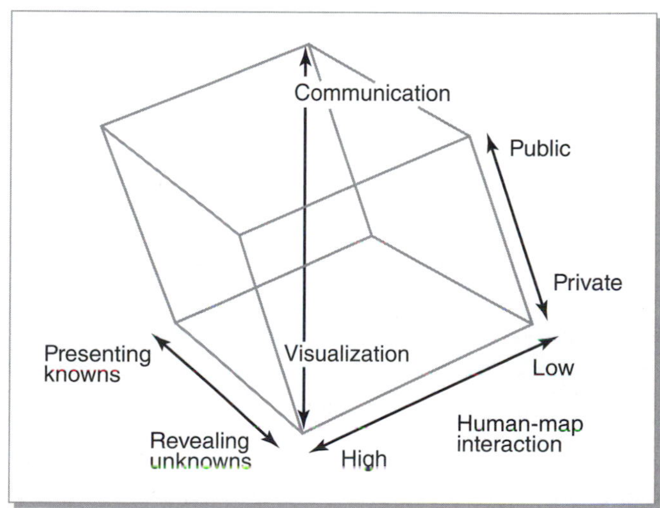

FIGURE 1.15 A MODEL OF MAP VISUALIZATION AND COMMUNICATION.
See text for explanation. Source: Redrawn from MacEachren 1994a, 3.

the last 35 years is recognizing that map readers are different, and not simple mechanical unthinking parts of the process, that they bring to the map reading activity their own experiences and cognition. *The view held in this book in general supports the idea that map communication is the component of thematic mapping the purpose of which is to present one of many possible results of a geographical inquiry, and that while the cartographer may want to convey a specific message, he/she cannot with certainty do this.* Maps, however, are useful in ways other than to convey (communicate) findings to others. They are also useful as tools to ". . . prompt insight, reveal patterns in data, and highlight anomalies" (MacEachren 1994b, 2). In this context, maps are seen as tools for the researcher in *finding patterns and relationships among mapped data,* not simply for the communication of ideas to others.

Cartographic design can be practiced, then, within a communication system framework. The kinds of questions asked by the designer and the steps he or she will take are better controlled by this approach as they best focus the design activity. In practice, maps are always made with a purpose; the map reader's abilities and needs and the limitations of the graphic media all influence design decisions (Robinson 1966). This is what makes cartographic design so interesting.

A very general graphic depiction of thematic map communication would be that represented in Figure 1.16. Two broad components recognizing the contributions of others are, first, "cartographic thinking," those activities in the private realm including visualization of data looking for patterns and relationships and using unencumbered and unstructured symbolization. This idea is not new, however, as mentioned previously. In this context, maps are seen as tools for the researcher in *finding patterns and relationships among mapped data,* not simply for the communication of ideas to others. The second major portion of the model is "cartographic communication," which resides in the public realm and involves the making of a final or optimal map or maps (within the context of the problem) with structured symbolization. Making the map includes those activities usually associated with map design—scale, symbolization, color choice, lettering, and so forth. Going from the private realm of unstructured symbolization to structured symbolization involves the cartographic processes of generalization and abstraction (discussed later).

In this picture of cartography and communication, *mapmaking* involves the conversion of the visualized data into a set of graphic marks (symbols) that are applied in rigorous ways and are placed on the map. Cartographic communication requires that the cartographer know something about the map reader so that the data are transferred to the reader. Communication error develops when there is a discrepancy between the message intended by the map author and the one gained by the map reader. The cartographer may never be certain that the intended message is conveyed.

A **map author** is someone who wishes to convey a spatial message. He or she may or may not be a mapmaker (or cartographer in the broadest use of the term) (Monmonier 1977). When a map author wishes to structure a spatial message most effectively, he or she may employ a trained cartographer to design the map. A cartographer who originates a message is his or her own designer. Going from the private world of visualization and exploration to a final map used for communication requires *cartographic generalization,* which involves *selection, classification, simplification,* and *symbolization* (Muehrcke 1978; Morrison 1975). Going from unmapped data to map form is sometimes referred to as the **cartographic process.** These topics will be more fully discussed later in this chapter.

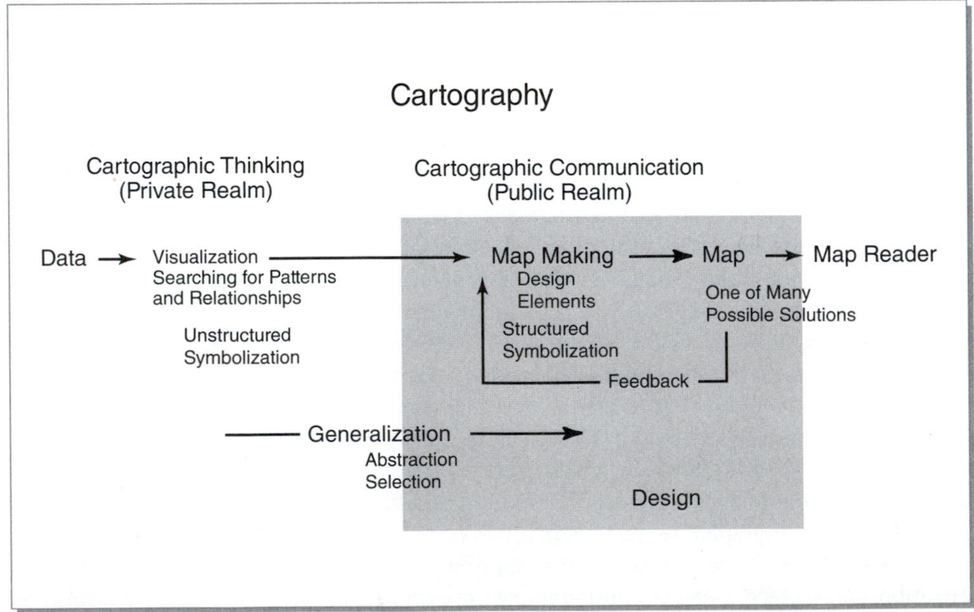

FIGURE 1.16 A MODEL OF MAP COMMUNICATION.
See text for explanation.

TABLE 1.1 SOURCES FOR THE INITIATION OF A MAP MESSAGE

1. A decision by intending map users or their representatives.
2. A decision by mapmakers in anticipation of a need.
3. A decision by a map author as a consequence of his or her own need for explanation.
4. A decision by a scientific body to contribute to the total information in a specialist field.
5. A decision by an individual to express himself or herself through making a map.

Source: Adapted with changes from Keates 1982, 103–4.

Several constraints are imposed on the map author or cartographer in developing the message. These include, especially, the purpose of the map, map format, scale, symbolization, graphic and printing limitations, and economic considerations (Morrison 1975). These may be considered to be *design constraints* in cartographic communication; they are the main subject matter of this book.

Within the context of mapmaking, initiation of a map can come from a variety of sources (see Table 1.1). Maps are made for aims other than pure communication, as the last item in Table 1.1 suggests. Some are constructed as decorative pieces of art, purely for visual appreciation. In such cases, map content plays a lesser role in the map's design.

In cartographic communication and design a fundamental idea is that the mapmaker and map percipient are not independent of each other (Robinson and Petchenik 1975). A **map percipient** (map reader) is one who gains a spatial knowledge by looking at a map (Robinson and Petchenik 1976, 20); most thematic maps are designed for percipients.

Cartographic researchers recognize that the map reading activity is very complex and involves the human nervous system. Muehrcke (1978) states that **map use** comprises reading, analysis, and interpretation. In **map reading,** the viewer looks at a map and determines what is displayed and how the mapmaker did it. On closer inspection, the map user begins to see different patterns; this initiates thoughtful **map analysis.** Finally, a desire to explain these patterns leads to **map interpretation.** In this last case, causal explanations (probably not displayed on the map) are sought by the map user.

Many complex reading tasks have been identified (see Table 1.2). It is no wonder that map communication is so difficult! Maps become especially difficult to read when the map author or cartographer incorporates several tasks in one design. Simplicity in design is a goal and can be achieved in part by reducing the number of map-reading tasks on a single map. Reader training has been shown to improve the efficiency of map communication (Olson 1975).

Older cartographic communication models (including the one used in previous editions of this book) usually contained an element called *feedback.* In the linear sense, feedback is helpful to the map author in that it allows him or her to alter the design by incorporating positive changes suggested by the map user. Unfortunately, most cartographic designers are separated from users by time and space, so it is difficult to make use of feedback. The wise designer will make use of it when possible to improve design. As a possible alternative to direct feedback, we strongly recommend at least running a map by some of your trusted colleagues to reveal needed changes or improvements.

TABLE 1.2 MAP-READING TASKS

1. Pre-Map-Reading Tasks
   - Obtaining, unfolding, and so on
   - Orienting
2. Detection, Discrimination, and Recognition Tasks
   - Search
   - Locate
   - Identify
   - Delimit
   - Verify
3. Estimation Tasks
   - Count
   - Compare or contrast
   - Measure
   - a. Direct estimation
   - b. Indirect estimation
4. Attitudes on Map Style
   - Pleasantness
   - Preference

Source: Morrison 1978, 106.

The Importance of Meaning

Discussion among professional cartographers has emphasized the importance of *meaning* in cartographic communication. The *needs* of the map user are very important in the transfer of knowledge between map author and user. It is essential to include only those objects of specific relevance to the context of the map's message. For example, Dornbach (1967) incorporated the users' requirements in his map design strategy during the development of aeronautical charts some years ago. The pilots of modern jet aircraft need maps with specific information that can be easily perceived. Not just any map will do.

In a given design task, then, *only information that is potentially meaningful to the context should be included on the map* (Guelke 1976). Because a map becomes meaningful only in relation to the previous knowledge the user brings to it (Petchenik 1975), the probable knowledge level of users should be taken into account before the final design of a thematic map is specified.

In sum, success in map communication depends on how well the cartographic designer has been able to interpret the requirements of the user (Guelke 1976). Cartography has been called "the science of communicating information between individuals by the use of maps" (Morrison 1978, 97). This statement implies that both map authors and map users are part of the process; the cartographer must pay particular attention to the needs of the user in designing each map.

Cartography and Geographic Information Systems

The use of geographic information systems, such as ArcGIS, Maptitude, MapINFO, or GRASS to produce thematic maps is now common practice. GIS is an effective tool for visualizing spatial data, and when the cartographer is ready to make a map for public viewing, there are now more tools than ever to make a quality cartographic product. In many government and private sector applications, it is common to produce map layouts as a one-time map production endeavor using GIS. Many GIS packages provide a number of thematic mapping tools, and with every software upgrade, these packages offer better mapping options, giving the cartographer a chance to design maps using well-established cartographic principles.

In addition to visualization and map production, the real strength in GIS is the capabilities of managing spatial data and for spatial analysis. As stated earlier, local, state, and federal agencies are making available spatial data sets that are specifically *created for GIS use*. The results of many GIS analyses (for example optimal site locations for a new business, detection of point sources of pollution, and identification of high risk areas for West Nile virus) are quite often thematic in nature. These activities agree with a summary definition of GIS quite succinctly: "Simply stated, GIS is a set of computer-based systems for managing geographic data and using these data to solve spatial problems" (Lo and Yeung 2002, 2).

GIS is also behind some exciting new developments in Internet mapping. Maps produced by products such as ArcGIS Server deliver interactive maps to the user's web browser. When the end-user specifies the map options, the GIS software (together with a server computer) generates the user-specified map, which is then displayed on the user's screen. The end-user does not even need to have GIS software on his or her computer. So far, GIS-server mapping seems to have been applied more to general purpose maps (for example, National Geographic's *MapMachine* 2007), but there is great potential in these technologies for thematic maps.

Artistic Drawing Programs

Many cartographers use GIS or other computer mapping packages, such as MicroCAM or MapViewer, which place less emphasis on analysis, in conjunction with a high end drawing software package, such as Adobe Illustrator, Freehand, or CorelDraw. Created material, such as base maps, can be edited and projected in the source program (the GIS or computer mapping package), and transferred to the drawing program for graphics enhancement (also called a finishing program). Finishing programs can also include packages that can be used to create animated and interactive maps, such as Adobe Flash, or layout programs such as Adobe InDesign and Quark Express, which are staples in the printing industry. High end drawing software as a finishing program gives the cartographer several important advantages in mapmaking:

1. These packages specialize in a certain function (for example, high end artwork, animation, interactivity, and so forth). Thus, by transferring maps from one software package to another, the strength of each package can be utilized.
2. For maps that will be published in high quality print form (for example, maps in books, atlases, brochures, and such), the high end drawing packages often provide more graphics options and file storage formats that are acceptable to a printing company or service bureau (always check with the printer about their file requirements before sending maps off to be printed).
3. High end drawing programs usually have the excellent text and color management capabilities and provide a greater number of artistic tools in which to practice the "art" in cartography. Although myriads of graphic options and functions do not in themselves ensure a better designed map, the *highest degree of artistic design control by the cartographer is found in the high end drawing packages.*
4. If more than one finishing software product is to be used (for example, a map in a high end drawing package is sent to animation or web software), the number of transfer file formats that can be sent and received from these packages are usually quite large. Depending on the source and destination software, some file formats transfer better than others, so a larger number of choices help ensure a quality product.

Of course, many finishing programs have overlapping functions. For example, Flash can be used for interactive maps, animated maps, or both. When possible, the cartographer should use the strength(s) of each package. Ultimately, the amount of work accomplished in the source program versus the finishing program(s) will depend on what combination of GIS and drawing program(s) are used and the purpose of the map.

CARTOGRAPHIC ABSTRACTION AND GENERALIZATION

Cartographic abstraction is that part of the mapping activity wherein the map author or cartographer *transforms* unmapped data into map form and *selects* and *organizes* the information necessary to develop the user's understanding of the concepts. "When we accept the idea that not all the available information needs to be presented, that instead information must be selected for particular purposes, then

> The possibility of placing data in inappropriate or misleading contexts is a real danger. Just as the meaning of statistics can be radically altered by judicious manipulation of their context, so too can the meaning of map data be changed according to the contextual information included. It is, therefore, essential that a cartographer have a broad background in a wide range of subjects to enable him or her to select appropriate contextual information against which the data mapped can be more readily comprehended. This ability implies that the cartographer has a knowledge of factors behind distributional patterns and some understanding of their causal interconnections (Guelke 1976, 117–18).

the mapping task becomes the identification of relevant elements" (Weltman 1979, 25). Of course, the selection is guided by the purpose of the map.

Selection, classification, simplification, and symbolization are each part of cartographic abstraction and are generalizing operations. Each results in a reduction of the amount of specific detail carried on the map, yet the end result presents the map reader with enough information to grasp the conceptual meaning of the map. Generalization takes place in the context of designing a map to meet user needs. The generalization processes lead to simple visual images, which are more apt to remain in the map user's memory (Arnheim 1976). Unless simplicity is achieved, the map will likely be cluttered with unnecessary detail. Appropriate generalization will result in a spatial message that is efficiently structured for the reader. On the other hand, excessive generalization may cause map images to contain so little useful information that there is no transfer of knowledge. A balance must be struck by the cartographer.

Selection

The **selection** process in the generalization operation of cartographic design begins the mapmaking activity. Selection involves early decisions regarding the geographic space to be mapped, map scale, map projection and aspect, which data variables are appropriate for the map's purpose, and any data gathering or sampling methods that must be employed. Selection is critical and may involve working very closely with the map author or map client (Ommer and Wood 1985). The selection activity requires the cartographer to be familiar with the map's content, especially the nature of the data being used in the mapping process.

Classification

Classification is a process in which objects are placed in groups having identical or similar features. The individuality and detail of each element is lost. Information is conveyed through identification of the boundaries of the group. Classification *reduces the complexity* of the map image, helps to *organize* the mapped information, and thus enhances communication.

In thematic mapping, classification can be carried out with qualitative or quantitative information. Qualitative data might include the identification of geographical regions; for example, the wheat belt, corn belt, or bible belt. It may be far simpler to communicate the concept of region by one large area than to show individual elements. Quantitative classification is normally in numerical data applications. Generally, an entire data array is divided into numerical classes, and each value is placed in its proper class. Only class boundaries are shown on the map. This process reduces overall information but usually results in a map that is more meaningful.

Simplification

Selection and classification are examples of **simplification,** but simplification may take other forms as well. An example might be the *smoothing* of natural or man-made lines on the map to eliminate unnecessary detail. In the selection process, the cartographer may choose to include in the map's base material a road classed as all-weather. Simplification would be the process in which its path is *straightened* between two points so that it no longer retains exact planimetric location (even though this might be possible at the given scale). It can be straightened because the purpose of this map is simply to show *connectivity* between two points, not to illustrate the road's precise locational features (see Figure 1.17).

Symbolization

Perhaps the most complex of the mapping abstractions is **symbolization.** Developing a map requires symbolization, because it is not possible to create a reduced image of the

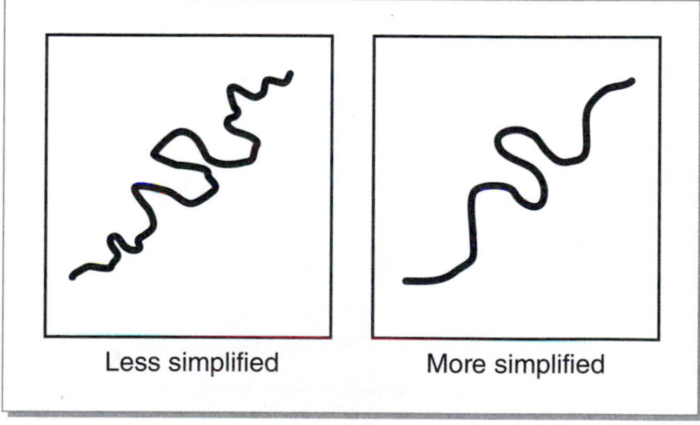

FIGURE 1.17 CLASSIFICATION AND SIMPLIFICATION IN CARTOGRAPHIC COMMUNICATION.

More effective cartographic communication may result if data are first classified and simplified.

real world without devising a set of marks (symbols) that stand for real-world things. In thematic mapping, if an element is mapped it is usually said to be *symbolized.*

Two major classes of symbols are used for thematic maps: **replicative** and **abstract.** Replicative symbols are those that are designed to look like their real-world counterparts; they are used to stand only for tangible objects. Coastlines, trees, railroads, houses, and cars are examples. Base-map symbols are usually replicative in nature, whereas thematic-overlay symbols may be either replicative or abstract. Abstract symbols generally take the form of geometric shapes, such as circles, squares, and triangles. They are traditionally used to represent amounts that vary from place to place; they can represent anything and require sophistication of the map user. A detailed legend is required.

By its very nature, the symbolization process is a generalizing activity shaped by the influence of *scale.* At smaller scales, it is virtually impossible to represent geographical features at true-to-scale likeness. Distortions are necessary. For example, rivers on base maps are widened, and cities that have irregular boundaries in reality are represented by squares or dots.

The cartographer's choices of symbols are not so automatic that every possible mapping problem will yield similar symbol solutions. The symbolization process in thematic mapping is more complex than that, owing to a variety of factors (see Figure 1.18). This topic will be discussed in fuller detail in Chapter 4. It will suffice to say here that the selection of map symbols should be based on the logical association between the level of measurement of the map's data and certain graphic variables (sometimes called "primitives"), conventions or standards, appropriateness, readers' abilities to use the symbols, ease of construction, and similar considerations. Although various attempts have been made, only some standardization exists in the realm of thematic map symbolization. There are still new and exciting symbolization schemes being developed, particularly with regard to GIS and Internet mapping. The student is encouraged to go to the Internet and browse through the many online atlases and map collections to see many of the symbolizations used today. This is a good laboratory exercise and often leads to very interesting classroom discussions.

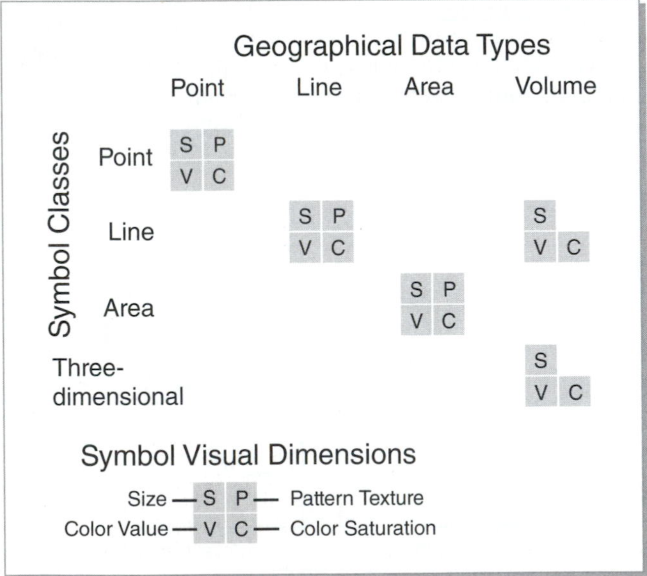

FIGURE 1.18 MAP SYMBOLIZATION VARIABLES.
The selection of thematic symbolization is compounded because of the complex nature of the world of symbols. Geographical data are found in one of four classes but must be symbolized in only one of three symbol classes. Furthermore, each symbol may or may not be varied, depending on four visual dimensions of symbols.

The Art in Cartography

Professional cartographers have long discussed whether cartography should be regarded as an art or as a science. Most would probably agree that it is not art in the traditional view of that activity—that is, an expressive medium. Most would certainly agree that it is a science in the sense that the methods used to study its nature are scientific. The idea that there is an art to cartography suggests that its practice as a profession involves more than the mere learning of a set of well-established rules and conventions. *The art in cartography is the cartographer's ability to synthesize the various ingredients involved in the abstraction process into an organized whole that facilitates the communication of ideas.*

Prominent British cartographer J. S. Keates believed strongly that art is a part of cartography. In his design process, there is a critical point that is "a series of design decisions embodied in the symbol specifications, the point at which design moves from any general ideas to the particular" (Keates 1984, 37). What is important here is that at that point the designer is not primarily interested in communication (the science of) but *representation,* and a desire for the map to be not only informationally effective but also aesthetically pleasing.

Each mapping problem is unique; its solutions cannot be predetermined by rigid formulas. How well the cartographic designer can orchestrate all the variables in the abstraction processes is the measure of how much of an artist he or she is. The reader must be considered, the final map solution must reduce complexity, and the map must heighten the reader's interest (Monmonier 1977; McCleary 1975). These considerations require an artist's skill, which can be acquired only through learning and experience.

THEMATIC MAP DESIGN

Now that cartography within the context of cartographic thinking and communication/visualization has been described, it is possible to develop the concepts of thematic map design. Cartographic design has been described as the "most fundamental, challenging, and creative aspect of the cartographic process" (DeLucia 1974, 83). But what is design? Is it a thing or an activity? We can say, "That map has a good design," or, "Don't bother me while I'm designing." Is there a way to relate the noun to the verb? Indeed, there is.

What Is Map Design?

What is design? There are almost as many definitions as there are designers. Design is the optimum use of tools for the creation of better solutions to the problems that confront us. The end result of design "is the initiation of change in man-made things" (Jones 1981, 6). Things are not just products, such as automobiles or toasters—they can be laws, institutions, processes, opinions, and the like—but they do include physical objects, including maps. Furthermore, design is a *dynamic activity.* "To design is to conceive, to innovate, to create" (Luzadder 1975, 13). It is a process, a sequential (and repetitive) ordering of events. To include the fashioning of everything from a child's soapbox racer to an artificial human heart, we can accept this definition: "The planning and patterning of any act towards a desired foreseeable end constitutes the design process" (Papenek 1971, 3).

Design operates in at least three broad categories: product design (things), environmental design (places), and communications design (messages) (Potter 1969). Cartographic design fits this classification most aptly in the category of communications, although it performs its functions with things (maps).

Map design is the aggregate of all the thought processes that cartographers go through during the abstraction phase of the cartographic process. It "involves all major decision-making having to do with specification of scale, projection, symbology, typography, color, and so on" (Robinson and Petchenik 1976, 19). Designers speak of the *principle of synthesis:* "All features of a product must combine to satisfy all the characteristics we expect it to possess with an acceptable relative importance for as long as we wish, bearing in mind the resources available to make and use it" (Mayhall 1979, 90). The cartographic designer seeks an organized whole for the graphic elements on the map, in order to achieve an efficient and accurate transfer of knowledge between map author and map user. Map design is the *functional relationship* between these two.

To reach the communication goals set forth for a given map design problem, the cartographer looks carefully at the intended audience and then defines the map's purpose with precision (Robinson 1975). This involves an investigation into the functional context of the map, its intended uses, and its communication objectives. Every manipulation of the marks on the map is planned so that the end result will yield a structured visual whole that serves the map's purpose.

Thus, cartographic design is a complex activity involving both intellectual and visual aspects. It is intellectual in the sense that the cartographer relies on the foundations of sciences such as communication, geography, and psychology during the creation of the map. Map design is visual in the sense that the cartographer is striving to reach goals of communication through a visual medium. When we speak of a map as having a design, we are actually looking at the product as the result of design *activity.*

A more detailed examination of the *design process* occurs in Chapter 12. There, such topics as design evaluation, creativity and ideation, visualization, and map aesthetics are dealt with.

Ethics in Cartography

Ethical questions regarding mapmaking are of increasing concern to professional cartographers—so much so that several sessions have been devoted to this topic at cartographic conferences. In fact, a published book titled *How to Lie with Maps* addresses this topic. Mark Monmonier, the author of the book, states in his introduction:

> *The purpose of this book is to promote a healthy skepticism about maps, not to foster either cynicism or deliberate dishonesty. In showing how to lie with maps, I want to make readers aware that maps, like speeches and paintings, are authored collections of information and also are subject to distortions arising from ignorance, greed, ideological blindness, or malice* (Monmonier 1991a, 2).

The important point is that because maps are made by humans they may as a consequence contain purposeful errors (lies) or errors of oversight and poor judgment, or both.

As you proceed through this book you will discover that there are no absolutely correct thematic maps, but only ones that are the best among several alternatives (Monmonier 1991b). Which to use? What should guide me? Monmonier addresses this problem as well, and we are inclined to agree with his philosophy when he remarks (on the subject of one-map solutions):

> *Any single map is but one of many cartographic views of a variable or a set of data. Because the statistical map is a rhetorical device as well as an analytical tool, ethics requires that a single map not impose a deceptively erroneous or carelessly incomplete cartographic view of the data. Scholars must look carefully at their data, experiment with different representation, weigh both the requirements of the analysis and the likely perceptions of the reader, and consider presenting complementary views with multiple maps* (Monmonier 1993, 185).

Dent's cartography students, both graduate and undergraduate, have looked at issues related to cartographic ethics for a number of years. The culmination of many discussions have led to several "codes of ethics" for the thematic cartographer, and includes these, in various forms:

1. Always have a straightforward agenda, and have a defining purpose or goal for each map.
2. Always strive to know your audience (the map reader).
3. Do not intentionally lie with data.
4. Always show all relevant data whenever possible.

5. Data should not be discarded simply because they are contrary to the position held by the cartographer.
6. At a given scale, strive for an accurate portrayal of the data.
7. The cartographer should avoid plagiarizing; report all data sources.
8. Symbolization should not be selected to bias the interpretation of the map.
9. The mapped result should be able to be repeated by other cartographers.
10. Attention should be given to differing cultural values and principles.

A good place to initiate any discussion of cartographic ethics may begin with these. It makes a great classroom exercise.

We strongly support the use of GIS and other mapping and artistic software in cartographic design. The capabilities of these programs are increasing each year, allowing the cartographer more flexibility in designing maps. However, the possibilities for the production of *maps without ethics* are compounded by the fact that the software allows non-cartography-trained persons to produce maps. In some cases they may *look* good, but often they are not consistent with any established professional standards or conventions. We also support maps being created for the Internet. If the goal is to reach as wide an audience as possible, Peterson (1999) argues that the Internet becomes the ethical choice of medium and mode of distribution. However, the possibilities for *maps without ethics* are further compounded by the ease in which these maps can be placed on the Internet. As McGranaghan (1999, 3) notes, "anyone with a modicum of technical savvy can 'publish' any content they wish on the internet …"

So, the question is what do we do about this? Becoming a trained map reader is one way that one can discern an unethical map. However, as far as map creation is concerned, at present no licensing or certification agencies oversee the cartography profession (as they do surveyors, for example). Certainly there are no governing bodies charged with guiding the ethical practices of thematic mappers. Thus the intent, honesty, and credentials of each cartographer must be dealt with on an individual basis, and his or her reputation must be considered. At the very least, cartographers should make it a practice to provide all base map and data sources, and as appropriate, methods of compilation, and appropriate dates on each product they design.

REFERENCES

Allen, J. 2008. *Student Atlas of World Geography,* 5th ed. Dubuque, IA: McGraw-Hill.

Arnheim, R. 1976. The Perception of Maps. *American Cartographer* 3: 5–10.

Bagrow, L. 1966. *History of Cartography,* rev. ed. Cambridge: Harvard University Press.

Bickmore, D. 1975. The Relevance of Cartography. In *Display and Analysis of Spatial Data,* ed. J. Davis and M. McCullagh. New York: Wiley.

Borchert, J. 1987. Maps, Geography, and Geographers. *The Professional Geographer* 39: 387–89.

Cuff, D., W. Young, E. Muller, W. Zelinsky, and R. Abler, eds. 1990. *The Atlas of Pennsylvania.* Philadelphia: Temple University Press.

DeLucia, A. 1974. Design: The Fundamental Cartographic Process, *Proceedings*, Association of American Geographers 6: 83–86.

DiBiase, D. 1990. Visualization in the Earth Sciences. *Earth and Mineral Sciences*. Bulletin of the College of Earth and Mineral Sciences, Pennsylvania State University, 59 (2):13–18.

Dornbach, J. 1967. *An Analysis of the Map as an Information Display System*. Ph.D. dissertation, Clark University, Worcester, MA.

Getis, A., J. Getis, and J. Fellman. 2008. *Introduction to Geography,* 11th ed. Dubuque, IA: McGraw-Hill.

Guelke, L. 1976. Cartographic Communication and Geographic Understanding. *Canadian Cartographer* 13: 107–22.

Haggett, P. 1990. *The Geographer's Art*. Oxford: Basil Blackwell.

Hearnshaw, H., and D. Unwin, eds. 1994. *Visualization in Geographical Information Systems*. New York: Wiley.

Hodler, T. 1994. Do Geographers Really Need to Know Cartography? *Urban Geography* 15 (5): 409–10.

Hodler, T., and H. Schretter, 1986. *The Atlas of Georgia*. Georgia: Institute for Community and Area Development.

Jenks, G. 1976. Contemporary Statistical Maps—Evidence of Spatial and Graphic Ignorance. *American Cartographer* 3: 11–19.

Johnson, H., and C. Duchon. 1995. *Atlas of Oklahoma Climate*. Norman: University of Oklahoma Press.

Jones, C. 1981. *Design Methods: Seeds of Human Futures*. New York: Wiley.

Keates, J. 1982. *Understanding Maps.* New York: Wiley.

———. 1984. The Cartographic Art. *Cartographica* 21: 37–43.

Lo, C., and A. Yeung. 2002. Clarke (ed.). *Concepts and Techniques of Geographic Information Systems*. Upper Saddle River, NJ: Prentice-Hall.

Loy, W., S. Allan, A. Buckley, and J. Meacham, 2001. *Atlas of Oregon, 2nd ed.* University of Oregon Press.

Luzadder, W. 1975. *Innovative Design, with an Introduction to Graphic Design.* Englewood Cliffs, NJ: Prentice-Hall.

MacEachren, A. 1994a. Visualization in Modern Cartography. In *Visualization in Modern Cartography,* eds. A. MacEachren and D. R. F. Taylor. 1–12. New York: Elsevier Science.

———. 1994b. *Some Truth with Maps: A Primer on Symbolization and Design*. Washington, DC: Association of American Geographers.

Mayall, W. 1979. *Principles of Design.* New York: Van Nostrand Reinhold.

McCleary, G. 1975. In Pursuit of the Map User. *Proceedings*, 1975 International Symposium on Computer-Assisted Cartography: 238–49.

McGranaghan, M. 1999. The Web, Cartography, and Trust. *Cartographic Perspectives* 32 (Winter 1999): 3–5.

Meynen, E., ed. 1973. *Multilingual Dictionary of Technical Terms in Cartography*. International Cartographic Association, Commission II—Wiesbaden: Franz Steiner Verlag.

Microsoft. 2007. *Encarta*. http://encarta.msn.com/encnet/features/mapcenter/map.aspx.

Moellering, H., 1984. Real Maps, Virtual Maps and Interactive Cartography. In *Spatial Statistics and Models*, ed. G.L. Gaile and C.J. Wilmott. 109–32. Boston: D. Riedel.

Monmonier, M. 1977. *Maps, Distortion, and Meaning*. Resource Paper No. 75–4. Washington, DC: Association of American Geographers.

———. 1991a. *How to Lie with Maps*. Chicago: University of Chicago Press.

———. 1991b. Ethics and Map Design: Six Strategies for Confronting the Traditional One-Map Solution. *Cartographic Perspectives* 10 (Summer 1991): 3–7.

———. 1993. *Mapping It Out: Expository Cartography for the Humanities and Social Sciences*. Chicago: University of Chicago Press.

Morrison, J. 1975. Map Generalization: Theory, Practice, and Economics. *Proceedings*, 1975 International Symposium on Computer-Assisted Cartography: 99–112.

———. 1978. Towards a Functional Definition of the Science of Cartography with Emphasis on Map Reading. *American Cartographer* 5: 97–110.

Muehrcke, P. 1978. *Map Use, Reading, Analysis, and Interpretation*. Madison, WI: JP Publications.

———. 1981. Maps in Geography. *Cartographica* 18: 1–37.

National Atlas of the United States. 2007. http://nationalatlas.gov.

National Geographic. 2007. *MapMachine*. http://plasma.nationalgeographic.com/mapmachine/.

Olson, J. 1975. Experience and the Improvement of Cartographic Communication. *Cartographic Journal* 12: 94–108.

Ommer, R., and C. Wood. 1985. Data, Concept and the Translation to Graphics. *Cartographica* 22: 44–62.

Papenek, V. 1971. *Design for the Real World*. New York: Pantheon Books.

Perry-Castañeda Library Map Collection (2006). http://www.lib.utexas.edu/maps/.

Petchenik, B. 1975. Cognition in Cartography. *Proceedings*, 1975 International Symposium on Computer-Assisted Cartography: 183–93.

Peterson, 1999. Maps on Stone: The Web and Ethics in Cartography. *Cartographic Perspectives* 34 (Fall 1999): 5–8.

Peterson, M. 2003. Maps and the Internet: An Introduction. In *Maps and the Internet*, ed. M. Peterson. 1–16. Oxford: Elsevier Science.

Potter, N. 1969. *What Is a Designer: Education and Practice*. London: Studio-Vista.

Robinson, A. 1954. Geographical Cartography. In *American Geography, Inventory and Prospect*, ed. P. James and C. Jones, 553–77. Syracuse, NY: Syracuse University Press.

———. 1966. *The Look of Maps*. Madison: University of Wisconsin Press.

———. 1975. Map Design. *Proceedings*, International Symposium on Computer-Assisted Cartography. 9–14.

Robinson, A., and B. Bartz Petchenik. 1975. The Map as a Communication System. *Cartographic Journal* 12: 7–15.

———. 1976. *The Nature of Maps: Essays toward Understanding Maps and Meaning*, 16–17 Chicago: University of Chicago Press.

Rooney, J., W. Zelinsky, and D. Louder, eds. 1982. *This Remarkable Continent: An Atlas of the United States and Canadian Society and Cultures*. College Station, Texas: Texas A&M University Press.

Sauer, C. 1956. The Education of a Geographer. *Annals* of the Association of American Geographers 46: 287–99.

Southworth, M., and S. Southworth, 1982. Maps: A Visual Survey and Design Guide. A New York Graphic Society Book. Boston: Little, Brown.

Weltman, G., ed. 1979. *Maps: A Guide to Innovative Design*. Woodland Hills, CA: Perceptronics.

Wood, D. 1992. *The Power of Maps*. New York: Guilford.

GLOSSARY

abstract symbols may represent anything; usually take the form of geometrical shapes such as circles, squares, or triangles

cartographic abstraction transformational process in which map author or cartographer selects and organizes material to be mapped; identification of relevant elements

cartographic generalization transformation process of abstraction involving selection, classification, simplification, and symbolization

cartographic process the transformation of unmapped data into map form involving symbolization and the map-reading activities whereby map users gain information

cartography somewhat broader in scope than mapmaking; the study of all subjects that are concurrent with maps including their history, reading and use, and collection

choropleth map a form of statistical mapping used to portray discrete data by enumeration units; area symbols are applied to enumeration units according to the values in each unit and the symbols chosen to represent them

classification placing objects in groups having identical or similar features; reduces complexity and helps organize materials for communication

dot map a map in which dots represent the distribution of phenomena; one dot will usually represent so many units of a certain phenomena

enumeration unit administrative areas to which attribute data is associated; examples of enumeration units at varying detail include countries, states, counties, and townships

flow map map on which the amount of movement along a linear path is stressed, usually by lines of varying thicknesses and/or color

general purpose maps also called reference maps; show natural and man-made features of general interest for widespread public use

isarithmic map a planimetric graphic representation of a three-dimensional volume

map a graphic representation of the milieu

map analysis map-use activity in which the map user begins to see patterns

map author anyone wishing to convey a spatial message; may or may not be a cartographer

map design aggregate of all thought processes that the map author or cartographer goes through during the abstraction phase of the cartographic process; activity seeking graphic solutions

map interpretation map-use activity in which the map user attempts to explain mapped patterns

mapmaking also called mapping; all processes associated with the production of maps; the designing, compiling, and producing of maps

map percipient anyone who gains spatial knowledge by looking at a map

map reading map-use activity in which the user simply determines what is displayed and how the mapmaker did it

map scale amount of reduction that takes place in going from real-world to map plane; ratio of map distance to Earth distance

map use comprises map reading, map analysis, and map interpretation

mental maps mental images having spatial attributes

proportional symbol map type of quantitative thematic map in which point data are represented by a symbol whose size varies with the data values; areal data assumed to be aggregated at points may also be represented by proportional point symbols

qualitative map thematic map whose main purpose is to show locational features of nominal data

quantitative map thematic map whose main purpose is to show spatial variation of amount

replicative symbols designed to appear like their real-world counterparts

selection a generalization process in which selections of items to be mapped are made, all within the context of map purpose and map scale

simplification reducing amount of information by selection, classification, or smoothing

symbolization devising a set of graphic marks that stand for real-world things

thematic map special-purpose, single-topic map; designed to demonstrate particular features or concepts

thematic overlay that part of the thematic map that contains the specific information (the map subject)

value-by-area cartogram name applied to the form of map in which the areas of the internal enumeration units are scaled to the data they represent

virtual map a map that is viewable but is impermanent, such as a map that is displayed on a computer monitor or a graphic in a Power Point presentation; the virtual map can be contrasted with a printed map product (sometimes called a "real" map)

2 BASIC GEODESY, COORDINATE SYSTEMS, AND SCALE

CHAPTER PREVIEW The Earth is an irregularly shaped body, roughly approximating a sphere but more precisely defined by a reference ellipsoid. The ellipsoid is often part of a map or spatial dataset's datum, which can also describe its origin and type of coordinate system used. Cartesian coordinate geometry provides the underpinnings of eastings and northings used in coordinate systems worldwide. Cartesian coordinate concepts are also incorporated in the geographic grid. Latitude and longitude positions may be reported in sexagesimal or DMS format, or as decimal degrees. An understanding of the deeper geometric relationships of the geographic grid can greatly assist the cartographer in choosing the map's projection and other design aspects. Scale and generalization, subjects discussed in Chapter 1, are essential concepts to finding and correctly utilizing the myriads of data that are available on the Web that can be used to design a thematic map. ■

In Chapter 1 it was noted that thematic maps are composed of three main structural elements: the base map, a thematic overlay, and a set of ancillary map elements, such as titles, legends, compilation credits, neatlines, and other objects. Chapters 2 and 3 focus on the issues pertaining to the first element: the base map. Historically, cartographers created their own base maps manually, obtained and associated the attributes to make the thematic overlay, and then created the ancillary elements to complete the thematic map. In the switch from manual to digital map compilation, obtaining source material for the thematic map (the base map and attribute data) is often just a click away on the Web.

Today, there is an ever increasing amount of digital base map data easily accessed in GIS and mapping software. These spatial data often come from the GIS or mapping software manufacturers, or are downloaded from federal, state, local, or other government, educational, or private institutions on the Web. Attribute data are often (but not always) included with the digital base map.

While the attributes determine *what* thematic topic is being mapped, such as population density, agricultural production, per capita income, or precipitation, the base map illustrates *where* those data occur. Are the data local, or are they from a different level, such as state or province, country, continent, or the world? If I download a state map containing counties, will the counties still look good if I zoom in on them? If I overlay a road or river network on top of this map, will the layers still line up? These are questions that are directly impacted by base map concepts that are foci of this chapter: *geodesy, datums, coordinate systems, the geographic grid, scale,* and *generalization* (the last two terms were also discussed in Chapter 1). These concepts also play a central role in another important related base map concept, the **map projection.** The map projection is the process by which we obtain our flat two-dimensional map from the three-dimensional Earth surface, and will be dealt with in Chapter 3.

BASIC GEODESY

Geodesy is the science of Earth measurement. Students are often surprised by the fact that there is more to describing the Earth's shape than as simply "round," and that many of the mathematical estimates of the Earth's size and shape can have an impact on the map's production. There are three important approximations of the Earth's shape: the sphere, the ellipsoid,

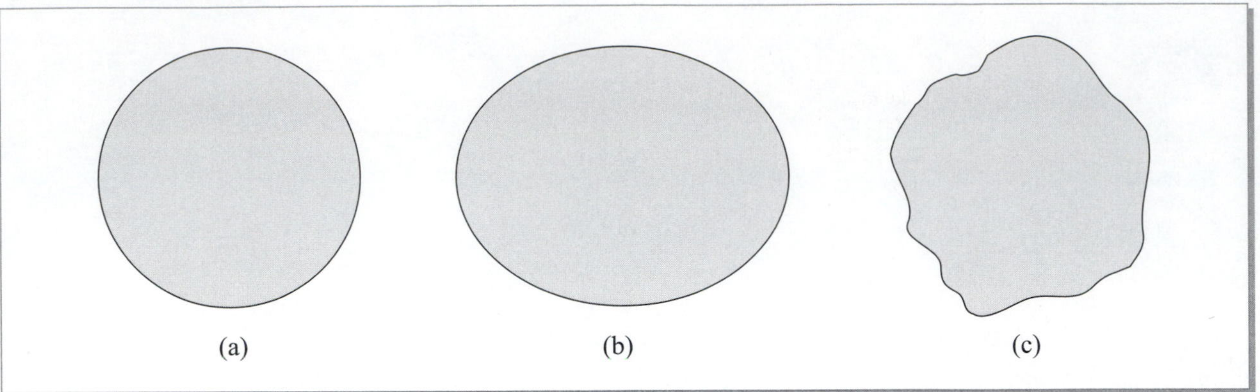

FIGURE 2.1 2-D COMPARISON OF THE SPHEROID (a), ELLIPSOID (b), AND GEOID (c) REPRESENTATION OF THE EARTH. Note that the shape of the ellipsoid and geoid are highly exaggerated for illustrative purposes.

and the geoid (see Figure 2.1). The next section will examine these important shapes and their impact on mapping.

The Size and Shape of the Earth

It is not known exactly when the Earth was first thought to be spherical in form, but Pythagoras (sixth century B.C.) and Aristotle (384–322 B.C.) are known to have determined that the Earth was a sphere. Aristotle based his conclusions partly on the idea, then widely held among Greek philosophers, that the sphere was a perfect shape and that the Earth must therefore be spherical. Celestial observations, notably lunar eclipses, also helped him reach this important conclusion. The idea of a spherical Earth soon became adopted by most philosophers and mathematicians.

Greek scholars turned their attention to measurement of the Earth. In fact, the Earth's size was measured quite accurately by the Greek scholar **Eratosthenes** (276–194 B.C.). Living in Alexandria, Egypt, Eratosthenes calculated the equatorial circumference remarkably close to today's measurement of 40,075 km (24,901 mi) (Campbell 2001).

Eratosthenes' ingenious method of measuring the Earth employed simple geometrical calculations (see Figure 2.2). In fact, the method is still used today. Eratosthenes noticed on the day of the summer solstice that the noon sun shone directly down a well at Syene, near present-day Aswan in southern Egypt. However, the sun was not directly overhead at Alexandria but rather cast a shadow that was 7°12' off the vertical. Applying geometrical principles, he knew that the deviation of the sun's rays from the vertical would form an angle of 7°12' at the center of the Earth. This angle is 1/50 the whole circumference. The only remaining measurement needed to complete the calculations was the distance between Alexandria and Syene; this was estimated at 5000 stadia (one stadia is the size of an athletic stadium of that time period; the *exact* value is unknown). By multiplying this number by 50, his estimate was 250,000 stadia, which most scholars place within about 15 percent of today's figure.

Not until the end of the seventeenth century was the notion of an imperfectly shaped Earth introduced. By that time, accurate measurement of gravitational pull was possible. Most notably, Newton in England and Huygens in Holland

TABLE 2.1 REFERENCE ELLIPSOIDS

Ellipsoid	Equatorial Radius (a) Statute Miles	Polar Radius (b) Statute Miles
Airy (1830)	3,962.56	3,949.32
Austrian Nat'l–South Am. (1969)	3,962.93	3,949.64
Bessel (1841)	3,962.46	3,949.21
*Clarke (1866)	3,962.96	3,949.53
Clarke (1880)	3,962.99	3,949.48
Everest (1830)	3,962.38	3,949.21
*Geodetic Reference System (1980)	3,962.94	3,949.65
International (1924)	3,963.07	3,949.73
Krasovskiy (1940)	3,962.98	3,949.70
World Geodetic System (1972)	3,962.92	3,949.62
*World Geodetic System (1984)	3,963.19	3,949.90

The statute mile is 5,280 feet. *Common in North America

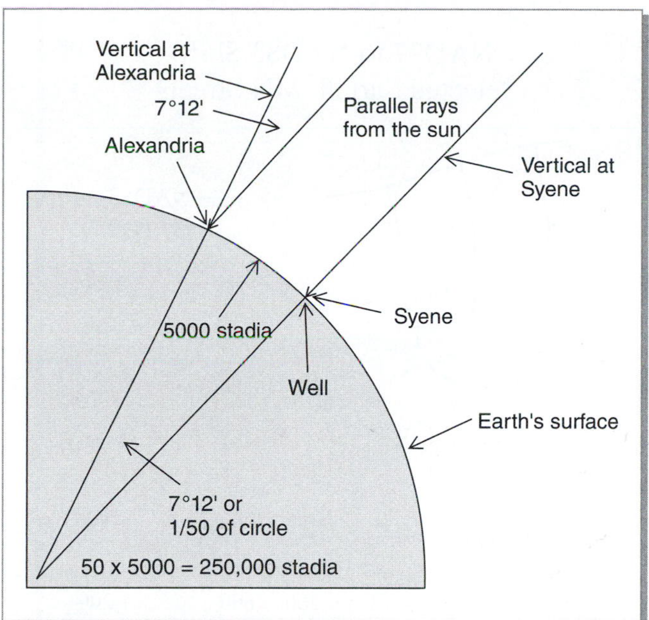

FIGURE 2.2 ERATOSTHENES' METHOD OF MEASURING THE SIZE OF THE EARTH.
See the text for a complete explanation.

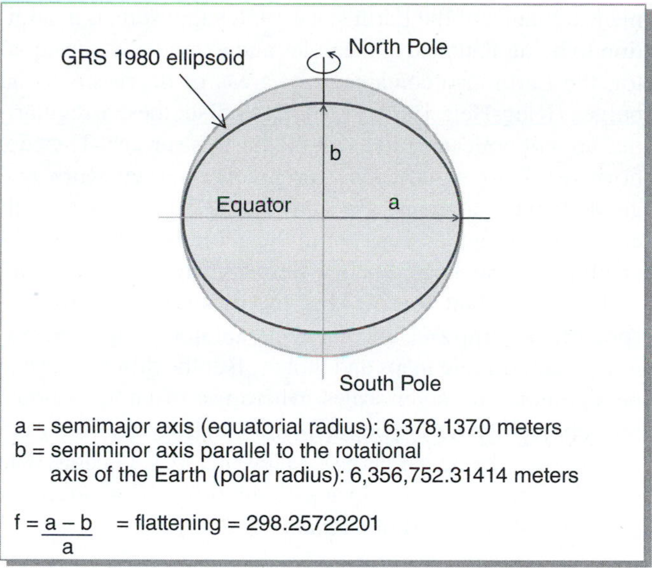

FIGURE 2.3 THE ELLIPTICAL SHAPE OF THE EARTH.
The ellipse is exaggerated for illustrative purposes. The GRS 80 ellipsoid is only one of many possible ellipsoids.

put forward the theory that the Earth was flattened at the poles and extended (bulged) at the equator. This idea was later tested by field observation in Ecuador and Lapland by the prestigious French Academy of Sciences. The Earth was indeed flattened at the poles! It is interesting to note that the first indication of this flattening came from sailors who noticed that their chronometers were not keeping consistent time as they sailed great latitudinal distances. The unequal pull of gravity, caused by the imperfectly shaped Earth, created different gravitational effects on the pendulums of their clocks.

The polar flattening and equatorial bulging of the Earth cause some cartographers to refer to the Earth as an oblate spheroid. The oblate qualities of the Earth can be modeled using a reference **ellipsoid** (see Figure 2.3). The geometrical solid is generated by rotating an ellipse around its minor axis and choosing the lengths of the major and minor axes that best fit those of the real Earth. Various reference ellipsoids have been adopted by different countries throughout the world for their official mapping programs, based on the local precision of the ellipsoid in describing their part of the Earth's surface (see Table 2.1). Some ellipsoids, such as the Geodetic Reference System 1980 and the World Geodetic System–1984 (WGS84) are designed for worldwide use. For example, Global Positioning Systems (GPS) technology is based on the WGS84 ellipsoid.

The most precise shape of the Earth is described by geodesists as a **geoid** (meaning Earth-shaped). Satellite

Mean Radius (2a + b)/3 Statute Miles	Ellipticity (Flattening) $f = \frac{a-b}{a}$	Where Used
3,958.15	1/299.32	Great Britain
3,958.50	1/298.25	Australia, South America
3,958.04	1/299.15	China, Korea, Japan
3,958.48	1/294.98	North America (especially NAD27), Central America, Greenland
3,958.49	1/293.46	Much of Africa, some countries of the Middle East
3,957.99	1/300.80	India, Southeast Asia, Indonesia
3,958.51	1/298.26	North America (especially NAD83)
3,958.63	1/297.00	Europe, individual States in South America
3,958.56	1/298.30	Russia (and former socialist states)
3,958.49	1/298.26	NASA; U.S. Department of Defense; oil companies, Russia
3,958.76	1/298.26	Worldwide, Global Positioning System

Sources: Defense Mapping Agency, Hydrographic Center, 1977, 117–20; Snyder, J. 1987, 12; Dana, P. 2003.

measurements of the Earth since 1958 suggest that, in addition to being flattened at the poles and extended at the equator, the Earth also contains great areas of depressions and bulges (King-Hele 1967, 1976). Of course, these irregularities are not noticeable to us on the Earth's surface. They do, however, have significance for precise survey work and geodetic measurements. For cartographers the most significant aspect of geodesists' use of the geoid is in the fitting of an ellipsoid and a coordinate system to distributed map data.

It is important to note that the differences between a spheroid, an ellipsoid, and the geoid are not going to be visible on small-scale maps and globes. But the differences can be significant at larger scales, where use of an appropriate reference ellipsoid is common (most cartographers do not directly use the actual geoid). Many of today's spatial data sets distributed for use in a GIS incorporate a reference ellipsoid, often as part of a *datum*.

Datums

A **datum,** or starting point, gives a context to locations and heights on the Earth's surface. At its most basic, a datum defines the size and shape of the Earth (usually a reference ellipsoid such as those described in Table 2.1) and some sort of tie-point that fixes the ellipsoid to the Earth's surface (see Figure 2.4a) or to the center of the Earth (see Figure 2.4b). For example, the North American Datum of 1927 (NAD27) uses the Clarke 1866 ellipsoid, which has its tie point located at Meades Ranch, Kansas, USA. The North American Datum of 1983 (NAD83) uses the Geodetic Reference System 1980 (GRS80) ellipsoid, which has its tie-point at the center of the Earth.

A datum will also describe the origin and orientation of the coordinate system used (Dana 2003). Coordinate systems will be described in more detail in the next section. Referencing the wrong datum can result in positional errors of hundreds of feet, depending on place and datum used. For example, the outlines from a section in central Minnesota (see Figure 2.5) illustrate the difference in the shift between NAD27 and NAD83 datums. The visual appearance of referencing the wrong datum is most pronounced at larger scales, but since public data distributed for GIS use are often distributed with a particular ellipsoid or datum, a mismatch in datum can cause a mis-registration among the layers, as in Figure 2.5.

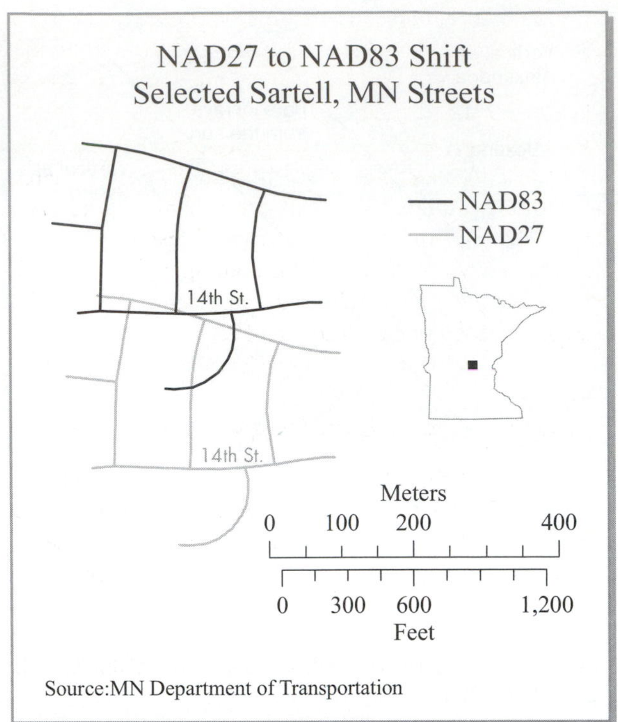

FIGURE 2.5 NAD27 TO NAD83 POSITIONAL SHIFT IN CENTRAL MINNESOTA.

For larger scale maps and downloaded GIS layers, datum selection can be very important. In this illustration, the *y*-displacement is over 200 meters (more than 650 feet).

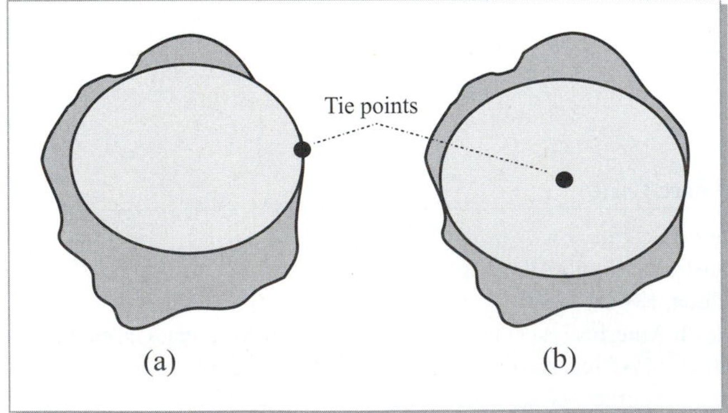

FIGURE 2.4 APPLYING THE REFERENCE ELLIPSOID TO THE EARTH.

In (a) the ellipsoid is tied to a point on the Earth's surface. In (b) the ellipsoid is tied to the center of the Earth. Again, both ellipsoid and geoid are highly exaggerated for illustrative purposes.

COORDINATE GEOMETRY FOR THE CARTOGRAPHER

Location was the key idea behind the historical development of the Earth's coordinate geometry. Ancient astronomers were naturally concerned with this question as they delved into debates related to the Earth's size and shape. During the fifteenth, sixteenth, and seventeenth centuries, when exploration flourished, exactness in ocean navigation and location became critical. Death often awaited mariners who did not know their way along treacherous coasts. Naval military operations in the seventeenth and eighteenth centuries also required precise determination of location on the globe, as is the case today.

Today, use of the **Global Positioning System (GPS)** is a common and useful means of determining location. GPS is a system of 24 or more orbiting satellites that transmit timing signals to ground-based receivers. The receivers' computer processor and software then calculate the location. The system can pinpoint one's location on the Earth within a few meters (a meter is about 3.28 feet) using consumer-grade receivers. However, land surveyors use higher precision devices and specialized methodologies that can place points within a centimeter (about 0.39 inch) of accuracy.

There are thousands of Earth coordinate systems available for cartographers, although the system of using latitude and longitude (also called the geographic grid, discussed in this section) is the most well known. GIS and mapping software allows for relatively easy conversion among these systems. This section focuses on the structure of coordinate systems in "Plane Coordinate Geometry" and "The Geographic Grid."

Plane Coordinate Geometry

Perhaps the best way to introduce the major coordinate systems is to examine *plane coordinate geometry.* **Descartes,** a French mathematician of the seventeenth century, devised a system for geometric interpretation of algebraic relationships. This eventually led to the branch of mathematics called analytic geometry (Van Sickle 2004). It is from his contributions that we have **Cartesian coordinate geometry.** This system of intersecting perpendicular lines on a plane contains two principal axes, called the *x*- and *y*-axes (see Figure 2.6). The vertical axis is usually referred to as the *y*-axis and the horizontal as the *x*-axis. The intersection of the *x*- and *y*-axes is referred to as the origin.

The plane of Cartesian space is marked at intervals by equally spaced lines. The position of any point (*Pxy*) can be specified by simply indicating the values of *x* and *y* and plotting its location with respect to the values of the Cartesian plane. In this manner, each point can have its own unique, unambiguous location. Relative location can easily be shown by plotting several points in the space.

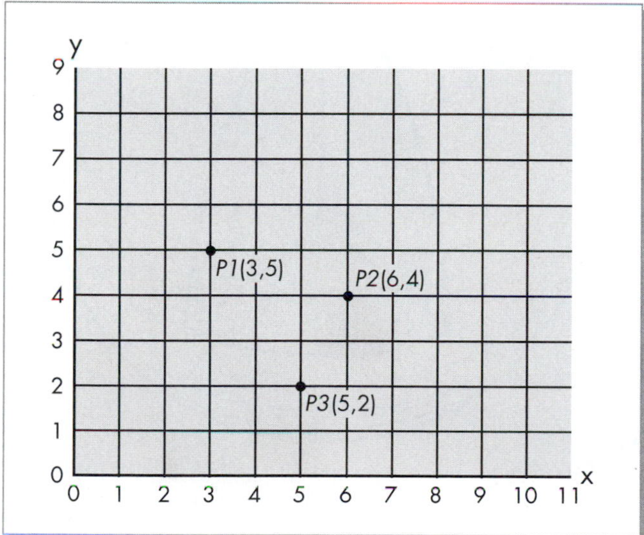

FIGURE 2.6 THE CARTESIAN COORDINATE SYSTEM.
Points (P1, P2, P3, and so on) can have their exact locations defined by reference to the grid. Absolute and relative locations are therefore easy to determine.

When describing real-world space, it is common among cartographers and land surveyors to refer to this system generically as eastings and northings. That is, a point can be defined by its easting (a measured distance along the *x*-axis from the origin) and its northing (a measured distance along the *y*-axis from the origin). The time-honored axiom of "read right up" can assist the map reader who is uncomfortable with which coordinate to list first.

In Figure 2.6, all of the coordinates have positive *x*- and *y*-values. But Cartesian coordinate geometry also allows for negative *x*- and/or *y*- values as well as positive ones by using quadrants (see Figure 2.7).

Many cartographic coordinate systems are set up so that the origin is always to the south and west of the coordinate space. That is, all *y* and *x* values will be positive (as in Figure 2.6). When a coordinate system is shifted so that everything is in quadrant one, the origin is sometimes referred to as a *false* origin. Alternatively, an arbitrary number can be added to the *x*-value and/or the *y*-value so that no values are negative. These numbers are sometimes called a *false* easting or *false* northing.

It is important to note that there are literally thousands of coordinate systems that exist throughout the world, at varying scales, often as part of a local, national, or global datum. Some use feet as their unit of measurement; many more use meters. The underlying principles of locating positions using eastings and northings work essentially the same way for every coordinate system.

There are a number of coordinate systems that are built on *map projections.* Two that are extremely important to cartographers are the State Plane and the Universal Transverse Mercator (UTM) coordinate systems. These are used throughout the United States—and for UTM, throughout much of the world—for storing and distributing spatial data. As these systems have an incorporated projection, most GIS and mapping software tend to include these coordinate systems in their

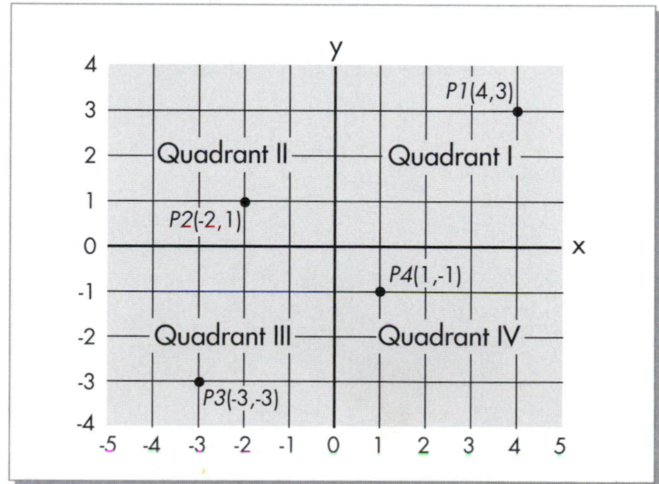

FIGURE 2.7 CARTESIAN SYSTEM QUADRANTS.
The four quadrants allow for positive and negative Cartesian coordinate values.

map projection options. Because they are built on specific parameters involved in the map projection process we will discuss these important systems in the next chapter.

The Geographic Grid

Concepts similar to those used in plane or Cartesian coordinate geometry are incorporated in the Earth's coordinate system, also called the **geographic grid** or the **graticule.** The Earth's geometry is somewhat more complex because of its spherical shape. Nonetheless, it can be easily learned. To specify location on the Earth (or any spherical body), angular measurement must be used in addition to the elements of the ordinary plane system. Angular measurement is based on a **sexagesimal** scale: division of a circle into 360 degrees, each degree into 60 minutes, and each minute into 60 seconds. This method of determining geographic grid values is sometimes referred to as a **DMS format** (for degrees, minutes, and seconds).

Our planet rotates about an imaginary axis, called the **axis of rotation** (see Figure 2.8). If extended, one of the axes approximately points to Polaris, the North Star. The place on Earth where this axis of rotation emerges is referred to as **geographic north** (the North Pole). The opposite, or **antipodal point,** is called **geographic south** or the South Pole. These points are very important because the entire coordinate geometry of the Earth is keyed to them.

If we were to pass an imaginary plane through the Earth perpendicular to and bisecting the axis of rotation, the intersection of the plane with the surface of the Earth would form a complete circle (assuming that the Earth is perfectly spherical). This imaginary circle is referred to as the Earth's **equator** (see Figure 2.8). Since the equator bisects the Earth into two halves, it is referred to as a **great circle.** The North and South Poles and the equator are the most important elements of the Earth's coordinate system.

Latitude Determination

Latitude is simply the location on the Earth's surface between the equator and either the North or the South Pole. Latitude determination is easily accomplished; it is a function of the angle between the horizon and the North Star (or some other fixed star; see Figure 2.9). As one travels closer to the pole, this angle increases. It can be demonstrated that the angle formed at the center of the Earth between a radius to any point on the Earth's surface and the equator is identical in magnitude to the angle made between the horizon and the North Star at that location. If an imaginary plane is passed through this point

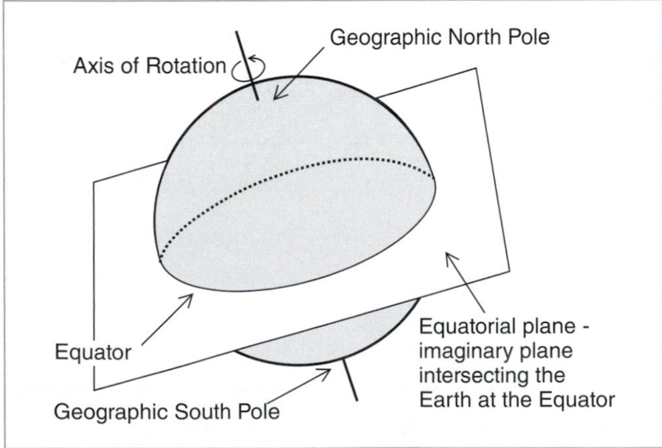

FIGURE 2.8 DETERMINATION OF THE EARTH'S POLES AND THE EQUATOR.

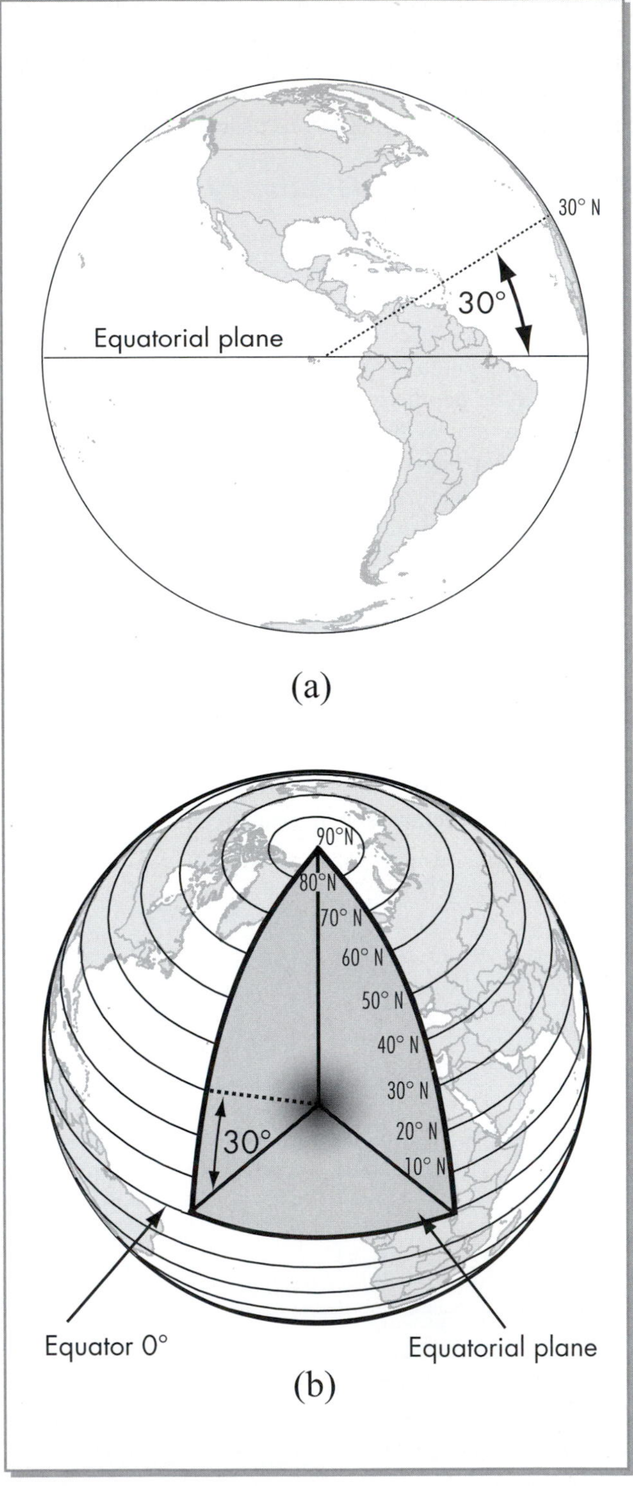

FIGURE 2.9 DETERMINATION OF LATITUDE ON THE SPHERICAL EARTH.
Two dimensional (a) and three-dimensional (b) illustrations of latitude.

parallel to the equatorial plane, it will intersect the Earth's surface, forming a **small circle,** or a **parallel of latitude** (see Figure 2.9). There are, of course, an infinite number of these parallels, and every place on the Earth can have a parallel.

Latitude is designated in angular degrees, from 0° at the equator to 90° at the poles. It is customary to label with a capital N or S the position north or south of the equator. Angular degrees are subdivided into minutes (60 minutes per degree) and seconds (60 seconds per minute) to provide a more precise latitudinal position. Thus common latitude designations would be 82° N, 16° 30' S, 47° 15' 47" N, and so on. The surface distance for each degree of latitude is about 69.2 miles (111.3 km).

Longitude Determination

Longitude on the Earth's surface has always been more difficult to determine than latitude. It baffled early astronomers and sailors, not so much for its concept, but for the instrumentation required to quantify it. Because the Earth rotates on its axis, there is no fixed point at which to begin counting position. Navigators, cartographers, and others from the fourteenth to the seventeenth centuries knew that in practice they would need a fixed reference point. They also knew that the Earth rotated on its axis approximately every 24 hours. Any point on the Earth would thus move through 360 angular degrees in a day's time, or 15 degrees in each hour. If a navigator could keep a record of the time at some agreed-upon fixed point and determine the difference in time between the local time and the point of reference, this could be converted into angular degrees and hence position.

The concept was simple enough, but the technology of measuring time was slow in coming. In 1714, the British Parliament and its newly formed Board of Longitude announced a competition to build and test such an accurate timepiece. John Harrison's famous **marine chronometer** was finally accepted in 1773, after a series of delays and changes in the Board's requirements (Campbell 2001). It was accurate to within 1.25 nautical miles (one nautical mile is 6,076.12 feet, as opposed to the land or statute mile, which is 5,280 feet). The puzzle of longitude was solved (Brown 1959; Sobel and Andrewes 1998).

At first, each country specified some place within its boundaries as the fixed reference point for calculating longitude (historically called "reckoning"). By international agreement, at the International Meridian Conference in 1884, the line passing through the British Royal Observatory at Greenwich, England, called the **prime meridian,** was designated as the origin for longitude.

If an imaginary plane is passed through the Earth so that it intersects the axis of rotation in a line, it will intersect the surface of the Earth as a complete circle. One half of this circle, from pole to pole, is called a **meridian of longitude** (see Figure 2.10). The meridian passing through Greenwich is referred to as the *prime meridian* and has the angular designation 0°. Longitude position is designated as 0° to 180° east or west of the prime meridian for a total of 360°. The 180° meridian provides the basis for the International Date Line, which is adjusted in places to keep countries entirely on one side or the other of the dateline, such as with the United States (Alaska) and Russia.

Unlike the surface distances between lines of latitude, lines of longitude are unequally spaced. At the equator, each

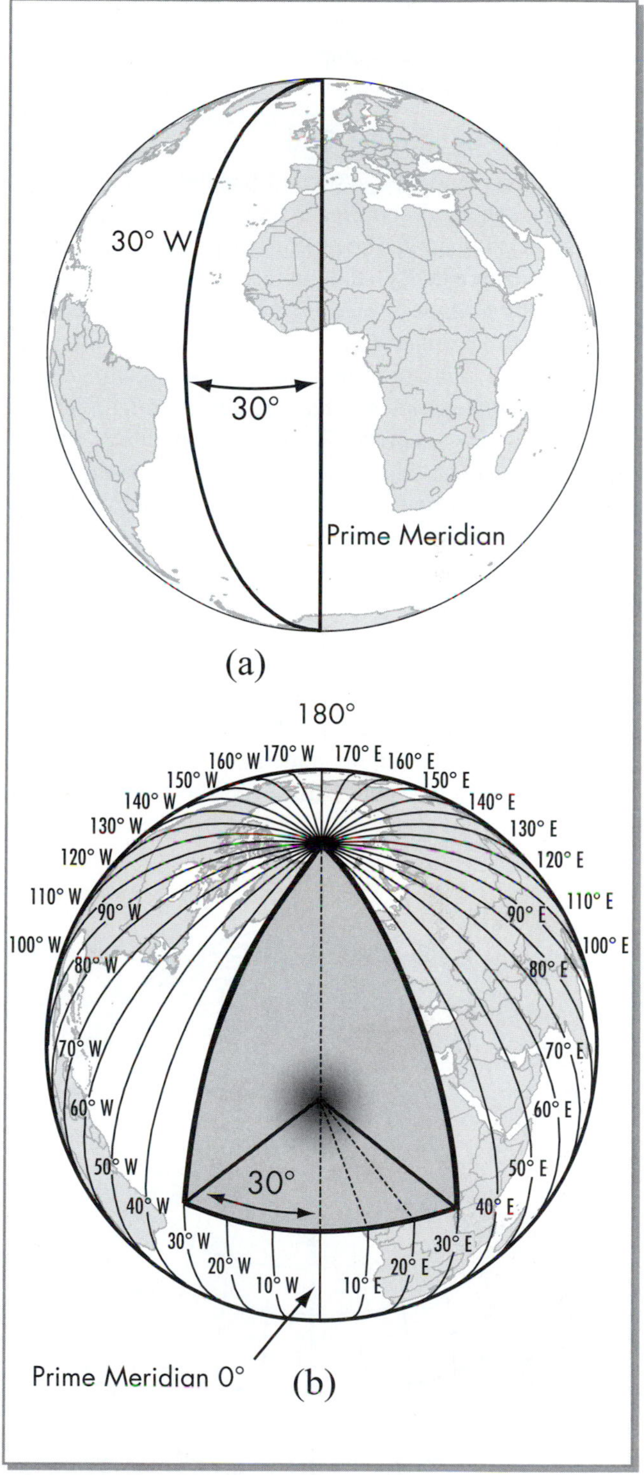

FIGURE 2.10 DETERMINATION OF LONGITUDE ON THE SPHERICAL EARTH.

Two dimensional (a) and three-dimensional (b) illustrations of longitude.

degree is 69.2 miles (111.3 km) apart. That distance narrows going away from the equator, until all the meridians converge at the poles. This phenomena is called the **convergence of meridians** (or meridional convergence).

Until the advent of radio after World War I and radar after World War II, navigation and calculating one's position on the Earth were accomplished by "shooting the stars" (celestial navigation which also includes latitude determination), by marine chronometer, compass, and sextant. Today, GPS has become foundational for most navigation.

The Complete Geographic Grid

With the aforementioned system of parallels and meridians, the Earth's spherical coordinate geometry is complete (see Figure 2.11). Its similarities to the plane Cartesian system should be apparent. By studying Figure 2.11 it should be clear that (1) *latitude is measured by counting the angular degrees north or south of the equator along a meridian* and (2) *longitude is measured by counting the angular degrees east or west of the prime meridian along a parallel.* Thus, the geographic grid is simply a spherical version of the Cartesian coordinate systems, where the equator serves as the x-axis, and the prime meridian serves the y-axis. The parallel to the planar coordinate system can be more clearly seen in the increasingly common use of decimal degrees instead of sexagesimal units to report latitude and longitude coordinates.

Decimal Degrees. The sexagesimal system described above is very common for teaching the basic principles of the geographic grid. It is also common in historic documents, books, and popular literature, and is sometimes a default setting for consumer-grade GPS units. However, the use of **decimal degrees** is becoming the standard in distributed data sets in which the coordinates of country borders, cities, features, and such are reported as latitude and longitude values. Decimal degrees are also used in describing certain aspects of map projections. Thus it is important to also understand the decimal degree format.

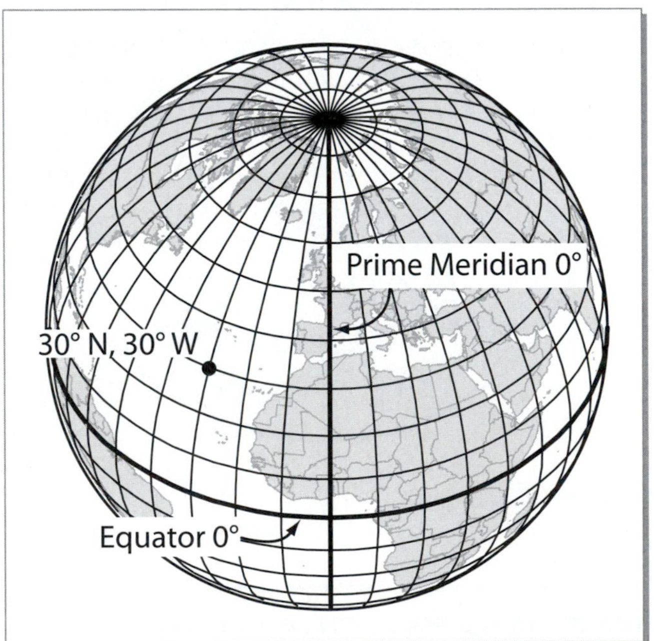

FIGURE 2.11 THE COMPLETE GEOGRAPHIC GRID.

Familiarity with the spherical geographic grid and the characteristics of the arrangement of meridians and parallels is important in estimating graticule distortion on the flat map. The student is encouraged to inspect a globe for this purpose. These important properties should be noted:

1. Scale is the same everywhere on the globe; all great circles have equal lengths; all meridians are of equal length and equal to the equator; the poles are points.
2. Meridians are spaced evenly on parallels; meridians converge toward the poles and diverge toward the equator.
3. Parallels are parallel and are spaced equally on the meridians.
4. Meridians and parallels intersect at right angles.
5. Quadrilaterals that are formed between any two parallels and that have equal longitudinal extent have equal areas.
6. The areas of quadrilaterals between any two meridians and between similarly spaced parallels decrease poleward and increase equatorward.

With decimal degrees, the degree measure stays the same, but the minutes and seconds are converted to a decimal format. For example, the point 45° 33' 37" N, 94° 9' 44" W in DMS format is represented as −94.162222, 45.560278 in decimal degrees. Besides the decimals, you may notice two things that happened in the conversion. First, longitude is reported before latitude, since it is following the eastings and northings convention described earlier. Second, there is also a negative sign in front of the 94. Decimal degrees use signed hemispheres instead of a letter designating N, S, E, or W, just like in the signed Cartesian coordinate quadrants described earlier and as shown in Figure 2.7. Figure 2.12 illustrates the sign that each hemisphere takes on in this system: positive numbers for the northern and eastern hemisphere, negative numbers for the southern and western hemisphere.

Principal Geometric Relationships of the Earth's Geographic Grid

The appropriateness of the final map depends in large measure on how well the cartographer knows the relationships of the elements of the geographic grid. This knowledge can be especially helpful when selecting a *map projection* (defined previously; was first used in Chapter 1, and is the focus of Chapter 3) because: (1) the projection process distorts and deforms the graticule in all sorts of ways—knowing how it really appears can help you assess the distortions in shape, area, distances, and direction that will occur, and (2) the projection process uses geographic grid positions and lines

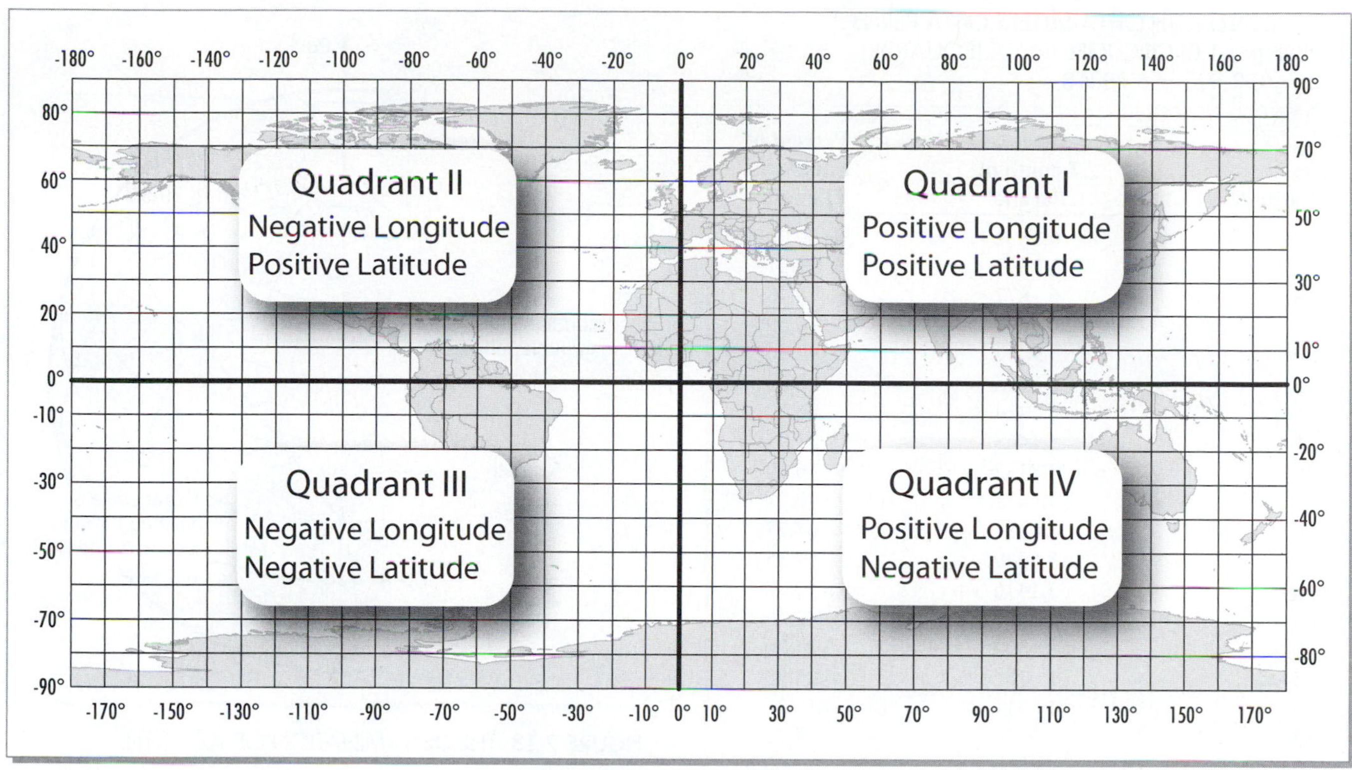

FIGURE 2.12 SIGNED HEMISPHERES FOR DECIMAL DEGREES.

As with Figure 2.7, the four hemispheres relate to the four Cartesian quadrants. Decimal degrees will be positive in the northern and eastern hemispheres, and negative in the southern and western hemispheres.

to describe some of the projection's qualities. This discussion is intended to provide a more in-depth look at the mathematical relationships of geographic grid.

Linear. Lengths of lines of the spherical grid have fixed relations to each other. The most important length is the magnitude of the radius (r); from this and easily learned formulas, most other line lengths can be calculated. For example, the diameter is $2r$. For the perfect sphere, the polar radius is identical to the equatorial radius. There is only one circumference, and it is equal to $2\pi r$ ($\pi = 3.1416$). On the real Earth, of course, the equatorial circumference does not equal the meridional circumferences because of flattening at the poles.

Meridional lengths are the easiest to handle. On the perfect sphere, a meridian is one-half the circumference of the globe. Normally, we want to deal only with the length of the degree along a meridian. On the perfect sphere, this length is simply the circumference divided by 360; every degree is equal to every other degree. On the real Earth, however, polar flattening causes the radius of the arc of the meridian to change, fitting it to an ellipse. Consequently, the length of the degree along the meridian is not constant. The meridian is equal to one-half the circumference—or half the length of the equator on a sphere. Knowing this is critical to the evaluation of projection properties.

Parallels of latitude also have fixed linear relationships. No parallel in one hemisphere is equal in length to any other in the same hemisphere on a perfect sphere. Parallels decrease in length at high latitudes. This relationship has a very definite mathematical expression, namely,

$$\text{length of parallel at latitude } \lambda = (\text{cosine of } \lambda) \cdot (\text{length of the equator})$$

The length of the degree of the parallel is determined by dividing its whole extent by 360. The cosine of 60° is 0.5. Thus the length of the degree along the 60° parallel is but one-half that at the equator (see Table 2.2). This information can be used intelligently in assessing map projections and consequent distortions. The lengths of the degree along the parallels on the "Earth" as defined by the GRS80 ellipsoid are listed on the text's website.

Angular. Inspection of a globe printed with a geographic grid will reveal important angular characteristics of the grid's elements. Most notably, it is easy to see that meridians converge poleward and diverge equatorward. Parallels, by definition, are parallel. What is more subtle is that meridians and parallels intersect at right angles. This is an important characteristic of the grid which can help in the evaluation of projection properties.

One line of note is a **loxodrome,** which has a constant compass bearing. The equator, all meridians, and all parallels are loxodromes. Other loxodromes, called oblique loxodromes, are special, and they too maintain constant compass

TABLE 2.2 LENGTH OF PARALLELS ON A PERFECT SPHERE ($r = 1.0$) CIRCUMFERENCE (EQUATOR) $= 2\pi r$ (OR πd) = 6.28318

Latitude (°)	Length of Parallel	Percent of Equatorial Length
0 (equator)	6.2831	100.00
5	6.2592	99.61
10	6.1877	98.48
15	6.0690	96.59
20	5.9042	93.96
25	5.6945	90.63
30	5.4414	86.60
35	5.1468	81.91
40	4.8132	76.60
45	4.4428	70.71
50	4.0387	64.27
55	3.6039	57.35
60	3.1416	50.00
65	2.6554	42.26
70	2.1490	34.20
75	1.6262	25.88
80	1.0911	17.36
85	.5476	8.71
90	.0000	.00

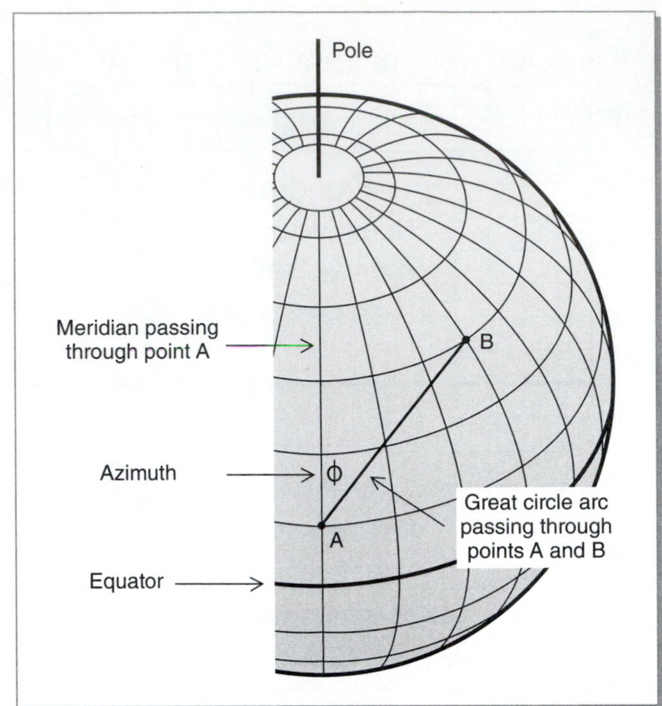

FIGURE 2.13 THE DETERMINATION OF AZIMUTH.

bearings, because they intersect all meridians at equal angles. Because meridians converge, oblique loxodromes tend to spiral toward the pole, theoretically never reaching it. Throughout history the loxodrome has always been important in sailing; mariners often wish to maintain the same heading throughout much of a journey. Unfortunately, loxodromes do not follow the course of the shortest distance between points on the Earth, which is a **great circle arc.** Navigators would usually approximate a great circle arc by subdividing it into loxodrome segments in order to reduce the number of heading changes during travel. Great circle arcs are followed very closely today, especially in airplane navigation. For example, the route travelled between New York City and London takes the traveler north near Greenland as a part of the great circle route.

Azimuth. Azimuth has a very specific definition in cartography (see Figure 2.13). Azimuth is always defined in reference to two points, for example A and B. *The azimuth from A to B is the angle made between the meridian passing through A and the great circle arc passing from A to B.* It is customary to specify azimuth as an angle counting clockwise from geographic north through 360 degrees. Most azimuths are from geographic north, although possible from the south. The angular measure is followed by either N or S, for example 49° N.

Area. The areas of quadrilaterals found between bounding meridians and parallels are important in understanding the areal aspects of the Earth's spherical coordinate system (see Figure 2.14). Between two bounding meridians, for identical latitudinal extents, the quadrilateral areas decrease poleward. Any misrepresentation of this feature during projection will have a profound effect on the appearance of the final land/water areas of the map. The relationship of these areas on a perfect sphere, based on change of latitude, is represented in

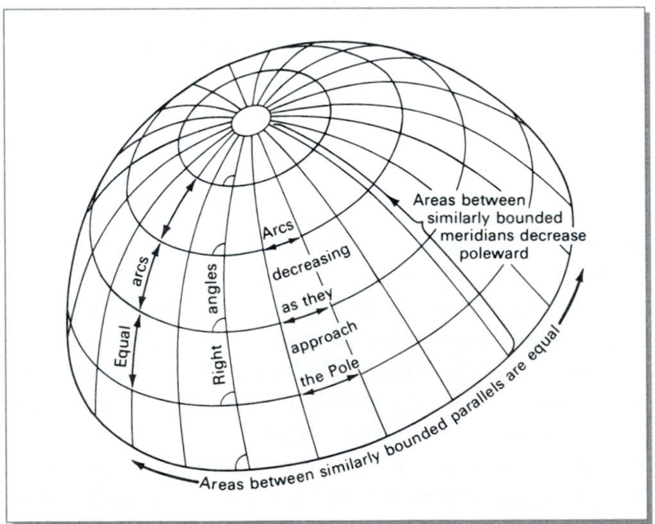

FIGURE 2.14 THE PRINCIPAL GEOMETRICAL RELATIONSHIPS OF THE EARTH'S COORDINATE SYSTEM.

TABLE 2.3 AREAL RELATIONSHIPS OF QUADRILATERALS USING THE GEOGRAPHIC GRID ON A PERFECT SPHERE ($r = 1.0$) 5° LONGITUDINAL EXTENT

Latitude (°)	Area	Percent of Lowest Quadrilateral
0–5	.007605	100.00
5–10	.007547	99.23
10–15	.007432	97.72
15–20	.007260	95.46
20–25	.007033	92.47
25–30	.006752	88.78
30–35	.006420	84.41
35–40	.006039	79.41
40–45	.005612	73.79
45–50	.005143	67.62
50–55	.004634	60.93
55–60	.004090	53.78
60–65	.003515	46.21
65–70	.002913	38.30
70–75	.002289	30.09
75–80	.001647	21.66
80–85	.000993	13.06
85–90	.000332	4.36

Table 2.3. The decrease in area relative to the lowest quadrilateral is more dramatic at the higher latitudes because of rapid meridional convergence. At lower latitudes, meridians converge slowly; as a result, the change in area is less marked.

Points. Perfect spheres are considered to be *allside surfaces,* on which there are no differences from point to point. Every point is like every other point, with the surface falling away from each point in a similar manner everywhere. In dealing with spheres and with the projection of the spherical grid onto a plane surface, we may think of points as having dimensional qualities. That is, they are commensurate figures, though infinitesimally small.

Circles on the Grid. Two special circles appear on the spherical grid. A *great circle* is formed by passing a plane through the center of the sphere. This plane forms a perfect circle where it intersects with the sphere's surface. Certain qualities about the circle are worth knowing: (1) The plane forming the great circle bisects the spherical surface. (2) Great circles always bisect other great circles. (3) An arc segment of a great circle is the shortest distance between two points on the spherical surface. Some of the elements of the geographic grid are great circles. All meridians are great circles; the equator is also a great circle.

Circles on the grid that are not great circles are called *small circles.* Parallels of latitude are small circles, except for the equator. On the Earth, to travel along a meridian (N–S) is to go the shortest distance. Traveling along a parallel (E–W) is *not* the shortest distance! Following the path of the equator, however, is an efficient way of travel.

SCALE REVISITED

In Chapter 1, we introduced the concept of map scale. These additional comments will also prove useful to a further understanding of scale. In cartography, scale is represented by a ratio of map distance to Earth distance:

$$\text{map scale} = \frac{\text{map distance}}{\text{Earth distance}}$$

It is generally expressed as a representative fraction (RF), and will always contain unity in the numerator. The denominator is referred to the RFD, or representative fraction denominator. The RF scale can be expressed as a traditional fraction (such as 1/25,000), but is most commonly written as a ratio using a colon, such as 1:25,000. RF scales are to be read, "One unit on the map *represents* so many of the *same* units on the Earth." The number in the fraction may be in any units, but both numerator and denominator will be in the *same* units. The cartographer should never say, "One unit on the map *equals* so many of the same units on the Earth." This is incorrect and logically inconsistent.

There are three customary ways of expressing scale on a map: the representative fraction, a graphic or bar scale, and a verbal scale. The RF is perhaps the most important expression of scale for three reasons:

1. It is unit independent: the other methods are designed for expression of scale in specific units. It is mathematically the most precise expression of map to real Earth distances.
2. Most GIS and mapping software allow you to specify a scale or scale range in which map data is to be displayed, which is expressed as an RF. This expression of scale is what cartographers use when interacting with the spatial data via GIS and mapping software.
3. Most distributed maps and map data have suggested scale ranges that are appropriate for display. Sometimes the scale at which the data was originally created is also provided. Provided scales or scale ranges are always in the RF form.

On many maps, a *graphic* (linear) bar scale is included. This bar is usually divided into equally spaced segments and labeled with familiar linear units, such as miles, kilometers, meters, or feet, depending on the scale of the map. This scale is read the same as an RF scale. This form of scale is very useful for the following reasons.

1. It has a fairly high communicative value when compared to the representative fraction. The average map reader probably does not understand what 1:100,000 means, but a graphic scale in miles or kilometers makes distances clear.

2. If the map is enlarged or reduced, the bar scale changes in correct proportion to the amount of reduction. For example, a virtual map will change its size from monitor to monitor, and also if displayed on a projector (for example, in a Power Point presentation). If a paper map is enlarged or reduced on a photocopier, the bar scale again changes. In each case, the bar scale will change with the map, but an RF will become incorrect.

Another common expression of scale is the *verbal* scale. This is a simple expression on the face of the map stating the linear relationship. For example, "one inch represents five miles" is an example of a verbal scale. This scale form is easily converted to an RF scale between map and Earth distances (for more on converting between scales and working scale problems, see "Appendix A"), but lacks the visual appeal of the graphic scale, and also is incorrect if the map is enlarged or reduced. This scale also locks you into *specific* units and numbers. For these reasons, some cartographers feel that the verbal scale is somewhat inflexible and therefore of less utility when compared with the RF and graphic scale expressions. Nonetheless, all three scale forms are common. Their inclusion in a thematic map, as with other design decisions, should rest on the purpose of the map, along with consideration of the audience who will view the map. If communication of distances is important to the map, then we recommend the use of the *graphic* (linear) bar scale.

Cartographers use the terms small, medium (or intermediate), and large scales quite often and somewhat casually. In Chapter 1 we saw that larger scales ("zooming in" in popular usage) mean that we have a smaller mapped Earth area, but more information detail with less generalized symbolization and vice versa (see Figure 1.14 in Chapter 1). But what constitutes large, medium, and small scales? We see a qualitative illustration of the basic principal in Color Plate 1.1, but is there a quantitative guide? Actually, there are many suggested ranges published from numerous sources as to what exactly defines small, medium, and large scales. For our purposes, we will use the guidelines of:

- Large Scale—1:30,000 or larger
- Intermediate Scale—1:30,000 to 1:300,000
- Small Scale—1:300,000 or smaller

Although these are relative terms, most texts provide ranges that are somewhat similar. But note that with continental, hemisphere, and world maps, which will all be small scale, one is better off saying "larger or smaller" scales relative to an existing map. For example, a world map in a typical printed atlas may be displayed at 1:100,000,000, but a map of the continental United States might be at a larger scale of 1:15,000,000. Likewise, local maps of varying scales may all fall in the category of large scale—1:1,000 scale mapping is a much larger scale than 1:24,000, but both are still considered large scale maps.

Scale and Line Generalization

While it has been noted that a majority of thematic mapping is done at smaller scales, we are seeing an upswing in large and intermediate scale thematic mapping. This is due primarily to the increases in availability of state and local data, including both base map and attribute data, which are available on the Web. To find some of these data, places such as the University of Arkansas Libraries web page of state, local, and national information can be a great place to start (University of Arkansas Libraries 2006). Another site that is illustrative of the situation is the Minnesota Data Deli, where large, medium, and small scale data can be found in abundance (Minnesota Department of Natural Resources 2004).

It is precisely this availability of base map and attribute data, created and distributed at multiple scales, that lead us to re-examine generalization levels (introduced in Chapter 1) when compiling base map data from different sources. As we will see, it usually best to mix and match data from similar scales. In other words, if we create a map that incorporates rivers and roads on a base map at 1:250,000, we are going to want to choose river and road data that was created or designed for 1:250,000 use; not 1:24,000 or 1:15,000,000. For most available spatial data sets, the map will have accompanying **metadata,** or data about data, that will describe the scale of the source data, and at what scale or scale range the data should be displayed.

Cartographers want a detail level that is appropriate for boundaries and other features at a particular scale. In geometric terms, we can say that the vertex density in the line work should be somewhat similar at similar scales. Of course, this will depend on the features being presented. A rectangular county will likely have four vertices for the corners, connected by lines to form the polygon area regardless of scale or generalization level. More complex figures are a different story altogether.

As an example, Figure 2.15a has a vertex density that is appropriate for its scale at 1:9,000,000. But if you zoom into a portion of the same map (see Figure 2.15b) at a larger scale of 1:850,000, the vertex density that worked well at the smaller scale is now too generalized, particularly if it is going to be matched with other data that was generated at 1:850,000. But Figure 2.15c is appropriate, since the data was generated with a higher vertex density that is more appropriate for 1:850,000.

If you are working with relatively large scale maps and data, your sources also have to be large scale and of similar vertex densities. In other words, at 1:850,000 in Figure 2.15b and c, 2.15c is from the better source. You cannot take small scale data (as in Figure 2.15a) and make it appropriate for a large scale map (as is attempted in Figure 2.15b). If you are creating your own data, perhaps by digitizing maps from aerial photographic imagery, the resolution of the imagery must be good enough for you to generate a reasonable vertex density.

If you are working with small scale data, you can either find similar small scale sources, or you can use larger scale data and perform line generalization to remove vertices from

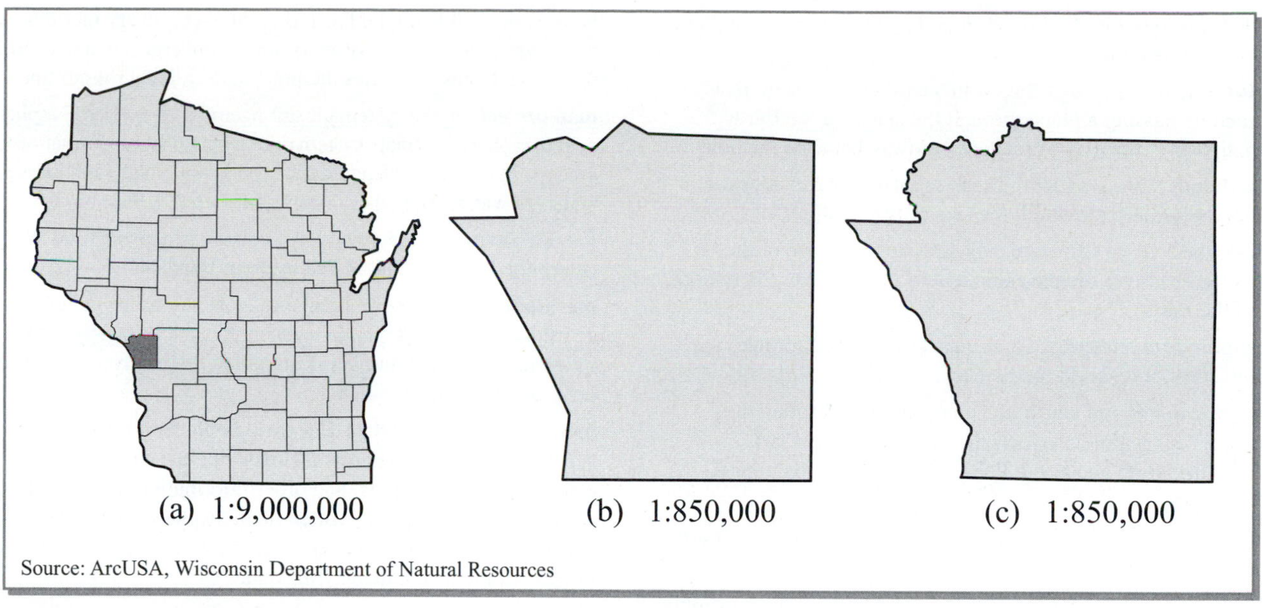

FIGURE 2.15 LINE GENERALIZATION LEVELS FOR DIFFERENT SCALES.
The first map (a) depicts Wisconsin at 1:9,000,000. The second map (b) shows La Crosse County at 1:850,000. Notice how the vertex density level, while appropriate for (a) is over-generalized for the larger scale. The third map (c) of the same area is also at 1:850,000, but has a vertex density that is appropriate for this zoom level. Notice that the real difference between (b) and (c) occurs in the curved lines.

the line work to make it appropriate for the smaller scale. GIS and mapping software usually provide a procedure to remove extraneous vertices from larger scale map data. In other words, the procedure will keep the most important points that best define the line's shape, and eliminates all others (Douglas and Peucker 1973). How generalized the line work should be will depend on the map's purpose, the medium for which the map is intended, and the audience for which the map is intended.

REFERENCES

Brown, L. 1959. *The Story of Maps*. New York: Bonanaza Books.

Campbell, J. 2001. *Map Use and Analysis*. New York: McGraw-Hill.

Dana, P. 2003. *Geodetic Datum Overview. The Geographer's Craft Project.* Department of Geography, The University of Colorado at Boulder. http://www.colorado.edu/geography/gcraft/notes/datum/datum.html.

Defense Mapping Agency, Hydrographic Center, 1977. *American Practical Navigator* 1. Washington, DC: Defense Mapping Agency.

Douglas, D., and T. Peucker. 1973. Algorithms for the Reduction of the Number of Points Required to Represent a Digitized Line or Its Caricature. *Canadian Cartographer* 10:2.

King-Hele, D. 1967. The Shape of the Earth. *Science* 183: 67–76.

———. 1976. The Shape of the Earth. *Science* 192: 1293–1300.

Minnesota Department of Natural Resources, 2004. *The DNR Data Deli*. http://deli.dnr.state.mn.us/

Snyder, J. 1987. *Map Projections—A Working Manual*. U.S. Geological Survey Professional Paper 1395. Washington, DC: USGPO.

Sobel, D., and W. Andrewes. 1998. *The Illustrated Longitude*. New York: Walker and Company.

University of Arkansas Libraries. 2006. *Starting the Hunt: Guide to Mostly On-Line and Mostly Free U.S. Geospatial and Attribute Data*. Second Edition. http://libinfo.uark.edu/gis/us.asp.

Van Sickle, J. 2004. *Basic GIS Coordinates*. Boca Raton, FL: CRC Press.

GLOSSARY

antipodal point point opposite; on the Earth, the North Pole is antipodal to the South Pole; 20° S, 60° E is antipodal to 20° N, 120° W

axis of rotation imaginary line around which the Earth rotates

Cartesian coordinate geometry system of intersecting perpendicular lines in plane space, useful in analytic geometry and the precise specification of location

convergence of meridians spacing of the meridians of longitude becomes less in a poleward direction from the equator

datum a starting or reference point that gives a context to things such as the size and shape of the Earth, and for any coordinate system that can be used for positioning (including elevation)

decimal degrees means of reporting minutes and seconds of latitude and longitude designations as a decimal (see also DMS format)

Descartes French mathematician whose early studies of algebra and geometry led to analytic geometry

DMS format the format for reporting sexagesimal latitude and longitude in degrees, minutes, and seconds (see also decimal degrees)

ellipsoid a geometrical solid developed by the rotation of a plane ellipse about its minor axis

equator imaginary line of the Earth's coordinate system that is formed by passing a plane through the center of the Earth perpendicular to the axis of rotation, midway between the poles

Eratosthenes (276–194 B.C.) Greek scholar living in Alexandria who first accurately measured the size of the Earth

geodesy the science that measures the size and shape of the Earth; often involves the measurement of the external gravitational field of the Earth

geographic grid spherical coordinate system used for the determination of location on the Earth's surface

geographic north and south the imaginary line forming the Earth's axis of rotation intersects the Earth's surface at two locations, the North and South Poles, referred to as geographic north or south

geoid term used to describe the shape of the Earth; means "Earth-shaped" and does not refer to a mathematical model

Global Positioning System (GPS) a system of 24 or more orbiting satellites that transmit timing signals to ground-based receivers which calculate location; used for navigation and mapping

graticule meridians of longitude and parallels of latitude on a map

great circle circles that result when a plane bisects the Earth into two equal halves (for example, the equator)

great circle arc segment of a great circle that is the shortest distance between two points on the spherical surface

latitude position north or south of the Earth's equator; designation is by identifying the parallel passing through the position; determined by the angle subtended at the center of a sphere by a radius drawn to a point on the surface

longitude position east or west of the prime meridian; designation is by identifying the meridian passing through the position; determined by angular degrees subtended at the center of a sphere by a radius drawn to the meridian and the position in question

loxodrome a line on the Earth that intersects every meridian at the same angle; because of meridional convergence, a loxodrome theoretically never reaches the pole; also called a rhumb line

map projection the systematic arrangement of the Earth's spherical or geographic coordinate system onto a plane; a transformation process that projects the geographic grid, along with land masses, bodies of waters, and other features onto a flat surface

marine chronometer relatively accurate timepiece used to determine longitude; perfected by John Harrison

meridian (of longitude) great circle of the Earth's geographic coordinate system formed by passing a plane through the axis of rotation; meridional number designation ranges from 0° to 180° E or W of the prime meridian

metadata data about data. For cartographers, information that accompanies spatial datasets that describe the dataset's spatial extent, attributes, lineage, and other information. In the context of scale and generalization, it often includes the scales of the source materials used to create the data set, as well as recommended scale ranges with which to display the data

parallel (of latitude) small circle of the Earth's geographic coordinate system formed by passing a plane through the Earth parallel to the equator; parallel number designation ranges from 0° at the equator (a great circle) to 90° at the Pole (either North or South)

prime meridian meridian adopted by most countries as the point of origin (0°) for determination of east or west longitude; passes through the British Royal Observatory at Greenwich, England

sexagesimal system of numbering that proceeds in increments of 60; for example, the division of a circle into 360 degrees, a degree into 60 minutes, and a minute into 60 seconds. Sexagesimal is also called DMS format

small circles any circles on the spherical surface that are not great circles; for example, parallels (except for the equator) are small circles

3 MAP PROJECTIONS

CHAPTER PREVIEW One of the most important aspects of thematic mapping is the map projection. The 3-D Earth is transformed to a 2-D flat surface by use of a geometric form, such as a plane, cylinder, or cone, or it may be derived mathematically. Important projection parameters such as standard points and lines, central meridian, latitude of origin, and projection aspect are created in the process. There are nearly an infinite number of possible projections that can be produced, but they generally fall into one of four projection families: azimuthal (planar), cylindrical, conic, and mathematical. Each family has their own characteristic grid and potential applications. The projection process involves map distortion. There are also four projection properties that can be maintained (but not at the same time) that help define what kind of distortion takes place. Projection distortion can be measured qualitatively by visual inspection and knowledge of a particular projection's parameters and distortion patterns, or it can be quantitatively measured by Tissot's indicatrix. Specific projections can be selected by asking questions about the map purpose, allowable distortion, and area of the world that is being mapped. Although there are thousands of named projections currently supported by most GIS and mapping software, knowledge of a fairly small number of projections and projected coordinate systems can fulfill a variety of mapping needs. When a projection does not exactly meet the requirements for a specific study area, the projection's parameters can often be adjusted to produce a more accurate and aesthetic map. ■

Chapter 2 introduced the concept of a datum, geoid, and reference ellipsoid, examined Earth's spherical grid, and revisited the concept of map scale. This chapter addresses the important related topic of map projections. The concept of a map projection originates from the idea that we have to *project* the imaginary lines of latitude and longitude from the Earth's surface onto a flat surface (for example, a piece of paper) as a grid. Using a map projection is the only practical way to portray the Earth's curvature on a flat surface.

Of course, the Earth can be portrayed using a globe. A globe provides a more accurate picture than a map of the Earth's surface in two important regards. First, on a globe the lines of latitude and longitude are positioned correctly. Parallels of latitude run parallel to each other; meridians of longitude converge at the poles. It can be said that there is no *deformation* of the graticule on the globe. Second, the relative size and shape of all the continents, oceans, and other area features are true. In addition, distances and directions between points are correct. In other words, the *properties* of area, shape, distance, and direction are true on a globe.

In the map projection process, however, distortion occurs. Distortion can refer to both the deformation of the graticule and to the loss of one or more of the properties (area, shape, distance, direction). The first part of this chapter, "The Map Projection Process," deals with the process of projecting the graticule and the Earth's features (landmasses, countries, bodies of water, and so on) to a geometric form, such as a plane, cylinder, or cone, or by other mathematical means. As we will see, four basic projection families result from this process, each with their own distinctive characteristics and distortion patterns.

The second part of the chapter, "Employment of Map Projections," focuses on matching the map projection to a particular application. There are hundreds of map projections within the four basic families, many of which are available in GIS and mapping software. Each one has its advantages and disadvantages. For example, a map projection that may be useful for the entire world may not be the correct projection for a continent, country, or a state. In order to select the optimum projection, it is important to understand not only the basic principles of the *projection properties* but also the *projection parameters*, which define how the projection is centered and the latitude or longitude values where distortion is minimized or eliminated. We will now take a look at the projection process to gain a better understanding of some of these important map projection principles and concepts.

THE MAP PROJECTION PROCESS

In thematic mapping, as in all cartography, we consider the production of a map a process of representing the Earth (or a part of it) as a *reduced model of reality.* Map projection is the transformation of the spheroid or elliptical surface to a plane surface. In most GIS and mapping software the cartographer simply selects the map projection that best suits the needs of the mapping project at hand. Technically, however, there is a three-step process that occurs. First, a spheroid or ellipsoid model is selected that best fits the geoid. For most distributed digital data, the incorporation of the ellipsoid or spheroid is already present, often as part of its datum (see Chapter 2 for a discussion of geoids, ellipsoids, and datums). Then there is the transformation of Earth coordinates to plane coordinates, often as eastings and northings, and the reduction in scale from Earth to map size. These last two steps are done mathematically by the GIS or mapping software.

Before computers and GIS were used in the projection process, when map projections were done manually, cartographers sometimes reduced the scale of the ellipsoid or spheroid model to a **reference globe** (also called a *nominal* or *generating globe*), from which the map projection was generated. As we will see, the reference globe is still a useful concept for illustrating basic projection principles.

Any map projection is the systematic arrangement of the Earth's or reference globe's meridians and parallels onto a plane surface. In addition to the graticule, the map projection process includes the transformation of other map features. For example, coastlines and boundaries are also transformed and, as a result, such map objects as landmasses and bodies of water may be distorted.

Developable Surfaces

The process of transforming and transferring the graticule and its features from a three-dimensional object onto a two-dimensional flat surface is the essence of a map projection. However, neither the Earth nor any of its three-dimensional representations (such as the geoid, ellipsoid, sphere, or globe) are **developable surfaces.** The Earth cannot be pulled or cut apart to lie flat the way a map does. Three geometric forms that have developable surfaces are the plane, the cylinder, and the cone (see Figure 3.1),

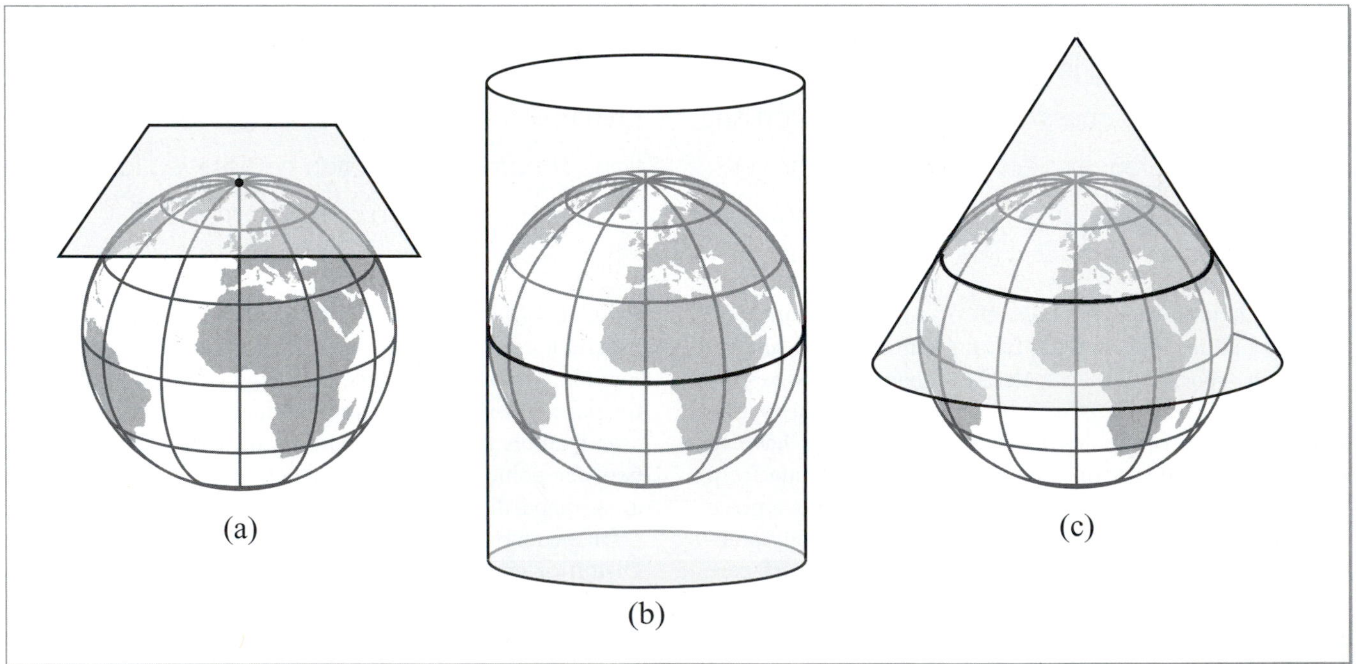

FIGURE 3.1 DEVELOPABLE PROJECTION SURFACES.
The plane (a), cylinder (b), and cone (c) are all geometrically developable surfaces. See the text for an explanation.

although many projections are simply mathematically derived graticules. These give rise to four overall families of map projections: azimuthal, cylindrical, conic, and mathematical. Each family has many individually named projections, where the name often contains one or more of the following: the name of the geometric figure, some of the projection's properties (discussed in the following sections), and the name of the individual identified as the originator of the projection.

Examining the three developable geometric surfaces is particularly useful in illustrating the concept of a map projection. As shown in Figure 3.1, each geometric surface is placed to a point (with the plane) or line of contact (with the cylinder and the cone). It may be helpful to conceptualize a light bulb at the center of the transparent reference globe in Figure 3.1, shining through the globe's surface, projecting the graticule and other features onto the new developable surface. The plane is already in its flat form, but the cone and the cylinder may be "taken off" from the 3-D surface and spread out without tearing, shearing, or distortion of the geometric surface (hence the surface is considered "developable"). However, as we shall see, the resulting maps will be far from distortion free.

The geometric surfaces' contact with a point or line, as in Figure 3.1, is called tangency, or the **tangent case** (or sometimes *simple form*). In the **secant case** (Figure 3.2), each of the surfaces is placed at a line (with the plane) or two lines (with the cylinder and the cone) of *intersection*. In the secant case, part of the geometric surface intersects and actually goes "into" the reference globe. The light bulb analogy breaks down a bit at this point, as the graticule and the features are projected up to or down onto the surface depending on the portion of the surface that is being examined.

Projection Parameters

Points and lines of tangency or intersection are called **standard** *points* and *lines*. If a standard line is also a parallel of latitude then the line is called a *standard parallel,* and if the standard line falls along a meridian of longitude then it is called a *standard meridian. Standard* points, lines, parallels, and meridians are one of the most important map **projection parameters,** because those corresponding places on the map will have no scale distortion. That is, the scale of the map along these lines will have the same scale as the reference globe. The farther away from the standard point or line(s), the greater the distortion (or deformation) that occurs. In Figure 3.1, the plane has the least distortion at the North Pole (its standard point); the cylinder has the least distortion at the equator (its standard parallel), and the cone has the least distortion at 30° N. The secant case helps minimize distortion over a larger area by providing additional control (see Figure 3.3).

The positioning of the plane, cylinder, and cone shown in Figures 3.1 and 3.2 are common positions, and are sometimes called their "normal" **projection aspect** (Hilliard *et al.* 1978). The projection aspect is the position of the projected graticule relative to the ordinary position of the geographic grid on the Earth, and can be visualized as the position of the developable geometric surface to the reference globe. In Figures 3.1 and 3.2 the axis of the cylinder and the cone run

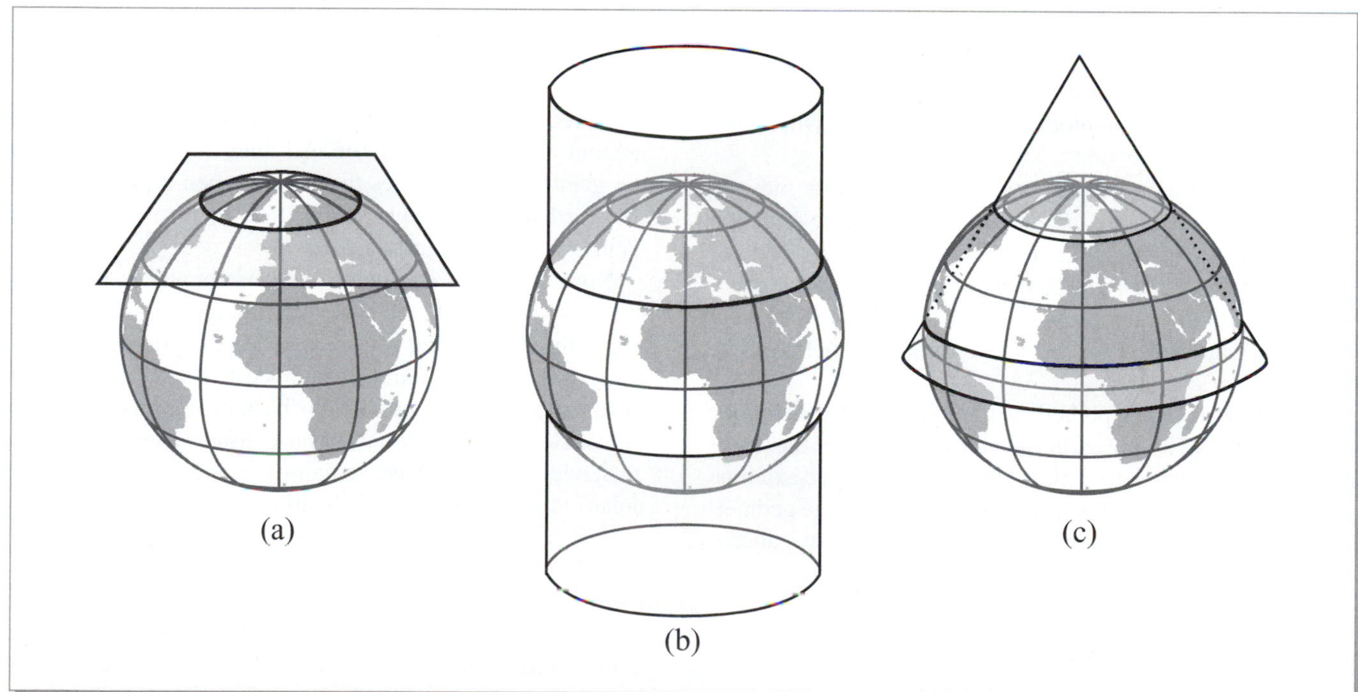

FIGURE 3.2 DEVELOPABLE PROJECTION SURFACES, SECANT CASE.
The plane (a), cylinder (b), and cone (c) intersect the Earth or reference globe. See the text for an explanation.

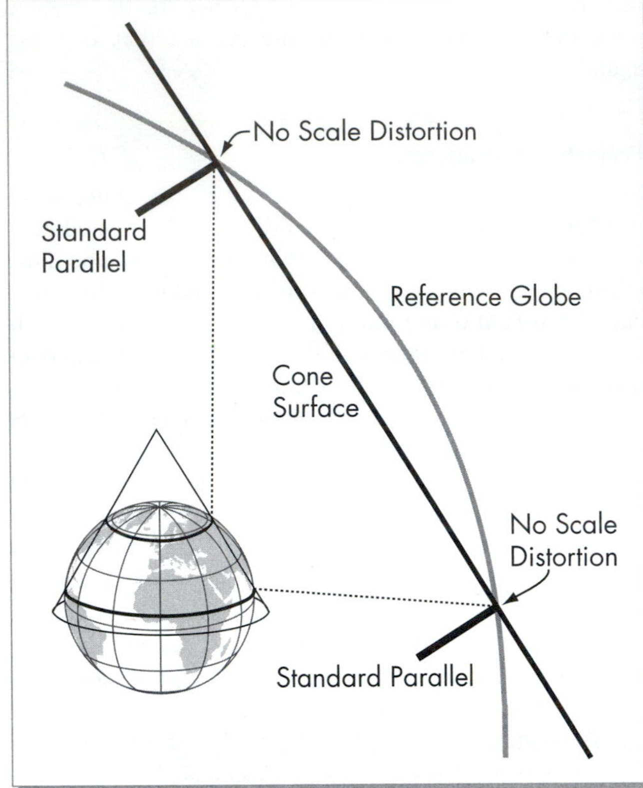

FIGURE 3.3 THE SECANT SURFACE.
The amount of scale distortion varies with the distance from the standard lines. In this example, those lines are standard parallels.

through the North and South Poles, and the plane is parallel to the equator. The *normal aspect* is the position that produces the simplest graticule. However, there is no rule that says these figures have to be placed in these particular positions. In particular, with azimuthal and cylindrical projections it is quite common to have other placement positions. Examples of other common projection aspects will be described in the next section.

Two other important *projection parameters* that indicate the projection aspect are the **central meridian** (the meridian that defines the center of the projection) and the **latitude of origin** (latitude of the projection's origin, such as the equator). Most GIS or mapping software, in addition to the selection of a map projection, allow for these parameters to be adjusted as necessary.

In addition, the hypothetical light source (for example, the "light bulb" in our earlier analogy) for any of these figures can be changed as well. In the next section we examine some of the most common projection aspects of the geometric forms and light sources, as well as the typical patterns and distribution of distortion for each of the families.

Projection Families

Selecting an appropriate projection involves a number of criteria. An understanding of the different projection families, along with typical **patterns of deformation** (the distribution of distortion over a projection) is one important aspect of this selection process. Although there is great diversity among projections, even within the projection families mentioned here, similarities in construction and appearance yield enough common elements to classify them into a few groups. The approach presented here is a conventional one.

Azimuthal Family

In the *azimuthal* or *planar* family, the spherical grid is projected onto a plane. This plane can be tangential to the sphere at a standard point (tangent or simple case; see Figure 3.1a), or pass through the sphere, making it intersect along its standard line, which is actually a **small circle** (secant case, see Figure 3.2a). A small circle is the circle that results when a plane intersects the Earth but does not go through its center, effectively dividing the Earth in two unequal parts. Patterns of deformation begin to emerge for the azimuthal class (see Figure 3.4a). Deformation increases with distance from either the standard point (tangent case) or the standard line (secant case). As with all projection families, scale distortion, and hence deformation, is nonexistent at standard points or lines. In the example shown in Figure 3.4a, the deformation increases outward in a series of concentric bands. In the secant case, however, there is also some distortion that occurs toward the center.

The plane may be tangent, of course, at any point on the spherical grid, depending on the projection aspect. Tangency at the pole is a *polar aspect;* at mid-latitude, an *oblique aspect;* and at the equator, an *equatorial aspect* (see Figure 3.5). Normal aspect for this family is the polar position when the plane is tangent at one of the poles, since it produces the simplest graticule. In this case, the meridians are straight lines intersecting the pole, and parallels are concentric circles having the pole as their centers (see Figure 3.4a). Directions to any point from the point of tangency (pole) are held true. All lines drawn to the center are **great circles** (circles that result when a plane bisects the Earth into two equal halves), as is also the case for equatorial and oblique aspects.

Normally, only one hemisphere is shown at a time on these projections. There are a few exceptions, however, in which the entire world is portrayed, but the grid departs radically from what we are accustomed to seeing. Azimuthal (also called *zenithal*) projections became quite popular during World War II, when there was considerable circumpolar air navigation; they have remained so today especially for polar mapping and other applications.

Light Source Variations. We mentioned earlier that the hypothetical light source can be from positions other than the center of the globe. Adjustments in light sources are most common in azimuthal projections, although they can occur in other families as well. Figure 3.6 depicts three primary positions for the light source. If the light is emanating from

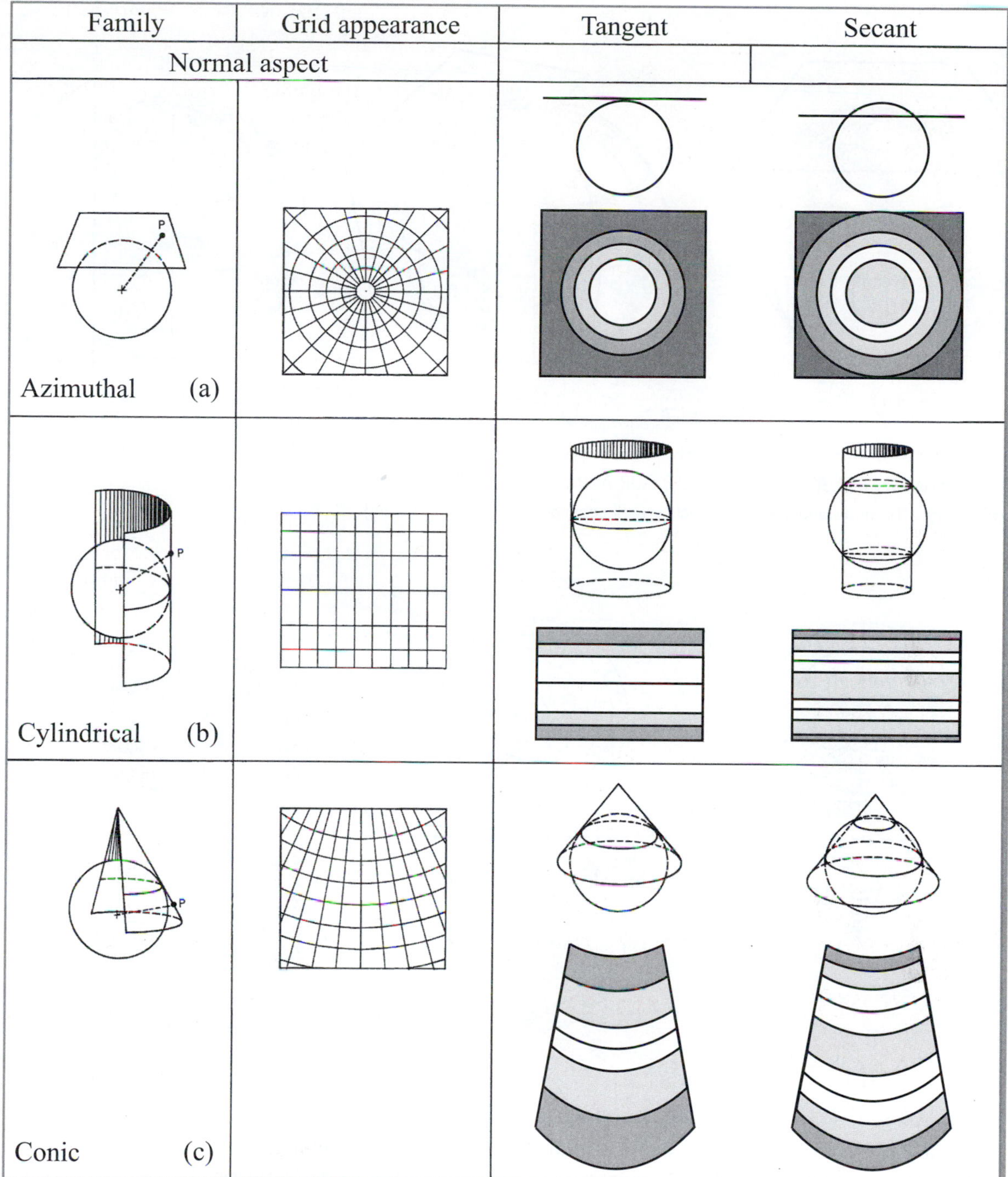

FIGURE 3.4 PROJECTION FAMILIES AND PATTERNS OF DEFORMATION.
The patterns are clearly related to the way the projection is devised for the azimuthal (a), cylindrical (b), and conic (c) projections. Darker areas represent greater deformation.

the center of the globe, it is a **gnomonic** projection. If the light source is at the point opposite the point of tangency (or antipode position), the projection is **stereographic;** if it is at a theoretical infinity (outside the generating globe, producing parallel light rays), an **orthographic** projection results. An examination of the graticule and continental features reveals the different deformation patterns that result due to changes in the light source positioning.

Cylindrical Family

Cylindrical or rectangular projections are common forms. They are sometimes seen in atlases and other maps

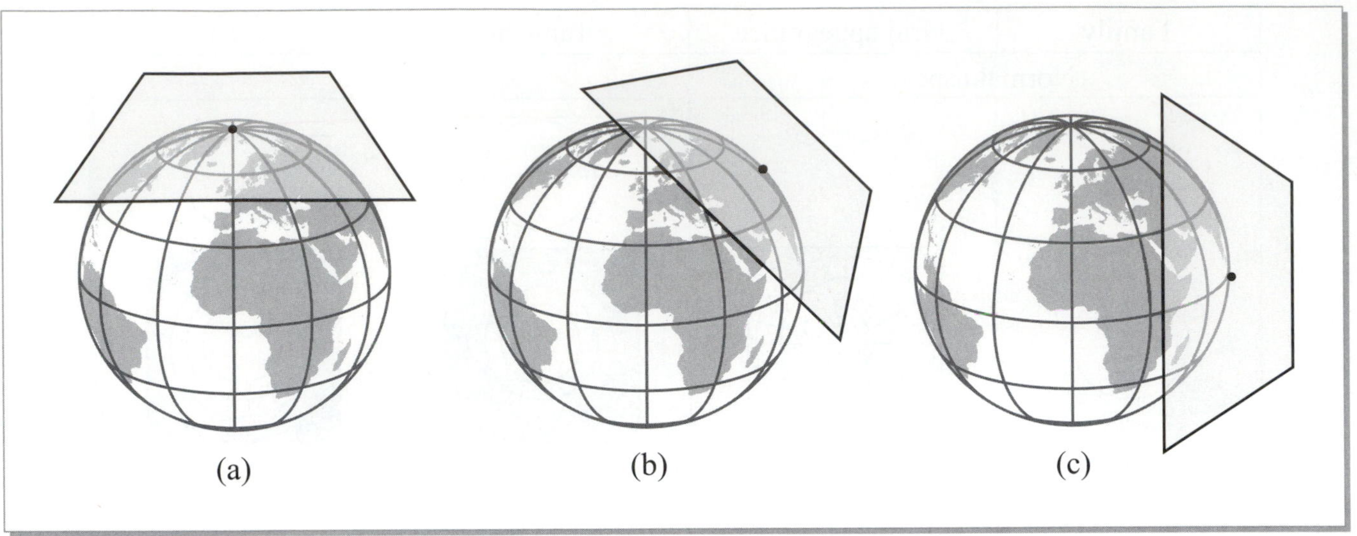

FIGURE 3.5 COMMON PROJECTION ASPECTS FOR AZIMUTHAL MAPPING.
The polar (a), oblique (b), and equatorial (c) aspects are all common in azimuthal mapping.

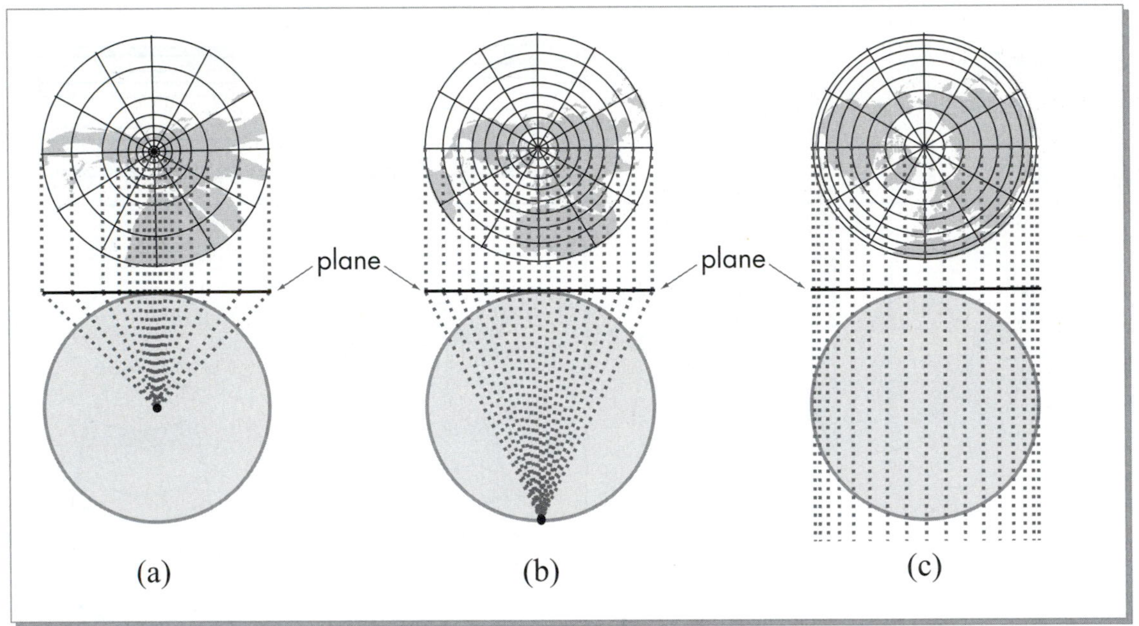

FIGURE 3.6 SOURCE OF ILLUMINATION VARIATIONS OF AZIMUTHAL MAPPING.
The light source and resulting graticule for the gnomonic (a), stereographic, (b), and orthographic (c) projections.

portraying the whole world, but are perhaps more typically used in medium- and large-scale mapping. The applications of such projections will be discussed in more detail later. They are developed (graphically or mathematically) by wrapping a flat plane or sheet into a cylinder and making it tangent along a line or intersecting two lines (secant case) on the sphere. Points on the spherical grid can be transferred to this cylinder, which is then "unrolled" into a flat map. The result is a rectangle-shaped map with parallels and meridians that intersect at 90° angles (see Figure 3.4b).

The normal aspect for these projections is the equatorial aspect, as shown in Figure 3.1b. In the tangent case, the equator is the standard parallel. Here too the standard line is also a great circle. In the secant case, the two standard parallels (*small circles*) will be located north and south of the equator (see also Figure 3.2b). The patterns of deformation in this case are not surprising; areas of least distortion are

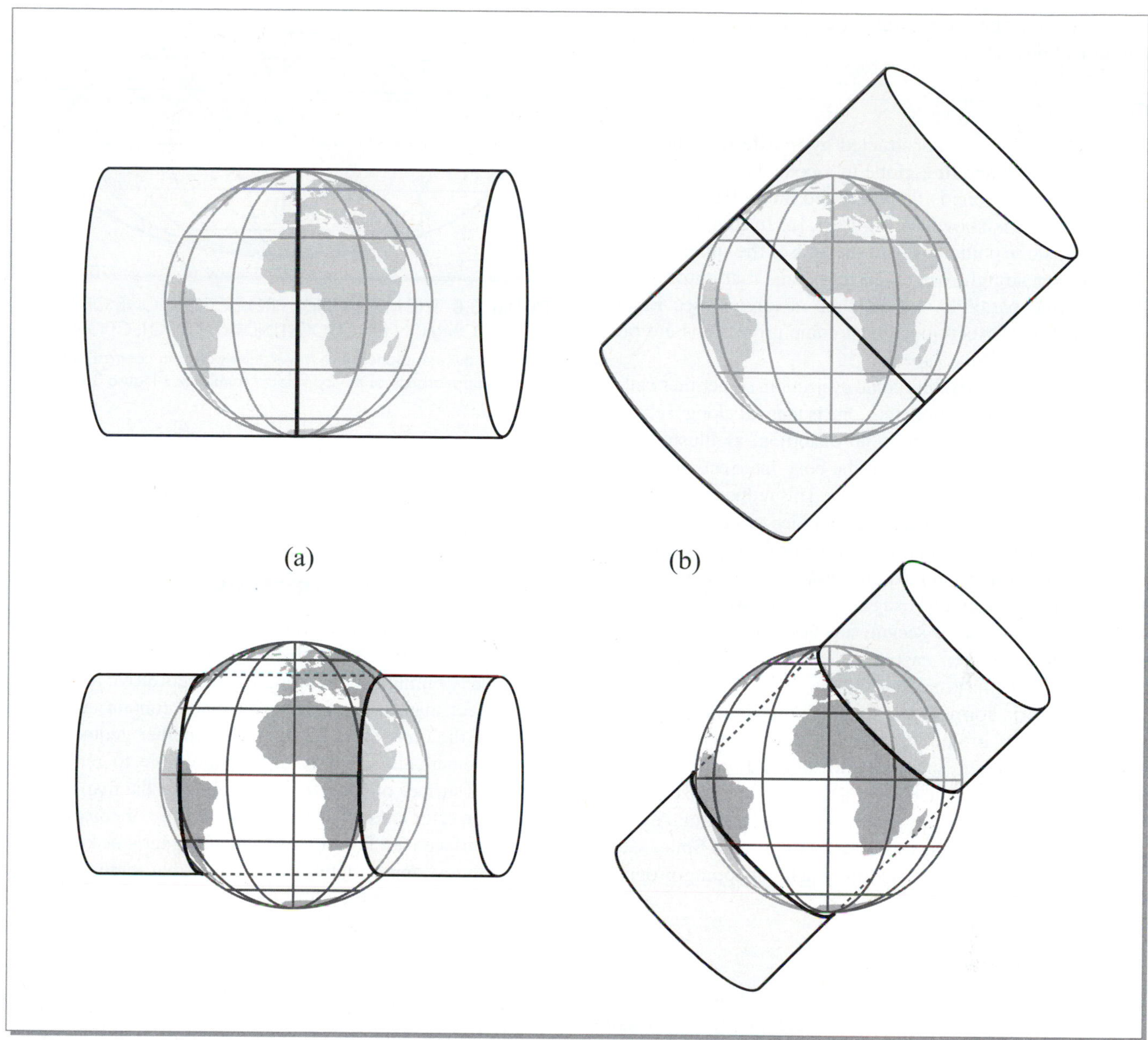

FIGURE 3.7 COMMON ALTERNATIVE PROJECTION ASPECTS FOR CYLINDRICAL MAPPING.
The transverse (a) and oblique (b) aspects in both tangent (above) and secant (below) cases are very common aspect forms for cylindrical mapping.

bands parallel to the standard parallel(s), with increasing exaggeration toward the outer edges of the map plane (see Figure 3.4b). Note that in this aspect, the scale preservation is in the east-west direction, parallel to the axis of the cylinder, and distortion progressively increases in a poleward direction.

Two other common aspects of this projection are the *transverse* (polar) and *oblique* cases (see Figure 3.7). Transverse means that the axis of the cylinder is turned parallel to the equator. In the tangent case, the standard parallel has now become a standard meridian (Figure 3.7a), and is thus preserving the least scale deformation in the north-south direction. In the secant case of the transverse cylindrical, the small circles produce two standard lines. Not surprisingly, this aspect is popular for enumeration units with elongated north-south directions.

In the oblique aspect, the cylinder is placed at any other position on the globe. It is often placed in such a way that the standard line(s) (a great circle in the tangent case and two small circles in the secant case) are at or near the area that is to be mapped. Remember that scale distortion increases as you move away from any standard lines. Political entities that are elongated in a manner other than a north-south direction, such as Japan, which has a pronounced southwest-northeast alignment, can be successfully mapped using an oblique cylindrical projection.

Conic Family

Conic projections are constructed by transferring the graticule from the generating globe to a cone enveloped around the sphere. This cone is then unrolled into a flat plane. The normal aspect is shown in Figure 3.1c. In this aspect, the axis of the cone coincides with the axis of the sphere, which yields either straight or curved meridians that converge on the pole and parallels that are arcs of circles (see Figure 3.4c). With few exceptions, most conic projections are presented in their normal aspect.

In the tangent case of the conic projection, sometimes called the *simple* conic projection, the cone is tangent along a chosen parallel, along which there is no distortion, as illustrated in Figure 3.1c. In the secant case, the cone intersects the globe along two parallels (see Figure 3.2c). This reduces distortion.

The pattern of deformation includes concentric bands parallel to the standard parallels of the projection (see Figure 3.4c). Secant conics tend to compress scale in areas between the standard lines and to exaggerate scale elsewhere. Conic projections, tangent or secant, are best for mapping Earth areas having greater east-west extent (by the standard parallel[s]) than north-south extent.

One historic form of the conic family is the polyconic projection. This projection uses several cones of development and consequently has several standard lines. Theoretically, each parallel is the base of a tangent cone. Historically, it has been used for mapping areas of great latitudinal extent; the polyconic projection was used by the United States Geological Survey (USGS) in its topographic mapping program until the 1950s.

Mathematical Family

The purely mathematical projections (those that cannot be developed by projective geometry) are in some cartographers' taxonomies simply classified into the geometric families on the basis of their appearances. A few projections bear striking resemblance to the developable ones but are different enough to be classed as pseudocylindrical, pseudoconic, and pseudoazimuthal. Pseudocylindrical projections are perhaps the most common in the mathematical family, with their meridians that curve toward the poles. A prominent example can be found in the **Mollweide projection** (see Figure 3.8). Other common mathematical projections that will be shown or discussed later in the chapter include the *Sinusoidal* and the *Hammer* projections.

Still others are so different that they are difficult to relate to the three geometric forms at all. As Strebe (2007) notes, sometimes the practice of distinguishing geometric projections from mathematically derived ones is not that important, since all projections can ultimately be mathematically described and are typically projected in a GIS or other mapping software. In the next section we consider the important

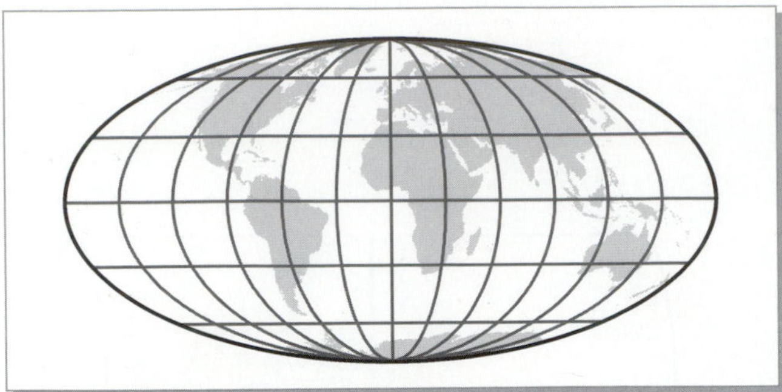

FIGURE 3.8 THE MOLLWEIDE PROJECTION, ONE OF THE MORE COMMON PSEUDOCYLINDRICAL PROJECTIONS.
Note that the meridians curve toward the poles, in contrast to the grid-like appearance of the cylindrical family. See Figure 3.4b for a comparison.

topic of describing types of deformation by examining projection properties.

Map Projection Properties

In the transformation process from the three-dimensional surface to a plane, some distortion occurs that cannot be completely eliminated. Although designers strive to develop the perfect map, free of error, *all* maps contain errors because of the transformation process, whether geometrically or mathematically derived. It is impossible to render the spherical surface of the reference globe as a flat map without distortion error caused by *tearing, shearing,* or *compression* of the surface (see Figure 3.9). The designer's task is to select the most appropriate projection so that there is a measure of control over the unwanted error.

These distortions and their consequences for the appearance of the map vary with scale. One can think of the globe as being made up of very small quadrilaterals. If each quadrilateral were extremely small, it would not differ significantly from a plane surface. For mapping small Earth areas (large-scale mapping), distortion is not a major design problem (although selection of datum can be very significant at larger scales; see Chapter 2 for discussion).

As the mapped area increases to subcontinental or continental proportions, distortion becomes an increasingly significant design problem for the cartographer. In designing maps to portray the whole Earth, surface distortions are maximized. At such scales, the map designer must contend with alterations of area, shape, distance, and direction, which are called a **projection's properties.** No projection of the globe's graticule can maintain all of these properties simultaneously; only on a globe are all four properties maintained.

Equal Area Mapping

Map projections on which area relationships of all parts of the globe are maintained are called **equal area** (or **equivalent**) **map projections.** The intersections of meridians and

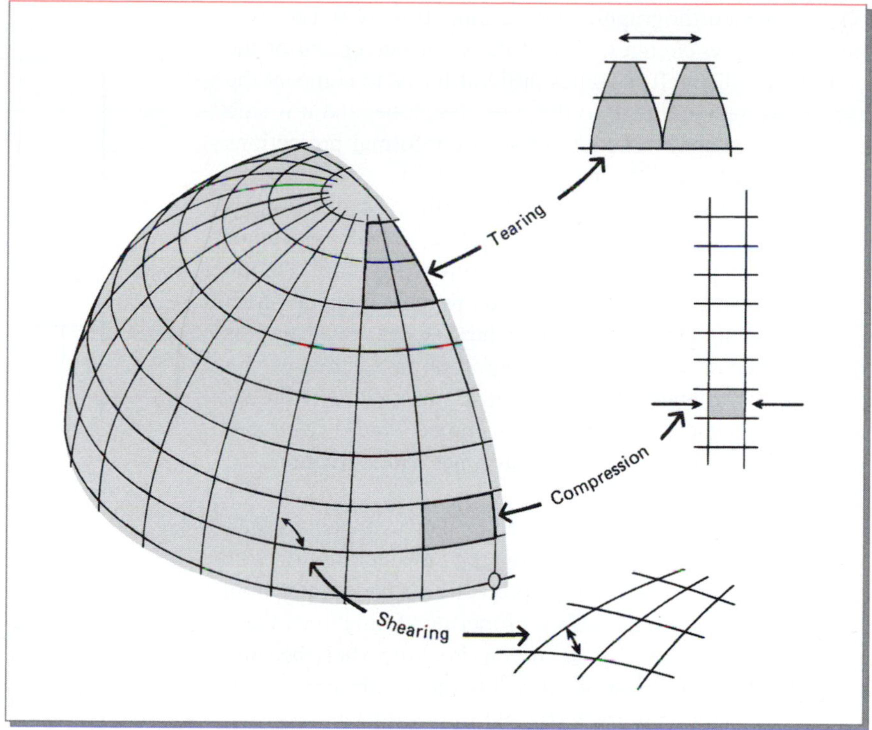

FIGURE 3.9 DISTORTION CAUSED BY THE PROJECTION PROCESS.
Transformation from sphere to plane may cause distortion brought about by tearing, shearing, or compression.

parallels are not at right angles, but areas of similarly bounded quadrilaterals maintain correct area properties. Linear or distance distortion often occurs in such projections. On equal area projections, therefore, shape is often quite skewed.

Equivalent projections are very important for general quantitative thematic map work, particularly at a global scale. It is usually desirable to retain area properties, particularly when enumeration units are compared or if area is *part* of the data being mapped (see Chapter 4); population density (that is, persons per square mile) is a prime example. Consider Marschner's classic argument for employing equal area projections:

> *It is hardly necessary to recapitulate here the arguments used in favor of equal area representation. There is no doubt but that the preservation of a true areal expression of maps, used as a basis for land economic investigation and research, is more important than a theoretical retention of angular values. Of the three geometrical elements that can be considered in mapping, the linear, angular, and areal, the last is the one around which land economic questions usually revolve. They do so for obvious reasons. Man does not inhabit a line, but occupies the area; he does not cultivate an angle of land, but cultivates and utilizes the land area. One of the principal functions of lines and angles is to define the boundary of the area. . . . They provide only the framework for controlling the relative position of features in the area, and with it a means for controlling the areal expression of the map itself* (Marschner 1944, 44).

Today many cartographers still agree that equivalent map projections can often be favorably selected for thematic mapping, *all other considerations being equal,* particularly at smaller scales. Note that some equal area projections may have deformation patterns that may be unacceptable to the map reader (or the cartographer). Depending on the map scale and purpose, other properties may be deemed important. As we explore the other projection properties, and later in the chapter, assess appropriate thematic map projections for different regions or tasks, we will see cases in which equal area maps are not as common, even for a thematic map. Nonetheless, as a generalization, thematic cartography tends to favor the use of equal area projections.

Conformal Mapping

Conformal (or orthomorphic) mapping of the sphere means that angles are preserved around points and that the shapes of small areas are preserved. The quality of conformality applies to small areas (theoretically only to points). On **conformal projections,** meridians intersect parallels at right angles, and the scale is the same in all directions about a point. Scale may change from point to point, however. It is misleading to think that shapes over large areas can be held true. Although shapes for small areas are maintained, the shapes of larger regions, such as continents, may be severely distorted. Areas are also distorted significantly at smaller scales.

The shape quality of mapped areas is an elusive element. If we view a continent on a globe so that our eyes are perpendicular to the globe at a point near the center of the continent, we see a shape of that continent. However, the shape of the continent is distorted because the globe's surface is falling away from the center point of our vision. We can view but one point orthographically at a time. If we select another point, the view changes, and so does our perception of the continent's shape. It becomes difficult for us to compare the shapes we see on a map to those on the globe, and it is safe to say that shapes of large areas on conformal projections should be viewed with caution (Dent 1987).

For large-scale mapping of small Earth areas, distortion is not significant. Indeed, the choice between an equivalent or a conformal projection becomes somewhat moot. At the smallest scales, the selection of the projection is more critical in the design process. Even at these scales, however, it is seldom necessary to specify a conformal projection, except in rare circumstances. Mapping phenomena with circular radial patterns may warrant such a choice. Radio broadcast areas, seismic wave patterns, or average wind directions are possible examples.

Most cartographers consider equivalence and conformality the two most important property considerations. Note that it is *impossible* for one projection to maintain both equivalency and conformality properties. Sometimes these properties are referred to as the *major* properties, because the properties can exist at all points on certain projections (Campbell 2001). Figure 3.10a and b illustrates the contrast between conformality and equivalence. The **Mercator projection** (3.10a) is the conformal map and distorts area, especially at the higher latitudes. This distortion is minimal near its standard parallel at the equator but increases greatly toward the poles. The **Hammer projection** (Figure 3.10b) is an equal area projection. The areas are equivalent, but conformality is nonexistent. Notice the shape distortion around its margin. The next two properties are considered *minor* properties, because the particular properties are not able to exist everywhere on the map.

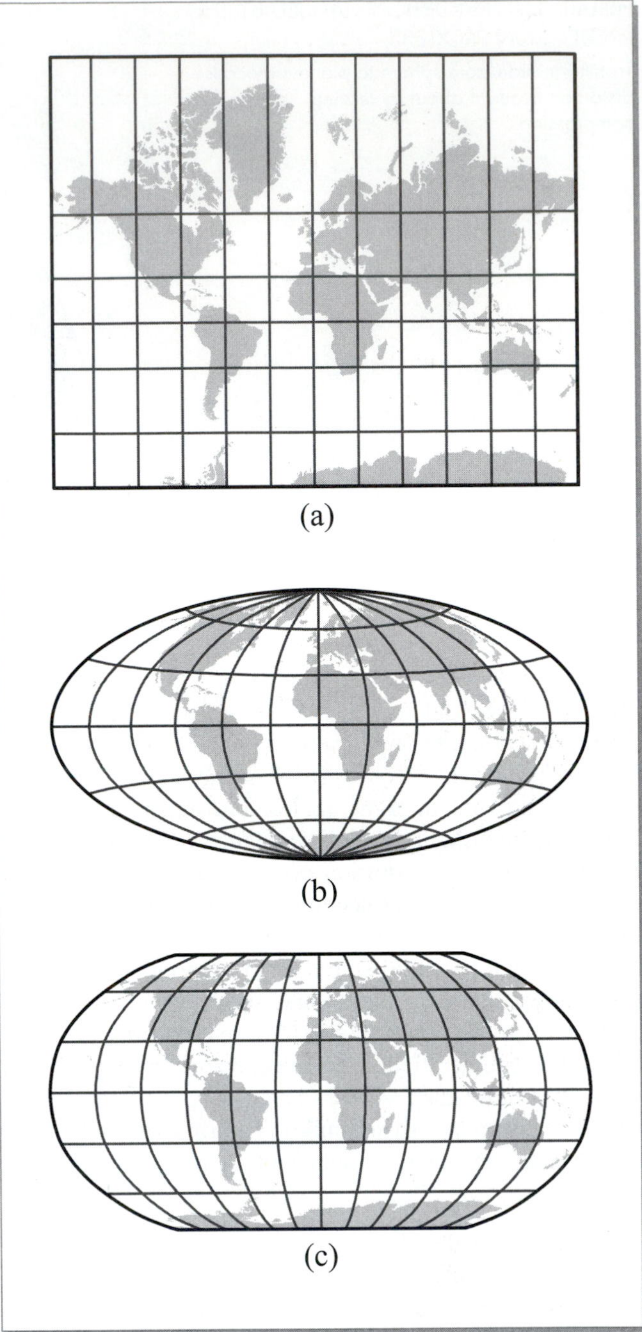

FIGURE 3.10 CLASSIC EXAMPLES OF A CONFORMAL, AN EQUAL AREA, AND A COMPROMISE PROJECTION.

The Mercator projection (a) provides an illustration of conformality. The Hammer projection (b), sometimes also referred to as the Hammer-Aitoff projection, illustrates an equal area projection, and the Robinson projection (c) illustrates a compromise or minimum error projection. See text for discussion.

Equidistance Mapping

The property of equidistance on projections refers to the preservation of great circle distances. There are certain limitations: distance can be held true from one to all other points, or from a few points to others, but not from all points to all other points. The distance property is never global. Scale will be uniform along the lines whose distances are true. Projections that contain these properties are called **equidistant projections.** Equidistant projections are sometimes used in general purpose maps in atlases because such projections are neither conformal nor equal area, and often have less distorted-appearing landmasses.

Azimuthal Mapping (Direction)

On **azimuthal projections,** true directions are shown from one central point to all other points. Directions or azimuths from the central point to other points are accurate, whereas from other (noncentral) points they are not. The quality of azimuthality is not an exclusive projection quality. It can occur with equivalency, conformality, and equidistance. Table 3.1 illustrates projection properties that can be combined. For example, azimuthal and equidistance properties can occur together in a map projection (a "yes" in the table), but conformal and equal area properties cannot exist simultaneously in a map projection (a "no" in the table).

TABLE 3.1 COMBINATIONS OF PROJECTION PROPERTIES THAT CAN OCCUR IN ONE PROJECTION

	Equal Area	Conformal	Equidistance	Azimuthal
Equal Area	—	no	no	yes
Conformal	no	—	no	yes
Equidistance	no	no	—	yes
Azimuthal	yes	yes	yes	—

Source: After Natural Resources Canada 2007.

Minimum Error Projections

Over the years, some cartographers have stressed the idea that **minimum error projections** (also called *compromise projections*) are best suited for general geographic cartography. These projections are essentially hybrids that attempt to control or minimize all four map projection properties in varying degrees. Often this is done to try to produce a better and more realistic depiction of the globe or parts of it. These projections are chosen by the designer on much the same basis as a projection having a uniquely preserved property: by selection of those total qualities that best suit the mapping task (Canters 1989). Figure 3.10c depicts the **Robinson projection,** which holds no property as true. When compared with the conformal and equal area (major property) figures, it is easy to see why some designers advocate for projections such as these. Indeed, for a decade the Robinson projection was the official map projection used by the National Geographic Society for their world maps.

Determining Deformation and its Distribution Over the Projection

Two chief methods are available for determining projection distortion and its distribution over the map. One is to depict a geometrical figure (square, triangle, or circle) or familiar object (such as a person's head) and plot it at several locations on the projection graticule (see Figure 3.11). Distortion on the projection is readily apparent. This method is very effective and quite sufficient for most general cartographic analyses; its weakness is the lack of a general quantitative index of distortion.

Another method, conceptually and mathematically more complex, uses **Tissot's indicatrix.** Tissot, a French mathematician working in the latter part of the nineteenth century, developed a way to show distortion at points on the projection graticule (Tissot 1881). Following his work, others have computed these indices and mapped them on several projections to show the patterns of distortion. The weakness of Tissot's method is that it is somewhat more complex mathematically than plotting simple geometrical shapes. Its strength lies in its quantitative ability to describe distortion.

The construct of Tissot's indicatrix consists of a very small circle, whose scale is unity (1.0), on the globe's surface. This small circle and two perpendicular radii appear on the plane map surface during transformation as a circle of the same size, a circle of different size, or an ellipse. For equal area mapping, the circle is transformed into an ellipse with the ratio of the new semimajor and semiminor axes such that its area remains the same (unity, or 1.0). Angular properties are not preserved (see Figure 3.12). On conformal projections, the small circle is transformed on the projection as a circle, although its size (area) varies over the map. Because the small circle is accurately projected as a circle, angular relationships are maintained. On some projections,

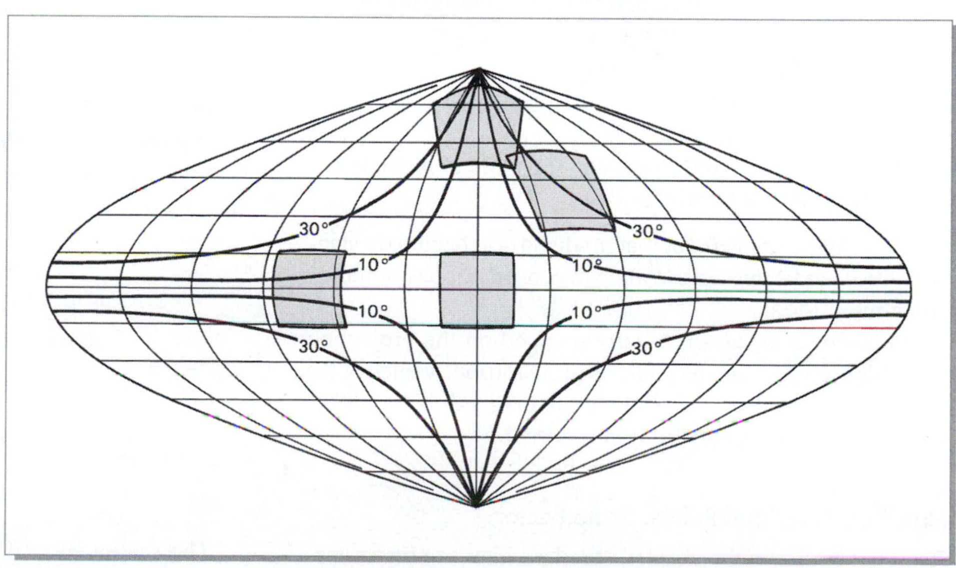

FIGURE 3.11 SHAPE DISTORTION ON AN EQUAL AREA PROJECTION.

On the globe, the squares in the figure are all the same size. On this Sinusoidal projection, the angular relations (and hence shape) are not preserved. Notice that distortion occurs most severely at the edges of the projection, as shown by the plot of 2ω values (10° and 30°). See text for explanation.

Source: Squares drawn from data in Chamberlin 1947, 97.

FIGURE 3.12 MAJOR AND MINOR AXES IN TISSOT'S INDICATRIX.

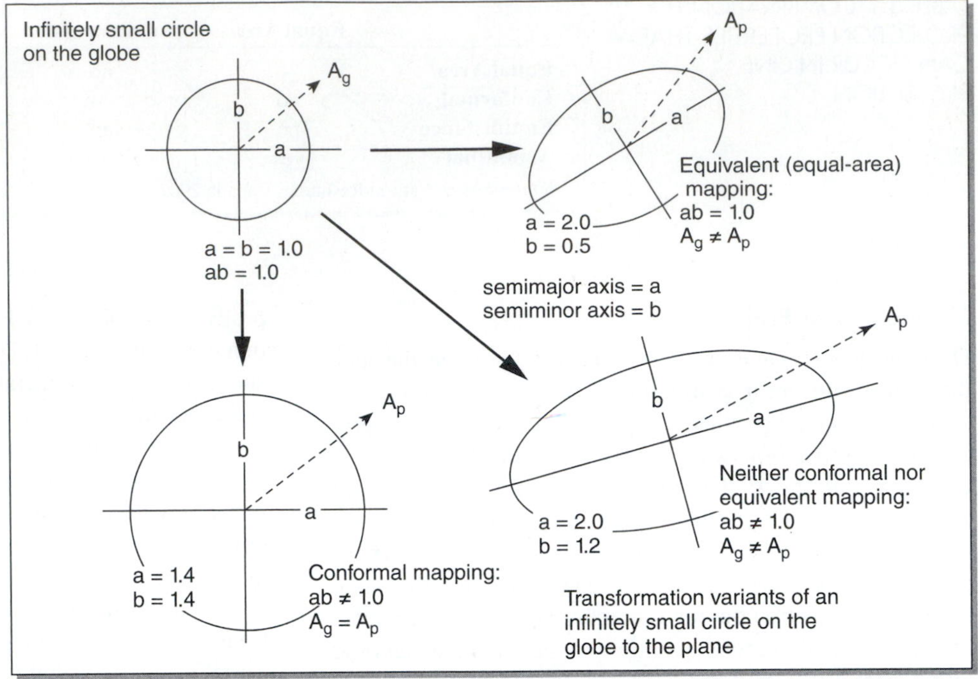

the small circle becomes an ellipse that preserves neither equivalency nor conformality. Such projections are not classified as either equivalent or conformal.

For purposes of computation and explanation, the indicatrix has these qualities:

- Maximum angular distortion = 2ω
- Scale along the ellipse semimajor axis = a
- Scale along the ellipse semiminor axis = b
- Maximum areal distortion = S

Every point on the new surface (map) has values computed for 2ω, a, b, and S. These conditions inform us of the distortion characteristics:

- If S = 1.0, there is no areal distortion, and the projection is equal area. If S = 3.0, for example, *local* features are three times their proper areal size.
- If $a = b$ at all points on the map, it is a conformal projection. S varies, of course.
- If $a \neq b$, it is not conformal, and the amount of angular distortion is represented by 2ω.

It should be pointed out that no distortion occurs on standard lines, and values of the indicatrix need not be computed along these lines.

Values of S or 2ω are usually mapped on the projection graticules to illustrate the patterns of distortion. Values of 2ω have been plotted on the **Sinusoidal projection** (a mathematical equal area projection) in Figure 3.11.

Standard Lines and Points, Scale Factor

As noted earlier in the chapter, standard lines are lines (usually meridians or parallels, especially the equator) on a projection that have identical dimensions to their corresponding lines on the reference globe. For example, if the circumference of the reference globe is 15 inches, the equator on the projection is drawn 15 inches long if it is a standard line. Scale can be thought to exist at points, so every point along a standard line has an unchanging or true scale when compared to the scale of the generating globe. On standard lines, scales are true. At all points on a standard line, using the indicatrix as a guide, $a = b = 1.0$, S = 1.0, and $2\omega = 0°$ (see Figure 3.13).

In practice, the lines on the projection can be compared to their corresponding lines on the reference globe by a ratio called **scale factor** (S.F.):

$$\frac{\text{projection scale fraction}}{\text{nominal scale fraction}}$$

In this equation, nominal scale is the scale of the reference globe. If the scale of a reference globe is 1:3,000,000 and the equator on the projection is drafted as a standard line, then

$$\frac{\text{projection scale fraction}}{\text{nominal scale fraction}} = \frac{\frac{1}{3{,}000{,}000}}{\frac{1}{3{,}000{,}000}} = \frac{3{,}000{,}000}{3{,}000{,}000} = 1.0 \text{ (S.F.)}$$

The scale factor of the equator will have a value of 1.0. For another example, suppose the nominal scale is

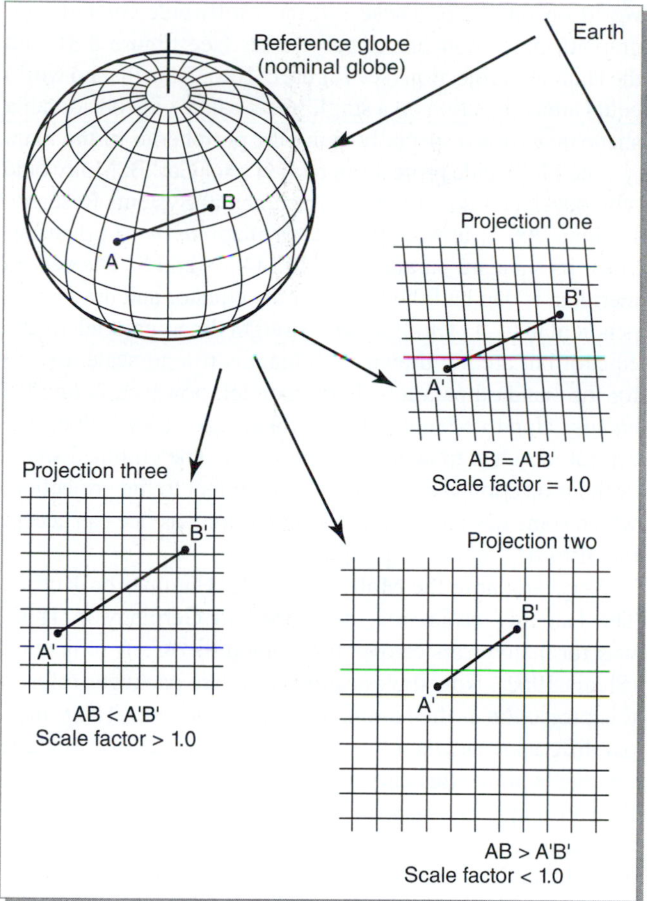

FIGURE 3.13 VARIATIONS OF THE STANDARD LINE AND THE DETERMINATION OF SCALE FACTOR.

1:3,000,000 and a line on the projection has a linear scale of 1:1,500,000:

$$\frac{\text{projection scale fraction}}{\text{nominal scale fraction}} = \frac{\frac{1}{1{,}500{,}000}}{\frac{1}{3{,}000{,}000}} = \frac{3{,}000{,}000}{1{,}500{,}000} = 2.0\ (\text{S.F.})$$

In this example, the line is two times *longer* on the map than it should be, because of stretching. In our final example, the nominal scale is 1:3,000,000 and a line on the projection is drafted at 1:6,000,000:

$$\frac{\text{projection scale fraction}}{\text{nominal scale fraction}} = \frac{\frac{1}{6{,}000{,}000}}{\frac{1}{3{,}000{,}000}} = \frac{3{,}000{,}000}{6{,}000{,}000} = 0.5\ (\text{S.F.})$$

Here we find the line to be one-half as long as its corresponding line on the globe. *Compression* has taken place in the transformation process.

Standard lines and scale factors are important to the overall understanding of projection distortion. The idea of scale factor can also be used in the assessment of areal scales. In this instance, small quadrilateral areas on the projection can be compared to the corresponding areas on the globe to determine the amount of *areal exaggeration* occurring on the projection. The bottom line is that the least distortion can be achieved not only by selecting the correct projection for the area being studied, but to the extent possible, having the standard points and lines run through the area of interest will produce a more accurate result.

Before we leave this section it is good to note that expressions of scale (for example verbal, graphic, or bar scale, or representative fraction), while often desirable as one of the basic elements of the map (see Chapters 1, 2, and 13 for a discussion of scale as a basic map element), are intrinsically inaccurate for world maps. They will be accurate only in places where there are standard points and lines. The wise designer will generally avoid an expression of scale on thematic world maps, unless there is a pressing need for the scale. If this is the case, a statement of where scale is accurate should accompany the expression of scale.

EMPLOYMENT OF MAP PROJECTIONS

In the previous section we focused on the map projection process. Considering all of the projection parameters that can be adjusted (geometric figure, aspect, standard points and lines, central meridian, and so on), there is a theoretically infinite number of potential map projections. There are hundreds or even thousands of named projections that currently exist. Indeed, most software today is able to work with many of the most common or popular projections.

In this section we provide some guidelines for selecting a projection from the myriads that are in existence, as well as survey a few of the most common map projections used for regions at different scales. These include ones that are commonly in use today or have had a major influence on modern cartography—even if it is a projection that we do not necessarily recommend for thematic cartography. Of course, no projection survey or list will satisfy every cartographic need or expectation. From an historical perspective, we recommend John Snyder's *Flattening the Earth: 2000 Years of Map Projections* (Snyder 1993) for a comprehensive look at many different map projections. From an applied perspective, we recommend examining the projections that are supported in your GIS or mapping software. Read about them in your software's help section, or better yet, take the (usually provided) world map data (or maps at other scales) and start projecting them. As you examine the projections visually, also try to examine the various projection properties to the extent that your software allows. If you do not have dedicated GIS or mapping software, there are also several web pages that provide a fairly extensive gallery of map projections. Some of these sites provide a link to free or low-cost projection software that can easily be downloaded to your computer (Anderson 2007; Geocart 2007).

Essential Questions

Several essential elements must be carefully considered in the selection of a map projection:

1. *Projection properties.* Are the properties of a particular projection suited to the design problem at hand? Are equivalency, conformality, equidistance, or azimuthality needed?
2. *Deformational patterns.* Are the deformational aspects of the projection acceptable for the mapped area? Is linear scale and its variation over the projection within the limits specified in the design goals? Do the characteristics of linear scale over the projection benefit the *shape* of the mapped area?
3. *Projection center.* Can the projection be centered easily for the design problem? Can the software accommodate experimentation with the re-centering of the projection, such as an adjustment of the projection's central meridian?
4. *Familiarity.* Will the projection and the appearance of its meridians and parallels be familiar to most readers? Will the form of the graticule detract from the main purpose of the map?
5. *Software support.* Is the particular projection supported in your software?
6. *Part of an existing map series or online digital map collection.* Does the map belong to a series that already has a projection? Do you want to continue to match that projection (and especially at larger scales, the datum and coordinate system)?

Again, there are literally hundreds of projections from which the cartographer may choose. Certain ones have proven more useful in mapping particular places (see Table 3.2). The ones included in our discussion are offered only as a guide.

World Projections

Many world projections may be selected for thematic mapping. World map projection selection can be both debatable and subjective at times. In some cases, they are the most controversial. We have grouped several types of world maps for convenience.

Mathematical, Equivalent Projections

As suggested earlier, the property of equivalency is often the overriding concern in thematic cartography. Two mathematically derived, pseudocylindrical equal area projections are presented here as good choices when depicting the entire world on one map. These are the Mollweide (or homolographic) projection mentioned earlier (see Figure 3.8), and the Hammer projection (see Figure 3.10b). Note that no world equal area projection on a single sheet can avoid considerable shape distortion, especially along the peripheries of the map.

The Mollweide projection, named after Carl B. Mollweide who developed it in 1805 (Environmental Systems Research Institute 2007), is widely used for mapping world distributions. Its standard parallels are 40° 44' N and S. The central meridian is one-half the length of the equator and drawn perpendicular to it. Parallels are straight lines parallel to the equator but are not drawn with lengths true to scale, except for the standard parallels. Each parallel, however, is divided equally along its length. The parallels are spaced along the central meridian to achieve equivalency. The elliptical shape of the projection gives a kind of global feel to the projection, which some designers find pleasing as long as one can accept the distortion along the peripheries.

Very similar to the Mollweide is the Hammer projection. The Hammer projection, developed in Germany in 1892, was for many years erroneously called the Aitoff projection (Steers 1962). Nonetheless some software packages refer to this projection as the Hammer-Aitoff projection. The principal difference between this projection and the Mollweide is that the Hammer has curved parallels. This curvature results in less oblique intersections of meridians and parallels at the extremities, and thus reduces shape distortions in these areas. The outline (that is, the ellipses forming the outermost meridians) is identical to the Mollweide.

Hammer's projection also is quite acceptable for mapping world distributions. A comparison of it with the Mollweide shows little difference. Because the parallels are curved, east-west exaggeration at the poles is less on the Hammer than on the Mollweide. This is most notable when comparing the Antarctica landmasses. Africa is less stretched along the north-south axis on the Hammer. Overall, however, these projections are very similar in appearance and attributes.

Also worth mentioning is the Sinusoidal equal area projection, also called the Sanson-Flamsteed projection (see Figure 3.11). This pseudocylindrical projection, with its straight parallels and curved meridians, was popular for mapping worldwide thematic distributions. The top-like shape of the outline makes it distinctive, although there are other similar projections. We have seen a slight falling off of its use in the twenty-first century. The unique shape could be a distraction to the thematic symbolization. There is also a rather extreme shape compression of the polar features, particularly at the North Pole. Cartographers seem to be favoring more elliptically shaped projections such as Mollweide or Hammer when equal area mathematical projections are desired.

Interrupted Projections. One solution to minimize distortion is to make an interrupted projection of the world. Many of the previously mentioned world projections can be turned into an interrupted projection. The most famous of the interrupted projections is perhaps the **Goode's Homolosine projection,** a

TABLE 3.2 GUIDE TO THE EMPLOYMENT OF PROJECTIONS FOR WORLD-, CONTINENTAL-, AND COUNTRY-SCALE THEMATIC MAPS

Principal Use	Suitable Projections	Notes
1. Maps of the world		
Equal area	Sinusoidal (Sanson-Flamsteed)	Awkard shape
Equal area	Mollweide	Pleasing shape
Equal area	Hammer	Sometimes called Hammer-Aitoff in software
Compromise	Robinson	Pleasing shape, balances extremes
Compromise	Winkel Tripel	May be most accurate compromise
2. Continental areas		
A. Asia and North America		
Equal area	Bonne*	Considerable distortion in NE and NW corners
Equal area	Lambert Azimuthal Equal Area	Bearings true from center
B. Europe and Australia		
Equal area	Lambert Azimuthal Equal Area*	
	Bonne*	
	Albers Equal Area Conic; ideal for United States	
Conformal	Lambert Conformal Conic	
C. Africa and South America		
Equal area	Lambert Azimuthal Equal Area*	
Equal area	Mollweide*	
Equal area	Sinusoidal*	
Equal area	Homolosine*	
3. Large countries in mid-latitudes		
A. United States, Russia, China		
Equal area	Lambert Azimuthal*	
Equal area	Albers Equal Area Conic	
Equal area	Bonne*	
Conformal	Lambert Conformal Conic	
4. Small countries in mid-latitudes		
Equal area	Albers Equal Area*	
Equal area	Bonne*	
Equal area	Lambert Azimuthal*	
Conformal	Lambert Conformal Conic*	
5. Polar regions		
Equal area	Lambert Azimuthal	
6. Hemispheres and continents		
Visual	Orthographic	View of Earth as if from space; neither equal area nor conformal

***Must take special care to scale (zoom in) and re-center projection parameters to the particular area of interest.**

Sources: Compiled from a variety of sources listed in the references, especially Raisz 1962; Steers 1962; Snyder 1987; Dana 1999; and Environmental Systems Research Institute 2007.

pseudocylindrical equal area projection created by J. Paul Goode (see Figure 3.14). Goode, the founder of the *Goode's World Atlas* took portions of the Mollweide and Sinusoidal projections and combined them in a manner that created six distinct lobes. Each lobe has its own central meridian, and the distortion that would occur in a continuous projection now is directed over water bodies (there are also versions that keep the water bodies intact and pull apart the land areas). Shape is not as distorted as some of the equal area projections that we have seen so far. The downside to Goode's and other interrupted projections like this is threefold. First, as with the Sinusoidal projection, the shape of the projection can be distracting. Second, the projection requires the map reader to understand that the gaps along a line of latitude do not exist. Third, if the thematic phenomena to be mapped are spatially continuous, such as climate data, then the projection is wrong for the thematic map type.

Minimum Error Projections

A number of cartographers strongly believe that minimum error or compromise projections (projections that do not

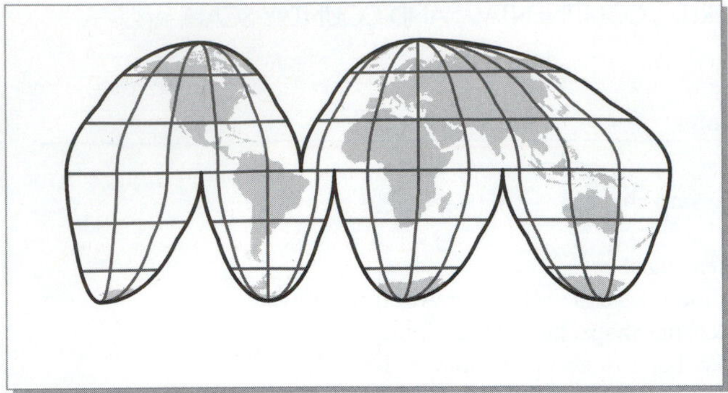

FIGURE 3.14 THE INTERRUPTED GOODE'S HOMOLOSINE PROJECTION.

Perhaps the most famous of the interrupted projections, this was used in the Goode's World Atlas series.

retain any specific property but try to minimize the worst distortions) are the best way to go. Two projections that we feel are also quite acceptable for world thematic mapping are the Robinson projection mentioned earlier (see Figure 3.10c) and the **Winkel Tripel projection** (see Figure 3.15).

The Robinson projection was developed in 1961 by Arthur Robinson for Rand McNally's world maps (Robinson *et al.* 1995). It is more famous, however, as being adopted in 1988 by the National Geographic Society for use as a world projection. It was replaced in 1998 by another minimum error projection, the Winkel Tripel projection.

The Winkel Tripel projection was developed in 1921 by Oswald Winkel (Environmental Systems Research Institute 2007). It has slightly less compression in the polar regions than does the Robinson projection. Physicists Goldberg and Gott (2007) have shown quantitatively, using Tissot's indicatrix and other measures, that the Winkel Tripel may be the best (lowest distortion) compromise projection yet.

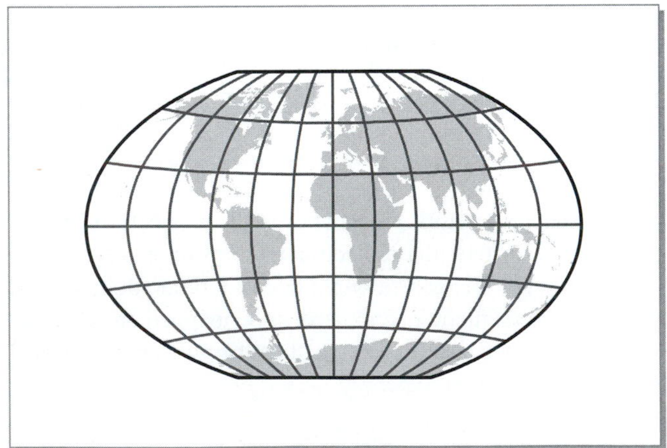

FIGURE 3.15 THE WINKEL TRIPEL PROJECTION.

Compromise (minimum error) projection used by the National Geographic Society for world maps.

Cylindrical Projections

This last category in World Projections is more of an acknowledgement that many cylindrical (also called rectangular) projections have been produced and used throughout cartographic history rather than a recommendation or endorsement for their use in world thematic mapping. The most famous projection in this category is doubtlessly the Mercator projection (see Figure 3.10a).

The Mercator projection was developed by Gerardus Mercator in 1569. It was developed for navigational use. All lines of constant compass bearings, called loxodromes or rhumb lines, are shown as straight lines. Navigators could therefore follow a compass heading using these maps (Snyder 1993). Over time, however, the map became popularized as a general reference map of the world. The problem with this use is that landmass areas such as Greenland and Russia (and Antarctica on those Mercator maps that include that area) are extremely distorted in the higher latitudes. Of course, as seen in some of the previous maps in this section, professional cartographers have provided alternatives for years, but the Mercator's popularity as a world projection did not start to wane until the latter half of the twentieth century. It is exactly this resilience of the projection as a world map and the fear that it had imbued a distorted "mental map" in the minds of many people that caused a fairly major controversy in the cartographic community.

Gall-Peters Projection. Special consideration is devoted here to what has become known as the **Gall-Peters projection,** primarily to stimulate the cartographic designer to look further into the literature, and partly because of the controversy surrounding its use (see Figure 3.16). In recent times, probably no other map projection has received as much attention in both the scientific and popular literature.

In 1972 Dr. Arno Peters of Germany published what he called a new map projection—the Peters projection (Robinson 1985; Snyder 1988). In fact, this projection had been devised earlier (mid-1880s) by Gall who called it the

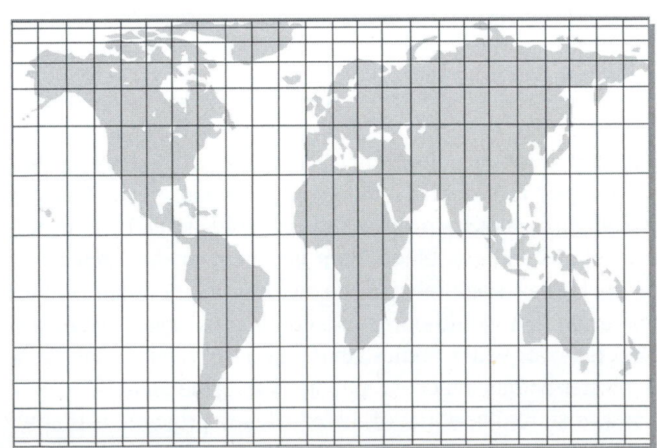

FIGURE 3.16 THE GALL-PETERS PROJECTION.

RESOLUTION REGARDING THE USE OF RECTANGULAR WORLD MAPS

WHEREAS, the Earth is round with a coordinate system composed entirely of circles, and

WHEREAS, flat world maps are more useful than globe maps, but flattening the globe surface necessarily greatly changes the appearance of the Earth's features and coordinate system, and

WHEREAS, world maps have a powerful and lasting effect on people's impressions of the shapes and sizes of lands and seas, their arrangement, and the nature of the coordinate system, and

WHEREAS, frequently seeing a greatly distorted map tends to make it "look right,"

THEREFORE, we strongly urge book and map publishers, the media, and government agencies to cease using rectangular world maps for general purposes or artistic displays. Such maps promote serious, erroneous conceptions by severely distorting large sections of the world, by showing the round Earth as having straight edges and sharp corners, by representing most distances and direct routes incorrectly, and by portraying the circular coordinate system as a squared grid. The most widely displayed rectangular world map is the Mercator (in fact a navigational diagram devised for nautical charts), but other rectangular world maps proposed as replacements for the Mercator also display a greatly distorted image of the spherical Earth.

American Cartographic Association
American Geographical Society
Association of American Geographers
Canadian Cartographic Association
National Council for Geographic Education
National Geographic Society
Special Libraries Association, Geography and Map Division

Source: Committee on Map Projections 1989, 223.

orthographic—a form of cylindrical projection. (It may be speculated that Peters had no knowledge of Gall's previous work) (Loxton 1985). In the conventional aspect, the Peters projection is an equal area rectangular projection with standard parallels at 45° North and South. Because the projection is not new, it is referred to as the Gall-Peters projection.

The purpose behind the invention was to counter the Mercator projection's areal exaggeration of the high latitudes and its pervasive use as a general purpose map. Peters believed that the Mercator projection accentuated European dominance over the Third World (Faintick 1986), because the projection's high latitude exaggeration visually minimizes countries located in the tropics. As is evidenced in Figure 3.16, the Gall-Peters projection emphasizes many portions of the tropics dominated by less developed countries. Because of the prevalence of the Mercator projection, Peters stressed that his projection be used exclusively. He also claimed that it portrays distances accurately, which is false (Snyder 1993; Monmonier 2004).

Proponents of the projection argue that the realigning of perceived areas and shapes shakes up our preconceived notions of Third World areas. Others argue that the shape distortion is too great to be of much utility (especially with other available mapping options); it especially distorts shape in the very underdeveloped areas that Peters was trying to portray more fairly (Stocking 2005). The real objection to Peters, however, is with the misconceptions he renders about the projection (the claim that Peters is the only projection that should be used, and that all distances on the projection can be measured correctly). Nonetheless, it has been widely adopted by three United Nations organizations (UNESCO, UNDP, and UNICEF), the National Council of Churches, and Lutheran and Methodist organizations (Loxton 1985).

One good thing has come from the controversy, namely, that the inadequacies of using the Mercator projection for world thematic mapping have finally been noted to the general population. Snyder, for example, has said:

> *Nevertheless, Peters and Kaiser (his agent) appear to have successfully accomplished a feat that most cartographers only dream of achieving. Professional mapmakers have been wringing their hands for decades about the misuse of the Mercator Projection, but, as Peters stresses, the Mercator is still widely misused by school teachers, television news broadcasters and others. At least Peters' supporters are rightly communicating the fact that the Mercator should not be used for geographical purposes, and numerous cartographers agree. The Gall-Peters Projection does show many people that there is another way of depicting the world* (Snyder 1988, 192).

One final comment may be made regarding the selection of projections for world mapping. Concern about the use of rectangular projections for world presentations reached the attention of the Committee on Map Projections of the American Cartographic Association, and the resolution it passed has the endorsement of most professional geography and cartography associations (see boxed text). They do not endorse the use of any rectangular world projection. Note that some cartographers feel that this may be an overreaction to the Peters projection. Nonetheless, none of the world projections recommended in this section are of the rectangular type; all have rounded or more globe-like margins.

Projections for Mapping Continents

There are a number of projections that can be used for mapping continental areas. Two that can be of great utility are the **Lambert Azimuthal Equal Area projection** and the Bonne projection, a favorite in earlier versions of this text. As with most of the projections discussed so far, both of these projections are well supported in GIS and mapping software packages.

The Lambert Azimuthal Equal Area projection is a versatile projection. In its equatorial aspect, this may be one of the best choices for mapping a hemisphere, and certainly a continent. The fact that its standard point can be placed anywhere (and most GIS and mapping software support oblique positioning for this projection) make it a truly useful projection for mapping continents (and countries, as we will see in a moment; see Figure 3.17). Since the distortion is radial outward from the center of the projection, it is critical that the latitude and longitude of the standard point be placed at the center of the continent or other area of interest.

The **Bonne projection** is named after its inventor, Rigobert Bonne (1727–1795) (Raisz 1962). It is an equal area conical projection, with a central meridian and the cone assumed tangent to a standard parallel (see Figure 3.18). All parallels are concentric circles, with the center of the standard parallel at the apex of the cone. Scale is true at the central meridian and also for each parallel (Environmental Systems Research Institute 2007). If the standard parallel selected is the equator, the projection becomes identical to the Sinusoidal. Map designers select the Bonne projection for a variety of continental mapping cases. It is commonly used to map Asia, North America, South America, Australia, and other large areas. Europe may also be adequately mapped with the Bonne projection. Caution should be exercised, however; although equivalency is maintained throughout, shape distortion is particularly evident at the northeast and northwest corners. Because of this, the Bonne projection is really best suited for mapping compact regions lying on only one side of the equator (McDonnell 1979). Because shape is best along the central meridian, the distortion becoming objectionable at greater distances from it, the selection of the central meridian relative to the important mapped area (and zooming to that area) is critical.

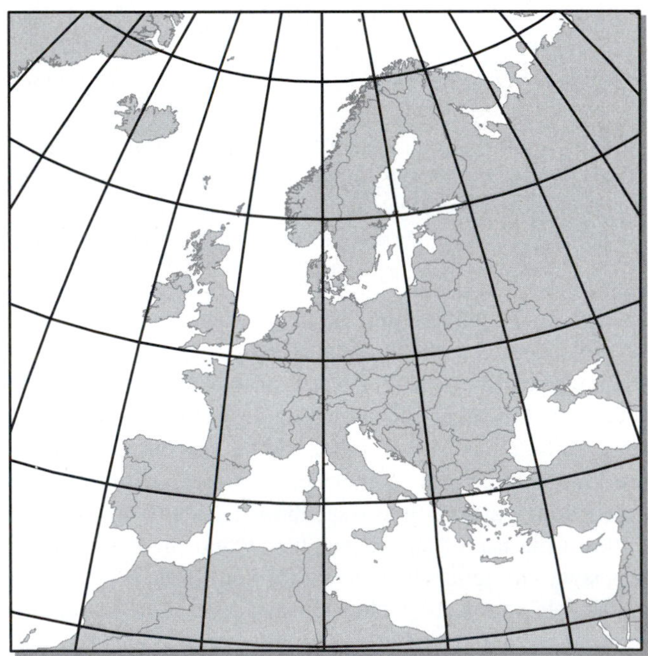

FIGURE 3.17 LAMBERT'S AZIMUTHAL EQUAL AREA PROJECTION, OBLIQUE ASPECT.

This projection can be an excellent choice for continents and countries that extend in all directions. The projection parameters for this particular map of Europe has the central meridian set at 10° E and the latitude of origin set at 30° N.

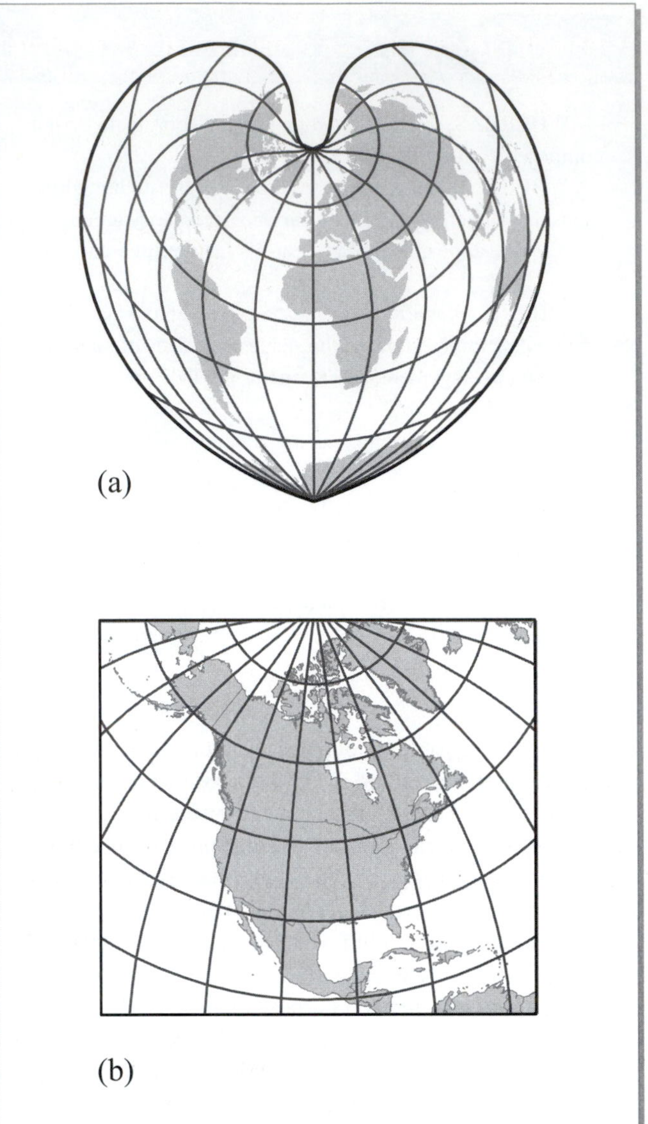

FIGURE 3.18 THE BONNE PROJECTION.

The Bonne projection depicting the entire world (a) and more appropriately re-centered and zoomed into North America (b). This projection is suitable for mapping continents, as in (b), but should never be used for mapping complete hemispheres. In (a) notice the severe shape distortion, brought about by shearing, at the northeast and northwest corners of the projection.

Some projections that are suitable for world maps are sometimes not the best for mapping continental areas, but others, such as the Sinusoidal, Mollweide, Goode's Homolosine, or even one of the compromise projections, can also be used effectively for a continent such as Africa or South America, as long as the central meridian is adjusted properly and you are zoomed into the area of interest appropriately. As always, you should let the purpose of the map and the "essential questions" presented earlier in this chapter help guide in your choices. If the projection that you choose for the continent does not meet your needs or hold to your expectations, try another projection.

Mapping Multiple Size Countries at Mid-Latitudes

Mapping larger countries at mid-latitudes can be handled in a variety of ways. The Lambert Azimuthal Equal Area or the **Albers Equal Area Conic projection** may be used. If conformality is desired, the **Lambert Conformal Conic projection** can be selected, and depending on scale, the visual appearance will often be somewhat similar to Albers. In general, a conic projection is usually adequate for mapping rather large countries or political units that have an east-west extent. Even the Bonne projection, with adjustments for the correct reference latitude, central meridian, and scale can work for some countries.

For countries or even groups of countries that extend in all directions, the Lambert Azimuthal Equal Area projection can be a great choice. Described in the previous section, the projection should have its standard point placed in the center of the area of interest (the oblique case), such as the center of the country. In addition to the equivalency property, the azimuth of any point on the map (as measured from the center of the projection) is correct. This makes this projection especially useful when mapping phenomena having an important directional relationship to the standard point chosen for the map.

For countries and political entities that have a pronounced east-west orientation, either the Albers Equal Area Conic projection or the Lambert Conformal Conic projection can be great choices. Both projections considered are the secant case, utilizing two standard parallels to lessen scale distortion.

We often recommend the Albers Equal Area Conic for thematic maps in these pronounced east-west cases. Besides maintaining the property of equivalency, the overall scale distortion is nearly the lowest possible for an area the size of the United States (see Figure 3.19). Not surprisingly, it is one of the best choices for mapping the continental United States. The Lambert Conformal Conic projection, however, does not produce appreciably different graticule in maps at this scale, as long as the standard parallels and the central meridian are set appropriately for the study area. Lambert Conformal Conic is used throughout the Atlas of Canada (Natural Resources Canada 2007), as well as in many projected coordinate systems (to be discussed) worldwide and in the United States.

While the Mercator projection is not recommended for world thematic mapping, the **Transverse Mercator** projection

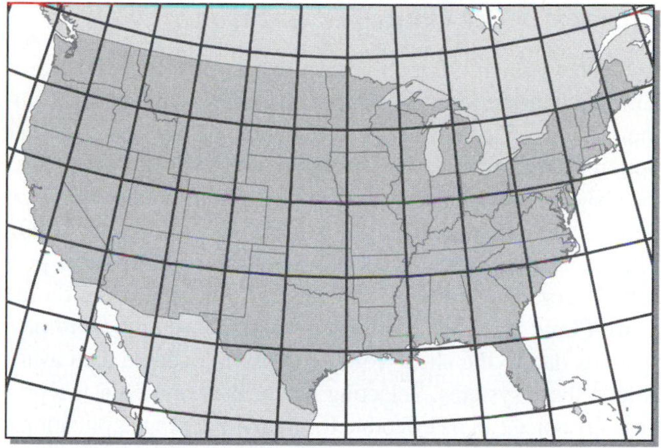

FIGURE 3.19 THE ALBERS EQUAL AREA CONIC PROJECTION. This projection is frequently used for mapping the continental United States. Angular distortion is minor, nowhere exceeding 2°.

is often used for countries and other political entities that have a pronounced north-south orientation. As described earlier in the section "Cylindrical Family," and illustrated in Figure 3.7a, the cylinder is rotated so that the standard parallel becomes a standard meridian, meaning there is no scale distortion in the north-south direction at the standard meridian. If the secant case is employed, then two standard lines (small circles) straddle the projection's central meridian, increasing the amount of relatively low distortion areas. This would be an appropriate choice for mapping a country such as Chile, and is also employed in projected coordinate systems such as the State Plane and UTM coordinate systems (to be discussed).

It is important to note that as mapped areas become smaller in extent, the selection of the projection becomes less critical; potential scale errors begin to drop off considerably. Using a localized projection and coordinate system might become an important consideration when this happens.

Mapping at Low Latitudes

Many of the projections already discussed are suitable for mapping countries, large and small, that are on or near the equator. The Lambert Azimuthal Equal Area projection, in its equatorial aspect, is one great choice. Cylindrical projections that have a standard parallel at the equator (see Figure 3.4b) or secant cylindricals that have standard parallels near the equator can also be good choices for low latitude mapping. A projection such as Mercator, while best avoided for world and other small scale mapping applications, has no distortion at the equator (in the tangent case). So mapping a country such as Ecuador would not be a problem using Mercator. But the Lambert Azimuthal Equal Area projection in its equatorial aspect could be equally as effective. As with other projections, make sure that you are zoomed into the area of interest appropriately and the central meridian is centered in the area of interest (for example, the center of Ecuador).

Projected Coordinate Systems

As we have seen so far, all projections involve coordinate systems, such as the decimal degree coordinates that are so common in mapping and setting projection parameters. Another option for mapping is to use a related but yet distinct concept of the projected coordinate system, which combines the projection process with the parameters of a *particular* grid (Iliffe 2000). There are a number of national, state, and county level projected coordinate systems. For this text, we will focus on the **State Plane** and **Universal Transverse Mercator** coordinate systems due to the sheer volume of downloadable data available in these systems. Selecting one of these systems as a projection choice can be effective for many mapping applications, particularly maps with scales at the state level and larger.

State Plane Coordinate (SPC) System

The original State Plane coordinate system was developed in the early 1930s by the then United States Coast and Geodetic Survey (USCGS), now the National Ocean Service (National Oceanic and Atmospheric Administration 1989). It was devised so that local engineers, surveyors, and others could tie their work into the reference then used, the Clarke ellipsoid of 1866, which was used in the North American Datum (NAD) of 1927 but has since been migrated to the North American Datum (NAD) of 1983. What they desired was a simple rectangular coordinate system on which easy plane geometry and trigonometry could be applied for surveying, because working with spherical coordinates was cumbersome.

Earlier we mentioned that if the area of the Earth being mapped is small enough, virtually no distortion exists. This is the principle behind the SPC system. To ensure stated accuracies of less than one part in 10,000, the states are partitioned into a series of zones (see Figure 3.20). In the continental United States, these zones are elongated either in the north-south direction or the east-west direction. Many states have two or three zones, a few states have only a single zone; Alaska, California, Hawaii, Texas, and Wyoming have four or more zones. Each zone can be referred to by its name; for example, Minnesota has a North, Central, and South zone (see the "Zone Examples" inset in Figure 3.20), or by a FIPS code (Federal Information Processing Standard, discussed in more detail in Chapter 4). Each zone is assigned its own coordinate systems with its own origin and its own projection.

There are three conformal projections used to map the states—the secant case of the *Lambert Conformal Conic* for zones with elongated east-west dimensions, the secant case of the *Transverse Mercator* for zones with elongated north-south dimensions, and the secant case of the Oblique Mercator for one section of Alaska. In each case, over small areas these projections essentially project as rectangular grids, with little or no areal and distance distortion. Because they are conformal, no angular distortion between the meridians and parallels is present.

The projection's parameters are then adjusted for each zone. For example, a zone employing the Lambert Conformal Conic projection will have its central meridian fit to the center of the zone, and the standard parallels are positioned near the northern and southern portions of the boundary (see Figure 3.21). In a zone employing the secant case of the Transverse Mercator projection (Figure 3.7a), the central meridian is again fit to the center of the zone, and its two standard lines (small circles) will straddle the meridian on either side (although much closer than is depicted in Figure 3.7a), near the western and eastern margins of the zone.

A particular zone's coordinate system is designed so that surveyors work only with positive (quadrant one) coordinates. The easting and northing measurements are in feet if NAD27 is used and in meters if NAD83 is used. A false origin (see Chapter 2 for discussion of this concept) is placed so that the *x*- and *y*- axis are the south and west of the zone, respectively (see Figure 3.21). The exact placement of the axes can vary greatly between states and zones. A typical configuration for SPC NAD83 would place the *y*-axis 600,000 meters west of the zone's central meridian in the Lambert Conformal Conic Projection (or 2,000,0000 feet west in NAD27), and 200,000 meters of the zone's central meridian in Transverse Mercator projection (500,000 feet west in NAD27). The *x*-axis position is more arbitrary but is still placed to the south of the zone in all cases (Van Sickle 2004).

In GIS and mapping software, selecting SPC is usually accomplished in a manner similar to that of other map projections. In many cases, local data are readily available in SPC. Because the distortion is small and the projection is centered, SPC works well for larger scale (up to state level) thematic maps.

If SPC is used to map an entire state, we recommend using as centralized a zone as possible. The final shape is fairly aesthetic (see Figure 3.21). For states with only two north-south elongated zones (such as Illinois, Indiana, or Georgia), we recommend selecting either the western or eastern zones, and then adjusting the central meridian west or east as necessary to provide a centered and balanced appearance to the state.

Universal Transverse Mercator (UTM) System

The Universal Transverse Mercator (UTM) system, along with the Universal Polar Stereographic System (for polar regions), was created after World War II by several allied nations in order to produce a unified and consistent coordinate system after years of attempting to trade information in disparate coordinate systems. The United States military soon followed with its own adaptation of the UTM system (Van Sickle 2004). We will be following the civilian use of the system.

The UTM system is not quite as accurate as the State Plane system, with accuracies as large as one part in 2,500 (Van Sickle 2004). But the increased coverage and overall acceptance of the system has had a major impact on the mapping community. Many U.S. states, for example, distribute much of their public domain spatial data in UTM format (usually NAD27 or NAD83 datum), which can be easily loaded into GIS and mapping software, where the projection, datum, and/or the projection's parameters can be adjusted if need be.

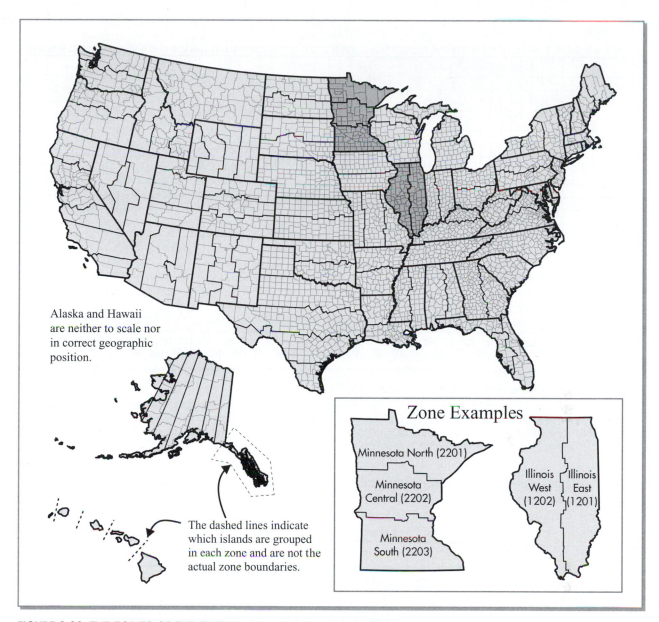

Alaska and Hawaii are neither to scale nor in correct geographic position.

The dashed lines indicate which islands are grouped in each zone and are not the actual zone boundaries.

Zone Examples

Minnesota North (2201)

Minnesota Central (2202)

Minnesota South (2203)

Illinois West (1202)

Illinois East (1201)

FIGURE 3.20 THE ZONES OF THE STATE PLANE COORDINATE SYSTEM.

This is the NAD83 configuration of zones for the SPC system, which follow state and county lines for the contiguous United States. The inset of zone examples includes a state with elongated east-west zones (Minnesota) and a state with elongated north-south zones (Illinois).

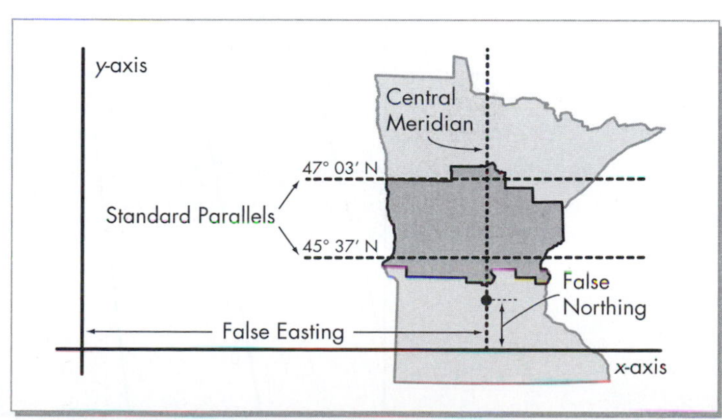

FIGURE 3.21 MINNESOTA CENTRAL ZONE OF THE SPC SYSTEM.

Positioning of the Lambert's Conformal Conic standard parallels, and the central meridian, as well as the SPC origin and axes for this zone. The exact positioning of the *x*- and *y*-axes varies from state to state and zone to zone, but the axes are always placed so that all coordinates are positive. (See text for discussion.) Note that the projection produces a reasonably aesthetic shape for Minnesota.

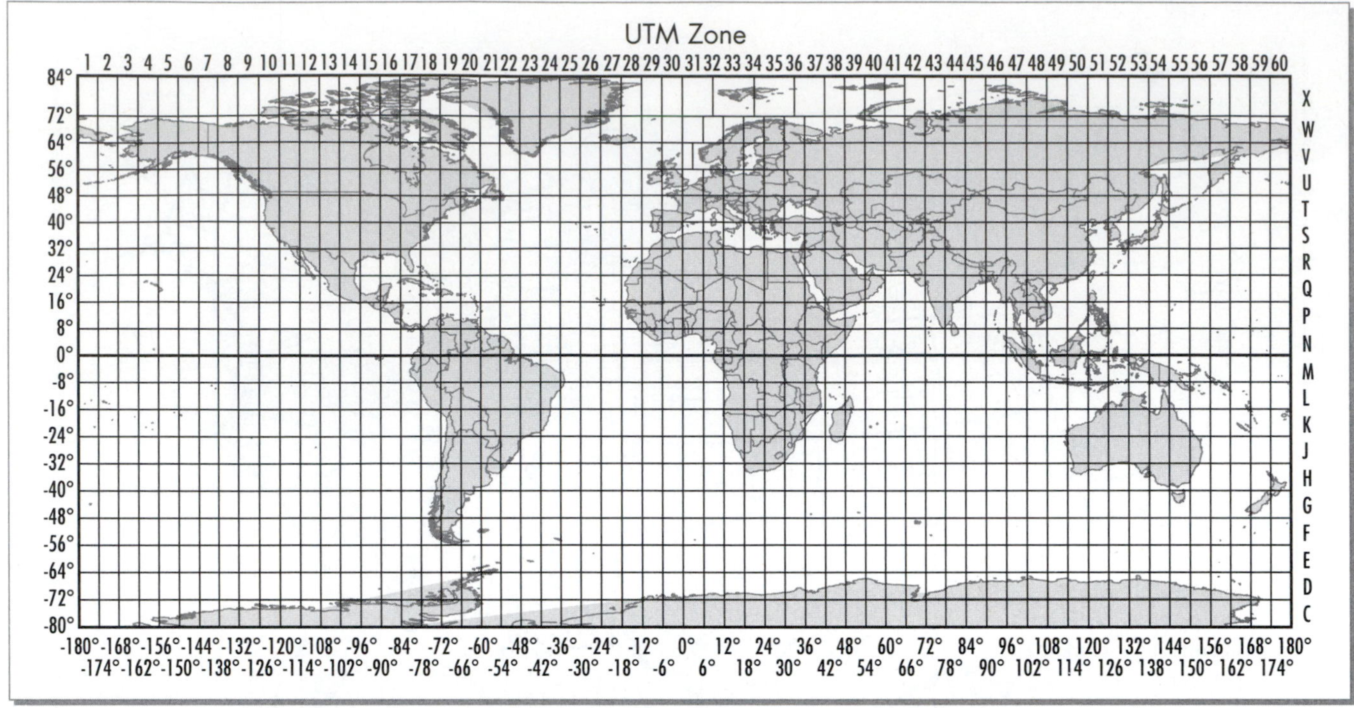

FIGURE 3.22 THE UNIVERSAL TRANSVERSE MERCATOR SYSTEM.

The UTM system is made up of sixty −6° zones that extend from 80° South to 84° North. The numbers on the left and bottom portions of the graphic are latitude and longitude values in decimal degrees. The letters on the right side of the graphic are referred to as zone designators (after Dana 1999). While zone designators are in the official description of the UTM system (Van Sickle 2004), almost all civilian uses of this system simply use the UTM zone(s), along with a designation of N or S to indicate the relevant hemisphere.

The UTM system is a projected coordinate system that covers the entire world from 80° South to 84° North. It is subdivided into 60 six-degree zones that are elongated in the north-south direction (see Figure 3.22). For example, zone one extends from 180° to 174° W as well as from 80° S to 84° N. The continental United States has ten zones (zones 10–19) that extend from 126° W to 66° W (see Figure 3.23).

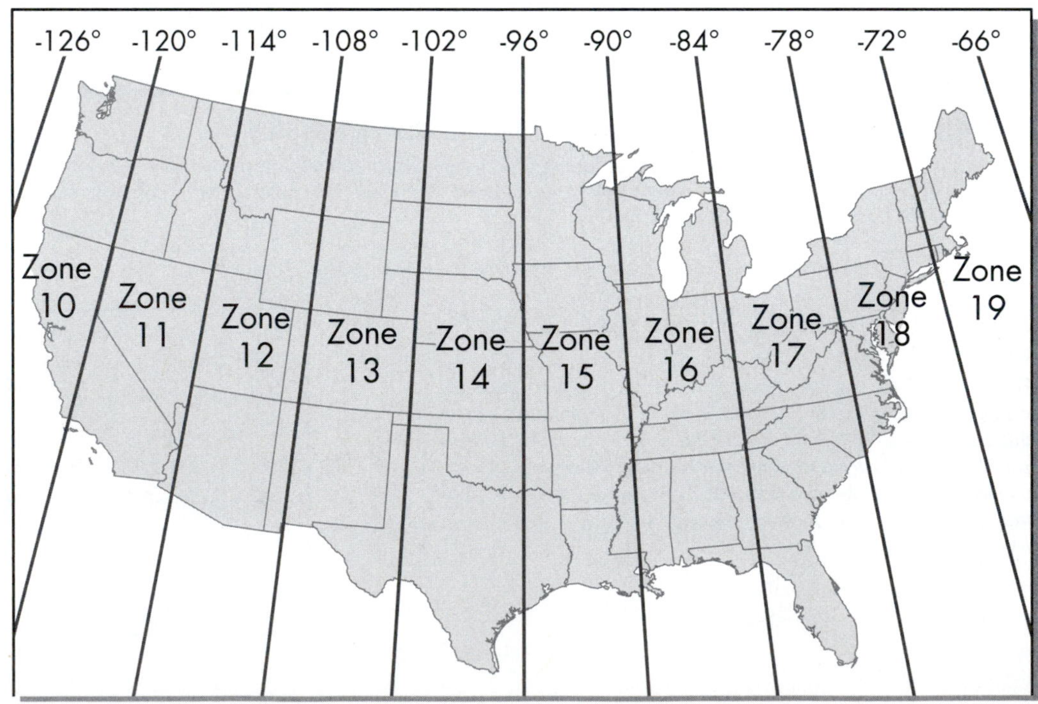

FIGURE 3.23 THE TEN UTM ZONES FOR THE CONTINENTAL UNITED STATES.

Zones 10–19 are used in the continental United States.

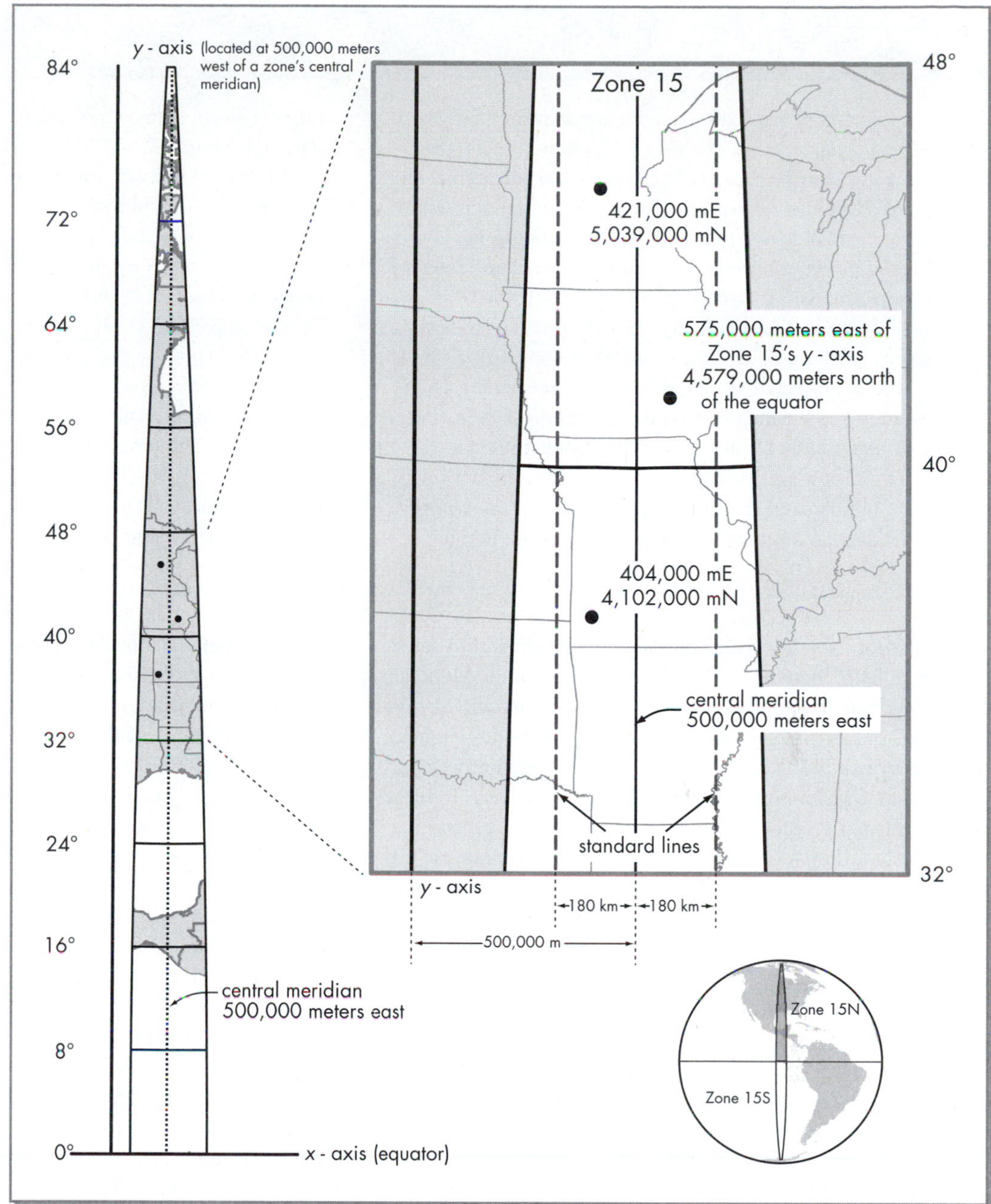

FIGURE 3.24 UTM ZONE 15 NORTH.
This diagram illustrates some of the most important aspects of a UTM zone. See text for discussion.

The projection for this system is the secant case of the Transverse Mercator (just like the north-south zones of the SPC). The transverse cylinder is positioned such that the central meridian runs north-south through the zones center. The central meridian for zone one is 177° W. The two standard lines (small circles) will straddle the central meridian at 180 kilometers (111.85 miles) to the meridian's east and west. Relatively low scale distortion is thus maintained in the north-south direction. The scale factor is about 0.9996 along the central meridian, and increases to one at the standard lines (Van Sickle 2004). The cylinder is repositioned 60 times, once for each zone.

The civilian coordinates are based on the equator, which serves as the *x*-axis. The *y*-axis is positioned at 500,000 meters west of a particular zone's central meridian. By generating this false origin, as with the SPC, all coordinates will now be positive. In the northern hemisphere, then, a point is measured as so many meters east (of the *y*-axis that intersects the false origin) and so many meters north (of the equator) (see Figure 3.24). In the southern hemisphere, the origin is moved south by 10,000,000 meters, which is not too far from the South Pole. But again, by having a false origin to the west and south of a particular zone's area, the coordinates will always be positive to the east and north of a particular zone's false origin.

As with SPC, UTM is often used to map an entire state, since there is relative accuracy of the coordinates, and so much data is readily available in this format. However, in Figure 3.23 it can be observed that many states are not centered or even entirely contained within each zone. Zone boundary

GIS "PROJECTION ON THE FLY"

Many students new to GIS mapping are unaware that when they load a map and select a map projection in their software, that the projection that they see on the screen may not necessarily be the same projection as the one in their original map data. When there is a difference between the coordinates or projection in the map data and the coordinates and projection being displayed in the GIS software it is known as "projection on the fly" (Environmental Systems Research Institute 2007). For example, if our original world map is in unprojected latitude-longitude coordinates, and we want to see the map in a Winkel Tripel projection, some GIS software allows you to make this projection on the fly without altering the original data's projection or coordinate system. In this case, the cartographer sees the map within the GIS display in the Winkel Tripel projection, but the original world map data *is still in unprojected latitude-longitude coordinates.*

While most software can effectively translate between projections and coordinate systems, the GIS software must "know" what projection and coordinate system is inherent in the data. In other words, it is not possible for a software to project or otherwise change coordinates for data that does not have its projection/coordinate system properly defined.

As a final note, working across two coordinate systems and projections can be confusing for some cartographers, and in some cases errors result when trying to edit the data across the two systems. Thus, many work environments require their employees to edit within the original projection and coordinate system or enact changes to the projection of the *original* data (and *not* project on the fly) before editing the map in the GIS.

determination was not based on "nice positioning" but rather on the regular 6° increments. States such as California, Montana, Kentucky, and Virginia (among others) are squarely in two zones. When people choose one zone or another, they find that the state's positioning does not look correct. The northern part of the state, for example, may tilt too far east or west. In these cases we usually recommend trying to move the central meridian eastward or westward to make the state look correct. If this is unsuccessful, change to an SPC coordinate system, or choose an appropriate projection in which the standard lines and central meridian can be adjusted to go through the state.

Some states, such as Wisconsin (which is also split by two UTM zones), have developed their own projected coordinate system, precisely to avoid such a split. This system is similar to UTM, except that they designated 90° W as the central meridian. The Wisconsin Transverse Mercator, as it is called, is available in both NAD27 and NAD83 datums (Wisconsin State Cartographer's Office 2004). This is one example among many of how states and other local governmental organizations respond to mapping needs on a more localized level.

A survey of state (or county or country) systems is beyond the scope of the text, but if creating thematic maps at a localized scale (even down to the city level) then it may be prudent to be aware of systems that are standard in that area. A local system may be required if you are making maps as a city or county employee. Even if it is not a *requirement*, if your state or other political unit has a projected coordinate system that is supported in the software being used, examine its properties and parameters. If these are acceptable, then you just may have found another viable projection option for mapping that area.

Adjustments in Projection Parameters

Many times, cartographers assemble maps that are groups of counties, states, countries, or other levels of areas and features that are sometimes outside a particular projection or projected coordinate system's "ideal" region. This final brief but important discussion centers on adjusting projection parameters to some of these specialized cases.

Perhaps the most important guideline in map projections is the centering of the projection of the map on the area of interest. As suggested in a number of places throughout the chapter, this means having the standard point, line or lines, as well as the central meridian, run directly through your study area or area of interest. Figure 3.18 is illustrative of this principle. What many students are often unaware of, however, is that if they are using a GIS or mapping software approach to projections, these parameters can be adjusted easily to suit the projection needs of a particular region.

For example, if the cartographer is going to make maps of the Mid-Atlantic and New England states region of the United States (see Figure 3.25a) he or she may start out selecting the Albers Equal Area Conic projection for the continental United States (see Figure 3.20). And this may not always be a bad choice, if the other states are going to be present to give a context to these states. However, if these states are extracted, as they are on so many thematic maps produced today, then the default central meridian (at 96° West or −96° in decimal degrees) and standard parallels (near which is the least scale distortion, at 29.5° and 45.5° North) no longer center the essential projection parameters on the newly created study area. Adjustments can be made to the standard parallels and central meridian, such that the projection appears better centered (on −47°), and will be more accurate by the adjustment of the standard parallels to 45° and 42°, respectively (see Figure 3.25b). For any other region, then, we suggest starting with the projection or projected coordinate system that most closely approximates the ideal projection properties and parameters, and then make the appropriate adjustments.

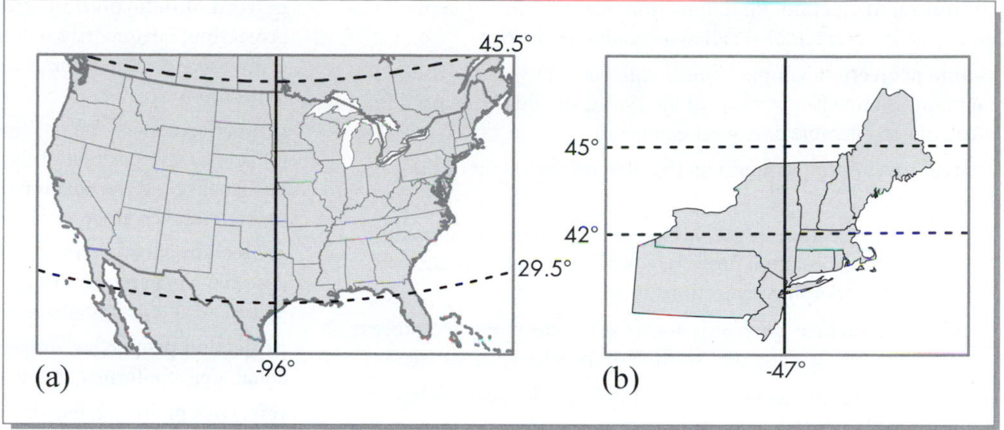

FIGURE 3.25
REPOSITIONING PROJECTION PARAMETERS FOR THE MID-ATLANTIC AND NEW ENGLAND STATES.
Default Albers Equal Area Conic projection for continental United States (a) can provide a great starting point. But adjusting the standard parallels and central meridian (b) can produce greater accuracy and an improved aesthetic look.

REFERENCES

Anderson, P. 2007. *A Gallery of Map Projections*. http://www.galleryofmapprojections.com

Campbell, J. 2001. *Map Use and Analysis*. 4th ed. Dubuque, IA: McGraw-Hill.

Canters, F. 1989. New Projections for World Maps: A Quantitative-Perceptive Approach. *Cartographica* 26: 53–71.

Chamberlin, W. 1947. *The Round Earth on Flat Paper.* Washington, DC: National Geographic Society.

Committee on Map Projections. 1989. Technical Notes. *American Cartographer* 16(3): 222–23.

Dana, P. 1999. *Coordinate Systems Overview*. *The Geographer's Craft Project.* Department of Geography, The University of Colorado at Boulder. http://www.colorado.edu/geography/gcraft/notes/coordsys/coordsys_f.html

Dent, B. 1987. Continental Shapes on World Projections: The Design of a Poly-Centered Oblique Orthographic World Projection. *Cartographic Journal* 24: 117–24.

Environmental Systems Research Institute. 2007. *ArcGIS Desktop Help*. Online Help for ArcGIS 9.2.

Faintick, M. 1986. The Politics of Geometry. *Computer Graphics World*. May: 101–4.

Geocart. 2007. *Geocart's Projections List: Simply Professional Map Projections*. http://www.mapthematics.com

Goldberg, D., and J. Gott. 2007. Flexion and Skewness in Map Projections of the Earth. *Cartographica* 42(4): 297–318.

Hilliard, J., U. Bosoglu, and P. Muehrcke. 1978. *A Projection Handbook.* Madison: University of Wisconsin Cartographic Laboratory.

Iliffe, J. 2000. *Datums and Map Projections*. Caithness, UK: Whittles Publishing.

Loxton, J. 1985. The Peters Phenomenon. *The Cartographic Journal* 22: 106–8.

Marschner, F. 1944. Structural Properties of Medium- and Small-Scale Maps. *Annals of the Association of American Geographers* 34: 44.

McDonnell, P. 1979. *An Introduction to Map Projections.* New York: Marcel Dekker.

Monmonier, M. 2004. *Rhumb Lines and Map Wars*. Chicago: University of Chicago Press.

National Oceanic and Atmospheric Administration. 1989. *State Plane Coordinate System of 1983,* Manual NOS NGS 5. Rockville, MD: National Geodetic Information Center.

Natural Resources Canada, 2007. *The Atlas of Canada—Map Projections*. http://atlas.nrcan.gc.ca/site/english/learningresources/carto_corner/map_projections.html

Raisz, E. 1962. *Principles of Cartography.* New York: McGraw-Hill.

Robinson, A. 1985. Arno Peters and His New Cartography. Views and Opinions. *American Cartographer* 12: 103–11.

Robinson, A., J. Morrison, P. Muehrcke, A. Kimerling, and S. Guptill. 1995. *Elements of Cartography.* John Wiley and Sons.

Snyder, J. 1987. *Map Projections—A Working Manual*. U.S. Geological Survey Professional Paper 1395. Washington, DC: USGPO.

———. 1988. Social Consciousness and World Maps. *The Christian Century,* February 24: 190–92.

———. 1993. *Flattening the Earth: 2000 Years of Map Projections*. Chicago: University of Chicago Press.

Steers, J. 1962. *An Introduction to the Study of Map Projections.* London: University of London Press.

Stocking, A. 2005. Everything is Somewhere: Two Views of One Planet. *The American Surveyor* 2 (7): 54–56.

Strebe, D. 2007. Map Projection Essentials. http://www.mapthematics.com/Essentials/Essentials.html

Tissot, A. 1881. *Memoire sur la Representation des Surfaces et les Projections des Cartes Geographiques.* Paris.

Van Sickle, J. 2004. *Basic GIS Coordinates*. Boca Raton, FL: CRC Press.

Wisconsin State Cartographer's Office. 2004. Regional Coordinate Systems. http://www.sco.wisc.edu/pubs/wiscoord/regional.php

GLOSSARY

Albers Equal Area Conic projection secant conical projection having equal area properties; useful for mapping areas of east-west extent

azimuthal projection directions from the projection's center to all points are correct; also called a zenithal projection

Bonne projection simple conical equal area projection, useful for mapping continent-size areas of the Earth; should not be used for areas of considerable east-west extent

central meridian the meridian that defines the center of the projection

conformal projection preserves angular relationships at points during the transformation process; cannot be equal-area; also called an orthomorphic projection

developable surface geometric form used in the projection process without tearing, shearing, or distortion of the geometric surface

equal area map projection no areal deformation; cannot be conformal; also called an equivalent projection

equidistant projection preserves correct linear relationships between a point and several other points, or between two points; cannot show correct linear distance between all points to all other points

equivalent map projection also the property of equivalence; see equal area map projection

Gall-Peters projection rectangular equal area projection emphasizing lower latitude countries; used by several UN organizations for world mapping

gnomonic projection the hypothetical light source that projects points onto a plane surface is placed at the center of the globe

Goode's Homolosine projection classic interrupted equal area projection of the world

great circle circle that results when a plane bisects the Earth into two equal halves (for example, the equator); the great circle arc is a segment of the great circle that is the shortest distance between two points on the spherical surface

Hammer projection equal area projection useful for world mapping; also called Hammer-Aitoff

Lambert Azimuthal Equal Area projection has equal area properties useful for mapping up to hemisphere areas (in its equatorial aspect) or up to continent-size areas (in its oblique aspect) on the Earth, especially if the area extends in all directions

Lambert Conformal Conic projection has conformal properties; used for areas with an east-west extent; one of projections used in the State Plane coordinate system (SPC)

latitude of origin latitude of the projections' origin, such as the equator

Mercator projection historical conformal map used for navigation; distortion in higher latitudes make it unsuitable for most world mapping applications

minimum error projection no equivalency, conformality, azimuthality, or equidistance; chosen for its overall utility and distinctive characteristics

Mollweide (or homolographic) projection equal area projection useful for world mapping

orthographic projection the hypothetical light source that projects points onto a plane surface is placed at a theoretical infinity

pattern of deformation distribution of distortion over a projection; customarily increases away from a standard point or line(s) (see also *standard points, lines, parallels, and meridians*)

projection aspect the position of the projected graticule relative to the ordinary position of the geographic grid on the Earth; can be visualized as the position of the developable geometric surface to the reference globe

projection parameters latitude and/or longitude values that describe a map projection's standard point, line or lines, its central meridian, and its latitude of origin

projection properties properties of a projection that make it an equal area, conformal, equidistant, or azimuthal

reference globe the reduced model of the spherical Earth from which projections are constructed; also called a nominal or generating globe

Robinson projection classic minimum error (compromise) projection used in world mapping

scale factor ratio of the scale of the projection to the scale of the reference globe; 1.0 on standard lines, at standard points, and at other places, depending on the system of projection

secant case in a map projection, each of the surfaces is brought to a line (with the plane) or two lines (with the cylinder and the cone) of intersection on the reference globe

Sinusoidal projection a mathematical equal area projection for world mapping

small circle any circle on the spherical surface that is not a great circle; for example, parallels (except for the equator) are small circles

standard points, lines, parallels, and meridians one of the map projection's key parameters; the point, line, or lines on a map that have no scale distortion. In regard to geometric figures, places where the plane, cylinder, or cone touches or intersects the reference globe during the projection process

State Plane Coordinate (SPC) system rectangular plane coordinate system applied to states; employs the Lambert Conformal Conic or Transverse Mercator projection for each zone

stereographic projection the hypothetical light source that projects points onto a plane surface is placed on the opposite side of the globe with respect to the plane's point of tangency

tangent case in a map projection, each of the geometric figures are brought to a point or line of tangency on the reference globe

Tissot's indicatrix mathematical construct that yields quantitative indices of distortion at points on map projections

Transverse Mercator projection has conformal properties; maintains little or no scale distortion in the north-south direction. One of the projections used in the State Plane coordinate system

Universal Transverse Mercator (UTM) system rectangular plane coordinate system applied to north-south zones worldwide; employs the Transverse Mercator projection for each zone

Winkel Tripel projection minimum error projection used in world mapping, may be least error in minimum error projections

4 THE NATURE OF GEOGRAPHIC DATA AND THE SELECTION OF THEMATIC MAP SYMBOLS

CHAPTER PREVIEW The presentation of spatial data using thematic maps assumes an understanding of both the nature of the data and the symbolization necessary for thematic map design. Spatial data is described by (1) its characteristics of location, form, and time, and (2) its level of measurement hierarchy. Characteristics of location tell us about the manner in which the data are distributed, that is, do the data distributions exist as points, lines, or areas. Form describes the data characteristics according to its inherent nature, such as are the data qualitative or quantitative, discrete or continuous, totals or derived. The characteristics of the spatial data dictate the type of thematic map used. Data measurement (nominal, ordinal, interval/ratio) categorizes the data according to a hierarchical structure.

Thematic maps are designed by matching the nature of the data to map symbols. Thematic map symbols use point, line, and area symbolization for displaying the spatial nature of the data. Such symbols are matched to visual variables that are components of the graphic system. Symbol placement positions the data in geographic space while the nature of the symbol provides communication clues that permit the understanding of the data distribution. The nature of symbols must match the nature of the data.

Before we begin the cartographic process, we should ask ourselves why we create a map in the first place. Normally, we have a reason to generate a map. Such a reason may be to communicate graphically something that we know (personal knowledge) to someone else (public knowledge), the reader of the map. We create maps for many of the same reasons that we take photographs, send emails to friends, or post a short movie on the Web. We want to show something or tell something to someone else about things we've seen or things we know, or we want to present a map of our area. We also create maps to visually display geographic phenomena. A map becomes a form of communication similar to the written or spoken word. ■

Cartography, like language, use rules so that both the presenter and the receiver of information understand what is being communicated. Both maps and language communicate using symbology. For example, the word "tree" is a term that we associate with a vegetative form. How the word is interpreted, for example, a broadleaf or needle-leaf tree, is based upon the experiences of the individuals. Cartographically we use symbols to represent a tree. Some of these symbols may resemble a tree, replicative symbols, on the landscape while others may be a more abstract form. Whether or not communication occurs is dependent upon an understanding of the symbology by both the cartographer and the map reader.

Karl Sauer's (1956) comment that maps are the language of geography suggests a strong link between language, cartography, and geography (see Chapter 1). Thus, the world of maps allows for the display of information about the milieu, our physical environment, whether it be above, on, or below the Earth's surface, and about our cultural environment, including socio-economic, demographic, political, and other human-induced activities. These maps may also pertain to statistical data that are collected about a specific location or region that can vary from a small area such as your backyard or a much larger area such as the entire Earth (see the discussion of scale in Chapter 1).

The subject matter and intent of the map will dictate the map type created. General purpose maps (see Chapter 1) include those topics that display a planimetric base on which we map the locations of cultural and physical features, for example, buildings, parks, lakes, forests, streets, or political boundaries. These maps are created for areas such as cities and towns, counties, recreational areas, and other locations where we want to communicate the distribution and location of such entities. State transportation departments and private mapping corporations create highway maps that focus on the transportation network of a particular area. The general purpose map is considered to display information qualitatively by using symbols to communicate distributions on the cultural landscape.

Thematic maps have traditionally involved the display of quantitative data that are easily incorporated into a GIS or advanced mapping software so that we can communicate the spatial variability of that data. As we will see as we progress through the remainder of the book, the characteristics of the data will guide us in the selection of which thematic map type to use. That is, the data dictate the map type for the display of the information.

Within a GIS database lies the numerical foundation from which mapping occurs. The data are presented in a row and column format in which the geographic location, whether it be for a specific location that can occur at a point, along a line, or within a polygon, is normally represented in rows and the many attributes collected for that location are presented in the columns. As an example, we may collect the total population, median family income, percentage of owner-occupied housing, or other such data for each county in Ohio. When we view the database, its contents represent nonspatial data, also referred to as attribute or characteristic data (Meyer 1997). That is, the data are independent of their geometric relationships. The maps we create from that database display the information in a spatial context such that we observe its distribution which permits us to visualize patterns and to draw inferences. The data, when mapped, become spatial data which display characteristics of location, size, or amount.

The remainder of this chapter will deal with the nature of spatial quantitative data. We will also investigate the statistical characteristics of the data and how they assist us in the selection of which thematic map to use. Presented too is a discussion of the visual variables used in creating thematic maps. We conclude with an examination of how statistics are used in data analysis.

THE NATURE OF DATA

Thematic maps are used to display the spatial qualities of geographic data. A distinction must be made, however, between **geographic phenomena** and **geographic data.** Data are facts observed or measured from which conclusions can be drawn. *Attribute* is another word that can be used in place of *data.* Geographic data are selected features (usually numerical) that geographers use to describe or measure, directly or indirectly, phenomena that have a spatial quality. This information is the attribute of a sampled location that is either qualitative or quantitative. For example, the phenomenon of climate can be described in part by looking at precipitation data. The use of thematic maps allows the user to understand the *where, what,* and *when* factors of spatial data. These three represent the nature of all data as they are tied to location, characteristics, and time. Data are gathered by geographical location, which we will see later in this chapter is directly tied to the symbolization form used to map that data. These locations may be at a point, along a line, or over an area. Their spatial components are generally considered as zero-dimensional (point), one-dimensional (line), two-dimensional (area), three-dimensional (volume), and four-dimensional (space-time continuum) (Harvey 1969; see Figure 4.1). Location characteristics can be directly influenced by and may vary according to the map scale and time. Most geographical study involves explanation of phenomena of the first four kinds, and to some degree the fifth.

An interesting feature of phenomena is that their form is intricately related to scale and may change with the level of

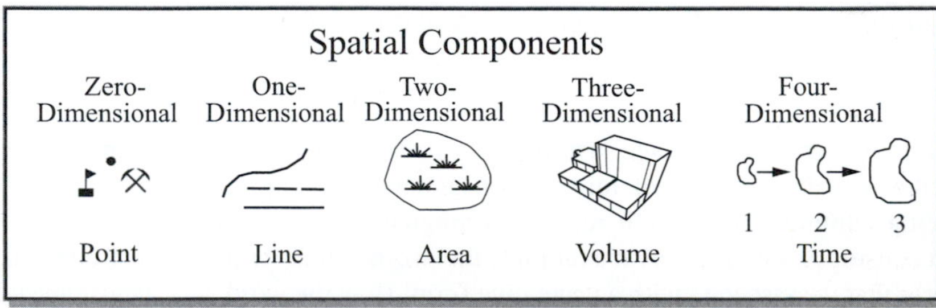

FIGURE 4.1 THE SPATIAL COMPONENTS OF MAP SYMBOLS.
Point symbols are considered as zero-dimensional as they are indicative only of an *x–y* location.

inquiry. A city, for example, may be a point phenomenon at the macroscale but can be considered to have two-dimensional qualities when examined at micro levels. At one level of investigation, a road may be a link between points (one-dimensional), but at micro levels the road can have the two-dimensional, areal qualities of length and width.

Examples of volume phenomena include landforms, oceans, and the atmosphere. Other geographic phenomena that are conventionally treated as three-dimensional because of their similarity to volumes are rainfall, temperature, growing-season days, and such derived ratios as population density and disease-related deaths. Rainfall collects in a glass and is volumetric; people piled close together make up a volume. Space-time phenomena are best exemplified by succession (for example, human settlement over time) and migration/diffusion. Location characteristics can be directly influenced by and may vary according to the map scale and time.

We observe these phenomena but we map data. Such data may take on various characteristics that include location, form, and time.

Data Characteristics

Location

Observations of data may be found as a point, line, or area feature. Attributes are assigned to these locations within a database and then mapped in order to depict the spatial variation of that data.

Point Data. The implication of point data is that it is applied to a specific location that has a unique geospatial coordinate set (*x*–*y*)—such as generic *x, y* positions, or latitude-longitude or eastings and northings (which are actually *y*–*x*). Within a particular topic, population for example, there will be only a single value for each point. However, within the attribute database there may be any number of these point-data relationships that have the potential to be mapped. The data value is referred to as the attribute, the location (point) as a node or location ID, and the number of individual points as the observations of the data set. Point data may be counted, measured, or estimated (see Figure 4.2). Examples of these are listed in Table 4.1.

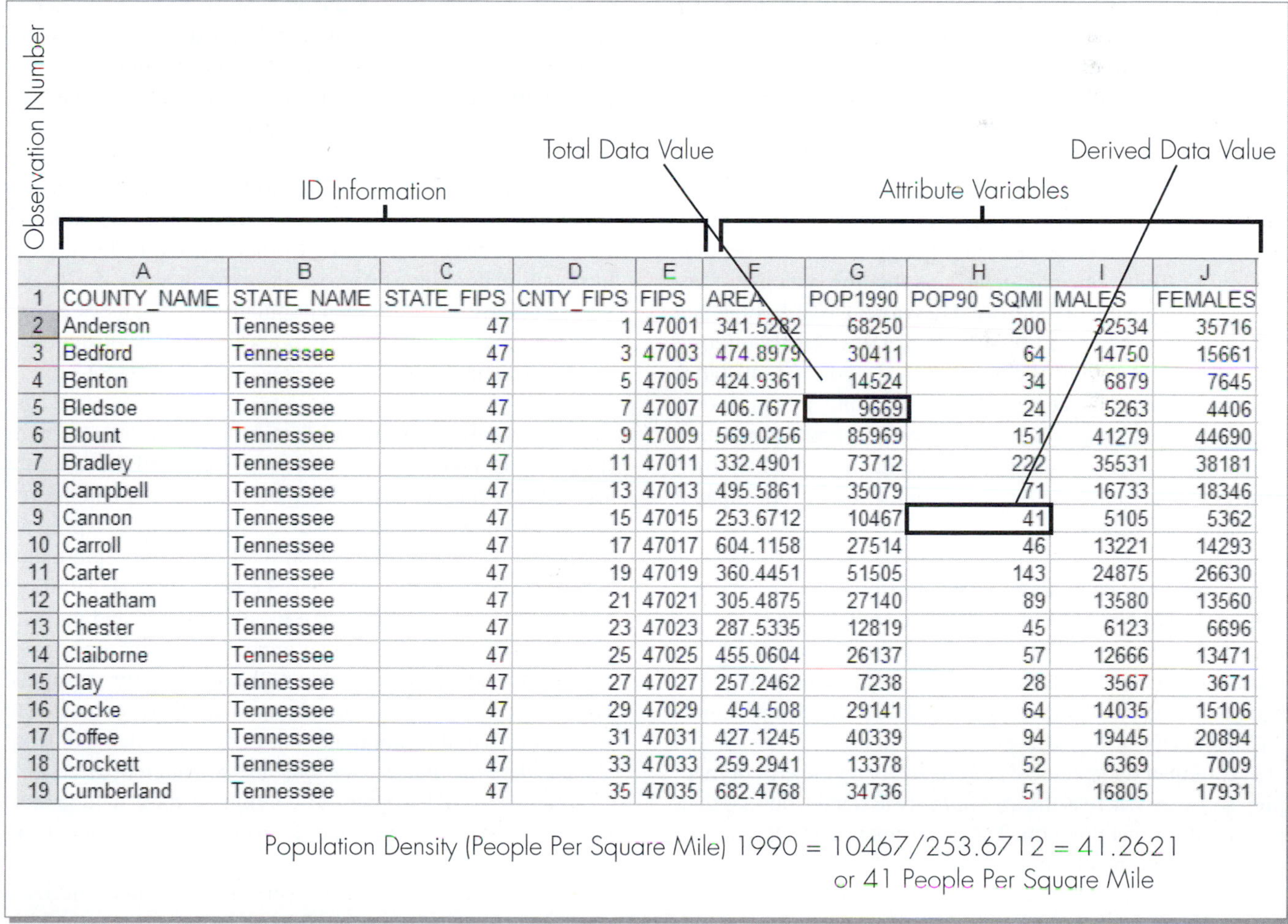

	A	B	C	D	E	F	G	H	I	J
1	COUNTY_NAME	STATE_NAME	STATE_FIPS	CNTY_FIPS	FIPS	AREA	POP1990	POP90_SQMI	MALES	FEMALES
2	Anderson	Tennessee	47	1	47001	341.5282	68250	200	32534	35716
3	Bedford	Tennessee	47	3	47003	474.8979	30411	64	14750	15661
4	Benton	Tennessee	47	5	47005	424.9361	14524	34	6879	7645
5	Bledsoe	Tennessee	47	7	47007	406.7677	9669	24	5263	4406
6	Blount	Tennessee	47	9	47009	569.0256	85969	151	41279	44690
7	Bradley	Tennessee	47	11	47011	332.4901	73712	222	35531	38181
8	Campbell	Tennessee	47	13	47013	495.5861	35079	71	16733	18346
9	Cannon	Tennessee	47	15	47015	253.6712	10467	41	5105	5362
10	Carroll	Tennessee	47	17	47017	604.1158	27514	46	13221	14293
11	Carter	Tennessee	47	19	47019	360.4451	51505	143	24875	26630
12	Cheatham	Tennessee	47	21	47021	305.4875	27140	89	13580	13560
13	Chester	Tennessee	47	23	47023	287.5335	12819	45	6123	6696
14	Claiborne	Tennessee	47	25	47025	455.0604	26137	57	12666	13471
15	Clay	Tennessee	47	27	47027	257.2462	7238	28	3567	3671
16	Cocke	Tennessee	47	29	47029	454.508	29141	64	14035	15106
17	Coffee	Tennessee	47	31	47031	427.1245	40339	94	19445	20894
18	Crockett	Tennessee	47	33	47033	259.2941	13378	52	6369	7009
19	Cumberland	Tennessee	47	35	47035	682.4768	34736	51	16805	17931

Population Density (People Per Square Mile) 1990 = 10467/253.6712 = 41.2621
or 41 People Per Square Mile

FIGURE 4.2 ATTRIBUTE TABLE.

Observations are county-based values. The ID information is used by software to match the values to the map locations. Examples shown here include both discrete and derived data values.

TABLE 4.1 NATURE OF DATA DEFINED BY LOCATION

Data Type	Examples
Point Data	Number of shoppers at a mall
	Number of students enrolled at an elementary school
	Temperature measured at the local airport
	Dollar value of your home
	Value of damage caused by a tornado that hit your town
Line Data	Traffic flow in vehicles per day
	Stream discharge
	Movement of a commodity exported to various locations
Area Data	Population density in people per square mile
	Crop yield in bushel per acre
	Acreage of various county land-use categories
	Precipitation from a storm event

Line Data. Quantitative line data are generated by the collection of information as applied along a linear feature or path. Thus a series of nodes are used to define the path and the attributes are used to define the characteristics of that path. For example, transportation flow maps are generated by a series of data collection points along a street or highway. These points are selected by the traffic engineer as representative of travel along a particular corridor. These can be identified by the rubber tubing stretched halfway across a street and connected to a counting device mounted alongside the street. By determining how many vehicles pass that point in a given time period and then associating that point to the other collection locations, one can determine the flow volume of vehicles per day along a roadway. Therefore, a set of point data attributes is used to create line data. Data can be inferred only along the length of the line based upon the attributes determined to exist at various points along that line (see Table 4.1). Data may also be assigned to a line representing the quantity of goods shipped from one location to another. The data would apply to the length of the line without spatial variation along the line.

Area Data. Attributes that exist over a two-dimensional extent (polygon) comprise area data. In certain instances we are able to determine the attribute from aerial photography or other remote sensing techniques. The area of the polygon in square kilometers of a lake or land in forest can be determined based upon the scale of the photograph or image. The polygon area is frequently generated automatically by the GIS or mapping software whereas the area *data* are assigned, observed, or measured for that polygon. The government collects statistics that tell us how much land area is associated with a particular land use. By knowing the total number of acres in farms and the total bushels of corn produced by those farms, we can calculate an attribute of bushels per acre. Other common examples of area data include population density of people per square mile (see Figure 4.2). Area data include attributes that may be applied to the entire area as measured by a series of point locations within that area or by a quantity thought to exist uniformly over the entire area (see Table 4.1).

A special case of area data is that of volume data comprised of attributes that exist over a three-dimensional extent of an interpolated surface. An example of these data is the total volume of precipitation that falls over a particular stream's drainage basin. There may be a network of precipitation gauges that are used to measure the amount of precipitation at a point and these data are extrapolated out to the entire surface area. The total volume of precipitation for a particular storm event can be calculated once the interpolated surface has been created (see Chapter 9).

Form

Data form establishes the contextual characteristics in which the data are found. Three context forms permit the user to understand the details of the data characteristics. These forms include either quantitative or qualitative, discrete or continuous, and total or derived data.

Qualitative/Quantitative Context. The form of data characteristics is comprised of three relationships. The primary form of the data is determined by the qualitative versus quantitative nature. The qualitative attribute of the data describes the inherent nature of the feature, for example, house, church, railroad, swamp (see Table 4.2a). These features may appear as point, line, or area locations and are described according to their cultural or physical form. Quantitative data utilize numerical values to indicate the differences in attributes according to some measurement scale, for example, feet above sea level, total population, or degrees Celsius. The quantitative data form also applies to any of the locational characteristics.

Spatial Context. This second data form is determined by its distribution. Point data is unique in form and represents a value that applies to a specific node within the spatial framework. Data may be considered as unique if the node represents a single *x–y* coordinate location or if the node represents a centroid identifying a line or area component. That is, both temperature data measured at a weather station and the total population of a county or other enumeration unit are considered as discrete data (see Table 4.2b). Frequently associated with data counts or totals, the data are uniquely associated to a singular spatial location, no matter how it is defined. Nonunique data are those data that are considered to be of a continuous form. Examples of continuous data include most weather-related variables and the topographic surface. Continuous data are areal and exist everywhere. It is impractical to measure these variables at every position so we utilize a sampling procedure to collect data at various locations within the areal extent. These data may then be displayed as a computed continuous variable using interpolation procedures (refer to Chapter 9).

TABLE 4.2 CHARACTERISTIC FORM OF DATA

a) Characteristic View

Qualitative	versus	Quantitative
Mine		Tons of coal
Airport		Number of aircraft
River		Volume of water
Forest		Board feet of timber
Farms		Acres of farmland

b) Spatial View

Discrete	versus	Continuous
Temperature at your home		Temperature across the United States
Precipitation at the airport		Precipitation in the Southeast
Elevation of the bridge		Topographic surface

c) Attribute View

Totals	versus	Derived
Total population		People per square kilometer
Ohio immigrants to Minnesota		Ohio immigrants to Minnesota as a percentage of all Minnesota immigrants
Employment in mining		Mining employment as a percentage of all employment

Attribute Context. This data form is characterized as either a total or derived form. Total data, similar to discrete used in the spatial context discussed above, may be observed, collected, or measured at a location so that it has a single value. That value is associated only with the location for which it was acquired. One can easily identify a total value by the fact that the data are counts or measurements identified by their descriptor, for example, 12 degrees Celsius or 9669 people. In this context, total data may be considered as a single observation of one variable among possibly many variables of a larger data set (see Figure 4.2). Derived data are represented by an attribute definition that indicates the data have been mathematically calculated. This is frequently done to normalize (standardize) the data so as to either adjust for the impact of area/size or to represent the data as a rate or percentage (see Chapter 5). Derived data are also used to make comparisons, for example, the variation in crime rates per county.

Area plays an important role in the potential quantity that can occur within its bounds. Logically, a county of larger area should be able to hold a larger population or larger farms should be able to produce a greater quantity of crops. In order to make these data comparable, we must normalize the data adjusting for the area differences. This is achieved by dividing the area into the total population or into the total quantity of crops produced. The resulting values are represented as people per square kilometer or in bushels per acre. Similar logic applies to data characterizing a sector of the population. Expressing data in the form of rates or percentages normalizes the data to a specific size of population or a number per 100. Table 4.2c provides examples of these data forms.

The data characteristic of form is utilized in order to define its spatial association and/or to indicate whether the data have been normalized to adjust for the bias of size. In order to communicate effectively the nature of the data, the data are described according to its form: qualitative or quantitative; discrete or continuous; total or derived. These characteristics also play an important role in determining the thematic map type used to display that data. Again, the data dictate the map type for the display of the information.

Time

It is also important to recognize that all data are time specific. Therefore, time is a necessary attribute to a data set. We display data on a map and we tell the map reader the subject matter being presented. It is also necessary to communicate the time period for which the data applies. Data may have a very specific time stamp, such as 1200Z, September 13, 1944. (The Z is used to specify time adjusted to Greenwich mean time in order to eliminate confusion between time zones or the application of daylight savings time.) We find this specificity applying to meteorological data. We may also find data applying to a specific month, year, or decade.

Temporal data are used either to graph sequential (longitudinal) changes in values for a location or to determine magnitude of change between time period one and time period two. Frequently, this magnitude is expressed in terms of a percentage change since the initial date. For example, the change in population between 1960 and 2005 may be mapped as a percentage change between the two time periods. Change values may be either positive or negative, depending on the population characteristics over time. Temporal data may include only a beginning and ending date as just described or the data may include a variety of time periods. Even though time is a continuum and thus uniform, the collection periods may not be uniformly spaced.

Whatever the time period, we are obligated to communicate this in the map's title. The date associated with the data source does not necessarily communicate the date of the data. For example, in many statistical publications one can find historical data that can be mapped. Only the data year and not the publication year should be communicated to the map reader. Temporal data may also cover a period of time, multiple years or multiple decades. Again, the communication of time sequence displayed is what is important.

Data Transformations

The cartographer will frequently modify the data prior to its display. Data collected or imported may be appropriate as nonspatial tabular data and yet not be ready for cartographic applications. In order to prepare the data for mapping, the cartographer will manipulate either the attribute data or the geographic space to which it applies. Three forms of transformations which may occur prior to mapping include the modification of scale, form, or geometric boundary.

Scale

The first is the modification of the data based upon changes in map scale. Data that may be appropriate for a large scale map, such as 1:24,000, may not be appropriate for a small scale map, such as 1:1,000,000. If we are converting the scale of display from the large-scale to the small-scale map, the data may need to be aggregated. Thus, data that originally was displayed according to census tracts may require aggregating to county values at the smaller scale. Neighborhood statistics may be combined to represent city statistics. The transformations of the data as a result of changes to a smaller scale are easily done by merging the data into a single value. The data transformations include a change in the definition of the data location. Data that originally was applied to a point location may now be aggregated and applied to an area location (see Figure 4.3).

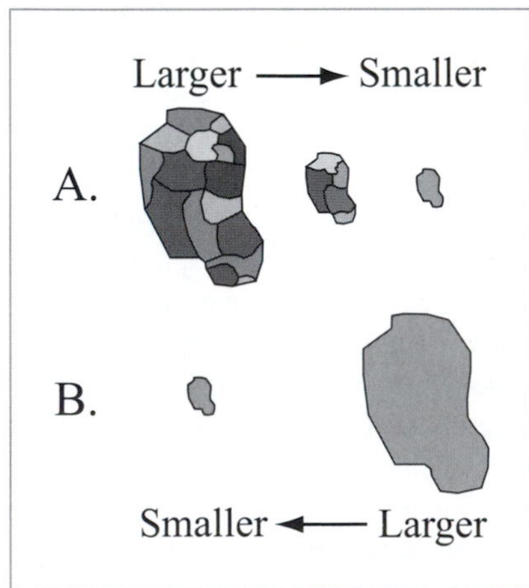

FIGURE 4.3 DATA AGGREGATION.
As the scale decreases, data should be aggregated in order to maintain clarity.

The transformation of data when changing to a larger scale presents different problems. Data that are collected for a larger area in order to produce the small-scale map will appear highly simplified when retained for the depiction at the larger scale. A disaggregation of the data requires a greater effort. Census data collected at the county level may also be available at the census tract level. If the data cannot easily be disaggregated, a return to the original data source may be required in order to acquire more appropriate data for the new map scale. This may require revisiting the original data source for acquisition of more refined data.

Form

A second type of transformation of the data may be associated with the data form. Data that are collected at a series of random sampling points may be used to transform the data from discrete point data to the display of the data as a continuous variable. Such a transformation is common when mapping surfaces. Data that occur at random locations will require a transformation to continuous data via interpolation (see Chapter 9).

Data form transformation may also involve the conversion of totals to rates or percentages. This transformation is important to adjust for the impact of size and area as explained above. In order for spatial comparisons to be made, this transformation process is frequently required depending upon the thematic map type being selected.

Boundary Changes

One problem frequently overlooked when comparing data in a temporal sequence is that the boundary of an enumeration unit may vary in different time periods. This is commonly referred to as the Modifiable Area Unit Problem (MAUP) and is a common occurrence when using U.S. Census tract boundaries (Openshaw 1984). Just as important as it is to compare data of the same data form, it is equally important to compare data covering comparable areas. The manner in which the Census Bureau defines census tracts and their boundaries often changes from one decennial census to another. Some boundary changes are minor, creating sliver polygons when the boundaries are overlaid. Other times the changes are significant. You will even find situations where the boundary remains the same but the census tract number is changed so that the tract number may appear in sequential censuses but applies to different polygons. As a cartographer, you must be aware that such boundary definitions and locations may be different between the dates of comparison. The simplest approach in addressing this problem is to aggregate the data to a smaller scale or county level in this example. Other more complicated approaches involve the use of spatial analysis modeling (Green and Flowerdew 1996) and/or remote sensing techniques for the adjustments of boundary changes (Holt *et al.* 2004).

DATA MEASUREMENT

S. S. Stevens, a noted scientist and psychologist, has said that measurement is the "assignment of numerals to things so as to represent facts and conventions about them" (Harvey 1969, 306). It has been customary in recent years for geographers to classify the ways they measure events into categories of data measurement. Measurement is an attempt to structure observations about reality. Ways of doing this can be grouped into four levels, depending on the mathematical attributes of the observed facts. A given measurement system can be assigned to one of these four levels: nominal, ordinal, interval, and ratio, listed in increasing order of sophistication of measurement. Methods of cartographic symbolization are chosen specially for representing geographic phenomena or data at these levels of measurement. The measurement scale will also play an important role in identifying the thematic map type to be used.

Nominal

Nominal scaling is the simplest level of data measurement (Taylor 1977), sometimes considered a qualitative measurement to be descriptive; answering the question of what is being mapped. An example would be the nominal identification of wheat regions, corn regions, and soybean regions. Each crop region is distinct; arithmetic operations between the regions are not possible at this level. Political party affiliation (Democrat, Independent, Republican), sex (male, female), and response (yes, no) are other examples. At this level of measurement, mathematical operations cannot be performed between classes. Equality/inequality between groups according to their classification or identification or the dominance spatially of one group versus another may be ascertained using this data measurement level.

Ordinal

The underlying structure of ordinal measurement is a hierarchy of rank. Objects or events are arranged from least to most or vice versa, and the information obtainable is of the "greater than" or "less than" variety. Ordinal measurement provides no way of determining how much distance separates the items in the array.

There are several types of ordinal measurement. In complete ordering, every element in the array has its own position, and no other element can share this position. This kind of ordinal measurement is considered relatively strong because statistical observations about the ranking are possible. This is not the case with the second main type, weak ordering, in which elements can share positions (called paired ranks) along the ordinal continuum. In the first instance there may be only one major seaport along a coast while in the second instance there may be several minor ports. The definition of major and minor may be determined by the number of ships serviced in a year or by the total tonnage of goods that pass through the port annually.

One interesting feature of the ordinal class is that we can attach any numerical scale to the ranking without violating the underlying structure. If we know that the order is K L M, we can have either K = 5, L = 3, M = 1, or K = 500, L = 300, M = 1, while retaining the original representation. Remember that in ordinal measurement we do not know how much difference separates the events in the array. For example, several geography students spend a summer touring 50 large cities throughout North America. After they return, their professor asks them to rank the cities based on their appeal, from best liked to least liked. In this array, a city's position is known in the overall ranking, but it is not possible to discern how much it differs from those ranking above and below it.

Other examples of the ordinal measurement are social class, social power (more, less), and agreement (strongly agree, strongly disagree). We may also utilize the ordinal class in refining a nominal description based upon importance of the category. Whereas, nominally we identify a linear feature as a road, we use the ordinal measurement to differentiate between a major road or minor road, an interstate highway or a national highway. Other examples would include the distinction between a seaport versus a major seaport or a forest versus a mixed deciduous forest. Again, the characteristic that defines ordinal apart from other approaches is that we cannot discern magnitude differences between the observations, only a hierarchical distinction of categories.

Interval

At this measurement, we can array the events in order of rank and know the distance between ranks. Observations with numerical scores at the interval measurement are important in geographical analysis because data at this level are needed to perform fundamental statistical tests having predictive power. Units on an interval scale are equal throughout; that is, one degree on the Fahrenheit scale is assumed to be the same regardless of whether it's at 22–23 or 78–79 degrees.

Magnitude scales at the interval level have no natural origin; any beginning point may be used. The classic example is the Fahrenheit temperature scale. There are no absolute values associated with interval measurement; they are relative. In the interval approach, units are agreed upon by researchers and are assumed to be standard from one set of conditions to another. Variables at the interval scale do not have absolute zero as a starting point.

Try this experiment for an example. Place a yardstick next to a tabletop, but not touching the floor. You can slide the yardstick up and down relative to the tabletop, but as it is not touching the floor, there is no absolute height for the table, only a relative height as shown by the sliding yardstick.

Ratio

Like interval, ratio measurement involves ordering events with known distances separating the events. The difference is that ratio magnitudes are absolute, having a known starting point. This scale of measurement has a zero (absence of a magnitude) as its starting point. The Kelvin absolute temperature scale is an example here. Other examples include weight (20 lbs is twice as heavy as 10 lbs), and distance (100 miles is twice as far as 50 miles). Elevation above sea level is another ratio scale where the average elevation of sea level is set as zero. The ratio approach is important to geography because more sophisticated statistical tests can be performed using this level of measurement.

In the yardstick/tabletop example, if the yardstick is placed on the floor, then the height of the tabletop can be measured relative to a zero starting point (the floor). If we chop off its legs, we know exactly how much shorter it has become.

The amount of information that can be obtained, statistical confidence, and predictive power increase as one progresses from nominal to ratio measurement. Cartographically, we view interval and ratio as being equal. The thematic map generated from interval data will utilize the same design techniques if it was created using ratio data.

DATA: THEMATIC MAP RELATIONSHIPS

The nature of the geographic data described above has a direct relationship on the thematic map type selected and thus the cartographic symbolization utilized. Table 4.3 identifies the data characteristics and their associated thematic map type. General reference maps utilize data that are qualitative with either a nominal or ordinal measurement that occur at discrete locations spatially. The thematic map types introduced in Chapter 1, which are presented in this table are addressed in detail within the chapters of Part II of this text. Thematic maps display quantitative data that primarily utilize ordinal or interval measurements of discrete values of totals. The exceptions to this generalization are both the choropleth and surface maps which display derived and continuous data, respectively. The characteristics of the data dictate the thematic map type that should be created.

Map Symbols

The selection of symbols is one of the more interesting tasks for the cartographic designer. The choice is wide, and no firm rules prevail. Symbol selection, however, is increasingly based on a compelling system of logic tied to both the type of geographic phenomenon mapped and certain graphic primitives or variables.

Symbols are the graphic marks used to encode the thematic distribution onto the map. From a vast array of symbols having different dimensions, the cartographer selects the symbol that best represents the geographic phenomena. Fortunately, the task is reduced somewhat by controlling factors, such as cartographic convention and the inability of most map readers to easily understand the more complex symbols that might be chosen.

The three generally recognized cartographic symbol types are directly tied to the data characteristics of location. These point, line, and area symbols continue to be the standard in thematic mapping. The use of GIS and advanced mapping software has given the cartographer the power to generate maps in either two- or three-dimensional design

TABLE 4.3 DATA AND THEMATIC MAP RELATIONSHIPS

Characteristic View		Measurement			Spatial View	
Qualitative	Quantitative	Nominal	Ordinal	Interval/Ratio	Discrete	Continuous
X		X	X		X	
X		X	X		X	
X		X	X		X	
	X		X	X	X	
	X			X	X	
	X		X	X	X	
	X			X		X
	X		X	X	X	
X	X		X	X	X	

1. Other data restrictions exist that should be observed and will be detailed in their constituent chapter.
2. Densities are inappropriate for this map type.

(see Figure 4.4). Maps that use the three-dimensional design display either volumetric data or surface phenomena.

There is a logical (and traditional) correspondence between geographic phenomena (point, line, area, and volume) and the employment of symbol types (point, line, area). Of course, the match is not a convenient one-to-one correspondence. Point data (for example, cities at the appropriate scale) are customarily mapped by point symbols such as dots, or scaled circles. Roads, which are linear phenomena, can be mapped by line symbols; geographic phenomena having areal extent (lakes, countries, nations) can be mapped with areal symbolization—solid fills of varying hue, intensity, or value, or sometimes patterns. Geographical landforms have been mapped by a form of area symbolization called hypsometric tinting: areas between selected elevation boundaries are rendered in various color shades. Elevation, also a 3-D form, is mapped by contours of elevation, which are 2-D line symbols. The use of three-dimensional mapping has become a standard form of map symbol in which to represent volume phenomena (see Figure 4.4).

The three symbol types may be matched to the nature of the data, both characteristics and measurement levels, to produce a valuable typology of map symbols (see Figure 4.4). The selection of map symbolization has been made easier by the inclusion of a wider array of symbols in GIS and mapping software. Clipart and other sources of pictoral symbols have widened the options of the cartographer significantly. Symbolization choice is enhanced through careful consideration of the visual variables.

Visual Variables

The way we use symbols to display spatial phenomena is to create a graphic scene in which tshe reader observes and reacts to the individual components. These components, whether observed individually or in concert with each other, serve to communicate the topic to the reader. These symbols serve as visual variables from which the reader gathers information and interprets the map. Bertin described the use of symbols as "components of the graphic system [to] be called 'visual variables'" (Bertin 1983, 42). Two of the variables are, in his words, the planar dimensions. That is the X and Y positioning in two-dimensional space. Such variables are used to represent the location of objects whether they be represented by point, line, or area symbols. In the nominal or ordinal measurement, they are used in a general reference map to identify location.

The remaining six variables are size, shape, orientation, texture, saturation, and value. These variables allow the cartographer to provide additional information that serves to display more complex spatial settings or quantitative data in the form of thematic maps (Bertin 1983; see Figure 4.5 and Color Plate 4.1).

Size

Size is used to imply relative levels of importance. Proportional symbols are used in which the size of the geometric form is scaled proportionally to the data. Proportional circles, squares, and other geometric forms as well as irregular shapes like cartograms are frequently used to portray variations in data through the use of varying size of the symbol. Line thickness is also a representation of size and thus communicates the nature of flow data.

Shape

Shape is used to "(1) reveal similar elements, and therefore, different elements and (2) to facilitate external identification,

Attribute View		Symbol			Chapter in
Totals	**Derived**	**Type**	**Dimensionality**	**Map Type**	**Text**
				General Purpose:	
		Point	0	Reference	
		Line	1	Reference	
		Area	2	Reference	
				Thematic:	
	X	Area	2	Choropleth	6
X		Point	0	Dot Density	7
X	X[1,2]	Point	0	Proportional	8
X	X	Area	2/3	Surface	9
				Cartogram:	10
X	X[1,2]	Area	2	Value-by-Area	
X	X[1,2]	Line	1/2	Flow	11

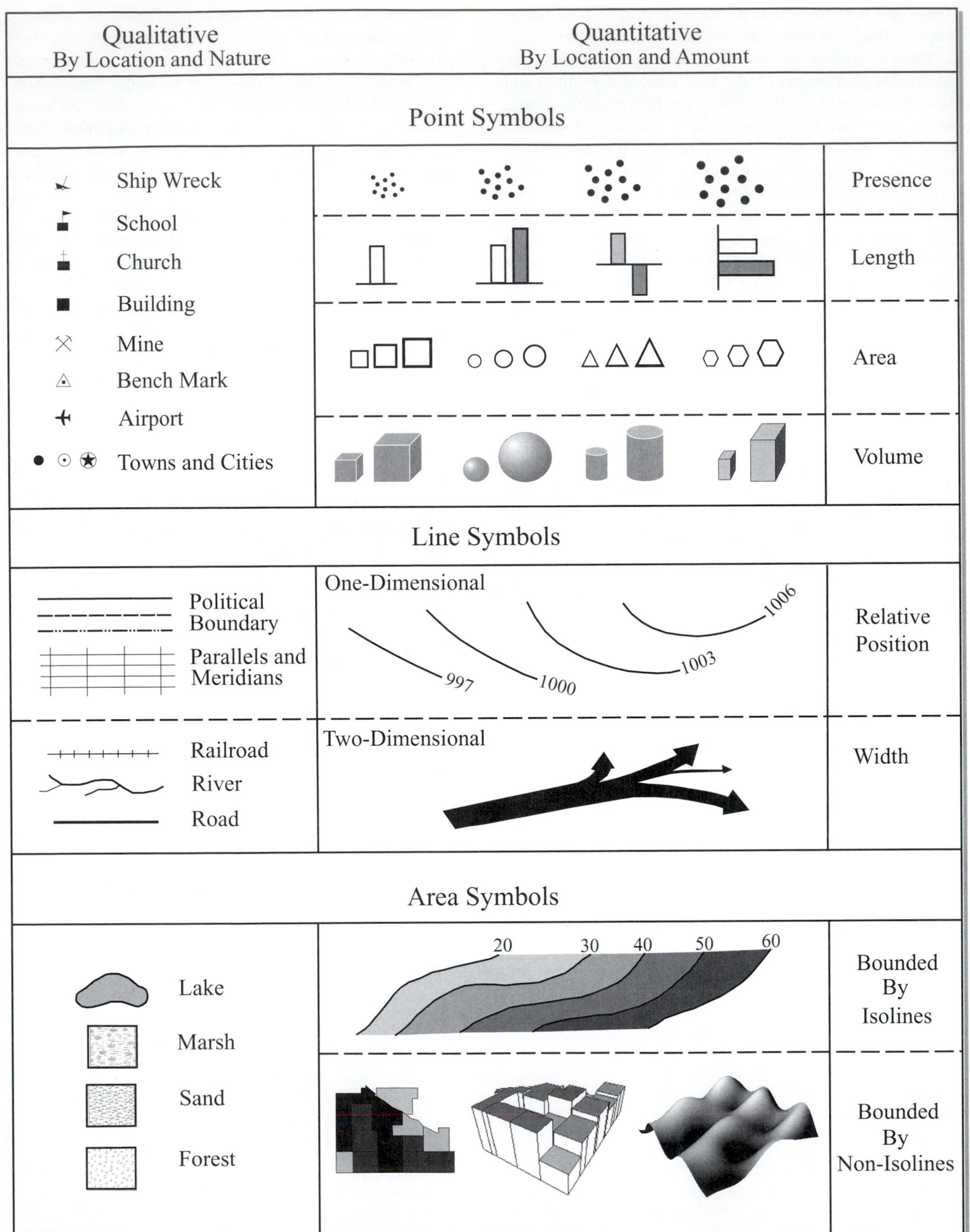

FIGURE 4.4 THE COMPLEXITY OF MAP SYMBOLS RESULTS FROM THE NATURE AND CHARACTERISTICS OF THE DATA.

Source: After Anonymous 1944.

FIGURE 4.5 VISUAL VARIABLES.
Source: MacEachren, 1994. Used by permission.

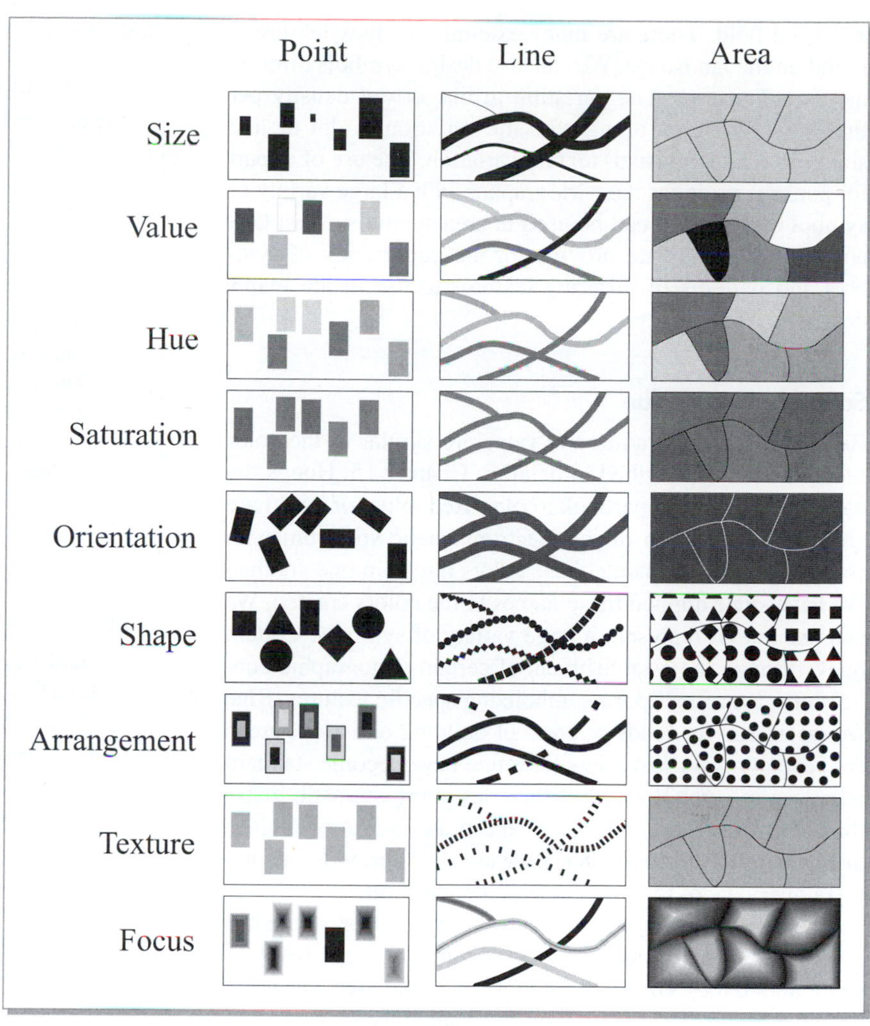

through shape symbolization" (Bertin 1983, 95). Thus, we select point symbols to represent a graphic display of the actual entity being mapped. For example, we use squares with a small flag on top to represent schools, those with crosses to represent churches, crossed pickaxes to represent mines, and countless other symbol shapes that we have grown to recognize by cartographic conventions. Cartographic traditions have established certain shape-object relationship. For example, the use of a star to represent a capital or squares to represent buildings is recognized by the most novice of map users. Linear and area symbols are also recognized by their inherent shape in the process of representing roads, lakes, and so on.

Orientation

The order of things on the landscape should not be changed to increase the interpretation of information. The orientation of building symbols and symbols of other structures should be such that it represents their actual position as closely as possible. However, when we are using symbols in a more general sense to represent an idea of the existence of objects, the orientation of those symbols creates a perception by the map reader that those objects are unique and belong together as a group. As similar orientation creates an essence of order and thus similarity, misalignment of symbols can achieve the opposite affect so that objects stand out on the map landscape.

Special care should be taken when designing graphs in which a fill is used to help differentiate the data variables, especially in bar graphs. Schultz (1961) found that the use of diagonal line patterns can actually create a perception that the bars tilt as opposed to their actual parallel construction. Thus the use of diagonal lines should be avoided, considering the other fill options available.

Texture

As the complexity of a map increases and we become challenged to find different visual designs to use as symbols, we can select different textures (or patterns) for the symbols or as an overlay to color in order to increase the number of symbol options. Frequently, texture is used in the area symbolization to help communicate areas where the landscape is somewhat coarse or smooth. The use of aerial photographs in a map layer helps to create this texture of the surface. The upper surface of a forest canopy is coarse when compared to a plowed field. There are many examples of how texture is found on the landscape. We can also design symbols through the use of texture. The variation in line or dot density per linear measurement along with the variation in dot or line size will create a (visual) texture variation. The use of clipart for patterns provides the cartographer with a large variety of symbol options that can be used in area symbols. Most GIS and mapping software provide for the importation of symbols and patterns of differing textures to use in the map's design.

Saturation and Value

Bertin's use of saturation and value are similar to the color model of HSV described in detail in Chapter 15. Hue is the name we apply to a particular color. Red, blue, or green are created by a portion of the electromagnetic spectrum with wavelengths that separate these colors out from one another. We have since applied these names to the colors created. We can use hue to represent a large variety of symbols. Certain hues, through the establishment of certain cartographic conventions, are reserved for symbolizing specific features. Blue for water, red for roads or areas of building omission, green for vegetation are just a few hues that have become standard practices in symbolization. When designing thematic maps, the selection of color is one of the most exciting activities available to the cartographer. Brewer (2005) provides a complete overview of the process of color selection.

Saturation is thought of as a level of brightness of the hue where value is considered as a sequence of steps between light and dark. The combination of these two variables provides the cartographer with the capability of designing quality maps using either grayscale or a color model. The software utilized provides the cartographer with the choice of millions of color options based upon hue, saturation, and value. These colors are easy to use in a virtual environment but may become more of a challenge in the hard copy or printed environment. The cartographer should not rule out the use of grayscale in the design of maps. Many professional journals are just now beginning to publish color graphics. The grayscale map is still a most effective means of map design.

The interpretation of these visual variables and how they are used to achieve a visual hierarchy in map design is discussed in Chapter 12 of this text. The fundamentals of figure-ground association are achieved by the careful selection of the visual variables and the manipulation of their components.

CARTOGRAPHIC ERROR

There are many places where inaccuracies can enter in the mapping process. The kinds of error that are encountered in creating a map include source error, processing error, and design error. These errors are discussed in this section and outlined in Table 4.4.

TABLE 4.4 SOURCES OF ERROR ON THEMATIC MAPS

SOURCE ERRORS
- Source Map Error
  - Existing Maps
    - Scale
    - Detail
    - Accuracy
    - Projection
    - Currency
- Data Entry Error
  - Boundary Data
    - Download of inaccurate boundary
    - Digitizing
      - Manual
      - Heads-up
  - Attribute
    - Incorrect entry
    - Inappropriate link to ID
    - Incomplete data

PROCESSING ERRORS
- Numerical Rounding in Computing
- Data Classification
- Data Transformation
- Interpolation Technique
- Inappropriate Use of Algorithm

CARTOGRAPHIC DESIGN ERRORS
- Thematic Map Type
- Scale
- Projection
- Generalization
- Symbolization
- Use of Color

Source Error

Source errors are those errors that are found in the data collection, compilation, and date entry procedures (Beard 1989). As a mapping project begins, a search is conducted to determine the existence of maps (both digital and paper) for the study area and data topic. Often these maps are used as reference or a base on which other information is compiled or placed. The use of base maps is quite common in cartography providing foundations for exploratory design and layout (see Chapter 1).

Caution must be taken in the selections of existing maps for use as a layer in the generation of a new map. These maps are limited in scale, detail, accuracy, projection, and the date to which the information applies. These limitations may inject data and location discrepancies created by the generalization level required by a particular map scale. Perhaps the map projection was inappropriate for the data being displayed or even created without a geographic base. Using maps created by others often perpetuates error and thus the map's lineage must be considered. Careful evaluation of the map's accuracy is essential before making use of such a foundation. The date of

SIR JOSIAH STAMP

When using data collected by others, we urge the reader to heed the words of Sir Josiah Stamp (1869–1941) of England's Inland Revenue Department, who observed:

The government are very keen on amassing statistics. They collect them, add them, raise them to the nth power, take the cube root and prepare wonderful diagrams. But you must never forget that every one of these figures comes in the first instance from the village watchman who just puts down what he damn pleases.

Source: Stamp 1929: 258–59.

the map may be too old for use in a current study. Both physical and cultural features may be left off the map as a result of its age. The caution applies to both hard-copy maps and those in digital form. Assuming that the map is correct doesn't necessarily make it so.

The federal government creates topographic maps at varying scales that include all these features plus a set of contour lines helping the map reader interpret and visualize the undulations in surface topography. These maps are created so that any location or elevation meets the National Map Accuracy Standards and thus gives us a base that is geometrically accurate. These maps have served as base maps on which cartographers have added other layers of information.

For data that is downloaded from government sites, educational institutions, or other online sources, checking the metadata (data about data) can be one way to ascertain the source map's qualities. Most correctly documented spatial metadata includes information such as its spatial reference (that is, map projection and coordinate system), an assessment of data quality (including the lineage of both spatial and nonspatial attributes that the data may contain), and contact information about the data set, as well as other information.

Error can also enter via the data entry process. This process includes the acquisition of digital boundary files and associated attribute data. The source of the boundary file must be carefully considered when downloading from the Web (see Federal Governmental Agencies, page 76). Only reputable sources should be considered. The production of new boundary/location files via manual or heads-up digitizing also generates the potential for errors in geometry of the map or subsequent creation of erroneous polygons that are nonexistent. The accuracy and precision of the instrument and the experience of the operator are also sources of error in the map geometry (Burrough and McDonnell 1998). Careful editing must be included in the project to be certain that errors of omission or commission are not included.

The keying of attribute data into a database can create errors that are carried throughout the data analysis and data display phase of any project. Entry errors that include entering the wrong attribute, transposing numbers in a data value, or incorrect attribute-ID association may cause the map to display spatial patterns that do not actually exist. Care must be taken so that values of zero are not interpreted the same as missing data.

Processing Error

Processing errors can result from the cartographic transformation of data as a result of changing scale, projection, or data form. Line simplification and data classification are techniques involved in generalization (see Chapter 1). Cartographic generalization based upon scale, such as the removal of islands or the straightening of sinuous lines, often produces errors that are unknown to the map user (Clarke 1990). Classification that overly generalizes the data will tend to reduce the spatial patterns that exist in that data. Careful consideration of classification methods is crucial (see Chapter 5).

When utilizing computer software for data transformation, the values of the raw data must be considered. Simple formulas used in spreadsheet software that permit the conversion from total to derived data often provide results with seven to ten decimal values. To include this level of precision will generate a false perception of accuracy of the data when used in mapping. Numerical rounding of such values to a single decimal equivalent permits the use of such calculations without falsely implying greater accuracy. We suggest that an increase of only a single decimal value over that of the original data be used.

Transformation from discrete to continuous data requires the use of interpolation in order to generate the database. Care must be taken to use an appropriate algorithm for that interpolation (Lo and Yeung 2007). A discussion of such conversions is provided in Chapter 9.

Cartographic Design Error

Once the data have been collected and processed, the cartographer should take time to examine the data attribute table for inconsistencies and completeness. Following that examination, the cartographic design process begins. The first potential for design error is in the selection of the wrong thematic map type based upon the data. As suggested, the data dictate the map type for the display of spatial data. Careful consideration of the nature of the data will serve as a guide in the selection of the appropriate thematic map (see Table 4.3).

The cartographer must make many decisions when it comes to map design. Each of those decisions has the potential for introducing map error whether it is factual on the map or one of incorrect perception by the map user. Choices in scale, projection, generalization, symbolization, and color are based upon traditions and conventions in cartography. To vary widely from such traditions may cause a decrease in the successful communication of the theme presented. An examination of the best choices to be made in these areas are presented in both Part I and II of this text.

Not all errors can be eliminated, but every conceivable means must be employed to reduce error to tolerable levels. Each mapping method discussed in this text has its own error sources, based on its unique way of mapping and symbolization. The thematic map designer must learn to take these error sources into account.

DATA SOURCES

Data (GIS) Clearinghouses

Traditionally, the cartographer has had to access hard copy of data by going to the library or by contacting local, state, or federal government agencies and then manually entering that data into the mapping program. Currently, GIS and mapping software come with a large data supply as a part of the program. These data are frequently acquired from the 1990 or 2000 U.S. Census of Population and may be used to produce a variety of thematic maps. With the rapidly expanding Internet, access to data is faster and easier than before. Most state governments provide GIS data clearinghouses from which data can be downloaded. Google searches provide many links to data for almost any desired topic. The U.S. federal government provides data access from agencies, such as the Bureau of the Census (population and housing, manufacturing and agriculture), the Geological Survey, and the Department of Agriculture, to name a few. Most data warehouses allow for free access to data from the public domain.

Federal Governmental Agencies

Not only are attribute data available via the Internet, one can also acquire digital boundary files for practically any location. Most GIS and mapping software contain basic boundary files with associated attribute data for the countries of the world, as well as many of their internal political units, such as states and counties. Frequently, these boundary files are of a proprietary nature or they are not readily available for a location to be mapped. Care should be taken when utilizing boundary files acquired from remote sources. The quality and reliability of the digital boundaries may not be of the highest standards. We recommend that you acquire boundary files from either reputable government agencies, such as the U.S. Census Bureau or the U.S. Geological Survey, who offer these products, or from the software vendors themselves.

U.S. Census Bureau

In February 1989 the U.S. Census Bureau released the first TIGER/Line (Topologically Integrated Geographic Encoding and Referencing) files, called the Prototype TIGER/Line files. The TIGER/Line files provided the first seamless nationwide street centerline coverage of the United States and Puerto Rico. Over the past 17 years, based upon user requests for additional data content, the TIGER/Line files have grown from a file containing six record types to a file containing 19 record types.

With the modernization of the Master Address File (MAF) and TIGER systems, the Geography Division will, in 2008, begin releasing TIGER spatial data in the following formats:

- Shapefiles
- TIGER/GML™
- The Census Bureau also will make available the TIGER spatial data over the Web:
  - WebTIGER™—A Web Feature Service (WFS) interface allowing requests for geographic features across the Web. It uses the XML (Extensible Markup Language)-based GML (Geographic Markup Language) for data exchange.
  - Web Map Server (WMS)—A WMS producing maps of spatially referenced data dynamically from TIGER in PNG, GIF, and JPEG formats or as vector-based graphical elements in Scalable Vector Graphics (SVG).

Available with the 2003 TIGER/Line files contain updated national ZIP Code Tabulation Areas (ZCTAs) reflecting the October 2002 U.S. Postal Service ZIP Codes (U.S. Census Bureau Website 2008 see Table 4.5).

U.S. Geological Survey

The U.S. Geological Survey produces a variety of digital vector and raster files that can be used in the generation of maps. The USGS Digital Cartographic Data include products such as DLG, DRG, DEM, DOQ, and NHD files. Table 4.5 provides a summary of files scale and content. These products are available from the USGS website (USGS 2008).

Federal Information Processing Standards (FIPS)

Most software utilizes a field (column) in the database to link the data to the location for which it applies. Frequently, this field will be one either containing the name of the associated observation or the associated Federal Information Processing Standards (FIPS) code. Every country, state, city, metropolitan

TABLE 4.5 EXAMPLES OF DIGITAL BOUNDARY AND DATA FILES AVAILABLE FROM THE U.S. CENSUS BUREAU AND THE U.S. GEOLOGICAL SURVEY

U.S. Census Bureau:

Vector/Raster:

Topologically Integrated Geographic Encoding and Referencing (TIGER®) System.

Content of Shapefiles:

- Blocks
- Block Groups
- Census Tracts
- Counties
- County Subdivisions
- Places
- Urban Areas
- Congressional Districts
- States
- And many more

U.S. Geological Survey:

Vector:

Digital Line Graphs (DLG) are digital vector representations of cartographic information derived from USGS maps and related sources.

Scales Available: 1:24,000, 1:100,000, and 1:2,000,000

Layers Available:

- Public Land Survey System (PLSS)
- Boundaries
- Transportation
- Hydrography
- Hypsography
- Non-vegetated Features
- Vegetation
- Survey Control and Markers
- Manmade Features

Raster:

Digital Raster Graphics (DRG) is a scanned image of a U.S. Geological Survey (USGS) standard series topographic map, including all map collar information. The image inside the map neatline is georeferenced to the surface of the Earth and fit to the Universal Transverse Mercator projection. The horizontal positional accuracy and datum of the DRG matches the accuracy and datum of the source map. The map is scanned at a minimum resolution of 250 dots per inch.

Digital Elevation Models (DEM) were comprised of scanned topographic map series. As of November 2006, the DEM was no longer offered by the USGS as it has been replaced by the NED.

National Elevation Data sets (NED) has seamless elevation coverage for the United States with a resolution of 1arc second. Elevation is provided in meters.

area, or other identifiable unit for which data are collected are assigned a FIPS code. Country codes are comprised of two-letter designation, U.S. states and territories are designated by a two-digit number, and U.S. counties are designated by a three-digit number.

Potential Problems

It is the cartographer's responsibility to utilize accurate data in a responsible manner. Caution is advised when copying columns of data into a database. The order in which the states appear within a column will depend upon the manner in which the alphabetical list is created. Table 4.6 provides a list of select states ordered alphabetically by their two-letter designation. If this is the order in which data are provided, the data will be in neither alphabetical order by state name nor by county FIPS code. Many databases are created using an alphabetical order by county name. The cartographer is cautioned that data errors can occur when combining data sets if the states contain counties that begin with "Mc." Table 4.7 displays a select list of counties for the state of Tennessee (state FIPS = 47) and those counties whose name begins with "M." The FIPS code order for these counties place McMinn and McNairy counties at the beginning of the "M" list (county FIPS of 107 and 109, respectively). Many data sets are produced alphabetically with these counties occurring farther down the list, after Maury County (119). Care must be taken when importing

TABLE 4.6 FEDERAL INFORMATION PROCESSING STANDARDS (FIPS) CODES FOR SELECT STATES.
The States Beginning with the Letter "A" in Order by Their Two-Letter Designation.

2-Letter Designation	State Name	State FIPS
AK	Alaska	02
AL	Alabama	01
AR	Arkansas	05
AZ	Arizona	04

TABLE 4.7 FEDERAL INFORMATION PROCESSING STANDARDS (FIPS) FOR SELECT COUNTIES.
Examples are from the State of Tennessee. (a) FIPS Code Order. (b) Alphabetical Order.

(a) State FIPS	County FIPS	County* Name	(b) Alphabetical List
47	107	McMinn	Macon
47	109	McNairy	Madison
47	111	Macon	Marion
47	113	Madison	Marshall
47	115	Marion	Maury
47	117	Marshall	McMinn
47	119	Maury	McNairy
47	121	Meigs	Meigs
47	123	Monroe	Monroe
47	125	Montgomery	Montgomery
47	127	Moore	Moore
47	129	Morgan	Morgan

*State alternatives for county: Alaska—Borough; Louisiana—Parish

data into an existing database. If the existing database and the one being imported are arranged differently, the combined database will have incorrect data associated with these counties.

REFERENCES

Anonymous. 1944. A Proposed Atlas of Diseases. *Geographical Review*, Vol 34 (4), 642–52.

Beard, K. 1989. Use Error: The Neglected Error Component. Auto-Carto 9 (Ninth International Symposium on Computer-Assisted Cartography) 808–17.

Bertin, J. 1983. *Semiology of Graphics*. Madison, WI: University of Wisconsin Press.

Burrough, P., and R. McDonnell. 1998. *Principles of Geographical Information Systems*. New York: Oxford Press.

Brewer, C. 2005. *Designing Better Maps: A Guide for GIS Users*. Redlands, CA: ESRI Press.

Clarke, K. 1990. *Analytical and Computer Cartography*. Englewood Cliffs, NJ: Prentice-Hall.

Green, M., and R. Flowerdew. 1996. New Evidence on the Modifiable Area Unit Problem. In P. Longley and M. Batty (eds), *Spatial Analysis Modelling in a GIS Environment* (41–54). New York: Wiley.

Harvey, D. 1969. *Explanation in Geography*. New York: St. Martin's Press, 293–98.

Holt, J., C. Lo, and T. Hodler. 2004. Dasymetric Estimation of Population Density and Areal Interpolation of Census Data to Compensate for Census Geography Changes, Metropolitan Atlanta, 1980–2000. *Cartography and Geographic Information Science* 31 (2): 103–21.

Lo, C., and A. Yeung. 2007. *Concepts and Techniques of Geographic Information Systems*, 2nd ed. Upper Saddle River, NJ: Prentice-Hall.

MacEachren, A. 1994. *Some Truth With Maps*. Washington, DC: Association of American Geographers, 129.

Meyer, T. 1997. NCGIA Core Curriculum in GIScience.

Openshaw, S. 1984. *The Modifiable Area Unit Problem*. Norfolk, VA: Norwick.

Sauer, K. 1956. The Education of a Geographer. *Annals* of the Association of American Geographers 46: 287–99.

Schultz, G. 1961. Beware of Diagonal Lines in Bar Graphs. *The Professional Geographer* 13 (4): 28–29.

Stamp, J. 1929. *Some Economic Factors in Modern Life*, 258–59.

Taylor, P. 1977. *Quantitative Methods in Geography: An Introduction to Spatial Analysis*. Boston: Houghton Mifflin.

U.S. Census Bureau. 2008. http://factfinder.census.gov. Accessed January 12, 2008.

U.S. Geological Survey. 2008. http://nationalmap.gov/gio/status.html. Accessed January 12, 2008.

GLOSSARY

area data attributes that may be applied to an entire area as a quantity that is assumed to exist uniformly (rightly or wrongly) over an entire area; examples include people per square kilometer and bushels per acre

attribute data values or characteristics associated with a point, line, or area

continuous form variables that are areal and exist everywhere, such as temperature or barometric pressure; these values are often interpolated from sample observations that lie within the area of study

data form contextual characteristics that include identification as either quantitative or qualitative, discrete or continuous, and total or derived data

data measurement an attempt to structure observations about reality; observations can be grouped into four levels, depending on the mathematical attributes of the observed facts, as either nominal, ordinal, interval, or rational measurements

derived data permits the comparisons of observations through data manipulation to normalize (standardize) the data so as to either adjust for the impact of area/size or to represent the data as a rate or percentage

design error mistakes made by the cartographer in the selection of thematic map type, their symbols, and other map characteristics

discrete data associated with data counts or totals, the data are uniquely associated to a singular spatial location, no matter how it is defined

FIPS code Federal Information Processing Standards (FIPS) used to identify every country, state, city, metropolitan area, or other identifiable units for which data are collected

geographical phenomena elements of reality that have spatial attributes; any spatial phenomena can be the subject of geographical analysis within the limits of scale

geographic data facts about which conclusions can be drawn; chosen to describe geographic phenomena; associated with a spatial dimension

interval measurement scales that have no natural origin; any beginning point may be used; the classic example is the Fahrenheit temperature scale; values are relative and do not have absolute zero as a starting point

line data generated by the collection of information as applied along a linear feature or path

metadata data about data; for cartographers, information that accompanies spatial data sets that describe the dataset's spatial extent, attributes, lineage, and other information

node a point defined in geographic space or a series of points used to define the limits of a line or polygon

nominal the simplest level of data measurement, sometimes considered a qualitative measurement to be descriptive; answering the question of what is being mapped

nonspatial data attribute or descriptive characteristics of data; these data are normally found as entries in a database or spreadsheet

observation a single entity or place normally displayed as a row in an attribute table

ordinal a measurement that is a hierarchy of rank; objects are defined in an order that permits comparison in a general sense; examples include an arrangement of least to most, greater than or less than, or some level of importance

point data a geographic location with a set of geospatial coordinates matched to a set of attributes

processing errors miscalculations or transformations of data through computer applications

qualitative descriptive characteristics that describe the inherent nature of a feature

quantitative numerical values used as attributes that establish measured positional sequences in an absolute or relative sense

ratio cartographically treated the same as interval; the measurement uses an absolute datum for numerical comparisons

source error found in the data collection, compilation, and data entry procedures

spatial data displays characteristics of location, size, or amount that allows for the visualization of patterns within the data

symbol types cartographic symbols are classed as point, line, or areal symbols; these symbols may be two- or three-dimensional and include pictoral counterparts

total data an attribute (or series of attributes) that is collected at a location

typology of map symbols description of thematic map symbols based on a cross-tabulation of measurement scales and symbol types

visual variables symbols used to communicate to the map reader; these include variables of location, size, shape, orientation, texture, saturation, and value

volume data attributes that exist over a three-dimensional extent of a continuous surface

5 DESCRIPTIVE STATISTICS AND DATA CLASSIFICATION

CHAPTER PREVIEW The data sets used in thematic mapping come in all sizes and characteristics. Prior to mapping, the data are evaluated in terms of their statistics, size, and forms. Data classification provides measures for simplifying large data sets to assist in analysis and display. The decisions facing the cartographer involve the number of classes and the data classification scheme to be used in generalizing the data. Nine classification schemes are examined along with assessment indices for evaluating the classification accuracy. The variations in spatial patterns based on the number of classes and data classification are presented in both numeric and graphic form. ■

Statistical data provide the cartographer with insight into a variety of issues and topics. The size of a database is directly dependent upon both the number of observations and the number of variables. Generally, we work with one variable at a time in an attempt to identify the spatial aspects of a single topic which is the approach of thematic cartography. Examining a single variable may be a simple or complex task depending upon the number of observations in the data set. If the attribute data are associated with political or governmental boundaries, the number of observations may be, for example, the 194 countries of the world or the fifty U.S. states. County data sets may be small, three counties in Delaware; moderate, 254 counties in Texas; or large, 3,219 counties in the United States. An even larger data set could include all 66,304 Census Tracts (see Table 5.1). The number of observations that are incorporated in a primary research project are defined by the researcher. Data are collected, stored in a database, and used for cartographic display of the spatial components of the data (see Chapter 4).

Working with large data sets is sometimes difficult and cumbersome. The advantage of course is that there is a unique attribute data value for each observation. In the purest sense, this allows for visualization of spatial patterns without data manipulation by the cartographer. Statistically, rarely does the number of observations in the study present problems. However, as we analyze the data and present findings in reports, graphs, and maps, large numbers of observations often become a problem. How we communicate the conclusions drawn from research and the visualization of results becomes a daunting task and presents an obvious need for simplification.

In an effort to handle large quantities of data, we tend to classify the data into smaller groupings. In doing so we are generalizing the data in order to simplify the process of analysis and mapping by placing the attributes into convenient categories. The result of classification enhances the understanding of the spatial patterns and information contained within the data. Cartographically there are many techniques available that permit the classification of data. Such classification simplifies the number of symbols used on the map and assists in the exploration of the spatial patterns.

TABLE 5.1 U.S. CENSUS SUBDIVISIONS, 2000

Subdivision	Number
States	50
Counties*	3,219
Highest: Texas	254
Least: Delaware	3
Census Tracts	66,304
Postal ZIP Codes	33,233

*Louisiana = parish
Alaska = borough
Source: U.S. Census Bureau, 2008.

OVERVIEW OF A DATA SET

For the purposes of discussion we will use the attribute data provided in Appendix B for all examples in the remainder of this chapter. The data are provided using the 159 counties in the state of Georgia as the enumeration (statistical) units. The data set found in Appendix B contains a column header in abbreviated form which identifies the data variable found within that column. Table 5.2 provides a full description of the header abbreviations used in Appendix B. The county name (observation) and the state-county FIPS code are displayed in the first two columns. Each may serve as an identifier (ID) that links the location in the data set to the related polygon in the map. The remaining twelve columns (variables) contain data values related to various attributes, such as demographic and housing characteristics. Each of these **variables** can be classified in order to obtain a better understanding of its spatial distribution. This data set was selected as a representative size that one frequently encounters in the mapping process.

TABLE 5.2 VARIABLE NAMES AND DATA SET HEADER ABBREVIATIONS FOR THE SAMPLE DATABASE FOR GEORGIA See Appendix B.

Header Abbreviations	Variables
POP1990	Total county population for 1990
AREA	Area of each county in square miles
POP/SQ MILE	Population density (people per square mile) for 1990
BIRTHS	Total births in 1990
FEMALE_15_44	Total number of females per county ages 15 to 44
GFR	General Fertility Rate
BIRTHRATE	Number of live births per 1,000 women of childbearing age
HOUSING UNITS	Total number of housing units in a county
H_U_OCCUPIED	Number of total housing units that are occupied

Ratio, Proportion, Percent, and Rate

Four of the simplest *derived* indices used by geographers and cartographers are ratios, proportions, percents, and rates. The first two are often used interchangeably, but in fact they differ. A **ratio** is a good way of expressing the *relationship* between two data entities. It is expressed as

$$\frac{f_a}{f_b}$$

where f_a is the number of items in one entity and f_b the number in a second entity. The number of items is referred to as the *frequency,* a term used in most statistical work. The result of the calculation is set so we compare a relationship between one of the first entity to some value of the second. The numerator is established to represent the unit value and the denominator the comparison value.

A familiar ratio in geography is *population density,* defined as the number of people per square mile or other areal unit. This statistic is used to allow for comparison of population data without the impact of size. Counties with larger area have, by the nature of size, a greater propensity to have more people within their boundaries. The conversion for total population to population density counteracts for the impact of size and thus allows the data to be comparable. Since a county is greater than one square mile, we must reduce the relationship of people-to-area ratio so that unity (one square mile) is in the denominator. The Georgia county population density can be calculated for each county using the formula:

$$\frac{\textit{Number of people in a county}}{\textit{Area of county in square miles}} = \text{People per square mile}$$

As an example, consider the ratio of the sexes enrolled in the population of a major southern university in the fall semester of 2006. There were 10,876 males and 14,248 females matriculating. The ratio is calculated as:

$$\frac{14{,}248}{10{,}876} = 1.31$$

This result produces a ratio expression of 1:1.31 or extrapolated to 100 males to 131 females, since we prefer to think of people in terms of whole numbers.

In Appendix B we find that Appling County's population (POP2000) is 17,419 and the county area (AREA) is 510.3680 square miles producing this calculation:

$$\frac{17{,}419}{510.3680} = \frac{34.13}{1}$$

Therefore the calculated population density for this county is 34.13 people per square mile. Again, since the data are whole numbers and represent people, and we normally don't think of a portion of a person we drop the .13 from the density value by rounding to the nearest whole number and specify that Appling County's population density is 34 people per square mile. This same operation can be applied to all counties as well as for state totals (the sum of all county population and area values). Georgia's state population density is 140 people per square mile.

Proportion is the ratio of the number of items in one group (class) to the total of all items. It is written:

$$\frac{f_a}{N}$$

where f_a is the number of items (frequency) in a class and N is the total number of items or total frequency. To determine the proportion of housing units that are owner occupied, again using Appling County's data, the calculation would be:

$$\frac{6{,}606}{7{,}854} = 0.8411$$

Typically, proportions are multiplied by 100, yielding a **percentage.** In this case,

$$0.8411 * 100 = 84.11\%$$

or 84.11 percent of the county's housing units are occupied. Here, since we are not dealing with the human form, we can retain the portion of a percent.

Percentage change is another frequently calculated variable based upon a single variable for two different time periods. The resultant value may be positive or negative change and is normally expressed as (Healy 1996):

$$\frac{\text{New Value} - \text{Old Value}}{\text{Old Value}} * 100 = \text{Percent Change}$$

The Georgia data set (Appendix B) provides total population for both the 1990 and the 2000 U.S. Census. Percentage population change is calculated by:

$$\frac{POP2000 - POP1990}{POP1990} * 100 = \text{Percent Change 1990–2000}$$

Using the data for Forsyth and Chattahoochee counties, the percentage change depicts one county with significant growth and one declining in population.

$$\text{Forsyth County } \frac{98{,}407 - 44{,}083}{44{,}083} = \frac{54{,}324}{44{,}083}$$
$$= 1.2322 * 100 = 123.23\%$$

This represents a change in population so that the county more than doubles its population in the decade of the 1990s from 44,083 to 98,407.

$$\text{Chattahoochee County } \frac{14{,}882 - 16{,}934}{16{,}934} = \frac{-2{,}052}{16{,}934}$$
$$= -0.1212 * 100 = -12.12\%$$

This negative percent (−12.12%) indicates that the county lost about one-fifth of its population during the decade and now has 2,052 fewer people living in the county in 2000 than in 1990. Percentage is a familiar and useful index in many geographical analyses. It is simply derived and, when coupled with location, can yield information about potential change in geographical concentration from time to time and/or place to place.

Rates are similar to percentages except that the relationship is a value per some much larger value. It is determined by the relationship of an observed number compared to a potential number of occurrences for a given time or place. The result is then multiplied by a power of ten, usually relative to the denominator, to make the result meaningful (Healy 1996).

$$\frac{\textit{Number of Occurrences}}{\textit{Number of Possible Occurrences}} \times 1000^{*}$$

(or other appropriate power of 10)

For example, the General Fertility Rate (GFR) for a given county is calculated using the formula:

(Number of Live Births in an age group / Female population ages 15–44) × 1,000

The resulting rate provides us with the number of live births to women of any age per 1,000 females ages 15–44. Again using Appling County, the data are calculated as:

$$\frac{278}{3{,}731} = 0.0745 * 1{,}000 = 74.5$$

or a GFR of 74.5.

With the data provided in the table, the death rate for that same county would be 11.19 deaths per 1,000 people in the year 2000. This was determined by dividing the number of deaths by the total population of the same year times 1,000.

Descriptive Statistics

Although this is not a textbook on statistics, an understanding of many of the basic terminologies employed in statistical analysis is important. We cover the fundamentals as they apply to cartography but recommend that you refer to a modern statistics reference for a more detailed examination.

Every variable possesses a set of descriptive statistics which provides general information about the data. The fourth column of the data set (labeled POP2000) contains the total population for each county in the census year 2000. Using the **sort** function of standard spreadsheet software, the data can be reorganized so that they are arranged in a descending order from largest to least values. This sorting places the data in a **rank order** of importance. The largest data value is traditionally given a rank of one and incremented sequentially so that the larger the rank number the lower it is in the numerical sequence. Before classification of data begins, the data should be sorted into its numerical order.

The summary of data can be thought of in three different ways. The first is a measure of the *central tendency* of the data. This includes characteristics that assist us in understanding the basic distribution of the data, such as the mean, median, or mode. The measures of *dispersion* depict the variability within the data such as the **range,** variance, and standard deviation. Lastly, measures of *shape* describe the nature of the distribution such as skewness or kurtosis (McGrew and Monroe 2000). This shape is easily visualized using a histogram or frequency distribution curve. Variance, standard deviation, skewness, and kurtosis will be defined in more detail later in this chapter.

A common set of statistics about a variable include the data's total, maximum and minimum values, data range, arithmetic mean, frequency of observations (number of counties), and the standard deviation. When presented in a formula, the individual observation (a county) is represented by the letter (symbol) n and the data value of that observation is identified as x. The total of all the observations is indicated by N. The Greek symbol sigma (Σ) is used to indicate that the data should be added together or summed. Thus, the total of the POP2000 column is 8,186,453. The formula would be:

$$\Sigma x = 8{,}186{,}453$$

The data maximum (x_{max}) is 816,006 and the data minimum (x_{min}) is 2,007. The data range is determined by subtracting the lowest from the highest value. Therefore, the data range is:

$$x_{max} - x_{min} = 813{,}999$$

The arithmetic **mean** is derived by adding all the values together and dividing by the number of values:

$$\overline{X} = \frac{\Sigma x}{N}$$

where $\overline{X}$ is the mean of the data set. In the case of the POP2000 data,

$$\overline{X} = \frac{8{,}186{,}453}{159} = 51{,}487.13$$

Again we are dealing with people, the .13 of the mean can be rounded, down in this case, to a mean of 51,487 people.

It is important to examine the maximum (x_{max}) and minimum (x_{min}) values of the sorted data in order to determine their potential for being an **outlier** value. An outlier is a value that is quite different from the remaining portion of the data. Such outliers will cause a shift in the data set mean higher or lower so that the value of that mean is falsely representative of the remaining values. For example, the students in a cartography class receive their percentage grade for a midterm exam. When the instructor announces the mean score for the exam, a greater majority of the students fall below the mean as a result of one individual who receives a much higher score than the rest of the class. The opposite case may also occur if the majority of students fall above the mean as a result of one student earning a very low failing grade which pulls the mean downward. In either case, the extreme grade is an outlier causing the mean to shift creating a potential erroneous view of the entire data set. This outlier, as we will see later in this chapter, may also create difficulties in data classification.

The concept of an average is often thought of as occurring somewhere in the middle of the data. The **median** is the midpoint of the data and should not be confused with the data mean. The median of the population data falls between observations 79 and 80 or a value approximating 21,100. Thus, the mean (51,487) is significantly larger than the median. If no outlier exists, larger data values impact the calculation of the mean more significantly than the value on the lower end of the order. This is easily identified in our example as the position of the mean occurs so that only 34 counties lie above the mean and 125 counties fall below the mean (see Figure 5.1). Again, the cartographer should observe the distribution of values within the data set when evaluating the central tendency of that data (see boxed text: "Mean and Median Example").

The variability of the data within the data is a measure of dispersion. One might think that a good way to express this dispersion is to calculate the average deviation about the mean. Doing this, however, yields no measure at all! If we calculate each x-value's deviation from the mean and sum these, they would total 0. The positive deviations above the mean will be offset by the negative deviations below the mean.

The **standard deviation** is the solution to this dilemma. We begin by first squaring each deviation. This eliminates the negative signs (minus times minus equals a plus). Now

MEAN AND MEDIAN EXAMPLE

The following two hypothetical data sets can be used to depict these differences:

	Set 1	Set 2	
	8	5	
	12	8	
	14	12	
Median = 16	16	14	Median = 15
Mean = 18.3	22	16	(14 + 16 = 30/2 = 15)
	26	22	Mean = 16.6
	30	26	
		30	
	128	133	

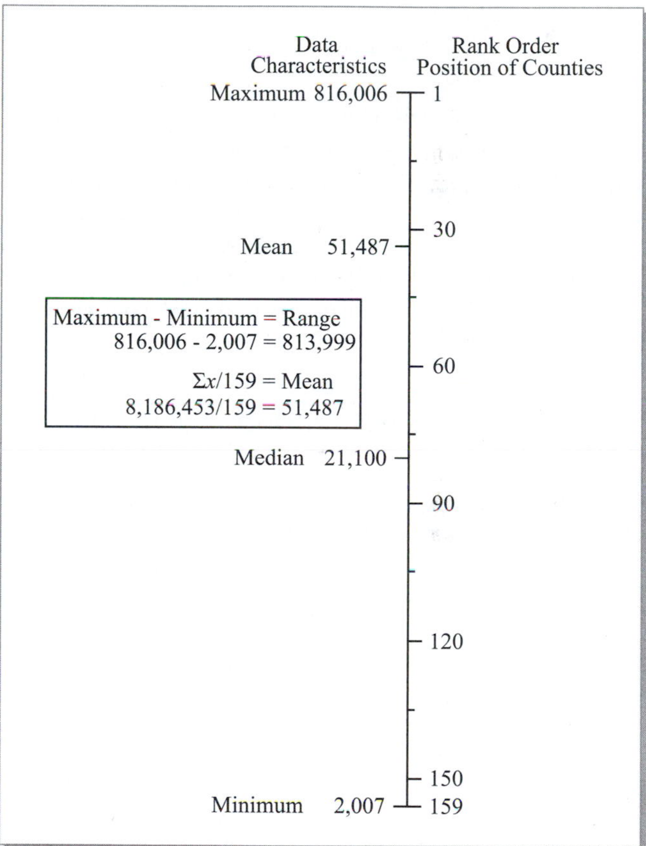

FIGURE 5.1 MEAN, MEDIAN, AND RANGE OF GEORGIA'S 2000 COUNTY POPULATION.
Distribution of Georgia's 159 county populations by value and rank order with the plotting of the mean and median locations.

we can sum them, divide by N, and compute the mean squared deviation. This is also called the **variance** and is symbolically written

$$\sigma^2 = \frac{\Sigma(x - \overline{X})^2}{N}$$

where σ^2 is the variance; x is an individual observation in the data set, $\overline{X}$ is the mean of the data set and N is the total number of observations. $(x - \overline{X})^2$ is used to compute each observation's (x) deviation from the mean, and Σ indicates to sum all deviation values.

There are two main advantages to using this method of determining variation. First, it is relatively simple, a mathematical function easily determined in the spreadsheet. Second, the units of the variance are identical to the original units of the variable. It is often more common to take the square root of the variance to obtain the measure of deviation in the same units as the data, and this is called the standard deviation:

$$\sigma = \sqrt{\frac{\Sigma(x - \overline{X})^2}{N}}$$

The standard deviation is especially useful when comparing two variables. The smaller of two standard deviations indicates values occur closer to the mean than a larger standard deviation. The two fictitious variables (X and Y) in Table 5.3 illustrate this principle. Variable X's σ of 14.03 indicates a wider dispersion than variable Y's 3.43. Their ranges (48 and 10 respectively) can also be compared. Of

TABLE 5.3 TWO DATA SETS WITH IDENTICAL MEANS (UNITS = TONS)

Variable X	Variable Y
6	28
39	24
11	26
22	29
23	23
44	27
49	21
1	20
31	21
19	29
28	30
27	22
$\Sigma x = 300$	$\Sigma y = 300$
$N = 12$	$N = 12$
$\overline{X} = 25$	$\overline{Y} = 25$
Range $= \lvert 49 - 1 \rvert = 48$	Range $= \lvert 30 - 20 \rvert = 10$
$\overline{X}^2 = 625$	$\overline{Y}^2 = 625$
$\Sigma x^2 = 9{,}864$	$\Sigma y^2 = 7{,}642$
σx^2 (variance) = 196.84	σy^2 (variance) = 11.76
σ (standard deviation) = 14.03 tons	σ (standard deviation) = 3.43 tons

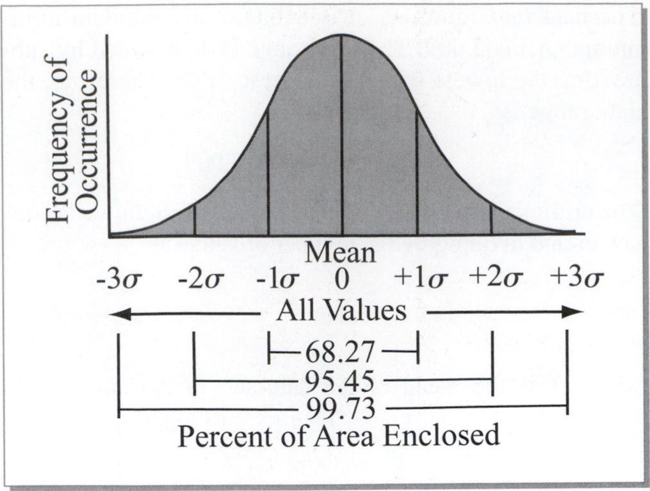

FIGURE 5.2 THE NORMAL DISTRIBUTION.
This figure illustrates the percentage of observations that fall within ±1, ±2, and ±3 standard deviations. Note that values above or below three standard deviations are very rare (less than 1 percent).

course, comparison of standard deviations is meaningful only with variables using the same units of measure.

Frequency or relative frequency distributions having certain specified shapes represented by bell-shaped curves are known as **normal distributions.** It has long been thought that a great many naturally occurring phenomena display such a distributional shape. This is not necessarily true, but the normal distribution remains at the heart of much statistical analysis. If data are distributed normally, *specific* percentages of observations occurring within spaces between $\overline{X}$ and σ can be specified (see Figure 5.2). 68.27 percent of all observations should fall within one standard deviation about the mean. As one progresses to three standard deviations, 99.73 percent of the observation should be found. Each step of a standard deviation above or below the mean has the same numeric data value. If a single observation is found in or beyond either tail of the distribution, then it is probably an outlier.

Central to the use of normal distributions is the calculation of the *probability* that certain observed values will occur. For normally distributed data, events that occur at further distances from the mean are less likely to occur. It is rare that a distribution will be purely normal. As the number of observations increase into the thousands, the greater the chance that the data will be of normal distribution. Distributions which are not *normal* can be described as having the property of skewness and/or kurtosis.

Skewness

When the peak or **mode** of a distribution is displaced to either side of the mean in a distribution, it is referred to as *skewed.* If the bulk of the frequencies are found to be left of the mean (and a long tail of low frequencies to the right), it is positively skewed. If the long tail is to the left of the mean, and the greater portion of frequencies greater than the mean, it is negatively skewed. **Skewness** is the measure of the displacement, and is calculated this way (Ebdon 1985):

$$\text{skewness} = \frac{\Sigma(x - \overline{X})^3}{n\sigma^3}$$

Negative values (or negative skewness) indicate a greater distribution of observations are on the right side of center (see Figure 5.3) and positive values (positive skewness) have the greater distribution to the left side of center (see Figure 5.3). The larger the numerical value of skewness, the greater the distribution is away from a normal distribution. Normal distributions have a skewness of 0.0.

Kurtosis

Kurtosis is a measure that describes the flatness or peakedness of a distribution. A flat distribution is one in which there is a nearly equal number of observations distributed throughout. A distribution having a high peakedness is one in which the greater bulk of the observations are in one cell. Kurtosis is calculated using the formula (Ebdon 1985):

$$\text{kurtosis} = \frac{\Sigma(x - \overline{X})^4}{n\sigma^4}$$

A normal distribution has a kurtosis of 3.0. Values greater than 3.0 indicate peakedness known as a leptokurtic distribution and values below 3.0 a flatness in the distribution or platykurtic (see Figure 5.4).

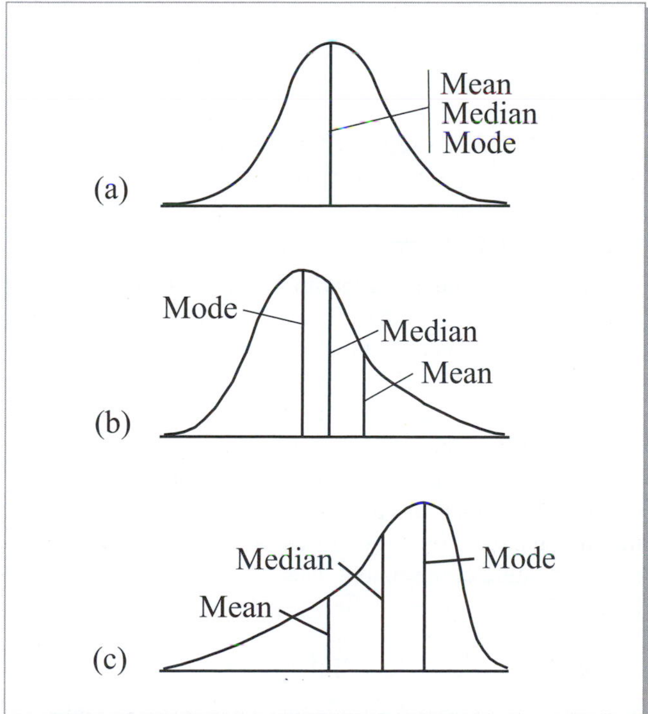

FIGURE 5.3 EXAMPLES OF SKEWNESS.
(a) symmetrical distribution; (b) positively skewed; (c) negatively skewed. Source: After McGrew and Monroe 2000, 40.

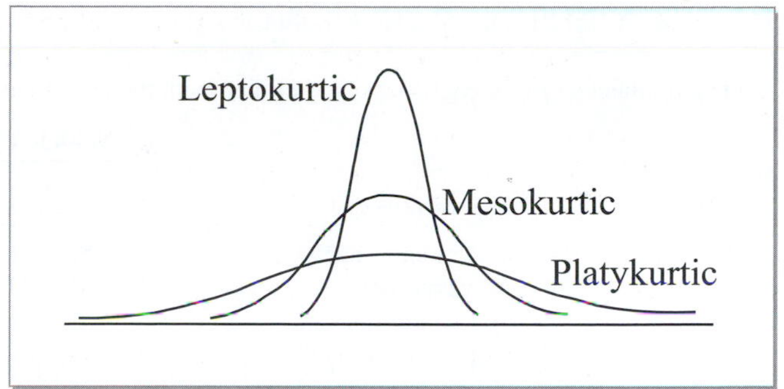

FIGURE 5.4 SKEWNESS OF DATA.
Data distribution with leptokurtic, mesokurtic, or platykurtic curves.
Source: After McGrew and Monroe 2000, 40.

DATA CLASSIFICATION

The classification of any variable, whether qualitative or quantitative in character, has a rudimentary purpose of grouping items that are alike. Geologists look for rock characteristics that identify a particular rock formation, soil scientists examine the pedons in defining soil series, and remote sensing specialists utilize spectral signatures in classifying land use and land cover categories.

The objective of classification is to group data in such a manner that not only are the observations within a class similar but also the classes themselves are dissimilar. There is no *one way* to create such groupings. In fact, classification can be done in a variety of ways. If we observe a set of 100 vehicles in a shopping center parking lot, we will see a large variety of makes, models, color, and size. Each of these descriptors can be used to classify the vehicles into five classes. Table 5.4 displays possible groupings so that within a group or class, the vehicles are similar in character but when comparing the different classes, they are different. This is similar to techniques discussed below in classifying quantitative numerical attribute data.

The classification of quantitative numerical data involves a series of steps that require decisions by the cartographer. These decisions will impact both the level of generalization of the data, ranging from very complex to very simple, and the degree of simplification accomplished in the process. Prior to classifying any data variable, the values must be sorted into rank order. This may be done manually within a spreadsheet or obtained through the automatic function of the classification schemes within GIS and advanced mapping software. The three steps taken in data classification include (1) the selection of the number of classes, (2) the classification procedure utilized, and (3) an analysis of classification accuracy.

Selection of the Number of Classes

How many classes to use has been a long-standing conundrum faced by many cartographers. How many are too many, too few, or just right? If we are talking about qualitative data, whether it be nominal or ordinal form, the number of

TABLE 5.4 POTENTIAL CLASSIFICATION OF AUTOMOBILES IN A PARKING LOT

The frequency (f) for each class is identified with the total frequency being 100 vehicles.

	Vehicle Characteristics			
Class*	Makes (f)	Models (f)	Colors (f)	Sizes (f)
1	Honda (32)	Sedan (33)	White (28)	Full-size (52)
2	Ford (21)	SUV (28)	Red (22)	Compact (29)
3	Chevrolet (20)	Mini Van (22)	Black (21)	Mid-size (10)
4	Dodge (17)	Pickup (14)	Blue (18)	Luxury (6)
5	Volvo (10)	Convertible (3)	Green (11)	Sub-compact (3)

*No order is implied in the use of class numbers.

classifications utilized will be dictated by the occurrence of categories in the data. Quantitative numerical data present the cartographer with a decision on just how many classes should be used in the classification process.

The more classes utilized, the more complex and often confusing the classification. Too few classes oversimplifies the data and can hide detail. The cartographer often selects four or five classes in which to group the data. There are no rules that state how many classes are required. Sturges (1926), however, provides a useful starting point based upon the log function of the number of observations. He suggests the following formula to determine the number of classes:

$$C = 1 + 3.3 * \log(n)$$

where C is the number of classes and n is the number of observations being classed. The results suggest the use of five or six classes when the n equals thirty and seven or eight when n equals 100. Again using Georgia as an example, n equals 159 with a log value of 2.20. Using Sturges' formula, eight classes (after rounding) are suggested. This approach should be used only as a starting point or rule of thumb when beginning the classification of data. Quite often, however, we fall back on old traditions of using either four or five classes when grouping quantitative data. These traditions essentially stem from design limits imposed by grayscale cartography. An understanding of the topic and the nature of the data should also be taken into consideration when selecting the number of classes.

Nothing Is Sacred

There is nothing sacred about grouping the data into four or five classes. The statistical classification of ordinal and interval/ratio data is unrestricted mathematically. Mathematically we have no limits on the number of classes. True statistical software allows the user to specify any number of classes up to the total number of observations, or no classes. Only when we attempt to map the data for visual display do we begin to set limits on the number of classes. GIS and advanced mapping software allow for a practically unlimited number of classes when classifying the data. The selection of the number of classes should not be made without forethought and consideration of the number of observations in the data set.

Values of Zero or No Data

A data variable may depict a value of zero for an observation or have a cell in the database be empty, possibly indicating *no data* or a null value. Care should be taken not to view these as being equal. A zero indicates that when the data count was taken, that observation was determined to have a value of zero. The concept of *no data* implies one of several possibilities: that the data were not collected for that observation; that the data were lost; that the data were corrupted in some manner; or that the statistical unit did not exist during that particular time period. The latter may occur when using temporal data sets in which the number of observations increase (or decrease) over time. For example, Georgia has not always had 159 counties.

No matter whether the variable contains observations with a zero or no data, these data should not be treated the same in the classification process. An observation with a value of zero should be included in the calculation of the data set's arithmetic mean whereas observations with vacant values, a designation of "ND," "null," or "−99," or similar marking should be excluded from the determination of the arithmetic mean or other descriptive statistics.

Impact of the Number of Classes

Beyond the mathematical manipulation of the data in order to classify the data is the impact of visualizing the spatial variability of the data. Without consideration of a classification scheme, Figure 5.5 displays the impact of differing number of classes on the spatial patterns. As the number of classes increase, the visual interpretation of the distribution changes as the patterns become more complex. The *art* of cartography comes into play as the cartographer determines the *best* number of classes to use in order to produce the inherent nature of the spatial data.

Data Classification Schemes

The selection of the appropriate data classification scheme is determined by the characteristics of the data and the desired level of generalization. The appropriate scheme maintains the character of the data. Jenks and Coulson (1963), in their

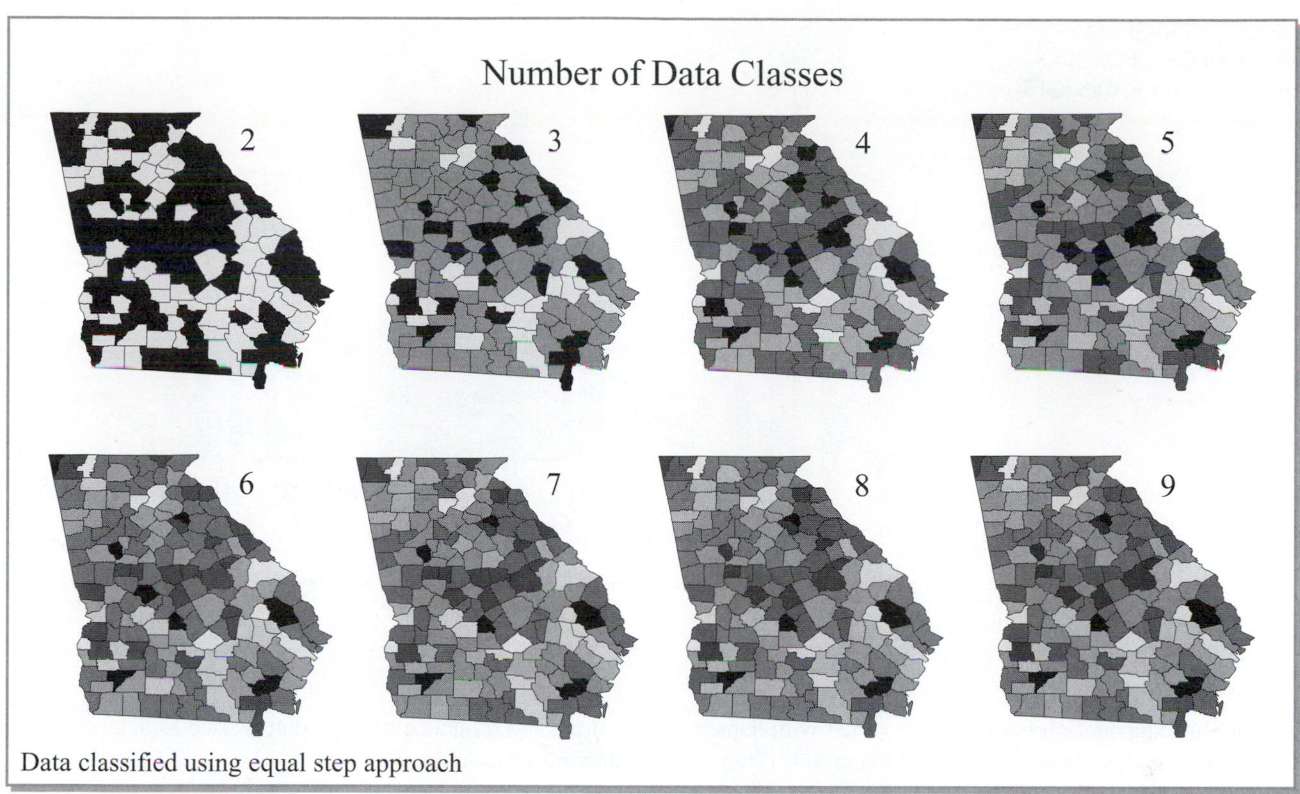

FIGURE 5.5 SPATIAL PATTERNS CREATED BY VARYING THE NUMBER OF DATA CLASSES USED.
As the number of classes increase, the visual interpretation of the distribution changes as the patterns become more complex. All maps were classified using the equal step approach.

analysis of the selection of class intervals, suggested the following five requirements should be met:

1. *Encompass the full range of the data.*
2. *Have neither overlapping values nor vacant classes.*
3. *Be great enough in number to avoid sacrificing the accuracy of the data, but not be so numerous as to impute a greater degree of accuracy than is warranted by the nature of the collected observations.*
4. *Divide the data into reasonably equal groups of observations.*
5. *Have a logical mathematical relationship if practical.*

These five requirements are laudable and every cartographer ought to attempt to use these as guidelines in determining the class limits in data classification. The means necessary to achieve these requirements were not fully presented by Jenks and Coulson. In fact, the techniques for data classification are quite varied.

Generally, when we classify data our goal is to group items together that are as alike as possible. In doing so, we generalize the data by grouping however many observations may be placed in the group and consider them as one. We also wish for the different groups to be as different as possible. The concept of minimum variance is a numerical process whereby we evaluate the difference between the attribute value of each class member and the mean of that class. This is known as within class variance. The desire is to have as small a value as possible which would indicate similarity.

There are nine common techniques used to classify data: natural breaks, optimization, nested means, mean and standard deviation, equal interval, equal frequency, arithmetic, geometric, and user defined. Some of these schemes are fairly simple, easily calculated, and somewhat arbitrary while others are very complex and rely upon the computer program to classify the data.

The discussion below will examine each separately but contain a common theme for comparison. Except for where the classification specifically indicates the number of classes, we will use five classes for each technique. A table will be generated in which the classification parameters of minimum class value, maximum class value, and the frequency of observations per class are identified. Using Georgia's General Fertility Rate (GFR) (Georgia DHR 2008) by county for 2000 we observe that the data range from a minimum of 41.14 (live births to women of any age per 1,000 females ages 15–44) to a maximum of 101.45.

Natural Breaks

When the numerical values of a rank order data set are examined, the lack of a numerical continuum is observed. The steps

FIGURE 5.6 HISTOGRAM DISTRIBUTION OF GEORGIA'S GENERAL FERTILITY RATES (GFR), 2000.

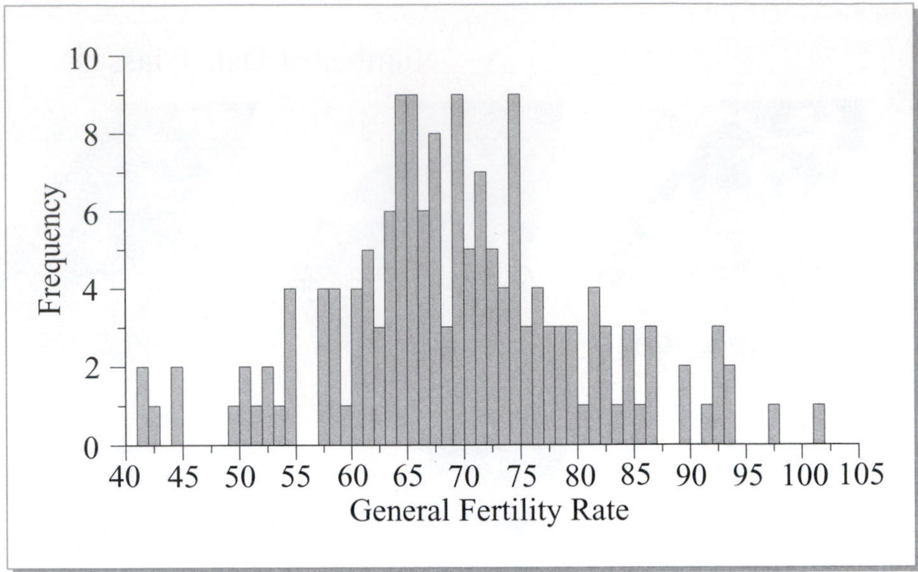

(gaps) between sequential observations are varied with some gaps being quite small and others of varying magnitude. These gaps are what make the data have a spatial non-uniform distribution. The cartographer can use the breaks in the data sequence as break points in the classification process. When the cartographer uses a graphic display of the data distribution, such as histograms or dispersion graphs, the steps are visually identified as **natural breaks** in the data. These breaks can also be identified mathematically using a standard spreadsheet program. When this approach is used the procedure is said to identify **maximum breaks** in the data. Normally, however, both processes are referred to as a natural break classification.

The breaks in the data permit the grouping of values that are thought to be alike by their nature of being closely packed. At the same time, groups are separated indicating their differences from one another. The separation of data into groups of like values and with gaps between the groups indicating their differences is the primary purpose of data classification. Fisher (1958) refers to this process as classifying for "maximum homogeneity." Many cartographers believe this to be the ideal classification technique for communicating the data distribution to the map reader.

The philosophy behind this technique is that we expect the data to cluster naturally, falling into small groups with gaps or breaks between the groups. The data can provide assistance in determining the number of classes to be used based upon the number and/or size of the gaps. Figure 5.6 displays the histogram of the 2000 GFR and the position of natural breaks. The determination of these breaks is done by an iterative process whereby the largest gap between two observations is identified, breaking the data into two classes. In the GFR data, a gap of 4.6 exists between the fifth and sixth observation. The size of each of the remaining gaps is determined and addressed in decreasing order. Table 5.5 identifies the sequence of gaps that is used to determine the following seven-class widths:

Class	Minimum–Maximum	Frequency
1	101.45–101.45	1
2	97.21–97.21	1
3	91.85–93.30	6
4	89.10–89.46	2
5	57.12–86.93	133
6	49.53–54.70	11
7	41.14–44.93	5

The problem frequently encountered with this approach occurs when the data are distributed such that all but 26 of the 159 observations are placed in the same class and the top two classes contain only a single observation each. This is a common consequence of this classification scheme.

Optimization. An algorithm for determining an optimal selection of natural breaks was developed by Walter Fisher (1958) and implemented by George Jenks in 1977. Although this is not another classification method as such, rather a more sophisticated method of pursuing natural breaks from a quantitative, algorithmic standpoint. The Fisher-Jenks algorithm has been implemented as a part of practically all GIS and mapping software. This classification approach is more popularly referred to as the "Jenks Optimization Method" or even shortened to the "optimal method." It uses an iterative procedure for examining all possible solutions based upon the sum of deviations about the median. Jenks' contribution to the field of cartography has occurred in many areas, not the least of which is data classification.

TABLE 5.5 DATA BREAKS USED FOR CLASSIFICATION: GEORGIA FERTILITY RATES, 2000

Value	Gap Value	Gap Order	Class
101.45			Class 1
	Gap = 4.24	2	
97.21			Class 2
	Gap =3.91	3	
93.30			
93.01			
92.98			Class 3
92.48			
92.40			
91.85			
	Gap = 2.39	5	
89.46			Class 4
89.10			
	Gap = 2.17	6	
86.93			
[131 closely packed values]			Class 5
57.12			
	Gap = 2.42	4	
54.70			
54.69			
54.50			
54.48			
53.99			
52.67			Class 6
52.21			
51.61			
50.46			
50.13			
49.53			
	Gap = 4.60	1	
44.93			
44.09			
42.98			Class 7
41.69			
41.14			

Using our GFR data and the Fisher-Jenks algorithm, the resultant classification will produce the following distribution:

Class	Minimum–Maximum	Frequency
1	89.10–101.45	10
2	77.06–89.09	25
3	68.03–77.05	49
4	57.12–68.03	59
5	41.14–57.11	16

The determination of the variance within a class requires an iterative process through an initial classification of natural breaks (Jenks and Caspall 1971). The absolute value of total within class variance is determined for each class and those values totaled to achieve a number which represents the total variance of that classification. A second iteration moves a single observation to a higher or lower class and a new set of total variance is determined. If the variance increases, that number is replaced and a subsequent number is moved and new calculations derived. This process continues until the lowest value of total variance is achieved. At that point, the distribution of observations within the class is determined to be as similar to the other class observation as possible. This "classification does the best job of evaluating how data are distributed along the number line" of interval data (Finn *et al.* 2006).

Nested Means

Nested means is a classification technique based on the arithmetic mean of the data (sometimes referred to as the grand mean) in order to group the data into two classes, one above the mean and one below the mean. The secondary means can be calculated using the observations of each class, thus dividing those classes into two more classes. This can be done a third time for tertiary means. Although not normally found in current GIS and mapping software, the method can easily be determined manually. The ease of calculation is the most noted advantage of this technique. The major drawback is that it is restricted to only 2, 4, and 8 classes (Scripter 1970). The general fertility rate (GFR) per county produces a grand mean is 69.56 and thus divides the data into two classes:

Class	Minimum–Maximum	Frequency
1	69.46–101.56	78
2	41.14–69.55	81

The secondary means are 78.26 and 61.19 for Class 1 and 2 respectively (see Figure 5.7). Using these secondary means, the data can be grouped into four classes. The distribution of the data is fairly well distributed. This is not always the case and when using totals, such as total housing units, the data become significantly skewed, stacking in the highest or lowest class.

Class	Minimum–Maximum	Frequency
1	78.26–101.45	33
2	69.56–78.25	45
3	61.19–69.55	50
4	41.14–61.18	31

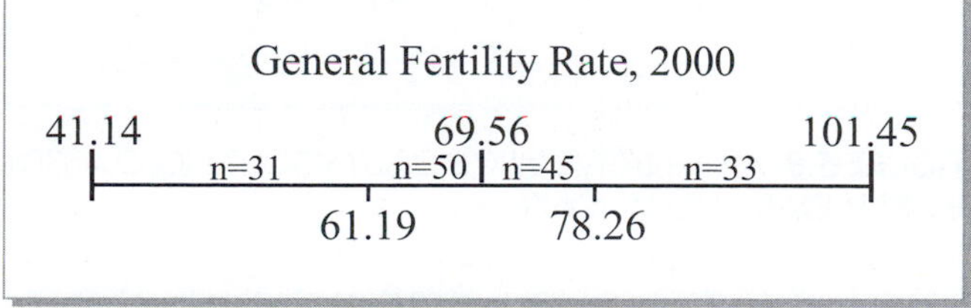

FIGURE 5.7 FOUR CLASS NESTED MEANS CLASSIFICATION OF GEORGIA'S GENERAL FERTILITY RATES (GFR), 2000.

Source: Georgia DHR Division of Public Health, 2008.

Scripter (1970) identifies several limitations of this technique including the restriction on the number of classes, the potential for the mean to coincide with a data value, and the possibility of highly skewed data producing several classes with one observation. Because this technique utilizes the grand and subsequent means of the data, deviation about the mean is minimized.

Mean and Standard Deviation

If the data set displays a normal frequency distribution, class boundaries may be established by using its standard deviation value. Class boundaries are compiled by comparing the **mean and standard deviation,** then determining the boundaries by adding or subtracting the deviation from the mean. Usually no more than six classes are needed to account for most values in a normal distribution (Figure 5.8). This method yields a constant class interval because the standard deviation is unchanging.

This technique assumes two things. First, a normal distribution of the data exists, producing the traditional bell-shaped curve. Second, the number of observations should be large enough in order to justify the classification/simplification of the data. The GFR data closely approximate such a distribution. The mean and standard deviation technique requires accessing three standard deviations both above and below the mean to include all data values. This classification and distribution using 6 classes is:

Class	Minimum–Maximum	Frequency
1	91.90–101.45	7
2	80.74–91.89	18
3	69.56–80.73	53
4	58.40–69.55	59
5	47.24–58.39	17
6	41.14–47.23	5

Comparing these values to what would be expected using normally distributed data, only minor differences occur (refer to previous discussion on standard deviation and Figure 5.2).

Number of Deviations above and below the mean	Percent Predicted per step	Percent Resulting in GFR data classification
1	68	70.44
2	95	92.25
3	99	100.00

A problem frequently encountered with this classification scheme results when the data are skewed, either positively or negatively. This results in unequal distribution in the number of standard deviation steps used. That is, only one standard deviation above the mean may use all the observations while three or more standard deviations are required to include all the observations below the mean. When this occurs, it is recommended that this technique be abandoned and a more favorable classification utilized. A major disadvantage in using this technique is that it requires a basic understanding of statistical concepts and may not be appropriate for all users (CDC 2007).

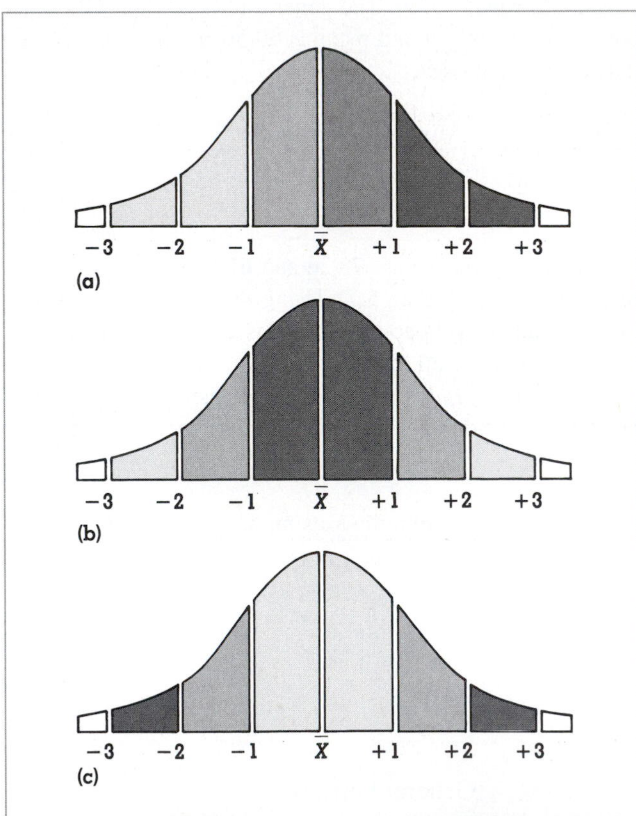

FIGURE 5.8 ALTERNATIVE WAYS OF SYMBOLIZING CLASSES BY STANDARD DEVIATIONS.

In (a), the classes range in visual importance from -3σ to $+3\sigma$ in a continuum. In (b), greater importance is given those values farthest from the mean. In (c), greater importance is assigned to those values farthest from the mean. The purpose of the map will dictate the choice of a symbolization method. Because of the bi-directional nature of the standard deviation, however, there appears to be little intuitive appeal for method (a).

Equal Interval

The **equal interval** classification assumes a desire for the data range of each class to be held constant. This is sometimes referred to as an **equal step** classification. The determination of that step is relatively simple. Simply divide the range of the entire data set by the number of classes being used. The GFR data has a data range of 60.31 (101.45−41.14). Using a five-class scheme, the class range should be maintained as steps of 12.06 in order to maintain equal class width. This produces the following classification where the range remains the same but the frequency of observations will vary considerably between classes.

Class	Minimum–Maximum	Frequency
1	89.39–101.45	9
2	77.33–89.37	25
3	65.27–77.32	71
4	53.21–65.26	43
5	41.14–53.20	11

A zero frequency class is to be avoided at all cost (Jenks and Coulson 1963) as it would imply a distribution that does not exist. This is a possible problem that can result when using data sets in which the largest value is far greater than the second ranked value. The problem is that a class may be generated in which no observations are found. To counter this problem, one treats the first observation as an *outlier,* placing it in a class by itself. The remaining data values are then considered as a new data set. A new data set range and class range is determined. However, if the final classification is to maintain a five-class map, only four classes are used for the remaining observations.

Using a hypothetical data set as an example, the data set range is a maximum of 1188 and a minimum of 88. A five-class equal interval would be comprised of steps of 220. No problem exists as long as the second ranked data value is within the step value of the largest data value. To help visualize the problem, assume that the second value is 741. This is how the classification would look:

Class	Minimum–Maximum	Frequency
1	968–1188	1
2	748–967	0
3	528–747	44
4	308–527	71
5	88–307	55

Because 741 falls in the third class, class two has zero observations and as such would invalidate the classification. The adjustment that should be made is to treat the 1188 value as an outlier, truncate it from the remaining data, and set it aside as a class by itself. This would result in a new data set range of 653 (741–88) that will be classified into four classes. The new class step would be 163 determined by dividing 653 by 4 (result is then rounded since we are dealing with whole numbers). The new classification would appear as:

Class	Minimum–Maximum	Frequency	
1	1188–1188	1	(named observation)
2	578–741	26	
3	415–577	57	
4	252–414	45	
5	88–251	42	

To help the user of the data understand that you have separated the outlier out of the original 171 observation data set, that outlier is named as a part of the classification display.

The outlier can occur on either end of the data. If the outlier were the data set minimum, it would be treated in the same manner as described here for the data set maximum value. This technique of truncation can be applied to any data set and any classification scheme when an outlier occurs.

Equal Frequency

Equal frequency classification distributes the number of observations equally among each of the classes. Frequently the cartographer divides the data into **quantiles.** This term is used to describe the assigning of total frequency (observations) into a set of equal proportions. Commonly, **quartiles** (four divisions) or **quintiles** (five divisions) are used. The first places twenty-five percent of the total observations into each class while the second disperses twenty percent into each class. The use of quartiles and quintiles is an accepted technique among scientists and researchers in the business world as earnings are reported on a quarterly basis. Teachers evaluate their students and identify an individual who is in the top twenty percent of the class. This level of data generalization also allows for the comparison between variables in order to establish potential or probability. For example, counties classified in the highest quartile in median household income, rate of unemployment, and percent poverty may indicate counties that have relatively high rates of nonviolent crime.

A quantile **equal frequency** classification will produce five classes with ranges of various widths in order to preserve the equal distribution of the observations. This technique will never produce the possibility of a zero observation class as we just observed with equal interval approach. This method works well when the number of observations is easily divisible by five (or however many classes are specified). This is not always the case however. In the Georgia example, there are 159 observations and when divided by five suggests 31.8 observations per class. Therefore, four classes with 32 and one with 31 observations will result. As a rule of thumb, overload the classes equally beginning with the lower class ranges. Also overload the lower class ranges, even if done manually, because the larger values have greater impact on statistical applications or determination of variance about the mean. The resulting classification distribution would be:

Class	Minimum–Maximum	Frequency
1	78.30–101.45	31
2	71.73–78.29	32
3	66.57–71.72	32
4	61.47–66.56	32
5	41.14–61.46	32

Arithmetic and Geometric Intervals

These mathematically defined interval systems produce class boundaries and intervening distances that change systematically. They should be used only when a graphic plot of the mapped values tends to replicate mathematical progressions (see Figure 5.9). From this plot, it is possible to see if an orderly mathematical function exists by comparing the shape of the curve to that of a typical arithmetic or geometric progression.

An *arithmetic progression* (identical in idea with the equal step method described above) is defined as

$$a, a+d, a+2d, a+3d, \ldots a+(n-1)d$$

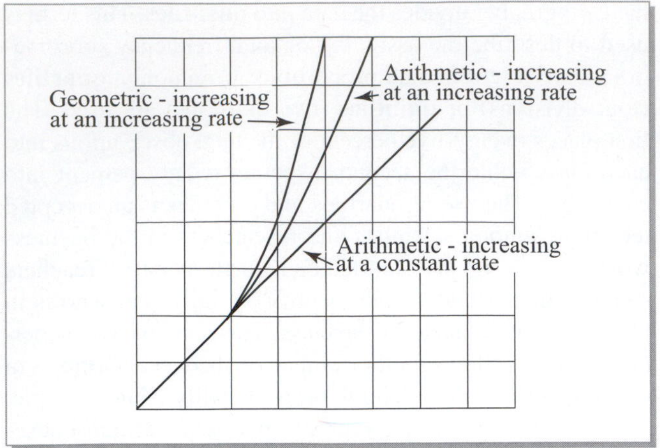

FIGURE 5.9 COMMON MATHEMATICAL PROGRESSIONS.
A graphic plot (or array) of the data values on arithmetic paper is compared visually to common mathematical progression plots. This can assist in selecting an interval plan.

where

- a = first term
- d = common difference
- n = number of terms
- l = last term = $a+(n-1)d$

(Note: The word *term* above refers to a given numerical value of the set of values in the entire progression.)

A *geometric progression* is defined as

$$a, ar, ar^2, ar^3, \ldots ar^{n-1}$$

where

- a = first term
- r = common ratio
- n = number of terms
- l = last term = ar^{n-1}

Different expressions of d and r will yield a variety of progression curves.

Arithmetic schemes are useful when classifying data with significant ranges. For example, when working with Minnesota cropland acreage by county, the attribute values range between zero and 597,000 acres. When examining global population by country, a geometric progression is useful for the dramatic range of population in 2004 of 11,468 for Tuvalu to over 1.3 billion in China.

User Defined

Although this method is not defined by a mathematical formula or rules for distributing the observations as with the previous classification schemes, it is included in the GIS and mapping software as a technique to permit maximum utility. The **user defined** method permits the cartographer to specify data values for indentifying class breaks. Under certain circumstances, the user may have specific needs to identify specific divisions in the interval data. If the results of the classifications presented here produce what may seem as odd break points within the data, the user can specify a new data range and implement the procedure as if it were an equal interval classification.

There are times when the classifications techniques produce distributions that create spatial patterns that are somewhat questionable. If one were to have a classification whereby a single enumeration unit of the next higher (or lower) class is found within a much larger grouping of observation, it would imply some difference that may be created only by the mathematical approach taken. Upon examining the data, the user defined method would allow for adjustment in the range of the class which would produce a more homogeneous spatial pattern. The user defined method permits the cartographer to apply personal knowledge or logic in order to produce a more meaningful visual display of the spatial data.

Assessment Indices

The decision of which classification scheme to use is often difficult. It coincides with the selection of the number of classes to use. Too often the cartographer simply relies on the default options of the software. The fact remains that the cartographer should consider all options of classification. Frequently we utilize equal-frequency or equal-step classifications as a result of a traditional scientific preference in presentation. Certainly the use of quantiles allows for standardization for comparison among multiple variables. However, if we are classing only one variable and not making comparison between variables, the use of classification indices allow for justification of using one over another classification scheme. Three indices that assist in classification evaluation are the Tabular Accuracy Index, Goodness of Variance Fit, and Goodness of Absolute Deviation Fit (Slocum *et al.* 2005).

These three indices evaluate the variance about either the median or mean of the data class and the data set. Such comparisons determine the degree to which the classification technique achieves a minimum variance result. Again the goal of classification is to place observations into classes that are as alike as possible and yet have the classes be as different as possible. The formulas for the three algorithms are quite similar, producing a decimal value less than one (see boxed text: "Goodness of Variance Fit (GVF) Assessment Index"). Only a data set which has not been classified could be thought of as having one observation per class. Each value represents the class mean and median; therefore, each value is considered as one containing no deviation. In this case, a mean (median) would be identical to the one value in the class producing a variance of zero. Across all classes, the index value would be 1.0 or perfect classification. Of course we do not have classes for each value in conventional classification approaches. Real solutions are less than 1.0 and the "best" solution will only tend toward 1.0.

These indices can be used to evaluate an individual classification or to compare two or more classifications. For the single classification scheme, a rule of thumb is to

GOODNESS OF VARIANCE FIT (GVF) ASSESSMENT INDEX

One of the more common assessment indices is called the **goodness of variance fit (GVF).** The procedure is a minimizing one in which the smallest sum of squared deviations from class means is sought. The steps in computing the GVF are as follows:

1. Compute the mean ($\bar{X}$) of the entire data set and calculate the sum of the squared deviations of each observation (x_i) in the total array from this array mean:

$$\Sigma(x_i - \bar{X})^2$$

This will be called SDAM (squared deviations, array mean).

2. Develop class boundaries for the first iteration. Compute the class means ($\bar{Z}_c s$). Calculate the deviations of each x from its class ($x_i - \bar{Z}_c$) mean, square these, and calculate the grand sum.

$$\Sigma(x_i - \bar{Z}_c)^2$$

This will be called SDCM (squared deviations, class means).

3. Compute the goodness of variance fit (GVF):

$$\mathbf{GVF} = \frac{SDAM - SDCM}{SDAM}$$

The computed difference between SDAM and SDCM is the sum of squared deviations between classes.

4. Note the value of GVF for iteration 1. The goal through the various iterations is to *maximize* the value of GVF.
5. Repeat the above procedures until the GVF cannot be maximized further.

Data classifications in which each value is a class unto itself are considered to be one containing no errors. In this case, a class mean would be identical to the one value in the class, so the squared deviation would be 0. Across all classes, then, SDCM = 0.0, and GVF = 1.0, the maximum value of GVF. Of course, we do not have classes for each value in conventional choropleth mapping. Real solutions for GVF will be less than 1.0, and "best" solutions will only tend toward 1.0. It should be apparent that applying an optimization technique requiring experimentation with class intervals is a timely process.

Earlier in this chapter, the measures of kurtosis and skewness were defined as numerical ways of describing the structural characteristics of data distributions. In one study conducted to determine the relative effectiveness of the Jenks optimization method, when compared to four other classing methods (quartile, equal interval, standard deviation, and natural breaks), the following results were achieved (Smith 1986):

1. For the quartile method, GVF scores decreased with increasing skewness, and decreased similarly with increasing kurtosis.
2. For the equal interval method, GVF scores showed little correspondence with skewness, that is, GVF scores remained little changed with increasing skewness. GVF scores did not change greatly with increasing kurtosis, but did show to be somewhat better with flat distributions.
3. For the standard deviation method, GVF scores generally declined with increasing skewness, but not always. The relationship between kurtosis and GVF was similar.
4. For the natural breaks method, GVF varied greatly with skewness and kurtosis. In general, distributions with high coefficients of skewness and kurtosis class the data more accurately than those which are close to normal.
5. For the Jenks optimization method, high GVF scores result for distributions ranging from low to high skewness, and from low to high kurtosis. This procedure produces consistently high GVF indices irrespective of the degree of skewness and kurtosis because the method selects classes on the basis of the GVF criteria.
6. None of the methods proved wholly reliable because the GVF did vary, but the Jenks optimization method proved best overall.

The implications for the designer are apparent. The structural characteristics of the data distribution to be classed must be considered before classing is begun. Skewness and kurtosis should be calculated, and GVF scores computed. *Here is the point: It appears abundantly clear that any numerical clustering procedure is better used than none at all, at least when the geographical objective of classing is of primary concern.*

achieve an index value greater than 0.70. It is preferred, however, to achieve an index value of 0.80 to be totally confident in the classification. Depending on the data set, even the lower value (0.70) is frequently hard to achieve. Care must be taken when using an index value when comparing different classification schemes. No comparisons can be made if the classifications contain different number of classes. A classification with a greater number of classes is one step closer to the initial unclassed (Peterson 1979) data set. When comparing classifications using assessment indices, each scheme must have the same number of classes and, at that time, the selection of the higher index value is advised.

Things to Watch Out For

Outliers and Data Truncation

Data sets often have one or more values that are unlike most of the other values. These observations make data classification very difficult, because such an observation defies the very purpose of classification—to group like phenomena. Extreme observation(s) can be treated as outliers and given their own class. The outlier or outliers are usually removed or separated from the data set (truncation), and then classification takes place on the remaining observations. If a five-class scheme is desired, the truncated values are treated as a class. After extracting the outliers, the remaining observations should be classed using only four classes.

The determination of what is an outlier, or how to handle multiple outliers, can require some creativity on the part of the cartographer. As suggested earlier, it is important for the cartographer to study the data table and make graphs to get a complete picture of the data's key characteristics. If an outlier can be measured in standard deviations from its counterparts, then treating the observation as an outlier is probably a good idea. Multiple outliers, if grouped relatively close together, may simply make another class of observations. Single observation outliers should be identified with the enumeration unit name and the specific value.

Zero-Observation Classes

One potential error to guard against in classification is producing a classification containing a zero-frequency class. Not only does this produce erroneous assessment index values but it communicates an erroneous perception of the data distribution. If the data were mapped using a zero-frequency classification, the map reader would assume that each class would be represented on the map when in fact it is not. If a classification method produces a zero-frequency class, the method should be discarded as a viable method.

There is an exception that can be made however. If you are comparing temporal data of the same variable, it would be possible to have a class with zero observations if a single legend is applied to a series of maps, either multiple static maps or in an animation sequence.

SUMMARY AND COMPARISON OF MAJOR CLASSIFICATION METHODS

Data classification provides the cartographer the opportunity to simplify large data sets and to generalize that data in preparation for displaying it thematically. Such classification is easily applied to numerical data in the form of interval or ratio form while some schemes may be applied to nominal or ordinal data. The determination of the number of classes initially is paramount to the selection of the method of classification. Although any number of classes is possible, cartographers tend to be most comfortable with using four or five classes. Some schemes are easily computed manually and provide for an easy understanding of the methodology involved. Other schemes, such as Jenks Optimal, require computer software to determine the class breaks and their methodology is not easily understood. We have presented both techniques included in GIS and mapping software as well as those not included but easily computed manually.

The results of data classification are displayed in a legend. Such legends include the minimum and maximum data values for each class and should include those values for the data set. The legend is used to communicate numerically the resulting classification of the data while the map displays the resulting spatial distribution based upon that classification. How well the data are classified may be determined using assessment indices such as the goodness of variance fit (GVF), tabular accuracy index (TAI), and the goodness of absolute deviation fit (GADF). While these values are not part of the final mapping process, they may be used to evaluate the validity of one classification scheme compared to another. Table 5.6 provides a summary of the classification scheme discussed, using Georgia's GFR data. When comparing the classifications using five classes (a, d, e, and f), the Jenks Optimal classification scheme produces the highest index values. Therefore, one can assume that this classification is the best among those compared. Again, no comparison can be made between schemes with a different number of classes.

The impact of the classification scheme is the resulting spatial patterns displayed on the map. Using the choropleth map to display the GFR data, the variations of spatial patterns are visualized (see Figure 5.10). The pattern variation created by the application of four of the classification techniques was presented in Table 5.6. Such spatial patterns are compounded by the number of classes selected as we saw in Figure 5.5.

A comparison of the advantages and disadvantages of the classification schemes is summarized in Table 5.7. What are advantages to some techniques appear as disadvantages to others. A recurring theme in this comparison is whether the classification scheme considers the distribution of the ordered data. The Natural Breaks (including its similar approaches of Maximum Breaks and the Jenks Optimal method) and the Mean and Standard Deviation methods consider such distribution. The latter, however, is applicable only if the data are normally distributed. Some methods are easily computed manually (Natural Breaks, Nested Means, Equal Interval, and Equal Frequency) while others (Maximum Breaks and Jenks Optimal) require computer software for their determination. This table incorporates Jenks and Coulson's (1963) five requirements for class interval selection in its analysis.

The User Defined method, while not a true classification technique, provides the cartographer maximum control over the class breaks in legend creation. Once the data are classified using any of the other classification methods, the user defined techniques may be used to "tweak" the data ranges per class. Once the data are mapped, spatial patterns may be examined and adjustments made to the classification based on personal knowledge of the data. Knowing the characteristics of the data is important for any data classification.

TABLE 5.6 CLASSIFICATION SCHEMES

(TAI and GVF values are relative to the data set, and cartographer's numbers will vary. Assessment indices are provided only for classifications using five classes.)

(a) Natural Breaks [Georgia Fertility Rate, 2000]

Class	Min–Max	Frequency	
1	97.21–101.45	2	TAI = 0.3797
2	89.10–93.30	8	GVF = .6333
3	57.12–86.93	133	Total Variance = 66.1462
4	49.53–54.70	11	
5	41.14–44.93	5	

(b) Nested Means (2 Classes) [Georgia: General Fertility Rate, 2000]

Class	Min–Max	Frequency
1	69.46–101.56	78
2	41.14–69.55	81

Nested Means (4 Classes) [Georgia: General Fertility Rate, 2000]

Class	Min–Max	Frequency
1	78.26–101.45	33
2	69.56–78.25	45
3	61.19–69.55	50
4	41.14–61.18	31

(c) Mean and Standard Deviation (6 Classes) [Georgia: General Fertility Rate, 2000]

Class	Min–Max	Frequency	
1	91.90–101.45	7	TAI = 0.7225
2	80.74–91.89	18	GVF = .9337
3	69.56–80.73	53	Total Variance = 45.8915
4	58.40–69.55	59	
5	47.24–58.39	17	
6	41.14–47.23	5	

(d) Equal Interval (5 Classes) [Georgia: General Fertility Rate, 2000]

Class	Min–Max	Frequency	
1	89.39–101.45	9	TAI = 0.6630
2	77.33–89.37	25	GVF = .9073
3	65.27–77.32	71	Total Variance = 61.1843
4	53.21–65.26	43	
5	41.14–53.20	11	

(e) Equal Frequency (5 Classes) [Georgia: General Fertility Rate, 2000]

Class	Min–Max	Frequency	
1	78.30–101.45	31	TAI = 0.6803
2	71.73–78.29	32	GVF = .8735
3	66.57–71.72	32	Total Variance = 77.8166
4	61.47–66.56	32	
5	41.14–61.46	32	

(f) Jenks Optimal (5 classes) [Georgia: General Fertility Rate, 2000]

Class	Min–Max	Frequency	
1	89.10–101.45	10	TAI = 0.7029
2	77.06–89.09	25	GVF = .9226
3	68.04–77.05	49	Total Variance = 58.4266
4	57.12–68.03	59	
5	41.14–57.11	16	

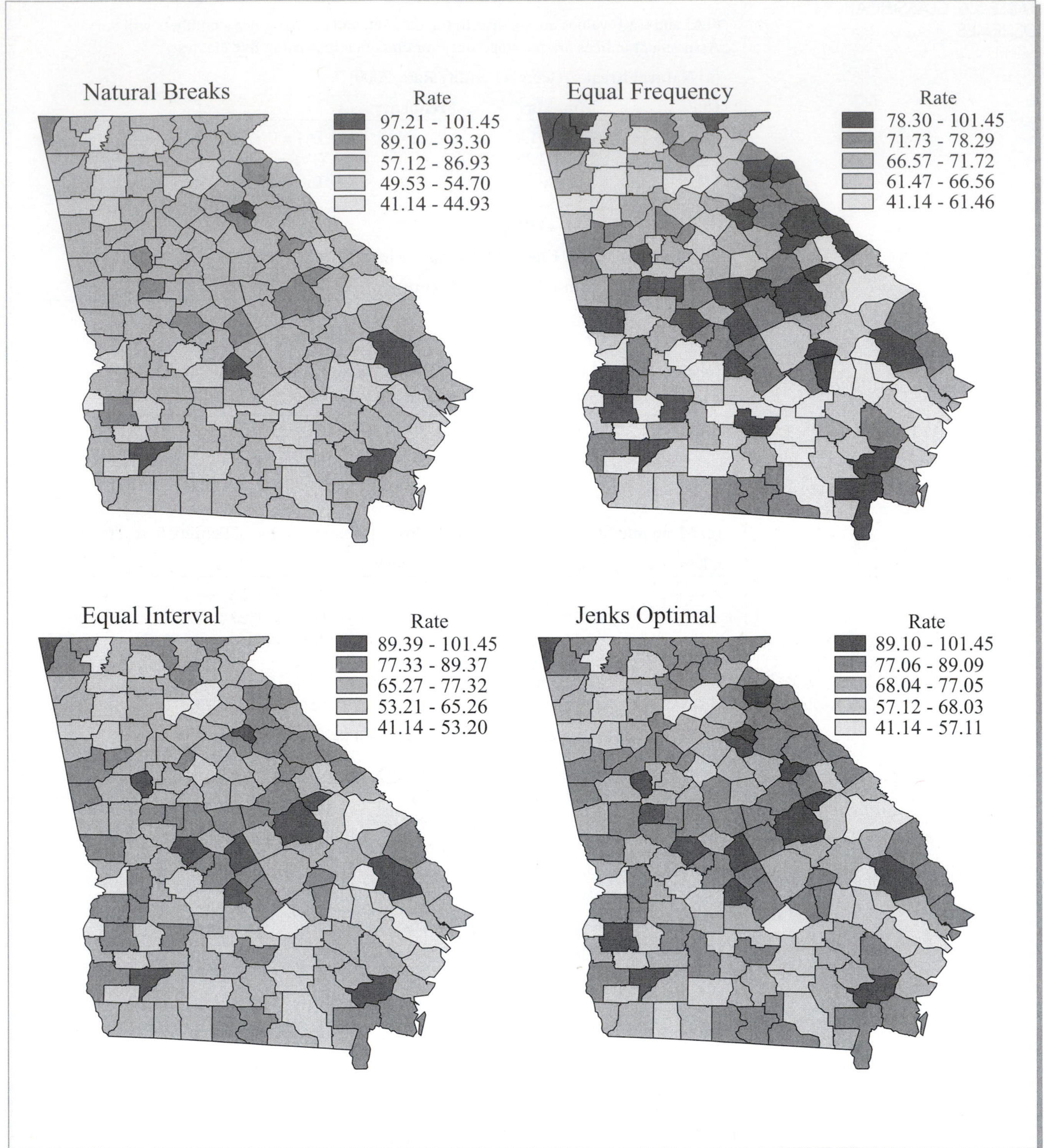

FIGURE 5.10 GEORGIA FERTILITY RATES MAPPED USING FIVE CLASSES AND FOUR DIFFERENT CLASSIFICATION SCHEMES.
Spatial patterns of change resulting from classifications listed in Table 5.6.

TABLE 5.7 COMPARISON OF CLASSIFICATION SCHEMES: ADVANTAGES vs. DISADVANTAGES

Natural Breaks
- Advantages
  - Considers the distribution of the data
  - Uses groupings based on break points visible graphically
  - Intuitive
- Disadvantages
  - Class breaks are subjective
  - Breaks are not necessarily obvious
  - Difficulty to determine breaks increases with larger data sets
  - May miss natural spatial clustering of data

Maximum Breaks
- Advantages
  - Considers the distribution of the data
  - Easily computed using a spreadsheet
  - Size of gaps can be identified using a hierarchy
- Disadvantages
  - May miss natural spatial clustering of data

Jenks Optimal
- Advantages
  - Computer program attempts multiple solutions to natural breaks
  - Attempts to minimize within class variance and maximize between class variance
  - Produces classification with high accuracy
- Disadvantages
  - Complicated
  - Difficult to understand the procedure for grouping

Nested Means
- Advantages
  - Easily computed
  - Mathematically intuitive
- Disadvantages
  - Limited to 2, 4, or 8 classes
  - Does not consider distribution of the data, possibly placing majority observations in one class
  - Not included as an option in GIS and mapping software
  - Requires the use of User Defined in order to apply within the software

Mean and Standard Deviation
- Advantages
  - Good for data with normal (bell-shaped) distribution
  - Considers distribution of the data
  - Produces constant class intervals
- Disadvantages
  - Most data are not normally distributed but skewed in some manner
  - Requires an understanding of basic statistics
  - Not easily understood by the map reader

Equal Interval
- Advantages
  - Easily understood by the map reader and straight forward
  - Simple to compute
  - No gaps in the legend
- Disadvantages
  - Does not consider distribution of the data
  - May produce classes with zero observations

cont.

TABLE 5.7 COMPARISON OF CLASSIFICATION SCHEMES: ADVANTAGES VS. DISADVANTAGES *CONTINUED*

Equal Frequency (Quantiles)
- Advantages
  - Easily calculated using rank-ordered data
  - Applicable to ordinal data
  - No empty classes
- Disadvantages
  - Does not consider distribution of the data
  - Variable class width and/or gaps in legend
  - Distribution unequal when division of observations by the number of classes does not result in a whole number
  - Duplicate data values at class break requires manual adjustments
  - Value of an observation could be closer to values in a different class than its own

Arithmetic and Geometric Intervals
- Advantages
  - Good for data with significantly large ranges
  - Break points determined by rate of change in the data
- Disadvantages
  - Not appropriate for data with small ranges or linear trends

User Defined
- Advantages
  - Provides complete flexibility for the user
  - Break points may be defined based on spatial distribution observed by classification methods
- Disadvantages
  - Logic of legend breaks not apparent
  - May not consider distribution of the data
  - Not easily repeated

REFERENCES

Center for Disease Control (CDC). 2007. http://www.cdc.gov/brfss/maps/faqs.htm

Ebdon, D. 1985. *Statistics in Geography,* 2d ed. Oxford, England: Basil Blackwell, 28–31.

Finn, M., M. Williams, and L. Usery. 2006. An Implementation of the Jenks-Caspall Algorithm for Optima Classification of Geographic Visualization. American Society of Photogrammetry and Remote Sensing. Annual Conference: Reno, NV (Poster).

Fisher, W. 1958. On Grouping for Maximum Homogeneity. *Journal of the American Statistical Association,* 53 (December).

Georgia Department of Human Resources, Division of Public Health. 2008. http://oasis.state.ga.us/oasis/qryMCH.aspx

Healy, J. 1996. *Statistics: A Tool for Social Research.* Belmont, CA: Wadsworth.

Jenks, G., and F. Caspall. 1971. Error on Choroplethic Maps: Definition, Measurement, Reduction. *Annals* of the Association of American Geographers, 61 (2).

Jenks, G., and M. Coulson. 1963. Class Intervals for Statistical Maps. *International Yearbook of Cartography,* 3: 119–33.

McGrew, J., and C. Monroe. 2000. *An Introduction to Statistical Problem Solving in Geography,* 2d ed. New York: McGraw-Hill Higher Education.

Peterson, M. 1979. An Evaluation of Unclassed Cross-lined Choropleth Maps. *The American Cartographer,* 6 (1).

Scripter, M. 1970. Nested-Means Map Classes for Statistical Maps. *Annals* of the Association of American Geographers, 60 (2), 385–93.

Slocum, T., R. McMaster, F. Kessler, and H. Howard. 2005. *Thematic Cartography and Geographic Visualization,* 2d ed. Upper Saddle River, NJ: Prentice-Hall.

Smith, R. 1986. Comparing Traditional Methods for Selecting Class Intervals on Choropleth Maps. *American Cartographer,* 38, 62–67.

Sturges, H. 1926. The Choice of a Class-Interval. *Journal of American Statistical Association,* 21, 65–66.

U.S. Census Bureau. 2008. http://www.census.gov/

GLOSSARY

equal frequency classification scheme where each class has the same number of observations

equal interval classification scheme where the range of each class is the same

equal step see equal interval

kurtosis used to describe the data distribution curve: symmetrical, flat, or peaked

maximum breaks similar to natural breaks only the divisions are identified using a spreadsheet program

mean the average of a class or data set determined by the summation of the data values divided by the number of observations

mean and standard deviation class ranges are determined for a normally distributed data set using the mean and standard deviation steps above and below the mean

median the middle value in a rank-ordered set of numbers

mode the most often occurring number or value in a data set

natural breaks division of a data set into classes based on breaks visible on a dispersion graph or histogram

nested means a classification which uses the mean of the data set to identify two classes; the means of these classes provide additional classes; the number of classes are restricted to 2, 4, or 8

normal distribution usually associated with larger data sets, the data are distributed symmetrically about the mean

outlier a value that is quite different from the remaining portion of the data and may be a data maximum or a data minimum value

percentage an associated amount based on a fraction of 100: also a proportion multiplied by 100

proportion the ratio of the number of items in one group (class) to the total of all items

quantile the division of a group into equal component parts

quartile the division of a group into four equal parts

quintile the division of a group into five equal parts

range the difference between the largest and smallest data values in a data set

rank order the sequencing of numbers arranged from high to low

rate a ratio of the number of items to a standardized value, for example, general fertility rate

ratio the number of one item compared to the number of another item

skewness the deviation from a normal distribution with the data being distributed asymmetrically

sort reordering of the data into numerical sequence

standard deviation a measure of dispersion of observations about the mean

user defined arbitrary determination of class breaks specified by cartographer

variable a column within a database which includes data values for a specific topic

variance the square of the standard deviation

PART II
TECHNIQUES OF QUANTITATIVE THEMATIC MAPPING

The chapters in Part II are devoted to the various quantitative thematic mapping techniques overviewed in Table 4.3 in Part I. Each chapter presents a different thematic technique, and each includes the rationale for its adoption along with a discussion of appropriate data and symbolization for each type. The type of geometric form (for example, point, line, or area) and the attribute data type (for example, totals or derived data) will be the primary factors in determining which mapping technique is selected. Part II begins with an examination of the choropleth map, perhaps one of the most widely used thematic map forms today. Area enumeration data are symbolized by graduated colors or shades in this form of map. Chapter 7 addresses the dot density map, in which area attributes are symbolized by small dots. This kind of map is used quite often in presenting agricultural data. Symbolizing quantities at points is presented in Chapter 8, which introduces the technique of mapping by proportional symbols. In this technique, larger data values are symbolized by larger symbols and vice versa.

Chapter 9 introduces surface mapping. This technique includes isarithmic mapping (similar to familiar contour maps) and 3-D mapping. Chapter 10 deals with value-by-area mapping, or mapping by transforming familiar geographical space into some other space based on attribute values. This form of mapping is more abstract than real, yet can yield interesting results and is often used to attract the reader's attention. It is a useful pedagogical device in geography. The design of flow maps concludes the chapters in Part II. The only method devoted to linear geometric form, this two-century-old technique is often used for mapping the movement of things or ideas, connections and interactions, and spatial organization, and frequently finds its way into economic presentations. Each of the map types covered in Part II presents interesting design challenges for the cartographer.

6 MAPPING ENUMERATION AND OTHER AREALLY AGGREGATED DATA: THE CHOROPLETH MAP

CHAPTER PREVIEW Choropleth mapping is a common technique for representing enumeration data. The **choropleth map** was introduced early in the nineteenth century. It was used by the Bureau of the Census in several statistical atlases in the last half of that century and has been a favorite of professional geographers and cartographers ever since. Its name is derived from the Greek words *choros* (place), and *plethein* (to fill). Choropleth mapping has also been called *area* or *shaded* mapping. For the most part, the rationale of the choropleth technique is easily understood by map readers. Major concerns of the cartographer are data classification, areal symbolization, and legend design. From a variety of classification methods discussed in Chapter 5, the designer is faced with selecting the one that best serves the purpose of the map and best depicts the spatial distribution of the data. One result of grouping data is the creation of different maps based on the classification scheme chosen by the designer. There is also the unclassed choropleth map, chosen when the map designer wants to eliminate attribute generalization error associated with classified maps. A form of map not used often today, the dasymetric map, is discussed because of its relation to choropleth mapping and GIS.

An entire chapter is devoted to choropleth mapping because of its widespread use and appeal, not only for professionals but for the general public as well. Because of this, the student must look at the choropleth map in considerable detail to learn its advantages and consider the standards for its use in a variety of mapping situations. ■

SELECTING THE CHOROPLETH TECHNIQUE

What guides cartographers in selecting one mapping technique over another when approaching a given design task? When is a given technique not appropriate? The following section describes under what conditions the choropleth map should be chosen.

Mapping Rationale

The choropleth technique is defined by the International Cartographic Association as follows: "A method of cartographic representation which employs distinctive color or shading applied to areas other than those bounded by isolines. These are usually statistical or administrative areas" (Meynen 1973, 123). Because this form of mapping is used to depict bounded areal classified or aggregated data (often defined by administrative areas), it is sometimes called *enumeration* mapping (see Figure 6.1). Examples of typical enumeration units used in choropleth maps include countries, states, provinces, counties, census tracts, or any other unit that has associated attribute data that correspond to the enumeration units.

Choropleth mapping may be thought of as a three-dimensional histogram or stepped statistical surface (see Figure 6.2). A choropleth map is simply a planimetric

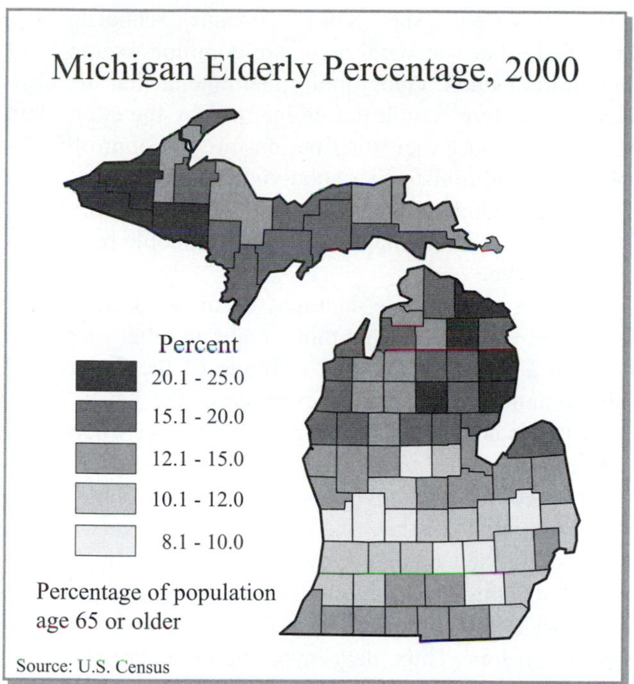

FIGURE 6.1 A TYPICAL CHOROPLETH MAP.
Each enumeration unit, in this case a county, has an areal symbol applied to it, depending on the class in which its data value falls. Over the entire map, it is possible to determine spatial variation of the data.

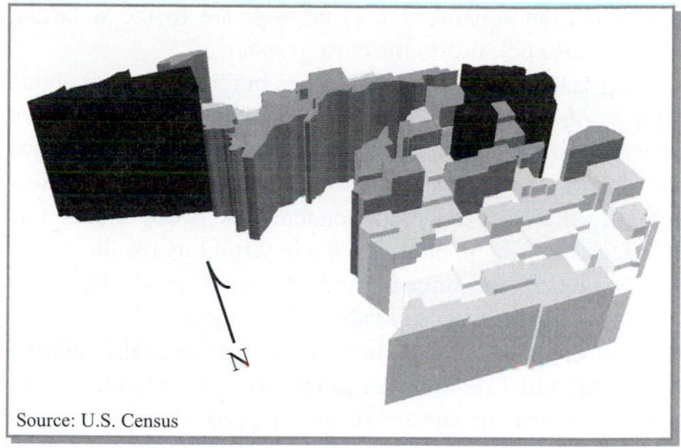

FIGURE 6.2 THE THREE-DIMENSIONAL HISTOGRAM OR STEPPED STATISTICAL SURFACE IN CHOROPLETH MAPPING.
In this conceptual model, each enumeration unit is a prism raised vertically in proportion to the value it represents. The attribute data is the same as in Figure 6.1.

representation of this three-dimensional prism model. In the model the height of each prism is proportional to the value it represents. The planimetric way of looking at the model incorporates areal symbolization to depict the heights of the prisms. It is helpful to think of this conceptual model when beginning a choropleth design task. In black-and-white mapping, the higher prisms are normally represented by darker-area symbols; conversely, the lower prisms are represented by lighter-area symbols. In color choropleth map production, generally darker or more saturated hues represent higher values, and lighter or less saturated represent lower values. See Chapter 14 for more detail on color terminology.

Choropleth mapping technique normally requires the cartographer to collect attribute data by statistical or administrative areas. These attribute data correspond observation for observation to the individual enumeration units. An areal symbolization scheme is selected or devised for these values, and symbols are applied to those areas on the map whose data fall into the symbol classes (see Figure 6.3). In GIS and other mapping packages, once the attributes have been associated, joined, or linked to the geographic base map, the process is highly automated. However, the responsibility for understanding and selecting a proper classification and

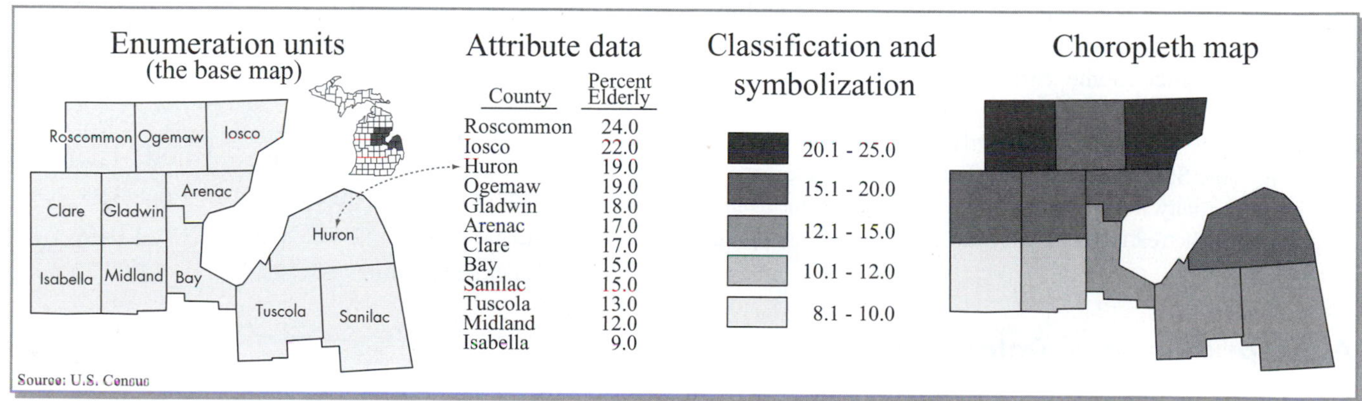

FIGURE 6.3 THE CHOROPLETH TECHNIQUE.
Cartographic designers begin with the enumeration units and associated attribute values. Class ranges are selected, as are the area symbols that will represent those ranges. The final map is the result of the application of appropriate area symbols to the enumeration units, based on each unit's attribute data value.

symbolization scheme that is appropriate to the data and map purpose belongs to the cartographer.

Map readers use choropleth maps in three ways: to obtain a sense of the overall geographical pattern of the mapped variable with attention to individual values; to compare one choropleth map pattern to another; and to ascertain an actual value (or the class range) associated with a geographic area. When using printed and some static virtual maps, the reader who wants to find only individual values would be better served to consult a table of values. With many online interactive choropleth maps and choropleth maps in a GIS environment, individual values often can be seen as the reader moves the cursor over the surface of the map, as a complement to visualizing the overall distribution.

Using two or more choropleth maps to compare geographical distributions is an acceptable application of choropleth mapping. This is often done to look for positive correlations between the maps. If the juxtaposed maps are relatively small in size, and have similar structure and context, they are sometimes referred to as **small multiples** (Tufte 2001). Similarly, using a sequence of two or more choropleth maps using the same attribute data but for different time periods allows for examining change through time. For example, county population densities in a particular state for 1950 and 2000 can be mapped and compared. More recently, animated choropleth maps, which occupy the same space on a display and have the enumeration units change symbolization with change in value (usually over a specified time period), have received attention in the cartographic community. All three of these extensions to basic choropleth mapping require careful attention to attribute data classification. Although our presentation throughout the chapter concentrates primarily on methods useful in the production of an individual choropleth map whose purpose is to portray a single geographical theme, we will touch on these, and other variations in choropleth mapping, as appropriate.

It must be pointed out that the choropleth map has come under close scrutiny by cartographers because for any given choropleth map solution there are also others that could be selected. Which is best? There is never a conclusive test to assist the designer. Some cartographers suggest that we should never offer only one map for a particular data set, but several, so that the reader has the advantage of seeing other map solutions. Still others suggest that an ethical alternative to providing many maps would be to add a statement to the map that tells the reader that this is but *one* of many mapping alternatives.

Appropriateness of Data

The choropleth technique should be selected only when the form of data is appropriate. Typically, and appropriately, choropleth maps are constructed when data occur or can be attributed to definite enumeration units that are areal in nature. Administrative political subdivisions used by the Bureau of the Census such as blocks, counties, school districts, states and other statistical areas are common examples of enumeration units. Geographic phenomena that are continuous in nature should not be mapped by the choropleth technique because their distributions are not controlled by political or administrative subdivisions. For example, to map average annual temperature by this method would not be appropriate, but to map the number of people per square mile would be.

Enumeration attribute data may be of two kinds: *totals* or *derived values* (rates or ratios). The number of people living in a census tract is an example of the former and average annual income is an example of the latter. Traditionally, it is not acceptable to map total values when using the choropleth technique. This convention is based on sound reasoning. In most choropleth mapping situations, the enumeration units are unequal in area. The varying size of areas and their mapped values will alter the impression of the distribution (see Figure 6.4). It has therefore become customary to use either *ratios involving area* or *ratios independent of area.* Thus, data involving areas are standardized over the map.

A familiar example of a ratio involving area is density, of which many kinds could be named. Population per square mile is often used, as is crop yield per acre. Ratios independent of area include per capita income, infant deaths per 100,000 live births, and so on. Proportions, including percentages, are also used. Much of the enumeration data available to cartographers is in the form of aggregated areal data, such as average value of farm products sold, median family

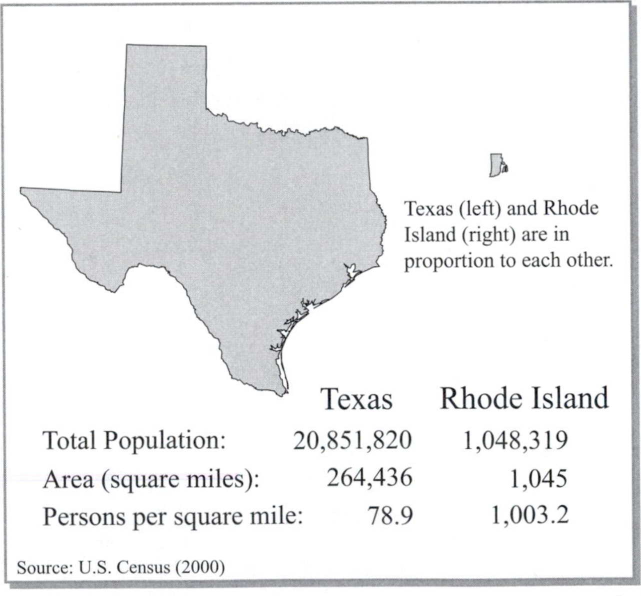

Texas (left) and Rhode Island (right) are in proportion to each other.
Texas
Rhode Island
Total Population: 20,851,820 1,048,319
Area (square miles): 264,436 1,045
Persons per square mile: 78.9 1,003.2
Source: U.S. Census (2000)

FIGURE 6.4 TOTAL VALUES SHOULD NOT BE MAPPED BY THE CHOROPLETH METHOD.

Although Texas has a greater population than Rhode Island, it also is much larger in area. Mapping total population masks the fact that the population density of Rhode Island is much greater than that of Texas.

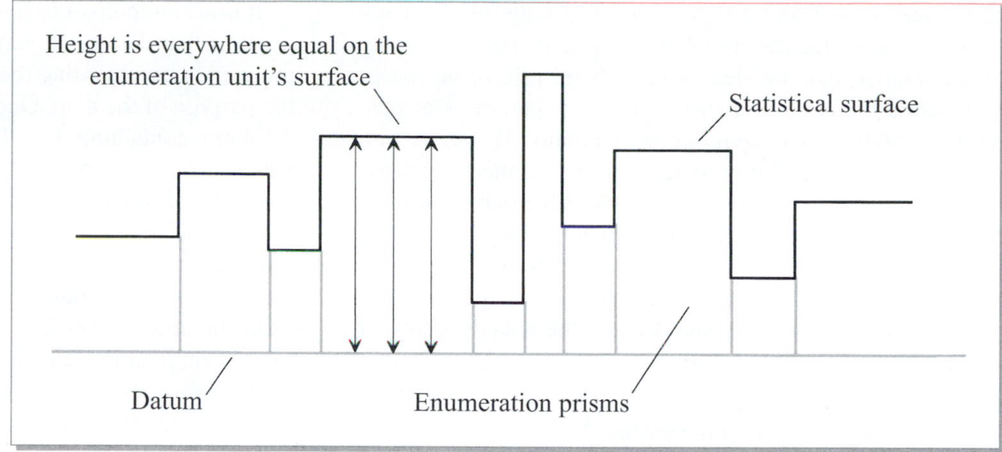

FIGURE 6.5 THE STEPPED STATISTICAL SURFACE.
The choropleth technique assumes a stepped statistical surface on which the value in each unit is constant. The view shown here is as if a vertical profile or a slice were taken out of a three-dimensional data model.

income, average annual income, and average number of people per household unit. These are treated in the same manner as ratios or proportions in choropleth mapping.

The most important assumption made in choropleth mapping is that the value in the enumeration unit is spread uniformly throughout the unit (see Figure 6.5). The top of each prism in the data model is horizontal and unchanging. Wherever one places a pencil point in the area, one finds the amount of the variable chosen to represent the entire area. Thus, the choropleth technique is insensitive to changes of the variable that may occur at scales larger than the chosen enumeration unit. If the variable is changing within the enumeration unit, the change cannot be detected on the choropleth map. (See the discussion about dasymetric mapping at the end of this chapter.) Geographic phenomena such as population density, agricultural variables, education data, or other economic data are often not consistent across each enumeration area. If the level of variation is too great within enumeration units, a smaller enumeration unit level could be considered.

In choropleth mapping, the boundaries of the polygons or the individual enumeration units (sometimes called statistical units, chorograms, or simply the *areal units*) have no numerical values associated with them. They function only to separate the enumeration areas and signify the geographic extent to which the enclosed area values apply. This is in contrast to the lines on the isarithmic map which do have values, and which will be discussed in Chapter 9.

When to Use the Choropleth Map

The choropleth technique is appropriate whenever the cartographer wishes to portray a geographical theme whose data occur within well-defined enumeration units. If the data cannot be dealt with as ratios or proportions, they should not be portrayed by the choropleth technique. Also choropleth (or any other mapping technique) should not be used if the interest is *only* to show actual, precise values within enumeration units. Choropleth mapping is simple and should be used only when its assumptions are acceptable to the cartographer and to the eventual reader.

Preliminary Considerations in Choropleth Mapping

Important considerations in the design of a choropleth map include thorough examination of the geographic phenomenon and its elements, map scale, number and kind of areal units, data processing, data classification, areal symbolization, and legend design.

Geographic Phenomena

All map design begins with careful analysis of just what it is that is being mapped. A careful designer assembles facts that will help in understanding the mapping activity. In a mapping problem to illustrate the geographical aspects of retailing, what measures should be used? Dollar sales, payrolls, and number of employees might be appropriate. What industrial or trade indices are commonly accepted and used by analysts? What surrogate measures might be used? What other geographical variables accompany the one being mapped? How does this phenomenon behave spatially or aspatially, with or without other geographic phenomena?

Cartographic designers must equip themselves with as much knowledge about the map subject (including the specific data set that will be mapped) as possible. Regardless of the individual steps taken, premapping research is necessary for effective map design, just as a product designer must measure the proportions of the human hand before designing an electric drill.

Map Scale

Map scale as it relates to choropleth design involves two considerations: necessity and available space (Muller 1974). Necessity dictates that the scale be sufficient to accommodate symbol recognition—the areal units must be large enough for the reader to see and differentiate areal patterns. In many cases, however, the cartographer is forced to operate in a map space smaller than the ideal, such as maps printed on spaces less than a typical 8 ½ inch-by 11-inch paper, or as a static virtual map that will be viewed on a monitor. However, with the increasing availability of large format plotters (such as, for printing a large poster), it is also possible that choropleth maps with less detail could be printed at too large a size. The increased size provides no additional detail or information, and simply occupies more space. Whether using a small or large format, the designer should seek the balance that best serves the purpose of the map.

Number and Kinds of Enumeration Units

For practically all choropleth mapping, the larger the number of enumeration units used for the entire study area, the more details of the geographical distribution the map can show. Scale is significant in this respect. Spatial detail is added as the number of enumeration units is increased; conversely, spatial coarseness increases as the number of units is decreased (see Figure 6.6).

Symbolization also has a direct effect on determining the number of enumeration units. As the number of units increase, their size decreases, sometimes making it more difficult to differentiate symbols.

The choice of how many enumeration units to use depends also on such variables as time, cost, map purpose, map size and scale, and symbolization. Each design task will have its own set of constraints.

The kind of enumeration unit is usually dictated by map purpose, level of acceptable generalization, data availability, and the scale considerations mentioned earlier. For much choropleth mapping, the kind of enumeration unit is determined before actual mapping begins—and most often is dictated by availability of data. State and county level enumeration units are two common levels (but certainly not the only ones) used in the United States for choropleth maps. Cartographers seldom have the option of specifying enumeration units, which are usually pre-established by local, state, or federal census sources. This is not always undesirable, however; standardization leads to easier comprehension.

Data Processing

Ideally, the designer would like to live in a world in which data are available in mappable form, and this is happening in increasing measure each year. Data from many online sites can easily be imported into a GIS or mapping program. However, this ease of obtaining and quickly creating a map tends to make the designer less aware of their data (see "Geographic Phenomena"). In keeping with the principle of "know your data," we strongly recommend creating histograms or other graphs of your data. A graph can also help tremendously when visualizing and selecting class boundaries, as seen in the data legend.

It is also important to remember that choropleth mapping requires that data be in derived form, such as ratio or rate, most often necessitating the processing of data according to the purpose of the map. One common derivation is to divide a column containing the total data (such as total population values per county) by a column containing the areas of each enumeration unit (for example, square miles), creating a new column of derived data (for example, people per square mile). Most mapping software allows for creating derived data from totals or other values. Some cartographers prefer using familiar spreadsheet software which can process and export enumeration data directly to the mapping software.

Data processing may require consultation with experts familiar with the purpose of the map, such as the map client. Data processing is an integral part of the total design activity and deserves careful attention.

DATA CLASSIFICATION REVISITED

Data classification, a major topic within Chapter 5, presented the concepts and methods behind this important topic. Indeed, there is perhaps no other issue so central to choropleth mapping as data classification. Values are grouped into classes to simplify mapped patterns for the reader. Assigning values to groups on the *choropleth map* is a form of data classification that leads to simplification and generalization (Jenks and Coulson 1963). Details may be lost, but as with all kinds of generalization, the results allow more information to be transmitted. Classification *and* symbolization of these resulting categories is what has been referred to as the **conventional choropleth technique.**

There are no absolutes when it comes to selecting the number of classes. However, the number of classes became somewhat standardized when it was learned that map readers could not easily distinguish between more than 11 areal symbol gray tones (Jenks and Knos 1961). A conventional recommendation that some cartographers adhere to is that no more than six classes be used (Monmonier 1977), and that a minimum of four are considered good practice. Since the reasons for classification have been better management of symbol selection and map readability, and not for any inherent advantages in grouping data (Brassel and Utano 1979), we recommend that the cartographer thoughtfully experiment with the number of classes (the conventional recommendation may be a good place to start) and to pay particular attention to the number of observations. See Chapter 5 for a more complete discussion on the relationship between the number of observations and selecting the number of classes.

It is also possible to symbolize each value by its own unique areal symbol (line, tone, or hue), thus producing an *unclassed* choropleth map (Tobler 1973). Such maps have not become widely used because not all cartographers approve of the result of not classifying the data: an unstructured or ungeneralized view of the mapped phenomenon. Differences between the conventional choropleth technique and the unclassed variety

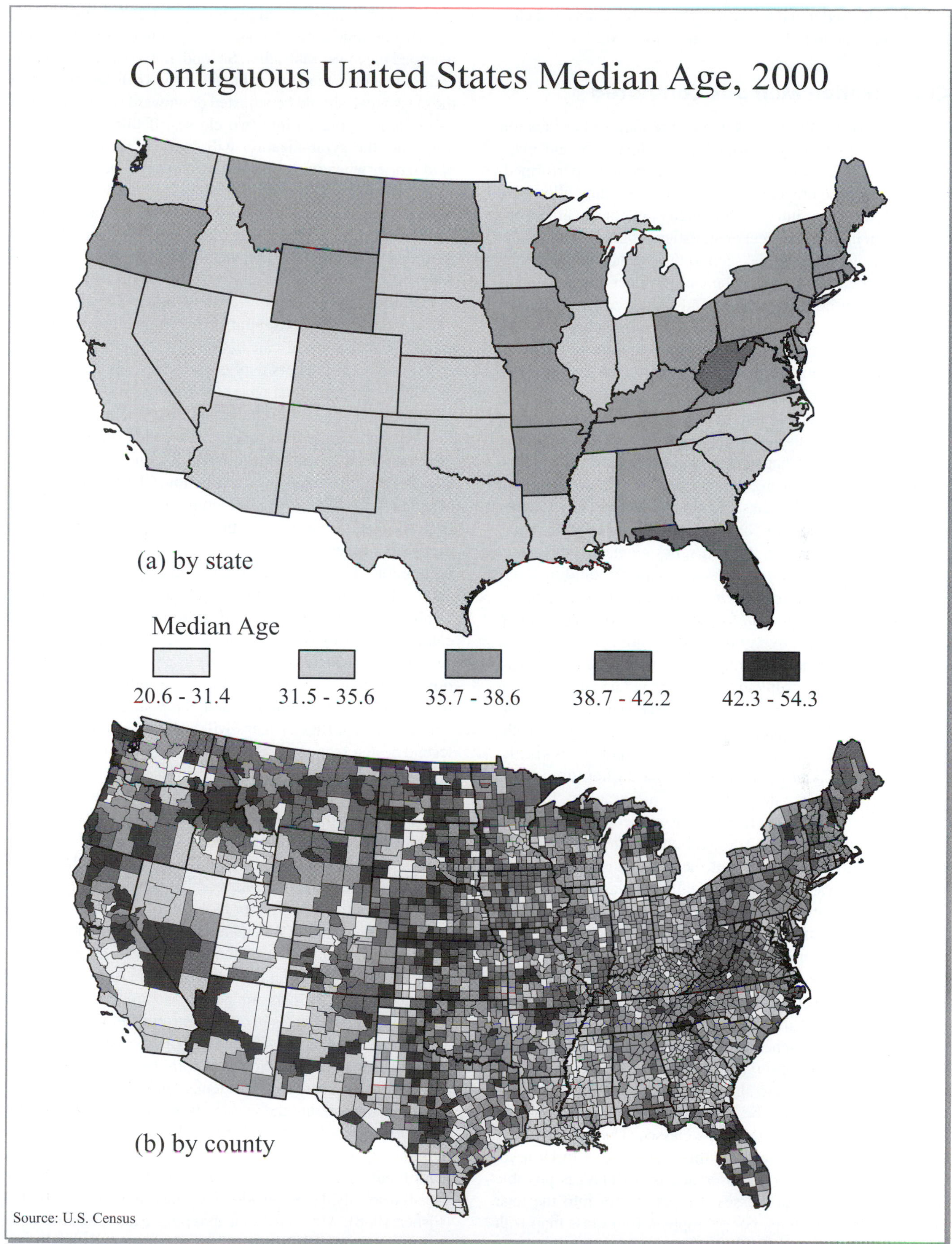

FIGURE 6.6 CHOROPLETH MAP DETAIL AND THE NUMBER OF ENUMERATION UNITS.
Greater distributional and attribute detail is possible with a larger number of enumeration units.

will be addressed in the following paragraphs. Unclassed choropleth maps are referred to by some as *tonal maps.*

Classification Methods Compared

In Chapter 5 we discussed some of the major classification methods employed by cartographers. Most GIS and mapping software allow for these classifications to be performed, or if one of these procedures is not provided, they allow you to specify the class intervals and breaks manually after your calculation in a spreadsheet or statistics program. The results of the classification will be seen in the legend. When selecting a classification method, the cartographer should always keep in mind the map purpose and the audience for which the map was intended. It is also extremely important for you to *know your data.* Study your data table and make graphs to get a complete picture of your data's key characteristics. It is important to keep in mind that while many general patterns may be somewhat similar across different classification schemes, there are also many differences, both in appearance (see Figure 5.10 in Chapter 5) and in other ways that will be discussed in the following paragraphs. We will consider and briefly review the nine classification schemes as they apply to choropleth mapping: *equal interval*, *equal frequency*, *arithmetic and geometric intervals, nested means*, *mean and standard deviations*, *natural break methods* (including traditional *natural breaks* and *optimization*) and *user defined.*

Equal interval and *equal frequency* (also called equal step and quantiles, respectively) historically are two of the most popular classification schemes. Prior to the automation of the conventional choropleth technique via GIS and mapping software, these schemes were the easiest to manually calculate, and are established, time-honored classification techniques (although we will see that they may not always be the best choices). *Equal interval* schemes are often used when constant intervals are desired. Some cartographers feel that irregular class intervals are harder to understand. Equal interval classification is one logical choice in this case. The class ranges may be equal, but the number of observations per class will vary. If the equal interval classification produces classes with no observations, then this classification method should *not* be selected. Equal interval classification is a great choice for layouts that include multiple choropleth maps that use the same legend for all maps. It also works well for animated choropleth maps, where enumeration units may change classes throughout a given time period.

Equal frequency schemes, in which the goal is to generate an equal number of observations per class, produce different class ranges from class to class. An equal number of observations in each class may be important if you are performing some statistical tests between the classes. There are two important issues that can arise if this technique is to be used. First, an equal number of observations per class is possible only if the number of classes divides evenly into the total number of observations. For example, a four class map with 16 observations means that exactly four observations will be placed in each class. When the number of observations is not exactly the same for each class, cartographic convention is to overload the lower ranks first. Second, if two (or more) observations that have the same value fall into two different classes, the class break should be adjusted downward so that identical values are not placed into two classes. If this adjustment is *not* done, the symbolization will *imply a difference to the map reader* when in reality the attributes are the same.

Arithmetic and *geometric* schemes are used for data that replicate mathematical progressions. Data that has a propensity for increase, or increase at increasing rates, may be good candidates for one of these methods. Currently, some GIS and mapping software packages do not support these schemes as classification options. Fortunately, most GIS packages allow the cartographer to input breaks manually, allowing a spreadsheet or manual calculation of the class breaks. Programming cartographers may wish to take advantage of the many GIS packages that have scripting or programming capabilities and add these schemes as classification choices.

Mean and *standard deviations,* also called simply standard deviations, can be used in choropleth mapping if the data set displays a normal frequency distribution. Class boundaries are calculated by adding or subtracting the standard deviation from the data set mean. Note that if this scheme is applied to data that are skewed (that is, not normally distributed) it may be possible, for example, to have three standard deviations above the mean and only two standard deviations below the mean (or vice versa), and thus create an uneven number of class ranges. This classification is often applied to data sets such as income increase or decrease as a percentage, or education levels above and below an average indicator. When applied with the right symbolization, a *bipolar choropleth map* may be created. Bipolar maps will be discussed further in the legend design section.

A somewhat related scheme, *nested means*, can be applied to similar data sets, but it does not have the requirement of a normal distribution as does the mean and standard deviation scheme. Class boundaries are determined by the mean of the data set to produce two classes. The means of these new classes provide additional class breaks. The number of classes is usually restricted to 2, 4, or 8. As with arithmetic and geometric schemes, most GIS and mapping software packages do not directly support this method. But since the breaks are fairly easy to calculate and GIS and mapping software support the manual entry of class ranges, it becomes another viable alternative in the choices in classification schemes.

Natural break methods, including both visual inspection of a graphed array (traditional "natural breaks") and a minimum variance scheme such as Jenks *optimization*, group like values by minimizing the within class variance (making sure that all values within a class are as alike as possible), and by maximizing the between class variance (making sure that class breaks fall where there are larger breaks in the data). This is also called classification for "maximum homogeneity" (Fisher 1958). Most GIS and mapping packages now have some form of natural breaks as a classification option. Of course, if visual inspection is used, the term "natural breaks" applies only if the cartographer is conscientiously trying to group like observations and find the larger class breaks.

A number of cartographers take the position that natural break methods may be the best choice for communicating classified data accurately to the map reader. The map reader quite naturally *expects that two enumeration units that have the same symbolization are more alike than two enumeration units with different symbolizations*. This expectation can best be met using a natural break method. In most cases, no other classification scheme achieves the level of class homogeneity: equal interval and equal frequency are particularly poor in this regard. Quantitative measures of class homogeneity, such as the goodness of variance fit (GVF) (Jenks and Coulson 1963), the tabular accuracy index (TAI) (Jenks and Caspall 1971), and the goodness of absolute deviation fit (GDF) (Slocum *et al.* 2005), when applied to natural breaks and the other classification schemes using an equal number of classes, will show natural breaks to be superior in establishing class homogeneity. The largest disadvantage to natural breaks classification schemes are the varying class intervals and varying number of observations per class.

There are proper applications for all of the classification schemes, depending on the nature of the data set and the purpose of the map. No one classification scheme is right for all circumstances. Sometimes if the cartographer wishes to identify specific divisions in the data, he or she will select a user-defined scheme. If no specialized mapping needs stand out for a given data set, we suggest using a natural break method due to its capacity for greater class homogeneity.

Data Truncation and Outliers

Data sets often have one or more values that are unlike most of the other values. These observations make data classification very difficult, because such an observation defies the very purpose of classification—to group like phenomena. Extreme observations can be treated as outliers and given their own class. The outliers are usually removed or separated from the data set (truncation), and then classification takes place on the remaining observations.

The determination of what an outlier is, or how to handle multiple outliers, can require some creativity on the part of the cartographer. As suggested above, it is important for the cartographer to study the data table and make graphs to get a complete picture of the data's key characteristics. If an outlier can be measured in standard deviations from its counterpart, then treating the observation as an outlier is probably a good idea. Multiple outliers, if grouped relatively close together, may simply make another class of observations. Single observation outliers should be labeled in the legend with the enumeration unit name and the specific value.

Different Maps from the Same Data

One consequence of having a variety of methods of classing is that different maps can result from the same data (refer to Figure 5.10). An alteration in either the number of classes or the class limits can cause such change. The designer has the task of mapping the data in the manner that best represents the distribution of the geographic phenomenon. This requires thorough knowledge of the mapped phenomenon and awareness that the map reader's perception of choropleth map patterns is the result of several conditions (Monmonier 1974):

1. The map reader's experience with cartographic materials
2. The cartographic method used to portray the distribution
3. The complexity of the map pattern

Pattern complexity has been studied by cartographic researchers, and most agree that complexity should be reduced. Pattern complexity is defined in these terms:

1. Number of regions (contiguous enumeration units within the same class)
2. Fragmentation index
3. Aggregation index
4. Contrast index
5. Size disparity index

Space does not permit a detailed explanation of these indices. However, at least one researcher has discovered that map readers tend to judge choropleth maps having different classing schemes as similar in all pattern-complexity indices. This has led him to conclude that only *one* index, perhaps the number of regions, can be used to compare pattern complexity among different maps (Chang 1978).

Cartographic designers need to examine closely the visual pattern that results from a particular classing method. Although no exact indices cover all cases, it is best to look for a pattern that results in a *simple* picture. The purpose of the map, the background and experience of the reader, and the mapping method must all be considered.

Unclassed Choropleth Maps

The **unclassed choropleth map** occurs when each observation is its own symbol. For example, in an unclassed choropleth map of the United States, each state will have its own symbol and the legend will usually have several representative shadings or colorings. In most GIS and mapping packages, the cartographer will simply generate a 50 class map, and modify the legend accordingly. The burden of deciding on a classification scheme is removed from the designer in this choropleth map variation (Muller and Honsaker 1978). Figure 6.7 illustrates the differences and similarities between a classed and unclassed choropleth map.

In such maps, however, the cartographer loses the ability to direct the message of the communication (Dobson 1973). Map generalization, simplification, and interpretation are left to the reader. There is evidence that readers can make reasonably good judgments of the mapped values by inspecting the legends that accompany the map bodies and, specifically, that the unclassed method can convey values more accurately to most map readers than classed maps having fewer than six classes (Peterson 1979).

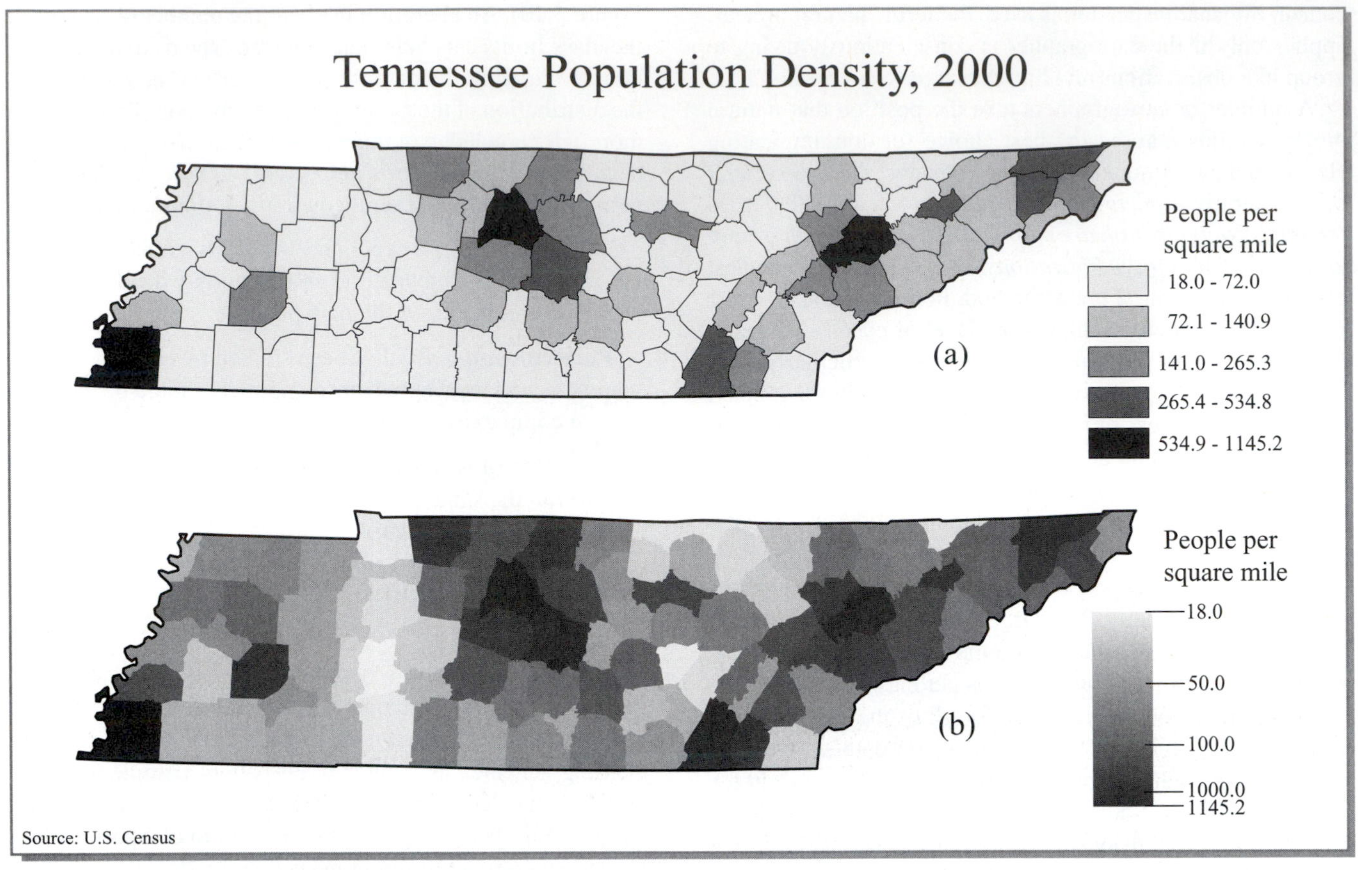

FIGURE 6.7 CLASSED AND UNCLASSED CHOROPLETH MAPS.
The five class map (a) gives a strikingly different appearance than the unclassed version (b). Note that the legends are treated differently. Enumeration boundaries are often not portrayed in unclassed maps.

The authors of one research effort tested the effectiveness of classed and unclassed choropleth maps, with both crossed-line (cross-hatched) and solid-fill (tinted) area symbols (Mak and Coulson 1991). They wanted to determine the relative effectiveness of each with respect to value discrimination (the ability to discern differences between patterns), value estimation, and ability to regionalize (from the patterns on the map). The results are summarized here:

1. In value estimation, classed choropleth maps were significantly better than unclassed, regardless of the symbol type used. The authors concluded, "classed maps have a distinct and statistically significant advantage over unclassed maps in value-estimation tasks" (Mak and Coulson 1991, 120).
2. In region-estimation tasks, the results were far less clear. "Overall, there is no significant difference between classed and unclassed choropleth maps, whether crossed-line or solid-fill symbology is used" (Mak and Coulson 1991, 120).
3. In value-discrimination tasks, crossed-line symbols are more accurate than solid-fill designs (both in and out of map contexts). "In value-estimation . . . crossed-line shading was significantly better within a map context on both classed and unclassed maps" (Mak and Coulson 1991, 121).

These authors also tested maps to determine if increasing the number of data units leads to different responses to both classed and unclassed maps. The answers to this question were not so unequivocal. In conclusion, the authors state:

> *If one intuitively prefers generalizing data into classes for choropleth maps and accepts the Jenks Optimal system as resolving the long-standing problem of what is the "correct" classification method, this study provides evidence to support continued classification in choropleth maps. If, however, one is less sanguine over the classification problem, the levels of accuracy attained with unclassed maps, and the failure of classed maps to perform significantly better overall in the region recognition task, provide evidence for the effectiveness of unclassed choropleth maps. Crossed-line symbology has been shown to perform better than the solid-filled method. What is now required is further empirical work to validate and expand present findings. In particular, extension of experimentation to include . . . color sequences* (Mak and Coulson 1991, 121).

Past versions of this text have questioned whether or not unclassed choropleth maps were even choropleth maps at all, and these versions noted that cartographers occasionally used the term "continuous-tone quantitative map" to describe an unclassed map. The reasoning for this view is that unclassed choropleth maps intentionally do nothing to preserve the integrity of the enumeration unit, and most often eliminate the discreteness of the units altogether—the quality that serves to define the choropleth map.

Although probably less controversial today than a couple of decades ago, it should also be noted that we see relatively few unclassed choropleth maps being produced—the norm seems to be toward classification. Perhaps we fall back to conventions to which we are accustomed. Ultimately, cartographic designers must look to the purpose of their design activity and make individual judgments regarding the use of unclassed maps. They may be used in conjunction with ordinary classed maps to provide the reader with an unstructured inventory of the mapped area. Other possible uses (other than as a main map) may yet be discovered.

LEGEND DESIGN, AREAL SYMBOLIZATION, AND BASE MAP DESIGN

Class-interval specification is not the only activity that occupies the designer's time while preparing a choropleth map. Other design considerations include legend design, symbol selection (with fill colors, shades, or patterns), and decisions regarding the amount of detail in the base map.

Sources of Map-Reading Error and the Need for Accurate Design Response

Reading and interpreting a single choropleth map are at best difficult assignments. The design must be planned carefully to reduce the reading task and eliminate as many sources of error as possible. One of the chief sources of error results from the classification of the map data. Effective classification and corresponding legend design will help reduce this source of method-produced error. The enumeration unit fills can also produce error whether the designer uses solid color fills, grayscale fills, or symbols such as hatch or line patterns. Yet another area that creates problems is the choice of areal symbols; some symbols are easier to read and less distracting to the eye. Finally, the addition of general reference features for the reader is an area that has not received much attention from cartographic researchers and will be addressed in this section.

Legend Design

One of the most important elements of a choropleth map is the legend. The following discussion highlights some of the more important design conventions that have been developed, and also provides some guidelines and options for constructing more effective choropleth map legends.

Box Shape, Size, Orientation, and Range Placement

Choropleth legend boxes are usually rectangular in form, and need to be large enough to provide a visual anchor for the map reader to correctly interpret the symbolization (fills) but not so large as to detract from the map. The size of the enumeration units can provide a starting point in determining box size. For maps that have smaller enumeration units, the boxes are drawn at approximately the same size as an average size polygon. When larger enumeration units are used, the boxes are drawn at approximately half to one-third of the area of an average size polygon. In both cases, the box dimensions can be adjusted upward or downward as necessary to achieve balance with the map body.

The boxes are most commonly displayed in a vertical presentation, although horizontal presentations are possible on maps where the map body has a longer east-west orientation. Class ranges (and outliers, if present) are usually placed on the right side of the boxes in a vertical presentation, and underneath the boxes in the horizontal presentation. The legend boxes may represent data from lowest to highest classes in a vertical presentation *or* highest to lowest from the top to the bottom box. In the case of a horizontal presentation, the boxes are placed from lowest to highest from left to right (see Figure 6.8).

Typically, there is some space between the legend boxes, but some cartographers feel that compressing the boxes gives a sense of continuity to the data. Compression also allows for

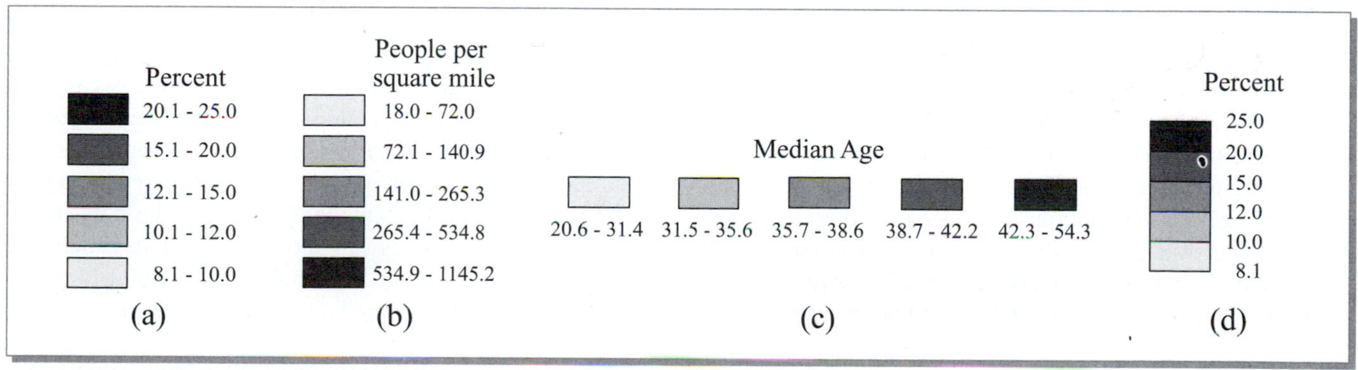

FIGURE 6.8 SAMPLE CHOROPLETH LEGEND DESIGNS USED IN THIS CHAPTER.
Vertical presentations may have class ranges increase from top to bottom (a) or the class ranges may decrease (b). Both forms are extremely common and acceptable. Some cartographers prefer (a), because it puts the emphasis on whatever is"most" in the map. Others prefer (b) for aesthetic reasons, since darker, "heavier" colors and shades are positioned at the bottom of the legend. Also, if the largest class contains an outlier or two (refer to Figure 1.8, Chapter 1), then (b) will provide room for labeling the outlier(s). With horizontal presentations (c) the class ranges should increase to the right. With compressed data boxes (d) the class breaks can be labeled with a single value.

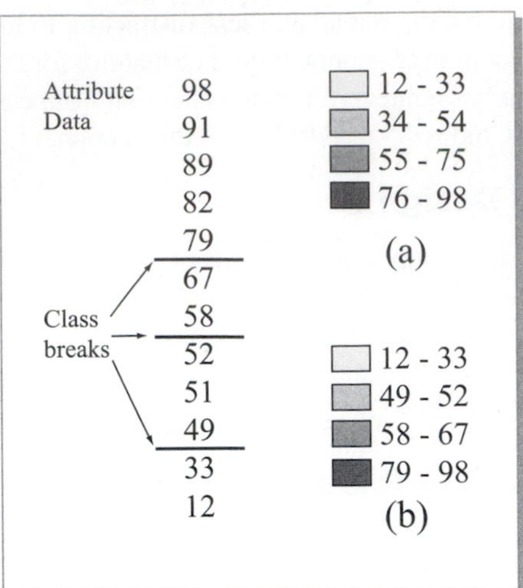

FIGURE 6.9 CONTINUOUS AND NONCONTINUOUS REPORTING STYLES FOR CLASS RANGES.

Two class range reporting styles, including the continuous style (a) and the noncontinuous style (b). See text for discussion.

other options in labeling class ranges, such as placing a single number to indicate a class break (see Figure 6.8d). If the boxes have been drawn as independent graphics, make sure that all of the boxes are congruent and that they line up appropriately.

Continuous and Noncontinuous Class Ranges

Class ranges may be presented using either **continuous** or **noncontinuous** styles. The continuous style (see Figure 6.9a) is the more traditional style of reporting the class ranges. In this style, there are no gaps or overlaps in the class ranges (Jenks and Coulson 1963). The class ranges as originally calculated during classification usually have to be expanded to achieve continuity. The noncontinuous style (see Figure 6.9b) configures the class breaks so that only the values actually existing in each class form the class ranges. If there are gaps in the data, then there will be gaps between class breaks. Although previous versions of this text favored the noncontinuous style over the continuous style for reducing map reading error, *both* forms are used extensively in choropleth mapping, and it is important for the thematic cartographer to be aware of the characteristics and potential uses for each presentation method.

The continuous style is used by map designers for two major reasons. The first is that "exhausting the data" is part of the original requirements of choropleth map classification. Before data automation with GIS and other software, "no gaps–no overlaps" in the class ranges ensured that no data fell through the cracks, especially if there were last minute adjustments in the map data. Some map designers also find a more complete aesthetic sense in this more conventional style. Second, and perhaps more importantly, is that the same legend is often applied to a series or sequence of maps (for example, three population density maps for a region from 1950, 1975, and 2000) in order for there to be an accurate comparison between the maps. In the same manner, animated choropleth maps also use only one legend—the enumeration units may change symbolization through time, but the legend is usually static. When single legends are used in map sequences or animated maps, *then the continuous style of reporting is required.*

The noncontinuous style is an option for cartographers who will be producing *only a single map*. The strongest argument for using noncontinuous class ranges has to do with reducing map reading error. When the class ranges are narrowed, the reader's estimate of the actual value is also narrowed, resulting in greater accuracy in map interpretation. Conversely, the artificial widening process in the continuous style increases the range of possible values in estimation, thus increasing reader error. A second benefit to the noncontinuous style is that it is actually easier to visualize the distribution of data, because the gaps in the class ranges are seen and the real class ranges are presented. It is important to remember that with each style, it is only the *reporting* that changes; the symbolization as determined by the classification choice will stay the same.

Class Range Formatting, Legend Titles, and Other Legend Information

There are several small but important formatting conventions that can help the legend become more readable and the map look more professional. One of the most important of these conventions is that the values for each class always increase left to right (for example, 501 to 925, *not* 925 to 501). Both a dash (for example, 501–925) and the word "to" (for example, 501 to 925) are used in depicting class ranges. Some cartographers hold that the word "to" should be used if there are negative numbers in the data (such as, −321 to −211) to avoid any confusion between the dash and the negative sign. Outliers are generally not associated with a data range, but as suggested previously in "Data Truncation and Outliers," single observation outliers should have the enumeration unit name and the specific value presented in the legend. Sometimes two outliers can be labeled (refer to Figure 1.8), but if the outliers are too numerous to list, the addition of another class should be considered to accommodate them.

Large and small numbers in the class ranges can be a challenge to work with. If numbers in the class ranges exceed four digits, it is good practice to use commas (for example, 16,789 is better than 16789). If larger numbers start to get unwieldy, consider presenting the data in another measure, such as "in thousands." If decimals are present in the attribute data, make sure enough decimal places are used to "capture" all of the data. For example, if two consecutive classes run from 10–20 and 21–30, and the data set has an observation of 20.3, that value is excluded. Class ranges of 10.1–20.0 and 20.1–30.0 will include the observation. Conversely, if more decimals are used in the class ranges than are in the data, then the legend is implying a greater precision than actually exists in the attribute data, possibly conveying a false impression of increased data quality.

Well-written legend titles can also greatly assist the map reader in interpreting the choropleth map. The legend title, also called the legend heading, should be succinct and match the data and the title or topic of the map (as with any quantitative map type). In a choropleth map entitled "Nebraska Population Density, 2005," a reasonable legend title would be "People per Square Mile." As with text in other parts of the map, do not use abbreviations, and symbols such as "$," "%," "etc.," and similar characters should be avoided, especially in the legend title. If the topic of the map is unemployment rate, for example, then the legend title could simply be "Percent Unemployed" or "Percent" as opposed to "% Unemployed or "%."

If the cartographer wishes to present additional information that will not fit into concise titles (or subtitles) or legend headings, ancillary textual information (text statements) can become an extremely valuable addition to the legend. For example, if the data are aggregated in any way, then there should be a statement about the original data enumeration unit, especially if not stated in the title or if it is different than what is implied on the map. Ancillary text is often placed below the legend boxes but can be placed in a manner which best achieves map balance and supports the purpose of the map (refer to Figure 6.1).

In some cases the cartographer may wish to provide even more information about the attribute data than is apparent with standard legend inclusions. For choropleth maps of average crime rates or median income, for example, some sort of delineation of the mean or median in the legend boxes is logical. Where class breaks are particularly uneven, the inclusion of *class* means or medians (usually labeled alongside class boundaries) can be beneficial. Although not common, some cartographers advocate the inclusion in the layout of a frequency histogram of the attribute data. This gives the reader additional information about the aspatial qualities of the attribute data, and provides background on how the class intervals and boundaries have been selected.

Map Sequences and Animated Maps Considerations

Proper legend design is crucial for a series of choropleth maps. The series, for example, could include three side-by-side choropleth maps of population density for the United States in 1950, 1975, and 2000. One legend should be created that will refer to all three maps in order to facilitate comparison between the maps. A single classification scheme must be used that will accommodate attribute data for all three maps. Because the data are changing from map to map, it usually is not practical to classify for maximum homogeneity. Rather, by convention, an equal interval classification scheme is most often used, in which the step ranges are calculated based on the lowest and highest values in the total data range for all three maps. In this case the continuous style of class range presentation should be used. With sequence mapping it is possible that not all of the class ranges in the legend will be depicted in all of the maps, especially if the variation between the maps' data values is relatively large. Because the overall data range changes for each map in the sequence, we also recommend placing text stating the minimum and maximum values alongside the map body for each map in the sequence.

For a choropleth map animation of the same data, the presentation is that one map will replace another map on top of each other (for example, animation *frames*), giving a sense of movement. (See Chapter 16 for further discussion of map animation terminology.) A single, static legend is used for the entire animation so that the class ranges are unchanging for the entire animation. The legend is designed with the same guidelines as with the map series (equal interval classification and continuous class range presentation). The only difference is that with an animated map, the greater the data variation between the animation frames the greater the chance that there becomes a class that does not have any observations (see equal interval guidelines, in this chapter) in one or more of those frames. While decreasing the number of classes could reduce this possibility, it is important to note that the decrease in classes will also lessen the amount of information that can be compared between maps or animation frames.

Symbolization for Choropleth Maps

Selecting symbolization for filling enumeration unit areas is one of the most important aspects of choropleth mapping. Automation via GIS, mapping, and other graphics software packages allow for visualizing many symbolization schemes in a relatively short period of time. How do we know what symbolization schemes will be effective? With what media will they work? In seeking answers to these questions, we need to examine black and white (or grayscale) and color symbolization for the class ranges.

Black and White Mapping

Black and white choropleth maps appear in many textbooks, journals, newsletters, brochures, pamphlets, and other venues where the higher costs of color printing is a factor. They also pose few problems for readers who have deficiency in perceiving color. In black and white mapping, patterns or shades are used to create the impression of a light-to-dark gradation. In pattern fills, dot, line, or hachure patterns have the greatest density (smallest spacing between elements, also called *texture*) in areas of highest values (see Figure 6.10a). With solid grayscale fills, the value or percent black changes so that darker tones are used to indicate areas of highest values (see Figure 6.10b).

The use of pattern fills was an extremely popular method of symbolizing choropleth enumeration units throughout most of the last century, with a marked decline in their use by the 1990s. Although patterns of all types are available in modern GIS and graphics software, they are rarely used anymore for choropleth fills because solid color or grayscale fills give the map a more aesthetic and less dated appearance.

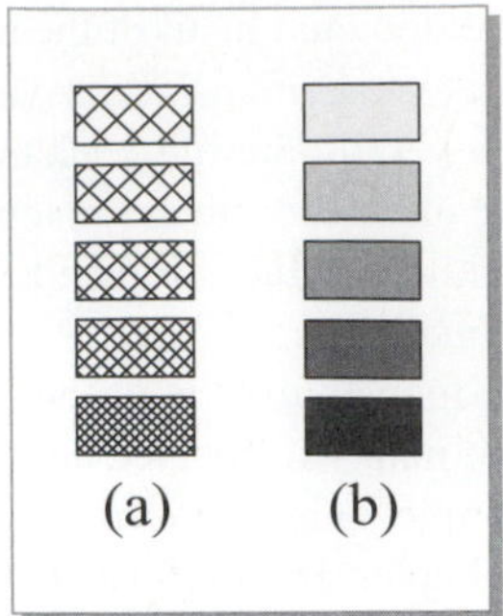

FIGURE 6.10 BLACK AND WHITE CHOROPLETH MAP SYMBOLIZATION.
Using pattern fills such as crosshatches (a) are not as common as grayscale fills (b). See text for discussion.

If pattern fills must be used, we recommend against using lines as enumeration fills. Lines as fills are harsh, and create undesirable patterns in the map that are very distracting. Dots and hachure patterns are usually preferred as they pose fewer visual problems for the map reader. Patterns that are meant to be used as nominal area fills (swamp symbols, water waves, bush symbols, and so on) should likewise be avoided.

It is more common to use grayscale fills for black and white choropleth maps. The value (the quality of lightness or darkness) is specified in percentage black (on a scale from 0–100) for printed graphics. For virtual maps, there are usually 256 levels of gray, but very few black and white choropleth maps are actually produced as virtual maps. These values are easily adjusted in GIS, mapping, and graphics software.

Empirical studies have demonstrated that perception of tones in graded series is not linear (Mak and Coulson 1991). In a series of printed tones, most map readers do not perceive the tones as varying from light to dark in perfect correspondence with the physical stimulus (the percentage of area inked, or simply percent black). Fortunately, sets of values have been developed and are available online from ColorBrewer (Brewer 2006), an online guide to map color selection (including grayscale for black and white choropleth maps). ColorBrewer has the advantage of giving numeric values not only for printing but for virtual environments and other software applications as well.

Many maps using solid grayscale fills unfortunately have total black or total white fills for the highest and lowest class ranges, respectively. Caution must be encouraged when using total black or total white fills. If the enumeration unit boundaries are black, as they often are in black and white cartography, the fill will merge with the boundary. Multiple contiguous enumeration units in the highest class will obscure the unit boundaries. With the *possible* exception of a single outlier, or perhaps a select few noncontiguous outliers, we recommend using a tone lighter than 100 percent black for the highest class ranges. Solid white fills for the lowest class ranges may not be the best choice either, as white fills often suggests that nothing is there. We recommend using some sort of fill, however light, for the lowest class range, and reserve white for enumeration units that do not have the quantity being mapped. The grayscales at ColorBrewer do not include total black or total white fills as long as there are not more than seven classes specified.

Color Map Symbolization

Color use in choropleth mapping is becoming fairly standard, owing to decreasing costs in color printing and increased use of virtual maps. Color gives more symbolization options than with black and white, but it also introduces a higher degree of complexity and subjectivity into the design process. With color choropleth mapping, the choice of hue, the color value, and color saturation all play an important role. Darker values and/or more saturated colors indicate more of a quantity. Sometimes the variation is along one hue (for example, light orange to dark orange); sometimes more than one hue is used (see Color Plate 6.1). In Chapter 4, Figures 4.4 and 4.5 illustrate the relationship between quantities and the visual variables.

The term **color ramp** seems to have been popularized with GIS software, in which the color variation is automatically applied by the software to each of the class ranges. Most mapping software provides quite a variety of color ramps that can be used as a starting point. The cartographer can change the color parameters (for example, hue, saturation, and value) for each class, to his or her liking. (Other common parameters and color models for printing and virtual maps are discussed in Chapter 14.) Almost all fills in color choropleth mapping are solid fills, so this discussion does not include color/pattern combinations.

How does one choose a good color scheme? What changes should be made to "default" color ramp in a GIS? The following is designed to give a relatively nontechnical list of suggestions in choosing and adjusting colors:

1. Make sure that there is enough differential between the symbolization in the class ranges so that the difference can be clearly seen in the map, and yet at the same time has a hue-value-saturation change that suggests more and less. One simple technique that we have used with great success is to cover up or temporarily delete the legend, and then ask yourself and/or trusted colleagues if the symbolization appropriately suggests "more" and "less" for each class range. The symbolization for the highest class range, for example, should suggest "more" of a quantity than other class ranges.
2. If the map is to be a virtual map, understand that "display monitor types and brands," their settings, and the graphics cards that drive the monitor cause a wide range of virtual viewing environments. Make sure that there is enough visual separation of class ranges so that slight variations in monitor brightness, for example, won't change the map reader's ability to distinguish classes.
3. If you will be making a number of printed maps that are going to the same printer or plotter, then printing a color table of selected color ramps or other hue/value/saturation combinations is often helpful, since the onscreen appearance often varies from the printed product.

4. Use ColorBrewer. As suggested earlier, this online guide allows the map designer to select color schemes. It is an excellent tool because it is based on established color plans for quantitative maps that follow the data, but one does not have to be trained in the physics of color to use the site. The explanations are concise and jargon free. The color scheme recommendations are provided in a number of (mostly) software-independent values. All that has to be done is to select the number of classes, select the legend type (*sequential* for most data sets; see bipolar symbolization in the following section) and color scheme, select the kind of color numbers you wish to obtain (for example, hex values for Web and artistic drawing programs), and the colors and values will appear, along with a graphic legend stating whether fill scheme is appropriate for various media, such as monitors, color printers, LCD projectors, and so on (Brewer 2006). Note that the color schemes are also available in hardcopy (Brewer 2005).
5. Avoid using qualitative discrete color fills that are normally reserved for categorical data. For example, on a four class map, blue–yellow–red–green is a horrible sequence for choropleth maps. Such a sequence does not in any way suggest "more" or "less," and is inappropriate for continuous quantitative data. Fortunately, most of today's GIS and mapping software does a fairly reasonable job of ensuring that continuous data are not accidentally mapped with discrete color schemes. In addition, cartographers obtaining their color values from the ColorBrewer will see that discrete color schemes are clearly labeled as "qualitative," minimizing the chance of selecting a poor color scheme.

Color also allows more flexibility for encoding special categories, such as "no data reported" or empty categories. Borrowing from black and white symbolization techniques, white can still suggest that "nothing is there" from the attribute that is being mapped. A light gray can effectively contrast with many color schemes in representing a "no data" category. No data reported is sometimes used in tables where values exist, but they have not been reported (which is *not* the same thing as "zero"). Here again, knowledge of what is being mapped, both with the topic and the data involved, is crucial in making the distinction.

While many designers stick with black for depicting enumeration unit boundaries, many other cartographers have effectively used varying grayscale values for the boundaries. Both black and grayscale values can be effective in color choropleth mapping, as their neutral quality *tends to accentuate the color fills*. If color is to be used in the boundaries, care must be taken that it does not take away, distract, or otherwise clash with the more important enumeration fills. Likewise, enumeration line widths (or weights) should be thick enough to distinguish the enumeration units but not take away from the fills or the density patterns in the overall map.

Bipolar and Bivariate Symbolization

There are two specialized types of choropleth maps that can be produced: **bipolar choropleth maps** and **bivariate choropleth maps.** Bipolar maps are created when a special color ramp scheme is applied to data that diverges from a central value (or sometimes a middle class range). The center (or pivot) can be as simple as the overall data average, but it can also be any other meaningful value or class range which logically splits the data. The scheme can be applied to any data that diverges in two directions from that center, such as percent increase/decrease (see Color Plate 6.2), profit/loss, or above/below average (see Color Plate 6.3). The data are symbolized so that in both directions from the central value or class, two different color schemes with increasing color value and/or saturation occur as the class ranges move farther from that central value. At the ColorBrewer website, selecting a legend type of *divergent* (a term used in some GIS software as well) will reveal a number of color schemes appropriate for bipolar choropleth maps. When the central value is a class mean, and if the data are distributed normally, then a mean and standard deviation classification scheme is often used (see earlier in this chapter).

Bivariate choropleth maps are ones that attempt to show two related variables on the same map. Line spacing combinations and/or creative use of color gradients (see Color Plate 6.4) are used to encode the data. Because the legend can become very unwieldy in this type of mapping, it is rare for there to be more than three or four classes for each of the two attribute data sets. Unlike bipolar maps, bivariate choropleth schemes are not universally supported in GIS and mapping software, so cartographic creativity and tenacity, or perhaps scripting or programming prowess, is required to create bivariate maps. Likewise, multivariate choropleth mapping (using more than two variables) has been the subject of research (Carr *et al.* 2005), but is beyond most current mapping software technology.

Adding Other Reference Features to the Map

Most cartographers would agree that proper data and enumeration unit selection, scale, classification, legend design, and symbolization processes are all important parts of choropleth map design. One item that has not yet been considered is whether or not the placement of select general reference features such as cities, roads, or rivers on the map will assist the reader in map interpretation (see Figure 6.11). In past versions of this text and in much of the literature, the term for selected reference features is "base information," or "base data." For the GIS-using cartographer, base data are simply other layers of information that may be added to the map.

Some cartographers point out that thematic maps, by definition, are supposed to be simple. If it is to be a single-topic map, then why try to turn it into a general purpose map? In many choropleth maps, particularly ones with small enumeration units and/or ones with vibrant color schemes,

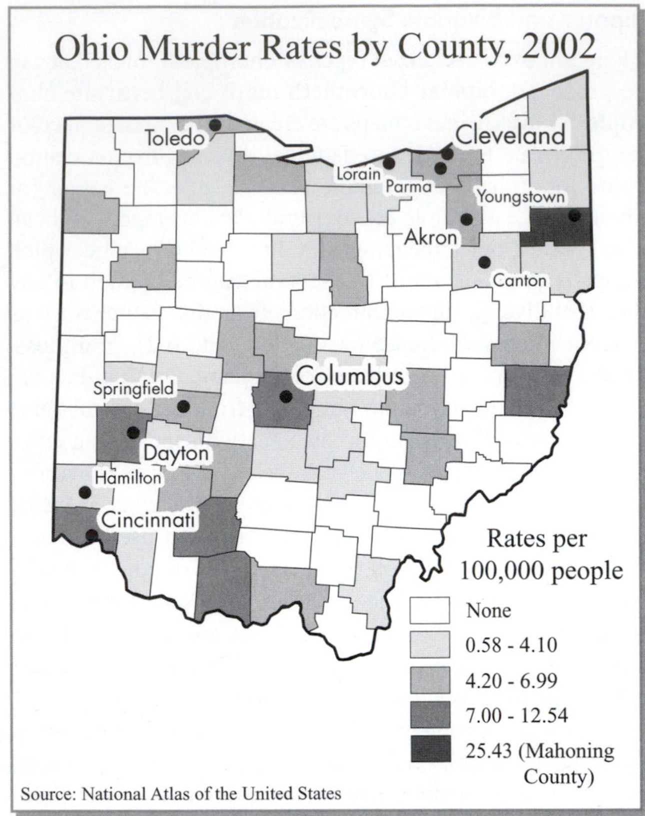

FIGURE 6.11 CHOROPLETH MAP WITH MAJOR CITIES INCLUDED.

The addition of carefully selected general reference features can assist the map reader in interpreting a choropleth map. See text for discussion.

adding features may unduly clutter the map, and in some cases obscure the data. Many cartography students often remark that adding too many general reference features takes away from the main point of the map.

Others, however, point out that many choropleth maps are made for the general public—people who want to be able to tie the choropleth map pattern to other features that can provide locational cues. Unless the reader is especially familiar with the mapped area, much of the information-carrying potential of the choropleth map is lost. Fitzsimons notes that:

> *The simplistic notion that base data [selected reference features] are merely elements of graphic design, subject to the whims of individual map makers, has reinforced the notion that these data are separate, unrelated map components. The cartographic solution to the problem of base-data selection must be to evaluate the relative appropriateness of all data to map purpose; the same criterion should be used in selection of both base and thematic data—their power to contribute to map purpose* (Fitzsimons 1985, 61).

As with most other decisions in thematic cartography, the map purpose, scale, and the experience and needs of the reader will dictate the choice of whether or not to include additional general reference features, or how many features to include if they are used. A choropleth map of the United States that uses counties as enumeration areas has the capacity to show regional variations of the mapped phenomenon. If this is the main purpose of the map, it can function quite well without ancillary general reference features. On the other hand, a marketing analyst working in an urban area might want political boundaries (other than the main enumeration unit), interstate highways, major thoroughfares, or the competition mapped as well. The designer must not make arbitrary decisions; the needs of users and map clients must be taken into consideration.

When ancillary general reference information is to be added to a printed or static virtual map, there are two important principles to consider. First, it is important to not provide any more features than are absolutely necessary. Many GIS data layers are too detailed, so the information must often be generalized. If large numbers of reference features are absolutely necessary, consider placing a reference map next to the choropleth map instead. Second, for the reference information to be visible, the choropleth fills must be somewhat subdued. Black and extremely dark hues should be avoided, favoring lighter values and/or less saturation for the fills instead. Making the reference features visible without sacrificing effective color ramp schemes for the class ranges can be extremely difficult, which may be why reference features are not included on more maps. Fortunately, GIS and mapping software can show the map designer what each color scheme will look like with or without the reference features.

Interactive maps (and visualization in GIS) have an advantage in this regard—ancillary general reference information can often be turned on or off by the map reader. The map reader can turn the reference information off to get a feel for pattern and quantity of the data, and then turn the reference information back on to see how the features relate to the thematic data. This type of visualization can be very effective, and it helps strike a philosophical balance between the view that thematic maps need to be simple with the need to include general reference features. In addition, interactive choropleth maps can be developed so that when the map reader clicks or holds the mouse over the enumeration units, the specific enumeration unit value and the name of the enumeration unit can be displayed. This strategy facilitates keeping the map simple (less information is actually placed on the map), assists the map reader in map interpretation, and further reduces the error that is inherent in classification.

DASYMETRIC MAPPING

A form of mapping somewhat linked to choropleth and enumeration mapping is called **dasymetric** mapping. The dasymetric map has been used as early as the 1820s, though not often (McCleary 1969), although the *term* dasymetric was first used by John K. Wright in 1936 (Chrisman 2002). This form of mapping has been mostly identified with mapping

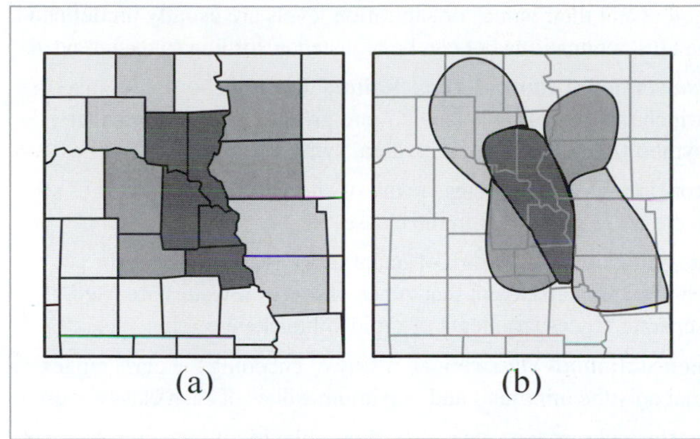

FIGURE 6.12 CHOROPLETH AND DASYMETRIC MAPPING.
Standard choropleth presentation in (a); dasymetric presentation in (b).

population density. Discussion of this method by cartographers and population geographers first reached an apex in the 1920s and 1930s but has now seen a resurgence in the research literature with help from remotely sensed imagery (for example, Holloway *et al.* 1997; Holt *et al.* 2004) and has seen attention in some recent textbooks as well (for example, Chrisman 2002; Slocum *et al.* 2005; Harvey 2008).

Choropleth enumeration units portray data uniformly throughout the enumeration unit, but the real distribution (for example, population density) often does not conform to those units. For example, a county that has a city or part of a city within its borders may be partially urbanized and partly rural, but the choropleth map does not make this distinction. Likewise, farm productivity will be nonexistent in places such as urban areas, lakes, or forested areas. In dasymetric mapping, the idea is to create zones of uniform statistical value (McCleary 1969) that may not necessarily follow enumeration unit boundaries. These zones are based on other data layers such as a land use map, land use imagery, or other relevant information. Once the zone boundaries are created, the (derived) data can be mapped in a manner similar to a choropleth map (see Figure 6.12).

Although dasymetric mapping has not been practiced often in thematic mapping in the last forty years, there is a resurgence of interest in the cartographic and GIS community. Most mapping and GIS software do not currently have a built in "dasymetric" procedure, but it is still possible to create this map type because the method simulates what the cartographer does when layers, serving as filters, are applied in GIS mapping. Much more work needs to be done in order for dasymetric mapping to be as widely used as other quantitative thematic mapping types.

REFERENCES

Brassel, K., and J. Utano. 1979. Design Strategies for Continuous-Tone Area Mapping. *American Cartographer* 6: 39–50.

Brewer, C. 2005. *Designing Better Maps: A Guide for GIS Users*. Redlands, CA: ESRI Press.

———. 2006. *Color Brewer—Selecting Good Color Schemes for Maps*. http://www.ColorBrewer.org

Carr, D., A. White, and A. MacEachren. 2005. Conditioned Choropleth Maps and Hypothesis Generation. *Annals* of the Association of American Geographers 95: 32–53.

Chang, K. 1978. Visual Aspects of Class Intervals in Choropleth Mapping. *Cartographic Journal* 15: 42–48.

Chrisman, N. 2002. *Exploring Geographic Information Systems*. New York: Wiley.

Dobson, M. 1973. Choropleth Maps Without Class Intervals? A Comment. *Geographical Analysis* 5: 358–60.

Fisher, W. 1958. On Grouping for Maximum Homogeneity. *American Statistical Association Journal* 53: 789–98.

Fitzsimons, D. 1985. Base Data on Thematic Maps. *The American Cartographer* 12: 57–61.

Harvey, F. 2008. *A Primer of GIS: Fundamental Geographic and Cartographic Concepts*. New York: Guilford Press.

Holloway, S., J. Schumacher, and R. Redmond. 1996. *People and Place: Dasymetric Mapping Using Arc/Info*. Wildlife Spatial Analysis Lab, University of Montana—Missoula.

Holt, J., C.P. Lo, and T. Hodler. 2004. Dasymetric Estimation of Population Density and Areal Interpolation of Census Data. *Cartography and Geographic Information Science* 31:103–21.

Jenks, G., and F. Caspall. 1971. Error on Choropleth Maps: Definition, Measurement, Reduction. *Annals* of the Association of American Geographers 61: 217–44.

Jenks, G., and M. Coulson. 1963. Class Intervals for Statistical Maps. *International Yearbook of Cartography* 3: 119–34.

Jenks, G., and D. Knos. 1961. The Use of Shading Patterns in Graded Series. *Annals* of the Association of American Geographers 51: 316–34.

Mak, K., and M. Coulson. 1991. Map-User Response to Computer-Generated Choropleth Maps: Comparative Experiments in Classification and Symbolization. *American Cartographer* 18: 109–24.

McCleary, G. 1969. *The Dasymetric Method in Thematic Cartography.* Ph.D. dissertation, University of Wisconsin—Madison.

Meynen, E., ed. 1973. *Multilingual Dictionary of Technical Terms in Cartography*. International Cartographic Association, Commission II—Wiesbaden: Franz Steiner Verlag.

Monmonier, M. 1974. Measures of Pattern Complexity for Choropleth Maps. *American Cartographer* 1: 159–69.

———. 1977. *Maps, Distortion, and Meaning*. Resource Paper No. 75–4. Washington, DC: Association of American Geographers.

Muller, J. 1974. *Mathematical and Statistical Comparisons in Choropleth Mapping*. Ph.D. dissertation, University of Kansas—Lawrence.

Muller J., and J. Honsaker. 1978. Choropleth Map Production by Facsimile. *Cartographic Journal* 15: 14–19.

Peterson, M. 1979. An Evaluation of Unclassed Cross-Line Choropleth Mapping. *American Cartographer* 6: 21–37.

Slocum, T., R. McMaster, F. Kessler, and H. Howard. 2005. *Thematic Cartography and Geographic Visualization*. Upper Saddle River, NJ: Prentice Hall.

Tobler, W. 1973. Choropleth Maps Without Class Intervals. *Geographical Analysis* 5: 262–65.

Tufte, E. 2001. *The Visual Display of Quantitative Information*. Cheshire, CO: Graphics Press.

GLOSSARY

bipolar choropleth map a kind of choropleth map where data class ranges diverge from a central value or class and are encoded with a divergent color scheme

bivariate choropleth map a choropleth map that shows two variables instead of one, for the purposes of showing a relationship between the two

choropleth map a form of statistical mapping used to portray discrete data by enumeration units; area symbols are applied to enumeration units according to the values in each unit and the symbols chosen to represent them

color ramp color symbolization that is applied to each class range in a choropleth map, usually with darker values and/or more saturated colors being associated with the larger numbers; the colors and their values or saturation levels are usually predefined in GIS applications but can be adjusted according to design needs

conventional choropleth technique form of choropleth map in which the data array is classed into groups and the groups are symbolized on the map with areal symbols

continuous class ranges a way of encoding the class ranges so there are no data gaps in the classes

dasymetric map a form of quantitative map similar to the choropleth map except that zones of uniform values are used, separated by escarpments of rapid value change

noncontinuous class ranges a way of encoding the class ranges so that only the minimum and maximum values of each class are used

pattern complexity measure of complexity of repeating elements in a visual image; used in studying choropleth maps

small multiples multiple maps or graphics that have a similar structure (such as two or more choropleth maps positioned next to each other for easy comparison)

unclassed choropleth map the data array is not classed into groups; the data are symbolized by continuously changing proportional area symbols

7 THE DOT DENSITY MAP

CHAPTER PREVIEW Mapping geographic phenomena can be accomplished by dot density mapping, also simply called dot mapping. Its main purpose is to communicate variation in spatial density. In use for over 100 years, the method is popular because of its simple mapping rationale: one dot represents so many items (such as one dot represents 500 bushels of harvested corn). The technique works best for data that are tabulated in enumeration areas as totals. GIS and mapping software highly automate the dot generation and placement pro-cess, usually placing the dots randomly within each enumeration unit. Significant design decisions include the selection of the enumeration unit, the dot value and size, and in some cases adjustments in the placement of the dots. Legend design is also important, especially because empirical research indicates that map readers do not perceive dot quantity or density in a linear relationship. ■

From the previous chapter's focus on choropleth mapping, a form of area mapping, we now proceed to dot-distribution maps. These are also quantitative maps; information pertaining to density and distribution is gained by visual inspection of the spatially arrayed symbols to arrive at relative magnitudes. Precise numerical determination of density is subordinate in this form of mapping. Although any point symbol can be used, it has become customary to use small dots—thus the name **dot density mapping.** A dot density map can also be simply called a dot map.

MAPPING TECHNIQUE

Dot maps were introduced as early as 1863 (Hargreaves 1961), and most current GIS and mapping software support this technique, so it is not surprising that their use and acceptance are widespread. In the simplest case, the technique involves the selection of an appropriate point symbol to represent a quantity of a geographically distributed phenomenon. The symbol form (size, shape, color, and so on) does not change, but the frequency of dots changes from area to area in proportion to the number of objects being represented. The technique works best for data that are tabulated in enumeration areas as totals, and has been used extensively for mapping agricultural production data and population data (see Figure 7.1). GIS and mapping software that support this technique randomly place the symbols within the respective enumeration units. Dot locations are approximate at best. If ancillary data are available to further position dots more closely to where the phenomena actually occur, then the map's quality improves.

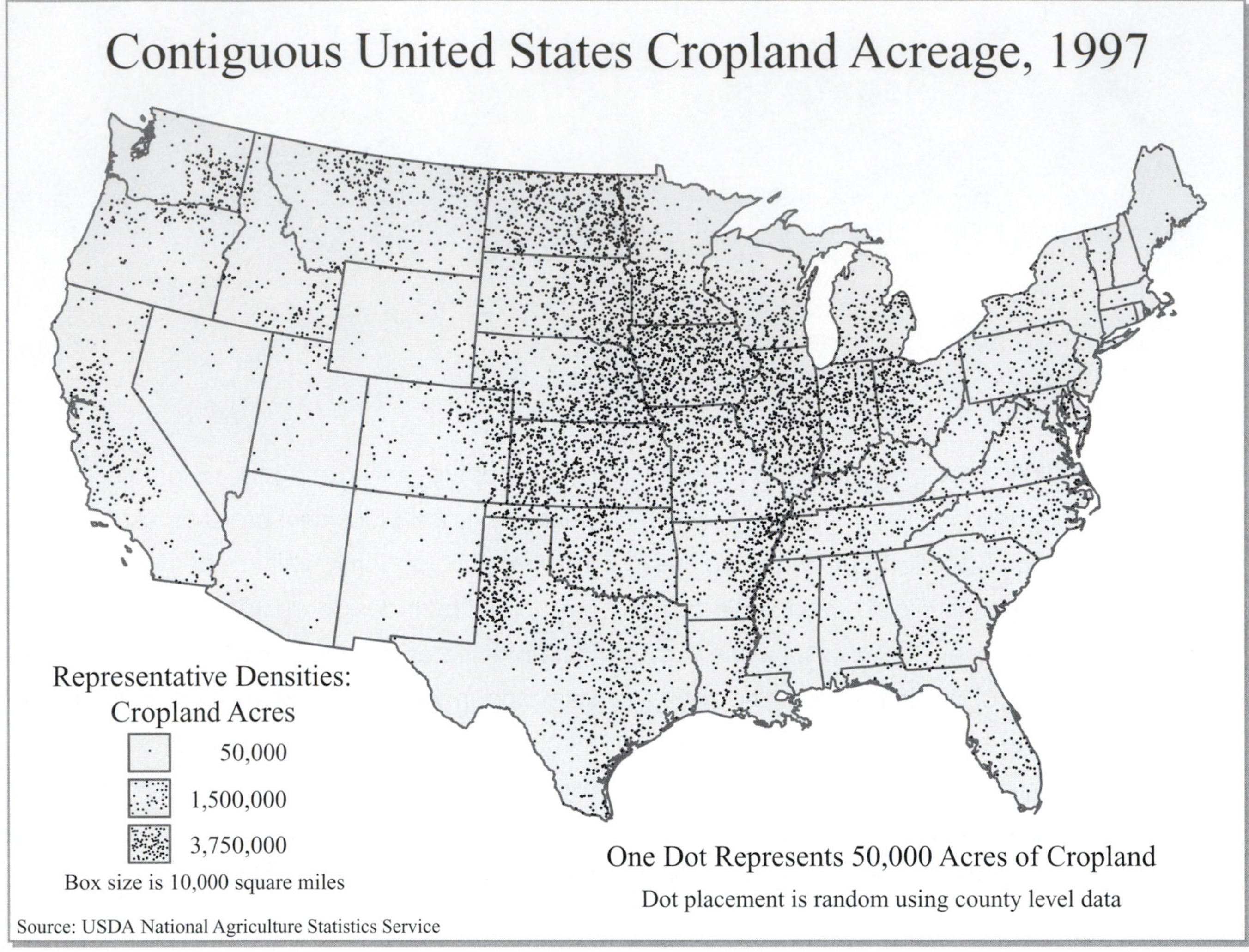

FIGURE 7.1 A TYPICAL DOT DENSITY MAP.
The elements of the typical dot map include the dots, reference political boundaries, and a legend that includes dot value and, if possible, representative densities taken from the map. The reader gets the idea that the amount of the item varies from place to place.

Advantages and Disadvantages of Dot Density Mapping

After a brief decline in the number of dot maps appearing in professional journals and other publications, there now seems to be a resurgence in popularity of this technique, largely due to the relative ease in generating this type of thematic map in GIS and mapping software. Not all of these recent GIS dot maps follow sound cartographic principles, as optimum dot map construction actually requires considerable research and effort. Therefore, it is important for the cartographer to have an understanding of the advantages and disadvantages of the method before selecting this map type.

The advantages of dot mapping include:

1. The rationale of mapping is easily understood by the map reader.
2. It is an effective way of illustrating variations in spatial density.
3. Original data may be recovered from the map if the map has been designed for that purpose.
4. More than one data set may be illustrated on the same map. As with any bivariate or multivariate map, there should be a distributional or functional relationship between the sets.
5. GIS and mapping software that support this technique allow the cartographer to quickly view and

evaluate many dot value and size combinations with relative ease.

Possible disadvantages of dot mapping would include:

1. Reader perception of dot densities is not linear. A person viewing an area with 10 times the number of actual dots compared with another area will usually not estimate values in those two areas in the same proportion as depicted by the dots on the map.
2. GIS and mapping software typically randomize dots within enumeration units, resulting in dots that may not be close to the phenomena they represent.
3. Ancillary data layers or imagery *should* be used in controlling dot placement, but in many cases this is not practicable. Possible workarounds are discussed later in the chapter.
4. Large ranges in data values make it very difficult to select a single dot value that is visually acceptable across areas of highest and lowest density.
5. When the map has been designed for optimum portrayal of relative spatial density, it is practically impossible for the reader to recover original data values.

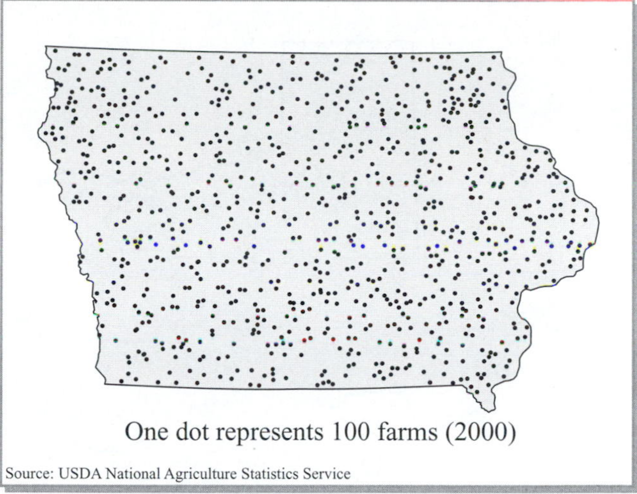

FIGURE 7.2 DOT MAPS DO NOT WORK WELL FOR ATTRIBUTE DATA THAT HAVE A SMALL RANGE OR UNIFORM DISTRIBUTION.

Data that have small attribute data ranges and are uniformly distributed should not be mapped, such as the number of farms in Iowa. See text for discussion.

Dot mapping is a valuable technique that is unique in purpose and appearance, and is likely to be included among thematic map types for the foreseeable future. Michael Coulson eloquently sums up the value of the dot mapping technique:

> *It is not the dot size, or value, or even placement, that give the real power to the dot map—assuming some reasonable decisions have been made. Rather, the power of the dot map is in the overall pattern of the distribution that is revealed* (Coulson 1990, 56).

In summary, map designers must weigh these advantages and disadvantages carefully before deciding to apply dot mapping to a given project. As in all design choices, the map purpose, readership, and appropriateness of the data are central concerns in making the decision.

Data Suitability

There are several data considerations that should be made before selecting this method. As with the choropleth map, this technique is used extensively for data that are tabulated in enumeration areas. Unlike the choropleth map, where some sort of derived data is generally desirable, totals or nonderived quantities are used in dot mapping. Common examples include agricultural production data (such as crops, crop productivity (in bushels or tons), numbers of livestock or farms), and population totals. In these cases, one dot may represent 500 bushels of harvested wheat, or 1,000 persons. Density is visually inferred by the distribution of the dots *instead of by using derived tabulated data*. Using derived or ratio data, particularly ratios involving area, cannot be interpreted accurately by the map reader, and should be avoided for this technique.

Data sets with extremely small or large attribute data ranges are often more difficult to portray effectively with dot maps than with other thematic map types. Dot maps use a single number to represent the value of the dot. Data with small attribute ranges will produce a spatial distribution that is uniform (see Figure 7.2). If similar numbers are uniformly distributed in the enumeration units, then construction of a dot map should not take place.

Data with extremely large attribute ranges make setting a dot value and dot size that is visually satisfactory for both highest and lowest density areas of the map quite difficult. For example, many state population maps are made from county level enumeration units with huge differences in rural and urban population totals. If the dot value is set so that the distribution is clearly seen in urban areas (a dense pattern of just coalescing dots), the rural areas will become vacant. If the dot value is adjusted to see the pattern in rural areas, the urban areas become a solid fill of indiscernible dots (see Figure 7.3). Of course, large data ranges can be problematic in other map types, but mechanisms not usually used in dot mapping, such as classification, or changes in visual variables such as color, can assist the cartographer in dealing with data extremes in other map types. Setting the dot value and dot size are two crucial components of dot mapping, and will be discussed in depth in the following section.

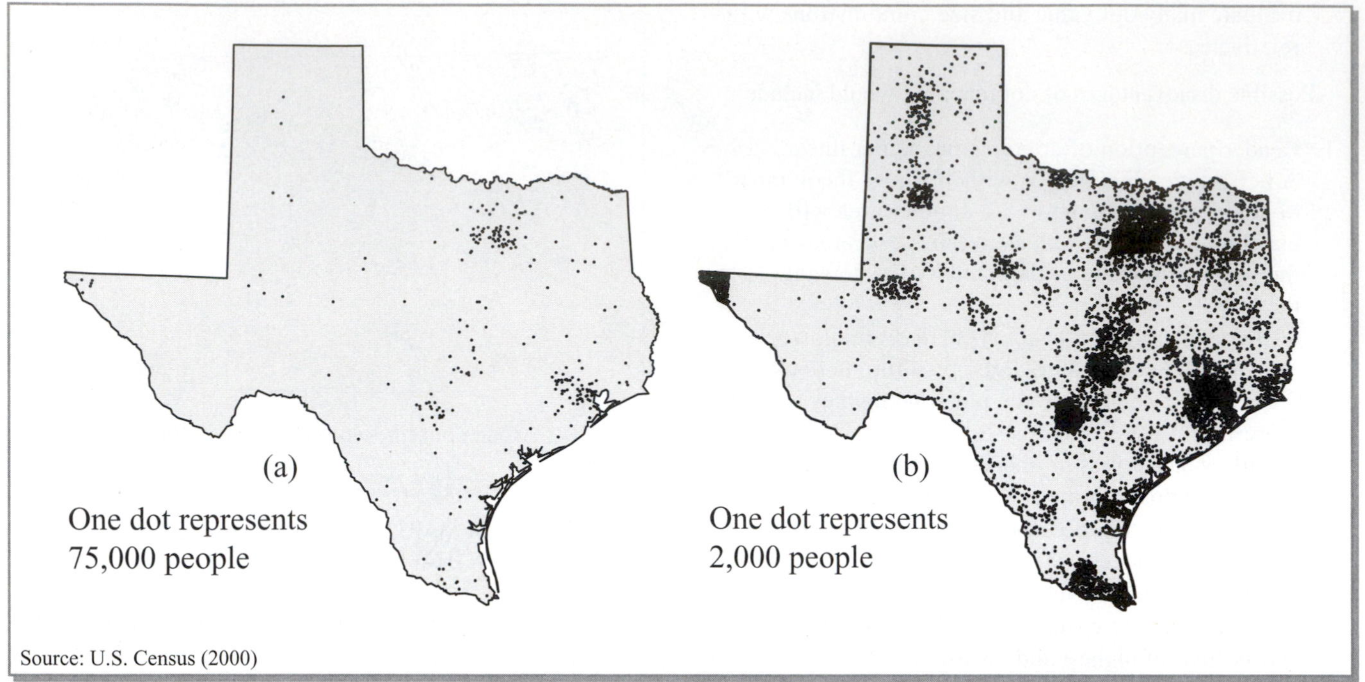

FIGURE 7.3 DOT MAPS DO NOT WORK WELL WITH DATA THAT HAVE AN EXTREMELY LARGE ATTRIBUTE DATA RANGE.
Data that have extremely large attribute data ranges, such as Texas county population, are nearly impossible to set a dot value that works for the entire map. In (a) the dot value is set so that more populated counties in and near urban areas are presented fairly well, but rural areas are completely left out. In (b) the dot value has been reduced so that more rural areas can be seen (although some counties are still left out even at this low ratio) but the urban and near urban areas become unreadable. A choropleth map of population density may be one possible alternative in this case. See text for discussion.

The Mapping Activity

Most dot maps are constructed at small or intermediate scales. At these scales, attribute data are normally collected in tabular form for enumeration areas. Therefore, the use of the term "dot map" is applied to a many-to-one situation: each dot represents more than one mapped element (but see the boxed text "One-to-One Mapping"). Each dot can be thought of as a **spatial proxy** because it represents at a point some quantity that actually occupies geographical space (Coulson 1990; see Figure 7.4). These geographical spaces are called **dot polygons** or territorial

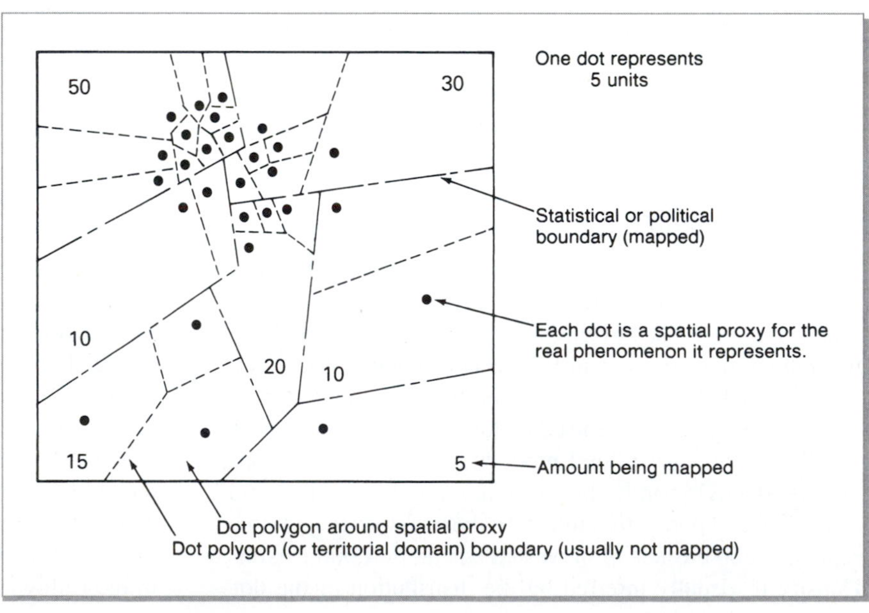

FIGURE 7.4 THE CONCEPTUAL ELEMENTS OF A DOT MAP.
Each dot functions as a spatial proxy for the item it represents. Around each spatial proxy is its polygon or territorial domain. Political or statistical units are normally included to provide the reader with spatial cues, but the boundaries of territorial domains are not. These domains help the mapmaker in conceptualizing and producing the map.

ONE-TO-ONE MAPPING

The dot is one of the most versatile symbols in cartography. Dots have been used for everything from uniform fill patterns to labeled cities on general reference maps to the location of specific point features. However, just using the dot symbol *does not make it a dot density map.* Dot density maps in quantitative thematic mapping signify mapping a continuous, statistical surface, using data collected in enumeration unit areas and mapped with a many-to-one correspondence. The resulting dots in this case are approximately positioned. But what about using dots with a one-to-one correspondence?

One-to-one mapping involves data that are collected at individual points and placing a symbol as precisely as possible to the location of occurrence of the phenomena. For example, a symbol could represent earthquake epicenters or the location of a new restaurant or store. In this type of map, the map reader can reasonably expect that an earthquake epicenter, a new restaurant location, or any other feature represented *is accurately placed.* For this reason, the value in the dot density procedure in most GIS and mapping software (which will randomize the dots within enumeration unit areas), should *never* be set to a one-to-one ratio.

One-to-one mapping is more accurately called "nominal point thematic mapping" because these types of data are collected at specific points (as opposed to being collected by enumeration unit areas), are qualitative in nature (if you scale the symbol by magnitude of the earthquake, you get a quantitative proportional symbol map), and should be mapped accurately (as opposed to being mapped as a spatial proxy in a statistical surface). The procedure in Golden Software's Map Viewer is called a "pin map." In ArcGIS, point placement of this type is simply referred to under a category called "single symbol."

To avoid confusion with the quantitative dot density maps discussed in this chapter, we strongly recommend using a symbol other than the dot for nominal point symbol placement. Other symbolization is becoming easier with the rapid proliferation of geometric and pictorial symbol sets available for use in GIS, mapping, and graphics software packages. If the data has a quantitative attribute associated with it (such as earthquake epicenters with the quake's magnitude), consider creating a proportional symbol map instead. See Chapter 8 for a discussion on making proportional symbol maps.

domains. They do not normally appear on the final map but are assumed to exist around each dot, making up the total enumeration area.

It is highly desirable in dot density mapping that associated spatial data showing functionally related variables be obtained prior to mapping, so that *somewhat* greater precision in dot placement can be achieved. Each mapping task has its own peculiar set of related materials. Some examples include satellite imagery; maps or data sets of physical relief; soil types; hydrographic features including lakes, ponds, rivers, and streams; land-use; national forests or parks; and military reservations. The related distribution maps or data sets operate as **locational filters** and serve to further position the placement of the dots (see Figure 7.5). Because the dots can vary in density within the various enumeration units, some cartographers see a relationship with *dasymetric mapping* (introduced in Chapter 6; see boxed text "Dot Density and Dasymetric Mapping" on the next page).

Once appropriate data and associated resource materials have been obtained, the designer can begin constructing the map. Major design decisions must be made regarding scale (of the overall map and the enumeration level used), dot value, and dot size. These are closely interrelated in the construction of the map; a decision regarding one affects the others. In this phase, the preliminary decisions are tested, largely by trial and error. Once these issues are resolved, issues of legend design and other dot map design issues can be addressed.

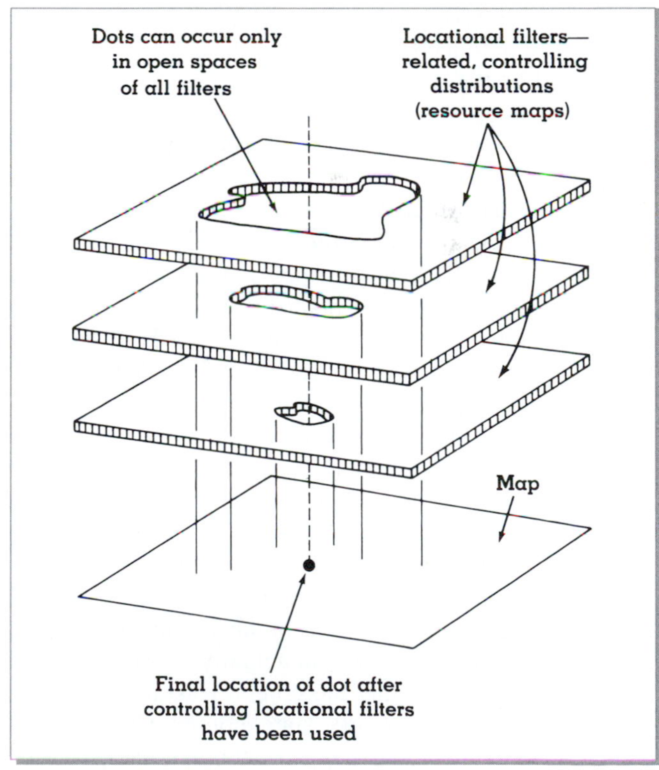

FIGURE 7.5 DOT PLACEMENT.

Locational filters (related distributions) determine the location of dots. Each dot's territorial domain can exist only in that part of the map which has not been influenced by other distributions.

DOT DENSITY AND DASYMETRIC MAPPING

The cartographic community often sees a close relationship between the dot density map and the dasymetric map. The dasymetric map was defined and discussed in Chapter 6 and is illustrated in Figure 6.12b. In dasymetric mapping, area fills are used to delineate zones of uniformity. The parallel to dot mapping is that the zones of uniformity (dasymetric) and the concentrations of dots (dot density) should not be constrained by enumeration unit boundaries. As noted in Chapter 6, farm productivity will be nonexistent in places within the enumeration that are dominated by urban landscapes, lakes, or forested areas, and these map types are capable of making this distinction. However, this parallel holds true only if ancillary spatial information in the form of locational filters are used in the dot mapping process (see Figure 7.5). If the dots are simply drawn randomly with the enumeration unit areas, then its tie to dasymetric mapping breaks down. Besides the obvious difference in physical appearance (dots as opposed to area fills), it is also important to note that dasymetric mapping generally requires derived data and dot mapping generally requires using totals (see discussions of data appropriateness and suitability in Chapters 6 and 7).

Size of Enumeration Unit

As previously mentioned, dot mapping usually involves the presentation of geographical quantities that have been collected by enumeration unit (see Chapter 1 for a discussion on enumeration units). Under most conditions, the smaller the enumeration unit in relation to the overall size of the map, the greater will be the relative accuracy of the final dot distribution (Birch 1964). Smaller enumeration units mean a smaller territorial domain for each dot, reducing the chance of locational error. Large-scale maps (small earth areas) require quite small statistical units. In fact, it is often found that common units used for enumeration data (block, tract, or county-sized units) are too large for dot mapping at large scales. Therefore, cartographers normally use intermediate to small scales for dot density mapping.

Cartographers know that scale is relative in all mapping activities. This is especially true for dot mapping. As stated previously, most GIS and mapping software place dots within enumeration units *randomly*. If the cartographer cannot use locational filters or manually move the dots, additional locational control is lost, *except that provided by small enumeration units*. To ensure that dot placements will be closer to the actual phenomena they represent, strive for the smallest enumeration unit that is practical.

The enumeration unit boundaries in which the data are collected should *not* be displayed on the map, especially if the dots are completely randomized within the enumeration unit. For example, in a state map of corn production, county level data might be used in the dot map procedure, but only the state outline would actually be shown. Some cartographers suggest that there be at least one level difference (for example, county, state, country) between the enumeration level of the mapped data and any enumeration units that are displayed on the map. Many United States agricultural distribution maps are created from county level data but may portray only the state or even country boundaries (see Figure 7.6). As discussed in "Legend Design," the enumeration unit from which the data are collected should be included in the legend or as ancillary text.

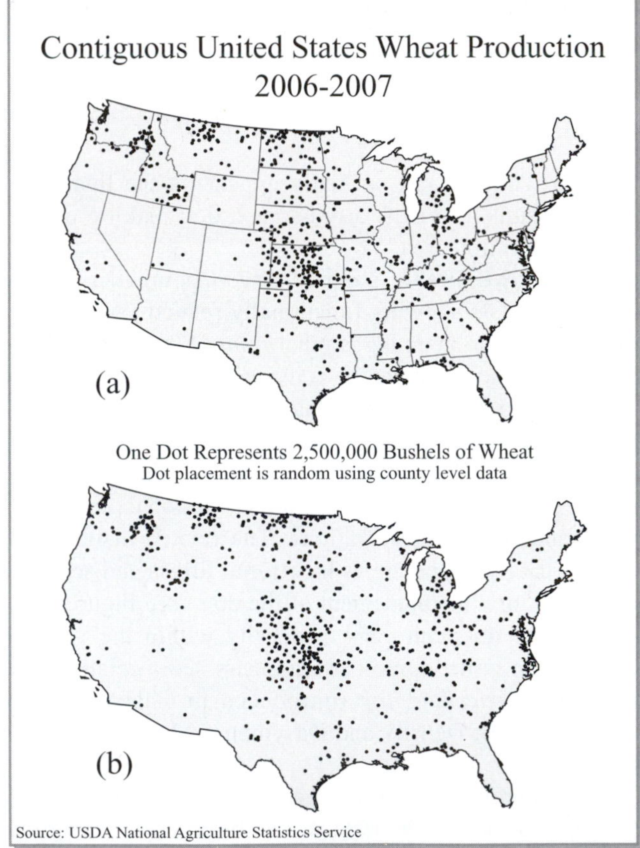

FIGURE 7.6 DEPICTED ENUMERATION UNIT LEVELS.
The boundaries of the enumeration units from which the data are collected are not shown. Some cartographers suggest at least one level of enumeration difference between the level of collection and depicted enumeration units. In this example, county data are used, and states (a) are shown, but some atlases also use only the country boundary (b).

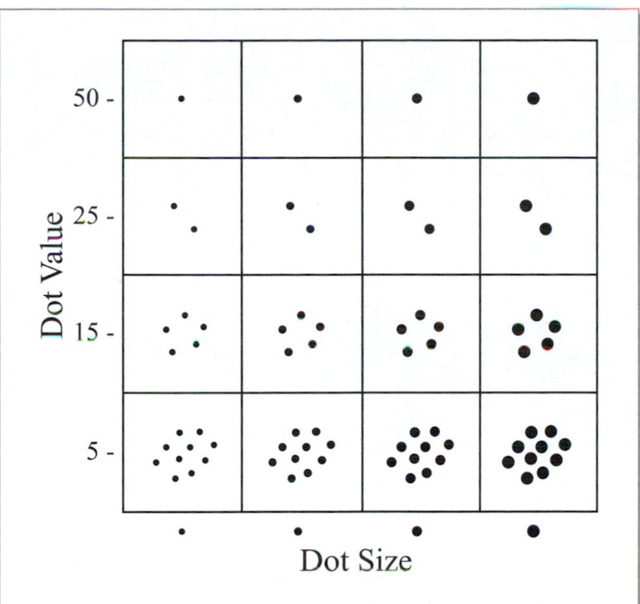

FIGURE 7.7 THE EFFECT OF CHANGING DOT VALUE AND SIZE.
The visual effects of changing dot value and size can be dramatic. For unusable extremes of emptiness and crowdedness, refer to Figure 7.3.

Dot Value and Size

Closely related to map scale is the determination of **dot value**—the numerical value represented by each dot (see Figure 7.7). Extreme care must be exercised in selecting dot value and size combinations, because they affect the map reader's impression and understanding of the map. Dots that are too small don't always stand out as proper figures relative to the rest of the map. When dots become too large the map takes on a crude appearance, especially if the dot value is too low. Dot values that are too high give the impression that the dots have been precisely placed, which is not the goal of dot density mapping.

It is difficult to list definite rules for selection of dot values and size, but here are some *general* guidelines:

1. Choose a dot value that results in two or three dots being placed in the statistical area that has the least mapped quantity.
2. Choose a dot value and size such that the dots just begin to **coalesce** in the statistical area that has the highest density of the mapped value (see Figure 7.8). When the symbols begin to overlap, the perceived concentration of symbols will be increased (Sadahiro 1997).
3. It is preferable to select a dot value that is easily understood. For example, 5, 500 and 1,000 are better than 8, 49, or 941.
4. Select dot value and size to harmonize with the map scale so that the total impression of the map is neither too accurate nor too general. This will require experimentation.

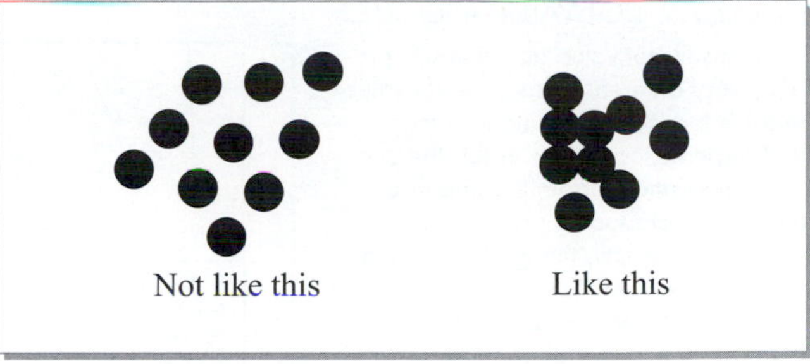

FIGURE 7.8 DOT COALESCENCE.
Dots should begin to coalesce in the densest part of the map.

5. Smaller dots may appear as squares on the monitor due to the shape of the pixels. This is not usually a problem in printed maps, as they will appear correctly on the printed page due to the higher resolution of the printer compared to the monitor. In virtual maps, the dots will have to be made larger or have zoom capability to avoid the squares.

It is important to consider that the data are generalized by the selection of the dot value. When selecting the dot value, a level of generalization occurs that is equal to one-half the dot value. For example, if the dot value is 500, one dot would be used for any data that fall between 250 and 749. Anything less than 250 would not have a dot representing that data. If one selects a dot value of 1,000, then any enumeration unit that has 499 will be displayed the same as the one that has a zero value.

At the same time, the designer has to consider dot size. What size dots, in combination with an appropriate dot frequency, will result in coalescence? Fortunately, trial and error testing using GIS and mapping software is relatively easy, as numerous dot value and size combinations can be created and viewed very quickly. Some GIS and mapping software packages also give previews of lowest, average, and highest density areas to assist in this task. In this case it is very easy to discern if the attribute data are too large or too small for dot mapping (see discussion in the section "Data Suitability"). Cartographers working entirely within graphic design software or using other manual techniques should develop representative (lowest, average, and highest density) enumeration areas at final map scale, or consider using mapping software to develop a mock-up.

Several different dot maps can result from identical data because the selection of dot value and size is subjective (see Figure 7.9). There is no quantitative index to tell the designer which is best. Knowledge of the distribution plus an intuitive sense of what looks right are the standards against which any dot map must be judged.

Counting Up Method. Traditional counting methods in dot mapping include first the assignment of a dot value (after the question of coalescence is considered), as just described, then this value is divided into each enumeration

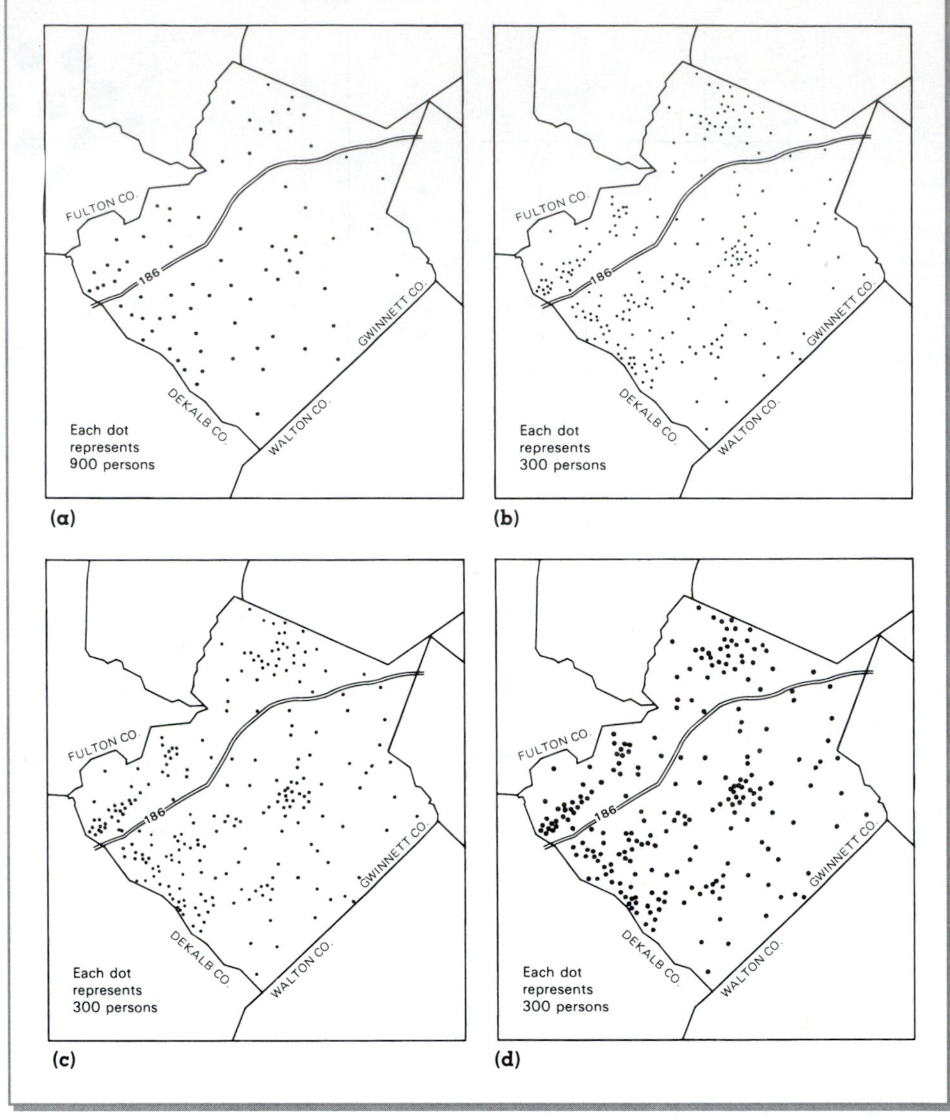

FIGURE 7.9 DOT VALUE AND SIZE. Changes in dot value and size will produce very different maps. The dot value in (a) is too large, leading to a map that appears too empty. In (b), the dot size is too small, again leading to a map that looks too empty. Dot size in (c) is still too small, though larger than in (b). Dot value in (d) remains at 300 persons, but dot size is now larger, giving a better overall impression for the map.

unit's total to determine the number of dots for that unit. Unfortunately, this often leads to error, as the *remainders* are never mapped. For example, if there are 4,500 people in an enumeration unit, and the dot value is 4,000, then that unit would get one dot, and the remainder of 500 people would never get mapped (the error). This error (or underestimation bias) is generally considered acceptable and even unavoidable when using some GIS and mapping software. Overestimation is avoided using this method. This might be thought of as "counting down."

Another method is one that "counts up." In this method, developed by the United States Census Bureau for their nighttime view of the United States 1990 population distribution map, a smaller enumeration unit is used, and the populations of neighboring units are added until the dot value is reached, and then the area gets a dot. The "remainder" is then added to other units until another dot value is reached (Hendrix 1995; see Figure 7.10). In this way, remainders are accounted for.

Dot Placement

Since dot maps are created from enumeration unit area totals, it has always been understood that the dots *do not* mark an *exact* spot where the geographic phenomena exist. The dots are simply a *spatial proxy* for data that exist somewhere in the vicinity. Of course the cartographic ideal, established back when all dots were placed manually, has been to locate the dots as close to the real distribution as possible, usually using the **center of gravity principle** (see Figure 7.11). Because each dot is a spatial proxy, it must be located so as to *best* represent all of its referents. This can require considerable knowledge of the real distribution, and relevant resource materials, acting as *locational filters*, are of invaluable assistance in this respect.

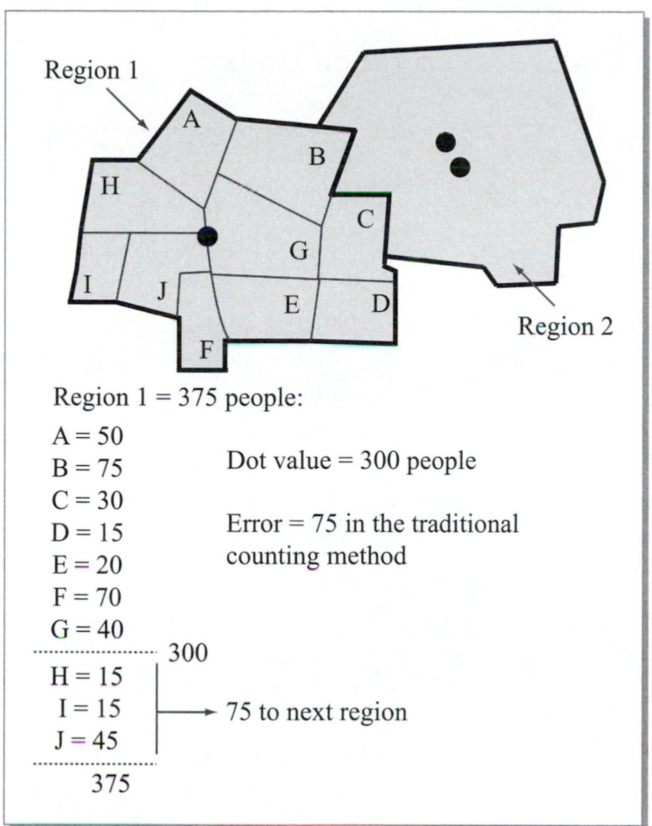

FIGURE 7.10 AN ALTERNATIVE METHOD OF DETERMINING DOT VALUE.

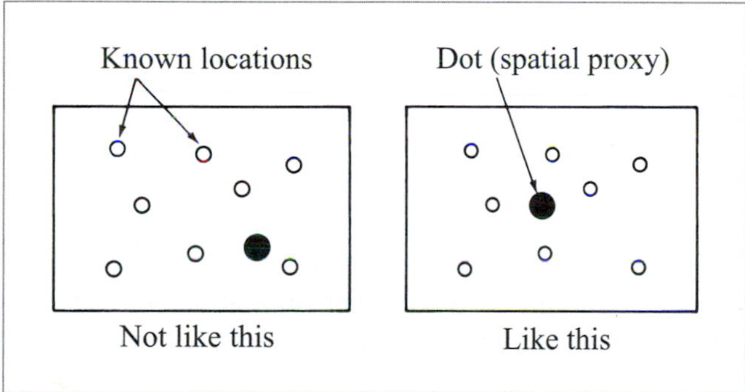

FIGURE 7.11 PROPER PLACEMENT OF DOTS.
Dots should be placed at the approximate center of gravity of the elements they represent.

If ancillary information is *not* used, then dots are simply randomized within the enumeration unit areas (see Figure 7.12). This is typically the way dots are placed in GIS and mapping software that support this technique. Dots representing people, for example, may be clustered in places such as the Mojave Desert or the Carson Sink. Likewise, dots representing harvested wheat may find their way into nonagricultural regions such as lakes or urban areas instead of farmland. As discussed earlier, the most common solution is to use the smallest enumeration unit possible to control the location of the dots.

Fortunately, there are a few other practical options that can help the GIS-using cartographer control dot locations. These following cases are considered:

1. *The software directly supports locational filters with a dedicated procedure*. Locational filters in the form of masks are now supported in some software, such as ArcGIS. These masks are created from another data layer of related information, and the placement of dots will be inside or outside of the masked area.
2. *The software does not directly support locational filters with a dedicated procedure*. New polygons can be created from raster imagery (such as land use) or from existing vector data using a GIS overlay procedure similar to the filters depicted in Figure 7.5. The attribute enumeration data can then be transferred to the newly created regions. The cartographer has effectively created newly shaped enumeration units. These new regions then serve as a proxy for the original enumeration units, and the dot density procedure is applied. The dots can be then be displayed on top of the original base map.
3. *The dots are transferred to an artistic drawing program*. The dot map can be initially created in the GIS or mapping software and digitally exported to ancillary graphic design software, where the dots can be manually moved into more desirable locations. The effectiveness of this option can vary greatly depending on what combination of GIS package, graphic design software, and transfer format is used.

Of course, dot maps can also be created entirely using graphic design software. Slocum *et al.* (2005) created a Kansas wheat harvest dot map by manually locating and placing the dots based on land use/land cover imagery in Freehand. Although time-consuming, manual dot placement by the cartographer ensures maximum control of dot location.

Geographic phenomena rarely occur uniformly over space, so a geometric arrangement of the dots within an enumeration area should be avoided (see Figure 7.13a–b). It is also good practice to not allow the political boundaries of the enumeration areas to exercise too much control over the location of the dots (see Figure 7.13c–f). Unintentional dot patterns are usually more of a problem in manual dot placement; political boundary effects can be a problem with both manual and GIS and mapping software. Refer to Figure 7.3b, where the county lines can be seen because of the large number of dots.

Legend Design

The most traditional dot map legend is a simple statement indicating the unit value of the dot. It is good practice for the legend to indicate that a dot represents a value but does not equal it. Including the dot symbol as part of the statement is not necessary. Also strongly recommended is the inclusion of

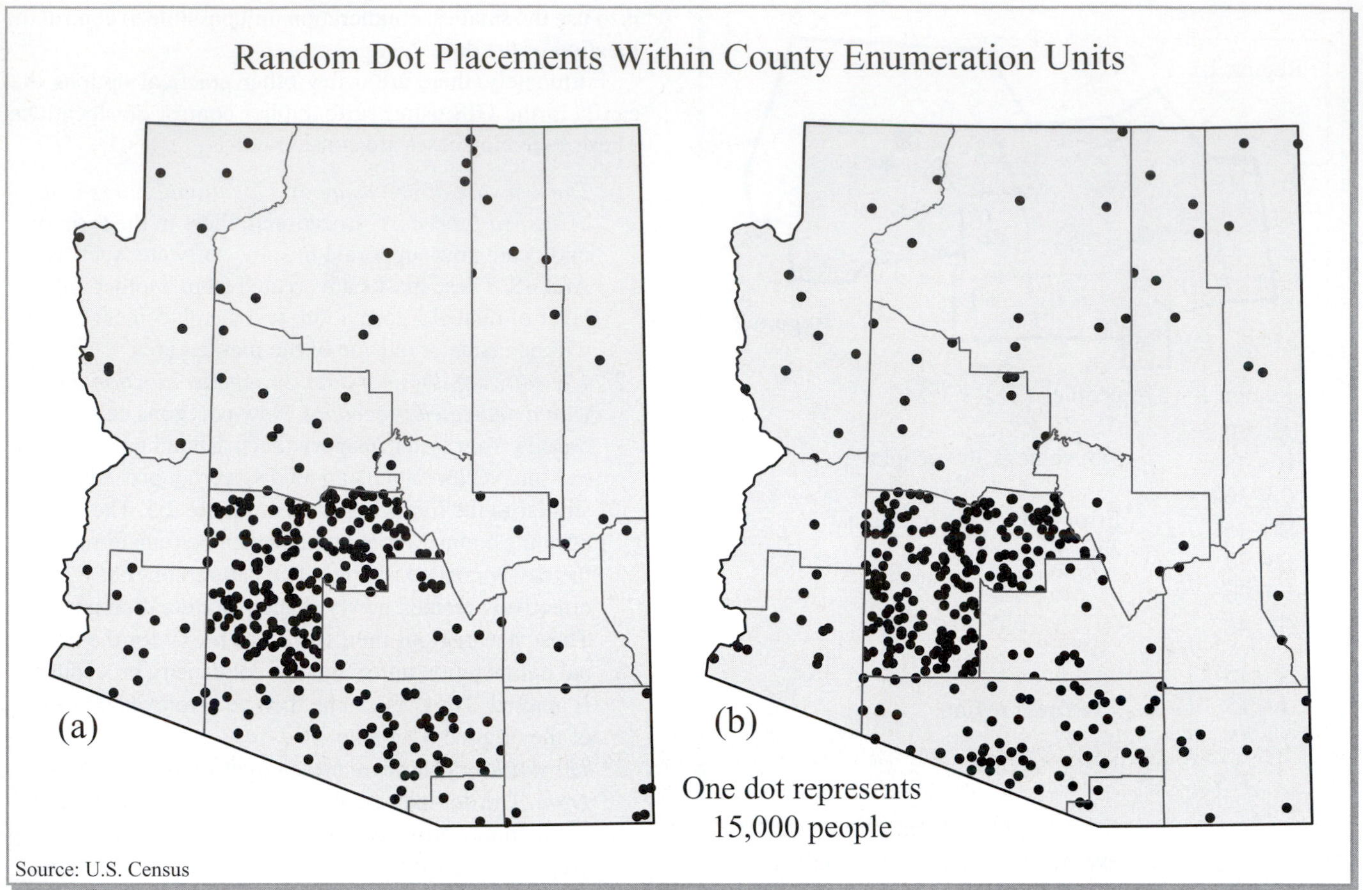

FIGURE 7.12 RANDOM PLACEMENT OF DOT SYMBOLS WITHIN ENUMERATION UNITS.
GIS and mapping software place dots randomly within each of the enumeration units. The same data and software is used for both (a) and (b), yet the placements are markedly different when comparing the same counties in (a) and (b). Normally removed or turned off, the county boundaries are left on in this figure for illustrative purposes. The effect of randomization is more noticeable with larger enumeration units, such as in the relatively large Arizona counties. This is why we recommend using the smallest enumeration unit possible and, if feasible, ancillary information to gain some control over dot location. Note that some software (for example, ArcGIS) allow you to keep an initially randomized dot placement constant, without re-randomizing every time the drawing page is refreshed.

a second component: a set of at least three squares (or other areas) that illustrate three different representative densities taken from the map (see Figure 7.14). The boxes usually contain a number of dots that illustrate a typical low, medium, and high density area on the map. These sample density boxes help readers interpret relative dot densities—an important concept that will be discussed in greater detail in the next section. The three squares may be arranged in any fashion that is unambiguous yet harmonious with the remainder of the map, but the legend must be designed for clarity. Finally, if the dots have been randomly placed, and/or the enumeration unit that is used to generate the map is not clear, we recommend that a statement clarifying these issues be included on the map, perhaps near the legend area.

Most GIS and mapping software generally do not have good automated procedures for generating complete (or always correct) dot map legends. Many cartographers simply add their own textual information for the unit value of the dot or for clarifying randomization and the enumeration unit. Density boxes can make a more effective map, but they are also time consuming to construct since at present there are no dedicated procedures for this process.

Other Dot Map Design Issues

One large challenge for many cartographers in creating a good dot map is establishing a figure-ground contrast to make the dots stand out appropriately without resorting to using overly large symbols. Two simple design strategies can assist the cartographer in this task. The first is to make sure that any displayed enumeration unit boundaries are supporting the dots by providing reference location, but at the same time are not visually competing with the dots. Following the principle discussed earlier of displaying the enumeration unit boundary at least one level greater than that of the original data set is one way to make sure boundaries are visually subordinate to the dots. In addition, de-saturation or lightening of the enumeration border lines' color or value causes

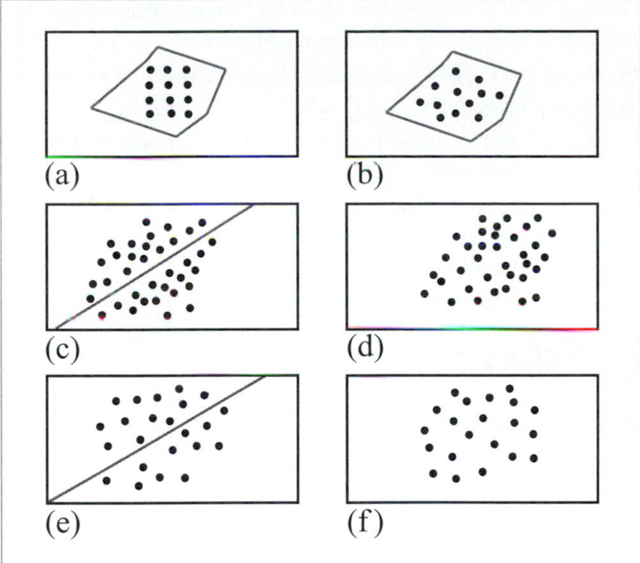

FIGURE 7.13 ACCEPTABLE AND UNACCEPTABLE METHODS OF DOT PLACEMENT.

The regular pattern in (a) should be avoided if manually placing or moving the dots. The irregular pattern in (b) is an improvement over (a). Enumeration unit boundaries should not exercise too much control over the location of the dots; when the boundary in (c) is removed or turned off, its effect is still felt (d). The dots in (e) have been located in such a way that if the boundary is removed or turned off, its control will not be noticed (f).

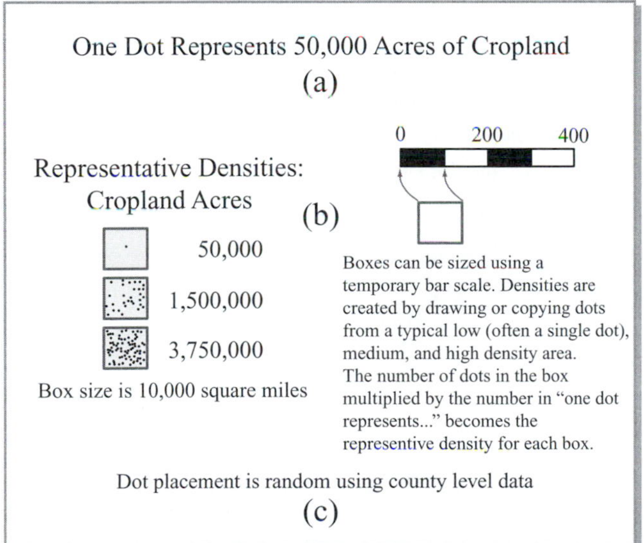

FIGURE 7.14 DESIGNING DOT LEGENDS.

A typical dot legend (from Figure 7.1) includes a notation of dot value (a). We also recommend including boxes with representative densities across the range of data values (b). The size of the density boxes and the densities represented are determined in accordance with the map scale. A temporary scale bar can be used to size the density boxes. Ancillary text for clarifying randomization and the enumeration unit is also recommended (c). This statement does not have to be part of the legend, and may be placed in whatever manner keeps the map visually balanced.

> Whatever dot value and dot size are used it should be borne in mind that an area full of dots on the map will carry a psychological impression that on the ground the same area is full of whatever is being mapped, and in the same way, too sparse a sprinkling of dots conveys an unmistakable impression of emptiness (Dickinson 1973, 49).

the dots to stand out more, as can decreasing the lines' widths. The second strategy is to use a very light or low saturation fill that distinguishes the background of the map from the map body, and the map body from the dots. These two strategies can work in conjunction with each other (see Color Plate 7.1 and refer to Figure 7.1).

Cartographers who wish to show a relationship between two distributions may want to create a bivariate dot map. The most conventional bivariate dot map is simply displaying the two distributions with different hues. Other variations include using two geometries, such as dots and squares, or overlaying dots on top of another area thematic map (such as a choropleth map), although the latter often produces figure-ground problems. Color Plate 7.2 illustrates placement of dots on top of several hues.

A multivariate dot map using three or more hue or hue/geometry combinations is also possible. If a bivariate or multivariate dot map is to be created, the cartographer should take care to make sure that the symbols and patterns are discernable. Note that with most GIS and mapping software, if multiple symbols are used, there is a chance that they may end up on top of each other. As with other bivariate or multivariate maps, if the result is becomes muddled or overly complex, then the distributions should be placed on two (or more) juxtaposed maps.

VISUAL IMPRESSIONS OF DOT MAPS: QUESTIONS OF NUMEROUSNESS AND DENSITY

It was noted earlier that reader perception of dot densities is not a one-to-one or linear relationship. The reader of a dot map is asked to judge relative densities and to make certain assessments about the spatial distributions of these densities. In most dot mapping we do not expect the reader to recover the original data—an impossibility if there is coalescence. The dot map is used to present data at the ordinal level in the sense that all the reader must do is to judge that there is more of the item in one place and less of it in another. **Apparent density** is the subjective reaction of the map reader to the physical stimulus of the actual density of dots per unit area (Olson 1977).

Numerousness is the subjective reaction to the physical number of objects in the visual field, without actually counting the objects during perception. Dot maps are usually selected as a way of illustrating the spatial distribution of continuous phenomena using discrete dot objects; the reader is not normally expected to judge the actual number. Cartographers select this mapping technique to convey the spatial variableness of density, and nothing more.

Cartographic research has demonstrated conclusively that most map readers underestimate the number of dots on a dot map. By extension, apparent densities are likewise guessed on the low side (Provin 1977). The evidence suggests that, when a person perceives an area as having 10 more dots than another with 1 dot, the actual ratio is probably closer to 15 or 16 to 1. This holds positive lessons for the cartographic designer. The following design ideas are suggested by this research.

1. Include a legend with examples of low, middle, and high densities. These act as visual anchors to counteract the map reader's tendency to underestimate relative densities (see "Legend Design").
2. When recovery of original data is important, an adjustment of the rule that dots should begin to coalesce in the densest part of the map will have to be modified. In this case, do not use coalescing dots; noncoalescing dots improve the estimation of dot number.

REFERENCES

Birch, T. 1964. *Maps: Topographical and Statistical.* Oxford: Oxford University Press.

Coulson, M. 1990. In Praise of Dot Maps. *International Yearbook of Cartography* 30: 61–91.

Dickinson, G. 1973. *Statistical Mapping and the Presentation of Statistics.* London: Edward Arnold.

Hargreaves, R. 1961. The First Use of the Dot Technique in Cartography. *Professional Geographer* 13: 37–39.

Hendrix, E. 1995. *Electronic Dot Map Preparation: An Examination of Design and Method,* unpublished Master's practicum research paper, Department of Geography, Georgia State University, Atlanta.

Olson, J. 1977. Rescaling Dot Maps for Pattern Enhancement. *International Yearbook of Cartography* 17: 125–37.

Provin, R. 1977. The Perception of Numerousness on Dot Maps. *American Cartographer* 4: 111–25.

Sadahiro, Y. 1997. Cluster Perception in the Distribution of Point Objects. *Cartographica* 34: 49–61.

Slocum, T., R. McMaster, F. Kessler, and H. Howard. 2005. *Thematic Cartography and Geographic Visualization.* Upper Saddle River, NJ: Prentice Hall.

GLOSSARY

apparent density the subjective response to the physical stimulus of the actual density of the objects in a visual field; density is underestimated in reading dot maps

center of gravity principle placement of the dot in its territorial domain in the way that best represents all the individual elements in the domain

coalescence merging of the dots in dense areas; the visual integrity of individual dots is lost

dot density mapping method of producing a map whose purpose is to communicate the spatial variability of density of discrete geographic data; also simply called dot mapping

dot polygon the area represented by each dot on a dot map; also referred to as the dot's territorial domain

dot value the numerical value represented by each dot in dot mapping; sometimes referred to as unit value

locational filter ancillary data layers or maps that influence placement of dots within the enumeration units

numerousness the subjective response to the physical stimulus of the actual number of objects in a visual field; not the result of the conscious activity of counting; number is underestimated in reading dot maps

spatial proxy the dot's function of representing geographical space

8 FROM POINT TO POINT: THE PROPORTIONAL SYMBOL MAP

CHAPTER PREVIEW Proportional or graduated quantitative point symbols have long been used in thematic mapping. The circle is one of the most popular forms, probably because of its compact size and ease of construction; many other forms are becoming popular. Cartographic and psychological research into how map readers estimate the size of symbols indicates that the physical elements of area and volume are underestimated during perception. This has led to apparent magnitude scaling methods. Research has concluded that underestimation can be controlled by careful legend design, whether or not absolute scaling is employed. A method that eliminates the need for any true direct scaling is to present symbols in range graded series, using point symbols of differential size, as in ordinal scaling. Attention to legend construction is paramount, since it plays a crucial role in symbol interpretation. Regardless of the method of scaling, proportional point symbols must be presented so that they appear as dominant, unambiguous figures in each map. ■

This chapter deals with one of the most flexible and popular thematic map techniques: *proportional point symbol mapping,* also called *graduated* or *variable* point symbol mapping. The conceptual basis for this technique is easily understood by most map readers. Its popularity among cartographers is likely to continue for some time but, like other processes in cartographic design, this method must be clearly understood if effective designs are to be created. Although much has been learned about this form of mapping in the last 40 years (notably in the area of symbol scaling), there is no single generally accepted design approach.

CONCEPTUAL BASIS FOR PROPORTIONAL POINT SYMBOL MAPPING

Proportional point symbol mapping is based on a fundamentally simple idea. The cartographer selects a symbol form (such as a circle, square, or triangle) and varies its size from place to place, in proportion to the quantities it represents. Shape and size are the two most important visual variables in this map type. Hue and/or value can also play an important role, particularly if more than one variable is shown on the map. Map readers can form a picture of the quantitative distribution by examining the pattern of differently sized symbols (see Figure 8.1). This simple approach can be spoiled by overloading the symbol form with too much information or by selecting scaling methods inappropriate to perceptual principles. The basic concepts of this form of quantitative thematic mapping are nonetheless easily grasped by most map readers, and most GIS and mapping software support this technique. We also find that, with a few notable exceptions, there is a greater flexibility in which data types can be used for this technique. This is no doubt why the method has reached its present level of popularity.

When to Select this Method—Data Suitability

There are two commonly accepted instances when the cartographer selects proportional point symbol mapping: when

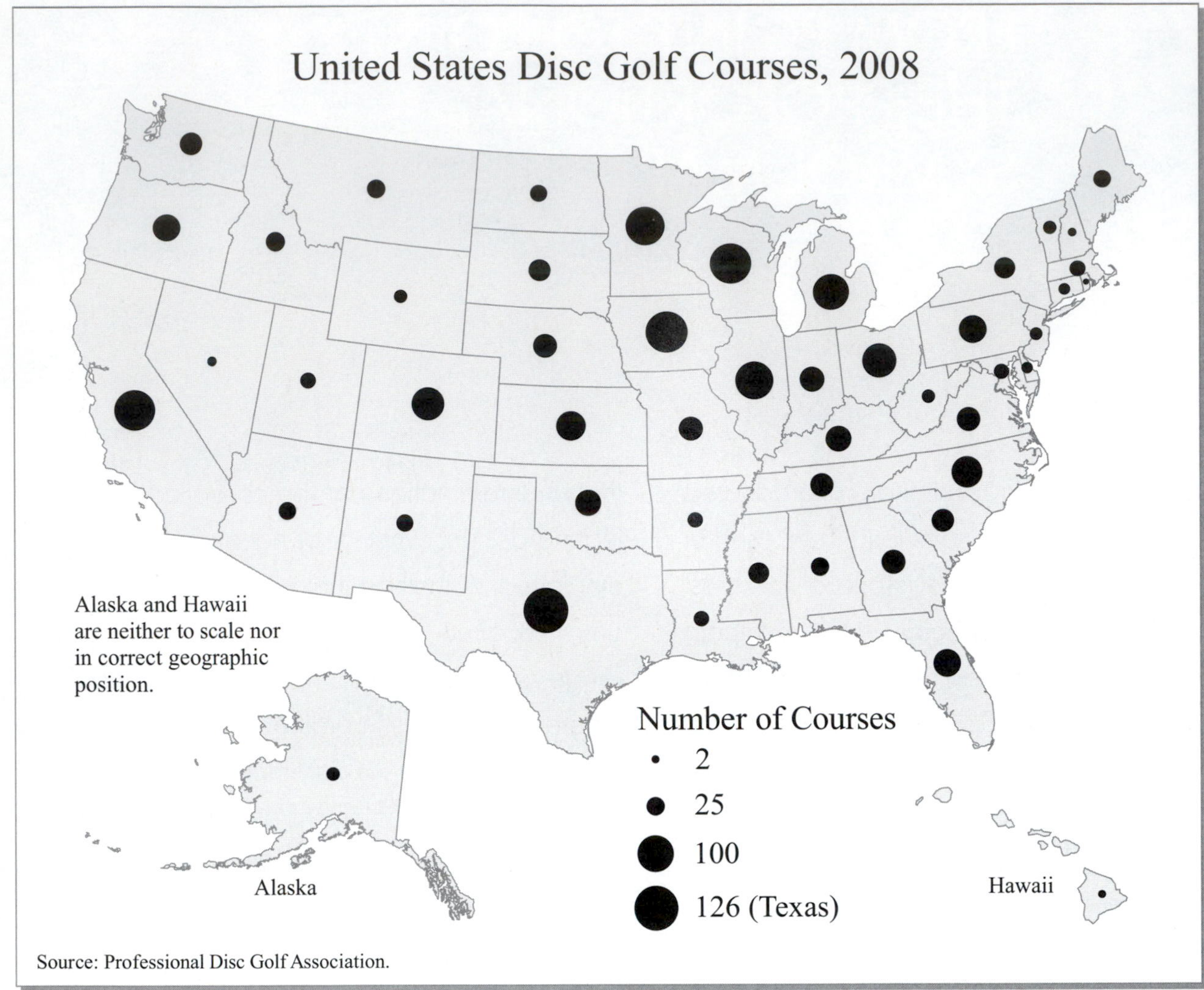

FIGURE 8.1 A PROPORTIONAL SYMBOL MAP.

data occur at points and when they are aggregated at points within areas (see Figure 8.2). Of course, data occurring at points are to be interpreted relative to map scale. Many attribute data magnitudes can be symbolized this way, including data such as totals, ratios (but see note on densities which follows), and proportions. This data flexibility is certainly a major factor in why proportional symbol mapping is such a popular technique.

However, there are three types of attribute data that should *not* be used with proportional symbols: interval level data, densities, and data with relatively small and unvarying attribute data ranges. The difficulty with interval data is that wherever there is no circle, a zero value is implied. For example, when mapping interval data such as average January temperatures in degrees Fahrenheit for U.S. cities, zero degrees is not the same thing as "no temperature," yet its value will not generate a symbol. Also, 60 degrees is not "twice as hot" as 30 degrees, yet the circle generated by such numbers would be twice its size. Densities (data normalized or derived by area) are normally symbolized by the choropleth technique. If the overall attribute data range is small and unvarying, the range of symbol sizes will also be small and reveal very little about the data (see Figure 8.3). The result will be a monotonous map that looks dull and undiscriminating (although it may be technically accurate). The intuitive appeal of the design can be a criterion of a map's appropriateness in such cases. With these important exceptions in mind, whenever the goal of the map is to show relative magnitudes of phenomena at specific locations, the proportional symbol form of mapping is appropriate.

The kinds of data that can be mapped are varied: total population; percentage of population; value added by manufacturing, retail sales, employment, any agricultural commodity, tonnage shipped at ports, and so on. Historically, proportional symbol maps have been used extensively to portray economic data, but magnitudes from the physical (such as earthquake magnitudes) and cultural worlds (such

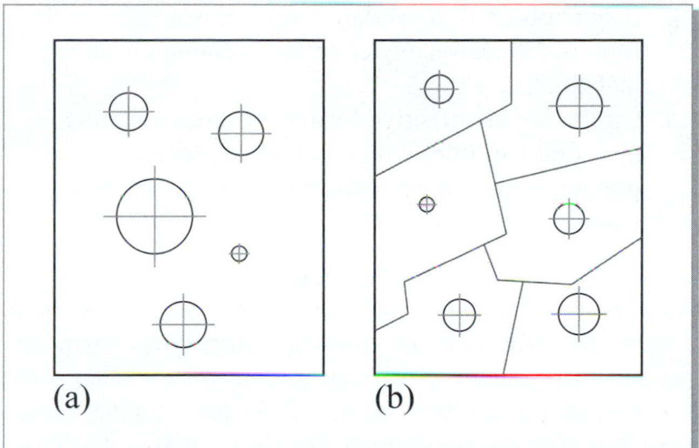

FIGURE 8.2 TWO KINDS OF LOCATION AND THE USE OF PROPORTIONAL SYMBOLS.

In (a), circles are placed at the points where quantities being mapped are located (cities, towns, stores, etc.). Area data assumed to be aggregated at points, as in (b), may also be represented by proportional point symbols. In each case, the symbols are used to show different quantities at distinct locations. The crosshairs in the circles illustrate positioning and are not mapped.

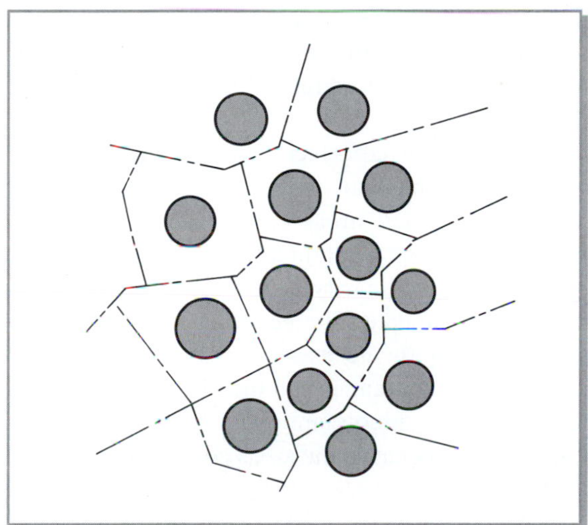

FIGURE 8.3 INAPPROPRIATE USE OF PROPORTIONAL SYMBOLS.

Maps in which unvarying geographic data are symbolized by proportional symbols are monotonous and reveal little. Designers should not use this mapping method with distributions of this kind.

as number of people speaking two or more languages) also may be illustrated.

A Brief History of Proportional Point Symbols

The earliest history of proportional point symbols on maps is really a history of the use of the proportional or graduated *circle;* other symbol forms (such as squares, triangles, spheres, and such) were not used until later. In 1801, William Playfair used graduated circles to depict statistical data, which were often compiled by geographical areas such as cities, states, or countries. However, even though the circles were scaled by data such as population or area, they were positioned along straight lines *within charts* rather than by location on maps (Tufte 2001; Spence 2006). Interestingly, Playfair used circle *areas* rather than scaling them to diameters or circumferences, a scaling practice still used today. In addition, the circle continues to be the form of proportional symbol most often used, although other choices are possible.

It was apparently not until 1837 that proportional circles were employed on maps. Harness used them to map city populations (Robinson 1955). In 1851, they were used by August Peterman in the mapping of city populations in the British Isles (Flannery 1956). They were also used in France by Minard by 1859 to portray port tonnages (Robinson 1967, 1982).

From these beginnings, the use of proportional symbols on thematic maps gradually increased throughout the twentieth century. Today, they are found in journals, magazines, atlases (for example, *Goode's World Atlas*), and in textbooks on mapping and GIS (for example, Chrisman 2002; Chang 2008), and can be created in most GIS, mapping, and artistic drawing software.

Although the circle is the most common form, some cartographers use bar graphs, line graphs, or other linear graphs as proportional symbols. The history of this form of point symbolization is unclear. We encourage caution with this method because the spatial distribution of phenomena becomes muddled, and in a few older software packages, the routines for making these graphs have been poorly implemented. It is virtually impossible to see relative magnitudes at points when, for example, a bar may stretch nearly halfway across the map. Proportional point symbol mapping requires compact geometrical symbols depicted at points.

A Variety of Symbol Choices

The steady increase of available symbols that are easily imported into GIS, mapping, and artistic drawing software to create proportional symbols has meant a corresponding increase in the choices of symbol forms available to the map designer. We can categorize these symbol types into two-dimensional geometric, three-dimensional geometric, and pictorial symbols.

Two-Dimensional Geometric Symbols

Two-dimensional symbols, such as circles, squares, and triangles, are the most common forms of proportional symbols, with the circle being the dominant form. In each case, *area* is the geometric characteristic that is customarily scaled to geographical magnitudes (see Figure 8.4). Circles have no

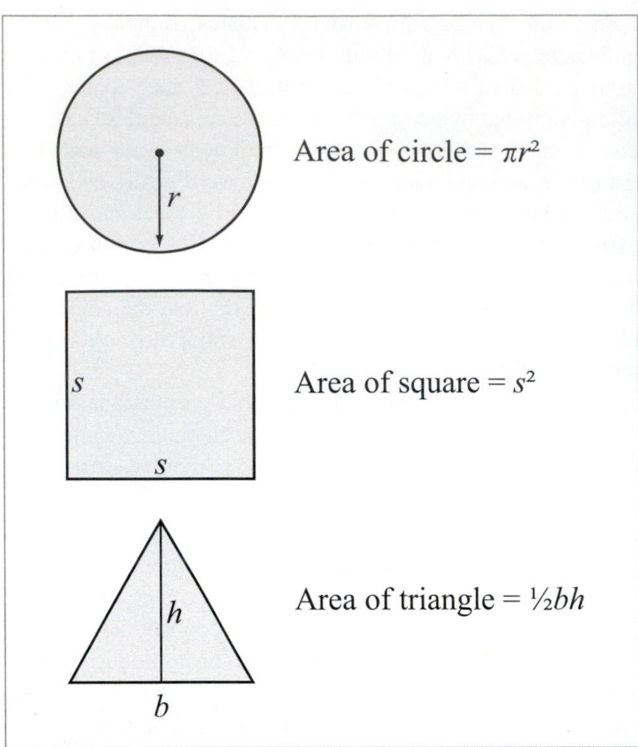

FIGURE 8.4 THREE TWO-DIMENSIONAL SYMBOLS (CIRCLE, SQUARE, AND TRIANGLE) AND THEIR AREA FORMULAS.

doubt become more common because of several advantages they have over the other symbols:

1. Their geometric form is compact.
2. Historically, circle scaling was less difficult than for other geometrical forms (with the exception of the square), because the square root of the radius can be used. Today, computers have eliminated any need for manual calculation of symbol sizes and areas.
3. Circles are more *visually stable* than other symbol forms and thus cause little eye wandering.
4. If symbols need to overlap (within reason), the circle form is still reasonably effective in communicating magnitudes.
5. Circles are an effective form for accommodating a second variable, which can be represented in a pie graph or with changes in the circle's hue or value.

Squares, along with pictorial symbols, are used increasingly because computer software can easily generate them. Before the widespread use of computer-generated symbols, squares were relatively difficult to construct, so they were not used as extensively as circles. The square is also considered less visually stable than the circle. It is not without advantages, however—these will be discussed later in the context of scaling. Triangles have disadvantages similar to those of squares; they are used even less.

Three-Dimensional Geometric Symbols

Cartographers and geographers have experimented with point symbols of three-dimensional appearance, including spheres, cubes, and other geometrical volumes (see Figure 8.5). The use of **three-dimensional point symbols** can result in very pleasing and visually attractive maps. However, most map readers cannot correctly gauge the scaled values of these maps. Potential solutions to this shortcoming will be discussed more fully on the following pages.

In addition to the visual effects, another important advantage of three-dimensional symbols lies in their scaling. With proportional circles, areas are scaled to the square root of the data (Table 8.1), but three-dimensional symbols, including the sphere, ordinarily have their volumes scaled to the cube root of the data. The net result is that in three-dimensional mapping, the necessary range of symbol sizes is reduced. Greater data ranges can thus be handled on the map when spherical symbols are used. Less crowding of symbols also results, because the areas covered by symbols are reduced.

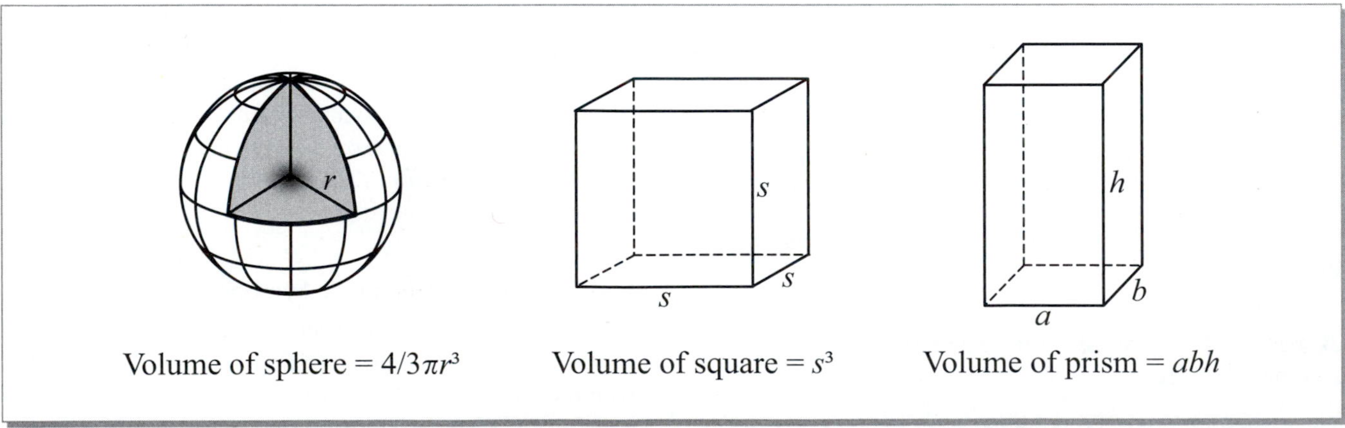

FIGURE 8.5 THREE THREE-DIMENSIONAL SYMBOLS (SPHERE, CUBE, AND RECTANGULAR PRISM) AND THEIR VOLUME FORMULAS.

TABLE 8.1 DIFFERENCES IN MAPPED AREAS OF CIRCLES AND SPHERES

Symbol	Formula	Quantity Represented: Small	Quantity Represented: Large	Radii: Small Symbol	Radii: Large Symbol	Ratio of Large Radius to Small Radius	Area on Map Covered by Small Symbol (sq units)	Area on Map Covered by Large Symbol (sq units)	Ratio of Area of Large Symbol to Area of Small Symbol
Circle	πr^2	100	1,000	.1	0.3162	3.162	0.0628	0.6278	9.99
Sphere	$4/3\pi r^3$	100	1,000	.1	0.2154	2.154	0.0628	0.2913	4.6

The difficulty is with reader perceptual scaling. Most map readers cannot correctly judge visually the relative sizes of quantitative three-dimensional symbols. We respond visually to the areas covered on the map by three-dimensional-looking symbols and do not "see" relative volumes. Cartographers have not fully explored the use of three-dimensional-*looking* symbols, especially spheres, scaled to the areas they occupy on the map sheet. As we will discuss in the section entitled "Proportional Symbol Scaling," one alternative may be to employ three-dimensional-looking symbols, but use *range graded* symbols, rather than having each symbol scaled to a unique value. Note that current software does not create true three-dimensional proportional symbols; they must be manually generated.

Pictorial Symbols

Pictorial symbols, also called pictographic, mimetic, or replicative symbols, are increasing in use for proportional symbol maps. This is due both to widespread availability of digital artwork (that is, clipart), and the ease with which the artwork can be imported into GIS and other software packages. Maps so produced are usually attention grabbing for the map reader and introduce an element of fun for the map designer. In the latter case, the novice designer sees the relatively mundane circle and square compared with a whole palette of interesting symbol choices (see Figure 8.6). Proportional cars, trees, or the outline of a human, can represent automobile production, timber harvested, or population, respectively. If the designer cannot find a symbol to their liking, they can design and incorporate their own symbol (Brewer and Campbell 1998). We have even seen proportional waterfalls representing average volume of water flow at the various waterfall locations.

Pictorial symbols, a long-time favorite of the graphic artist, have a couple of potential drawbacks. The more irregular the shape of the symbol, the harder it will be for the map reader to estimate magnitudes, or compare quantities from place to place. This is also true if highly irregular symbols overlap, and too many irregular symbols give the map a cluttered appearance. Mapping a second variable is difficult and usually not advised, with the possible exception of changing color or hue of the object. Because of these potential drawbacks, we usually recommend using a more conservative and studied form such as the circle. Of course, the map designer can choose pictorial symbols that have simpler geometric dimensions, such as a proportional coin (circle) or paper currency (rectangle). As with 3-D geometric symbols, range grading pictorial symbols may be the best solution for symbols that have a highly irregular shape.

It is not uncommon for two (or more) proportional forms to illustrate two (or more) distributions on the same map. For example, power plants could be symbolized with different shapes representing various production methods (for example, nuclear, hydroelectric, and so on) and their symbols scaled for amount of power generated. However, be aware that too much complexity can undermine the communicative effort. If the map becomes too cluttered or complex, the map designer is encouraged to consider developing two (or more) separate maps. Further discussion of bivariate and multivariate maps occurs later in the chapter.

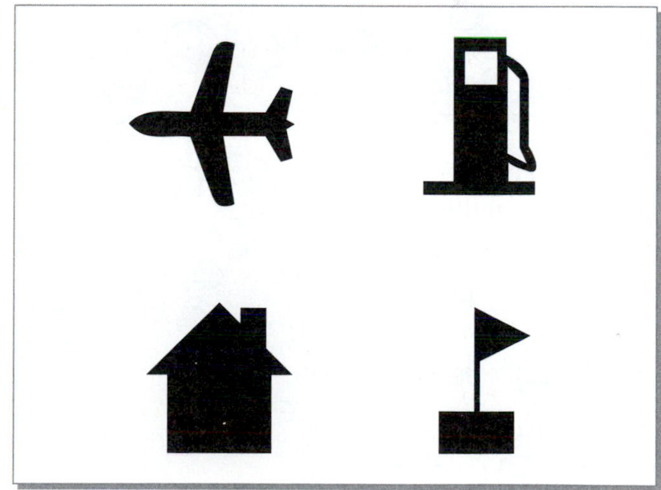

FIGURE 8.6 PICTORIAL SYMBOLS.
Pictorial symbols are attention grabbing and fun to use, but the more irregular the shape of the symbol, the harder it will be for the map reader to estimate magnitude, or compare quantities from place to place.

PROPORTIONAL SYMBOL SCALING

The scaling technique chosen is one of the most important aspects of proportional symbol mapping. Scaling circles, in particular, has been the subject of much study and debate. There are three common scaling techniques: **absolute scaling, apparent magnitude scaling,** and **range grading.** Absolute and apparent magnitude scaling are closely related techniques that are most synonymous with the term **proportional symbol** (see Figure 8.7a). Maps that employ range grading are constructed differently, take on a different appearance, and have communication goals that are different from those of maps with proportional symbol (absolute or apparent magnitude) scaling (see Figure 8.7b). Many GIS and mapping software programs use the term **graduated symbol** when range grading is applied. The following are working definitions for these important terms:

1. *Absolute Scaling*. In this method, symbols are scaled proportionally to their data values, and are therefore also in proportion to each other (see Figure 8.7a). For example, a symbol that is twice as large in area as another symbol will represent a value that is twice the data amount as the other symbol. To some extent this method is analogous to an *unclassed choropleth map* (described in Chapter 6), in that each observation in the attribute data has a unique symbol size assigned to it (as opposed to a unique hue, value, or pattern in the unclassed choropleth map). Each symbol will be unique in size, unless there is a repeated value in the attribute data source. With this type of scaling, magnitude estimation of the symbols is the communication goal for the map reader, based on the visual *anchors* (the symbols in the legend). This type of scaling is most conventionally associated with the term "proportional symbol."
2. *Apparent magnitude scaling*. This method is also known as perceptual scaling, psychological scaling, or Flannery compensation. Map readers tend to underestimate areas and volumes of objects, and this underestimation gets worse as symbol sizes become larger (such as, with larger data values). Apparent magnitude scaling is a variation of absolute scaling as correction factors are applied to compensate for the underestimation of a symbol's area or volume. In every other aspect, the resulting map is the same as a map that has been

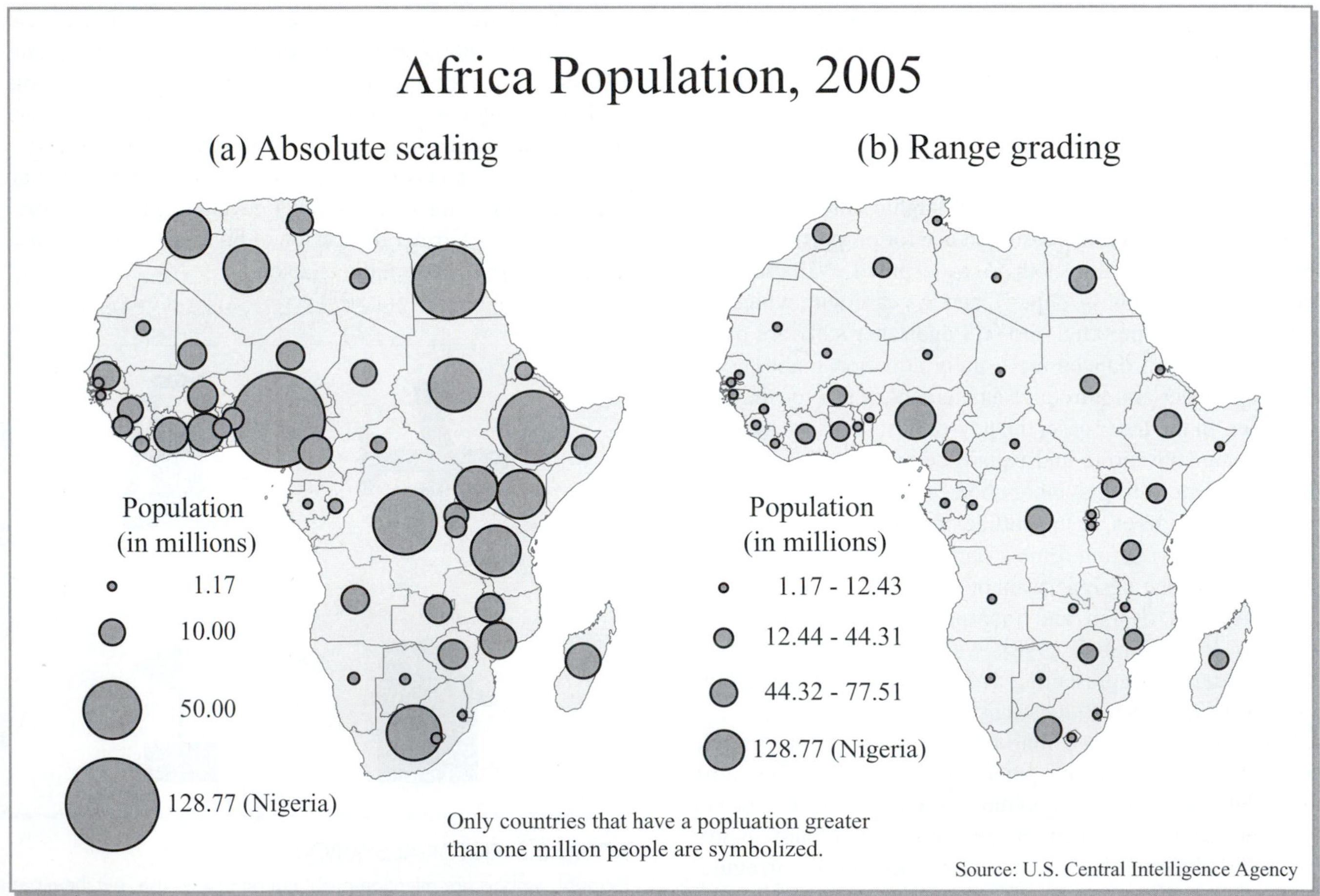

FIGURE 8.7 TWO MAPS FROM THE SAME DATA. SYMBOLS ARE SCALED VIA ABSOLUTE SCALING (a), AND RANGE GRADING (b). The same symbol size is used for the smallest data value (a) or range (b). A symbol outline of some kind is used if the symbols overlap (a).

created using absolute scaling, including the communication goal of magnitude estimation.

3. *Range grading.* In this scaling method, a symbol represents a range of data values (see Figure 8.7b). A certain number of ranges or classes are chosen (for example, four or five), and the data are then classified according to one of the classification schemes discussed in Chapter 5. The symbols are not usually in proportion to the attribute values, although larger symbols will imply "more" of whatever is being mapped, in the same manner that a classed choropleth map has darker or more saturated hues that signify higher values. Symbol size discrimination, then, is the primary goal in creating maps that are range graded. Many GIS and mapping software often use the term "graduated symbol" map when range grading is employed.

With these scaling definitions in mind, we may now more formally introduce several ideas concerning how map readers perceive quantitative symbols. For several decades, psychologists, cartographers, and graphical method statisticians have researched the mechanisms whereby map readers perceptually scale quantitative symbols with particular regard to the circle form and the absolute and apparent magnitude scaling techniques. The researchers study response patterns to stimuli—the symbols and their characteristics such as length, area, and volume. The research attempts to document carefully how stimulus and response interact, usually by way of mathematical description. These studies are usually referred to as *psychophysical investigations.*

Absolute and Apparent Magnitude Scaling: Psychophysical Examination of Quantitative Thematic Map Symbols

The general psychophysical relationship between visual stimulus and response can be described by this formula (Chang 1980):

$$R = K \cdot S^n$$

where R is response, S is stimulus, n is an exponent describing the mathematical function between S and R, and K is a constant of proportionality defined for a particular investigation. This relationship is sometimes referred to as the **psychophysical power law,** because S is raised to a power, n.

The results of the research efforts suggest overall that most people do not (or cannot) respond to the geometric properties of quantitative symbols in a linear fashion (see Figure 8.8). A linear response pattern means that if 10 units of stimulus are viewed, 10 units of response will be measured, and this one-to-one relationship will hold throughout the range of all stimuli. It has been found, overall, that length is correctly perceived, but area and volume are not. The geometrical property of area is increasingly underestimated as higher magnitudes of stimulus are judged. Volume is underestimated to a greater degree than area.

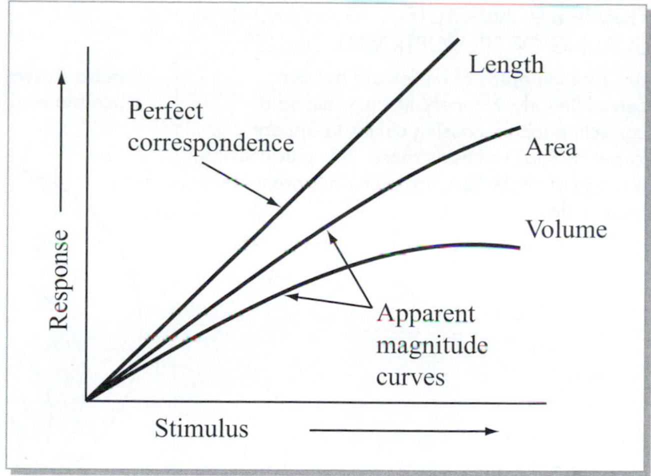

FIGURE 8.8 APPARENT MAGNITUDE CURVES FOR LENGTH, AREA, AND VOLUME.

In the visual psychophysical world, the area and volume of geometric figures are underestimated. Length is usually correctly perceived. Underestimation means that for every unit of physical stimulus perceived, less than unity is reported in response.

These findings have had considerable impact on map designers. The initial results, especially in the case of proportional circles, caused quick adoption of methods to adjust symbol sizes to compensate for the underestimation. This led to perceptually based *apparent magnitude scaling,* also called perceptual or psychological scaling (see Figure 8.9). The pioneering work of James Flannery must be cited in this context; his contributions to scaling adjustments became the standard for nearly 35 years (Flannery 1956).

Psychophysical studies have had a different impact on designers. They suggest that apparent magnitude scaling might be abandoned in favor of *absolute scaling,* but with more careful attention given to legend design (Chang 1980).

Absolute Scaling with Circles

Absolute scaling is simply the direct proportional scaling of magnitudes to the symbol's area. For example, if we have two values, 6,400 and 1,600, that are to be symbolized and scaled by proportional circles, we must first set up a proportion, remembering that the area of a circle is πr^2 (r = radius of the circle), thus:

$$\frac{\pi r_1^2}{\pi r_2^2} = \frac{\textbf{data value 1}}{\textbf{data value 2}}$$

This reduces to

$$\frac{r_1^2}{r_2^2} = \frac{\textbf{data value 1}}{\textbf{data value 2}}$$

and then, taking the square root of both sides of the equation, we get:

$$\frac{r_1}{r_2} = \frac{\sqrt{\textbf{data value 1}}}{\sqrt{\textbf{data value 2}}}$$

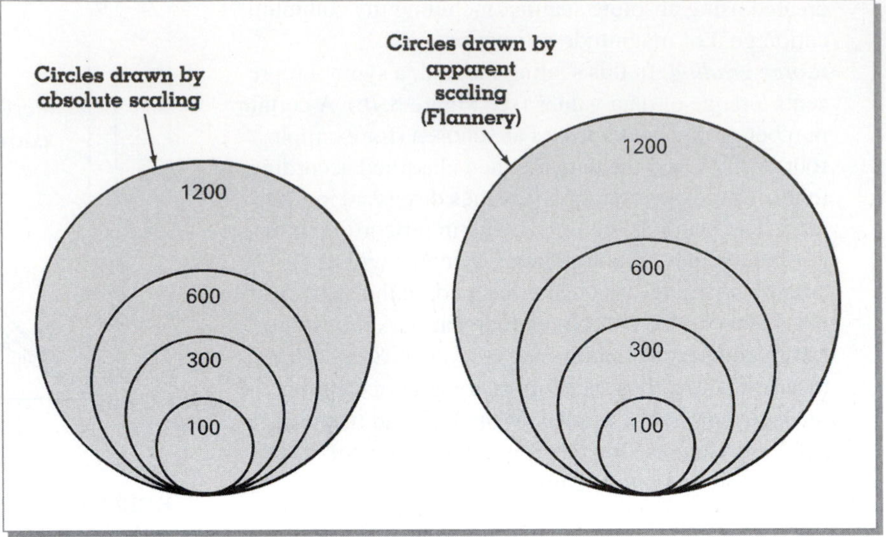

FIGURE 8.9 ABSOLUTE AND APPARENT SCALING OF PROPORTIONAL CIRCLES.
Because the areas of circles are not perceived linearly, Flannery has introduced a correction factor, causing circles to appear larger as their values increase. This produces a range of circle sizes scaled by apparent magnitude.

For this example, we next have

$$\frac{r_1}{r_2} = \frac{\sqrt{6,400}}{\sqrt{1,600}}$$

and then,

$$\frac{r_1}{r_2} = \frac{80}{40} = \frac{2}{1}$$

or the radius of data value 1 is twice that of data value 2.

In practice a radius is selected to represent the smallest data value that yields a circle that looks good on the map; then a radius is calculated for the largest data value to determine if the circle size is appropriate and, if so, all intermediate circle sizes for the remaining values are calculated. If not, the process is begun again and a final set of circle sizes is computed and drawn.

For example, suppose you had a range of data values with 10 being the smallest and 150 the largest, and decided to render the smallest circle with a radius of .1 inch:

$$\frac{\pi r_1^2}{\pi r_2^2} = \frac{10}{150}$$

$$\frac{r_1^2}{r_2^2} = \frac{10}{150}$$

$$\frac{(.1)^2}{(r_2)^2} = \frac{10}{150}$$

$$\frac{.1}{r_2} = \frac{\sqrt{10}}{\sqrt{150}}$$

$$\frac{.1}{r_2} = \frac{3.16}{12.25}$$

$$3.16\, r_2 = 1.225$$

$$r_2 = .387$$

The resulting 0.387 is the size of the largest circle representing the data value 150. If you like the computed size of the largest circle size, then you will compute the sizes of all intermediate circles. If not, you must adjust the smallest circle size upward or downward and recalculate the rest of the circle sizes. This process is repeated until you arrive at a circle-size set that looks good on your map.

Not too many years ago, textbooks on cartography included graphic devices to compute circle sizes. Today circle sizes are either scaled using GIS or mapping software, or are calculated using spreadsheet programs if the cartographer is creating the symbol sizes using artistic drawing software.

While most graphics programs allow the designer to work with their unit of choice, including inches, many GIS and computer mapping software packages use point sizes for circle scaling. The concept of point size originated with print typography, and has migrated to modern fonts that will be detailed in Chapter 13. One point is approximately 0.013889 inches (1/72 of an inch), and one inch is approximately 72 points. Therefore, if we wanted to convert the above calculations into point sizes:

.1 inch · 72 points per inch = Point Size: 7.2

.387 inch · 72 points per inch = Point Size: 27.864

You will have to round the results to the number of decimal digits that your software supports. This direct conversion does not work well for other geometric or pictorial forms, as these forms do not exhaust the entire space designated by the point size. For example, a 72-point object (for example, text or other miscellaneous symbols) does not take up the entire one-inch space allocated for the object. Even worse for size estimation, the actual amount varies depending on the type of object (geometric shape or letter) between objects. To see if the circle's diameter is both 72 points and one inch in your particular software package, you can usually test this by drawing a 72-point circle in the software's layout or drawing page. By obtaining the object's properties (usually by right clicking the object in some programs; many programs have rulers or a background grid in inches so that you can estimate the size), you can tell very quickly if the circle's diameter is one inch.

Apparent Magnitude Scaling with Circles

Scaling for circles has received more attention from psychologists and cartographers than scaling for any other form of symbol, because of the circle's popularity as a proportional symbol. When Flannery made clear that circle sizes need to be adjusted to compensate for underestimation, a standard was adopted based on his findings.

The psychophysical relationship, expressed as $R = K \cdot S^n$, is usually transformed into logarithmic form:

$$\log R = \log a + b \cdot \log S$$

This is an equation for a straight line where a is the constant K (and the intercept on the R-axis), b is equivalent to the slope of the line (and the n of the original power equation), and S and R are stimulus and response, as before. By regression methods, a line will then be fitted to the experimental data, and a value for n is produced.

In the original equation $R = S^n$ (omitting K for simplicity), if a perfect relationship exists between R and S, then the exponent n would be 1.0. Flannery's experimental data yielded a value of $n = 0.8747$. For example, if a circle of area value 2 is judged experimentally, it would be seen as having an apparent area of

$$2^{.8747} = 1.833$$

This underestimation of circle size is the rationale for adjusting such a circle's size upward. Flannery arrived at his *adjustment* factor by transposing the equations. In logarithmic terms, the square root of number N is $(\log n)(.5)$. To compensate for underestimation, the symbols are enlarged somewhat by using 0.5716 (known as Flannery's constant) instead of .5. This value is tied directly to the value of 0.8747 for n.

The procedure to scale circles this way includes the following steps (the method used by Flannery):

1. Obtain logarithms of data values.
2. Multiply these log values by 0.5716 (or rounded to .57).
3. Determine the antilogarithms of each.
4. Scale all values to the new set of antilogarithms in ordinary fashion.

With the availability of computer spreadsheets that easily raise any number to any power, it is unnecessary to use logarithms in these calculations. See Table 8.2 for an example of circles scaled by absolute methods and those by apparent-scaling methods. In this table, the reason for the calculated difference between the two final circle-size outcomes is that for linear scaling, the exponent used is .5, but for the apparent magnitude scaling, it is 0.5716.

In the psychophysical power law, n expresses how the response can be described relative to the stimulus. We might assume that n is a constant for circle judgment tests, but it is not. Research indicates that n fluctuates because of these factors: testing method, instructions, standard (or anchor) stimulus used, and the range of stimuli used in the experiments (Chang 1977; Cox 1976; Flannery 1971). These n values range from 0.58 to 1.20. It is now evident that no one correction factor developed for an apparent magnitude scale is the answer. This is especially true in light of parallel findings which show that nearly correct responses can be achieved with carefully designed legends.

Available studies also indicate that apparent magnitude scaling is not that much more efficient than absolute scaling if the range of required circle-size judgments is 10 or less (Griffin 1985). One final caution, too, is worthy of note. Some map readers simply cannot make proportional symbol judgments, even if apparent scaling is used. Map designers will need to weigh all these factors when making decisions regarding this form of quantitative map.

It must be pointed out, finally, as already shown in Figure 8.9, that depending on the magnitude of the data,

TABLE 8.2 WORKED EXAMPLE OF LINEAR AND APPARENT MAGNITUDE SCALING OF CIRCLE SIZES

Data Value (n)	Linear Scaling, Exponent = .500	Apparent Magnitude Scaling, Exponent = .5716	Linear Problem	Apparent Scaling Problem
20	4.47	5.54	$\frac{\pi r_1^2}{\pi r_2^2} = \frac{20}{200}$	$\frac{\pi r_1^2}{\pi r_2^2} = \frac{20}{200}$
30	5.47	6.99	$\frac{r_1}{r_2} = \frac{\sqrt{20}}{\sqrt{200}}$	$\frac{r_1}{r_2} = \frac{\sqrt{20}}{\sqrt{200}}$
40	6.62	8.24	$\frac{(.1)}{r_2} = \frac{4.47}{14.141}$	$\frac{(.1)}{r_2} = \frac{5.14}{20.488}$
50	7.06	9.30	$4.47(r_2) = (.1)(14.141)$	$5.54(r_2) = (.1)(20.488)$
.			$4.47(r_2) = 1.414$	$4.54(r_2) = 2.048$
.			$r_2 = .316$	$r_2 = .369$
.			$d_2 = .63$	$d_2 = .738$
200	14.14	20.66		

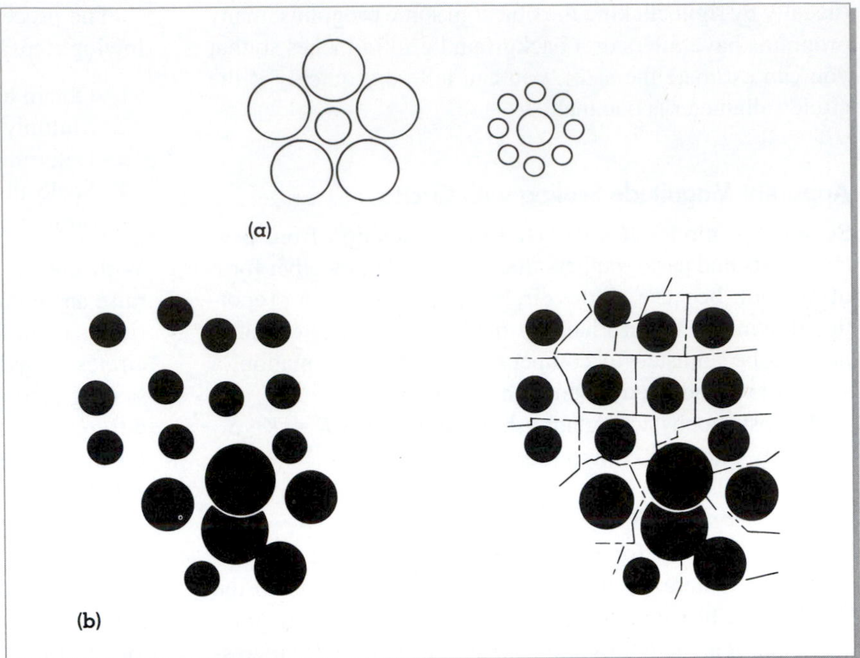

FIGURE 8.10 THE EFFECT OF NEIGHBORING SYMBOLS ON THE JUDGMENT OF CIRCLE SIZE.

In (a), the inner circles in the right and left drawing are of the same size. Because of the contrast effects produced by neighbors, the inner circle on the right appears larger than its counterpart on the left. In (b), the provision of internal boundaries reduces the contrast effect of neighboring symbols. Unfortunately, cartographic designers have little control over variables of this kind.

selecting the apparent magnitude scaling method requires greater map space, and this is often a critical factor in design, as space is usually limited.

Perception of Circles Among Circles. Most cartographic research on the judgment of circle size has measured rather abstractly how people perceive circle size in non-map settings. Moreover, and more troubling, the influences of neighboring circles on size judgment have been largely overlooked. However, at least one study has investigated this phenomenon and has produced these interesting results (Gilmartin 1981).

1. When a circle is seen among circles that are smaller, it is seen on the average about 13 percent larger than it would ordinarily be seen if judged among circles that are larger (see Figure 8.10a).
2. Circles surrounded by circles of the same size were judged both high and low relative to the surrounding circles.
3. The effect of surrounding circles can be reduced if internal borders are used on the map (see Figure 8.10b).

Thus circle judgment is affected by the size of neighboring circles on the map. At present no generally applicable solution is available to control these undesirable effects. If repeated design alternatives show that this problem cannot be solved, perhaps a different form of map should be constructed. Designers need, at least, to know that this problem can arise so that troublesome areas on the map can be treated with greater care.

One very important aspect of proportional symbol maps is their ability to show distribution and location in addition to magnitude (see Figure 8.11). This feature is extremely effective when symbol sizes are carefully selected for the chosen base map. Although overlapping symbols make it

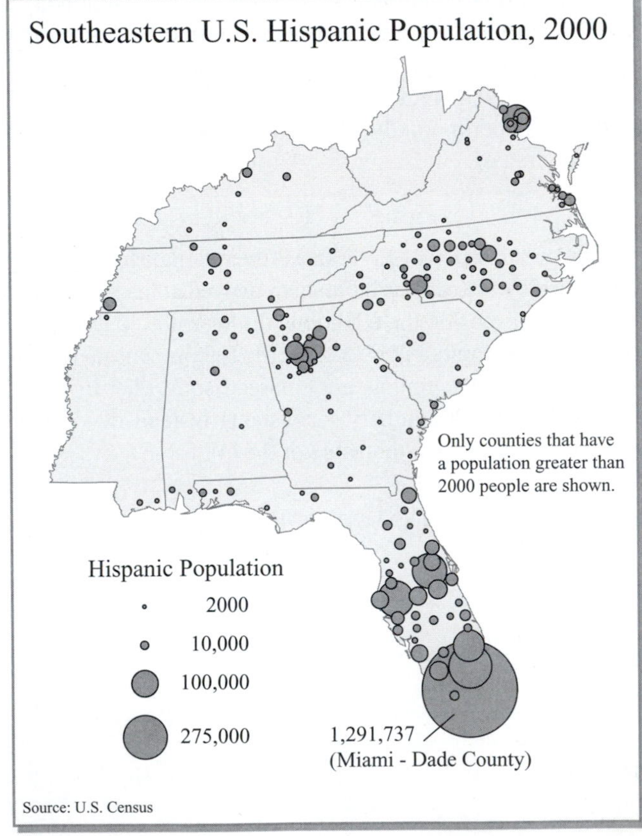

FIGURE 8.11 PROPORTIONAL SYMBOL MAPS SHOW SPATIAL DISTRIBUTION.

In this map the variability of the spatial distribution (in addition to magnitude) of Hispanic population in the southeastern United States is emphasized by the location of the proportional symbols, but overlapping symbols make it difficult to estimate symbol sizes. See text for discussion.

difficult to estimate individual symbol sizes, a sense of visual cohesiveness results, and this may lead to memorability, certainly a goal in cartographic communication (Peterson 1985; Slocum 1984). This dilemma, to show overall pattern versus specific symbol quantity, usually can be dealt with by careful examination of map purpose. If the two cannot be reconciled, either by finding a new scaling scheme, or by dealing with extreme outliers in the data, then another quantitative map type should be considered.

The Square Symbol

Graduated squares are used in quantitative point symbol mapping in much the same fashion as circles. The areas of the squares are scaled to data amounts yielding simple proportional relationships between the square roots of data and the sides of the squares:

$$\frac{s_1^2}{s_2^2} = \frac{\text{data value 1}}{\text{data value 2}}$$

$$\frac{s_1}{s_2} = \frac{\sqrt{\text{data value 1}}}{\sqrt{\text{data value 2}}}$$

Squares have never received quite the attention by researchers that circles have, probably because they are not as commonly used, and prior to computer production, they were more difficult to produce. Visually, they introduce a rectangularity to the map's design which is often difficult to coordinate with other map elements (see Figure 8.12). In addition, the square is often used for qualitative symbols for houses and buildings, so especially at large scales, squares are often avoided to prevent symbology confusion. Nonetheless, they can show geographical distribution—clustering, dispersion, or regularity—as well as circles.

One important consideration for the selection of squares is that their proportional areas are nearly perfectly perceived. In one study, the exponent value n in the psychophysical power law was experimentally determined to be 0.93 (Crawford 1973). Other psychological and cartographic studies suggest that the areas of squares are more accurately judged than circle areas. This does not mean that circles should be abandoned in favor of squares. With appropriately designed legends, both symbols can communicate well. The selection should be made in concert with other design elements and mapping goals.

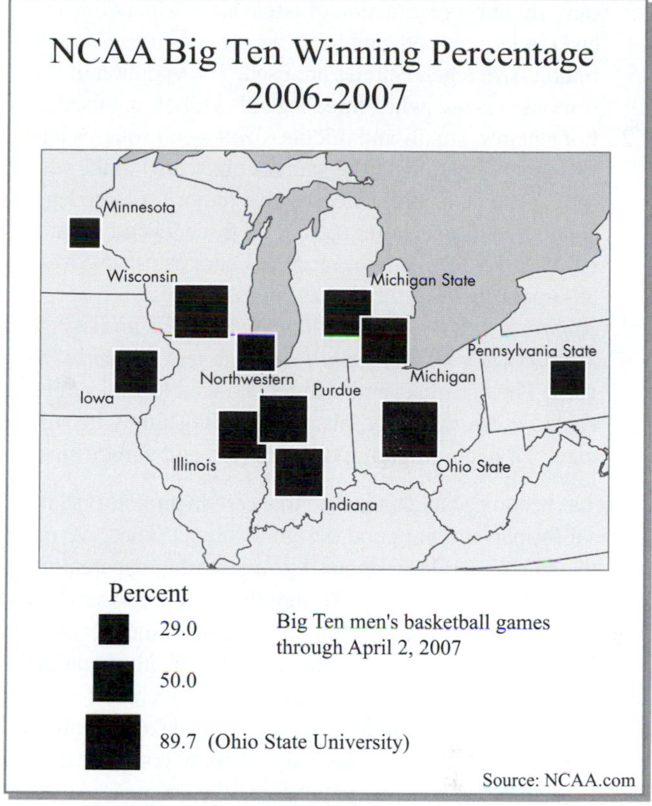

FIGURE 8.12 PROPORTIONAL SYMBOL MAP USING SQUARES.

Some cartographic designers dislike using squares because they introduce "rectangularity" on the map. However, squares are perceived more accurately than circles.

Absolute and Apparent Magnitude Scaling Design Implications

With some of the latest releases of GIS and mapping software, we are starting to see functions that allow for both absolute and apparent magnitude scaling of symbols. The map designer should be aware that apparent magnitude scaling methods (the introduction of corrective factors into absolute scales) are based on the "average" map reader who does not exist outside mathematical tables. Moreover, the landmark corrective factor of Flannery, although quite important in the early assessment of symbol scaling, is not applicable in all cases because it was developed from a single psychophysical study with one experimentally derived exponent. Subsequent investigations have demonstrated that the exponent itself varies depending on test conditions.

Legend designs have also been examined closely. This avenue of research is based on the theory of **adaptation level.** This theory holds that perceived judgments are based on an adaptation level (a neutral reference point) and that the perceptual anchors from which judgments are made affect this point (Cox 1976). Two **anchor effects** have been shown to exist: *contrast* and *assimilation.* The contrast effect is displacement of perceptual judgment away from the anchor; by assimilation, perceptual judgments are displaced toward the anchor. The size of the anchor has an effect on how perceptual judgments of the symbols on the map are made. In the case of proportional circles and squares, judgments are displaced toward the anchor, whether it is a large or a small anchor. This is evidence that assimilation is taking place.

One study has produced important findings regarding anchor effects in proportional circles and squares. There are important design implications (Cox 1976):

1. When graduated circles are scaled according to the apparent value (after Flannery), there is greater underestimation of symbol sizes when a single small legend value is used, less overestimation when a single large symbol is used,

only slight overestimation of circle sizes when *three* legend circles are used, and fairly accurate estimates when middle-size legend circles are used. The variation in responses is less when three legend symbols are used.

2. For squares, small- and middle-size legend squares lead to underestimations. (The squares are scaled to the square root of the data, with no corrective factor.) A large legend square yields overestimations, and three legend squares (small, middle, and large) lead to better estimates overall—principally at the higher end of the scale.
3. *Apparent scaling does not compensate for underestimation when a single small or large legend symbol was used.* The use of several differently sized legend symbols, scaled conventionally (absolute scaling), may be sufficient for accurate judgment of circle and square size.

It has become abundantly clear that certain proportional point symbol formats are not good design solutions. Thus, we make the following recommendation: If you are using apparent magnitude *or* absolute scaling, make sure that you use several circle sizes that represent well the entire data range in the legend. The smallest, largest, and intermediate symbols will all appear in the legend. Never use only one legend symbol (anchor) for either case. We also recommend that you do not scale any symbols to their volumes (cube roots of the data). A more complete discussion of legend design is provided later in the chapter.

Range Grading

Range grading is achieved in a manner similar to the development of classes for choropleth maps. The attribute data is divided into groups; each group is represented by a symbol that is clearly distinguishable from other symbols in the series (refer to Figure 8.7b). As with classification in the choropleth map, the presentation of circles is now on an ordinal scale from the classification of continuous (ratios or totals) attribute data, and therefore the symbols are usually not proportional to the data values. In this scaling method, symbol-size discrimination is the design goal, rather than magnitude estimation. In ArcGIS, the *graduated symbol* option means using ordinally scaled range graded symbols.

Meihoefer (1969) first suggested range grading of proportional symbols in the 1960s. Since then, use of range grading has been employed by some cartographers as an alternative to proportional (absolute and perceptual) symbol scaling. Research by Heino (1995) suggests that range grading will lessen mistakes made by the map reader. Unlike with proportional symbol maps, map readers can easily make a connection between the legend and the map symbols using range grading. Some cartographers also like this technique for attribute data with extremely high ranges and/or wildly varying enumeration unit spacing (which often results in maps with excessive symbol overlap). Therefore, getting symbols to "fit" on the map is easier with range grading.

Other cartographers disagree with range grading, arguing that the process introduces *too much* generalization into the map. Map readers *expect* a symbol that is twice the size of another symbol to represent data accordingly, but range grading *generally* does not maintain proportionality (see the alternative in "Proportional Midpoint Range Grading"). Another related concern is that attribute data with a low range (which we suggested earlier should not be used in proportional symbol maps) could be classified and symbolized to imply differences that are simply not there. Perhaps for both better *and* worse, range grading of symbols is now an accepted part of this thematic mapping technique, and is a default setting in some (but not all) GIS and other mapping packages that support this technique.

Since the goal in range grading is symbol discrimination, it is important to find symbol sizes that will not be mistaken for one another. Miehoefer (1969) devised a set of 10 circle sizes that were easily and consistently discriminated by his subjects and were applicable to most small-scale thematic maps (see Figure 8.13). An alternative array of circles useful for a range graded series is presented in Figure 8.14. After repeated use of the Meihoefer series, Dent determined that the circles with radii of 0.38 and 0.44 inches (9.77 and 11.28 mm), and to some degree the one with a radius of 0.54 inches (13.82 mm), are not that easily distinguished on actual small-scale maps. The circles presented in Figure 8.14 have not been tested rigorously in an experimental setting but have proved useful in practice. You might wish to explore using them, utilizing the given values for the radius, diameter, area, or point sizes as is appropriate in your particular software.

The number of circles (or other symbols) is dependent on the number of classes. We generally recommend from three to six classes, with four or five classes being typical choices. As with choropleth mapping, the choice of classification technique will determine the intervals with which each class is symbolized. (See Chapter 5 for the classification descriptions and Chapter 6 for their application to choropleth maps.) All of the qualities of each classification technique, along with the pros and cons for each type, are applicable to classification for range grading symbols. The classification scheme should be chosen based on the nature of the attribute data set and the map's purpose. Just as in determining choropleth class intervals, the goal is usually to devise classes that have the greatest degree of internal homogeneity within classes and a high degree of heterogeneity between classes (Dobson 1974). Therefore, the most appropriate method for a single range graded map in most circumstances is one of the natural breaks classification schemes. If one legend is to be used for several range graded maps, then an equal-step classification scheme based on the combined attribute data range is recommended.

As suggested earlier, range grading is strongly recommended for three-dimensional and pictorial symbols, as magnitudes are difficult or nearly impossible to estimate from volumes or irregularly shaped objects. It will be easier for the map reader to match range graded symbols in the legend with the symbols on the map for these symbol forms.

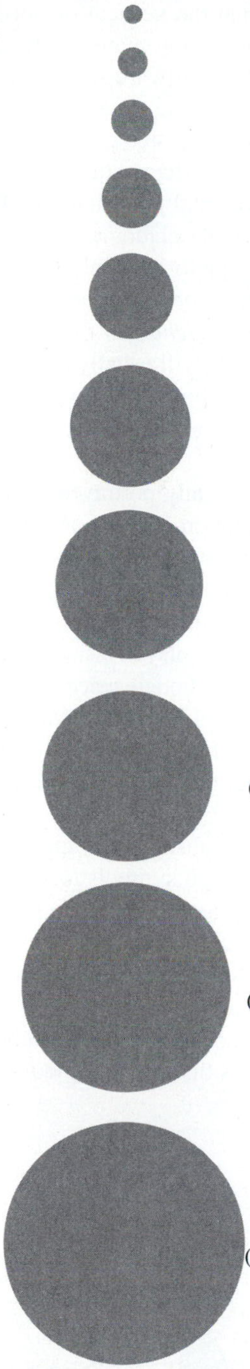

Meihoefer Circle Sequence	Radius inches (mm)	Diameter inches (mm)	Area in^2 (mm^2)	Point size
	0.05 (1.27)	0.10 (2.54)	0.01 (5)	7.20
	0.08 (1.99)	0.16 (3.98)	0.02 (12.5)	11.52
	0.11 (2.82)	0.22 (5.64)	0.04 (25)	15.84
	0.16 (3.99)	0.32 (7.98)	0.08 (50)	23.04
	0.22 (5.64)	0.44 (11.28)	0.15 (100)	31.68
	0.31 (7.98)	0.62 (15.96)	0.30 (200)	44.64
	0.38 (9.77)	0.76 (19.54)	0.45 (300)	54.72
	0.44 (11.28)	0.88 (22.56)	0.61 (400)	63.36
	0.54 (13.82)	1.08 (27.64)	0.92 (600)	77.76
	0.63 (15.96)	1.26 (31.92)	1.25 (800)	90.72

FIGURE 8.13 MEIHOEFER CIRCLE SIZES USEFUL IN A RANGE GRADED SERIES FOR SMALL-SCALE THEMATIC MAPS.

Each circle is easily differentiated from its neighbor—the fundamental criterion in a range graded series. No more than five classes are advisable for this form of scaling. When using this chart, any adjacent set of circles may be selected from the array.

Source: Data from Meihoefer 1969.

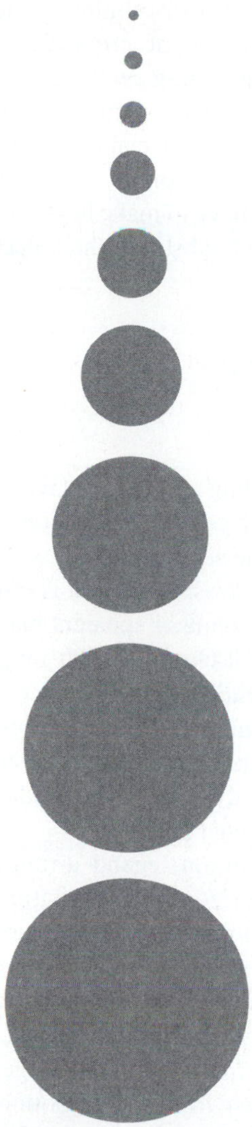

Dent Circle Sequence	Radius inches (mm)	Diameter inches (mm)	Area in^2 (mm^2)	Point size
	0.025 (0.65)	0.05 (1.3)	0.002 (1.3)	3.60
	0.045 (1.15)	0.09 (2.3)	0.006 (4.2)	6.48
	0.065 (1.65)	0.13 (3.3)	0.013 (8.6)	9.36
	0.11 (2.8)	0.22 (5.6)	0.04 (25)	15.84
	0.17 (4.3)	0.34 (8.6)	0.09 (58)	24.48
	0.245 (6.2)	0.49 (12.4)	0.19 (121)	35.28
	0.38 (9.65)	0.76 (19.3)	0.45 (293)	54.72
	0.51 (12.95)	1.02 (25.9)	0.82 (527)	73.44
	0.595 (15.1)	1.19 (30.2)	1.11 (716)	85.68

FIGURE 8.14 EMPIRICALLY DERIVED CIRCLE SIZES USEFUL IN A RANGE GRADED SERIES ON A SMALL-SCALE THEMATIC MAP.

Although useful in practice, these circles have not been experimentally tested. Any set containing three to six adjacent circles may be used.

Proportional Midpoint Range Grading

As pointed out earlier, one of the main criticisms of using ordinally scaled graduated symbols is the potential for overgeneralization. Dickinson (1973) argues that in range grading, the symbols should still maintain some sense of proportionality. In **proportional midpoint range grading,** the symbols are scaled proportionally using absolute scaling techniques earlier with regard to the midpoint of each class. This midpoint is usually the class mean or median.

Symbols scaled with midpoint range grading are more generalized than in a true proportional symbol map, but in this alternative the symbols are *more closely* in proportion to the data they represent. This method also tends to balance the communication goals of symbol size discrimination in range grading and magnitude estimation of proportional symbols. However, the relative ease in getting symbols to fit on the map using ordinally scaled graduated symbols is often lost. Since the default on most GIS and mapping software (graduated symbol) is the application of ordinally scaled symbols, the cartographer will have to make her or his own absolute symbol calculations and substitute the values for the software defaults.

PROPORTIONAL SYMBOL LEGEND DESIGN

From the foregoing discussions on scaling methods and reader perception of symbols, it is clear that legend design is crucial in proportional symbol mapping. The legend serves as the visual anchor for interpreting symbol sizes, whether it is estimating symbol magnitude (as with absolute or apparent magnitude scaling) or determining symbol class (as with range graded symbols). Major issues in legend design are choice of symbol arrangement and style, selecting the number of symbols to include in the legend when scaled absolute or apparent magnitude scaling has been used, and how to handle class ranges when the symbols have been range graded.

Figure 8.15 represents some common legend arrangements using circles. Vertical or horizontal presentations (Figure 8.15a and 8.15b) are the most typical, with the vertical presentation being default for many GIS and mapping software. Both are equally acceptable; select the arrangement that best balances the map layout. Vertical presentations may have the symbols progress from smallest to largest (most common) or largest to smallest from top to bottom. Horizontal presentations are smallest to largest from left to right. For the circle form, nested circles (Figure 8.15c) and nested semi-circles (Figure 8.15d) provide for a unique design when there may not be enough space for the more elongated vertical or horizontal legends. Nested presentations are somewhat less common than the vertical or horizontal arrangements, mostly because automated legend procedures in most GIS and mapping software do not directly support nesting. Thus, it takes some creativity and extra effort with the software to generate nested circles.

For symbols other than circles or perhaps squares, we usually recommend a vertical or horizontal presentation. Of course it is technically possible to create elaborate legends that nest complex geometric forms or pictorial symbols. But keep in mind that as the legend gets more complicated, it may *lose its effectiveness as the visual anchor for interpreting symbol sizes*. Keeping things simple will usually make it easier for the map reader to interpret the map. Ultimately, this is why we usually recommend circles over more complicated forms.

The numbers or data values are typically portrayed along the right side in a vertical presentation. In a horizontal presentation, they usually occur underneath the circles or symbols. As with choropleth map legends, outliers may be labeled with its specific value and/or the name of the enumeration unit. Occasionally, the numbers or ranges are displayed inside larger circles or other large symbols *if* there is enough space so that the numbers are legible *and* there is enough *figure-ground contrast* between the text and symbol fill. For consistency, if the number is placed inside of one circle, they should be placed inside all the circles in the legend.

For symbols that are scaled using absolute or apparent magnitude scaling, there must be enough symbols to represent effectively the data range in the legend. Typically high, medium, and low values in the data are shown (see section "Absolute and Apparent Magnitude Scaling Design Implications"). One approach that we strongly recommend is to choose the highest and lowest values, calculate those symbol sizes, and then calculate symbol sizes that are intermediate.

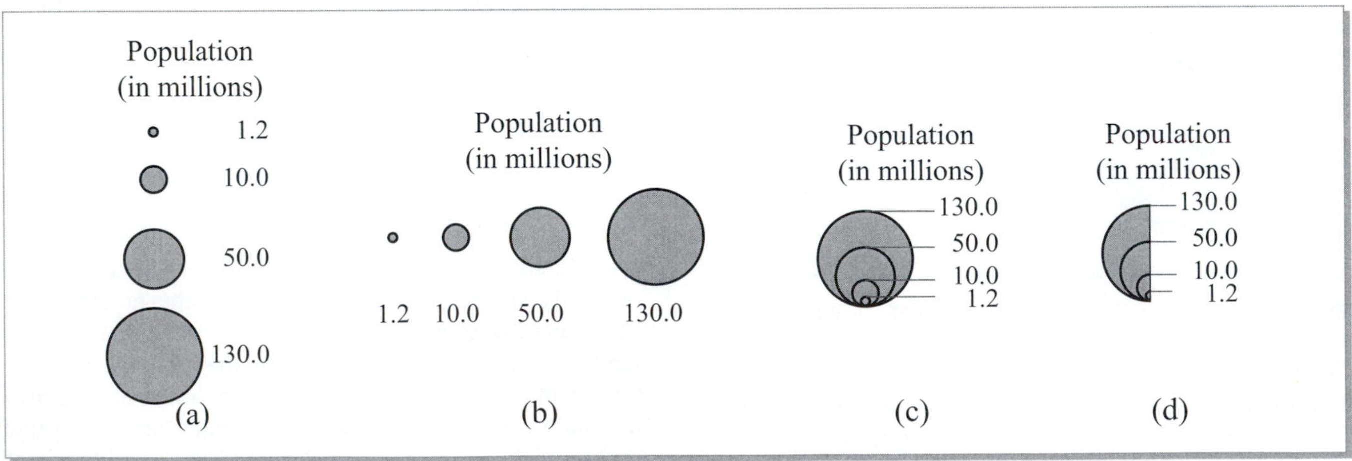

FIGURE 8.15 VARIOUS PROPORTIONAL SYMBOL LEGEND DESIGNS.
Common presentations include vertical (a) and horizontal (b). Other circle arrangements include nested circles (c) and nested semicircles (d).

GIS and mapping software that do "true" proportional symbols automate the symbol scaling process, but it is important to note that with some older software, the symbols sometimes do not stay in true proportion to their original data values, especially when the lowest and/or highest symbol sizes are changed. If apparent magnitude scaling is used, some cartographers feel that a statement that informs the reader that the symbols have been adjusted for underestimation is not only good practice but is also the ethical choice, since the symbols are no longer in true proportion to the data.

For range graded symbols, the number of classes will determine the number of symbols, and four or five classes are typical. It must be clear to the reader that the circles are indeed range graded and do not represent individual values. This may be best accomplished by listing the data ranges with the symbols (refer to Figure 8.7b). As with choropleth maps, data ranges always increase left to right, and the word "to" is better than a dash if negative numbers are used. Some cartographers include a statement that the data have been classified, and that the symbols are not in proportion to the data. For structuring the rest of the legend, review the section "Class Range Formatting, Legend Titles and Other Legend Information" in Chapter 6.

It is important to note that proportional symbol maps that portray a second variable require additional legend information. For example, proportional circles that contain pie charts require a legend to explain each part of the pie. Additional legend information is also required if the second variable is represented by changing each symbol's hue or value. In many cases, there will have to be two legend titles (see Color Plate 8.1).

GRAPHIC DESIGN CONSIDERATIONS FOR PROPORTIONAL POINT SYMBOL MAPS

Several issues relating to the design of proportional point symbols have been presented, focusing on the conceptual basis of their use, scaling methods, and legend design. Other design concerns deal more with their graphic treatment, discussed in this concluding section.

Graphic Treatment of Proportional Symbols

It is very common for designers to fill the symbols or circles, whether in black and white mapping (using grayscale or gray-tone) or more commonly today, with color. Although most of the testing on perception of circle sizes was done with black and white symbols, using different hues apparently does not affect the estimation of circle-size differences (Lindenberg 1986) any more than using grayscale fills. The following list, modified from Crawford (1971), highlights several advantages in using hue or grayscale fills for the symbols:

1. The amount of information can be expanded by showing two distributions, one with black symbols and one with gray in black and white mapping, or two hues or hue progressions for color maps. (See "Bivariate and Multivariate Proportional Symbols" for further discussion of handling two variables.)
2. Gray-tone symbols, or hues changing in saturation or value, allow for better ordering of information on the map. That is, symbols can be rendered in tones of gray in black and white mapping, or the hue's saturation or value in proportion to the values they represent (see Figure 8.16). This **redundant coding** can reinforce the information presented in the proportional symbols (Dobson 1983; Nelson 2000). However, some cartographers caution that darker or more saturated fills for larger symbols can direct *too much* attention to those symbols. It also gives the *appearance* of a bivariate map (discussed later), so careful legend design is critical if redundant coding is employed.
3. Some symbols can be de-emphasized (gray or lighter hue) relative to others (black or darker, more saturated hue).

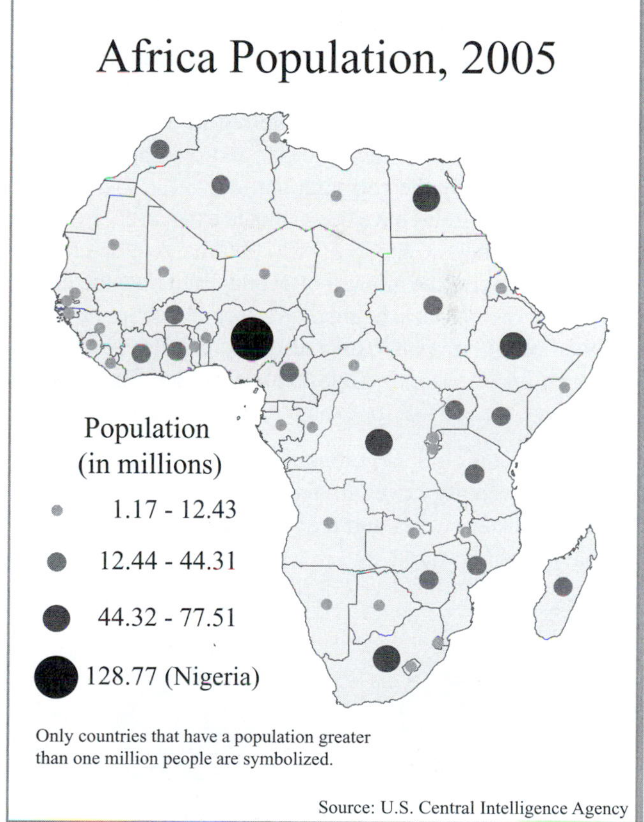

FIGURE 8.16 REDUNDANT CODING.
While redundant coding *can* reinforce the information in the symbol sizing, it can also direct too much attention to the largest figures, and it takes on the appearance of a bivariate map. See text for discussion.

In using color or grayscale progressions, remember that generally darker or more saturated hues (darker grayscale values) represent higher values, and lighter or less saturated hues (lighter grayscale values) represent lower values. As discussed in Chapter 6, we recommend use of ColorBrewer (Brewer 2006) to select color or grayscale schemes that match the attribute data.

Overlapping Symbols

One of the biggest challenges in proportional symbol design is producing a set of symbols that visually stand out (that is, are not too small) but at the same don't have excessive overlap (when largest symbols are *too* large). Even after adjusting the smallest and/or largest symbol, or switching from one scaling method to another, many cartographers often wonder if the amount of symbol overlap is acceptable. Although there is general cartographic agreement that some overlap is acceptable, there are no set rules to say how much overlap should be allowed. As an approximate guide, if the circles are obscured by more than about 25 to 33 percent in the most congested area on the map body, then it is possible that further adjustments in circle sizes are necessary (refer to Figure 8.7a). With more complex symbol forms, even less overlap is recommended.

Where symbols overlap, it is important that smaller symbols "cover" larger symbols (see Figure 8.17). Note that in some GIS and mapping software packages, the capability of controlling the drawing order or amount of overlap of symbols directly within a proportional or graduated symbols procedure is not supported. However, the symbols can often be converted to graphics or transferred to artistic drawing software, where both drawing order and overlap can be directly controlled by the cartographer.

Overlapping circles have been studied extensively by cartographic researchers. Cabello *et al.* (2006) have researched algorithms to control the amount of overlap, and how the circles will "stack" on top of each other. Overlapping circles have also been studied with regard to map reader perception, particularly with open and "cut-out" circle treatments (see Figure 8.18). Groop and Cole (1978), for example, made these discoveries:

1. Transparent circles (rendered so that edges of both are visible even after overlap) are more accurately perceived than cut-out circles and are seen similarly to circles having no overlap.

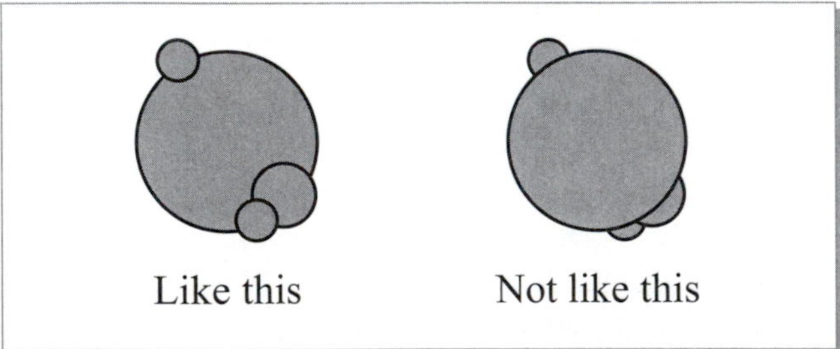

FIGURE 8.17 OVERLAPPING CIRCLES.
It is important for smaller circles or symbols to cover larger ones.

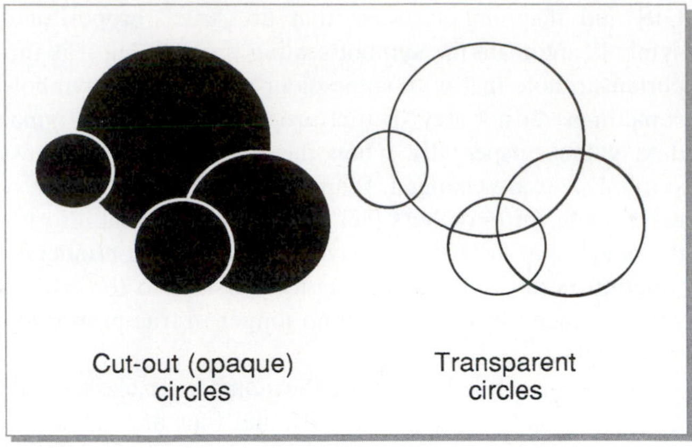

FIGURE 8.18 CUT-OUT AND TRANSPARENT CIRCLES.
Readers tend to make errors of perceptual judgment with cut-out symbols, in proportion to the amount of the circle obscured by the overlap. Transparent circles, on the other hand, are perceived about as well as circles that do not overlap. The designer must weigh these results against other design elements and the purpose of the map. Cut-out circles add a three-dimensional quality to the map, whereas a map containing many transparent circles looks flat and uninteresting.

2. With cut-out circles (in which the overlap obscures part of one circle), there is a strong relationship between the amount of overlap and estimation error.

Because of the complex nature of many proportional symbol maps, the effects of overlapping, tightly clustered circles on readers' estimations of circle sizes are not known with certainty. In some cases, the reader is asked to estimate the magnitudes of individual symbols; in others, range graded series. The designer must be aware that the reader's judgment is influenced by the graphic treatment of the symbols.

Another interesting research project suggests that map readers have preferences regarding the way circles are rendered on the map (Griffin 1990). For the map readers tested in the study, maps having circles with little contrast to their surroundings, or that are open with line work showing through, proved least popular (see Table 8.3). Black-filled circles were judged best, and gray-filled symbols also did well. The author of the test remarks that the map readers seemed to be more influenced by the features of transparency and opacity of the circles and less by whether the circles were black or gray. Note that the study also found that these types did not have any differences on symbol size judgments.

The results of the Griffin study tend to support the viewpoint later in this book (Chapter 12) that readers respond favorably to symbols with high contrast that appear as strong visual figures in perception. We have observed the same to be true with both color and black and white mapping. The proportional point symbol should be clear and unambiguous in meaning, its edges sharp and not easily confused with the background, and its surface character easily differentiated from other surfaces and textures on the map. Its scaling

TABLE 8.3 PREFERENCES FOR GRAYSCALE PROPORTIONAL CIRCLES IN DIFFERENT MAP ENVIRONMENTS

	Map Environment (Percent)					
Judged	**F**	**C**	**B**	**E**	**D**	**A**
Best (rank 1 of 6 ranks)	30	26	26	16	1	1
Worst (rank 6 of 6 ranks)	4	5	0	0	39	51

A = Transparent circles with line work showing through
B = Gray-filled circles with line work showing through
C = Black-filled circles with line work (in white) showing through
D = Opaque white-filled circles
E = Opaque gray-filled circles
F = Opaque black-filled circles

Source: Data from Griffin 1990.

should be precise, not confusing for the reader. The symbol should be made to stand out from its surroundings, both graphically and intellectually. If these simple guidelines are followed, the final map will perform well in the communication of spatial concepts.

Labeling Symbols

With some proportional symbol maps, it is possible or in some cases even advisable to label the proportional symbols, without sacrificing the simplicity that characterizes thematic maps. For example, in a map that has circles scaled in proportion to the population of cities, the map reader would benefit in knowing which city is which. Labeling symbols is possible where symbol overlap and congestion is minimal and there is enough space to add the labels. As a general rule, labels look best when placed in one of the first four point feature label location priorities that will be discussed in Chapter 13. (Refer to Figure 13.17 for those positions.) For online interactive maps, and maps in a GIS environment, individual cities and their values often can be seen as the reader moves the mouse cursor over the surface of the map, as a complement to visualizing the overall distribution.

Bivariate and Multivariate Proportional Symbols

The proportional symbol map is one of the most common quantitative mapping techniques for bivariate and multivariate mapping. The symbol form allows for a second (or more) variable to be portrayed on the map. For example, it is common to change each symbol's hue or value by a second variable (refer to Color Plate 8.1) while the symbol's size is scaled by the first variable. As mentioned earlier in the chapter, different symbols can represent two or more symbol forms as well. With careful selection of related data sets and attention to design principles, bivariate proportional symbol mapping can be an effective means of illustrating a functional

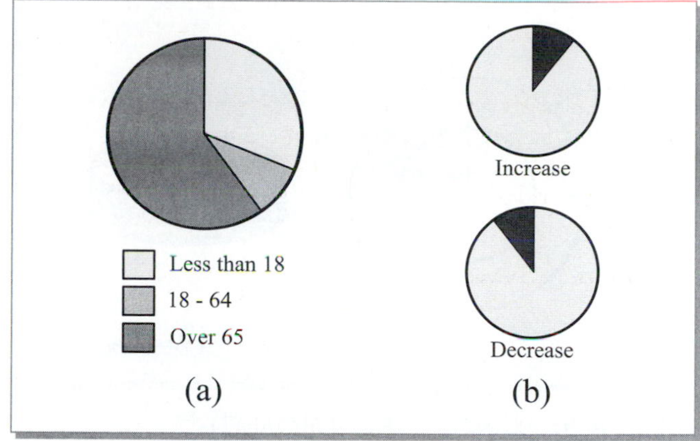

FIGURE 8.19 PIE CHART IN A CIRCLE (a) AND CIRCLES WITH INCREASE OR DECREASE INDICATORS (b).

In (a), pieces of the pie indicate proportion of the whole. In this case, it is age cohorts but could be anything from land use to energy production type. In (b), starting from the 12 o'clock position, phenomena can show increase (top) or decrease (bottom) of phenomena.

relationship between related variables (Nelson 2000). Unthinking design, as we will see, makes for a map that is potentially confusing to the map reader.

The circle form is particularly advantageous for portraying a second variable. The form allows for embedded pie charts within the proportional circles (see Figure 8.19a and Color Plate 8.2), as well as increase/decrease indicators (see Figure 8.19b). When pie charts are used, it is important to note that each segment of the pie must add up to a whole, or 100 percent of the expected value. If each pie piece represents a city's ethnicity, for example, and one-quarter of the pie represents Hispanic population, then the map reader will expect the Hispanic population to be 25 percent (or proportion of 0.25) of the city's population. Cartographers using GIS solutions must be especially cautious at this point, as the software will allow you to add *any* numeric variables to the pie, and the software will treat the added pieces as 100 percent. As noted earlier, bivariate map legends will need to include components for both circle sizes and for the pie segments.

Overloaded Proportional Point Symbols

The increasing availability and variety of symbol forms, coupled with increasingly more powerful and flexible design software, affords the cartographer a chance to be ever more creative in using the proportional symbol form, particularly with bivariate and multivariate symbolization. The map designer should be aware, however, that *multivariate* proportional symbols have not been researched extensively. All too often, cartographic designers fall into the trap of pushing the system beyond its capacities. Because it is possible to cram a lot of information into one symbol, the designer often tries for too much of a good thing. **Symbol overload** results when it is no longer easy for the reader to make judgments about the quantitative nature or spatial pattern of the distribution.

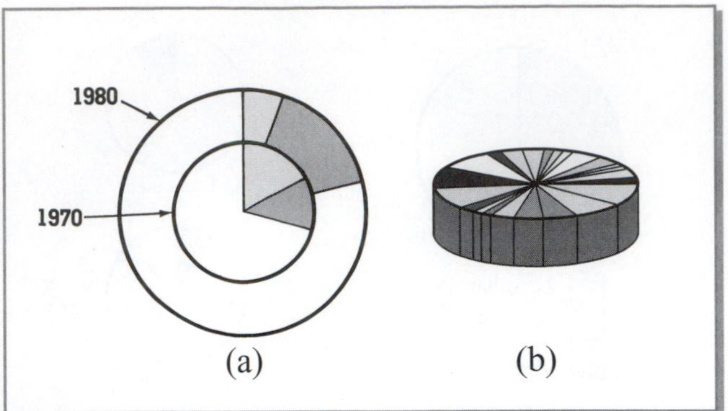

FIGURE 8.20 EXAMPLES OF OVERLOADED PROPORTIONAL SYMBOLS.

We strongly suggest not symbolizing a map like either of these examples. In (a), the map reader is expected to make an easy and quick assessment of the change in the geographical distribution of each part, at several locations. In (b), there are too many sectors in the embedded pie graph. Note that the 3-D view catches the eye, but magnitude estimations become more difficult.

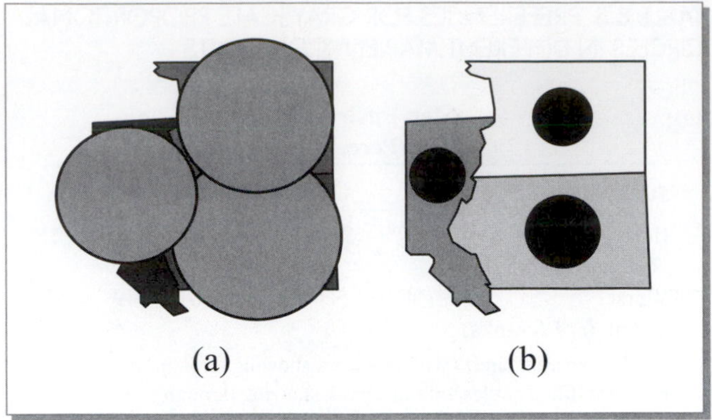

FIGURE 8.21 PROPORTIONAL SYMBOLS AND CHOROPLETH MAPS.

In (a), the proportional symbols are too large and the choropleth map fills are too dark. In (b), the symbols are small enough to see the choropleth background, and there is a better contrast between the symbols and choropleth map fills. See text for discussion.

Even the simple circle form can be carried to an extreme. Because it can be sized, segmented, colored, sectored, made into a 3-D model, and so on, some feel compelled to do all of these things on one map (see Figure 8.20). Unfortunately, the map reader can no longer see the distribution or its intent clearly and easily. These maps lose their thematic goals and exhibit complex graphic displays that almost totally destroy their communicative efforts.

Therefore, we suggest limiting the number of variables symbolized by proportional point symbols to one or possibly two, but rarely three or more. When two or more variables are being represented, consider keeping only one variable at the ratio level and the other at the nominal or ordinal level (refer to Color Plate 8.2). Two variables at the ratio level can possibly lead to excessive complexity and symbol overload.

Proportional Symbols and Choropleth Maps

With many GIS and computer mapping software packages, it is easy to overlay point symbols on top of area fills. A method that is increasing in popularity is overlaying proportional symbols on top of choropleth maps to illustrate relationships between variables. We caution the map designer that there is the same potential for reader confusion as with overloaded proportional symbols, and generally suggest positioning two such maps side-by-side instead. *If* this technique is used for multivariate or bivariate mapping, we recommend using highly contrasting schemes for the proportional symbols and the choropleth enumeration unit fills in order to create adequate figure-ground development. For example, if darker or more saturated hues for the choropleth symbolization are used, then the symbols should be filled with a lighter color (or vice versa). Limit the number of classes for the choropleth map, so that you are not forced to use fills that do not adequately contrast with the proportional symbols. We also recommend a simple geometric form—circles—and that the circles not represent more than one variable. Also, the circles cannot be so large as to obscure the enumeration units in the choropleth map (see Figure 8.21). Sometimes the enumeration units are just too small to include the proportional symbols, and thus the map reader will be unable to understand where each symbol is applied. Remember that simplicity is the hallmark of effective thematic maps, and as always, intuitive appeal of the design can be a criterion of a map's appropriateness.

REFERENCES

Brewer, C. 2006. *Color Brewer—Selecting Good Color Schemes for Maps*. http://www.ColorBrewer.org

Brewer, C., and A. Campbell. 1998. Beyond Graduated Circles: Varied Point Symbols for Representing Quantitative Data on Maps. *Cartographic Perspectives* 29 (Winter 1998): 6–21.

Cabello, S., H. Haverkort, M. van Kreveld, and B. Speckmann. 2006. Algorithmic Aspects of Proportional Symbol Maps. In *Lecture Notes in Computer Science*. 720–31. Berlin: Springer.

Chang, K. 1977. Visual Estimation of Graduated Circles. *Canadian Cartographer* 14: 130–38

———. 1980. Circle Size Judgment and Map Design. *American Cartographer* 7: 155–62.

———. 2008. *Introduction to Geographic Information Systems*. 4th ed. Dubuque, IA: McGraw-Hill.

Chrisman, N. 2002. *Exploring Geographic Information Systems*. 2nd ed. New York: Wiley.

Cox, C. 1976. Anchor Effects and the Estimation of Graduated Circles and Squares. *American Cartographer* 3: 65–74.

Crawford, P. 1971. Perception of Gray-Tone Symbols. *Annals* of the Association of American Geographers 61: 721–35.

———. 1973. The Perception of Graduated Squares as Cartographic Symbols. *Cartographic Journal* 10: 85–88.

Dickinson, G. 1973. *Statistical Mapping and the Presentation of Statistics.* London: Edward Arnold.

Dobson, M. 1974. Refining Legend Values for Proportional Circle Maps. *Canadian Cartographer* 11: 45–53.

———. 1983. Visual Information Processing and Cartographic Communication: The Utility of Redundant Stimulus Dimensions. In *Graphic Communication and Design in Contemporary Cartography,* edited by D.R.F. Taylor. New York: John Wiley.

Flannery, J. 1956. *The Graduated Circle: A Description, Analysis, and Evaluation of a Quantitative Map Symbol.* Ph.D. dissertation, Department of Geography, University of Wisconsin-Madison.

———. 1971. The Effectiveness of Some Common Graduated Point Symbols in the Presentation of Quantitative Data. *Canadian Cartographer* 8: 96–109.

Gilmartin, P. 1981. Influences of Map Context on Circle Perception. *Annals* of the Association of American Geographers 71: 253–58.

Griffin, T. 1985. Groups and Individual Variations in Judgment and Their Relevance to the Scaling of Graduated Circles. *Cartographica* 22: 21–37.

———. 1990. The Importance of Visual Contrast for Graduated Circles. *Cartography* 19: 21–30.

Groop, R., and D. Cole. 1978. Overlapping Graduated Circles: Magnitude Estimation and Method of Portrayal. *Canadian Cartographer* 15: 114–22.

Heino, A. 1995. The Presentation of Data with Graduated Symbols. *Cartographica* 32: 43–50.

Lindenberg, R. 1986. *The Effect of Color on Quantitative Map Symbol Estimation.* Ph.D. dissertation, Department of Geography, University of Kansas.

Meihoefer, H. 1969. The Utility of the Circle as an Effective Cartographic Symbol. *Canadian Cartographer* 6: 105–17.

Nelson, E. 2000. Designing Effective Bivariate Symbols: The Influence of Perceptual Grouping Processes. *Cartography and Geographic Information Science.* 27: 261–78.

Peterson, M. 1985. Evaluating a Map's Image. *American Cartographer* 12: 41–55.

Robinson, A. 1955. The 1837 Maps of Henry Drury Harness. *Geographical Journal* 121: 440–50.

———. 1967. The Thematic Maps of Charles Joseph Minard. *Imago Mundi* 21: 95–108.

———. 1982. *Early Thematic Mapping in the History of Cartography.* Chicago: University of Chicago Press.

Slocum, T. 1984. A Cluster Analysis Model for Predicting Visual Clusters. *The Cartographic Journal* 21: 103–11.

Spence, I. 2006. William Playfair and the Psychology of Graphs. *Proceedings* (2006 American Statistical Association), Section on Statistical Graphics: 2426–36.

Tufte, E. 2001. *The Visual Display of Quantitative Information.* Cheshire, CT: Graphics Press.

GLOSSARY

absolute scaling directly proportional scaling of area or volume of symbols to data values

adaptation level in psychological theory, a neutral reference point on which perceived judgments are based; affected by the perceptual anchors from which judgments are made

anchor effects the size of the visual anchor affects the estimate of magnitude of an unknown symbol; the contrast effect causes estimation to be away from the anchor, and the assimilation effect causes judgments to be toward the anchor

apparent magnitude scaling scaling of proportional symbols that incorporates correction factors to compensate for the normal underestimation of a symbol's area or volume

graduated symbol in the literature and classic sense, refers to any of the scaled point symbol mapping schemes; with GIS implementation implies range grading using ordinally scaled symbols

proportional midpoint range grading variation of range grading in which symbols are in proportion to the midpoint of a class range

proportional point symbol mapping type of quantitative thematic map in which point data are represented by a symbol whose size varies with the data values; areal data assumed to be aggregated at points may also be represented by proportional point symbols

proportional symbol traditional cartography and modern GIS term that describes symbols that are in true proportion to their attribute data; in GIS usually associated with absolute scaling, and in some GIS software, apparent magnitude scaling

psychophysical power law the mathematical expression that describes the relationship between stimulus and response; general form is

$$R = K \cdot S^n$$

where R is response, S is stimulus, n is exponent describing the relationship between R and S, and K is a constant

range grading a symbol represents a range of data values; differently sized symbols are chosen for differentiability for each of several ranges in the series of values being mapped; larger symbols indicate more of a quantity but are not in true proportion to the data they represent

redundant coding expressing the same variable by more than one visual variable; in proportional symbol mapping, the symbol's fill will change in saturation and/or value in proportion to the data that scale the symbol's size; larger symbols will have a darker or more saturated fill, smaller symbols have a lighter or less saturated fill

symbol overload too much information in a symbol, so that readers have difficulty making assessments about the quantitative nature of the data

three-dimensional point symbol any point symbol made to appear three dimensional, such as a sphere or cube

9 MAPPING GEOGRAPHIC SURFACES: ISARITHMIC AND THREE-DIMENSIONAL MAPS

CHAPTER PREVIEW Mapping the surface of a real or conceptual three-dimensional geographical volume may be achieved using either quantitative line symbols or three-dimensional models of that surface. The isarithmic quantitative thematic map dates back to the mid-sixteenth century, when isobaths were first charted. Today, two forms are recognized: isometric and isoplethic. Each involves the planimetric mapping of the traces of the intersections of horizontal planes with the upper surface of the volume. The isarithm is placed by "threading" it through a series of data points at which magnitudes are assumed to exist. Data points are locations where measurements are taken or places are chosen to represent unit areas. From these data points, three-dimensional surface models display the undulating surface from a variety of viewpoints. All surface maps contain error, as do other quantitative thematic maps, but the designer can learn to recognize potential sources of error and reduce their effects on the overall map. In the total map design, isarithmic lines should be placed at the top of the visual/intellectual hierarchy and made to appear as figures in perception. Three-dimensional models should be designed to maximize the display of the total volume's surface. Legends should be clear and unambiguous and should specify the units of measurement. ■

Six distinct kinds of quantitative thematic maps are treated in this book: choropleth maps, dot density maps, proportional symbol maps, isarithmic maps, value-by-area cartograms, and flow maps. Of these, the isarithmic map may be the most difficult conceptually. Isarithmic mapping requires three-dimensional thinking for surfaces that vary spatially. The three-dimensional model assists the reader in the understanding of that surface and should be used in conjunction with the isarithmic map. Isarithmic mapping is an important part of the cartographer's repertoire and should therefore be mastered. This chapter will first examine the nature and concepts of isarithmic mapping and will follow with a discussion of three-dimensional mapping.

THE NATURE OF ISARITHMIC AND THREE-DIMENSIONAL MAPPING

The basic concepts, diversity, and history of isarithmic mapping are first introduced so that the full range of the activity and its product, the isarithmic map, will be better understood. An understanding of three-dimensional surface models assists the map user in envisioning the variability of the surface.

Fundamental Concepts

A **surface** represents the uppermost layer of a real three-dimensional volume, or it may be an abstract mental construct

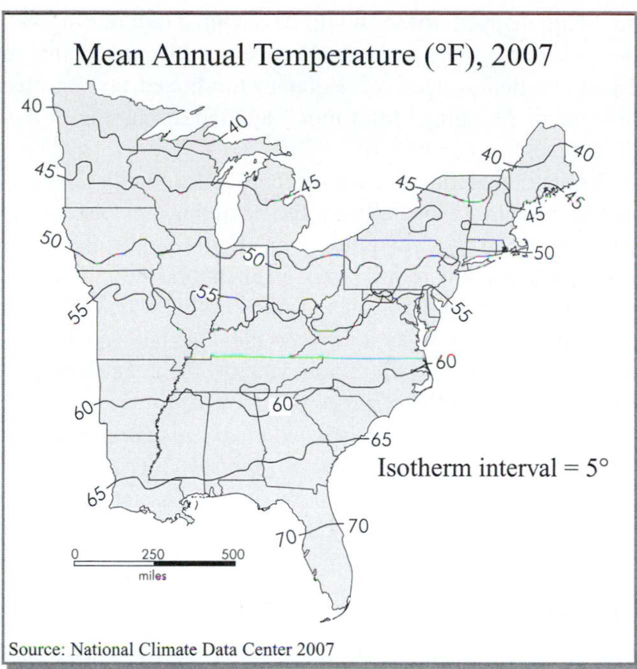

FIGURE 9.1 ISARITHMIC MAP.
The well-designed isarithmic map includes the following critical map elements: title, statement of isoline interval, linear scale, and source statement. (Climatological data are frequently mapped in unprojected latitude-longitude.)

representing some varying geographical distribution. On the one hand, for example, the lithosphere is a real geographical volume whose top forms what is called the *topographic surface.* It is possible to make a scaled-down version of that surface, or a portion of it in the form of either a map or model. An attempt to map the surface, using quantitative line symbols, results in an isarithmic map. On the other hand, the mind can form a construct of a geographic quantity that has volume—for example, precipitation—which is high in some areas and low in others. This abstract mental construct also can be mapped isarithmically. Either real or abstract models can provide the content for isarithmic maps, but the third dimension must be present or assumed. That third dimension can be displayed in a model that can be created from a variety of view locations.

An **isarithmic map** is a planimetric graphic representation of the surface of a three-dimensional volume (see Figure 9.1). The graphic image or map that results from **isoline mapping** is a system of *quantitative line symbols* that attempt to portray the undulating surface of the three-dimensional volume. The three-dimensional model displays that surface as a continuous image depicting the undulations. Regardless of the interval data being mapped or any complexities associated with the map content, a third dimension must exist or be assumed to exist if a mapping technique is to be called *isarithmic* (Hsu and Robinson 1970).

Another requirement of the isarithmic technique is that the volume's surface be continuous in nature, rather than discrete or stepped (Hsu and Robinson 1970; Jenks 1963). Geographic phenomena such as the locations of factories are discrete—values do not occur between points. Temperature, however, exists everywhere, both at and between observation points. Some geographic phenomena, such as population density, can be assumed to exist everywhere and thus can be mapped isarithmically (see Figure 9.2).

Isarithmic Categories and Terminology

Isarithmic maps fall into two distinct categories: the **isometric map** and the **isoplethic map.** The construction of these is achieved in similar fashion, but the nature of the data from which they are generated is quite distinct. Isometric maps are generated from data that occur or can be considered to occur at points; isoplethic maps result from mapping data that occur over geographic areas.

FIGURE 9.2 DATA TYPES AND ISARITHMIC FORM.
Notice that total values are never mapped isoplethically.

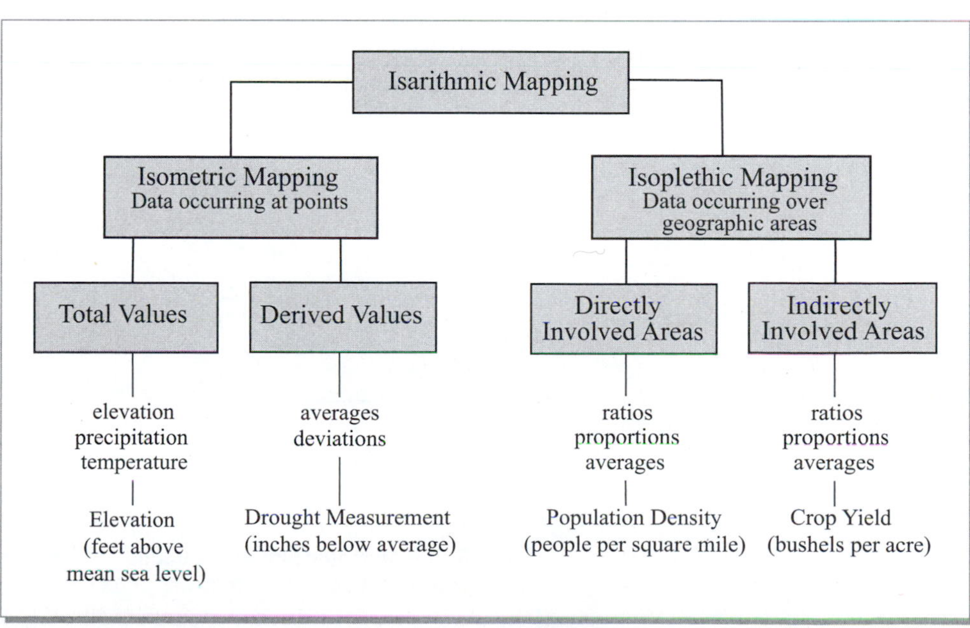

A distinct organization of data types and isarithmic categories has developed (Robinson *et al.* 1978; see Figure 9.2). Data that can occur at points, to be mapped isometrically, can be divided into *total* and *derived* values. Total data values (Robinson referred to these as actual data) include such things as temperature, precipitation, and elevation. Such values are generally obtained by recording instruments or by other means of point sampling in the field. Derived values are subdivided into two groups. One group includes such statistical measures as means and measures of dispersion; the other deals with such magnitudes as proportions and ratios.

Isometric maps are created from a series of **sampling points** where the data are measured and recorded. We know that the data exists everywhere, such as temperature, but we are limited by the means of collecting such data. From the sampling points, we can infer or approximate a data value at every location and thus display the spatial variations in that data.

Isoplethic maps (also *isopleth* maps) are generated from data that occur over geographic area. The values that can be represented include ratios that involve area either directly, such as population density, or indirectly, such as crop yield per acre. Professional cartographers generally do not illustrate totals that are applied to an area isoplethically. If data are being used that represent areas, they must first be converted into ratios or proportions that involve the areal magnitude. For example, if population totals by census tract are being mapped, these should be changed into density values before mapping as isodems (lines connecting points of equal population density). Refer to the boxed text entitled "Isoplethic Mapping" for a more detailed discussion of isopleth maps.

Isopleth mapping is more difficult conceptually than isometric mapping, especially for the beginner. It is not easy to conceive of an areal magnitude existing along a line. How can people per square mile exist in such a fashion? It is perhaps best not to think of these values in this way, but rather to visualize the line as a *surface element* below which is found the volume of population density. *Once the data have been converted to the ratios or proportions for an isoplethic map, the techniques of construction are the same whether the map category is isometric or isoplethic.* The following discussion of construction methods may assist in understanding the conceptual basis behind this form of mapping.

The Basis of Isarithmic Construction

The methods of isarithmic construction are straightforward but require a grasp of the basis behind the technique. Isarithmic mapping depicts the surface of a volume by quantitative line symbols. The construction of lines begins by imagining a series of pins erected at the **data points** so that they extend vertically in proportion to the magnitude they represent at each point (see Figure 9.3). This is the vertical scale. The

FIGURE 9.3 THE CONCEPTUAL DEVELOPMENT OF AN ISARITHMIC MAP.

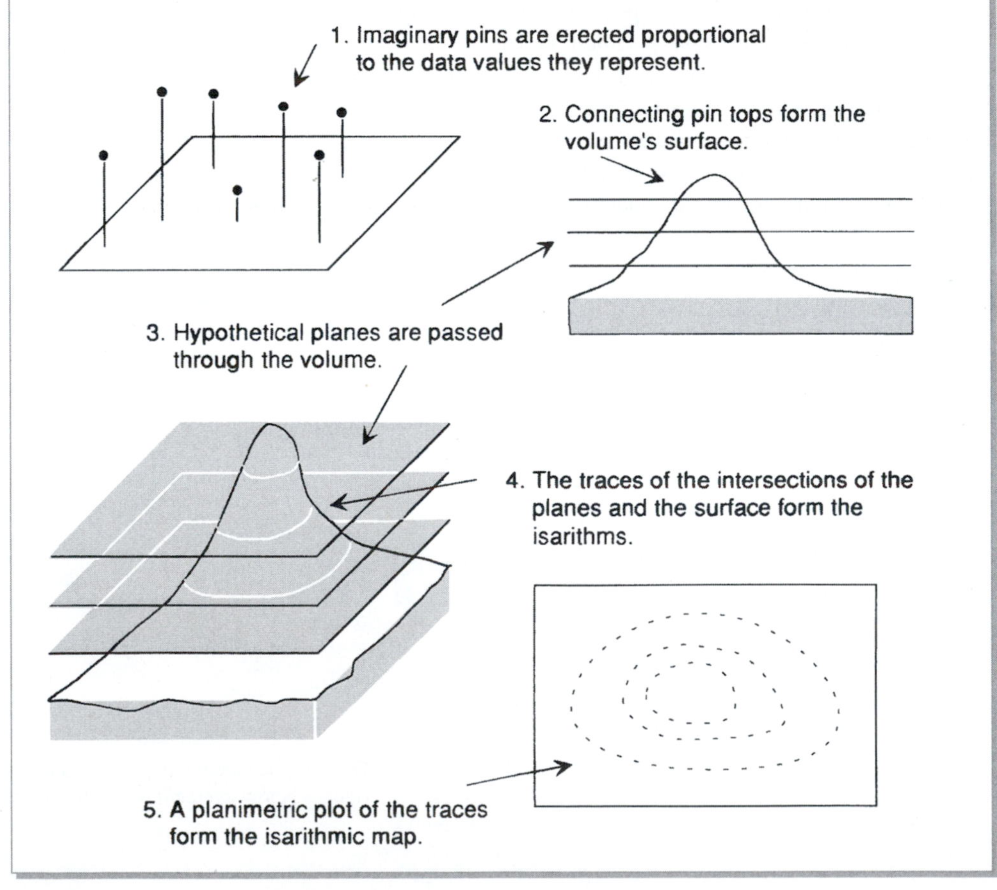

1. Imaginary pins are erected proportional to the data values they represent.
2. Connecting pin tops form the volume's surface.
3. Hypothetical planes are passed through the volume.
4. The traces of the intersections of the planes and the surface form the isarithms.
5. A planimetric plot of the traces form the isarithmic map.

ISOPLETHIC MAPPING

An isopleth map is one of the two categories of isarithmic mapping. This map is unique in two ways. First, the data are of the derived form in which the value represents a standardized number per unit area. Second, the data are applied to a point assigned to represent the area.

To begin with, the data are collected as totals for some predefined area or enumeration unit. In the United States, civil enumeration districts are states, counties, minor civil divisions, and census tracts. The data to be mapped are converted from total to derived data, such as people per square mile. This is easily done using a spreadsheet to divide the total population for a given time period by the area of its associated location. Crop yield per acre is another derived data example that is commonly mapped using this technique. GIS and other computer mapping packages identify the association between the enumeration link and the derived data through the use of FIPS codes. A data point, assigned by the software linking it to the geographic area, has the derived data value assigned to it. The location of this point within this area is a matter of concern, because it will affect the accuracy and appearance of the whole map. The data point serves as a representative location for the entire enumeration unit. This situation conceptually is a typical symbol transformation often used in cartography, that is, area being represented by a point.

The cartographer has little control over the selection of the point assigned by the software to represent the area. Usually, this point is determined by the geographic center of the polygon and is commonly referred to as its **centroid.** The centroid locations serve as the data point for the isoplethic mapping. If you have unique knowledge about the distribution of the data within the areal unit, you may elect to change the location of the centroid to better represent that distribution (Box Figure 9.1). Most software allow you to view the location of this point and in some instances alter its location. Such modification to the centroid should be carefully considered. Depending on the scale of the map, one could accidentally move the centroid to a location outside of the polygon boundary.

The distribution of these data points is affected by the irregular size of the enumeration units. As the size and shape vary, the centroid distribution may be widely spaced. If the area units are more compact in size, the centroid distribution may be more uniform (Box Figure 9.2). The cartographer should consider this distribution when selecting the interpolation method.

The isopleth map presents an interpretation problem for most map users. The data are areal data that are represented by a point. These points serve as coordinate locations used in the interpolation process. The mapping of isopleth isolines is handled in exactly the same manner as with isometric isolines. The isoline representing this area data passes through the enumeration unit. The interpolated positions of the isoline are depicting the surface variation representing a volume of people per square mile. Since isarithmic mapping implies a continuous variable, the implication is that the data change along some slope between the isolines. This then becomes a conundrum for the map user in interpreting the isarithmic map.

We recommend that you pay particular attention to the titling of your map and the definition of the isoline values. It would be beneficial if an explanatory statement be included on the map to assist the map user in understanding the distribution being displayed.

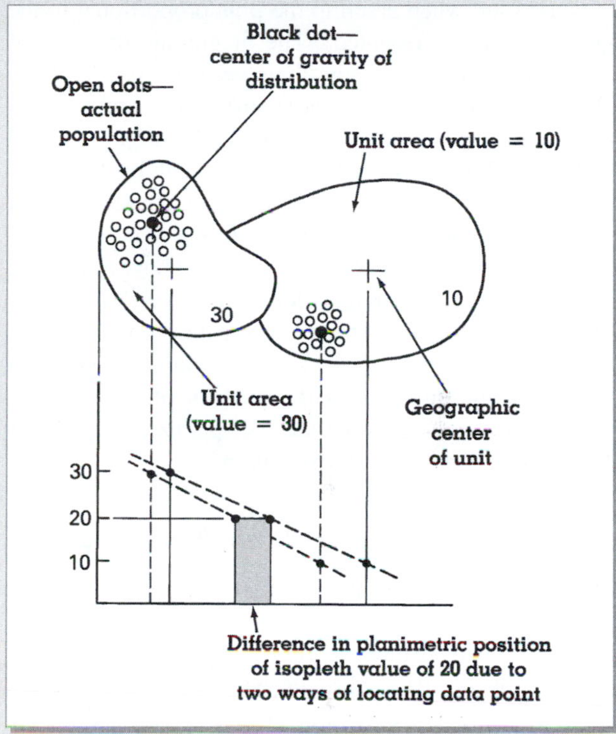

BOX FIGURE 9.1 DATA POINT PLACEMENT AFFECTS THE FINAL MAP.

Planimetric displacement may result from alternate ways of locating data points on isopleth maps. Over an entire map, the discrepancy can be considerable.

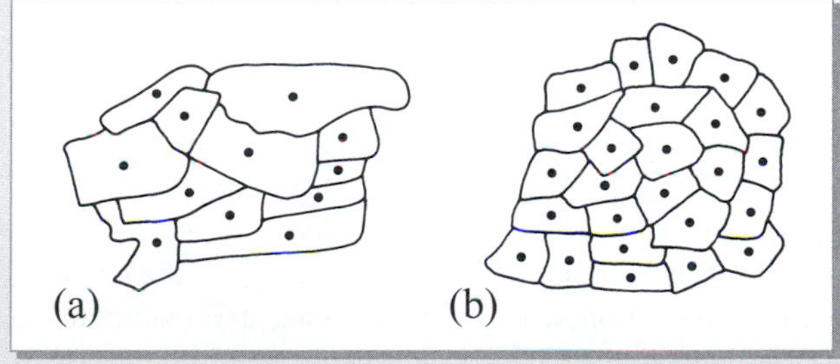

BOX FIGURE 9.2 DIFFERENT UNIT AREA PATTERNS.

Irregularly sized and shaped enumeration areas used as unit areas, as in (a) are less desirable than uniformly sized and shaped ones as in (b). (a) Irregular polygons produce a more dispersed pattern of data points whereas (b) polygons of similar size and shape produce a more uniform pattern.

tops of all the pins will form the surface of the new volume. The line symbols are the planimetric traces of the intersections of hypothetical planes with this three-dimensional undulating surface. The lines represent the surface, not the volume.

The vertical positions of the hypothetical planes are selected relative to a datum adopted for the particular map. The value of the datum is usually zero and the range of data values represented at the points will dictate the variability in the surface. Each plane has an assumed value associated with it depending on its placement relative to the vertical scale (the same scale used when erecting the pins proportional to the magnitudes they represented). The magnitude or value of the isarithmic lines represents their vertical distance from the datum. Because the planes are constructed parallel to the datum, each isarithm will maintain an *unchanging magnitude* or vertical distance from the datum. Actually, these isarithms are traces (isolines) of locations on the surface that have a value equal to the distance from the datum. For example, an isarithm that represent 40 feet above mean sea level connects places with that elevation.

The total effect of these traces or isarithms on the surface is to show the varying amounts of the volume beneath the surface. The interpretation of the **pattern of isarithms** is the critical element in reading these maps. Pattern elements are *magnitude, spacing,* and *orientation.*

A Brief History of Isarithmic Mapping

Isarithmic mapping in each of its two categories has had a fairly long history, at least in relation to thematic mapping in general. **Isobaths** (isometric lines showing depth of the ocean floor) were first used as far back as 1584 (Robinson 1971, 1982). This early use was no doubt the result of the urgent need for reliable map information for commercial and military navigation.

In 1777, the **isohypse** line was proposed by Meusnier as a way of depicting surface features. An actual map using the isohypse was made by du Carla-Dupain-Triel in 1782. The isohypse is an isometric line; isopleths were not used until somewhat later. **Isogones,** lines showing equal magnetic declination, were first used around 1630 by Borri, an Italian Jesuit (Robinson 1971, 1982). They were used on a thematic map by Edmond Halley in 1701. The renowned naturalist-scientist-geographer Alexander von Humboldt mapped equal temperatures using isometric lines called **isotherms.**

Arthur Robinson, in tracing the genealogy of the isopleth, placed its inception with Léon Lalanne, a Frenchman, in 1845 (Robinson 1971, 1982). Although based on the use of isometric lines, such as isobaths, isohypses, and isotherms, this marked a clear departure because point data representing area were part of the method. Lalanne's description of the method is still accurate today.

From these beginnings, the isometric and isoplethic techniques gained wide usage. Isopleth mapping came later and is clearly based on isometric examples. Isopleth mapping was adopted by American professional geographers late in the nineteenth century and earned a place of prominence during the early decades of the twentieth century (Porter 1957–1958). Its use peaked just before World War II. During these early decades, the isopleth technique was most often used in agricultural mapping, showing intensity of crop production and rural population density (Jones 1930). The number of differently named isolines has reached sizable proportions (see Table 9.1).

Isarithmic mapping has experienced variable usage among professional geographers during the last few decades. Prior to GIS and other mapping packages, the maps were prepared manually and were labor intensive in their design and construction. With the introduction of Golden Software's SURFER, ESRI's ArcGIS, and other application software, the creation of isarithmic maps is faster and easier to prepare. Proper design of the map still requires a wealth of knowledge about the area being mapped. This knowledge is especially important in order to recognize erroneous surface features

TABLE 9.1 LIST OF ISOLINE NAMES

Isobath	Depth below a datum (for example, mean sea level)
Isogonic line	Magnetic declination
Isocline	Magnetic dip (inclination) or angle of slope
Isohypse (contour)	Elevation above a datum (for example, mean sea level)
Isodynamic line	Intensity of the magnetic field
Isotherm	Temperature (usually average)
Isobar	Atmospheric pressure (usually average)
Isochrone	Time
Isohyet	Precipitation
Isobront	Occurrence of thunderstorms
Isanther	Time of flowering of plants
Isoceph	Cranial indices
Isochalaz	Frequency of hailstorms
Isogene	Density of a genus
Isospecie	Density of a species
Isodyn	Economic attraction
Isohydrodynam	Potential water power
Isostalak	Intensity of plankton precipitation
Isovapor	Vapor content in the air
Isodynam	Traffic tension
Isophot	Intensity of light on a surface
Isoneph	Degree of cloudiness
Isochrone	Travel time from a given point
Isophene	Date of beginning of a plant species entering a certain phenological phase
Isopectic	Time of ice formation
Isotac	Time of thawing
Isobase	Vertical earth movement
Isohemeric line	Minimum time (freight transportation)
Isohel	Average duration of sunshine in a specified time
Isodopane	Cost of travel time

Source: Thrower 1972, Appendix B.

that may result from input or computational errors. However, any thematic map is a form of generalization, so the isarithmic method is surely as conceptually correct as others.

When to Select the Isarithmic Method

Isarithmic mapping should be selected only if the advantages of its use contribute to achieving the goals of the mapping task. Certain additional requirements must be met before adopting this method:

1. The mapped data must be in the form of a geographical volume, or must be assumed to be voluminous, and must have a surface that bounds the volume.
2. It must be feasible to consider the mapped phenomena continuous in nature; discrete phenomena cannot be mapped isarithmically.
3. The cartographer must fully understand the distribution being mapped in order to verify the resulting map generated by the software. A casual acquaintance with the phenomenon will not suffice to develop a sensitive and accurate isarithmic solution to the mapped data.

Various advantages to the isarithmic technique must be weighed in the selection process:

1. Isarithmic mapping shows the *total* distribution of a spatially varying phenomenon.
2. It is flexible and can easily be adapted to a variety of levels of generalization or degrees of precision.
3. The technique is easily rendered by using computerized cartographic methods.

> *The cartographic designer chooses the isarithmic technique on the basis of these advantages, as weighed against those of other methods. Of course, such matters as data availability, base-map availability, and scale influence the ultimate selection. Beyond these considerations, other constraints are imposed by the method. Erwin Raisz, a noted cartographer and strong influence in the discipline during his life, once said, "Making true and expressive isoplethic maps is something of an art and requires the best geographical knowledge of the region."* (Raisz 1962, 201)

It is important for the cartographer to design the isarithmic map with the intended user in mind. The isarithmic map is a viable resource for communicating the variable surface, whether it be real or abstract.

ISARITHMIC PRACTICES

Like other forms of quantitative thematic mapping, the isarithmic variety contains elements and design strategies that are unique to the method. Although the computerized applications are most common, this section discusses the philosophy of generating isolines from a manual perspective in an effort to establish a basic understanding of the mechanics of the isarithmic software.

Elements of Isarithmic Mapping

The cartographer must master all elements of the isarithmic process, because they directly influence the quality of the finished map. The elements take on different significance in each mapping activity. Like with any computerized cartographic activity, the individual must understand the processes and procedures involved in the design and construction of the map.

Concepts in Isarithm Placement

In the construction of an isarithmic map, it is not usually necessary to erect pins or draw horizontal planes. The pin-and-plane model is, however, important conceptually. It should be thoroughly understood before beginning the planimetric design of the isarithms with respect to the array of data points.

Methods of projective geometry form the basis of isarithm placement (see Figure 9.4a). The method assumes that some generalized surface plane is interposed between adjacent data points. Where a hypothetical horizontal plane intersects this new surface plane, a trace is formed. The orthogonal projection of this trace is made to the map plane, intersecting the surface proportionally between the data points. In actuality the cartographer deals only with the map surface, the array of data points, and their values (see Figure 9.4b). The trace of the isarithm of a certain magnitude is manually placed by assuming linear distances between the data points.

First, all values of the horizontal planes are chosen. The cartographer then completes the rough map by "threading" the isolines through the array of data points. In the next step, these sharp, straight-line segments are *smoothed*—generalized—to give the appearance of a continuously varying, undulating surface (see Figure 9.5).

Locating Data Points

Data points have their locations specified by grid (xy) notation. Exact position can be determined either by an *x–y* rectangular coordinate reference or by geographic coordinates. Coordinate systems of Universal Transverse Mercator (UTM) and latitude-longitude in decimal degrees are frequently used.

In isarithmic mapping, the positions of data points can usually be specified exactly, because the positions of recording instruments are well known. In terms of meteorological data, weather station locations are precisely known. If the researcher is collecting primary data in the field, global positioning systems (GPS) are used to provide specific location data. In most cases, the data points used in isarithmic mapping exist in a non-uniform distribution. In any case, interval data collected at these specific points are used in the interpolation process. Problems usually relate to the uneven spacing of these points, because the lack of uniform spacing leads to varying precision in interpolation. In cases where

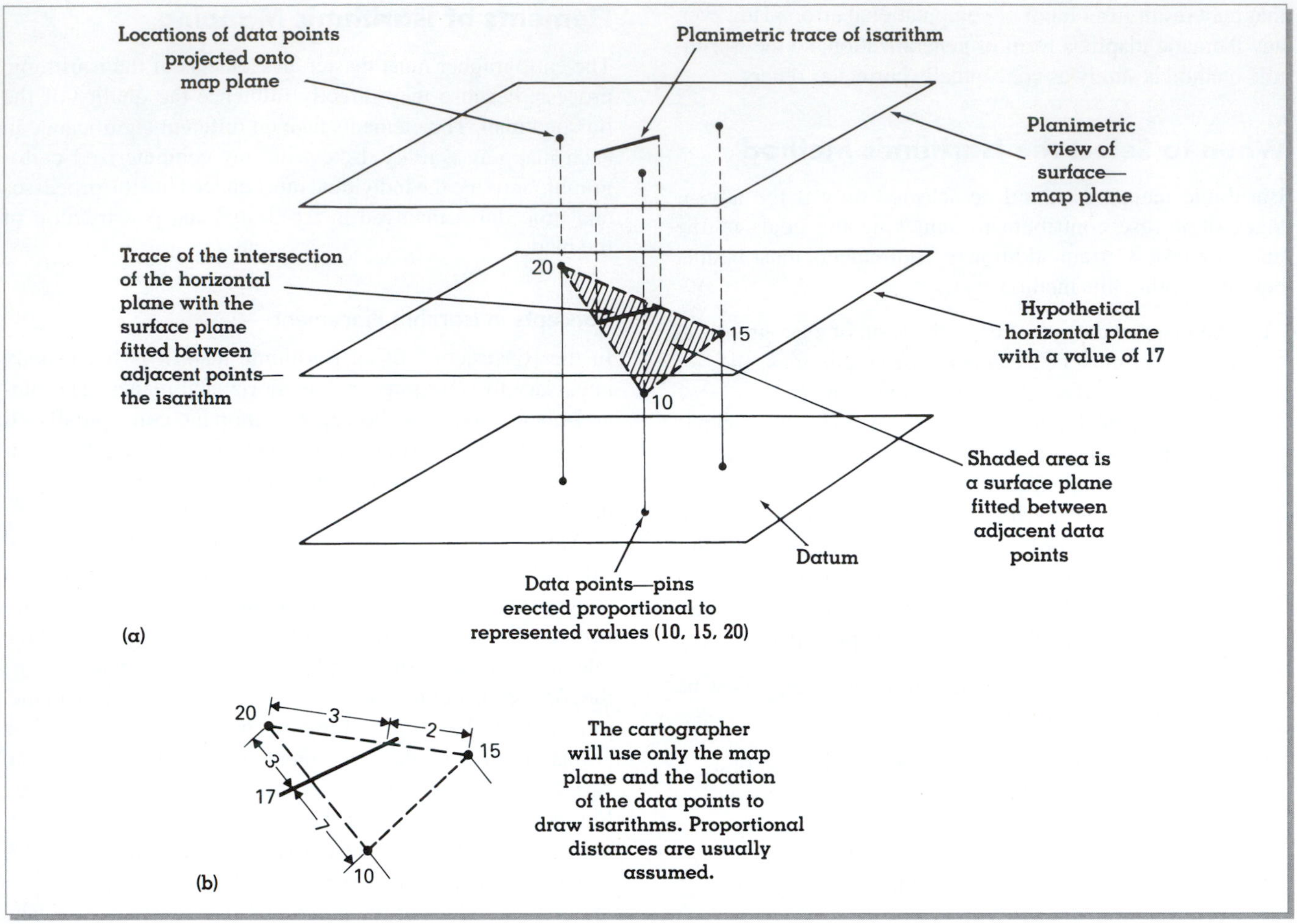

Locations of data points projected onto map plane
Planimetric trace of isarithm
Planimetric view of surface—map plane
Trace of the intersection of the horizontal plane with the surface plane fitted between adjacent points—the isarithm
20
15
10
Hypothetical horizontal plane with a value of 17
Shaded area is a surface plane fitted between adjacent data points
Datum
Data points—pins erected proportional to represented values (10, 15, 20)
(a)
20
3
2
15
3
17
7
10
The cartographer will use only the map plane and the location of the data points to draw isarithms. Proportional distances are usually assumed.
(b)

FIGURE 9.4 ISARITHMS.
Methods of projective geometry provide the basis for the planimetric location of isarithms.

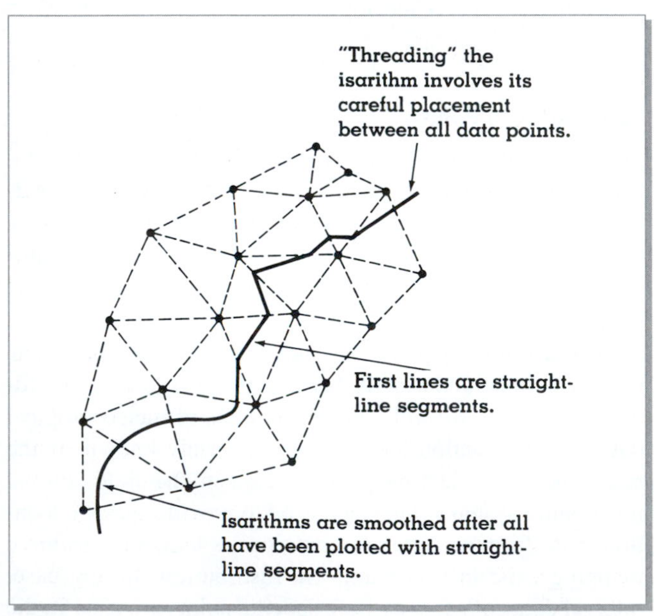

FIGURE 9.5 CONCEPTUAL BASIS FOR PLOTTING ISARITHMS.
Isarithm locations are first approximated as straight-line segments and then smoothed for the final rendering.

the cartographer does have control over the pattern of spacing, the selection of a point pattern that yields a regular *triangular* net is most desirable.

Concept of Interpolation

The key activity in isarithmic mapping is **interpolation** for determining isoline placement—a procedure for the careful positioning of the isolines in relation to the values of the data points.

Through projective geometry, the trace of the intersection of a hypothetical plane with a surface segment results in a line that falls proportionally between the data points. This is true, but the case described in Figure 9.4 was a linear fit. In *linear interpolation* between data points on the map plane, the cartographer positions an isarithm with a certain value between adjacent data points at proportionate distances from each (see Figure 9.6a). The linear fit assumes that the gradient of change in magnitude between adjacent data points is even and regular. Equal change in the unit over equal horizontal distances produces a uniform gradient.

Adoption of this method assumes that the distribution being mapped changes in linear fashion, with a uniform

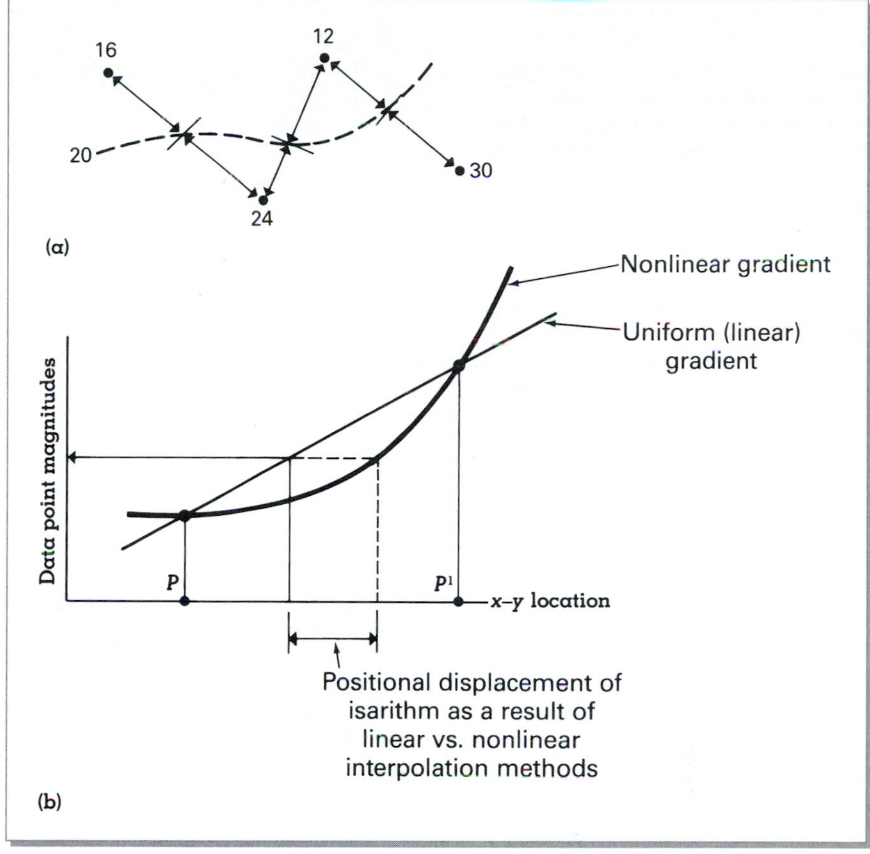

FIGURE 9.6 THE INTERPOLATION MODEL AND THE PLANIMETRIC LOCATION OF THE ISARITHM.

Linear interpolation is the traditional method of isarithm placement without the benefit of digital techniques as is shown in (a). Planimetric placement can result from the gradient determined by the gridding process and data point distribution.

gradient between data points. This may not be the case, however, as many geographic phenomena do not behave this way. Adopting a non-uniform gradient in preference to a linear one can result in considerable local displacement of the isoline (see Figure 9.6b). However, because of the underlying complexity of interpolating isoline position using a non-uniform gradient between all data points, and because the exact pattern of change from place to place is often unknown prior to mapping, most manual methods of interpolation assume uniform or linear change between data points.

Automated Isarithmic Mapping

The automated approach to the creation of an isarithmic map begins with the collection or acquisition of the data in *XYZ* format. Such data represent sampled locations of a continuous variable with the data points occurring as a non-uniform distribution. From these randomly spaced points, a regularly spaced grid of projected data will be generated. This interpolation process is referred to by a variety of names, such as **gridding** or **surface modeling.** The grid dimensions will extend to the maximum and minimum values of *XY*. The resolution of the grid is user specified and is accomplished by indicating the number of grid lines desired on the longest side. The number of grid lines is automatically calculated for the shorter side. A second approach is to specify the size of the grid cell and allow the software to determine the number of grid lines on each side. The cell size and/or the number of grid lines chosen will determine the size of the grid file generated.

Z-values, generated through an interpolation process, represent the continuous surface area. The known values from the data points are used to predict the *z*-values at the grid intersections (nodes) (see Figure 9.7). The automated gridding interpolation is not linear and therefore more complicated than can be determined manually.

The number of data points used in the interpolation process is of less importance to the accuracy of the final grid surface than is the distribution of those points. Although logic would suggest that more data points would produce a better interpolation result, this is generally not the case. Research has shown that interpolations produced from a larger number of data points were less accurate than those from smaller data sets. A data distribution with areas of clustering and areas of few data points produces a surface that is unevenly generalized and, thus, unevenly accurate (Yang and Hodler 2000). Robinson *et al.* (1995) found that increasing the number of data points may actually diminish the overall grid accuracy by producing a clustering effect. Wang (1990) determined that a sample density of approximately 10 points per square kilometer produced the best results. "The distribution of sampled points and their geometric and attribute accuracy may outweigh the effect of absolute number and therefore a larger sample size does not always ensure higher accuracy of products" (Yang and Hodler 2000, 173).

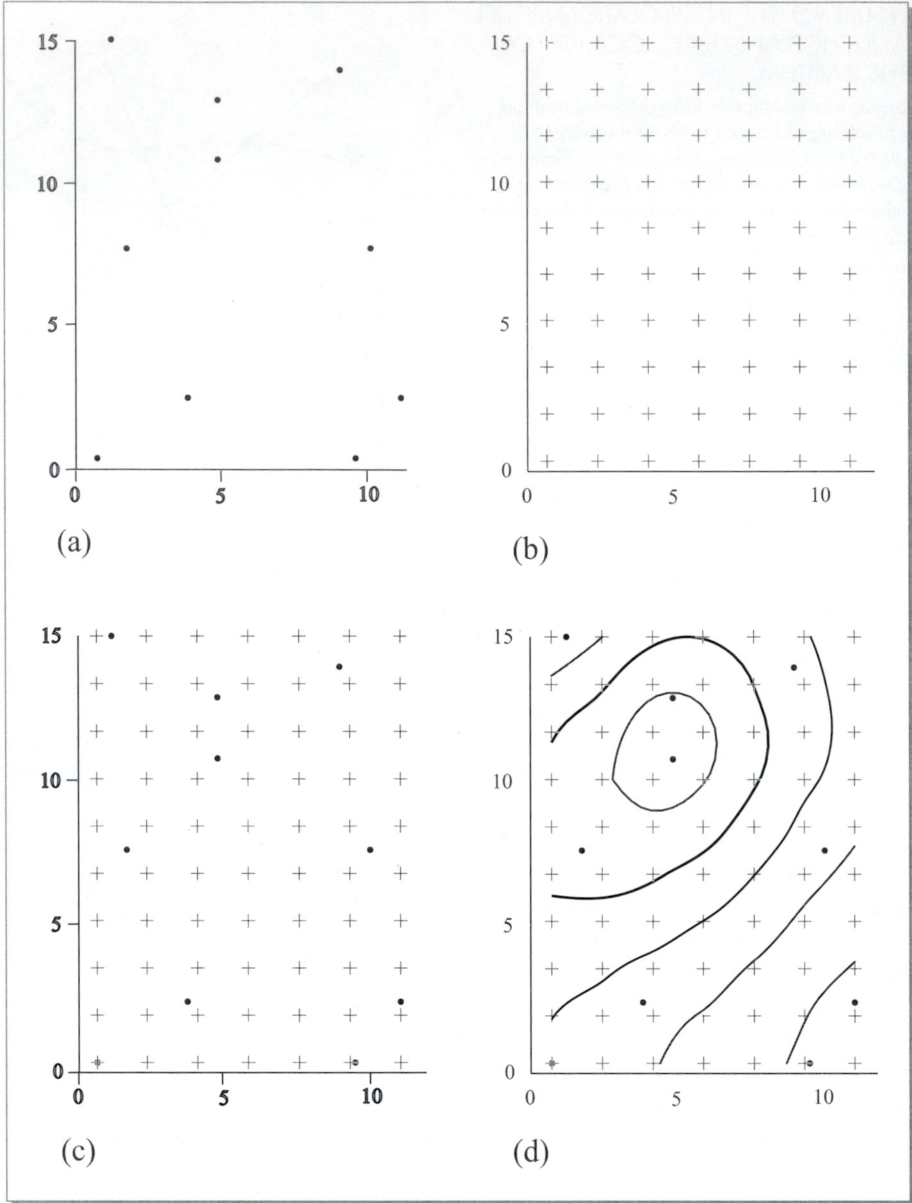

FIGURE 9.7 CONCEPTUALIZATION OF THE GRIDDING PROCESS.

Depicted here are the steps taken in going from an original data set (comprised of ten fictitious data here) to the generation of isarithms. (a) Data are collected at *xy* locations with some attribute *z-value*. (b) Within the software the grid size and frequency can be set, thus generating grid intersections (nodes). (c) Gridding assigns a new data value for each node which may or may not include the original input data as the original points would have to coincide with the grid nodes. (d) The software threads the isarithms based on the grid spacing and node values.

The gridding interpolation methods, discussed in the following paragraphs, are used to produce a regularly spaced grid of data values by which isarithmic and surface models can be generated. Known values of observation points are used to predict values of grid intersection points through an iterative process, one intersection at a time. It is logical to think that points closest to an intersection will have a greater impact on defining the value of a grid intersection than those points farther away. Such a relationship was first voiced as the first law of geography by Tobler (1970) when he indicated that everything is related to everything else, and near things are more related than distant things. How these distributions are utilized in the gridding process is what helps to identify their differences.

Gridding Methods of Interpolation

There are a variety of interpolation methods available to the cartographer (Franke 1982; Watson 1992; Yang and Hodler 2000). Some are known as **exact interpolators** and include the value of the data points when they coincide with one of the grid nodes, while others are **smoothing interpolators** which do not necessarily include the data point value in order to smooth the overall surface. Even when using an exact interpolator, it is possible not to honor the input data values if they are not coincident with a grid node. In order to increase the probability that your data point will coincide with a grid node, increase the number of grid lines used. Surfaces generated by exact interpolators may have greater variability in the range of surface values. If you are not certain about the surface being generated, use one of the smoothing interpolators for the gridding of your data.

Inverse Distance Weighted and Kriging interpolation methods can selectively produce either exact or smoothed results depending upon options specified. These common interpolation methods can be selected for either of the two gridding approaches, while Nearest Neighbor and Triangulation with Linear Interpolation approaches are always exact interpolators.

Inverse Distance Weighted (IDW). This method is one of the most widely used techniques for generating a surface. Data points that are close to a particular grid intersection will have greater influence in determining the predicted value than points farther away. That is, points that are more closely positioned will have more similar values than points that are farther away. The influence of a particular data point value in determining the grid node value will be inversely proportional to its distance from that node. Thus Inverse Distance is a weighted average interpolator whether exactly or smoothly defined. For each node in the grid, every point will be used to predict the value of the node. The software progresses sequentially through each node of the desired grid until all nodes have an assigned value.

A general formula used to calculate the z-values at a given node is

$$Z_{x,y} = \frac{\sum_{p=1}^{R} Z_p d_p^{-n}}{\sum_{p=1}^{R} d_p^{-n}}$$

where:

Z_p = Z variable at point p;

R = the number of points in the neighborhood;

d = distance from the grid node to point p; and

$Z_{x,y}$ = Z variable at the grid node.

The value of n is the weighting factor that causes distant points to have less influence on the predicted node than closer values. For a weight factor of 2, the method is commonly referred to as Inverse Distance Squared.

Kriging. Kriging was developed by Matheron in 1963 and named in honour of D.G. Krige. Kriging has been highly recommended as a method of interpolation for GIS (Oliver and Webster 1990, Burrough and McDonnell 1998). It has proven useful in geostatistical applications and is accepted in many fields. It is particularly noted for its generation of visually appealing maps. This method attempts to include information about the surface through the use of an underlying variogram. It treats the data as a regionalized variable which creates an unbiased estimate of the predicted surface (Lam 1983). Kriging has several advantages because it generates a surface that passes directly through the data points, and because it attempts to express trends found in the data (Lam 1983; Yang and Hodler 2000). It is similar to IDW by utilizing a weighting function of the surrounding data points in order to predict each node value. "However, the weights are based not only on the distance between the measured points and the predicted points (grid nodes) but also on the overall spatial arrangement among the measured points (ESRI 2007)."

Nearest Neighbor. This interpolator can be used for data sets that already exist in a uniform or gridded manner. If there are missing values or holes in the data, this method assigns a value to a node based upon the value of the node nearest its location. For example, if a particular node has no z-value specified, the interpolator will assign a value to that node based upon the values of the nodes immediately surrounding it. This adjustment for missing data does little to the resultant surface configuration as a result of z-value averaging.

Linear Triangulation Interpolation. The linear triangulation (LT) method makes it possible to join the data points with lines to form a patchwork of triangles in the grid. Delaunay triangulation produces triangles that are as close to being equilateral and well shaped as possible. Each triangle is treated as a plane surface. Once the surface is defined, the values for the interpolated nodes can be calculated. The data distribution should be evenly distributed as areas with sparse distribution will result in distinct triangular facets on the map. This surface can be either raster-based or vector-based (TIN). For a good discussion of this method and the resulting triangulated irregular network (TIN), refer to Lo and Yeung (2002).

We have found that the use of the Kriging interpolator generates the most reliable data surface in the majority of cases. This technique has become widely accepted as the interpolator of choice. Figure 9.8 depicts the results of the four methods gridded using the data points depicted in (e) while maintaining all other factors constant. While the IDW method is the simplest to understand, it is easily affected by an uneven distribution of data points (Lam 1983). A discussion of grid error in the following section provides for techniques to quantitatively evaluate the gridded surface when compared to the original data points.

Evaluating Grid Error

The chief sources of errors for isarithmic mapping relate to (1) the quality of the data, which may be affected by observational bias or sampling errors, (2) the number of sampled data points used, and (3) the interpolation method used. Observational error can be caused by either humans or machines. Faulty instrumentation or poor instrument-reading practices can lead to erroneous z-values. Sampling error is of considerable concern. Because the observations (z-values) used represent only one set from a much larger universe, the manner of selection (sampling) can become a source of error. The distribution of the sampled data may not be under the control of the cartographer. While the sampled data should be evenly distributed throughout the area, all too frequently the data are clustered with greater concentrations in some locations and sparse in others. This will cause the overall accuracy of the predicted data values to vary based upon that spatial pattern.

An evaluation of the accuracy of the final grid can be achieved by comparing the values of the original data points to their predicted values. Error results when the predicted

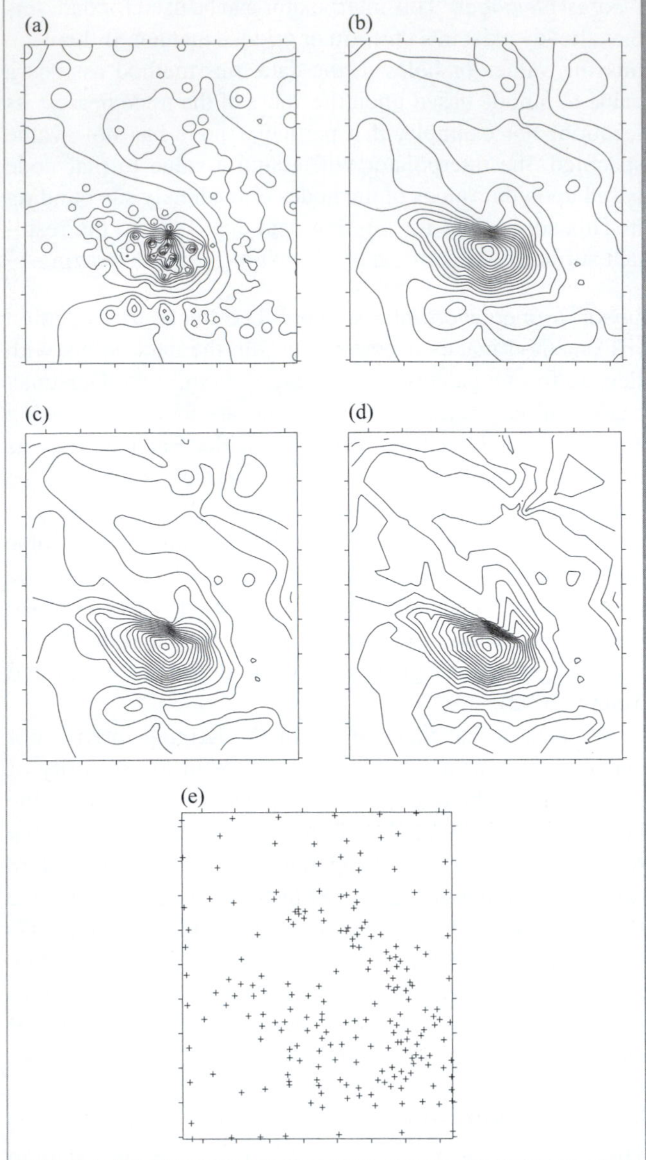

FIGURE 9.8 ISARITHM COMPARISON DETERMINED BY DIFFERENT GRIDDING METHODS.

(a) Inverse Distance Squared, (b) Kriging, (c) Nearest Neighbor, (d) Linear Triangulation Interpolation; (e) depicts the spatial distribution of data points used in the gridding process.

value is different from the observed value. The statistical accuracy can be evaluated in both the global (absolute) range of errors and their relative variations across the surface. The global error is based on the index of the root mean square error (RMSE) calculated for different gridded models. The RMSE has been considered as the best measure of goodness of fit in judging the suitability of models (Wang 1990). The RMSE is computed:

$$RMSE = \sqrt{\frac{\sum_{i=1}^{|n|} (Z_i^{dat} - Z_i^{grd})^2}{n}}$$

where

Z_i^{dat} is the surface property of interest, that is, the elevation value of point i in the original sample file;

Z_i^{grd} is the elevation value of the interpolated surface at the X,Y coordinate of point i;

Z_i^{dat} and Z_i^{grd} is the residual; and

n is the total number of sample data points.

In other words, the Z values of the interpolated model are compared with those of the original sample file. If the difference between these values is small, the RMSE is small, implying that the interpolated model is statistically more accurate, and vice versa.

The use of relative errors can provide information on how error varies across the surface, which is believed to be more useful than the absolute error values in many cases (Wood 1994). If the global errors are identical, a surface model with evenly distributed errors is much more reliable than one with highly concentrated errors. The relative errors were based on the residual error values for individual cells. The residual error (RE) is computed:

$$RE = abs(Z_i^{dat} - Z_i^{grd})$$

where abs stands for absolute value.

Voltz and Webster (1990) utilized a combination of RMSE and a form of residual error for validating various gridding methods for predicting soil properties from sampled data. They used both mean error and mean square error (MSE) for validating the gridding methods. Mean error sums the values of actual residuals and divides the total by the number of data points. The result measures the bias of the prediction and a value approaching zero is desired. They also used the mean square error which summed the square of deviation values divided by the total number of observations. The MSE measures the precision of the prediction.

When the distribution of the sampled data points maintains a reasonable density, their attribute accuracy may outweigh the absolute number of observations. It is safe to say that a larger sample size does not ensure a higher accuracy of the final product (Yang and Hodler 2000). The IDW method tends to be more sensitive to changing sample size than the Kriging method.

The Selection of Isarithmic Intervals

It is impossible to map all isarithms because of space constraints. The cartographer is faced with the task of choosing the more appropriate ones and a suitable number of them. An **isarithmic interval** must be selected for the map.

The only reasonable and logical approach is to select a uniform interval. Isoline values of 2, 4, 6, 8, and 10 are appropriate, but 2, 10, 15, and 20 are not. To put it another way, the hypothetical horizontal planes that pass through the three-dimensional surface should be *uniformly spaced vertically*. Only through this approach can the isolines show the total or integrated form of the three-dimensional surface

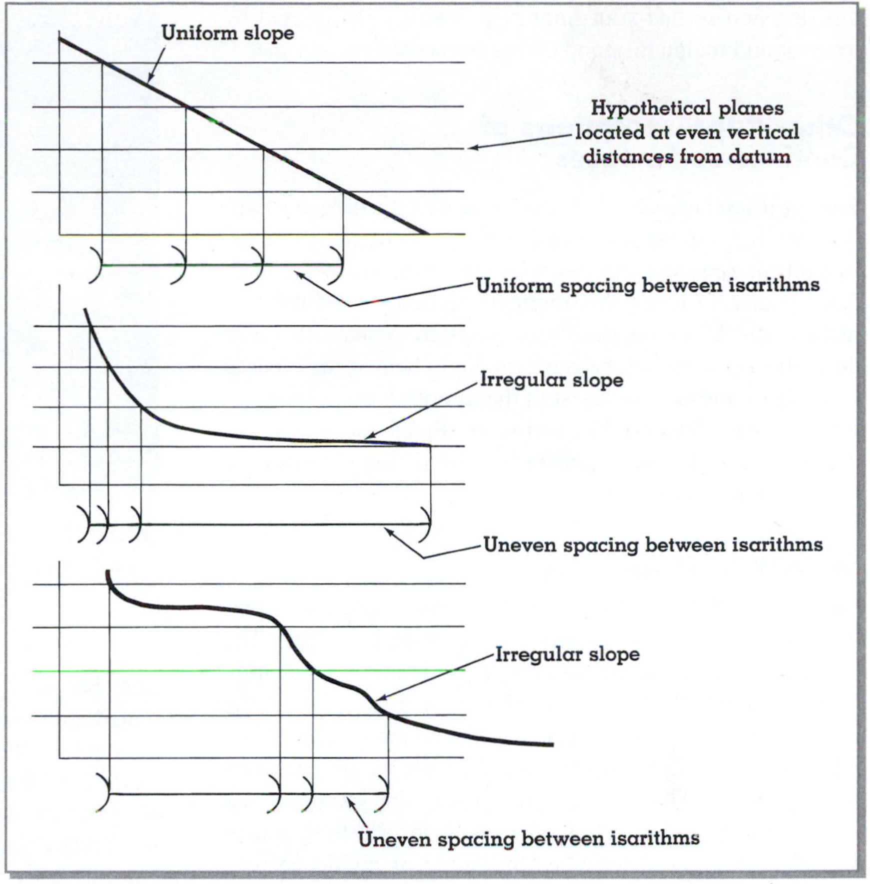

FIGURE 9.9 ISARITHM SPACING.
The spacing of planimetrically viewed isarithms determines the gradient of changes of the surface. Isarithms spaced close together indicate rapid change (steep slope), and those spaced farther apart suggest a relatively gradual slope. Experience in reading isarithmic maps is required.

(see Figure 9.9). This results in a particularly knotty problem at times, because most geographical distributions are skewed. The natural flow of a continuous surface is not planar or smooth. Rather, peaks and valleys of data exist that are difficult to display when abrupt changes occur. At best, the designer can choose map scale and isoline interval carefully to avoid crowding; the result will be a compromise.

Another aspect of interval selection is the determination of the lowest isoline value. Because of the uniform interval, the lowest isoline greatly influences where subsequent isolines fall in relation to the whole map. The overall range of the data must be considered. The datum of the data may not be a part of the actual data being mapped. Therefore, the isoline with the lowest value will be the first in the sequence of the isarithmic interval at or just above the minimum data value. For example, if the lowest data value in the interpolated grid is 104 and the selected interval 10, the lowest isoline will have a value of 110. Will the chosen interval show the distribution best? No universal rules apply here. As in so many other areas of thematic cartography, experimentation, geographical knowledge, study, and experience are necessary.

The relative relief of the surface, its highest value minus its lowest value, has a direct impact on the number of isolines displayed on the map. A surface that has large relief will have more isolines plotted than a surface with lower relief when using the same interval. Therefore, the selection of the isoline interval should take the amount of relief into consideration. Surfaces that have an abrupt change in slope will have the isoline very closely spaced and the reverse is true for surfaces with limited slope variation. An interval that provides too many isolines generally results in a map that looks either "too accurate" or too congested. A wide interval appears too generalized and reveals little. A contour interval of 10 feet is fine for central Iowa but would not be appropriate in western Colorado. This is especially true in cases where a large number of isolines have been interpolated from a small number of data points.

In most cases, a wise choice of interval requires experimentation with the choice of intervals. Current software provide for near instantaneous construction of the map in virtual form. The final selection of the isarithmic interval is achieved by multiple iterations that allow the cartographer to experiment with a variety of intervals. The final map represents only one of many possible solutions; it should therefore be based on a variety of views that provide the most information with the least congestion.

We often begin with the total relative relief divided by ten. The resultant number is a good starting point for determining your interval. This number should be adjusted using a logical approach suggested earlier. For example, if your resulting number is 12, you would begin the iteration process using an interval of ten. A number of 22 would suggest an interval of 25 and so on. The resulting interval determines the overall look of your map. Care should be taken in making this decision so that your final map is easily interpreted by the user and maintains good cartographic design.

Other Representations of Continuous Surfaces

The isarithmic map displays the surface of the data in a two-dimensional approach utilizing isolines to depict the data. Not all map readers can easily interpret the surface of that data accurately using this thematic technique. Of the thematic maps discussed, isarithmic maps are perhaps the most difficult to understand, because the logic behind the method is the most abstract. To assist in the interpretation of the isarithmic map, shaded relief and three-dimensional views of the surface can be used alone or in conjunction with its two-dimensional counterpart.

Shaded Relief Maps

Shaded relief maps are created so that a pseudo three-dimensional surface is displayed using a shadow effect. The design of such images has long been used in the presentation of surface data. The terrain is depicted using variations in the grayscale in order to imply relief. In the generation of shaded relief maps, the light source should be placed in the northwest quadrant. This projects shadows to the southeast and creates an illusion of relief. The height of the light source above the horizon determines the angle of incidence with which the light strikes the surface. Just as in the natural setting, a higher angle of incidence will produce shorter shadows while a lower angle produces longer shadows. Although this is quite an effective tool for visualizing the terrain in areas of moderate relief, mountainous areas may generate such long shadows that the detail of information within the shaded portion may be lost. The cartographer must be aware that a light source positioned in the southeast creates a phenomenon of relief inversions, whereby the hills appear as valleys and vice versa (Patterson 2002; see Figure 9.10). The generation of a shaded relief image is a functional component of automated mapping programs.

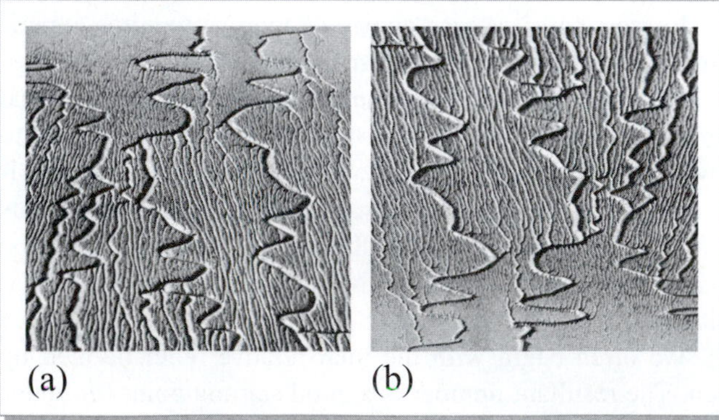

FIGURE 9.10 SHADED RELIEF.
(a) The light source placed in the upper left quadrant created a visualization of relief as the shadows are cast to the lower right. (b) Relief inversion results when the light source comes from the lower right causing the channels to be viewed as ridges. (Source: Hidore 1974. Used with permission.)

FIGURE 9.11 SHADED RELIEF MAP OF THE PACIFIC NORTHWEST.
(Source: National Geophysical Data Center 2008)

Shaded relief surfaces can be generated from digital data made available from the U.S. Geological Survey (USGS). National Elevation Dataset (NED) can be downloaded from the USGS Seamless Data Distribution Systems (SDDS) website. This elevation data replaces the Digital Elevation Models (DEMs) that have been available in 7.5 Minute Topographic Quadrangle format. The data are downloaded in raster format and the area desired is either specified by latitude-longitude coordinates or by selecting an area from an interactive map (USGS 2006). The elevation data are in meters with a seven-meter RMSE accuracy. The shaded relief images can be generated from 1-, 1/3-, or 1/9-arc second resolution (see Figure 9.11). Traditional DEMs in topographic map format continue to be available from USGS but are no longer being updated.

Wireframe and Surface Maps

Both the wireframe and the surface map allow the user to visualize the undulations of the three-dimensional surface. The wireframe represents the skeleton structure on which a continuous skin can be draped which hides the wireframe base. Whether you use the wireframe alone, the maps utilize the grid generated in the interpolation process to display the continuous surface in a three-dimensional view. A matrix of lines of the *x*- and *y-axes* are used in conjunction with the *z-axis* values (interpolated data values at each node) to create the wireframe. The cartographer can manipulate both the compass direction and view angle indicating the viewing position of the user. Compass direction can be established by selecting a particular angular rotation east or west of the traditional view direction of south. One of the more common view directions is 45° west of south.

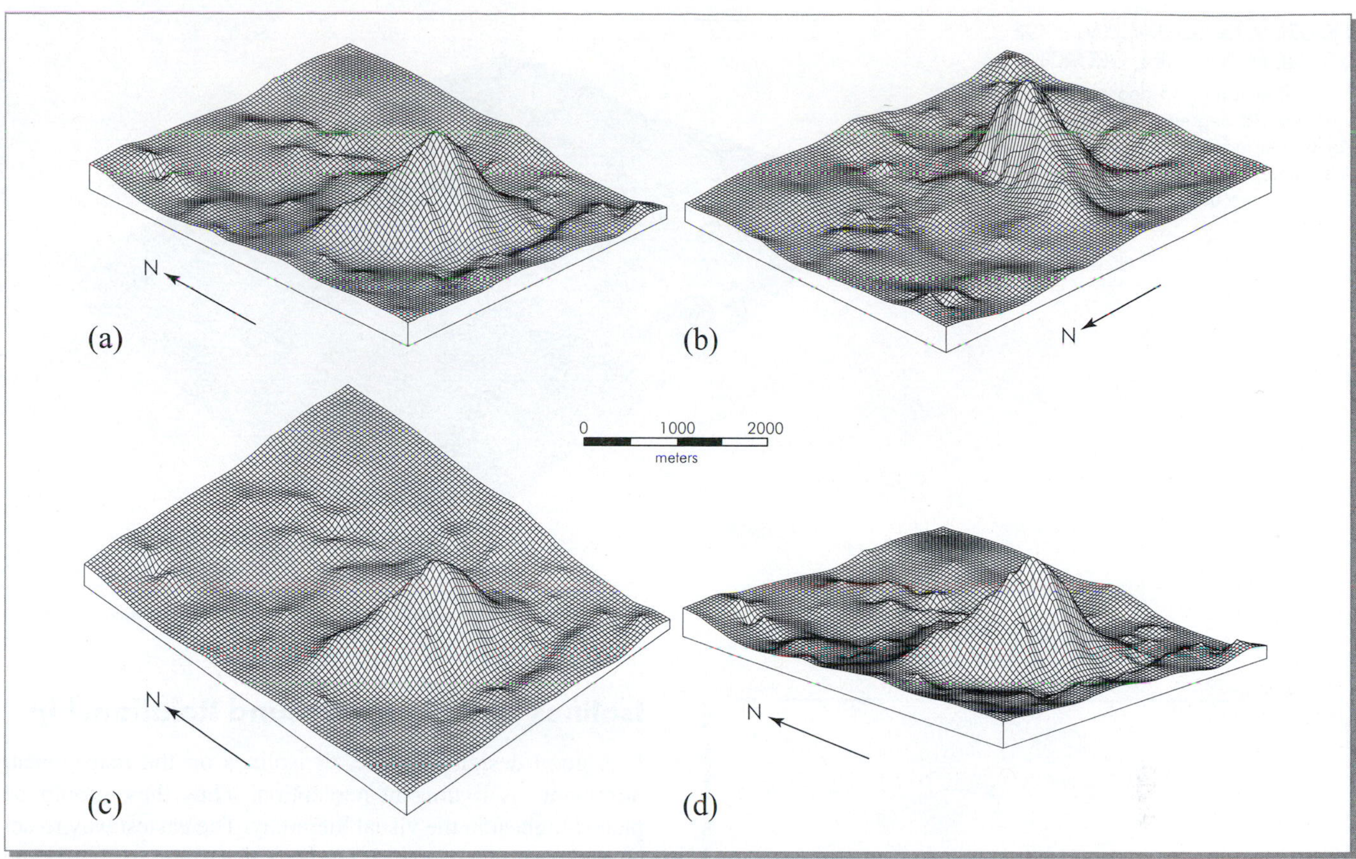

FIGURE 9.12 WIREFRAME MAP OF STONE MOUNTAIN, GEORGIA.
These four maps depict different view direction and view angle: (a) the standard view direction of 45 degrees and a view angle of 30 degrees above the horizon, (b) view direction of 135 degrees, (c) view direction of 45 degrees and view angle of 45 degrees, and (d) view angle of 20 degrees.

The normal view angle of 30° above the horizon creates the three-dimensional relief necessary to represent the variations in the surface. Lower than this angle and the surface may take on a profile view of the surface and higher than 30° reduces the amount of relief. As the view angle approaches 90°, or directly overhead, only the *xy* lines of the grid show in a planimetric manner.

GIS and mapping software allow for manipulation of these variables which in turn rotates and tilts the image for a new viewing position. This is done in order to maximize the visible surface. A general rule of thumb is to position the areas of higher elevation (*z*-values) to the rear of the image. This permits the viewing of the maximum surface area and minimizes the shadow effect created by taller objects hiding shorter ones (see Figure 9.12).

The map scale can be modified by changing the ratio relationship of *x*- and *y*-axes. The level of vertical exaggeration may be altered by modifying the scaling of the *z*-dimension. This will alter the display of the surface so that the surface undulations are either over- or underexaggerated. Care should be taken when modifying the vertical exaggeration so as not to alter the perception of that surface.

Surface maps are another form of three-dimensional mapping that combine the view angle and view direction of the wireframe map with the hill shading of the shaded relief map. GIS and mapping software build the surface without first displaying the wireframe (see Figure 9.13 and Color Plate 9.1). From the digital surface, new maps can be generated within a GIS in order to display characteristics of viewshed, slope, aspect, and other spatial components of the data. These surfaces are used in spatial analysis and are common components of a GIS software. The surface mapping functions share a commonality with 3-D fly through software whereby the surface is rendered to represent a different view location.

Communicating Using Multiple Map Displays

The communication of the variability of a surface can be enhanced through the use of multiple maps viewed simultaneously. This helps overcome the misinterpretation of the surface displayed by a single isarithmic map. Surfaces are difficult to understand and it is helpful for the cartographer to provide multiple views of the surface to help improve that understanding. These maps may be positioned side-by-side or stacked vertically for improved visualization (see Figure 9.14).

FIGURE 9.13 SURFACE MAP OF STONE MOUNTAIN, GEORGIA.
View direction of 45 degrees and a view angle of 30 degrees. (See Figure 9.11a for a comparison to the wireframe with the same view.) (See Color Plate 9.1 for this figure using color in place of grayscale.)

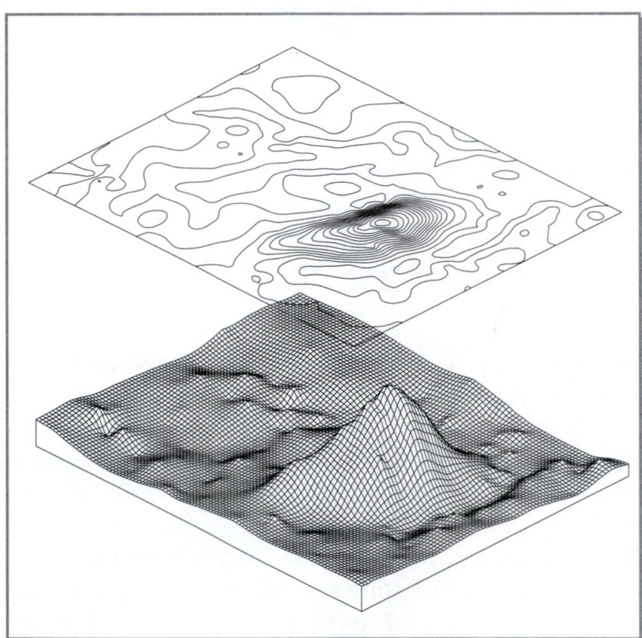

FIGURE 9.14 MAP COMPOSITE WITH CONTOUR MAP SUPERIMPOSED OVER THE WIREFRAME OF STONE MOUNTAIN, GEORGIA.
The use of both the isarithmic and three-dimensional view may make it easier to visualize the surface.

DESIGN ASPECTS FOR ISARITHMIC AND CONTINUOUS SURFACE MAPS

Several design considerations related to the appearance of the completed isarithmic and surface maps are addressed in this final section.

Isolines and Figure-Ground Relationship

It is good design to make all isolines on the map appear dominant, as figures in perception. Thus they should be placed highest in the visual hierarchy. The easiest way to accomplish this is to render the lines as solid dark lines, and lighten much or all of the remaining map information (see Figure 9.15a). In this way, the total form of the distribution becomes immediately recognizable, and the less important but necessary base-map material does not interfere. Figure 9.15b does not establish the figure-ground relationship and therefore renders the map ineffective. In some circumstances, it may be necessary to provide data point locations, and even their values, on the map. The points and values may then be screened to reduce their contrast and thereby lessen their figural strength.

The use of symbol size and color contrast also helps to establish the isolines as the figure and the remaining information as the ground. Heavier solid lines of a darker color hue and a lighter gray or pastel color will heighten the importance of the isolines above the background location. As in other forms of thematic mapping, contrast should be developed to emphasize the visual and intellectual hierarchy.

Isoline Labels

Isolines should be labeled periodically to make reading easy. Do not over label as this reduces the statistical surface shown by the lines. Labels should be easy to read but not so large as to dominate the map. Beginning students often make an easily avoidable mistake: placing labels upside down. Labels at the ends of isolines are appropriate, but each should be placed consistent with the flow of the line to which it is associated. The extension of the isoline beyond the map boundary is a possible solution appropriate as a result of mapping

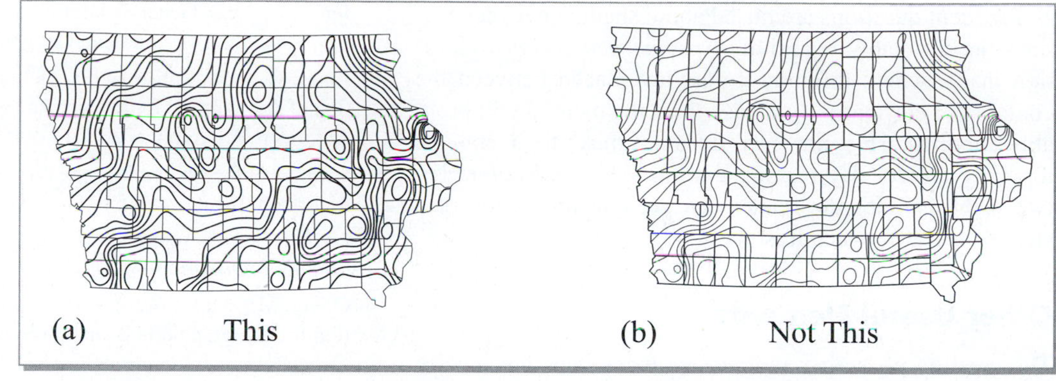

FIGURE 9.15 ISARITHMS AND FIGURE-GROUND RELATIONSHIP.

Increasing figure-ground differences allows for better visualization of the surface; (a) promotes better figure-ground relationship; (b) the figure-ground relationship is lost among the county lines.

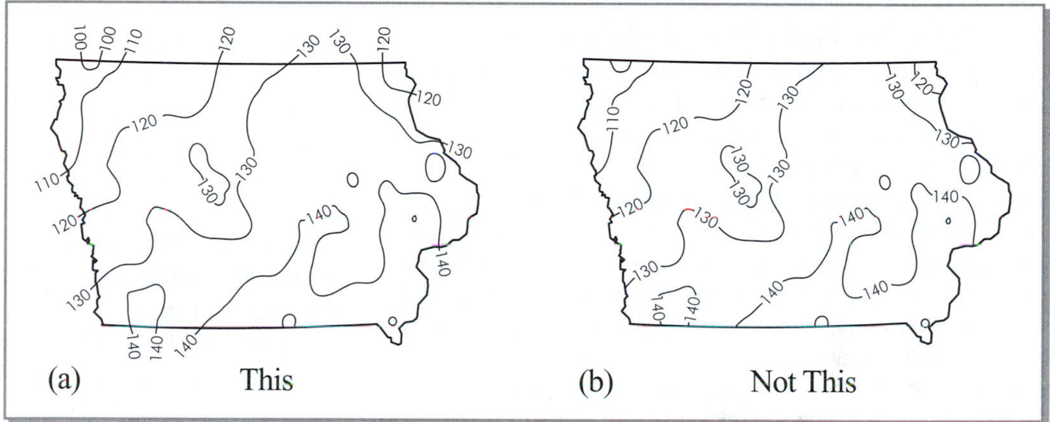

FIGURE 9.16 LABELING THE ISOLINES.

(a) Labels are placed at the end of the isolines or at a midpoint segment within the line; (b) too frequent placement clutters the map.

continuous data. All isolines would eventually be closed given a map large enough to cover near continental dimensions. Figure 9.16b is an example of too high a frequency of line labeling as well as having some labels placed in an upside down position. A better choice of the number of labels per line is displayed in Figure 9.16a, and the labels have been adjusted so that they are all correctly oriented.

Three variables control the placement of labels in automated systems: curve tolerance, label-to-label distance, and label-to-edge distance. Curve tolerance examines the curvature of the isoline and determines a threshold value at which a label may or may not be placed. Tight curves make it difficult to display the label along the line and still maintain the visibility of that line. Therefore, some lines that are highly curved should not be labeled. Label-to-label and label-to-edge measurements affect the frequency that a line is labeled. It is not wise to have a label placed close to the map boundary within the body of the map. The line should be extended slightly beyond the map boundary and then apply the label to the end of the line (refer to Figure 9.1).

Legend Design

Legends should be designed for clarity. The legend material most often contains only verbal statements. These must describe at least (1) the units of the isolines (for example, frost-free days, days of sunshine, average annual precipitation in inches, persons per square mile) and (2) the isoline interval. In those instances when shading is applied to areas between isolines, the tints, the isoline interval, and the representative values can be combined in one legend (see Figure 9.17 and Color Plate 9.2).

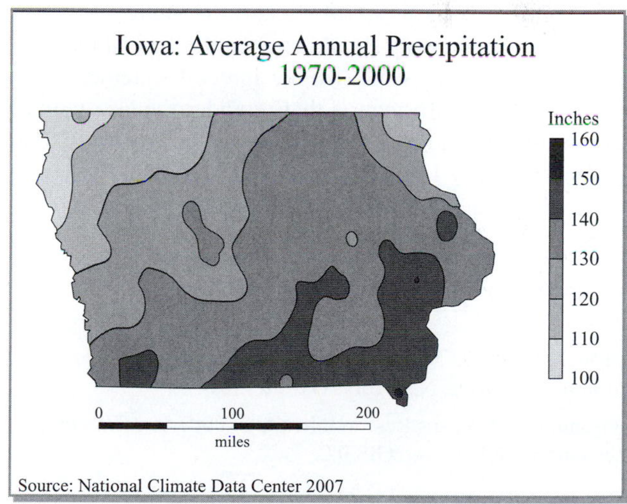

FIGURE 9.17 SHADING BETWEEN ISARITHMS.

The addition of shades of gray between isarithms identifies areas with similar value range. The scale of the map is the vertical identification of grayscale to data relationship. (See Color Plate 9.2 for this figure, using color in place of grayscale.)

Frequent questions are raised about shading between isolines, most notably regarding *hypsometrically tinted* elevation maps, where different colors are placed between the contours of elevation. By all accounts the method will probably continue. This, however, does not make the technique logically correct. By placing the *uniform areal colors* between the isolines the reader can get the impression of a stepped surface, which it is not.

Other Useful Elements

The creation of isoline and surface maps does not involve particularly difficult or unique circumstances. The use of grid interpolation and contouring software provides the cartographer with rapid evaluation of the accuracy of the mapped surface and the design of the final map product. Using isarithmic maps to display surfaces is further enhanced with the inclusion of other surface map products such as shaded relief, wireframe, or surface maps. A better understanding of the surface variation is achieved through the use of a combination of these maps whether placed side-by-side or stacked vertically. If the maps are stacked, the isarithmic map must also be rotated using the same view angle and view direction used to produce the three-dimensional maps (refer to Figure 9.13).

In order to achieve a viewing perspective when using a wireframe or surface map, the map is rotated to some position where north is no longer at the top of the page. It is imperative that a north arrow be included with these maps in order to allow the user to orient the map when compared to its planimetric version. Rotation of these maps should be made so that the greatest surface area is revealed. Frequently, a peaked surface positioned in the foreground of the figure will hide in its shadow valuable surface information. This effect can be lessened by modifying the view direction so that the higher peaks are to the back of the figure.

Every isarithmic map should contain the critical map elements of title, linear scale, isoline interval statement, data source, and other information that provides the user with an accurate understanding of the map display (refer to Figure 9.1). Chapters 1 and 13 provide a useful discussion of these critical map elements for all the thematic map types.

REFERENCES

Burrough, P., and R. McDonnell. 1998. *Principles of Geographic Information Systems*. New York: Oxford Press.

Environmental Systems Research Institute. 2007. *ArcGIS Desktop Help*. Online Help for ArcGIS 9.2.

Franke, R. 1982. Scattered Data Interpolation: Tests of Some Methods. *Mathematics of Computation* 38 (138).

Hidore, J. 1974. *Physical Geography: Earth Systems*. Glenview, IL: Scott, Foresman and Company.

Hsu, M., and A. Robinson. 1970. *The Fidelity of Isopleth Maps*. Minneapolis: University of Minnesota Press.

Jenks, G. 1963. Generalization in Statistical Mapping. *Annals* of the Association of American Geographers 53:1553:15–26.

Jones, W. 1930. Ratios and Isopleth Maps in Regional Investigation of Agricultural Land Occupance. *Annals* of the Association of American Geographers 20: 117–95.

Lam, N. 1983. Spatial Interpolation Methods: A Review. *The American Cartographer* 10 (2): 129–49.

Lo, C.P., and A. Yeung. 2002. *Concepts and Techniques of Geographic Information Systems*. New Jersey: Prentice Hall.

Oliver, M., and R. Webster. 1990. Kriging: A Method of Interpolation for Geographic Information Systems. *International Journal of Geographical Information Systems* 4(3).

Patterson, T. 2002. Getting Real: Reflecting on the New Look of National Park Service Maps. *Cartographic Perspectives* 43: 43–56.

Porter, P. 1957–1958. Putting the Isopleth in its Place. *Proceedings of the Minnesota Academy of Science* 25–26: 372–84.

Raisz, E. 1962. *Principles of Cartography*. New York: McGraw-Hill.

Robinson, A. 1971. The Genealogy of the Isopleth. *Cartographic Journal* 8: 49–53.

———. 1982. *Early Thematic Mapping in the History of Cartography*. Chicago: University of Chicago Press.

Robinson, A., J. Morrison, P. Meuhrcke, A. Kimerling, and S. Guptill. 1995. *Elements of Cartography*, 6th ed. New York: Wiley.

Robinson, A., R. Sale, and J. Morrison. 1978. *Elements of Cartography,* 4th ed. New York: Wiley.

Thrower, N. 1972. *Maps and Man*. New York: Prentice Hall.

Tobler, W. 1970. A Computer Movie Simulating Urban Growth in the Detroit Region. *Economic Geography* 46: 234–40.

U.S. Geological Survey (USGS). 2006. *National Elevation Dataset*. http://ned.usgs.gov/

Voltz, M., and R. Webster. 1990. A Comparison of Kriging, Cubic Splines and Classification for Predicting Soil Properties From Sample Information. *Journal of Soil Science* 41: 473–90.

Wang, L. 1990. *Comparative Studies of Spatial Interpolation Accuracy*. Master's thesis, University of Georgia.

Watson, D. 1992. *Contouring: A Guide to the Analysis and Display of Spatial Data*. Oxford: Pergamon.

Wood, J. 1994. Visualizing Contour Interpolation Accuracy in Digital Elevation Models. In *Visualization in Geographic Information Systems,* ed. H. Hearnshaw and D. Irwin, 168–80. New York: Wiley.

Yang, X., and T. Hodler. 2000. Visual and Statistical Comparisons of Surface Modeling Techniques for Point-based Environmental Data. *Cartography and Geographic Information Science* 27 (2).

GLOSSARY

centroid a location representing the geographic center of a polygon; if that center is outside the polygon, most GIS software place the centroid within the polygon

data points points at which magnitudes occur, or are assumed to occur, and from which isarithmic maps are constructed; each point's magnitude is referred to as its *z*-value

exact interpolator creation of a grid matrix whereby the original data points are included

gridding process of interpolation in which the data for the generation of a surface occur in the mapping software Surfer

interpolation procedure for the careful positioning of isolines between adjacent data points

isarithmic interval the vertical distance between the hypothetical horizontal planes passing through the three-dimensional geographic model; selection of the interval determines the degree of generalization and detail of the map

isarithmic map a planimetric graphic representation of a three-dimensional surface

isobaths isometric lines showing the depth of the ocean

isogones isometric lines showing equal magnetic declination

isohypse isometric line showing the elevation of land surfaces above sea level

isoline mapping isarithmic mapping

isometric map one form of isarithmic map; made from data that occur at points

isoplethic map one form of isarithmic map; constructed from data that occur over geographic areas

isotherms isometric lines showing equal temperature

map reading and analysis errors caused by the inability of the reader to interpret map accurately

method-produced errors result from the cartographic technique

pattern of isarithms arrangement of isarithms, especially their magnitude, spacing, and orientation, to reveal the surface configuration of the geographical volume

production errors result from either manual or machine rendering of the final map

sampling points location of observations where data are collected

smoothing interpolator creation of a grid matrix whereby do not necessarily include the original data points

surface the top-most boundary of a three-dimensional volume whether an actual surface, such as topography, or a mental construct of a varying geographical distribution, such as temperature distribution

surface modeling process of interpolation in which the data for the generation of a surface occur in the mapping software ArcGIS

10 THE CARTOGRAM: VALUE-BY-AREA MAPPING

CHAPTER PREVIEW Erwin Raisz called cartograms "diagrammatic maps." Today they may be called cartograms, value-by-area maps, anamorphated images, or simply spatial transformations. Whatever name one uses, cartograms are unique representations of geographical space. Examined more closely, the value-by-area mapping technique encodes the mapped data in a simple and efficient manner with no data generalization or loss of detail. Two forms, contiguous and noncontiguous, have become popular. Mapping requirements include the preservation of shape, orientation, contiguity, and data that have suitable variation. Successful communication depends on how well the map reader recognizes the shapes of the internal enumeration units, the accuracy of estimating these areas, and effective legend design. Complex forms include the bivariate map. Cartogram construction may be accomplished using manual or automated approaches. In either method, a careful examination of the logic behind the use of the cartogram must first be undertaken. ■

We are accustomed to looking at maps on which the political or enumeration units (for example, states, counties, or census tracts) have been drawn proportional to their geographic area. Thus, for example, Texas appears larger than Rhode Island, Colorado larger than Massachusetts, and so on. The areas on the map are proportional to the geographic areas of the political units. (Only on non-equal-area projections are these relationships violated, but the distortion is greater with a typical cartogram.) It is quite possible, however, to prepare maps on which the areas of the enumeration units have been drawn so that they are proportional to some space other than the geographical. For example, the areas on the map that represent states can be constructed proportional to their population, aggregate income, or retail sales volume, rather than their geographic size. Maps on which these different presentations appear have been called *cartograms, value-by-area maps, anamorphated images* (Tikunov 1988), and *spatial transformations.*

This chapter introduces this unique form of map. In these abstractions from geographic reality, ordinary geographic area, orientation, and contiguity relationships are lost. The reader is forced to look at a twisted and distorted image that only vaguely resembles the geographic map. Yet cartograms are being used more and more by professional geographers to uncover underlying mathematical relations, general models, and other revealing structures (Haggett 1990). Cartographers likewise use them for communication of these ideas. The eventual success of cartograms as a communication device rests on the ability of the map reader to restructure them back into a recognizable form. Regardless of these complexities, cartograms are popular. Their appeal no doubt results from their attention-getting attributes.

THE VALUE-BY-AREA CARTOGRAM DEFINED

All **value-by-area maps,** or **cartograms,** are drawn so that the areas of the internal enumeration units are proportional to the attribute data they represent (see Figures 10.1 and 10.2). This method of encoding geographic data is unique

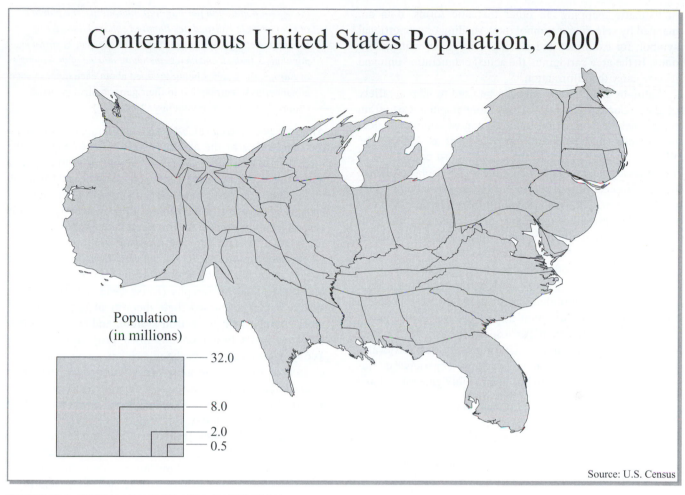

Conterminous United States Population, 2000

Population
(in millions)
32.0
8.0
2.0
0.5

Source: U.S. Census

FIGURE 10.1 TYPICAL VALUE-BY-AREA CARTOGRAM.

FIGURE 10.2 ELVIS CONCERTS ATTENDANCE PER STATE, 1970–77.

A contiguous value-by-area cartogram showing unique data. This map reveals that unique and rarely mapped data can be the subject of cartogram mapping and can attract unusual attention. Source: Map compiled by Andrew Dent and Linda Turnbull, Georgia State University. Used by permission.

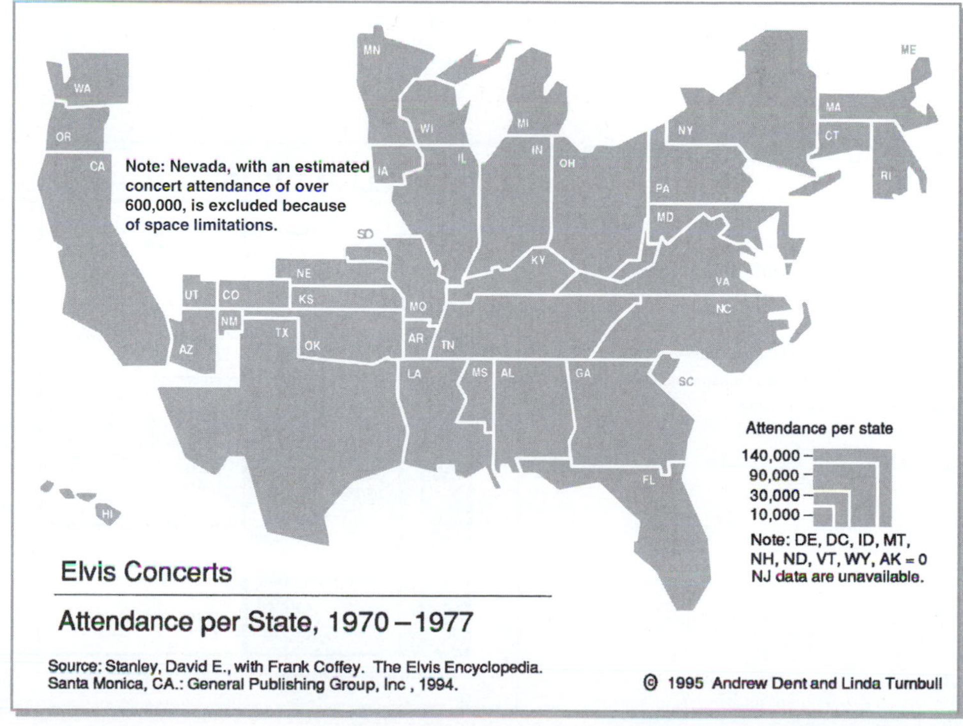

in thematic mapping. In other thematic forms, data are mapped by selecting a symbol (area shading or proportional symbol, for example) and placing it in or on enumeration units. In the area cartogram, the actual enumeration unit and its size carry the information.

Value-by-area cartograms can be used to map a variety of data. Currently, some of the more popular cartogram topics include population totals (the most common), election results, and epidemiology (Keim *et al.* 2004). The Worldmapper website has one of the largest collections of world cartograms to date, where the maps display a diverse range of topics from demographics to environmental issues (Dorling *et al.* 2008). Cartogram topics can be extremely diverse, but note that there are some important data constraints for this method (which will be discussed later, in the section entitled "Data Limitations").

Because of the method of encoding, there is no attribute data generalization. No data are lost through classification and consequent simplification. In terms of data encoding, the value-by-area cartogram is perhaps one of the purest forms of quantitative map, because no categorization is necessary during its preparation. Unfortunately, data retrieval is fraught with complexity, and readers may experience confusion because the base map has been highly generalized and distorted.

Brief History of the Method

As with so many other techniques in thematic mapping, it is difficult to pinpoint the beginning of the use of value-by-area maps. An early version was apparently used by Levasseur in his textbooks in both 1868 and 1875. To quote Funkhouser:

> *These include colored bar graphs showing the number of inhabitants per square kilometer of the countries of Europe, the school population per hundred inhabitants, the number of kilometers of railroad per hundred square kilometer of territory, etc.; squares proportional to the extent of surfaces, population, budget, commerce, merchant marine of the countries of Europe,* the squares being grouped about each other in such a manner as to correspond to their geographical position. *(Author's emphasis)* (Funkhouser 1937, 355)

Although not called a value-by-area cartogram by Levasseur, the appearance of the actual graph seems to support the idea that it was indeed such a cartogram. Others have traced the idea of the cartogram to both France and Germany in the late nineteenth and early twentieth centuries, respectively (Hunter and Young 1968). Erwin Raisz was certainly among the first American cartographers to employ the idea with the **rectangular cartogram** (Raisz 1934; see Figure 10.3). Cartogram construction techniques were treated by Raisz through several editions of his textbook on cartography (Raisz 1948, 1962). In 1963, Waldo Tobler discussed their theoretical underpinnings, most notably their projection system, and concluded that they are maps based on unknown projections (Tobler 1963). Cartograms have been used in texts and in the classroom to illustrate geographical concepts; their role in communication situations has been investigated (Dent 1975). Their popularity in magazines, in atlases, and on the Web continues today.

There has also been research in the last 35 years into designing computer **algorithms** to automate the cartogram construction process. An algorithm is simply the set of instructions for accomplishing a task, such as generating a cartogram. Typically, algorithms become software programs or scripts. (See the boxed text entitled "Software, Programs, and Scripts.") This research includes a number of algorithm and software developments by numerous

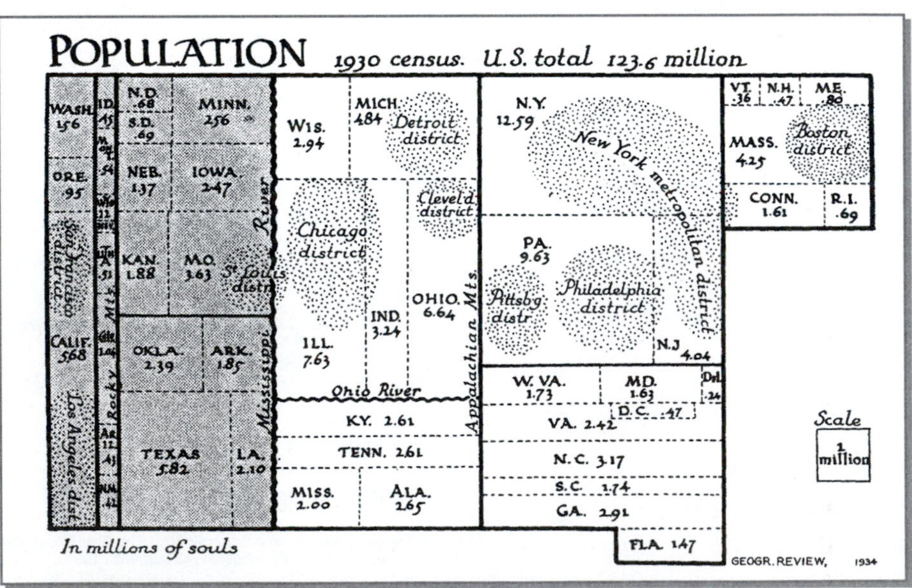

FIGURE 10.3 RAISZ RECTANGULAR CARTOGRAMS.

Enumeration units are highly generalized in this early version of the cartogram. Most cartograms made today attempt to retain some vestige of the enumeration unit's shape. Source: E. Raisz, "The Rectangular Statistical Cartogram," *Geographical Review* 24 (1934): 293, Figure 2. Used by permission of the American Geographical Society.

SOFTWARE, PROGRAMS, AND SCRIPTS

The discussions surrounding algorithms and automated approaches in generating cartograms often involve terms such as *software, programs,* and *scripts*. These related and overlapping terms can be confusing at first, as they are often used interchangeably in much of the popular literature. It may be helpful to think of a script as a type of program, and a program as a kind of software. Software includes the operating system, such as Windows or Linux, as well as the programs that run on those operating systems.

A program is simply a set of instructions, written in a computer programming language, that are understood by the computer. A computer language and programs written in that language are either compiled or interpreted. When a program is compiled, the instructions are turned into code that only the computer understands and are processed directly by the computer processor. Microsoft Excel, Adobe Illustrator, and ArcMAP are familiar examples of compiled software programs that are available commercially, but one can write his or her own program in a language such as C++ as well. A script is an interpreted program, in which the set of instructions are carried out by another secondary program. Lines of code are often closer to human instructions, so scripts are often easier to write. JavaScript and Perl are two of many scripting languages.

In GIS and mapping applications, scripts and programs are often written to expand the capabilities of the program beyond its original design. For example, if the GIS software does not support the creation of cartograms, a programmer could write a script, often based on an algorithm, to generate cartograms. Cartogram scripts and programs are often made available on the Web, such as those on ESRI's Web site (Environmental Systems Research Institute 2008).

cartographers and engineers beginning with Tobler's pioneering work in the 1960s and '70s (Tobler 1973, 2004) through modern innovative approaches by Inoue and Shimizu (2006). Particularly notable is Jackal's (1997) adaptation of the Dougenik *et al.* algorithm (1985). It was the first to be implemented in a major commercial GIS software (as a script in ArcView), bringing new possibilities to cartographers who don't write their own programs. Since then, others have followed suit, allowing cartographers to use several of the more popular algorithms in several leading GIS software or as an online application on the Web. More automated approaches are discussed in the section entitled "Cartogram Construction."

Ultimately, since their introduction, cartograms have been used in atlases, books, and magazines, and have made their appearance on the Web to illustrate geographical facts and concepts. This chapter treats area cartograms only. Linear transformations are also possible but are not discussed here.

Two Basic Forms Emerge

Two basic forms of the value-by-area cartogram have emerged: contiguous and noncontiguous (see Figures 10.1 and 10.4, respectively). Each has its own set of advantages and disadvantages, which the designer must weigh in the context of the map's purpose.

Contiguous Cartograms

In **contiguous cartograms**, the internal enumeration units are adjacent to each other. Although no definitive research exists to support this position, it appears likely that the contiguous form best suggests a true (that is, conventional) map. With contiguity preserved, the reader can more easily make the inference to continuous geographical space, even though the relationships on the map may be erroneous. Making the cartogram contiguous, however, can make the map more complex to produce and interpret.

Several advantages may be listed for the contiguous form:

1. Boundary and orientation relationships can be maintained, strengthening the link between the cartogram and true geographical space.
2. The reader need not mentally supply missing areas to complete the total form or outline of the map.
3. The shape of the total study area is more easily preserved.

The disadvantages of the contiguous form include:

1. Distortion of boundary and orientation relationships can be so great that the link with true geographical space becomes remote and may confuse the reader.
2. The shapes of the internal enumeration units may be so distorted as to make recognition almost impossible.
3. Although rapidly changing, the relative lack of dedicated GIS and commercial software procedures (like a proportional symbol or dot density function from a menu or list of thematic map choices) makes it more difficult for the cartographer to produce contiguous cartograms than other thematic map types.

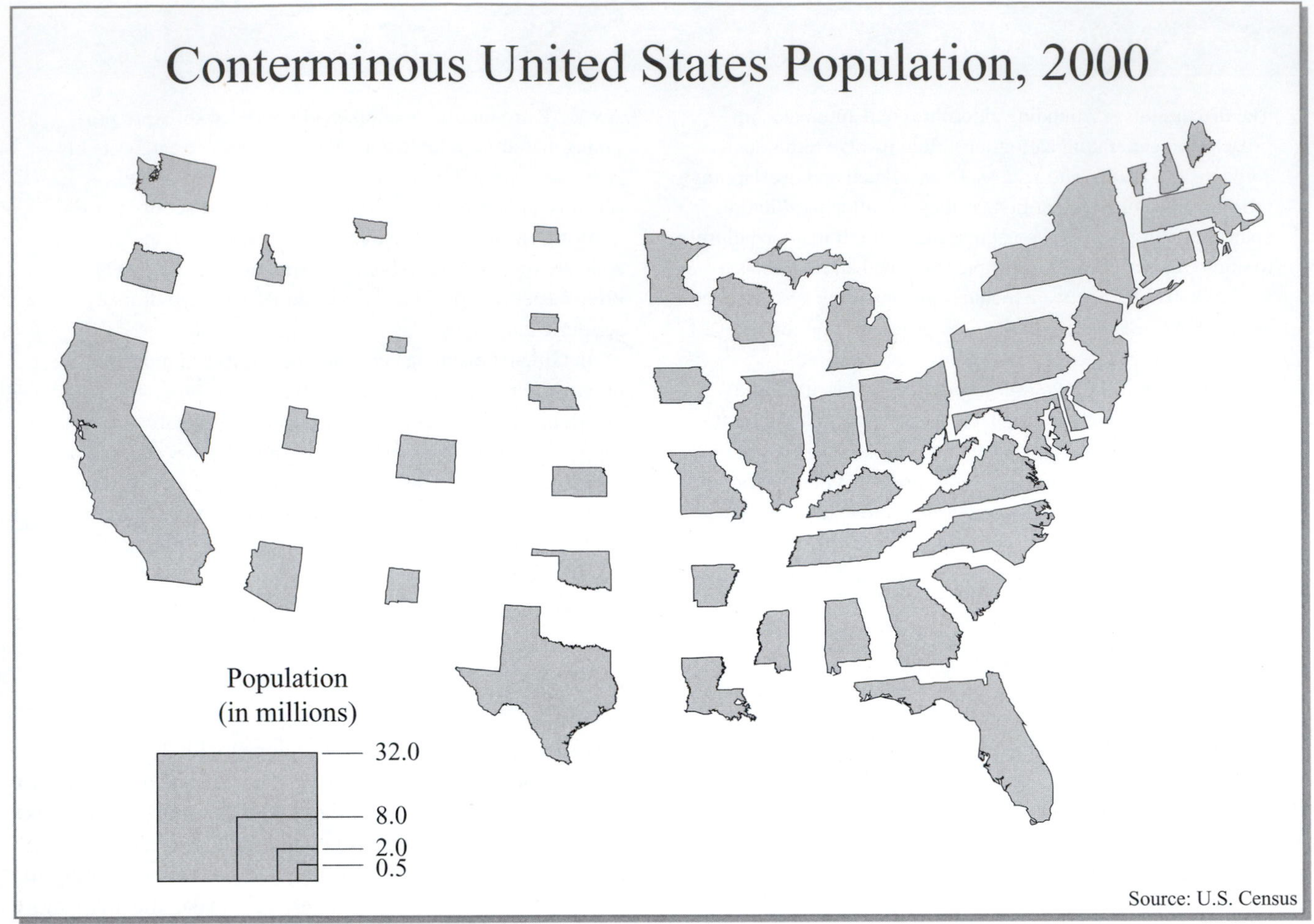

FIGURE 10.4 NONCONTIGUOUS CARTOGRAM.

Contiguous cartograms like Figure 10.1 are compact, and boundary relationships are attempted. In noncontiguous cartograms, enumeration units are separated and positioned to maintain relatively accurate geographic location without causing overlap.

Noncontiguous Cartograms

The **noncontiguous cartogram** does not preserve boundary relations among the internal enumeration units. The enumeration units are placed in more or less correct locations relative to their neighbors, with gaps between them. Such cartograms cannot convey continuous geographical space and thus require the reader to infer the contiguity feature.

There are nonetheless certain advantages in using noncontiguous cartograms:

1. They are relatively easy to scale and construct even in software that does not have a dedicated noncontiguous cartogram procedure.
2. The true geographical shapes of the enumeration units can be preserved.
3. Areas lacking mapped quantities (gaps) can be used to compare with the mapped units, for quick visual assessment of the total distribution (Olson 1976).

The disadvantages of noncontiguous cartograms include:

1. They do not convey the continuous nature of geographical space.
2. They do not possess an overall compact form.
3. Unless the transformed enumeration units are very small, there will be a trade-off between maintaining relative position of the enumeration units and having the symbols overlap, or moving the units so they do not overlap, and the shape of the entire study area will become distorted (Bortins and Demers 2002; see Figure 10.4 and caption).

One important variation of the cartogram form was developed by Daniel Dorling (1996). In the Dorling cartogram, (also called a **circle cartogram**), each enumeration unit is replaced by a circle. Each circle is then scaled to the appropriate size via the attribute value, while keeping the circular enumeration units in contact with each other without overlapping

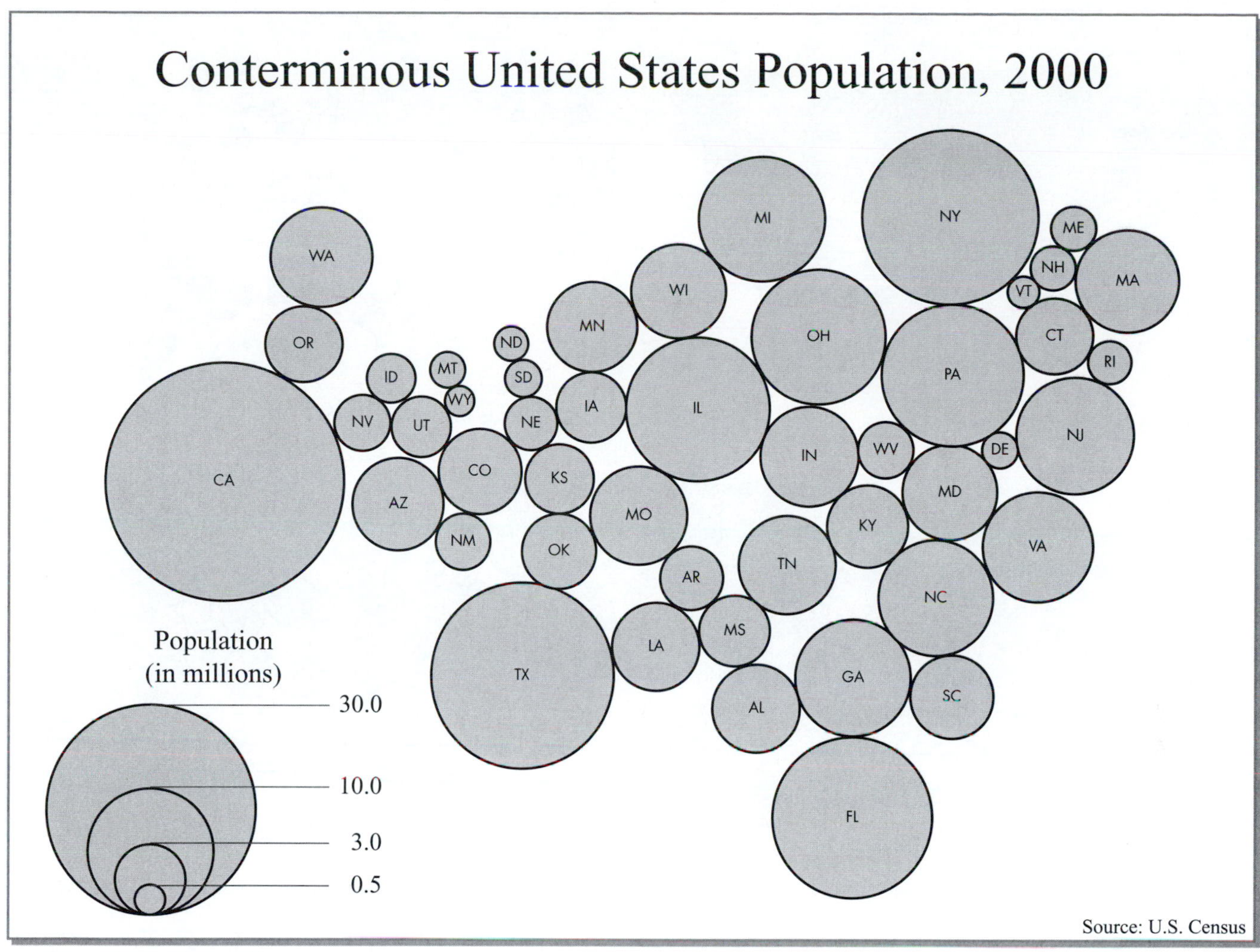

FIGURE 10.5 THE DORLING CIRCLE CARTOGRAM.
Enumeration units are generalized to a circle. Because there is no shape cue, we recommend labels whenever possible in this type of cartogram.

each other (see Figure 10.5). Some cartographers see this as a variation of the contiguous cartogram, since the circles are touching, at least partially meeting the adjacency requirement of contiguity. Others, such as Gastner and Newman (2004), see them as noncontiguous, since they touch at only one point along the circle, and leave gaps everywhere else.

Why circles? In the early 1990s there were centroids but no boundary data available for small enumeration units in Britain (Tobler 2004). Dorling developed this form to better represent British census statistics (Dorling 1993). The circle cartograms became a mainstay in his *New Social Atlas of Britain* (Dorling 1995), and his algorithm is implemented in several applications today (discussed later, in "Automated Solutions"). This circle cartogram solution is quite unique and the final image provides a startling view of the data. Figure 10.6 depicts a circle cartogram with a contiguous cartogram, a noncontiguous cartogram, and an undistorted base map for easy side-by-side comparison.

Mapping Requirements

Communication with cartograms is difficult at best, because it requires the reader to be familiar with the geographic relationships within the mapped space, which includes the total form of the study area as well as the shape, size, orientation, and contiguity of the internal enumeration units. The task may not be too difficult for students in the United States when the mapped area is their homeland and the internal units are states, but how many students in this country are familiar with the shapes of the Mexican states or those of the African nations? Likewise, are European students that knowledgeable about the shapes of the Canadian provinces or the states of the United States? On the other hand, by the very fact that they are unfamiliar with the mapped areas, map readers may pay more attention to the map than they otherwise would.

The situation can even be complex when mapping close to home. How many Tennessee residents know or could

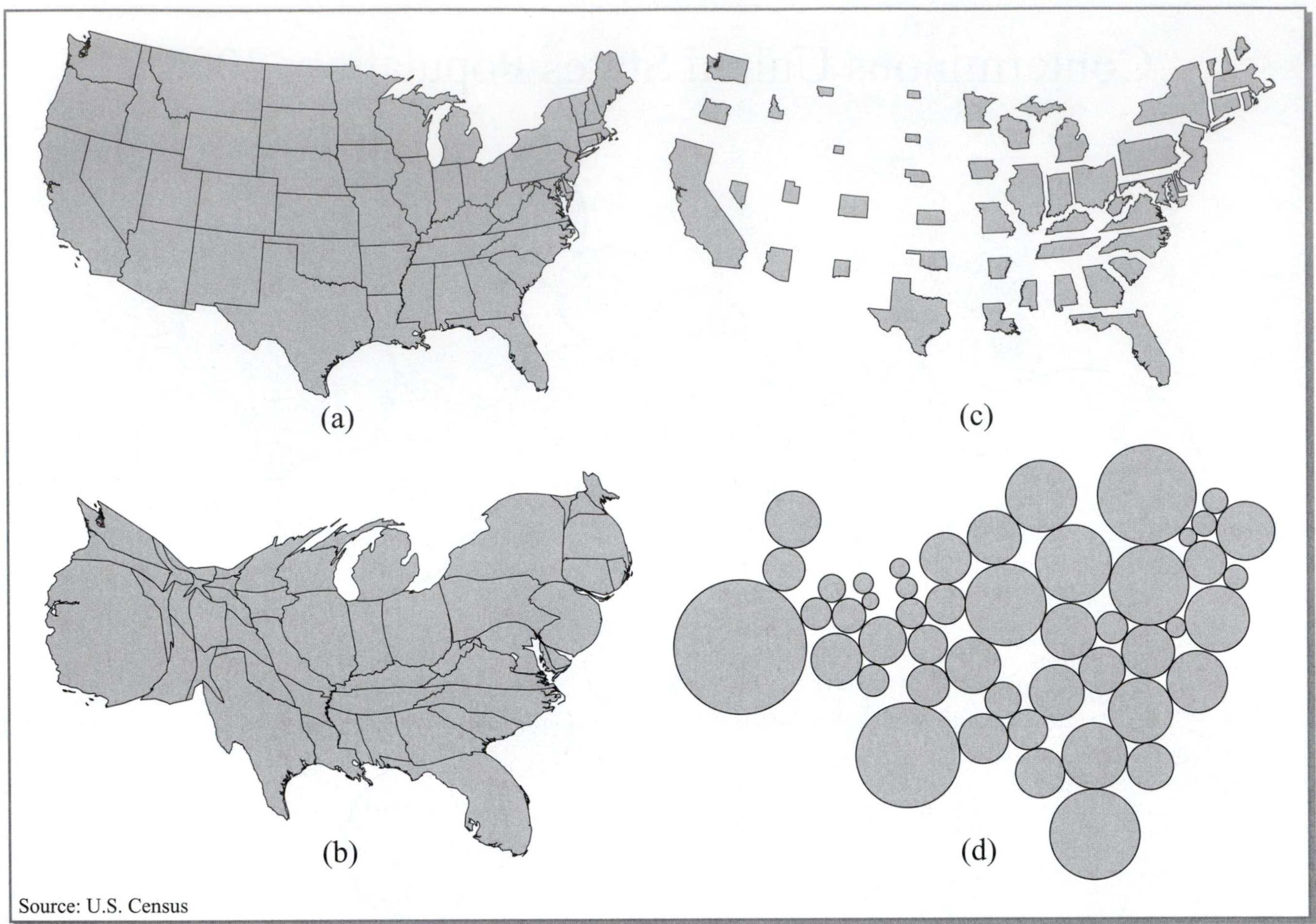

FIGURE 10.6 SIDE-BY-SIDE COMPARISON OF CARTOGRAM FORMS.
In (a), we see an undistorted base map. The contiguous (b), noncontiguous (c), and circle cartogram (d) forms can be visually compared with each other and the base map.

recognize the shapes of the counties in Tennessee? Georgia has 159 counties, Texas more than 200. Even worse, many counties are of similar shape. In some parts of the United States, rectangular and square counties are lined up row by row. Once distortion of the counties takes place, it will be nearly impossible to differentiate or identify the counties by shape or relative position. Fortunately, most professional cartographers realize the futility of mapping little-known places with cartograms.

Cartograms can present a unique view of geographical space (Dorling 1993; Gastner and Newman 2004). Tentative evidence indicates that map readers can obtain information from value-by-area maps as effectively as from more conventional forms. For this to happen, however, certain qualities of the true geographic base map must be preserved during transformation. The first of these is the **shape quality.** Preservation of the general shape of the enumeration units is so crucial to communication that the cartogram form should not be used unless some approximation of true shape can be achieved.

> The main disadvantage [of cartograms] is that they are unfamiliar, but we do not learn from familiarity (Dorling 1993, 167).

Conventional thematic maps are developed by placing graphic symbols on a geographic base map. Regardless of the form of the thematic presentation, the symbols are tied to the geographical unit with which the data are associated. Thus, for example, graduated symbols are placed at the centers of the states. Value-by-area maps, however, are unique in that the thematic symbolization also forms the base map. In a way, the enumeration units are their own proportional symbols, in addition to carrying the information of the conventional base map. On an original geographic base map of, for example, the United States, each state contains four kinds of information—size, shape, orientation, and

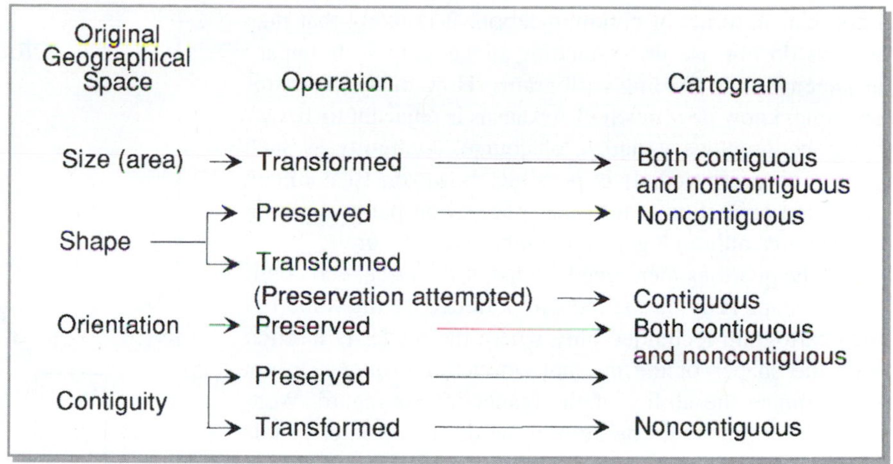

FIGURE 10.7 IDEAL CARTOGRAPHIC OPERATIONS IN VALUE-BY-AREA MAPPING.

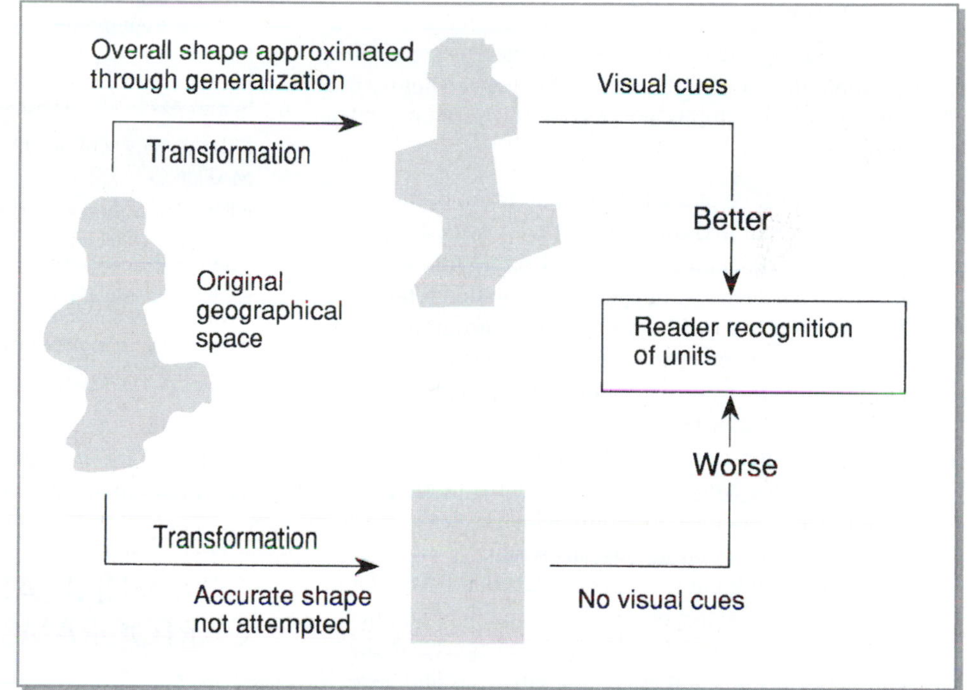

FIGURE 10.8 THE IMPORTANCE OF SHAPE IN CARTOGRAM DESIGN.
Shape preservation provides necessary visual cues for efficient reader recognition of original spatial units.

contiguity (see Figure 10.7). In value-by-area mapping, only size is transformed; the other elements are preserved as nearly as possible. Contiguity is somewhat special and may not be as important as the others in map reading.

Individual unit shapes on the cartogram must be similar to their geographical shapes. It is through shape that the reader identifies areas on the cartogram. Shape is a bridge that allows the reader to perceive the transformation of the original (see Figure 10.8). If the reader cannot recognize shape, confusion results and comprehension is difficult, if not lost altogether. The designer's problem is deciding how far it is possible to go along a continuum between shape preservation and shape transformation before the enumeration unit becomes unrecognizable to the majority of readers.

Geographical **orientation** is another important element in value-by-area mapping. Orientation is the internal arrangement of the enumeration units within the transformed space. Because the reader must be familiar with the geographic map of the study area to interpret a cartogram properly, the cartographer must strive to maintain recognizable orientation. When distortion of internal order occurs, communication surely suffers. How frustrating it would be to see Michigan below Texas!

Contiguity as an element in cartogram development relates, of course, only to the contiguous form. When producing this kind, it is desirable to maintain as closely as possible the original boundary arrangement from true geographical space. Of the elements mentioned thus far—shape, orientation, and contiguity—it appears that contiguity is the least important in terms of communication. It is likely that map readers do not use understanding of geographic boundary arrangements in reading cartograms. How many of us, for example, know how much of Arkansas is adjacent to Texas? On noncontiguous varieties, of course, contiguity as such cannot be preserved. It is possible, however, to maintain loose approximation of the total space by proper positioning of the units, although gaps remain between the units.

Of the qualities mentioned (shape, orientation, and contiguity), shape is by far the most important. Use the value-by-area cartogram technique only where the reader is familiar with the shapes of the internal enumeration units. Do not overestimate the ability of the reader in this regard. Well-designed legends can be helpful, as discussed later in this chapter.

Data Limitations

Although value-by-area maps present numerous possibilities for the communication of thematic data, they are not without their limitations. The limits are dictated by the data and their variability:

1. The data must be at least at the ratio level, because a real zero is required for each unit to be scaled appropriately. Total data that have a real zero work well for this method, such as in mapping population totals. Most derived data will work well for cartograms, but data derived by area should not be used (for example, population density—use population totals or create a choropleth map instead).
2. Negative values cannot be mapped.
3. Zero will effectively eliminate the enumeration units, creating a gap in the cartogram, which is particularly undesirable for a contiguous cartogram.
4. Extremely small numbers in data sets that have a large range can also be trouble for many computer applications. One possible solution may be to aggregate the enumeration units that are both contiguous and have small data values (such as combining the mountain states in a contiguous population cartogram).
5. Common sense must prevail. If a cartogram becomes so distorted as to be completely unrecognizable, even with a legend and/or an inset map of the original undistorted areas, use a different thematic map type, such as a choropleth map. Also, if the data so closely correlate to original map space, then the cartogram will be very uninteresting. This latter point requires some discussion:

It would be fruitless to map data that are exactly proportional to the areas of the enumeration units of the geographic base (see Figure 10.9). The cartogram would then replicate the original. It would be as if a cartogram was created using area as the variable. At the other extreme, there could be a single enumeration unit having the same area as the entire "transformed" space, in which case no internal variation would be shown. No cartogram (or any other map) would be needed. If the map does not illustrate the distribution in a visually dramatic but appropriate way, it is best abandoned.

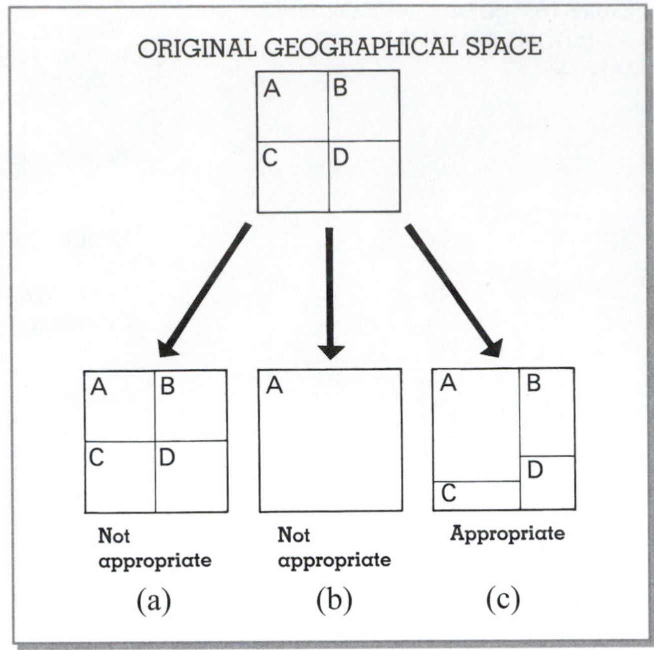

FIGURE 10.9 DATA LIMITATIONS AND VALUE-BY-AREA MAPPING.
If the original data lead to spatial transformation that is unchanged from the original (a), the value-by-area technique is inappropriate. Also inappropriate would be those cases resulting in only one enumeration unit remaining after transformation (b). Most suitable would be those instances when original data are transformed into new spatial arrangements dramatically different from the original (c).

COMMUNICATING WITH CARTOGRAMS

Success in transmitting information by the value-by-area technique is not guaranteed. There are at least three problem areas: shape recognition, estimation of area magnitude, and the stored images of the map reader. The designer should be familiar with the influences of each on the communication task.

Recognizing Shapes

It is by the shape of objects around us that we recognize them. We often identify three-dimensional objects by their silhouettes, and we can label objects drawn on a piece of paper by the shapes of their outlines. This holds true for recognition of outlines on maps. For example, South America can be seen as distinct from the other continents. The shape qualities of objects that make them more recognizable are simplicity, angularity, and regularity (Dent 1972). Simple geometric forms such as squares, circles, and triangles are easily identified. Shapes to which we can attach meaning are also easy to identify.

In the production of value-by-area maps, the cartographer ordinarily attempts to preserve the shapes of the enumeration units. How this is done is crucial to the effectiveness of the map. Many of the elements that identify the shape of the original should be carried over to the new generalized shape on the cartogram. The places along an outline where direction changes rapidly appear to be those that carry the most information about the form's shape (Dent 1972). Therefore, such points on the outline should be preserved in making the new map. These points can be joined by straight lines without doing harm to the generalization or to the reader's ability to recognize the shape (see Figure 10.10).

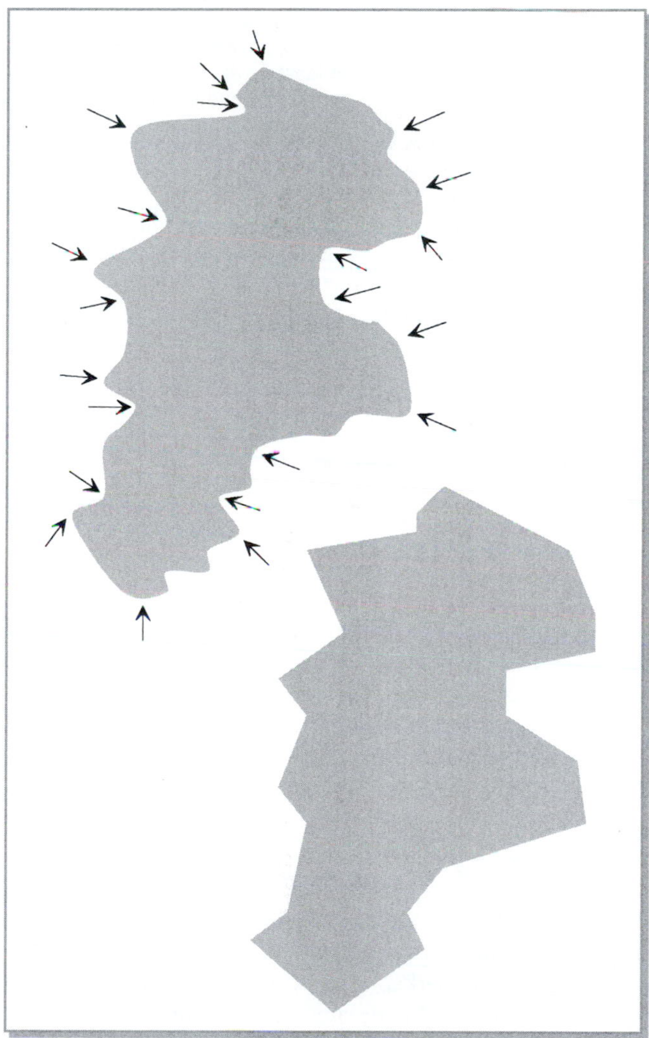

FIGURE 10.10 STRAIGHT-LINE GENERALIZATION OF THE ORIGINAL SHAPE.

Important shape cues are concentrated at points of major change in direction along the outline, as indicated here in the upper drawing. These points should be retained in transformation as a guide in the development of a reasonable straight-line generalization to approximate the original shape, as done here in the lower drawing.

Estimating Areas

Because each enumeration unit in a cartogram is scaled directly to the data it represents, no loss of information has occurred through classification or simplification. If any error results, it is to be found somewhere else in the communication process—most likely in the reader's inability to judge area accurately. The psychophysical estimation of area magnitudes is influenced by the shapes of the representative areas used in the map legend.

Research suggests that for effective communication of area magnitudes, the shapes of the enumeration units should be irregular polygons (not amorphous shapes) and that at least one square legend symbol should be used at the lower end of the data range (Dent 1975). It is best to provide three squares in the legend, one at the low end, one at the middle, and one at the high end of the data range. Some cartographers also suggest using one or two actual outlines (for example, state outlines) as legend entries. With proportional symbol maps we recommended symbol representation of the highest and lowest values, but that is usually not practical with cartograms. Of course, the overall communication effort may fail because the distortions from true shapes brought about by the method can interfere with the flow of information.

A Communication Model

It has been stressed thus far that communicating geographic information with cartograms is difficult unless certain guidelines are followed. First, shape-recognition clues along the outline of enumeration units must be maintained. Second, if the cartographer cannot assume that the reader knows the true geographical relationships of the mapped area, a geographic inset map must be included. Third, the cartographer should provide a well-designed legend that includes symbol representation at the low end (the most crucial), middle end, and high end of the value range.

These three design elements are placed in a generalized communication model of a value-by-area cartogram in Figure 10.11 (Dent 1975). In this view, design strategies should accommodate the map-reading abilities of the reader. In Step 1, all the graphic components are organized into a meaningful hierarchical organization so that the map's purpose is clear.

Accurate shapes of the enumeration units are provided in Step 2 by retaining those outline clues that carry the most information—the places where the outline changes direction rapidly.

In the United States, people are exposed from early childhood to maps of the country through classroom wall maps, road maps, television, and advertising. Satellite photographs have added to the already clear images of the country's shape in the minds of the population. How well these images are formed varies from individual to individual. Some people have well-formed images not only of the shape of the United

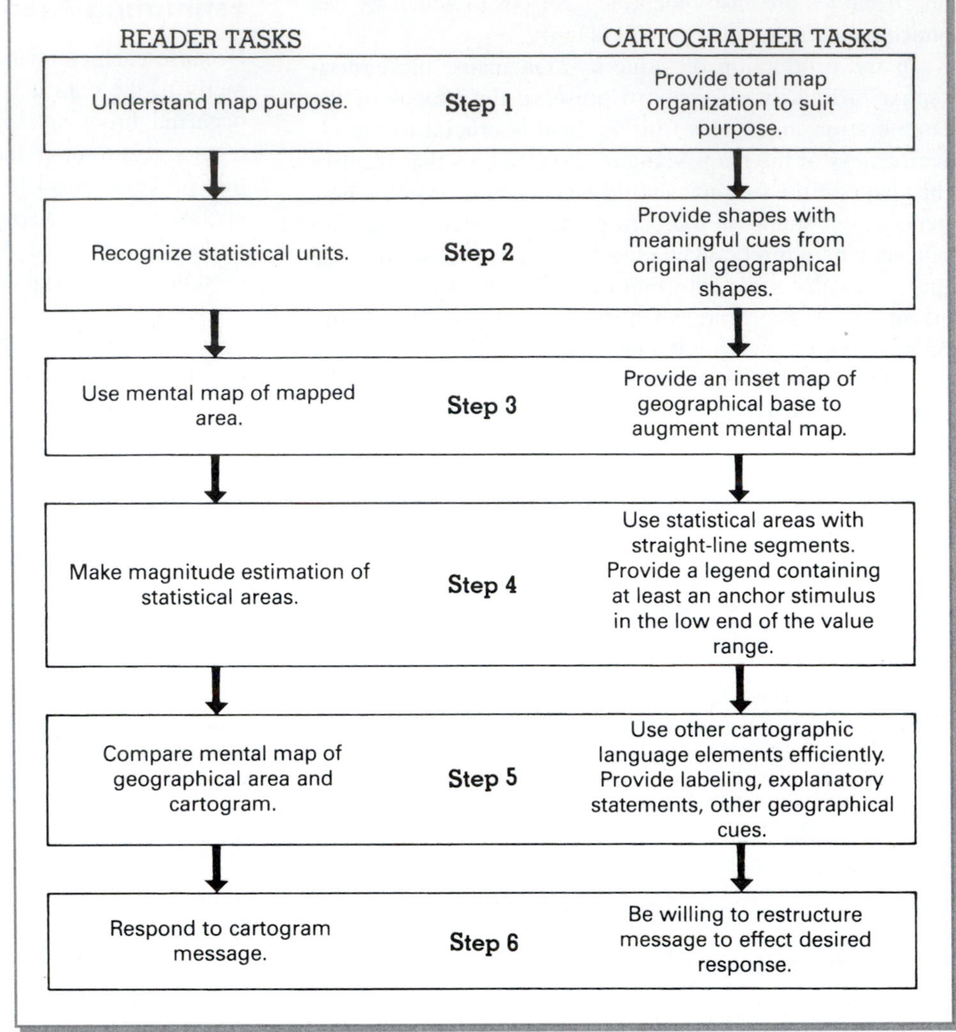

FIGURE 10.11 CARTOGRAPHER AND READER TASKS IN A GENERALIZED VALUE-BY-AREA CARTOGRAM COMMUNICATION MODEL.

Many of the steps are likely to occur simultaneously, not sequentially—especially Steps 2 through 5.

Source: Dent 1975.

States, but also of the individual states; others have difficulty choosing the correct outline from several possible ones. Successful cartogram communication may well rest on the accuracy of the reader's image of geographical space. Without a correct image, the reader cannot make the necessary match between cartogram space and geographical space. Confusion results if this connection is not made quickly.

In Step 3, the readers search through the represented geographic areas in an attempt to match what they see with their stored images (Dent 1976). Because the reader's stored images may be inaccurate, the designer should include a geographic map of the cartogram area in an inset map.

The map reader in Step 4 estimates the magnitudes of the enumeration units by comparing them with those presented in the legend. Effective legend design makes this task easier. Anchor stimuli in the legend should be squares, including at least one at the low end of the value range.

In Step 5, written elements, such as labels and explanatory notes, are included to assist the map reader in identifying parts of the map that may be unfamiliar at first. Finally, the designer should be willing to restructure the message to make the communication process better (Step 6). Inasmuch as the cartographer may not know what the reader thinks, because the cartographers and reader are usually separated in time and space, the first five tasks become even more important.

Advantages and Disadvantages

Preference-testing research has discovered that cartograms do communicate spatial information, are innovative and interesting, display remarkable style, and present a generalized picture of reality. Value-by-area maps are often stimulating, provoke considerable thought, and show geographical distributions in a way that stresses important aspects. On the other hand, they are viewed as difficult to read, incomplete, unusual, and different from readers' preconceptions of geographical space. It can be difficult to assess the communicative aspects of cartograms, since no two people manually devise identical cartograms of the same area, and different algorithms will also produce different looking cartograms. (This may be considered a strength rather than a drawback.) For the untrained map reader, the new configurations can cause visual confusion, detracting from the purpose of the map rather than adding to it.

The advantages of this thematic mapping technique are (Griffin 1980):

1. To shock the reader with unexpected spatial peculiarities.
2. To develop clarity in a map that might otherwise be cluttered with unnecessary detail.
3. To show distributions that would, if mapped by conventional means, be obscured by wide variations in the sizes of the enumeration areas.

Disadvantages include:

1. Some map readers may feel repugnance at the "inaccurate" base map that results from the study.
2. Map readers may be confused by the logic of the method unless its properties are clearly identified.
3. Specific locations may be difficult to identify because of shape distortion of the enumeration areas.

Design Strategies Recap—Legends, Inset Maps, and Labeling

From the preceding discussion, it is clear that constructing a good legend, preferably with several representations from the attribute data value range, is crucial in effective cartogram design. A number of examples of legend design can be found throughout the chapter, including nested and vertical presentations. A number of principles involved in proportional symbol map legends (see Chapter 8) can also apply to cartograms.

It is also clear that inset maps can help the reader interpret the cartogram, which can assist the reader in learning something about the distribution (Rittschof *et al.* 1996). As long as the cartographer is not trying to portray more than one variable, some map designers advocate using nominal level fills that match the enumeration units in the inset map to those in the cartogram (for example, North Dakota gets one color that will be the same in both the inset map and the cartogram). Again, we caution that inset maps can only go so far; if the cartogram becomes so drastically distorted as to be completely unrecognizable, use a different thematic map type.

One design element that has not been discussed is the labeling of the individual enumeration units in contiguous cartograms. Although not always used or required by the method, placing the names within the enumeration unit areas or within a reasonable proximity using a connector line can greatly assist the map reader in interpreting places on the cartogram. In Figure 10.12, for example, the labels work in conjunction with shape recognition to make an effective cartogram. With circle and rectangular cartograms, labeling the unit areas becomes an even more important consideration (refer to Figure 10.5), considering these forms rely heavily on geographical orientation (as opposed to shape). So as shape quality of the enumeration units is diminished, the need for labeling increases proportionally. Although we generally recommend using abbreviations as little as possible, standard and common contractions such as U.S. postal codes can be effective and are not usually distracting to the visual impact that cartograms have. If it is not practical to

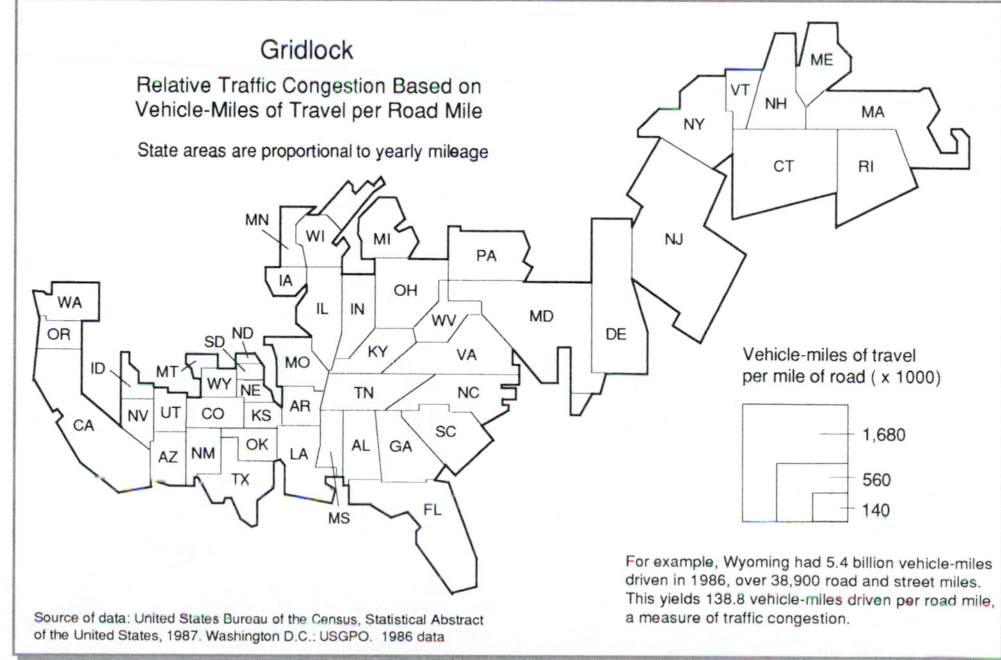

FIGURE 10.12 USE OF LABELS IN A VALUE-BY-AREA CARTOGRAM.

Postal code abbreviations are placed inside the unit areas where possible, and placed with connector lines for smaller units. In this example, there is an impression of discontinuity to the map, changing some polygons' orientation with others. The labels work in concert with shape to mitigate this effect, and assist the reader in cartogram interpretation. Source: Cartogram designed by Bernard J. vanHamond. Used by permission.

label the enumeration units on the cartogram, then labeling the units on an inset map can be considered.

Bivariate Cartograms

The discussion thus far has concerned only the use of a single attribute data set (variable), but as with other thematic map types it is possible to illustrate two (a **bivariate cartogram**) or more data sets on a single cartogram. Cartographers who view the cartogram as a unique projection rather than as a thematic map as such will see the cartogram as a base map that will be used to map another attribute. Either way, the result is the same. For example, on a cartogram of the United States in which the states are represented proportional to their populations, the cartographer can render individual states by gray tones or changes in color value or saturation according to a different variable like average income, as on a choropleth map. Other variable combinations are also possible (see Color Plate 10.1). This is the most common design approach for bivariate cartograms.

Other second variables can be accommodated on cartograms by proportional point symbol schemes. The second distribution can be represented by placing a proportional symbol within each enumeration unit of the cartogram. The reader must make the visual-intellectual comparison between the size of the enumeration unit and the size of the scaled symbol. As with "Proportional Symbols and Choropleth Maps" (see section in Chapter 8), we recommend a few cautions in this relatively unstudied approach. The most important of these is that the proportional symbols cannot obscure the enumeration unit with which the symbol is associated. If this happens, the cartogram variable will be lost in every unit that is obscured by the proportional symbol. There must be enough of the cartogram base showing for the symbols for the reader to form the association. If this technique is attempted, simple symbol forms such as circles should be used, and the shapes of the cartogram's enumeration units should not be too distorted—cartograms are often difficult to read even without a second variable! It is also important for the proportional symbols and the cartogram base to use fill colors or shades that will ensure adequate figure-ground development.

Most of the above discussion on bivariate cartograms applies to both contiguous and noncontiguous forms. In some cases the scaled enumeration units in noncontiguous cartograms are placed over their original units, to allow the reader some perspective on the orientation and the degree of scaling that is occurring. Cuff *et al.* (1984) used nesting of two or more symbols of the same enumeration unit to show proportions (such as proportion of land designated to uses such as forest, agriculture, and so on). A simple example of comparing area proportions is found in Figure 10.13, where the cartogram shows how much of a total area is occupied by internal geographic divisions. The sizes of the internal areas are drawn proportional to the data being mapped. Shape preservation is perhaps less of a central concern in such applications.

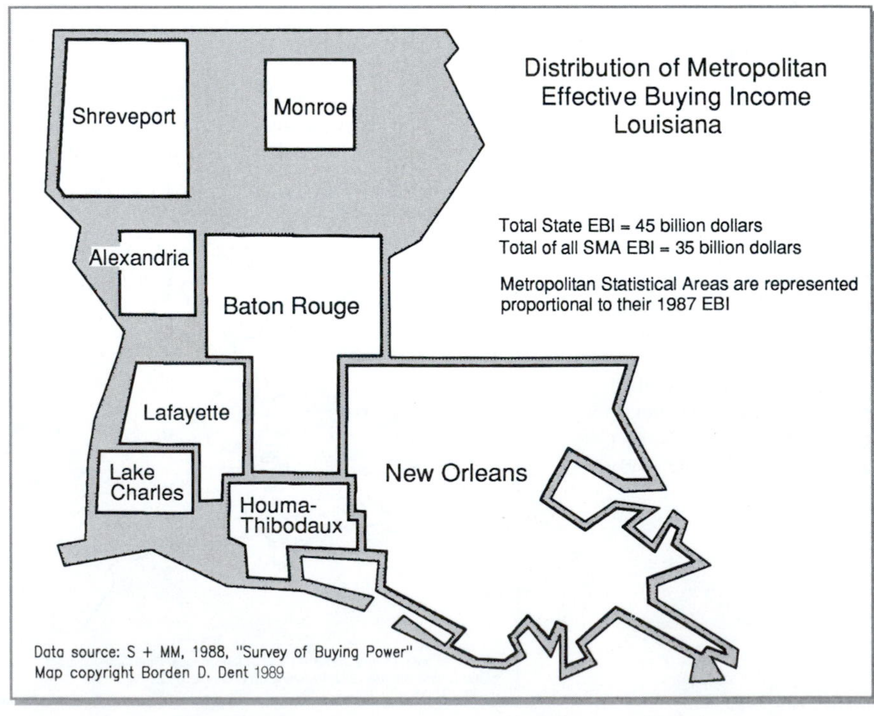

FIGURE 10.13 NONCONTIGUOUS CARTOGRAM TO SHOW GEOGRAPHICAL PROPORTION.
In this presentation SMAs are drawn proportional to their buying power and are shown relative to the total buying power of the state. Shapes of the SMAs are not as important in this form of cartogram, although relative location is.

CARTOGRAM CONSTRUCTION

There are two ways of producing value-by-area cartograms. One method is to use one of the algorithmic solutions that automate most of the construction process. The other is to manually create the cartogram, *usually by emulating the traditional, manual process* in GIS, mapping, CAD, or artistic design software. We begin with the latter, since the manual approach exemplifies the basic cartogram principle.

Manual Methods

Manual techniques for the construction of value-by-area maps are relatively simple, and they provide a viable alternative to those wanting more design control in their cartograms than is afforded in the automated methods. While the mathematics behind the algorithmic solutions is not always intuitive, studying the manual approach can be quite illustrative of the basic premise. In addition, the manual approach can be emulated in GIS, mapping, CAD, or artistic design software. The following is a description of the construction of a contiguous value-by-area cartogram for United States population.

The first step is to acquire the total population for each state. The cartographer must then decide what the total area for the transformation is to be, and what proportion of the total population is represented by each state. Then the area for each state is computed on the basis of its share (see Table 10.1). The cartographer must draw each state, preserving the shapes of the states while making their areas conform to the values computed. Of course, exact shapes are not preserved in contiguous cartograms.

To facilitate the drawing of the states, one must determine the total area, in square inches, of all states combined. This can be done using ten-lines-per-inch graph paper as an underlay to the map. Each small square, therefore, is .01 square inch. First, count all full square-inch blocks (for example, 25 or 25 square inches) that lie inside the boundary, followed by counting all of the small squares of the remaining area that also lie inside the boundary (for example, 1000 or 10 square inches). The sum of these two counts provides you with the total area of the base map (that is, 35 square

TABLE 10.1 DATA SHEET FOR A POPULATION CARTOGRAM OF THE UNITED STATES

State	2000 Population	Number of Counting Units	State	2000 Population	Number of Counting Units
Alabama	4,447,100	55	Montana	902,195	11
Alaska	626,932	8	Nebraska	1,711,263	21
Arizona	5,130,632	64	Nevada	1,998,257	25
Arkansas	2,673,400	33	New Hampshire	1,235,786	15
California	33,871,648	422	New Jersey	8,414,350	105
Colorado	4,301,261	54	New Mexico	1,819,046	23
Connecticut	3,405,565	42	New York	18,976,457	236
Delaware	783,600	10	North Carolina	8,049,313	100
Florida	15,982,378	199	North Dakota	642,200	8
Georgia	8,186,453	102	Ohio	11,353,140	141
Hawaii	1,211,537	15	Oklahoma	3,450,654	43
Idaho	1,293,953	16	Oregon	3,421,399	43
Illinois	12,419,293	155	Pennsylvania	12,281,054	153
Indiana	6,080,485	76	Rhode Island	1,048,319	13
Iowa	2,926,324	36	South Carolina	4,012,012	50
Kansas	2,688,418	33	South Dakota	754,844	9
Kentucky	4,041,769	50	Tennessee	5,689,283	71
Louisiana	4,468,976	56	Texas	20,851,820	260
Maine	1,274,923	16	Utah	2,233,169	28
Maryland	5,296,486	66	Vermont	608,827	8
Massachusetts	6,349,097	79	Virginia	7,078,515	88
Michigan	9,938,444	124	Washington	5,894,121	73
Minnesota	4,919,479	61	West Virginia	1,808,344	23
Mississippi	2,844,658	35	Wisconsin	5,363,675	67
Missouri	5,595,211	70	Wyoming	493,782	6

Total population (excluding District of Columbia and Puerto Rico) = 280,849,847. Total map area adopted in cartogram = 35 sq in. Counting unit size adopted for project = .01 sq in. Total number of counting units = 3,500. For each state, a ratio of the state's population to the national population was determined. The ratio was applied to the 3,500 total counting units to compute the number of units assigned to the state. For computation in this table, population figures were rounded to the nearest thousand.

Source: U.S. Census

inches or 3,500 **counting units**). The resulting cartogram will contain this same area (35 square inches) but redistributed based on the attribute data. By dividing the total population (280,849,847) by the total number of counting units (3,500), the value of data per .01 square inch can be derived. In this case, one counting unit represents 80,243 people. Now, by dividing the individual state's population by this value, the total number of counting units allocated per state is determined. For example, Iowa's population (2,926,324) is divided by the counting unit value (80,243) to give us 36 counting units that are assigned as the area for mapping Iowa's population. The cartographer need only arrange these small counting units until the shape of the state is approximated (see Figure 10.14). After the shape is achieved, the cartographer may wish to check the relative accuracy of the state's area.

Each state's shape is adjusted and fitted to adjacent states until the cartogram is completed. The shape of the entire study area must be roughly preserved throughout. This is not difficult but it is time consuming and often frustrating. It is wise to construct the larger enumeration units first, then the smaller ones. If odd shapes result, the noncontiguous cartogram may be selected. Note that cartograms produced in this manner will often take on some rectangular characteristics (refer to Figures 10.2 and 10.12).

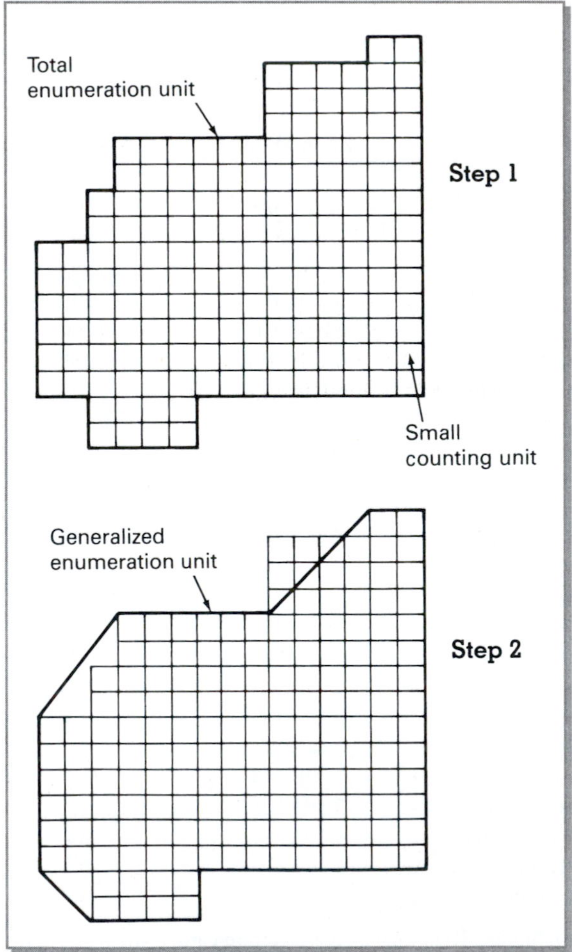

FIGURE 10.14 MANUAL CARTOGRAM CONSTRUCTION.
Small counting units are used to "build" the size and shape of the enumeration units (such as countries, states, counties) in Step 1. Step 2 involves smoothing to the approximate final shape.

A question is often raised about how to treat enumeration areas with zero value. As stated earlier, zero values effectively eliminate the enumeration unit, which is not usually desirable in a contiguous cartogram. If the eliminated units are/were on the periphery, the cartogram's contiguity will not suffer to the degree that will happen with internal zero value units. As an alternative, the cartographer could consider making a noncontiguous cartogram. In any case, zero value areas that will be omitted from the cartogram *should be listed in a note at the bottom of the map as having zero values so that they could not be mapped.* This informs the map reader that they were not *forgotten.* In a sense, this "other" space of the cartogram areas with zero value has simply collapsed.

For noncontiguous cartograms, computer use is standard. After the computations are made, most mapping and drawing programs allow you to scale polygons by a percentage or value (such as the number of counting units). The cartographer then positions the state outlines to form the shape of the total study area. Relative geographical position of each state is sought. The newly sized states, to the degree possible to avoid overlapping polygons, may be positioned in accordance with the centers of the states on a conventional map. Note that as the polygons are moved farther away from their original centers, the shape of the total study area becomes compromised.

Automated Solutions

Developing new algorithms to automate the production of contiguous value-by-area cartograms represents one of the most important avenues of research in cartogram theory. Scholars from a number of related science and engineering disciplines, both past and present, have joined geographers and cartographers in contributing to contiguous cartogram algorithm development (Tobler 2004). These automated approaches almost always introduce error into the cartograms, and the enumeration units' shape can become distorted in unaesthetic ways. In most algorithm development, authors strive to reduce cartogram error, improve the quality of shape, and in some cases improve the speed of the algorithm (traditionally contiguous cartograms take a long time to process). They do this by either improving upon existing algorithms, or by creating their own approach. Table 10.2 highlights some of the major algorithmic developments and their importance to the nonprogramming cartographer.

Contiguous cartogram algorithms require a base map and a set of attributes. The base map is often in an equal area

TABLE 10.2 MAJOR CONTIGUOUS CARTOGRAM ALGORITHM DEVELOPMENTS.

Algorithm Author (Major Publication Date)	Description
Tobler (1973)	Pioneering work in 1960s and 1970s began the algorithm development process
Dougenik *et al.* (1985)	Refinement of Tobler's work; algorithm is basis for several of the scripts and programs for current GIS software
Gausein-Zade and Tikunov (1993)	New mathematical algorithm; pushed algorithm research in new directions
Dorling (1996)	Circle cartograms originally developed in 1991; algorithm appears in mapping software and on the Web
House and Kocmoud (1998)	Introduced constraints to control wild fluctuations in shape that can occur with algorithmically generated cartograms
Keim *et al.* (2004)	Another algorithm developed to better control shape; runs relatively fast; available to cartographers as CartoDraw
Gastner and Newman (2004)	Available to cartographers on the Web as a C program; also has been developed as a Java applet, a Windows executable program, and an ArcGIS script; used in many of the 2004 election maps seen on the Web
Inoue and Shimizu (2006)	Algorithm based on triangulation of regions to help preserve polygon shape
van Kreveld and Speckmann (2007)	Algorithm to create rectangular cartograms in the tradition of Raisz

Sources: Tobler 2004; Inoue and Shimizu 2006; van Kreveld and Speckmann 2007.

projection (recommended when possible) or unprojected latitude-longitude coordinates. Geographic space in the base map is then transformed according to the mathematical procedures associated with a particular algorithm. With most automated contiguous cartogram generation, there is some error. That is, the enumeration unit areas are not perfectly in proportion to the attribute data they represent. For some of the algorithms, the transformation of the enumeration units is an **iterative** process, meaning that the enumeration unit areas become progressively closer to being true proportion with the attribute (see Figure 10.15).

Prior to about 1997, computer solutions consisted of published algorithms and distributed computer codes that laid a foundation for a new generation of contiguous spatial transformations. Algorithms by Tobler (1973), Dougenik *et al.* (1985), Gusein-Zade and Tikunov (1993), and Dorling (1996) are particularly notable for their longstanding impact on the cartographic literature. Since 1997, algorithm research and development has steadily continued, with notable contributions from House and Kocmund (1998), Keim *et al.* (2004), Gastner and Newman (2004), Inoue and Shimizu (2006), and van Kreveld and Speckmann (2007). The fact that some of these algorithms are now being distributed as programs that run in or with commercial GIS and mapping software or can be downloaded from the Web to work with widely used data structures is a large benefit to the cartographic community.

GIS and mapping software generally do not support automated contiguous cartogram generation in the same way that they do other thematic map types, such as with dedicated procedures used for making choropleth maps, dot maps, and proportional symbol maps. In 1997, Charles Jackal's implementation of the Dougenik *et al.* (1985) algorithm was the first cartogram program that became widely available for use in ArcView (Jackal 1997). The importance of this development cannot be understated, since prior to that time, one needed to be adept at scripting or programming to be able to successfully implement contiguous cartogram algorithms. With a little creativity, it always has been possible to create noncontiguous cartograms or emulate a manual approach without programming, but not so with the automated variety.

Since Jackal's contribution, we are seeing a continuing increase in availability of solutions to automated contiguous cartogram production. Adrian Herzog's Java implementation of the Dougenik and Dorling algorithms is available free on the Web (Herzog 2003, 2007). The Keim *et al.* (2004) algorithm is implemented in the program CartoDraw. As of version 6.0, MapViewer includes the Dorling circle cartograms.

Two of the most widely used algorithms today are the Dougenik *et al* (1985) algorithm, which was used in Figures 10.1 and 10.15, and the Gastner and Newman (2004) diffusion algorithm (see Figure 10.16 and refer to Color Plate 10.1). The Dougenik algorithm has been implemented in a variety of scripts and programs over the last 10 years, and the Gastner and Newman algorithm was popularized with the 2004 United States elections (Gastner *et al.* 2004) and in the previously mentioned Worldmapper site. Recent implementations of these two algorithms by Eric Wolf (2005) and Tom Gross (for the Dougenik *et al.* and Gastner and Newman algorithms, respectively) are available for use

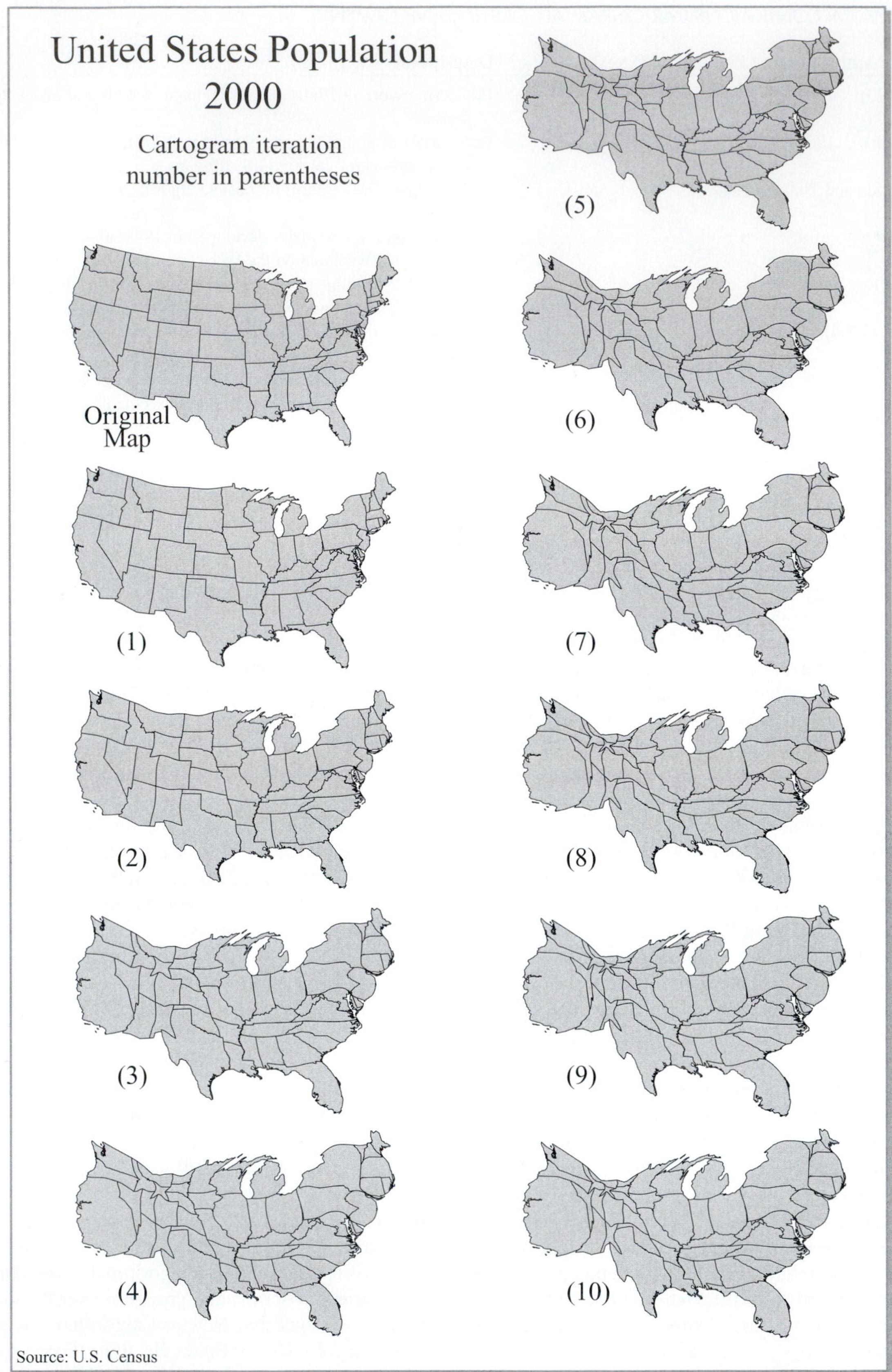

FIGURE 10.15 THE ITERATIVE METHOD.

Many cartogram algorithms are iterative in nature. In this approach, the enumeration unit areas become closer to being true proportion with the attributes with each iteration, such as in this implementation of the Dougenik *et al* (1985) algorithm. Note that after about eight to ten iterations, the amount of change is not very noticeable.

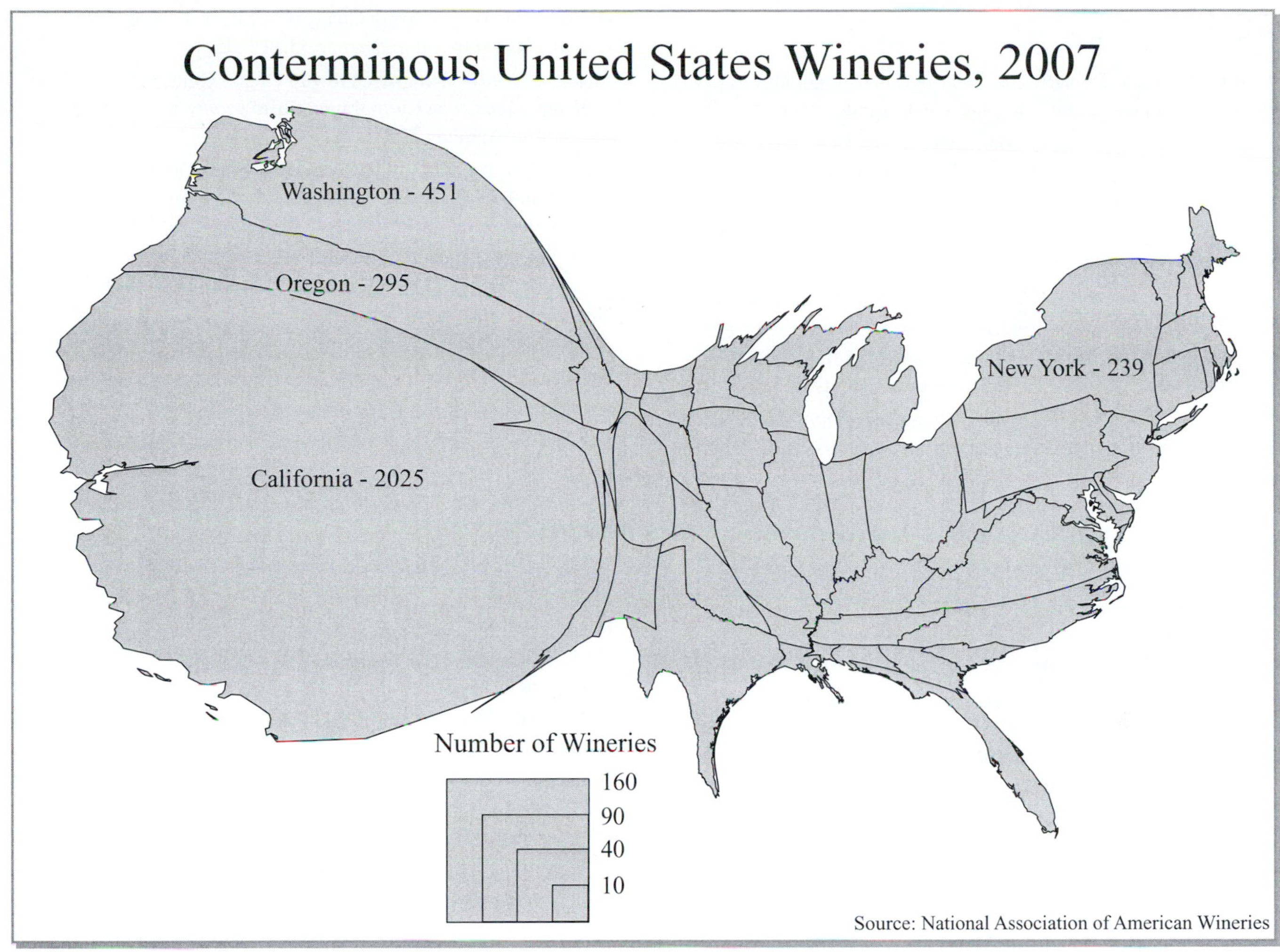

FIGURE 10.16 CONTIGUOUS CARTOGRAM USING THE GASTNER AND NEWMAN DIFFUSION ALGORITHM.

in ArcGIS (Environmental Systems Research Institute 2008). While these examples are not exhaustive, nor are they likely to stay current with the increased proliferation of such programs, the main point is to show that it is possible for the nonprogramming cartographer to create contiguous value-by-area cartograms using these automated approaches.

There are also a number of noncontiguous cartogram options as well. MapViewer supports noncontiguous in a dedicated thematic map procedure. Several scripts are available for ArcView and ArcGIS (Environmental Systems Research Institute 2008) as well. Since the size of polygons can be easily scaled and moved using virtually any GIS, artistic design software and spreadsheet software combination, some cartographers choose to bypass these scripts and programs altogether. Regardless of whether or not dedicated programs are used, noncontiguous cartograms preserve the original shapes, and they do not experience the scaling errors that occur in enumeration units in the automated contiguous cartograms.

One of the largest advantages that noncontiguous cartograms have over the automated contiguous cartogram is the increased input by the cartographer into the design process. For example, geographic orientation of the individual units can easily be adjusted or repositioned if necessary. This might occur if two or more polygons end up overlapping each other, or if the cartographer wishes to place the scaled polygons over the original enumeration units. In the latter case, it may be prudent to scale the polygons *downward*—that is, retaining the original size of the unit with the largest value and then reducing the size of the other units. This strategy was advocated by Nelson and McGregor (1983) to avoid overlapping polygons in the first place, and we have found this sensible design strategy in some of the current scripts and programs that support noncontiguous cartograms.

REFERENCES

Bortins, I., and S. Demers. 2002. Cartogram Central. http://www.ncgia.ucsb.edu/projects/Cartogram_Central/index.html

Cuff, D., J. Pawling, and E. Blair. 1984. Nested Value-By-Area Cartograms for Symbolizing Land Use and Other Proportions. *Cartographica* 21:1–8.

Dent, B. 1972. A Note on the Importance of Shape in Cartogram Communication. *Journal of Geography* 71: 393–401.

———. 1975. Communication Aspects of Value-by-Area Cartograms. *American Cartographer* 2: 154–68.

———. 1976. Postulates on the Nature of Map Reading. Unpublished paper presented at the Georgia Academy of Science.

Dougenik, J., N. Chrisman, and D. Niemeyer. 1985. An Algorithm to Construct Continuous Area Cartograms. *Professional Geographer* 37: 75–81.

Dorling, D. 1993. Map Design for Census Mapping. *The Cartographic Journal.* 30:167–83.

———. 1995. *A New Social Atlas of Britain.* Chichester, U.K.: Wiley.

———. 1996. Area Cartograms: Their Use and Creation. In *Concepts and Techniques in Modern Geography* (CATMOG), No. 59.

Dorling, D., M. Newman, G. Allsopp, A. Barford, B. Wheeler, and J. Pritchard. 2008. *Worldmapper—The World as You've Never Seen it Before.* http://www.worldmapper.org/index.html

Environmental Systems Research Institute. 2008. *ArcScripts Home – ESRI Support.* http://arcscripts.esri.com/

Funkhouser, H. 1937. Historical Development of the Geographical Representation of Statistical Data. *Osiris* 3: 269–403.

Gastner, M., and M. Newman. 2004. Diffusion-Based Method for Producing Density-Equalizing Maps. *Proceedings of the National Academy of Sciences of the United States of America* 101: 7499–7504.

Gastner, M., C. Shalizi, and M. Newman. 2004. Election Results Maps. http://www-personal.umich.edu/~mejn/election/

Griffin, T. 1980. Cartographic Transformation of the Thematic Map Base. *Cartography* 11 (1980): 163–74.

Gusein-Zade, S., and V. Tikunov. 1993. A New Method for Constructing Continuous Cartograms. *Cartography and Geographic Information Systems* 20: 167–73.

Haggett, P. 1990. *The Geographer's Art.* Oxford, England: Blackwell.

Herzog, A. 2003. Developing Cartographic Applets for the Internet. In *Maps and the Internet*, ed. M. Peterson. 115–28. Oxford: Elsevier Science.

———. 2007. *MAPresso: Choropleth Maps—Cartograms as Java Applet.* http://www.mapresso.com

House, D., and C. Kocmoud. 1998. Contiguous Cartogram Construction. *IEEE Visualization Proceedings of the Conference on Visualization* North Carolina: 197–204.

Hunter, J., and J. Young. 1968. A Technique for the Construction of Quantitative Cartograms by Physical Accretion Models. *Professional Geographer* 20: 402–6.

Inoue, R., and E. Shimizu. 2006. A New Algorithm for Continuous Area Cartogram Construction with Triangulation of Regions and Restriction on Bearing Changes of Edges. *Cartography and Geographic Information Science* 33:115–25.

Jackal, C. 1997. Using ArcView to Create Contiguous and Noncontiguous Area Cartograms. *Cartography and Geographic Information Systems* 24:101–9.

Keim, D., S. North, and C. Panse. 2004. CartoDraw: A Fast Algorithm for Generating Contiguous Cartograms. *IEEE Transactions on Visualization and Computer Graphics* 10: 95–110.

Nelson, B., and B. McGregor. 1983. A Modification of the Non-Contiguous Area Cartogram. *New Zealand Cartographic Journal*, 12: 21–29.

Olson, J. 1976. Noncontiguous Area Cartograms. *Professional Geographer* 28: 371–80.

Raisz, E. 1934. The Rectangular Statistical Cartogram. *Geographical Review* 24: 292–96.

———. 1948. *General Cartography.* 2nd ed. New York: McGraw-Hill.

———. 1962. *Principles of Cartography.* New York: McGraw-Hill.

Rittschof, K., W. Stock, R. Kulhavy, M. Verdi, and J. Johnson. 1996. Learning from Cartograms: The Effect of Region Familiarity. *Journal of Geography* 95: 550–58.

Tikunov, V. 1988. Anamorphated Cartographic Images: Historical Outline and Construction Techniques. *Cartography* 17: 1–8.

Tobler, W. 1963. Geographic Area Map Projections. *Geographical Review* 53: 59–78.

———. 1973. A Continuous Transformation Useful for Districting. *Annals of the New York Academy of Sciences* 219: 215–20.

———. 2004. Thirty-Five Years of Computer Cartograms. *Annals* of the Association of American Geographers 94: 58–73.

van Kreveld, M., and B. Speckmann. 2007. On Rectangular Cartograms. *Computational Geometry* 37(3):175–87.

Wolf, E. 2005. Creating Contiguous Cartograms in ArcGIS 9. *ESRI International User Conference Proceedings.* Paper 1155. http://gis.esri.com/library/userconf/proc05/index.html

GLOSSARY

algorithm a step-by-step method or set of rules for accomplishing a specific task in a finite number of steps; algorithms often end up as computer code, such as the algorithms for constructing contiguous cartograms

bivariate cartogram a value-by-area map on which a second, related variable is mapped, usually using area fills as is done with a choropleth map

cartogram name applied to the form of map in which the areas of the internal enumeration units are scaled to the data they represent; used synonymously with value-by-area map

circle cartogram a type of contiguous cartogram developed by Daniel Dorling in which the enumeration units are drawn as circles prior to scaling

contiguous cartogram a value-by-area map in which the internal divisions are drawn so that they join with their neighbors

counting unit small spatial unit used in the manual preparation of value-by-area cartograms

iterative cartogram algorithms that transform the enumeration units incrementally closer to being in true proportion with each iteration

noncontiguous cartogram a value-by-area map in which the enumeration unit boundaries do not join each other; units appear to float in mapped space

orientation the internal arrangement of the enumeration units within the total transformed region; cartogram communication relies heavily on the map reader's knowledge of the geography of the study area and that those units are oriented correctly (for example, South Dakota is still south of North Dakota after geographic space has been altered by the cartogram method)

rectangular cartogram a type of contiguous cartogram employed by Erwin Raisz in which the enumeration units are generalized to rectangular figures

shape quality a bridge allowing the reader to perceive the new value-by-area transformation of the original geographic base map; shape recognition is critical—without it, confusion results and communication fails

value-by-area map name applied to the form of map in which the areas of the internal enumeration units are scaled to the data they represent; used synonymously with cartogram

11 DYNAMIC REPRESENTATION: THE DESIGN OF FLOW MAPS

CHAPTER PREVIEW Maps that show linear movement between places are called flow maps and, because of this quality, are sometimes referred to as dynamic maps. Symbols on quantitative flow maps are lines that usually vary in width according to their attribute data, sometimes with arrows to show direction. In some modern presentations, color may be used in place of or in addition to scaling line widths. Flow mapping began with maps done by Henry Drury Harness in 1837. With the impetus supplied by the study of economic geography, flow mapping became common in geography textbooks in the early decades of the twentieth century. Particular attention must be paid to total map organization and figure-ground principles when designing flow maps because of their complex graphic structures. The projection for the flow map, line scaling and symbolization, and legend design are the chief design elements in flow mapping. ■

THE PURPOSE OF FLOW MAPPING

Maps showing linear movement between places are commonly called **flow maps** (also *dynamic maps* in older literature). Flow line symbolization is used when the cartographer wants to show what kind of (qualitative) or how much (quantitative) movement there is between two or more places. For the quantitative variety, the widths of the flow lines connecting the places are drawn in proportion to the quantity of movement represented. Quantitative width adjustments are the most common form for this map type, and most of the discussion in this chapter centers around this form. In some modern presentations, the line's hue, saturation, or value is altered to indicate magnitude of the flow *instead of* or *in addition to* varying line widths. In the latter case, if only one variable is used, then the **redundant coding** is used for visual emphasis (these maps can become visually complex). If two variables are used in this manner (that is, one variable scales the widths of the lines and one variable determines hue parameters), a **bivariate flow map** is created to examine the relationship between two variables.

Anytime the cartographer wishes to show movement between places, and has the appropriate attribute data to support this theme, a flow map is appropriate. Its purpose may be to show movement of actual items, or movement of relatively intangible things such as volume of Internet traffic, or ideas. In most instances, except for the *origin-destination* case (discussed below), the cartographer attempts to show the actual route taken by the movement, although this may be difficult because of map scale and the level of generalization selected for the map. Because of the nature of the symbolization process, and the inherent complexity of these maps, most flow maps become highly generalized.

Flow line symbols may be used on maps to show *qualitative* movement, as mentioned earlier. The lines are unscaled and generally have arrowheads to indicate direction of movement. Lines are usually of uniform thickness and color or shade. These lines are referred to as *streamlines*. Early qualitative flow maps date back to the late eighteenth century, but are still common today for showing shipping routes, airline service, ocean currents, migration flows, and other similar presentations. It is possible to map different types of flow data on one map by using different line characteristics, such as using blue and red lines for cold and warm ocean currents.

Quantitative Flow Maps

In quantitative flow maps, the line's width (see Figure 11.1) or hue parameters (see Color Plate 11.1) indicate the magnitude of the flow. In the former and more common case, thicker lines indicate greater flow magnitudes. In the latter case, darker and/or more saturated hues (or darker values if grayscale is used) usually indicate greater flow amount. A notable exception to this rule is that many online traffic flow maps use signal light colors of red, yellow, and green to indicate stopped or very heavy congestion, moderate or heavy traffic, and clear traveling conditions, respectively. Whether line widths or hue parameters are used, there are several important data and form considerations that must be carefully considered before construction of flow maps can begin.

Data Suitability

The flow map is a fairly flexible form for a variety of different attribute data. In the case of scaling line widths, the data requirements are similar to those of proportional symbol maps. Data that work well for this method include raw data such as totals, as well as ratios and proportions. For proper scaling of the lines there has to be a real zero value. Interval data will not ensure proportionality and should not be used (see discussion in Chapter 8, "When to Select This Method—Data Suitability). Zero will effectively eliminate the flow line, and negative values cannot be mapped. Note that densities (ratios of quantities to area) cannot be used since the spatial dimension is linear, and not areal. Weight, volume, value (dollar), amount, and/or frequencies are often the units used on quantitative flow maps. International commodity flows—such as overseas movements of grains, ores, and produce—are frequently mapped by quantitatively scaled lines.

If color is used to portray the primary variable, then the data requirements are not so stringent, particularly with regard to zero or negative numbers. If color is used to depict a secondary variable for bivariate mapping, then data may be used from any level of measurement, qualitative or quantitative.

Directed and Undirected Flows

Flow maps may or may not use arrows in their design. Flow maps with arrows are sometimes called **directed flow maps,** because the flow is traveling along the line only in the direction of the arrowhead. In some online applications, the arrowheads are replaced by animated flow lines that "move" in the direction of the flow.

In most printed and virtual static maps, flow maps without arrows imply movement in both directions. These types

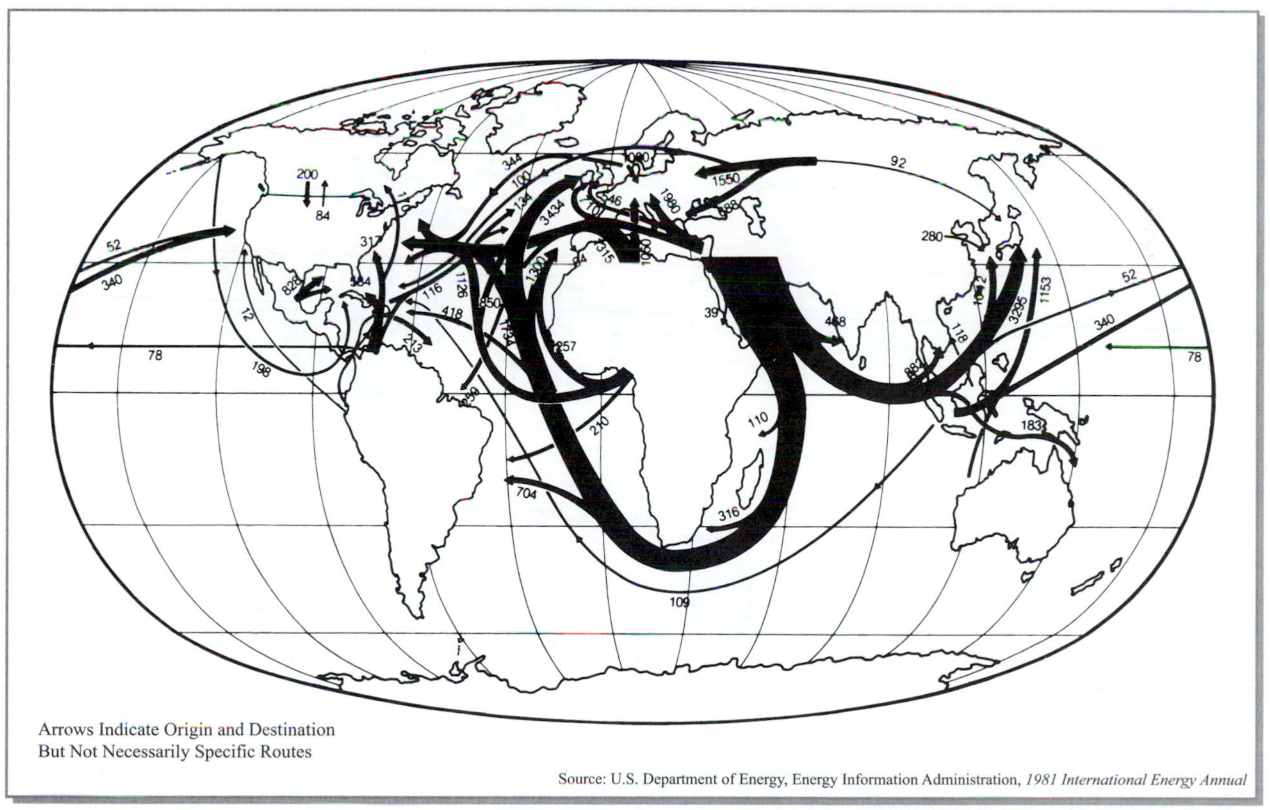

Arrows Indicate Origin and Destination
But Not Necessarily Specific Routes
Source: U.S. Department of Energy, Energy Information Administration, *1981 International Energy Annual*

FIGURE 11.1 INTERNATIONAL CRUDE OIL FLOW, 1980 (THOUSAND BARRELS PER DAY).
No legend is used on this map, but the lines are labeled to indicate the magnitude of flow.

of flow maps are sometimes referred to as **undirected flow maps.** This form of data generalization perhaps shows less information than do directed flow maps, but undirected flow maps *are* used, for one of three reasons:

1. Much of the current vehicular traffic data is already aggregated for both directions.
2. It is often impractical to portray both directions on many flow maps (such as a flow map depicting highway traffic at a small or medium scale). Again, flow maps can get very complex.
3. It is easier to generate undirected flows in most GIS and mapping software.

The Relevance of Flow Routes

Flow maps may or may not attempt to follow the actual routes of movement. In the case of many **traffic flow maps,** the actual routes (or some close approximation) show the organizational and hierarchical nature of the state or urban (Birdsall and Florin 1981) road systems. For example, varying line widths can be used to symbolize the number of vehicles passing over portions of a state's major highways (see Figure 11.2). The actual route and the magnitudes of flow are both important parts of the theme. Since most roads accommodate traffic in both directions, it is common to see traffic flow maps without directional symbols. The traffic flow concept can be expanded to include rail, water, and air transportation, or even bird flyways. Of course, the precision with which the routes are presented will vary; highway locations are generally fixed, but bird flyways are usually approximated.

There are also flow maps in which actual routes are not portrayed. In this case, the actual route may not be known or is deemed unimportant by the map designer. The locational emphasis shifts from the route to the origin and destination of the flows. **Origin-destination maps,** also called *desire*

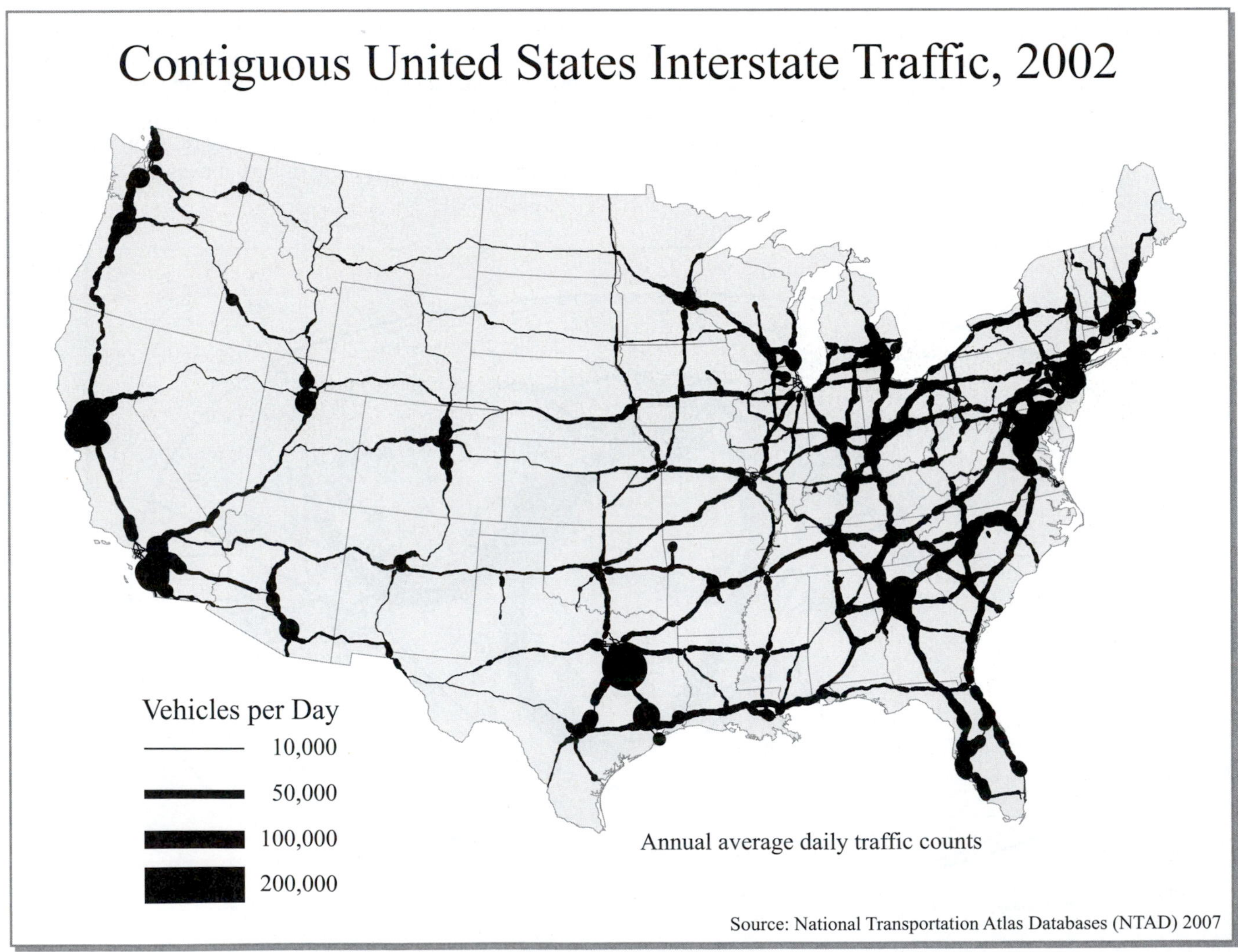

FIGURE 11.2 A TYPICAL TRAFFIC FLOW MAP.
In many instances such as this, direction of flow is not mapped. Width of lines shows number of vehicles passing in both directions for a specified time period.

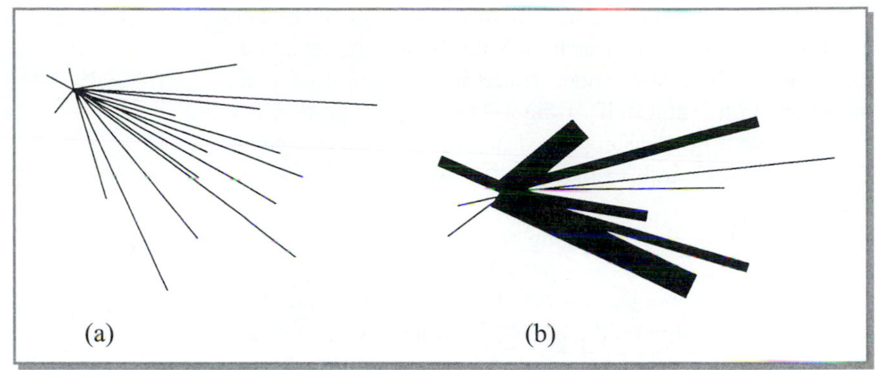

FIGURE 11.3 ORIGIN-DESTINATION MAPS. In (a), the lines are of uniform thickness (often joining origin and destination of the traveler). In (b), the lines vary in thickness based on aggregate interaction. They are depicted as straight lines here, but may also be portrayed as arcs.

line maps (Davis 1974), often depict a one-to-many (or many-to-one) relationship (see Figure 11.3). For example, a store owner may wish to map customers by zip code, and connect the store to the centroids of the zip code areas via flow lines. The application of origin-destination maps is best found in those instances where nodal geographical patterns are to be focused on (such as depicting air traffic between major hubs), or possibly where urban hierarchies are being stressed (Berry 1967; Berry *et al.* 1976). The origin-destination maps have been used frequently in this latter instance to show shopping or commuting structures.

Mapping Web traffic and other telecommunication data is another common example in which actual routes are not portrayed. These types of maps are fairly common on the Web, and are depicted in the Atlas of Cyberspace (Dodge and Kitchin 2001). These maps can be difficult to create, as in some cases a geographical origin and destination of the information flow may not be known or deemed important. More is said of this type of flow mapping later in the chapter in "Innovative Solutions."

HISTORICAL HIGHLIGHTS OF THE METHOD

Our discussion here is limited by space and coverage, but a few remarks about the history of flow mapping are necessary to place this form of mapping in the correct historical perspective. Quantitative flow mapping began in the "golden age" of statistical cartography in Western Europe, in the two decades preceding the middle of the nineteenth century. This form of map was utilized heavily in economic geography textbooks throughout the first half of the twentieth century.

Early Flow Maps

The earliest quantitative flow maps apparently were done by Henry Drury Harness when he prepared the atlas to accompany the second report to the Railway Commissioners of Ireland in 1837 (Robinson 1955). On one map the relative number of passengers moving in different directions by regular conveyance was shown, and on the other, the relative quantities of traffic in different directions. On the traffic conveyance map, the width of the lines is proportional to the average number of passengers weekly (derived data). These maps did not show exact routes, but showed straight lines of varying thicknesses connecting points (cities and towns). The maps of Harness remained unknown for nearly a century (Robinson 1982).

Within 10 years of the production of the flow maps of Harness, Belpaire of Belgium and, more notably, Minard of France began publishing flow maps of designs essentially similar to those of Harness. Charles Joseph Minard was more productive than Belpaire, and Minard's interests were primarily in the areas of economic geography. Minard apparently had no contact with geographers or cartographers, although he was instrumental in popularizing the flow line technique among statisticians. By his own account, he was interested in showing quickly, by visual impression, numerical accuracy, so much so that he often overgeneralized other portions of his maps so that the flow lines themselves would command attention (Robinson 1967). This design aim is still relevant today. Although Minard did not invent the flow map, Robinson has noted that he did bring "that class of cartography [flow maps] to a level of sophistication that has probably not been surpassed" (Robinson 1967, 105).

Minard produced some 51 maps, most of which were flow maps (Robinson 1982). The mapped subjects of the flow maps were varied, and included such topics as people, coal, cereal, mines, livestock, and others (see Color Plate 11.2). He mapped the distribution or flow of these commodities not only in France but worldwide as well. His style, especially on the maps showing movement of travelers on principal rail maps in Europe (1862), is the same as that used today. Perhaps the most unique and provocative of his illustrations is the flow chart showing the demise of Napoleon's army in Russia; as suggested by Tufte, "It may well be the best statistical graph ever drawn (Tufte 2001, 40)."

Flow Maps in Economic Geography

From the time of Harness, Belpaire, and Minard, quantitative flow maps have been used to map patterns of distribution of economic commodities, people (passengers), and any number of measures of traffic densities. As suggested earlier, the flow maps of Minard reached a sophistication of technique never really matched since. There was a period of time, however, in the first half of the 1900s, when a great many quantitative flow maps appeared in college economic geography textbooks and, in many cases, with laudable design techniques.

In a study of flow mapping in college geography textbooks by M. Jody Parks, several hundred flow maps were found among 71 books published between 1891 and 1984 (Parks 1987). Although the author did not consult all books in this category in this time period (an enormous task), the study is remarkable for its breadth and attention to detail, summarizing carefully the findings and including numerous examples of this kind of mapping. Although used extensively in this publishing medium, in general flow mapping lagged behind other forms of thematic mapping techniques.

It is interesting to note from the study done by Parks that qualitative flow maps appeared in these books as early as 1891, and that this form outnumbered the quantitative form by three to one (Parks 1987). Qualitative flow maps are used to illustrate migration routes, explorer routes, and transportation networks. Origin-destination maps as a category of qualitative flow maps are used also, especially after 1960. The earliest quantitative maps, from those textbooks studied by Parks, appeared in 1912. Transportation themes are most often illustrated by the quantitative flow maps, especially international import and export of agricultural commodities (Parks 1987).

The study by Parks also yields a classification of flow map design beyond those discussed above in "Quantitative Flow Maps," including these three distinct patterns: radial, network, and distributive (Parks 1987). **Radial flow maps** are easily distinguished by a radial or spoke-like pattern, especially when the features and places mapped are nodal in form. Present-day traffic volume maps fit into this category (see Figure 11.4a). **Network flow maps** are those used to reveal the interconnectivity of places, especially evidenced by transportation or communication linkages, such as intracounty commuting to work (see Figure 11.4b).

Flow maps in the third class include those that present the distribution of commodities or migration flows. These are called **distributive flow maps.** Trade flows, such as shipments of wheat among countries, is a good example of this form (see Figure 11.4c). Maps that show diffusion of ideas or things are included in this class. Maps illustrating diffusion are often found in textbooks on cultural geography.

Topics mapped by the flow line technique are quite varied, and suggest the innovation often employed by cartographers and geographers in using this method. Railway and airline route maps are common (showing interconnectivity between places). Shipments of natural gas, wheat, animals, ore, coal, cotton, and migration of people are just a few examples of the topics commonly represented on flow maps. Migration routes, such as those taken by American Indians, and the French, Spanish, and English people in settling the New World, and other topics have been mapped by the flow map technique. As long as the data requirements discussed earlier are met, there are few subjects that contain a from-to relationship that cannot be mapped by flow symbolization.

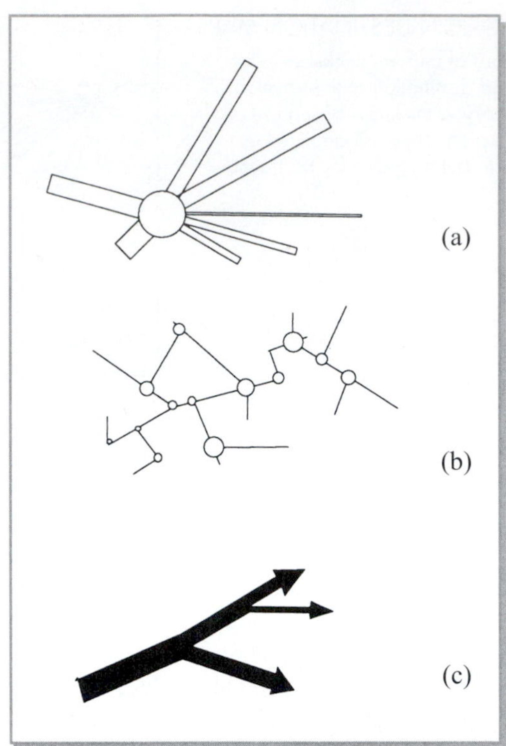

FIGURE 11.4 FURTHER CLASSIFICATION OF FLOW LINE PATTERNS IN GEOGRAPHY TEXTBOOKS.
A radial type is illustrated in (a), the network type is represented in (b), and the distributive type in (c).

This brief examination of flow maps in textbook cartography is intended to provide only a backdrop to the fascinating study of this form of mapping. Space will not permit a detailed presentation of hundreds of examples and the rich variety of design found among these maps. The student of thematic map design is urged to explore examples from the actual textbooks themselves. A worthwhile design activity can be found in such an experience.

DESIGNING FLOW MAPS

Creating effective flow maps through a careful and thoughtful design plan represents one of the more difficult challenges for the map designer. Since many GIS and mapping software do not have a dedicated "flow map" procedure, most map designers will simply use a graduated or proportional symbol function, and apply it to the lines' attributes. If the lines do not yet exist, they can be created or drawn via **heads-up digitizing.** The flow map procedure in Golden Software's Map Viewer allows the cartographer to input the "from" and "to" locations, and then modify the arc of the line between the two places. If the flows are accomplished using artistic drawing packages, the calculations of the line widths are usually done with the assistance of spreadsheet software. Regardless of production method, three aspects of design must be considered: map organization and figure-ground (including the selection of the projection), line symbolization and data scaling, and legend design.

Map Organization and Figure-Ground

The map's hierarchical plan must be carefully considered. The flow lines are almost always going to be the most dominant features on the map. As with other forms of thematic mapping, they should be placed high in the hierarchy so that they clearly stand out as strong figures. (The topic of the visual hierarchy and map design is dealt with in greater detail in Part III of this text, "Designing Thematic Maps.") Several figures in this chapter, notably Figure 11.1, illustrate this idea. However, achieving this hierarchy is sometimes difficult because the flow lines may stretch over several different levels on the maps. For example, lines may first be over land, then water, then land again, and so on. They may also intersect other flow lines, potentially confusing the map reader.

As with other thematic symbol types, flow lines should have strong edge gradients and be rendered so that visual conflict with other symbols does not result (see Figure 11.5). In Figure 11.5b, rendering the flow lines black, along with lightening the fills on the background, is a way to improve the thematic symbols on this map.

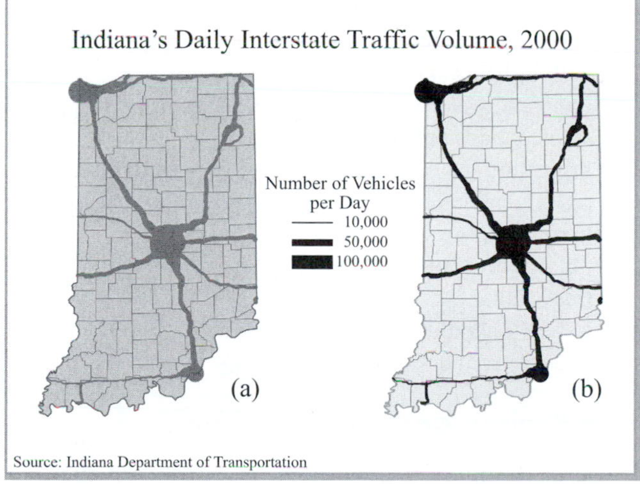

FIGURE 11.5 FLOW LINES SHOULD HAVE STRONG FIGURE CHARACTERISTICS.

The flow lines in (a) could be stronger figures by rendering them as black symbols, as in (b). Also, the county fills have been lightened in (b), producing a better figure-ground contrast.

Flow maps are ordinarily quite complex visual graphics. On most thematic maps, the organization of the graphic components, land and water, symbols, titles, legends, and other marks on the maps tends to fall neatly into an easily followed plan. Land and water contrasts are developed by following figure-ground principles, and symbols usually occupy space over land areas. Titles and legends are dealt with similarly. But on many flow maps, especially those illustrating international movements, the thematic symbols are likely to occupy spaces over land or water, or both, and many times the lines themselves are intertwined and appear to rest in different visual levels. The nature of the visual complexity on many flow maps, then, creates unusual and challenging design problems for the cartographer.

When applicable, the cartographer should provide clear land and water distinctions on flow maps. The flow symbols must be dominant figures in perception, with strong edge gradients and clear continuity. Labeling flow lines with their values, often useful to assist the reader, should not interfere with the symbol's visual integrity. Using color or shade fills on flow symbols should not lead to confusion with other areas on the maps; pattern fills should almost always be avoided for this reason. The scaling of the flow lines should not cause them to be too large for the maps, which can result in too little base-map information showing through. Attention to design details such as these will assist the cartographer in reaching successful results.

Projection Selection

In addition to achieving a good visual hierarchy on the flow map is the necessity of selecting an appropriate projection. Placement of the center of the flow, if there is a center, must be strategically planned and this may require careful consideration of the projection, its center, and aspect. Placement and design of the flow lines should be done so that the map does not become an incomprehensible mesh of confusing lines (see Figure 11.6). For flow mapping, the equal-area and conformal attributes of projections may not be as important as other factors, such as continental shapes (Dent 1987).

Selecting a projection, adopting its central meridian, choosing an aspect for it, and placing it in the map frame must all be considered when developing a flow map. The final map that results from these choices should be one that connotes organization and control of the image, and deftly satisfies the purpose of the map. It is the responsibility of the cartographer to know the flow pattern he or she wishes to portray, and then make decisions regarding the projection that best illustrates this pattern. For example, if the flows are single origin to multiple destinations, or if the pattern reflects multinodal origins and a single destination, the employment of the projection should complement the pattern.

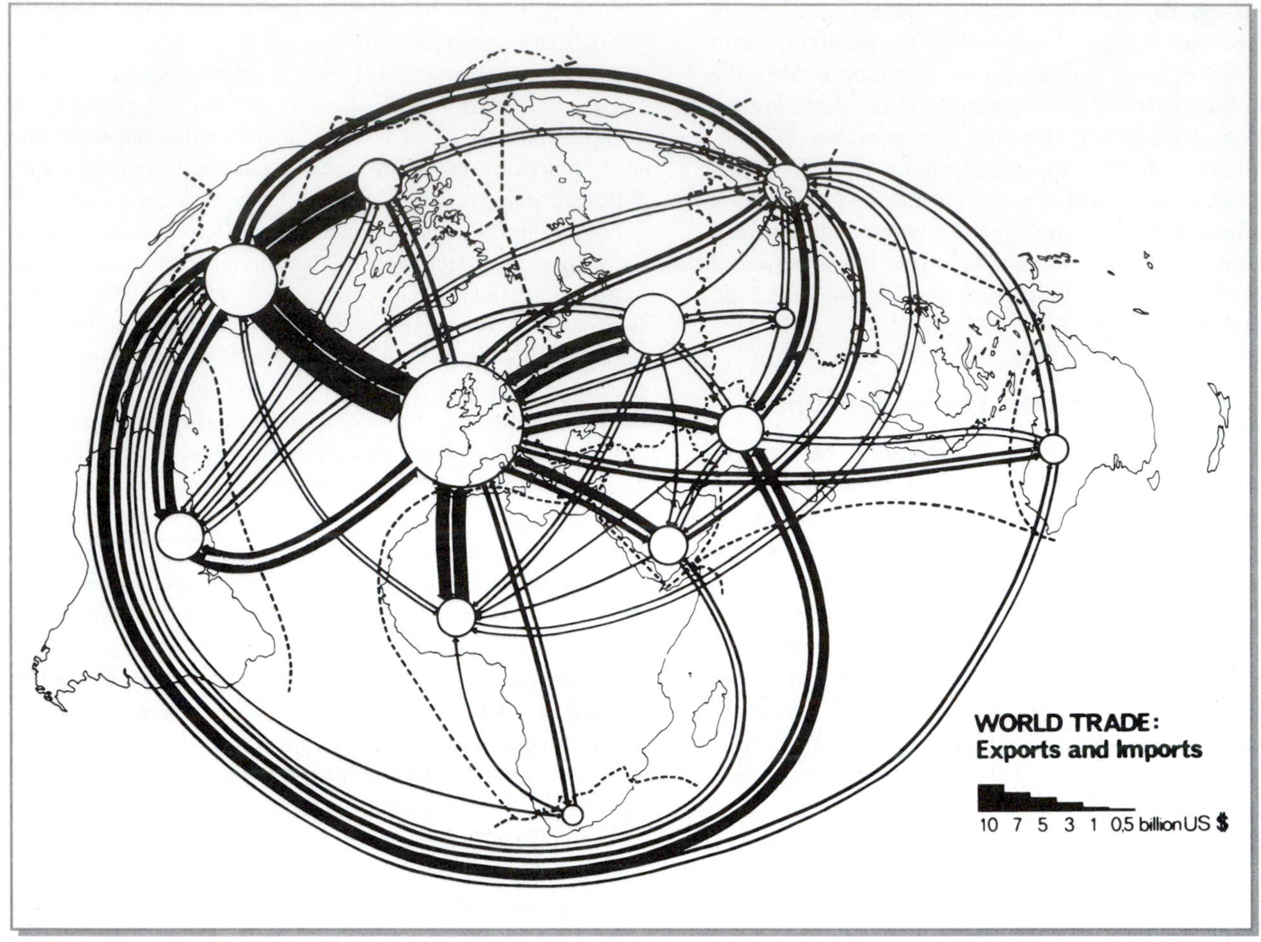

FIGURE 11.6 WORLD TRADE NET EXPORTS, 1967.
The selection of the projection (uncommon in this case), its orientation in the map frame, and the nature of flow distribution all contribute to a confusing array of lines. *Source:* Reprinted by permission of Van Nostrand Reinhold Company, Ltd.

Balance and layout of the map's elements conclude the design activities related to achieving the visual and intellectual plans for the map. Placement of map objects and utilization of space are dealt with so that a pleasing result is reached, which is defined as one in which no other solution seems merited. The map's elements, as with other thematic map forms, are placed in intellectual order and treated graphically to satisfy the plan. Titles, legends, scales, source materials, and other elements are therefore treated accordingly. However, the thematic symbols—the lines themselves—remain the most important features of the map, and all other elements are second in importance.

Essential Design Strategies

A summary of the essential design strategies for flow maps should include these principles:

1. Flow lines are the highest priority in visual/graphic importance.
2. Smaller flow lines should appear on top of larger flow lines *if* they must cross or overlap.
3. Arrows are necessary if direction of flow is critical to map meaning.
4. If the data permit the map designer control in line placement, lines should be placed in a manner that balances the entire map (that is, not too top-heavy, bottom-heavy, and so on).
5. Land and water contrasts are essential (if the mapped area contains both).
6. Projection, its center and aspect, are used to direct readers' attention to the flow pattern important to the map's purpose.
7. All information should be kept simple, including flow line scaling.
8. Legends should be clear and unambiguous, and include units where necessary.

Line Scaling and Symbolization

On most quantitative flow maps, the widths of the flow lines are proportionally scaled to the quantities they represent. Thus, a line representing 50 units will be five times the width of one symbolizing 10 units. This practice has been employed since the time of Harness, Belpaire, and Minard. Perceptually, the reader is being asked to make this visual judgment, and ordinarily most readers can do this in a linear fashion (McCleary 1970). Scaling, therefore, appears to be straightforward and should not pose severe problems for the designer.

Chief among the concerns for the cartographer is the overall attribute data range that must be accommodated by the widths selected. In many instances, this imposes considerable restraint on the project. Ordinarily, the best plan is to select the widest line that can be placed on the map (and still preserve the integrity of the base map), and then determine, by looking at the data range, the narrowest line on the map. If the widest line is 1.27 cm (0.5 inch), and the data range is 5,000 units, then the smallest line would be 0.000254 cm (0.0001 inch), obviously too small to be of practical use. Either the widest line would have to be enlarged, already determined to be unacceptable, or some other solution would need to be reached. It is important to note, too, that as the linear symbols become wider, they cease to appear as such, and may take on the qualities of area symbols. This limit should be avoided.

This problem is similar to what sometimes happens in proportional symbol mapping—what happens if the data range is so large that no combination of smallest or largest symbolization produces a reasonable result? One possibility is to range grade the symbols. Range grading involves classifying the attribute data according to one of the classification schemes described in Chapter 5 and assigning a set of ordinally scaled symbols to each class. There must be enough of a difference in line widths so that the reader can discriminate between the classes. This type of range grading is supported in most GIS and mapping software. Alternatively, the lines can be scaled proportionally to the midpoint of the classes (a complete discussion of range-grading technique and its advantages and disadvantages can be found in Chapter 8).

In other cases, where there are extreme outliers in the data set, it is possible to provide some sort of specially symbolized line to represent values above or below a certain critical value (and symbolize the remaining values by proportional scaling or range grading). In Figure 11.7, both a specialized symbol and range-grading techniques are combined on the same map. We recommend this approach only as a last resort, after every reasonable attempt has been made to scale the symbols proportionally (often by further reducing the width of the smallest line, which can often have quite an impact on the other symbol widths) or by range grading the symbols.

As mentioned earlier, color is also used to symbolize flow maps. With the exception of the "signal light" color schemes that are often applied to online traffic flow maps, darker and/or more saturated hues usually indicate greater flow amount; lighter or less saturated hues indicate less flow. In this case, it is common to classify the data according to one of the classification schemes mentioned in Chapter 5. Most GIS and mapping software allow for easy classification and color symbolization via "graduated color" or some similar procedure. Even though the flow lines' *widths* are not associated with data in this case, it is important that the lines must be of sufficient width for the map reader to perceive color correctly.

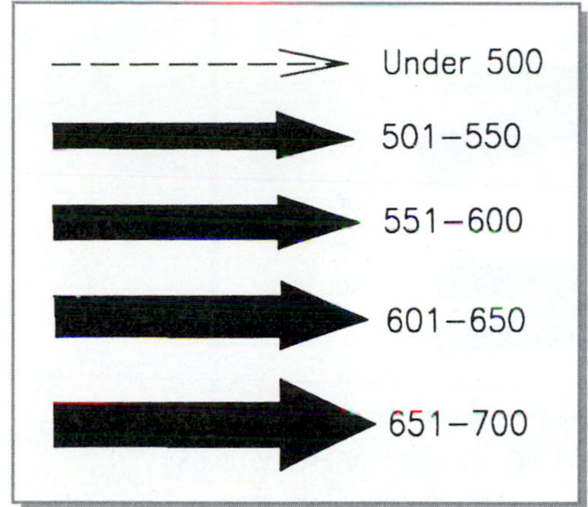

FIGURE 11.7 USE OF RANGE GRADING AND A SPECIALLY SYMBOLIZED LINE IN SCALING FLOW LINES.

When the data range is too great to scale all flow lines proportionally, range grading and/or the specially symbolized line may be selected to symbolize several values. See text for discussion.

In many modern flow maps, color is used in addition to scaling the lines. One strategy is to use the same variable for scaling line widths and for adjustment of the hue's value or saturation. In other words, the larger the value, the wider the line and the darker or more saturated its color. Because flow maps can become very complex, this type of *redundant coding* is used by some cartographers to help the map reader interpret the map. There are two potential downsides of redundant coding in this manner. First, the map takes on the appearance of a bivariate map. Assuming that the legend has accounted for both width and color, it may take a few moments for the reader to realize that both visual variables are tied to the same attribute data. Second, and perhaps more importantly, larger symbols already stand out more in the figure-ground contrast. This scheme can inappropriately add too much emphasis on the larger symbols. (Refer to Figure 8.16 for a similar situation with proportional symbols.) Inverting the color scheme so that the smaller symbols are darker or more saturated may perhaps be more aesthetically pleasing, but is a wrong choice because the *data do not match the symbolization*. A mismatch between data and symbolization should be avoided for flow maps or any other thematic map type.

It is also possible to make a *bivariate flow map* to illustrate the relationship between two variables. For example, an automobile traffic map could illustrate the relationship between traffic volume and number of accidents for a particular state's major highways during a given year. The flow line widths can be scaled by the average daily traffic count, and the line color saturation or value can be adjusted by number of fatalities.

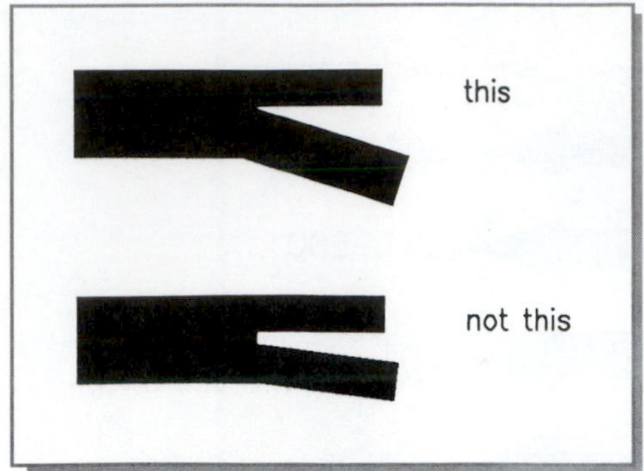

FIGURE 11.8 SYMBOLIZATION OF FLOW LINES.
When quantitative flow lines branch or unite, their widths are drawn proportionally.

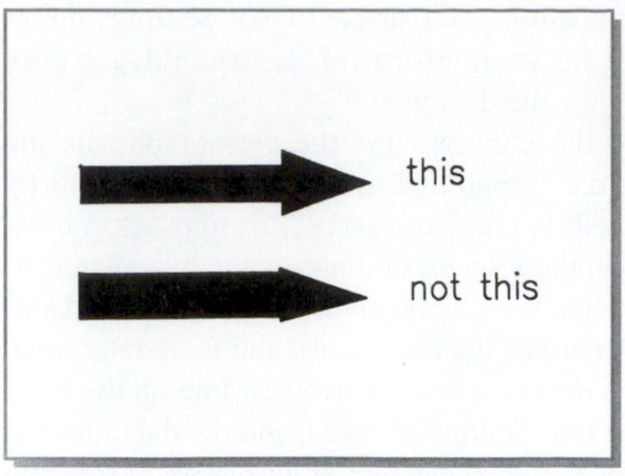

FIGURE 11.9 PROPER ARROWHEADS IMPROVE THE APPEARANCE OF FLOW LINES.

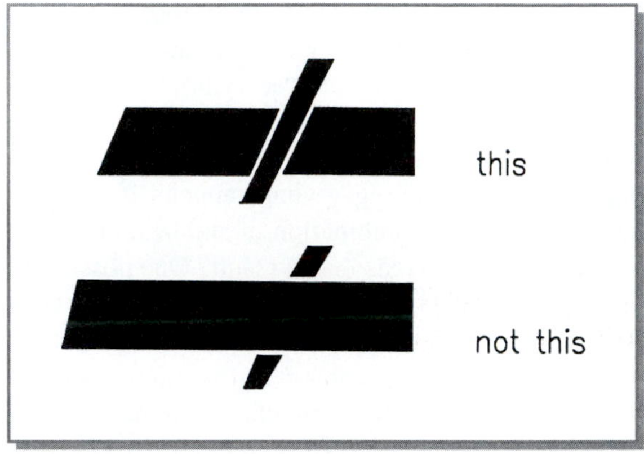

FIGURE 11.10 INTERPOSITION IS USEFUL IN FLOW LINE SYMBOLIZATION.
Interposition is a technique useful in making smaller lines appear to rest on top of larger ones.

Treatment of Symbols

No rules have emerged that govern how flow lines should be treated graphically in all maps. The only convention appears to be the way distributive flow lines are treated (see Figure 11.8). When mapping flows that separate into smaller flows, the widths of the individual branches should add up to the width of the trunk. This makes intuitive sense. The same applies when smaller branches come together to make a larger one.

In most GIS approaches, branching of this nature is difficult to accomplish. If there are not too many flows, we recommend that each origin to destination flow be its own line. If branching is desired, then the portions of lines that are to make up the trunk will have to be aggregated. In other words, a new line is created (for example, via *heads-up digitizing*) as a replacement for the segments that are to be combined. The attribute value for the new line will be the combination from the old lines.

When new lines are created for the map, such as in cases where exact routes are unknown or of less importance (for example, Internet traffic volume between continents), there is an opportunity for the cartographer to balance the flow lines between all parts of the map. In other words, where possible, place the lines so that the map does not become visually unbalanced (a top-heavy or bottom-heavy map).

If arrowheads are used, they should be clear and should be scaled proportionally to the lines to which they belong. Arrowheads with small shoulders should be avoided (see Figure 11.9). In many software applications, the arrowheads will have to be manually added after the flow lines are created and scaled. In such cases, the designer will have to take care in choosing a matching arrowhead form, as well as make sure that proper scaling, rotation, and superimposition of the arrow on the flow line occurs.

If flow lines overlap, smaller ones should be made to appear on top of larger ones (see Figure 11.10). This is accomplished if the cartographer has control of the drawing order so that the smallest flow can be brought to the top. To avoid the blending of the overlapping figures, some designers use a shadow, silhouette, or some sort of border line for the smaller flow. This gives the effect of breaking the larger flow line, and having the smaller one continue uninterrupted. More will be said of interposition in Part III of this text.

Legend Design

One of the more difficult tasks in the design of the flow map is the preparation of the legend. The legend is the crucial link in cartographic communication between cartographer and map reader, as it serves to explain carefully the symbols on the map. The legend must above all be clear and unambiguous. Units of measurements must be prominently displayed and it must be obvious how the lines are scaled, and the flow lines in the legend must appear exactly as they are on the map. If the data have been classed, the class boundaries should be clearly represented.

Of the myriads of possible legend designs, there are four arrangements that are discussed here. Specified key values

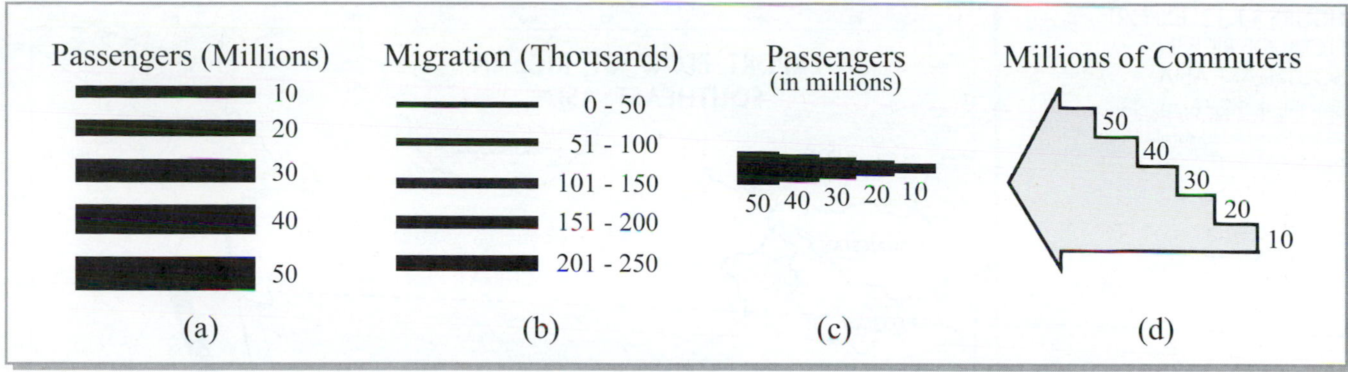

FIGURE 11.11 LEGEND DESIGNS FOR QUANTITATIVE FLOW MAPS.
A variety of design options exist for flow maps when the lines' widths have been scaled. These include specified key values (a), range graded (b), telescoping (c), and stairstep designs (d). In directed flow maps, arrows are sometimes placed on the lines in the legend. See text for discussion of color-based flow maps.

for flows that are proportional to their volumes and graduated lines for range graded values are common, and are usually the most practical to implement in GIS and mapping software (see Figure 11.11a and 11.11b). The telescoping design (Figure 11.11c) is a variation of the classical stairstep design (Parks 1987) (see Figure 11.11d). The latter two designs facilitate comparison of line widths. Some cartographers feel that if there are arrows on the flow map there should also be arrows on the scaled lines in the legend. There have been no exhaustive studies on flow map legend designs. Cartographers often must rely on their best judgments, perhaps borrowing legend design principles for proportional symbols, or, when possible, perhaps simply asking a respected colleague or representative reader about his or her legend design choices.

In maps that employ color as the primary vehicle for illustrating flows, legend design will follow those guidelines suggested for choropleth mapping (see Chapter 6), except that line segments will normally replace legend boxes. Some map designers retain the box design in order to provide a stronger visual anchor to the hue's value and saturation levels, although there is now a mismatch in spatial dimension between line and area symbolization. This mismatch can be avoided by making sure that on both the map and the legend, sufficient line width is used in order for the map reader to perceive the color correctly and make the connection between legend and map body.

For maps that combine line width scaling and hue parameter change, two legends will be needed if a *bivariate flow map* is produced: one for flow widths as discussed, and one for colors. For maps employing *redundant coding,* we recommend the range-graded approach for the width symbolization (see Figure 11.11b), and then applying color as appropriate to the flow classifications.

Legends have not always been employed on all flow maps. In some cases, the values of the lines on the map may be labeled for greater precision (refer to Figure 11.1). The overall pattern of the lines shows the organization of the movement, and the labeled values provide detailed information for the reader seeking tabular data. In other cases, although the lines are scaled and drawn proportionally, no labels at the lines are provided, and no legend is included (see Figure 11.12). Presumably, the cartographer is wishing to provide an approximate visualization of the data, emphasizing geographical organization and pattern. It would seem imperative that these drawings be accompanied by written narrative. Although there is some historic precedent for this latter style, the inclusion of a legend (or line values) *should normally be standard practice* for modern quantitative flow maps.

Innovative Solutions

A number of innovative and unique solutions for flow mapping have been used by cartographers and geographers. Up to this point in our discussion, the majority of examples have been drawn from a rather conventional pool. Other solutions do exist and are mentioned to stimulate experimentation. One such example illustrates global Internet connectivity by using flow lines that appear in the form of arcs, either above a tilted map base (Becker *et al.* 1995) or a globe (Munzner *et al.* 1996) (see Figure 11.13a-b). These types of maps began to emerge in the 1990s with the growth of the Web. Arc distance above the map or globe, line hue or value, or line width can be used to illustrate amount of data being transferred or type of connection used.

Mapping Internet traffic volumes as cartographic flows has continued with the increasing prevalence of the Web and related communications technologies. In most of these maps, the routes are highly generalized. One innovative approach taken by TeleGeography Research is to combine flow lines

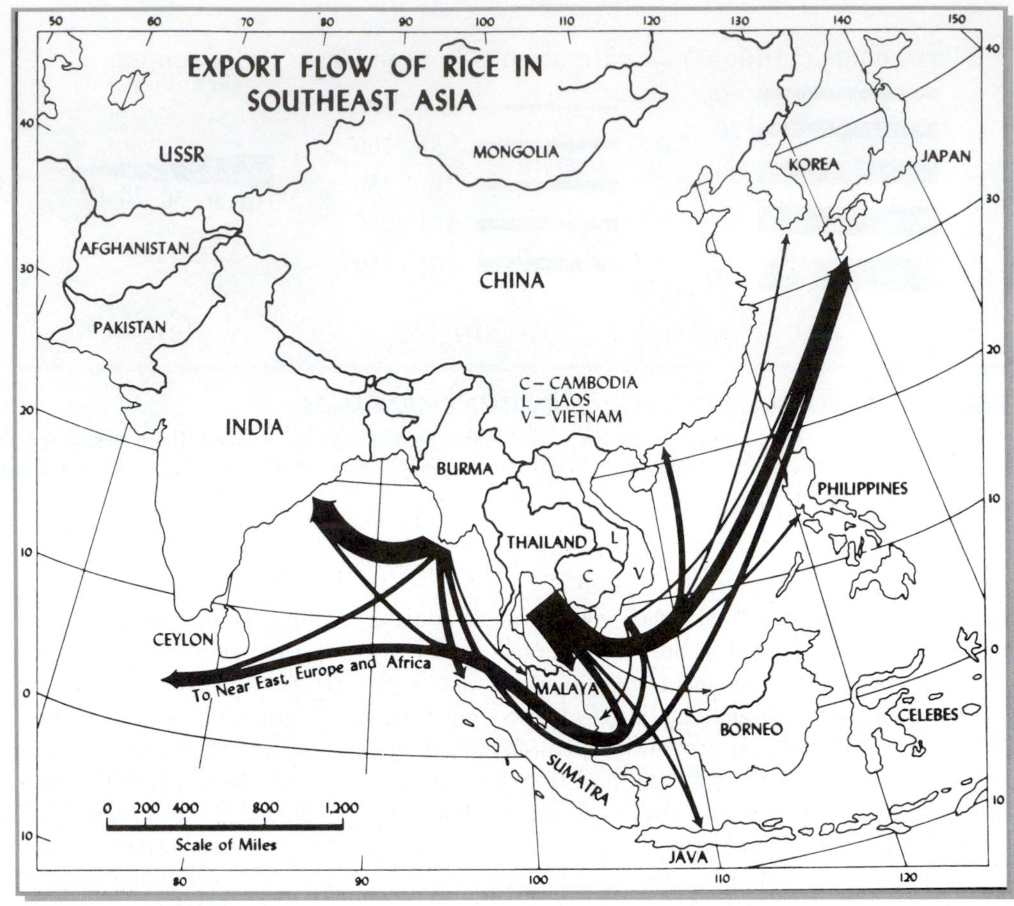

FIGURE 11.12 EXPORT FLOW OF RICE IN SOUTHEAST ASIA.
See the text for an explanation. *Source:* Reprinted by permission of Prentice Hall, Inc.

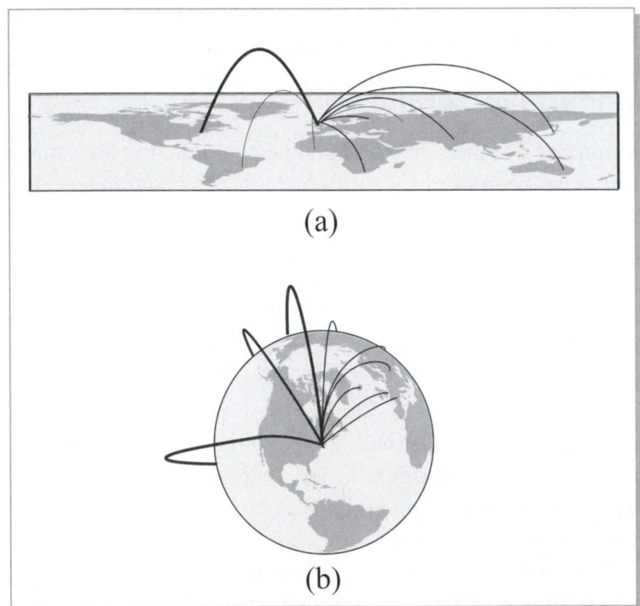

FIGURE 11.13 FLOW LINE REPRESENTATION AND SCALING USING ARCS.
See the text for an explanation. *Source:* After Becker *et al.* 1995 (a) and Munzner *et al.* 1996 (b).

(amount of data transfer between countries) with proportional symbols (depicting total data transfer for a particular country) (see Figure 11.14).

A number of other interesting Web visualizations depicting information flows can be found on the Web, such as the ones found at VisualComplexity (Lima 2008). The cartographer should be cautioned that, unlike Figure 11.14, most of these graphics do not have a geographic basis to them—they simply show connectivity between nodes on the Internet. In other words, the routes, the origin, and the destination are unknown, unimportant, or nonexistent in terms of real-world coordinates. As network display **spatializations** (making a graphical view or visualization of non-spatial data), these types of visualizations are beyond the scope of this chapter. However, it is interesting to note that some of the cognitive research related to these graphics seems to reinforce the effectiveness of conventional cartographic use of line width, hue, and value in the creation of the flow lines (Fabrikant *et al.* 2004).

One of the other innovative areas of research in flow mapping has been algorithmic control of flow line placement in order to minimize the number of line crossings. Tobler (1987) developed one of the first such algorithms to control

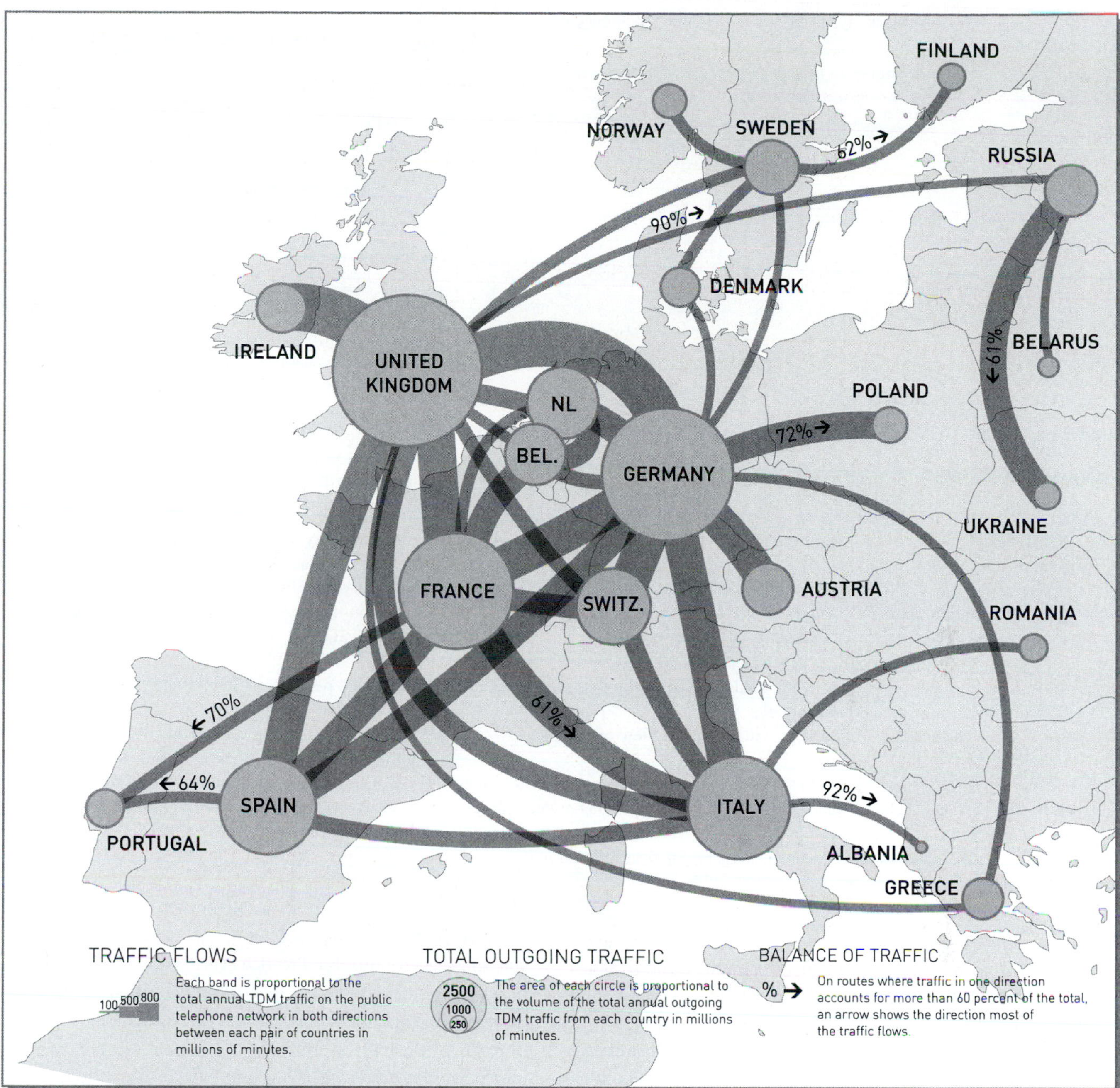

FIGURE 11.14 EUROPEAN TDM TRAFFIC FLOW, 2008.
In this map, flow lines are combined with proportional symbols. Notice that the semi-transparency used in the flow lines helps keep the lines' distinction when they cross each other, while still appearing as the dominant figures on the map. *Source:* The underlying data and the map were generated by TeleGeography Research (www.telegeography.com). Used by permission.

interposition of lines. Phan *et al.* (2005) developed a method that automatically generates flow maps that distort the positions of the nodes (the origins and destinations of the "from" and "to" positions) but hold their relative positions to each other while minimizing line crossings (see Figure 11.15). Enterprising designers will no doubt create new ways of creating new algorithms and symbolizing cartographic flows. We encourage such experimentation.

SUMMARY OF MAPPING TECHNIQUES

Six different mapping techniques have been presented in this part of the book: choropleth mapping, dot-density mapping, proportional symbol mapping, isarithmic and surface mapping, value-by-area or cartogram mapping, and flow mapping. This group accounts for the majority of thematic *quantitative* maps

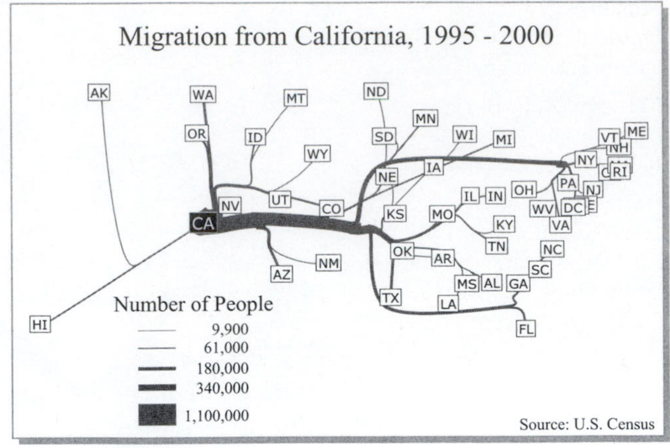

FIGURE 11.15 ALGORITHMIC SOLUTION TO FLOW MAPPING.
In this algorithm, the goal is to maintain the relative position between from and to nodes and also minimize the number of flow line crossings (Phan *et al.* 2005). *Source:* Map courtesy of Doantam Phan. Used by permission.

TABLE 11.1 SUMMARY OF THEMATIC MAPPING TECHNIQUES WITH EXAMPLES OF APPROPRIATE DATA

Technique	When Appropriate	Data Types: Example
Choropleth map (Chapter 6)	Portraying a geographical theme within areal enumeration units	Derived values, ratios, proportions: violent crime rate by county; population density
Dot map (Chapter 7)	Used to show spatial numerousness of geographic phenomena; usually area-based data	Nonderived variables such as totals: bushels of corn harvested by county
Proportional symbol map (Chapter 8)	Displaying discrete phenomena at points, or to represent aggregate data at points	Totals, non-density ratios and proportions; cannot be interval or unvarying data: population of cities or states
Isarithmic or Surface map (Chapter 9)	Used when total form of continuous phenomena is warranted	Often totals: population, temperature, or precipitation
Value-by-area cartogram (Chapter 10)	Used to provide a unique view of areal data	Variables that do not have negative numbers: state population
Flow map (Chapter 11)	Used to show movement of goods or ideas between places; used to show interaction between places	Totals, non-density ratios and proportions, but not at the interval level: automobile traffic counts; Internet volume

For more detailed explanations and examples, consult the individual chapter.

rendered today. It is a good idea at this juncture to review *when* each is to be used, and to give examples, before you continue through the remainder of the chapters. The review may be assisted by examining Table 11.1.

It is important to note that even though we have studied these techniques and have identified appropriate data for each, selecting the mapping technique should ultimately follow the data, not vice-versa. (Refer to Table 4.3 in Chapter 4.) For example, the cartographer makes a flow map because there is a need to map traffic data, and the flow map is the most appropriate technique choice. It is also a good idea to remember that matching map technique with the data continues to be somewhat subjective and will usually vary with the scale and purpose of the study or map.

REFERENCES

Becker, R., S. Eick, and A. Wilks. 1995. Visualizing Network Data. *IEEE Transactions on Visualization and Computer Graphics* 1:16–21.

Berry, B. 1967. *Geography of Market Centers and Retail Distribution.* Englewood Cliffs, NJ: Prentice Hall.

Berry, B., E. Conkling, and D. Ray. 1976. *The Geography of Economic Systems.* Englewood Cliffs, NJ: Prentice Hall.

Birdsall, S., and J. Florin. 1981. *Regional Landscapes of the United States and Canada.* New York: Wiley.

Davis, P. 1974. *Data Description and Presentation.* London: Oxford University Press.

Dent, B. 1987. Continental Shapes on World Projections: The Design of a Poly-Centered Oblique Orthographic World Projection. *Cartographic Journal* 24: 117–24.

Dodge, M., and R. Kitchin. 2001. *The Atlas of Cyberspace.* Addison-Wesley.

Fabrikant, S., D. Montello, M. Ruocco, and R. Middleton. 2004. The Distance-Similarity Metaphor in Network-Display Spatializations. *Cartography and Geographic Information Science* 31: 237–52.

Lima, M. 2008. *VisualComplexity.com: A Visual Exploration on Mapping Complex Networks.* http://www.visualcomplexity.com/vc/

McCleary, G. 1970. Beyond Simple Psychophysics: Approaches to the Understanding of Map Perception. *Proceedings of the American Congress on Surveying and Mapping*: 189–209.

Minard, C. 1862. *Emigration Map*. Paris: Charles Joseph Minard. From Library of Congress, *Map Collections: 1500–2004*. http://hdl.loc.gov/loc.gmd/g3201e.ct 000242

Munzner, T., E. Hoffman, K. Claffy, and B. Fenner. 1996. Visualizing the Global Topology of the MBone. *Proceedings of the 1996 IEEE Symposium on Information Visualization*: 85–92.

Parks, M. 1987. "American Flow Mapping: A Survey of the Flow Maps Found in Twentieth Century Geography Textbooks, Including a Classification of the Various Flow Map Designs." Unpublished Master's thesis, Department of Geography, Georgia State University, Atlanta.

Phan, D., L. Xiao, R. Yeh, P. Hanrahan, and T. Winograd. 2005. *Proceedings of the 2005 IEEE Symposium on Information Visualization* (INFOVIS'05).

Robinson, A. 1955. The 1837 Maps of Henry Drury Harness. *Geographical Journal* 121: 440–50.

———1967. The Thematic Maps of Charles Joseph Minard. *Imago Mundi* 21: 95–108.

———1982. *Early Thematic Mapping in the History of Cartography.* Chicago: University of Chicago Press.

Tobler, W. 1987. Experiments in Migration Mapping by Computer. *American Cartographer* 14: 155–63.

Tufte, R. 2001. *The Visual Display of Quantitative Information.* Cheshire, CT: Graphics Press.

GLOSSARY

bivariate flow map a flow map in which two variables are mapped on the flow lines; usually one variable is mapped to the line's widths, the other to changes in the line's hue, value, or saturation

directed flow map flow map that uses arrows to indicate direction of flow

distributive flow map flow map on which the distribution of commodities or migration is the principal focus

flow map map on which the amount of movement along a linear path is stressed, usually by lines of varying thicknesses and/or changes in hue, value, or saturation

heads-up digitizing creation of features interactively on screen. In GIS, these features can have attribute data. Required in flow mapping if the lines do not yet exist (such as in origin-destination maps)

network flow map flow map that reveals the interconnectivity of places

origin-destination map type of flow map in which actual routes between places are not stressed, but interaction is; direction of flow often not shown

radial flow map class of flow map that is characterized by a radial or nodal pattern

redundant coding expressing the same data by more than one visual variable; perhaps more common in flow mapping, the line's hue will change in saturation and/or value to the same data that scales the line's width. Wider lines will have a darker or more saturated fill; narrower lines will have a lighter or less saturated fill

spatialization creating a graphic display from non-spatial data, often used in Internet data visualizations

traffic flow map particular kind of flow map in which movement of vehicles past a route point is shown by scaled lines of proportionally different thicknesses

undirected flow map flow map which does not use arrows; a two-way flow is implied

PART III
DESIGNING THEMATIC MAPS

The text preceding Part III dealt with forming the base map, techniques of symbolization, data processing, and mapping techniques. All of these subjects are important in developing the final map. Other design considerations are required, however, before one completes the mapping assignment. This part of the book presents many of the techniques regarding the visual design aspects of thematic mapping.

Cartographic designers face two-dimensional design problems much as artists do. The elements of two-dimensional design can be studied carefully and techniques learned. Basic and advanced approaches, and advanced concepts and techniques, are presented in Chapter 12, which provides ideas for total map organization and figure-ground relationships. Of all the visual design ideas discussed in this book, the most fundamental—and necessary—is having a visual hierarchical plan. This plan steers the page design activity. Typographics, the study and application of language labels to maps, is a basic part of map design. Chapter 13 provides current ideas about this subject and briefly includes modern techniques of producing type by laser printing. Chief among the design variables for the thematic cartographer is color. Color essentials, color specifications, applications, and standards are subjects that cartographers need to know. These topics are presented in the last chapter of Part III.

12 THE MAP DESIGN PROCESS AND THE ELEMENTS OF MAP COMPOSITION

CHAPTER PREVIEW This chapter describes the design process and the appearance of the final map product. The design process is viewed as a series of stages beginning with problem identification and ending with implementation. Evaluation, creativity, experimentation, and map aesthetics are part of the design activity. The basic concepts of graphic composition as they relate to maps are presented, and they should encourage experimentation. No one best way to a design solution can be predetermined for all maps—only principles and general approaches can guide the cartographer. The map's design elements must be arranged into a graphic composition suitable to the map's communication purpose. The elements include the map's title, scale, neatline, symbols, and other components. By using visualization and experimentation techniques, the composition elements are arranged to satisfy the design goals. These elements include the planar organizational elements of balance, focus, internal organization, figure and ground, and a number of contrast elements. Visual acuity plays an important part in design; cartographers must be aware of the limitations of the human eye as they plan and develop map specifications. ■

Chapter 1 presented a brief explanation and several descriptions of map design. There, it was established that map design is a complex activity involving both intellectual and visual aspects. This chapter resumes the topic of design by looking in particular and in more detail at the design process and examining the elements of map composition—the first of the visual assignments that must be treated by the designer.

THE DESIGN PROCESS

Most designers would agree that all design takes place in sequential *steps,* ordered in a way that eventually yields the planned result. Identifying these steps or stages in design is helpful in learning how design takes place.

The **map design process,** like any act of designing, includes six essential stages: problem identification, preliminary ideas, design refinement, analysis, decision, and implementation (Hanks *et al.* 1978; see Figure 12.1). Students of cartography encounter these steps of the design process but not always in the order presented. However, the professional cartographer in a mapping firm oversees the mapping process that may involve a number of individuals. Thus, the design process must be more formalized.

Needs and design criteria are established in the first stage. Limitations are usually set in this stage. In the case of mapping, this stage includes the identification of map purpose, map audience, and such factors as cost and technical considerations. The most creative step in the design process occurs in the second stage, where preliminary ideas are formulated. Brainstorming or having problem-solving sessions involving creative thinking is helpful here. Many solutions to design problems are found unconsciously, and then "pop" into the mind's eye. Sorting through one's visual memory often takes place in this stage and is especially helpful in cartographic design.

In the third stage, called design refinement, all preliminary ideas are evaluated—and may be accepted or rejected. Those ideas that are retrieved are refined and sharpened, and

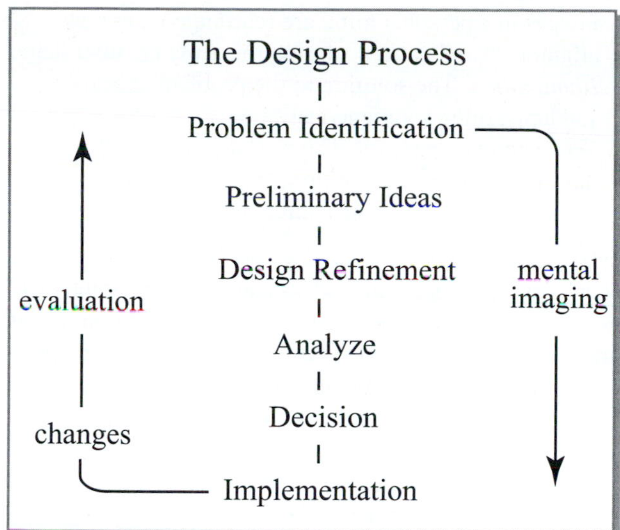

FIGURE 12.1 THE DESIGN PROCESS.
See text for an explanation.

decisions are made that will affect the whole process. For cartographers, this stage usually involves setting down in writing details of the mapping project. For example, critical data needs are reviewed, construction specifications are established, and a production schedule is finalized.

Models are often created in the analysis stage. For product designers, the models serve to make the drawings and sketches come alive and assist in the visualization process. Today, we use the computer to create models in graphic form, and real models are not made. Nevertheless, this remains a prototype stage. Map designers use this stage to develop detailed versions of their maps and to work out problem areas (such as tricky printing problems or unique symbol schemes). Cartographers may wish to test their preliminary designs on sample readers or trusted colleagues when this is possible.

The decision stage is as the name suggests. Changes, if any, are made on the prototype based on research and fact-finding from the previous stage. Ideas are rejected or accepted, and the final stage, implementation, begins. For cartographers this stage signals the beginning of the final map product.

Evaluation and modification (also called feedback) in the design process is continuous. Each design teaches us something about future problems and processes. Evaluation is a critical element that helps designers become efficient and recognize that each design process may be unique, and that not every design problem will utilize the design stages in exactly the same manner.

Mental imaging in the process assists the designer in anticipating solutions and problem areas. Mental imaging also involves visualization, especially in cartographic design. This imaging allows the designer to "see" in his or her mind's eye the end product, and thereby helps decision making along the way. Being able to visualize is essential, and comes with experience, regardless of the design area. This is easy enough to say but difficult to do or teach—experience is the only teacher. Nonetheless, the ability to conceive the final solution before it is physically mapped will reduce the number of wrong choices in design.

The designer may cycle through all stages of the design process as many times as required to reach an acceptable solution to the problem. Repetition is to be expected.

Design Evaluation

It is indeed difficult to evaluate map design, usually because the one doing the evaluating was not intimately involved in the design process. The outsider is never sure what compromises the cartographer had to make to balance the decisions in the design process. We do not know the relationship between map author and designer, and what sacrifices and learning had to take place to get anything in the map. One cartographer does suggest, however, that a map's design should be judged only with regard to the map's purpose and intended audience (Muehrcke 1982). The cartographer has a good idea as to who will be using their map. Frequently, people who were not a part of the intended audience will use the map, thus making design even more of an important factor.

There are a few guidelines that may be followed. Southworth and Southworth (1982), for example, list these design characteristics of successful maps:

1. A map should be suited to the needs of its users.
2. A map should be easy to use.
3. Maps should be accurate, presenting information without error, distortions, or misrepresentation.
4. The language of the map should relate to the elements or qualities represented.
5. A map should be clear, legible, and attractive.
6. Many maps would ideally permit interaction with the user, allowing change, updating, or personalization.

Creativity and Visualization

Creativity is the ability to see relationships among elements (regardless of the design arena). Although there is no recipe for creativity, certain activities appear to be shared by people considered to be great thinkers, scientists, or artists (The Burdick Group 1982).

1. Challenging assumptions—daring to question what most people take as truth.
2. Recognizing patterns—perceiving significant similarities or differences in ideas, events, or physical phenomena.
3. Seeing in new ways—looking at the commonplace with new perceptions; transforming the familiar into the strange, and the strange into the familiar.
4. Making connections—bringing together seemingly unrelated ideas, objects, or events in ways that lead to new concepts.
5. Taking risks—daring to try new ways, with no control over the outcome.

6. Using chance—taking advantage of the unexpected.
7. Constructing networks—forming associations for the exchange of ideas, perceptions, questions, and encouragement.

Cartographic designers can learn from this list. Conscious effort to participate in new ways of thinking during the transformation stage should become an integral part of the design procedure. New ways of thinking and taking risks should not occur unless they are balanced with established design principles.

The **visualization process,** or thinking by incorporating visual images into thought, occurs to the fullest when seeing, imagining, and graphic ideation come into active interplay (Samuels and Samuels 1975; Muehrcke 1981).

For the creative person, *seeing* is integral to all thought processes, and willingness to restructure visual images into new configurations is essential. Looking at a map inside out or upside down may yield solutions never thought possible. Mental imagining is also helpful in developing design experience. The ability to foresee in the design process calls for the strengthening of imagining talents.

There are many descriptions of how new ideas are formed in the minds of creative people. A composite pattern might include these four stages in the creative process (Samuels and Samuels 1975; see Figure 12.2).

1. *Preparation.* At this stage, a person consciously "files away" into memory visual images that can be useful for a problem at hand. This may be referred to as an **image pool.**
2. *Incubation.* The person releases all conscious hold on the problem and turns to other tasks. It is theorized that images in a person's mind are rearranged into new alignments and patterns during this truly creative stage.
3. *Illumination.* The solution to the problem appears suddenly, often spontaneously.
4. *Verification* or *revision.* The person consciously works out the details of the solution, bringing all efforts and skills to bear. Formal structures result.

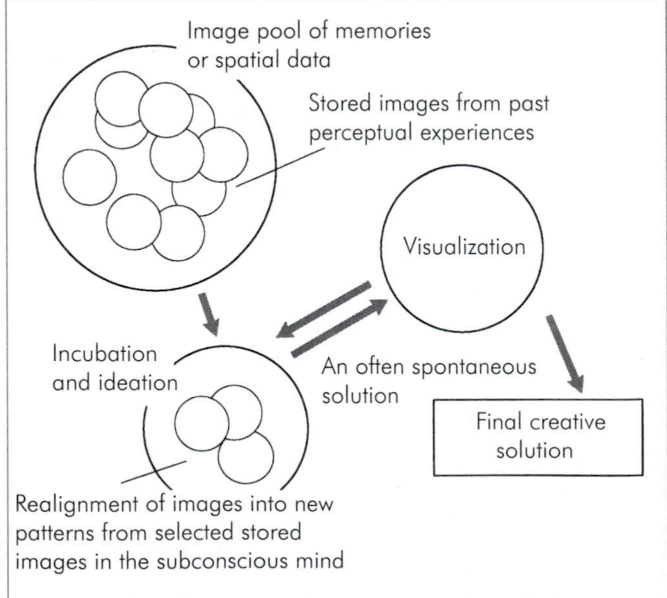

FIGURE 12.2 THE VISUALIZATION PROCESS.
It is theorized that creativity is the result of new arrangements of stored visual images in the subconscious mind. Visualization is the ability to "see" these arrangements, and is an illumination process leading to final solutions.

Map designers should try to develop this image pool in the subconscious, so that creative ideas can form during the incubation stage. Some psychologists believe that ideation results from realignment of images stored from earlier perceptual events (Samuels and Samuels 1975). *Visualization* is the process of experiencing or seeing these new creations—an illumination stage, or as described earlier, placing visual images into thought (DiBiase 1990). Cartographers should therefore experience as much graphic art, art, and cartography as possible. It is worthwhile to spend time exploring an art museum, seeing an animated film exhibit, or going through old atlases. From such experiences, the designer builds an image inventory that can later provide creative design solutions.

There are two views of visualization. One is the view just presented, where visualization begins from within and leads to new mental formations (creations) from previously stored images. The other view is called **scientific visualization,** a method using computers to transform data into geometric models that can be presented on-screen, rotated in three-dimensional space, or animated to portray longitudinal information. Both operations are instrumental in good cartographic design.

Graphic Ideation

Graphic ideation, or *the formation of mental maps,* is often practiced by creative persons. The goal of this activity is to bring vague images into clear focus (Muehrcke 1981). Designers can improve their creative potential by practicing seeing, imagining, and graphic ideation.

One real contribution of rapidly generated GIS virtual maps is their potential as "what if" images. Hundreds of maps with alternative designs can be created in a short time so that the designer can explore alternatives. Scale, projection, color, typography, classification, and other elements of design can be varied and checked almost instantly. Yet, do not be lulled into thinking this is a panacea for good design; the computer should serve as a *tool* to facilitate the individual's imaging abilities, providing for rapid graphic display of the creative process.

Experimentation

Experimentation is necessary in testing the new idea. Sometimes ingenious solutions do not work when the final map is produced. Creative solutions must be attempted first using the GIS or graphic mapping software. Map design in virtual form allows the designer to try many different possible scenarios of layout, color, type fonts, scale, and other design elements that help create the well-designed map. We are able to evaluate each possible iteration of ideas in order to find that best, if not perfect, solution. A willingness to explore alternative solutions is essential to developing the best graphic design for a map.

Unfortunately, many map designs are first solutions. Relying upon default setting of graphic software frequently creates a poorly designed map. All too often the novice user of GIS and other mapping software for analyzing complex data fails to communicate his or her research solutions as a result of a poorly designed map. It is altogether too easy to drift into simple solutions. The process from visualization through experimentation to final solution often involves much repetition and backtracking.

Map Aesthetics

Although maps may not be objects of aesthetic concern, there is a trend toward giving greater consideration to the quality of appearance of thematic maps. Writing several decades ago, map critic John K. Wright addressed this issue:

> *The quality of a map is also in part an aesthetic matter. Maps should have harmony within themselves. An ugly map, with crude colors, careless line work, and disagreeable, poorly arranged lettering may be intrinsically as accurate as a beautiful map, but it is less likely to inspire confidence* (Wright 1942, 542).

John S. Keates, a British cartographer, remarked, "The 'art' of cartography . . . is not simply an anachronism surviving from some prescientific era; it is an integral part of the cartographic process" (Keates 1982, 127).

Three elements have been identified as forming the basis for the evaluation of map aesthetics: harmony, composition, and clarity (Karssen 1980). *Harmony* is viewed as the relationship between different **map elements** (that is, how do the elements look *together?*). *Composition* deals with the arrangement of the elements and the emphasis placed on them. In other words, how does the structural balance of emphasis appear? Finally, *clarity* deals with the ease of recognition of the map's elements by the map user. "A map which lacks one or more of these three main elements, lacks beauty" (Karssen 1980, 124).

Cartographic designers have a certain degree of freedom in the design process. Of course, map function is the overriding concern, along with the needs of the user, but beyond that the designer is working in a subjective realm. How well he or she performs in the creative, aesthetic realm will more than likely depend on intuitive judgments, conditioned by fundamental training and experience.

The subjective elements of design have been listed as follows (Karssen 1980):

- Generalization—beauty of simplified shapes
- Symbolization—beauty of graphic representation
- Color—beauty of color accent and balance
- Layout—beauty of composition
- Typography—beauty of typographic appearance

This brief excursion into map aesthetics is not intended to be exhaustive. It has been included only to suggest to the reader that map design is subjective as well as objective. Beyond the *science* of cartography lies the *art* of cartography—the intuitive, artistic, and aesthetic world of maps where designers can exercise their expressive talents.

THE MAP'S DESIGN ELEMENTS

Thematic maps are instruments of visual communication. The marks that make up a map are visual elements, and transfer of information takes place through them. The map designer arranges the visual elements into a functional composition to facilitate communication. This functional approach to design was first expressed by Arthur Robinson: "Function provides the basis for design" (Robinson 1966, 3). That something is functional means that it has been "designed or developed chiefly from the point of view of use" (Robinson 1966, 3). Design decisions regarding the map's elements should be made on the basis of how each element is to function in the communication. The challenge is to make the map aesthetically pleasing as well as functional.

Thematic maps contain most of these map elements: titles, legends, scales, mapped areas, map insets, credits, graticules, borders, symbols, and place names (see Table 12.1). The task of the designer is to arrange these into a meaningful, aesthetically pleasing design—not an easy task. From this perspective, map design is like a series of filters in which selections are made at each filter (see Figure 12.3). This analogy directs attention to the fact that map design is a complex affair involving many decisions, each of which

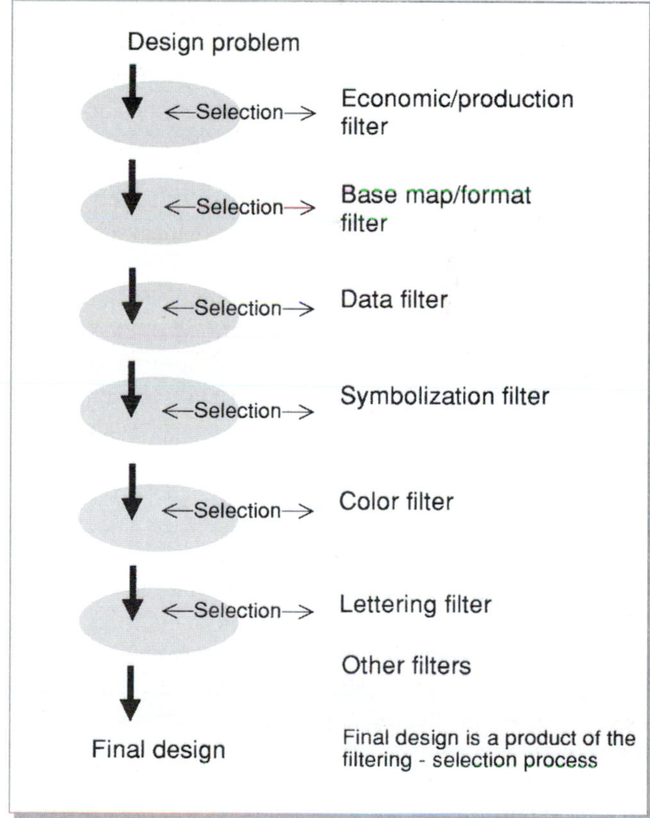

FIGURE 12.3 MAP DESIGN AS A FILTERING-SELECTION PROCESS.

In this view of design, a series of filters must be rotated (selection), allowing design activity to continue until an appropriate final solution is reached.

TABLE 12.1 TYPICAL ELEMENTS OF THE THEMATIC MAP

Name of Element	Description and Primary Function
Title (and subtitle)	Usually draws attention by virtue of its dominant size; serves to focus attention on the primary content of the map; brevity is desired; usually includes where, what, and when components; may be omitted where captions are provided but are not part of the map itself
Map legend	The principal symbol-referent description on the map; subordinate to the title, but a key element in map reading; serves to describe all unknown or unique symbols used; often has an associated legend title
Map scale	Sometimes included on a thematic map; it provides the reader with important information regarding linear relations on the map; can be graphic, verbal, or expressed as an RF
Mapped body	Objects, land, water, and other geographical features important to the purpose of the map; make the composition a map rather than simply a chart or diagram
Inset maps	Small ancillary map generated at a larger scale and including more detail; the outline of the inset should be identified on the mapped areas
North arrow	Often not included on maps, they are necessary when north is not at the top of the map—like many 3-D models, or the mapped area is unfamiliar or nonintuitive to the map reader
Location maps	Small ancillary map generated at smaller scale that identifies location of the map body; again used when the mapped area is unfamiliar or nonintuitive to the map reader
Credits	Can include the map's data source, an indication of its reliability, dates, and other explanatory material
Graticule	Often omitted from thematic maps today; should be included if their locational information is crucial to the map's purpose; usually treated as background or secondary forms; the graticule may be composed of latitude-longitude, UTM, state-plane, or other systems
Neatlines	Neatline is usually a relatively thin and unobtrusive line that surrounds the map body and all other map elements; serves to balance the map
Map symbols	Wide variety of forms and functions; the most important elements of the map, along with the geographic areas rendered; symbolization in the quantitative thematic map types in Part II of the text follows the type of data being used
Place names and labeling	The chief means of communicating in general purpose maps; also important in thematic cartography when reference features are used to orient the reader on the map
Ancillary text	Can be additional information included to provide a greater understanding of the map topic; often used to indicate special occurrences, missing data, or data manipulation pertinent to the interpretation of the map

affects all the others. Good design is simply the best solution among many, given a set of constraints imposed by the problem. The best design will likely be a simple one that works well with the least amount of trouble. The optimum solution may not be achievable, and what is good design today may be ineffective in the future. We are constantly learning more about the map user, and this will modify our future decisions as designers.

Not all elements occur on all maps as the inclusion of each element is dependent on the map purpose, readership, and context. Most thematic maps, for example, actually do not have a graticule. Inset and locators are not needed for the myriads of U.S. level choropleth maps, and the inclusion of place names occurs most frequently on general purpose maps.

Important design principles include simplicity, appropriateness in a functional context, pleasing appearance, and considerations of economy. The designer's tools of creativity, visualization, ideation, and problem solving are used to sift through the map elements in order to bring these principles into a proper balance.

The graphic solution should seek to enlighten the reader, as mentioned in Chapter 1. In viewing art, the reader may believe that nothing is required other than mere contemplation. When looking at effective graphic advertising, the reader responds by buying the product. In effective cartographic design, the reader should gain spatial knowledge and understanding.

DESIGN LEVELS ON THE MAP

It is useful at the outset to imagine the thematic map as composed of different planes or levels, just as in GIS where we work with data layers (see Figure 12.4). Usually the levels are differentiated by visual prominence. Each component of the map belongs to a specific level. More than one map element can be placed on a particular level, but a *single element should never be assigned to more than one level.* Maintaining a separation of these elements in this way will facilitate the map's overall design.

Map composition, the arrangement of the map's elements, takes place at each level and between levels. The arrangement at a given level may be called **planar organization,** and that between levels **hierarchical organization.** The cartographer ordinarily approaches design solutions by simultaneously manipulating all elements *at* and *between* all levels.

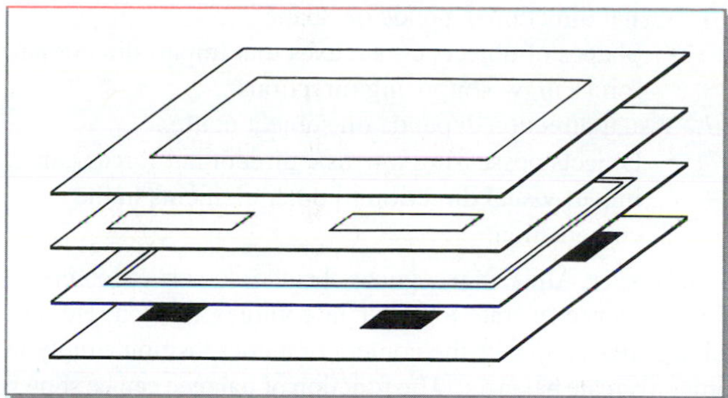

FIGURE 12.4 THE ORGANIZATIONAL LEVELS OF THE THEMATIC MAP.

A typical thematic map can be considered to be made of several distinct levels or planes. In planning a map, the designer assigns the various map elements to these levels. This causes the designer to think of each element in its proper role, thus leading to a more organized design.

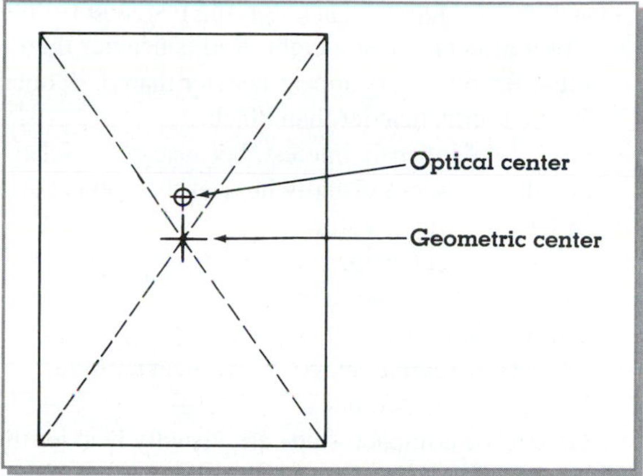

FIGURE 12.5 THE TWO CENTERS OF AN IMAGE SPACE.

The designer should arrange the map's elements around the natural (optical) center rather than around the geometric center.

ELEMENTS OF MAP COMPOSITION

The map's graphic composition is the arrangement or organization of its elements. The composition principles introduced in this chapter include the purpose of map composition, planar organization, figure and ground organization, contrast, and visual acuity. Knowledge of these principles and their application assists the cartographer in seeking better design solutions.

Purpose of Map Composition

Map composition is much more than layout, which is simply a sketch of a proposed piece of art, showing the relationships of its parts. The term *composition* is used here because it indicates the intellectual dimension as well as the visual. Map composition serves these ends:

1. Forces the designer to organize the visual material into a coherent whole to facilitate communication, to develop an intellectual *and* a visual structure
2. Stresses the purpose of the map
3. Directs the map reader's attention
4. Develops an aesthetic approach for the map
5. Coordinates the base and thematic elements of the map—a critical factor in establishing communication
6. Maintains cartographic conventions consistent with good standards
7. Provides a necessary challenge for the designer in seeking creative design solutions

Planar Organization of the Visual Elements

The three aspects of planar visual organization are balance, **focus of attention,** and internal (intraparallel) organization. Each is important to the designer's language, and their visual possibilities and effects must be explored.

Balance

Balance involves the visual impact of the arrangement of image units in the map frame. Do the units appear all on one side, causing the map to "look heavy" on the right or left, top or bottom? An image space has two centers: a geometric center and an **optical center** (see Figure 12.5). The designer should arrange the elements of the map so that they balance visually around the optical center.

Rudolf Arnheim (1965), a noted author on the psychological principles of art, has suggested in his writings that **visual balance** results from two major factors: weight and direction. Objects in the visual field (for example, within the borders of a map) take on weight by virtue of their location, size, and shape. Direction is also imposed on objects by their relative location, shape, and subject matter. Arnheim stresses that balance is achieved when everything appears to have come to a standstill, "in such a way that no change seems possible, and the whole assumes the character of 'necessity' in all its parts" (1965, 12). In Arnheim's view, unbalanced compositions appear accidental and transitory.

Arnheim's (1965) observations on balance resulting from visual weight and direction can be summarized as follows:

1. Visual weight depends on location.
   - Elements at the center of a composition pull less weight than those lying off the tracks of the structural net (see Figure 12.6a).
   - An object in the upper part of a composition is heavier than one in the lower part.
   - Objects on the right of the composition appear heavier than those on the left.
   - The weight of an object increases in proportion to its distance from the center of the composition.
2. Visual weight depends on size.
   - Large objects appear visually heavier than small objects.

3. Visual weight depends on color, interest, and isolation.
   - Color affects visual weight. Red is heavier than blue. Bright colors appear heavier than dark ones. White seems heavier than black.
   - Objects of intrinsic interest, because of intricacy or peculiarity, seem visually heavier than objects not possessing these features.
   - Isolated objects appear heavier than those surrounded by other elements.
4. Visual weight depends on shape.
   - Objects of regular shape appear heavier than irregularly shaped ones.
   - Objects of compact shape are visually heavier than those not so shaped.
5. Visual direction depends on location.
   - Weight of an element attracts neighborhood objects, imparting direction to them (see Figure 12.6b).
6. Visual direction depends on shape.
   - Shapes of objects create axes that impart directional forces in two opposing directions.
7. Visual direction depends on subject matter.
   - Objects possessing intrinsic directional forces can impart visual direction to other elements in the composition.

Of course, Arnheim recognizes that the elements of compositional balance operate together in complex fashion. He also advises not to forsake the content of a composition simply in order to create balance: "The function of balance can be shown only by pointing out the meaning it helps to make visible" (Arnheim 1965, 27). Once again, this underscores the idea that map content is more important than the map's design.

It is often difficult to achieve balance on the map. Many of the shapes and their locations are imposed by geographical or locational facts. The key is to find a way to use Arnheim's guidelines such that balance is achieved within the formal constraints inherent to the map. The guidelines presented by Arnheim should nevertheless be applied whenever possible.

To illustrate how balance can affect the impression one has when viewing a map, several different locations of the shape of Africa are included in Figure 12.7. The shape of the mapped

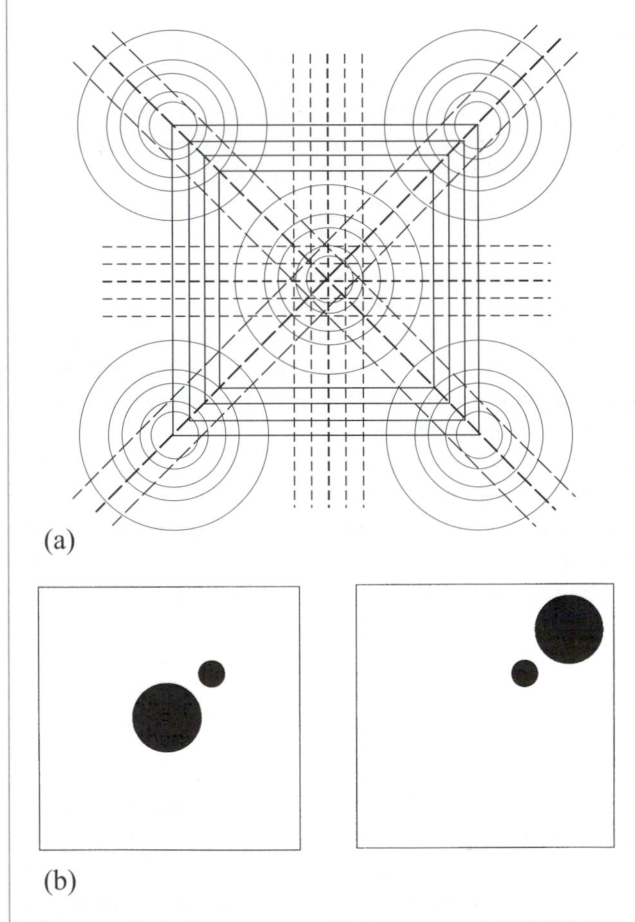

FIGURE 12.6 BALANCE IN THE VISUAL FIELD.

Arnheim stresses that a structural net, as in (a), determines balance. Objects on the main axes or at the centers will be in visual balance. An object is given direction by other objects adjacent to it. In (b), the small disc's directional element is shifted as the large disc's position is changed. Each thematic map will have a unique structural net created by the locational patterns of its elements.
Source: After Arnheim 1976.

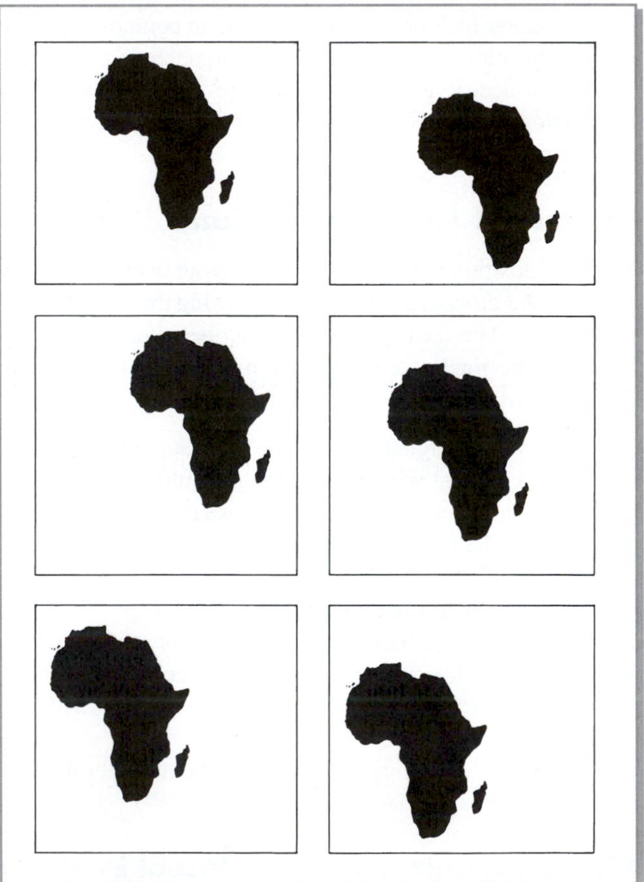

FIGURE 12.7 MAP BALANCE.

Position of map elements in the image space affects the balance of the map. The difference can be visually subtle, as this illustration shows. In which image does a natural visual equilibrium appear to exist?

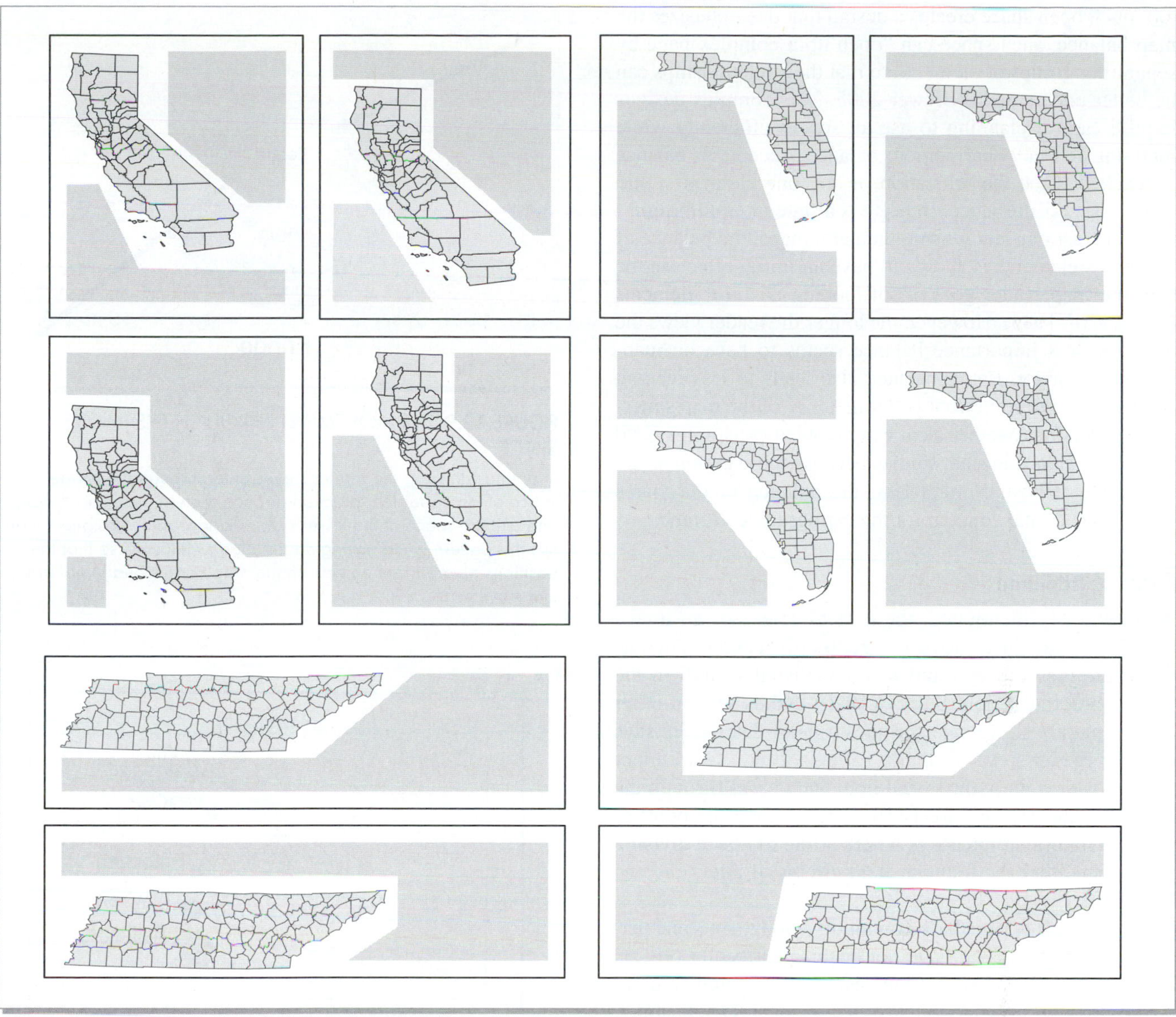

FIGURE 12.8 MAP BALANCE AND MAP WORK SPACE.

The positions of California, Tennessee, and Florida represent potential orientations and positions within the map neatline. The gray areas are displayed to highlight the possible locations for the various map elements.

area will frequently play a role in its positioning on the map. The Africa map example indicates how shape influences balance. The more rectangular the shape of the map body, the more impact it will have on overall balance. Take the states of California, Tennessee, and Florida as examples. California's elongation north-south creates a challenge in its positioning in terms of map balance. Tennessee presents an equal challenge as a result of its east-west elongation. Florida, with its truncated "T" shape, naturally provides a weight factor that pulls the reader to the upper right quadrant. Figure 12.8 depicts possible positioning of these three states within their map neatline. No single position is recommended here as that is a decision made by the cartographer when considering the entire map and its components. The maps are not presented at the same scale since this is not the norm when mapping several states separately on page-size media. The gray areas suggest usable work space of titles, legends, scales, reference, and ancillary information. Careful consideration must be made when adding the additional map elements in order to achieve a balanced map.

Achieving visual balance, of course, is not always as simple as the case just illustrated. Normally, thematic maps contain most of the elements mentioned earlier, and all must be handled in terms of balance. Visual weight caused by texture, solid black and white areas, and other elements must figure in the planning. Open spaces take up "balance space" and must be used effectively in the overall design. Brewer (2005) discusses the balancing of empty spaces. Although too much open space creates a design that deemphasizes the map balance, such space can "open up a complex page by separating groups of elements so that their relationships can be better understood" (Brewer 2005, 20). Complex designs require careful planning to use all spaces efficiently while retaining a visually harmonious balance. Acceptable balance is reached when the relocation of any one element would cause visual disturbance. Balance is a state of equilibrium.

In at least one study using thematic maps, the balance of the map's elements is shown to have an initial effect on the way the map reader goes about looking at those elements (Antes *et al.* 1985). However, the longer the reader views the map, the less importance balance seems to have on map-reading behavior. Better balance also leads to less reading difficulty and to somewhat better memory of the map's message. It is not altogether clear exactly what constitutes good and poor balance in such studies because these extremes are subjective at best. Nonetheless, the balance of the map's elements is a vital concern for the cartographic designer.

Focus of Attention

As previously mentioned, the optical center of an image area is a point just above the geometric center. This attracts the viewer's eye, unless other visual stimuli in the field distract attention. Surrey (1929) makes several other points that are significant for questions of design. He says that the reader's eye normally follows a path from upper left to lower right in the visual field and passes through the optical center (see Figure 12.9). Furthermore, the point of greatest natural emphasis is where a line of space division intersects either the focus or field circles of attention (see Figure 12.10).

Surrey's ideas were based on intuitive judgments and personal observations and have not been scientifically proven. Yet they do have appeal for the designer. An examination of recent print advertisements attests to the general applicability of his ideas. We can learn from these and other graphic designs. The map is a visual instrument, so the designer must learn what works in the visual world.

Internal Organization

The internal organization of the map's visual field relates to visual or perceptual order. Arnheim (1965) defines order as "a wealth of meaning and form in an overall structure that clearly defines the place and function of every detail in the whole" (45). Order implies an underlying structure, graphic or intellectual, that binds the parts of the whole together. In an ordered map, the graphic elements are arranged into a composition that develops a clear visual expression of the meaning of the communication and that shows an underlying structure of the graphic elements.

Brewer (2005) indicates that the internal elements of the map should maintain a linear balance. The positioning of the top of these elements should fall along a common linear position. The same can be true of elements that can

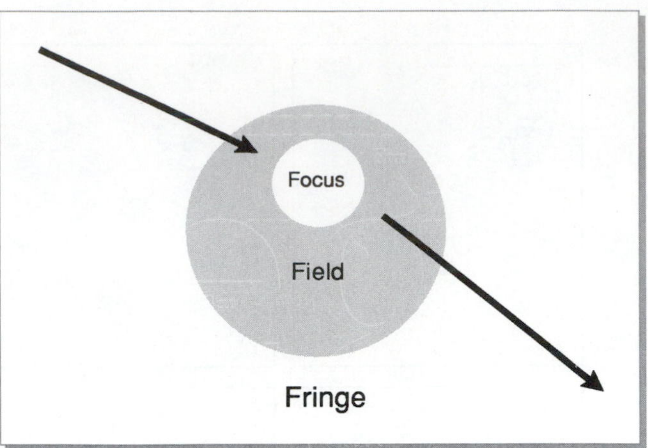

FIGURE 12.9 EYE MOVEMENT THROUGH THE IMAGE SPACE.

In normal viewing, the reader's eyes enter through the image space at the upper left, proceed through the visual center (focus), and exit the space at the lower right. Cartographic designers may use this pattern when arranging the map's elements, so that the positions of important objects on the map correspond to natural eye movements.

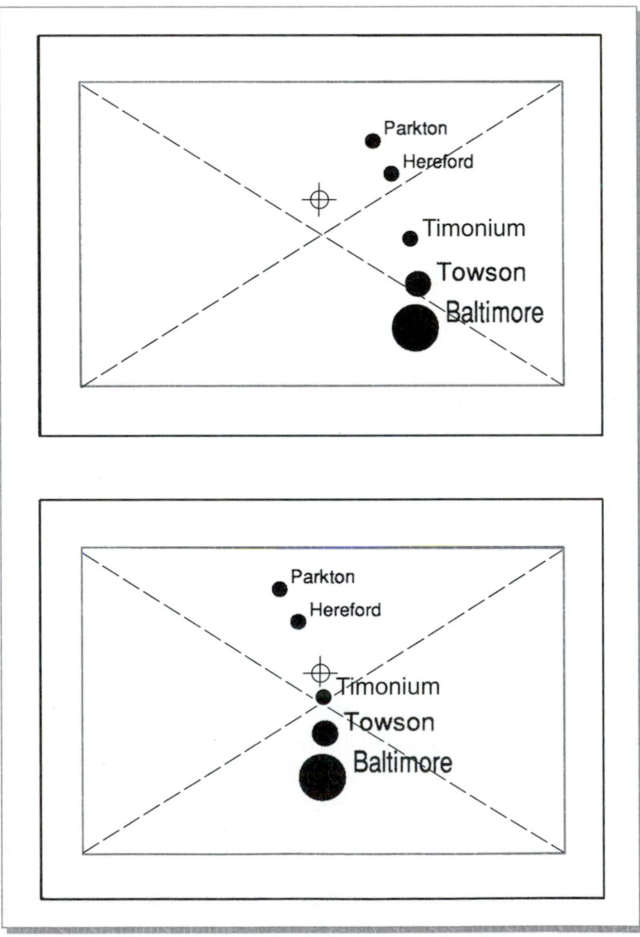

FIGURE 12.10 RECENTERING FOR GREATER CLARITY.

If possible, it is a good idea to recenter the map to place the central focus (in this case, the town of Timonium) closer to the optical center.

align vertically. This alignment brings an orderly appearance to the map and thus produces a harmonic balance to the layout design. These alignments can be achieved in mapping software through the use of guidelines and background grids.

Contrast and Design

Closely associated with figure and ground organization, and of nearly equal importance, is the feature called **contrast.** Contrast is fundamental in developing figure and ground but can be considered a design principle in its own right. Visual contrast leads to perceptual differentiation, the ability of the eye to discern differences. A lack of visual contrast detracts from the interest of the image and makes it difficult to distinguish important from unimportant parts of a communication. Map elements that have little contrast with their surroundings are easily lost in the total visual package. Contrast must be a major goal of the designer.

Contrast can be achieved through several mechanisms: line, texture, value, detail, and color. All of these could be used in one design, but the result might be visual disharmony and tension—potentially as unrewarding as having no contrast at all.

Line Contrast

Lines may be put to a variety of uses on maps. They can function as labels, borders, neatlines, political boundaries, quantitative or qualitative symbols, special symbols to divide areas, or graphic devices to achieve other goals. Line contrast can be of two kinds: character and weight. **Line character** derives from the nature of the line and its segments, or its value or color (see Figure 12.11a). The order of visual importance of various line characters has not been well established. The subject and purpose of the map very often restrict choice of line character. On some maps, there may be no lines. For example, some recent designs use edges rather than lines to evoke a response.

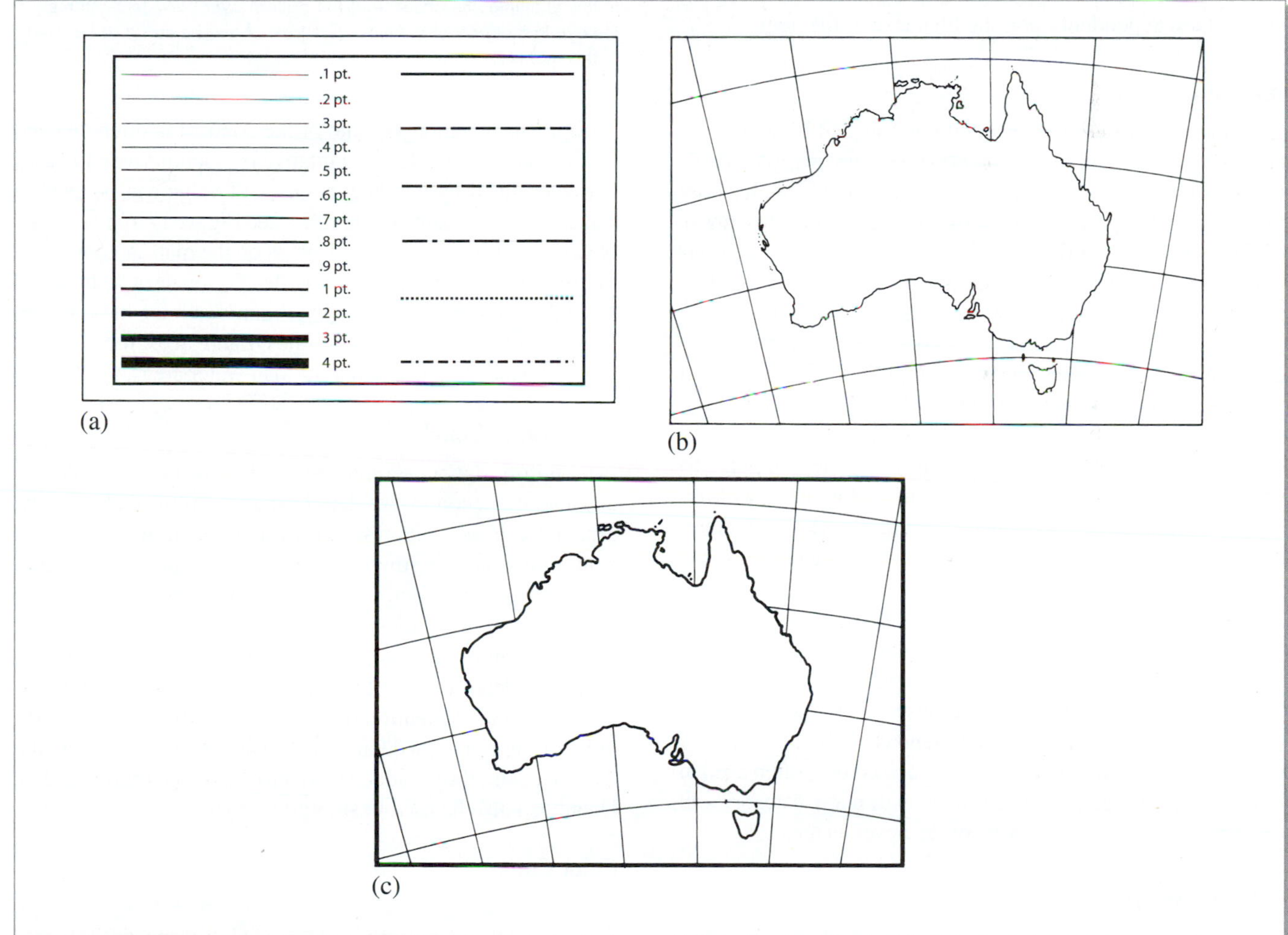

FIGURE 12.11 LINE CONTRAST.

The relationships between point sizes and their actual line weights are presented in (a) along with variable line characteristics. The visual effect of varying line weights on a map is illustrated by comparing (b) and (c). More visual interest is achieved with greater contrast of weight. In this case, the figure and ground organization is strengthened.

The thickness of a line is its **line weight,** although no clear-cut relationship exists between thickness and visual or intellectual importance. Although a broader line generally carries more intellectual importance, very fine lines also can be visually dominant. Strike a balance, keeping the map's purpose firmly in mind.

Contrast of line character and weight introduces visual stimulation to the map. A map having lines of all one weight is boring and lacks potential for figure formation (see Figure 12.11b). On the other hand, a map with lines of several weights and characters focuses attention, is lively, and aids the map reader's perceptual organization of the material (see Figure 12.11c). Most GIS and mapping software allow the cartographer to specify line weights of any width. Line weights are specified by point size, very similar to specifying type size to be discussed in Chapter 13. There are 72 points per inch. Most software offer a minimum default line weight of 0.5 point (.014 inch). Although differences of one-half point are discernible by many map users, increments larger than this are recommended to assure for better line weight discrimination. The actual line weights selected for creating contrast are dependent upon the final size of the map.

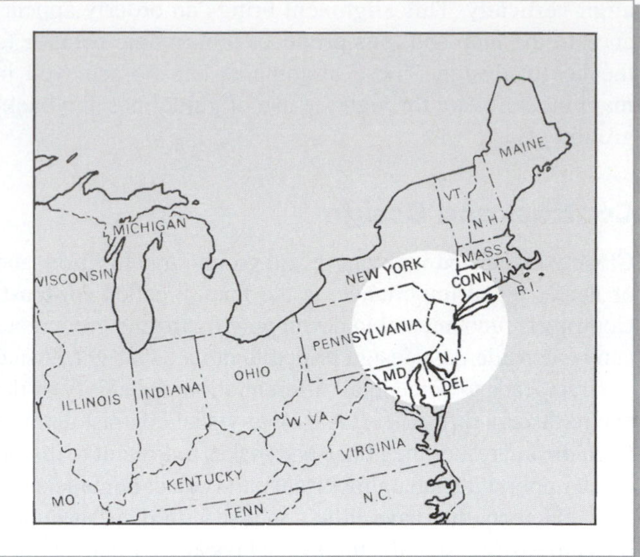

FIGURE 12.12 USE OF DIFFERENCES IN VALUE TO FOCUS THE READER'S ATTENTION.

In this example, the absence of the pattern over New Jersey forces the eye to that part of the map. Contrast of texture can also be used in this manner.

Texture Contrast

Contrast of texture involves areal patterns and how they are chosen for the map. In this context, texture is a pattern of small symbols (such as dots or lines) repeated in such a way that the eye can perceive the individual elements. Texture is often determined by the selection of quantitative or qualitative symbols for the map. Contrast considerations should be part of symbol selection. In some instances, patterns are selected and applied to the map solely to provide graphic contrast (for example, in the differentiation of land and water). Texture is sometimes applied in order to direct the reader's attention to a particular part of the map.

Texture is used more frequently in grayscale mapping. To achieve a differentiation of features, the cartographer will use grayscale values, patterns of lines at various angles, or pictoral patterns to display a diverse set of qualitative features. These patterns are included in most GIS and mapping software. Care should be taken when selecting these patterns as their interaction and distribution often create a map that is aesthetically poor. The texture can become a visual irritant. Texture becomes less of an issue when mapping because we are more likely to use solid color fills, but we recommend caution in using color pattern fills. A combination of texture and color is often used in multivariate cartography or on maps with many categories in order to create additional layers of data over an area.

Value Contrast

Texture is observable because the individual dots or other elements of the pattern are easily seen. Reducing such a pattern to the point where the elements are below the threshold of resolution acuity results in the perception of a visual tone or value. Contrast of value is another design technique used by cartographers, although some of the contrast is often dictated by the nature of the data (qualitative or quantitative). In cases not determined by the data, contrast of value can be used in ways similar to contrast of texture (see Figure 12.12). Contrast of value leads to light and dark areas on the map. A good place to use this contrast type is in the development of figures and grounds. To stand out strongly, figures should have values considerably different from grounds. Land areas, for example, should be made lighter or darker than water areas.

Variation of Detail

Although designers seldom think of it as a positive design consideration, contrast of detail can be employed effectively, especially in combination with other techniques. Along a continuum ranging from little detail at one end to great detail at the other, the reader's eye will be attracted to those areas of the map with the most detail.

This feature can work against the designer, however. Exquisite detail rendered to an unimportant feature can distract the reader's attention from the communication effort. By judicious use of extra detail in important areas of the map, the designer can subtly lead the reader to them (see Figure 12.13). Detail can also be used to strengthen figure formation.

Color Contrast

Employment of color is one of the chief techniques in the development of contrast in design. Color can differentiate areas on the map for a variety of purposes. In general purpose and other qualitative maps, multiple hues are selected to differentiate between features. In choropleth and other quantitative maps where data are classified, we use a single

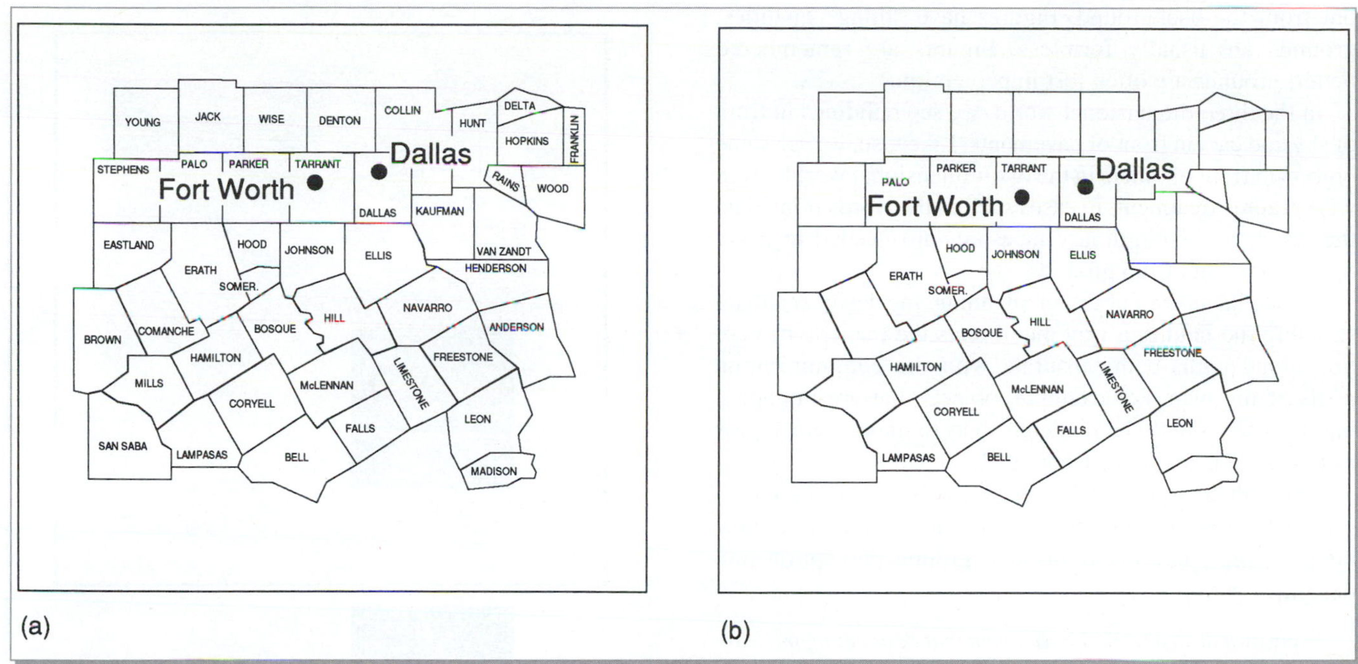

FIGURE 12.13 PROVISION OF DETAIL CAN DIRECT THE READER'S EYE.

In (a), the even distribution of place names does not focus the reader's attention. The eye is drawn to that part of the map in (b) that has the most detail—in this case, lettering.

hue with differing values to represent the various classes of data for a single variable. The cartographer makes use of differing levels of saturation, or brightness, in order to increase the design of the map. Color contrast can be easily achieved for virtual maps displayed on-screen using a red-green-blue, or RGB, color model. Maps that are designed to be printed will use the color complements of cyan-magenta-yellow, or CMY, color model. In both instances, careful selection of color greatly enhances the contrast of map features or data. Color as a major design ingredient is treated in detail in Chapter 14.

Vision Acuities

The map designer works in a visual medium, so all elements must be visible to the map reader. If the elements cannot be seen, the map's communicative attempt will be lost, no matter how well designed the map may be in other respects. There are two important measures of the human ability to see visual elements: **visual acuity** and **resolution acuity.**

Visual Acuity

So far it has been assumed that the map reader can see the map and all its design elements, but visibility must not be taken for granted. Fortunately, it is seldom a problem unless the designer overlooks it when preparing art for reduction.

Visual acuity is a measure of a size threshold (Potts 1972). One's visual acuity determines one's ability to discriminate between small objects or symbols. As with any process involving vision, individuals may have different acuity thresholds. Whether or not the map reader can identify the various line weights or circle sizes used in the generation of the map is determined by his or her individual acuity.

Resolution Acuity

Resolution acuity is somewhat different from visual acuity. Resolution is a measure of the detectable separation between objects in a visual field. When two objects are seen apart, the reader is said to resolve them. Again, the threshold is the point at which this occurs accurately. The average threshold separation of two black dots on a white background is approximately .003 inch (.076 mm) (Potts 1972). This measure can become critical in map design when patterns are specified in design, especially if art is to be reduced. If the elements in a pattern are closer than this, the observer does not see the pattern but begins to perceive only a continuous tone. If the difference in patterns is important for differentiation in such a case, the design has failed in its task of facilitating communication. Map elements are not functional at all if they cannot be seen. The very best map designs will falter if minimum thresholds are not maintained.

THE VISUAL HIERARCHY OF MAP ORGANIZATION: FIGURE-GROUND RELATIONSHIP

There is probably no perceptual tendency more important to cartographic design than **figure and ground organization.** A person's underlying perceptual tendency is to organize the visual field into categories: figures (important objects) and grounds (things less important). This concept was first introduced by Gestalt psychologists early in this century. Figures become objects of attention in perception, standing out from the background. Figures have "thing" qualities; grounds are usually formless. Figures are remembered better; grounds are often lost in perception.

In the three-dimensional world, we see buildings in front of sky and cars in front of pavements. Likewise, we see some objects in front of others in the two-dimensional world, given wise graphic treatment. In cases such as the words printed on this page, no extraordinary measures are needed to make figures stand out from ground.

In the planar level of design planning, the designer should structure the field in a way that directs the reader's perception along paths commensurate with the communication goals of the map. For example, objects that are important intellectually should be rendered so as to make them appear as figures in perception (Dent 1972).

Deborah Sharpe, a color designer not from the cartographic profession, has stated very succinctly the importance of incorporating figure and ground perception into design:

> *In my own design work, I have found that designating the features that are to represent figure and those that are to represent ground as a first step on the job eliminates the trial and error inherent in the beginning stages of most creative tasks* (Sharpe 1974, 102).

Visual Hierarchy Defined

Public speakers arrange their material to emphasize certain remarks and subordinate others. Professional photographers often focus the camera to provide precise detail in certain parts of the picture and leave the remainder somewhat blurred. Advertising artists organize ads to accentuate some spaces and play down others. Choreographers arrange the dancers so that some will stand out from the rest on the stage. Professional cartographers must go through similar activities in designing the map.

The **visual hierarchy** (or *organizational hierarchy*) is the intellectual plan for the map and the eventual graphic solution that satisfies the plan (Dent 1972). Each design activity should contain such a hierarchy. In this phase of design, the cartographer sorts through the components of the map to determine the relative intellectual importance of each, then seeks a visual solution that will cast each component in a manner compatible with its position along the intellectual spectrum. Objects that are important intellectually are rendered so that they are visually dominant within the map frame (see Figure 12.14).

Customary Positions of Map Elements in the Hierarchy

Each map has a stated purpose that controls the planning of the visual hierarchy. Mapped objects and their relative importance assume a place in the hierarchy. Although identical objects may vary in relevance, depending on the map on which they are placed, there are general guidelines to follow in developing the hierarchy (see Table 12.2).

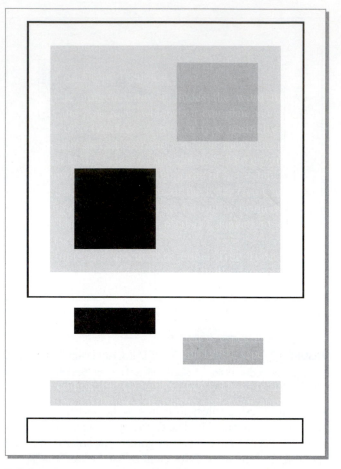

FIGURE 12.14 THE VISUAL HIERARCHY.

Objects on the map that are most important intellectually are rendered with the greatest contrast to their surroundings. Less important elements are placed lower in the hierarchy by reducing their edge contrasts. The side view in this drawing further illustrates this hierarchical concept.

TABLE 12.2 TYPICAL ORGANIZATION OF MAPPED ELEMENTS IN THE VISUAL HIERARCHY

Usual Intellectual Level*	Object	Visual Level
1	Thematic symbols	I
1	Title, legend material, symbols, and labeling	I
2	Base map—land areas, including political boundaries, significant physical features	II
2	Base map—water features, such as oceans, lakes, bays, rivers	II
3	Other map elements—scale graticule, inset map, north arrow	III
4	Important explanatory materials—map sources and credits	IV
5	Framing the map—neatline	V

*A map object with a rank of 1 has a greater intellectual importance to the map's message than one with a rank of 5. Visual levels I through V roughly correspond with the intellectual levels 1 through 5.

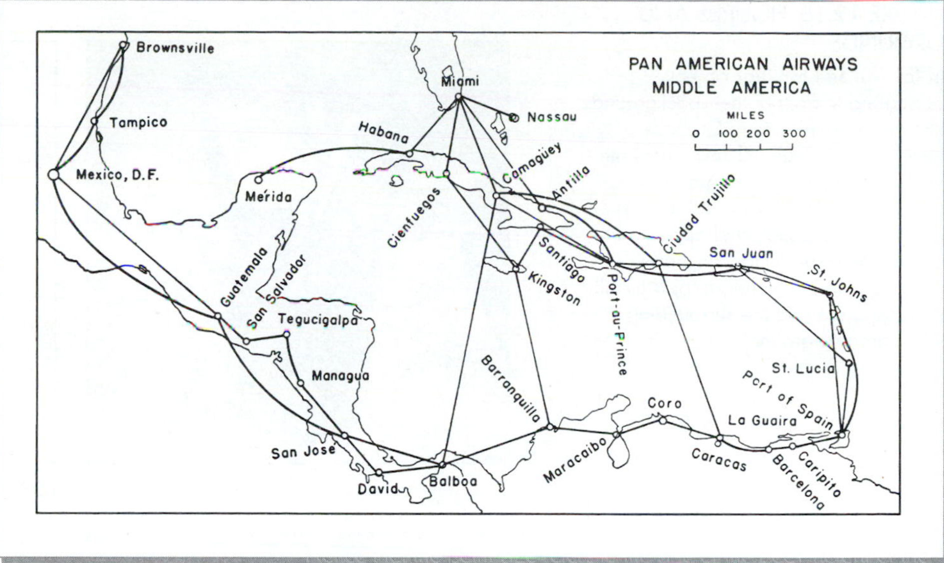

FIGURE 12.15 POOR ORGANIZATIONAL MAP PLAN.

In this case, it is difficult to see which areas on the map are water and which are land. Further difficulty is introduced by failure to place the thematic symbolization on the most dominant visual level. The map as a whole suffers from poor figure-ground contrast. *Source:* Carlson 1952. Used by permission.

It is highly unlikely that the symbols on a thematic map will assume any rank other than the topmost. Those map objects customarily toward the bottom of the hierarchy may fluctuate more. For example, water is ordinarily placed beneath the land in the order, but might assume a more dominant role if the purpose of the map has to do with marine or submarine features. The design activity calls for a careful examination of each element and its proper placement in the hierarchy.

An interesting activity for students is to analyze thematic maps with a view to their organizational schemes. One noticeable result will be that those maps without a visual plan are the least successful in conveying meaning and are visually confusing (see Figure 12.15). Unfortunately, too many such maps exist. On the other hand, a map with a carefully conceived and executed hierarchy pleases the reader, is visually stable, and does not require redesigning.

The oft-stated axiom that the best designs are not even noticed operates in cartography as well as in other disciplines.

ACHIEVING THE VISUAL HIERARCHY

Ordering the intellectual importance of map elements is a relatively simple task, especially when guided by a clearly stated map purpose. Making the hierarchy work visually is another matter, involving knowledge of the perceptual tendencies of map readers. The designer must learn these tendencies if effective results are to be achieved.

Fundamental Perceptual Organization of the Two-Dimensional Visual Field: Figure and Ground

The **figure-ground phenomenon** is often considered to be one of the most primitive forms of perceptual organization; it has even been observed in infants (Dember and Warm 1979). We tend to see objects having form because they are created using solid lines and brighter colors. Their background is subdued through the use of grayscale values or pastel colors. The use of contrast in the map design creates a visual hierarchy. Objects that stand out against their backgrounds are referred to as *figures* in perception, and the map base as *grounds.* The segregation of the visual field into figures and grounds is a kind of automatic perceptual mechanism. Figures will not emerge from homogeneous visual fields, however.

In the everyday three-dimensional environment, we see a table on top of the floor (with the floor continuing beneath the legs of the table), buildings in front of the sky, pictures in front of the walls on which they are hung, and so on. When we lay a pencil on a piece of paper, we see the paper continuing unbroken behind the pencil and the pencil as a complete object on the top of the paper (see Figure 12.16a).

Figure formation is possible in two-dimensional spatial organization as well. Figures perceived in this way are seen separately from the remainder of the visual field, have form and shape, appear to be closer to the viewer than the amorphous ground, have more impressive color, and are associated with meaning (Haber and Hershenson 1973). The ground usually appears to continue unbroken behind the figure, just as in the three-dimensional case. A simple example of figure from the two-dimensional world is a black disc placed within a frame (see Figure 12.16b).

The cartographic designer should use this perceptual tendency of figure-ground segregation as a positive design element in structuring the visual hierarchy, so that the most important map elements appear as figures in perception. With careful attention to graphic detail, all the elements can be organized in the map space so that the emerging figure and ground segregation produces a totally harmonious

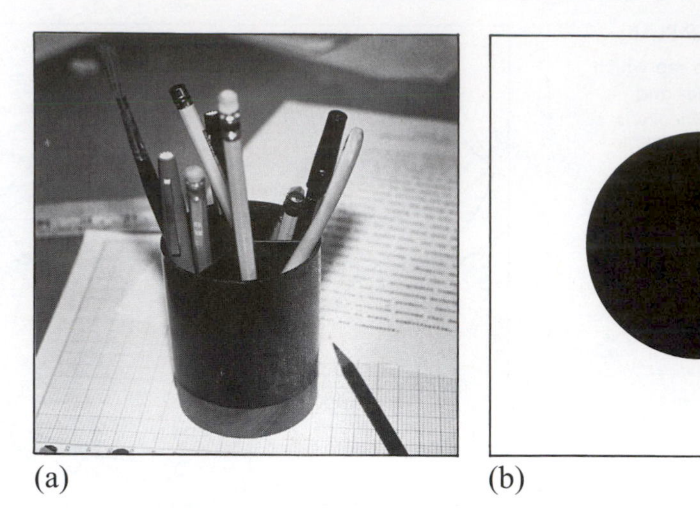

FIGURE 12.16 FIGURES AND GROUNDS.

In (a), we see familiar objects appearing in front of their backgrounds (the paper in this case). Grounds appear to continue unbroken beneath the figure objects. In the graphic two-dimensional world, as in (b), figures dominate perception and appear to rest on top of seemingly unchanging grounds. The black disc in (b) is usually seen as figure and the surrounding white area as ground.

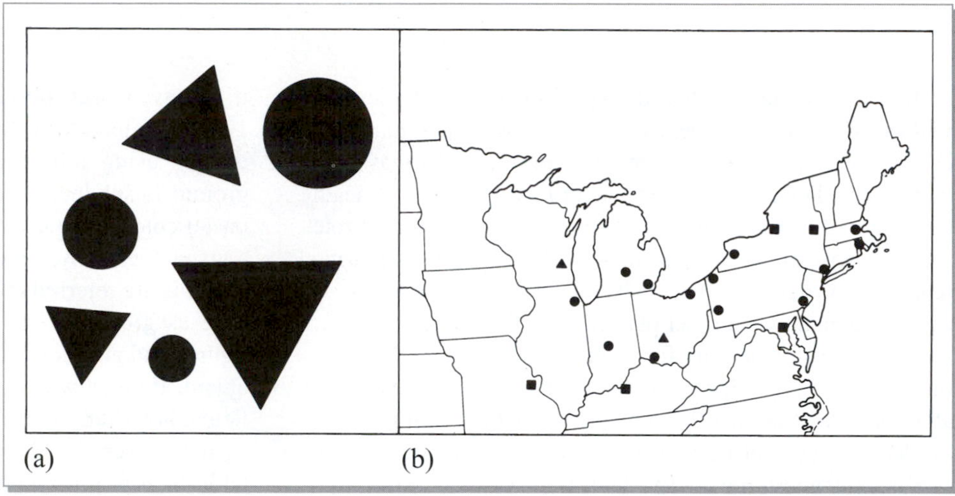

FIGURE 12.17 GROUPING BY SIMILARITY.

In perception, we tend to group similar objects. In (a), the triangles belong visually to a group distinct from the group formed by the circles. This perceptual tendency is fundamental in some cartographic situations, such as (b). We often use this perceptual grouping phenomenon to communicate thematic messages on maps.

design. Visual confusion is eliminated, and the intent of the message becomes clear. Fortunately, psychological researchers have examined many of the mechanisms that lead to figure formation, so designers have guidelines for organizing the map's graphic elements.

Perceptual Grouping Principles

Several perceptual grouping principles have been found to be primary mechanisms for figure formation. In **perceptual grouping,** the viewer spontaneously combines elements in the visual field that share similar properties, resulting in new forms or "wholes" in the visual experience. From the designer's point of view, these groupings can act in two opposing ways: as positive mechanisms to use if the map's elements will permit it, or as mechanisms to avoid if the map's elements are arranged in such a fashion that the spontaneous grouping will detract from the planned hierarchy. Their potential as design elements is explained and exemplified below.

Grouping by Similar Shape. Objects in the visual field possessing similar shapes are usually combined into a new group that appears distinct from the remainder (see Figure 12.17a). This perceptual feature undoubtedly comes into play when map readers view a map containing several different qualitative map symbols (see Figure 12.17b). In fact, if we could not visually combine identical symbols, it would be difficult to "see" the geographical pattern of one symbol type as distinct from others.

Grouping by Similar Size. Viewers tend to group similarly sized objects in the visual field into new perceptual structures (see Figure 12.18a). This tendency is especially important to cartographic design in at least two areas: the reading of different type sizes on maps and the visual assimilation of geographical patterns from maps containing range-graded proportional symbols (explained in Chapter 8). Designers, although they may not be aware of it, depend on this perceptual grouping feature when they work with these cartographic elements (see Figure 12.18b).

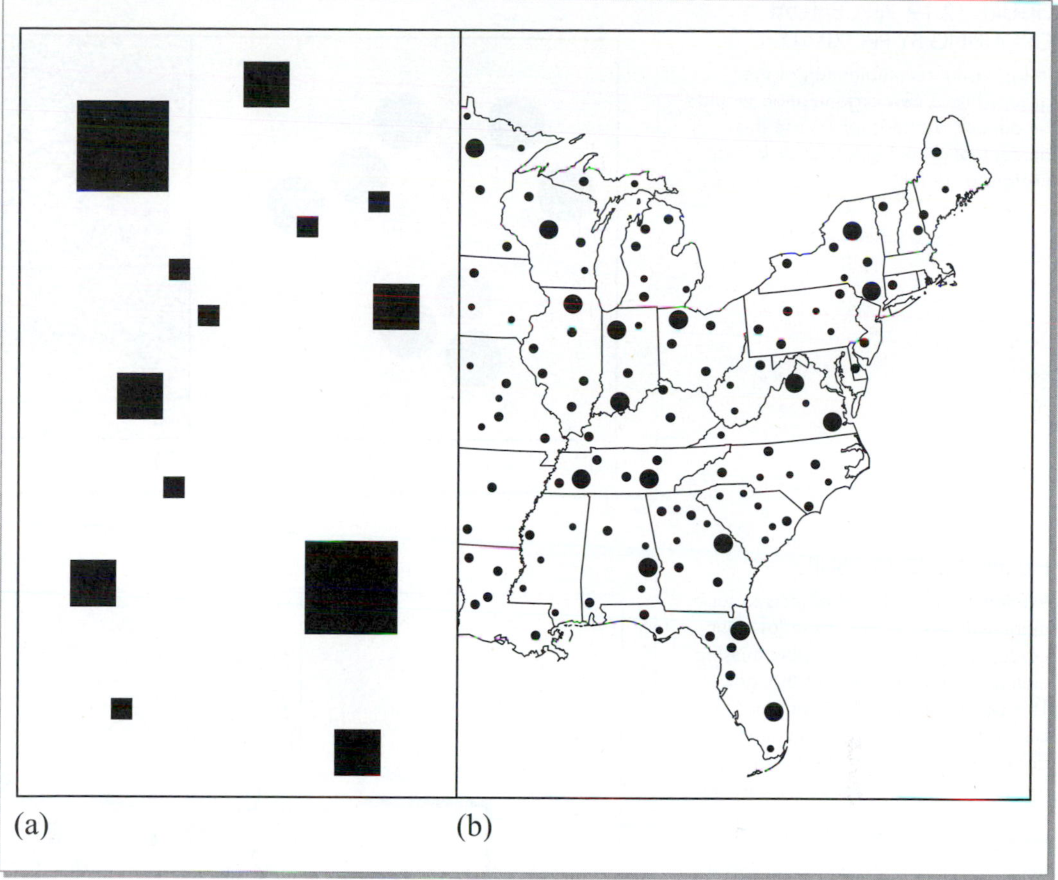

FIGURE 12.18
GROUPING BY SIMILAR SIZE.
In (a), objects in the visual field that are similar in size tend to be grouped together in perception. It is difficult to make the experience take on a different visual structure. Try to place a small square and a large square together into a coherent new group. This perceptual tendency to group similarly sized objects is used frequently by cartographers in the design of maps, as in (b).

Research on the perception of graduated point-symbol maps (discussed in Chapter 8) has uncovered a perceptual characteristic that is relevant to the matter of grouping by similar size. It appears that the perceived size of symbols is affected by their immediate environments. Consequently, the perceptual grouping of similarly sized symbols could possibly be affected by this condition. Patricia Gilmartin (1981) concluded her study with these findings:

1. When a circle (the point symbol tested) is among circles smaller than itself, it appears larger (an average of 13 percent) than when it is surrounded by circles larger than itself. Isolated circles not surrounded by others (larger or smaller) were judged to be of an intermediate size.
2. The effect of larger or smaller circles on a surrounded circle can be reduced if internal borders (boundaries around the area represented by each graduated symbol) are used on the map.
3. There is some evidence that smaller circles are more susceptible to being judged differently because of their environments than large circles.

It is clear from such perceptual studies that the designer's task is not easy, or even altogether clear. Perceptual tendencies can be confusing and difficult to manage in cartographic design. However, if the designer strives to create patterns that are unambiguous in perception, the design is likely to succeed.

Grouping by Proximity. Another strong perceptual grouping tendency is that of proximity. Elements in the visual field that are closer to other elements tend to be seen as a unit that stands out from the remainder (see Figure 12.19a). In fact, the words on this page are formed by the proximity principle—letters are grouped visually in words. The cartographer relies on this visual tendency when depicting geographical distributions, especially those containing clusters (see Figure 12.19b). If the eye did not combine elements in this way, it would be exceedingly difficult for designers to accomplish their task.

Figure Formation and Closure

Closure is an important perceptual principle. It refers to the tendency for the individual to complete unfinished objects and to see as figures objects that are already completed. A contour or edge is usually associated with closure (see Figure 12.20a). When viewing a contour line which is interrupted by the contours elevation value, the contours are spontaneously completed so that we see the "whole" object. Therefore, map designers need to provide strong, completed contours around figure objects. Sometimes these edges are "broken" by the diagram to provide for lettering or other map elements (see Figure 12.20b). Unless skillfully executed, this design solution can have negative effects on figure areas.

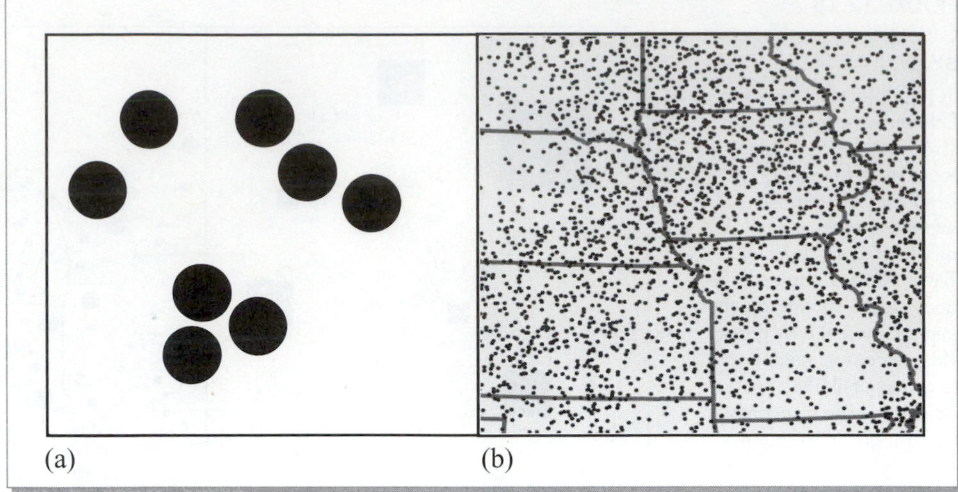

FIGURE 12.19 PERCEPTUAL GROUPING BY PROXIMITY.
In (a), visual recombination of the groups into a new organization would be difficult. Cartographers use this perceptual grouping tendency in map design, as in (b).

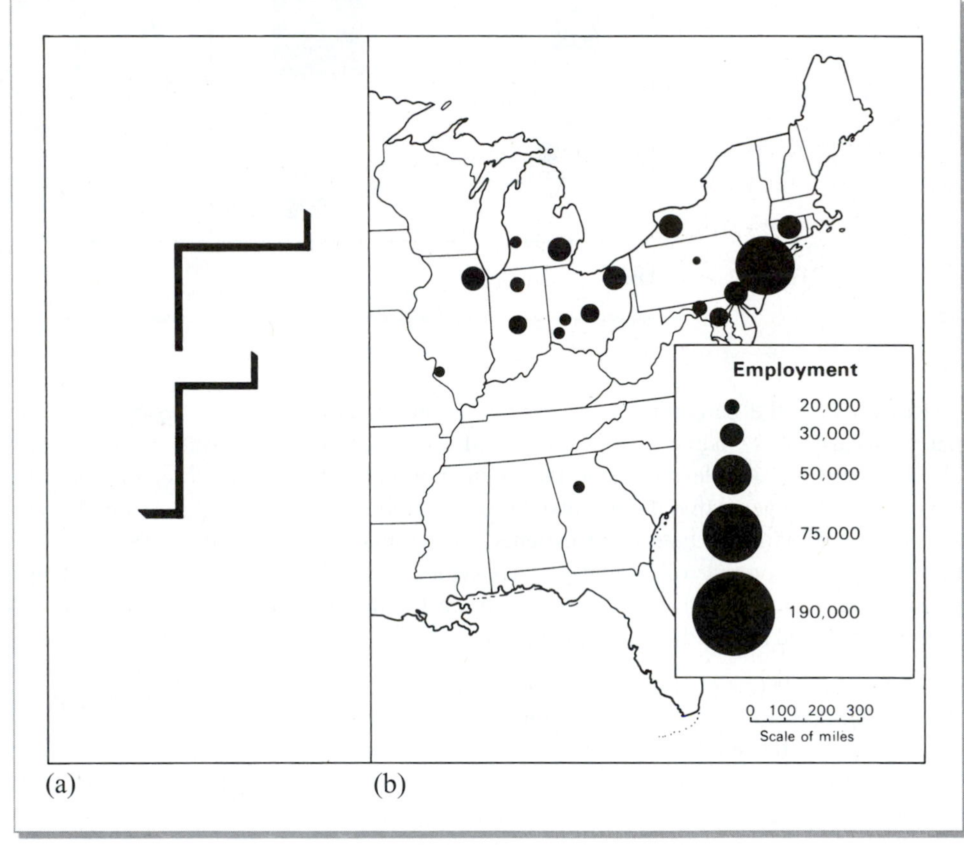

FIGURE 12.20 CLOSURE.
We tend to close figure objects to form more simple structures, as in (a). Our perceptual mechanisms supply missing elements without conscious thought. This occurs in many cartographic situations, and the communication does not suffer. In (b), the person reading the map provides the missing coastlines and is not really bothered by their physical absence. Normally the placement of the legend so that it hides the body of the map is not acceptable.

Strong Edges and Figure Development

One of the principal ways of producing a strong figure in two-dimensional visual experience is to provide crisp edges to figure objects. Conversely, figure dominance can be weakened by reducing edge definition. Edges result from contrasts of brightness, reflection, or texture. These characteristics have special significance and utility in cartographic design, particularly in coordinating the graphic elements in the planned visual hierarchy. For example, thematic symbols and other elements high in the hierarchy can be rendered in solid hue (black or color), and features of less importance can be displayed using a lighter color value or grayscale level, thus reducing the sharpness of their edges. Lines of lighter value reduce the intensity of the less important elements (see Figure 12.21).

Important work on the contributions that the figure-ground phenomenon can make on design continues. Wood, for example, remarks clearly on this subject by saying:

> *Cartographic communication can only be successful if sound map design principles are fully understood and used properly.*

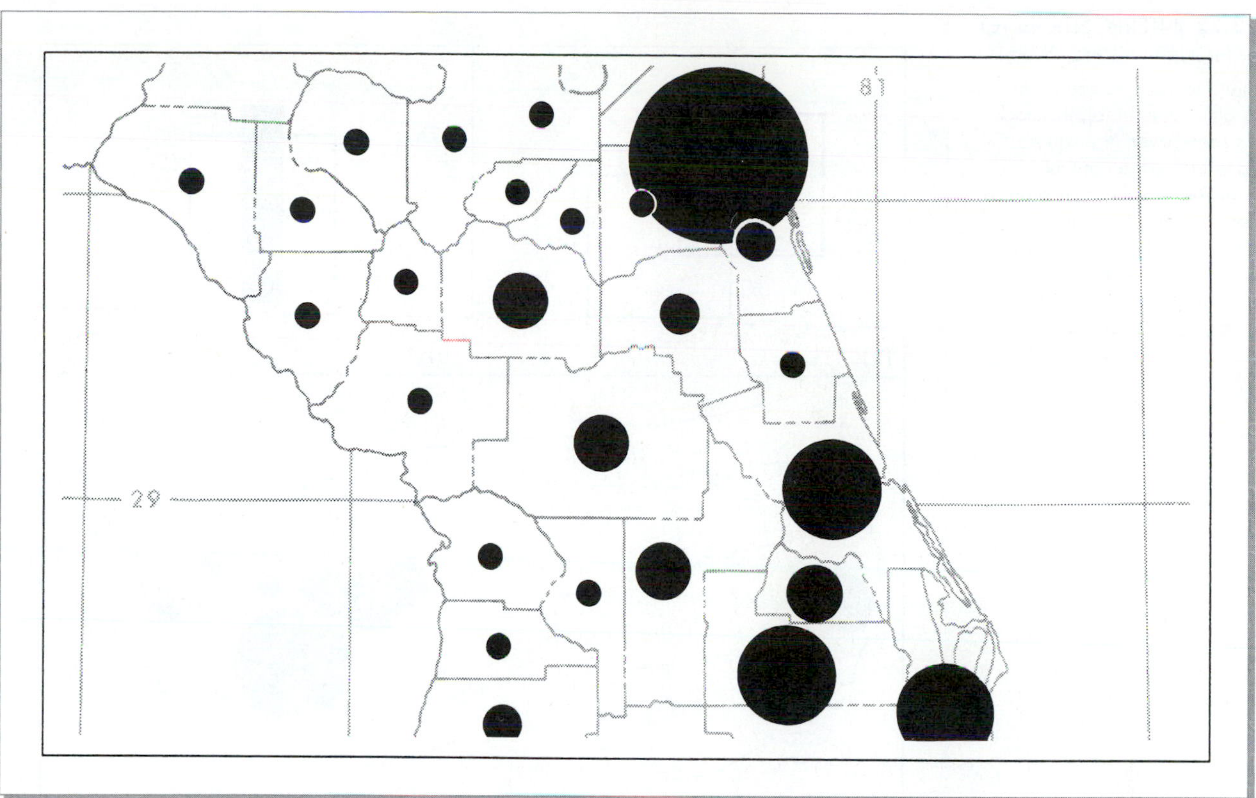

FIGURE 12.21 THE USE OF GRAYSCALE TO DEVELOP VISUAL HIERARCHY.
Less important elements (coastlines, political boundaries) have been presented in a shade of gray to reduce the shape of their edges. The graticule has been created using a lighter value, further reducing its contrast and thereby placing it on an even lower visual level. Important elements (symbols) are rendered in total black, which tends to emphasize their importance in the overall organization.

> *It is obvious that many cartographers still do not understand the fundamental principle of map design, the figure-ground relationship* (Wood 1992 128).

The figure-ground relationship formed the basis of a research effort that has important contributions for designers. Fundamentally, Wood researched the idea that designs incorporating figure-ground divisions would have an effect on the way readers read maps. The results showed that without a figure-ground inspired design, map readers have visual processing difficulties. Wood further remarks, "It is clear that figure-ground differentiation is a vital element in successful graphic communication (262)."

We believe that the figure-ground phenomenon, as a basic visual and perceptual tendency among map readers, must be accommodated for in *every* thematic map design project. To not do so would be a serious oversight. Figure-ground relationships are perhaps *the* most important design aspect of the map.

The Interposition Phenomenon

A most useful way of causing one object in the two-dimensional visual field to appear on top of or above another is to interrupt the edge or contour of one of the objects. This is usually referred to as the **interposition phenomenon** and is frequently cited as a depth cue in perception (Hochberg 1964). In cartographic design, this technique can be used to strengthen the dominance of certain objects in the visual hierarchy (see Figure 12.22). The result is an impression of depth on the map—an interesting, dynamic, and fluid solution.

It is possible to use interposition cues to produce stacking effects of clustered proportional symbols (these symbols were discussed in Chapter 8). Although this method results in a map that appears three-dimensional (thus adding interest to the design), it makes it difficult for the map reader to see the quantitative differences of the scaled symbols (Groop and Cole 1978). On the average, errors in symbol reading increase with the amount of the symbol obscured by an overlapping one. Because of this, the use of interposition is not recommended in cases where quantitative point symbols overlap by one-third for simple figures such as circles, probably less for more complex symbol forms. Otherwise, interposition can be used whenever it enhances the overall hierarchical plan for the map.

Figures and Grounds in the Map Frame

Within a bounded space in the two-dimensional visual field, areas that are smaller and completely enclosed will tend to be viewed as well-defined figures. Cartographic designers have

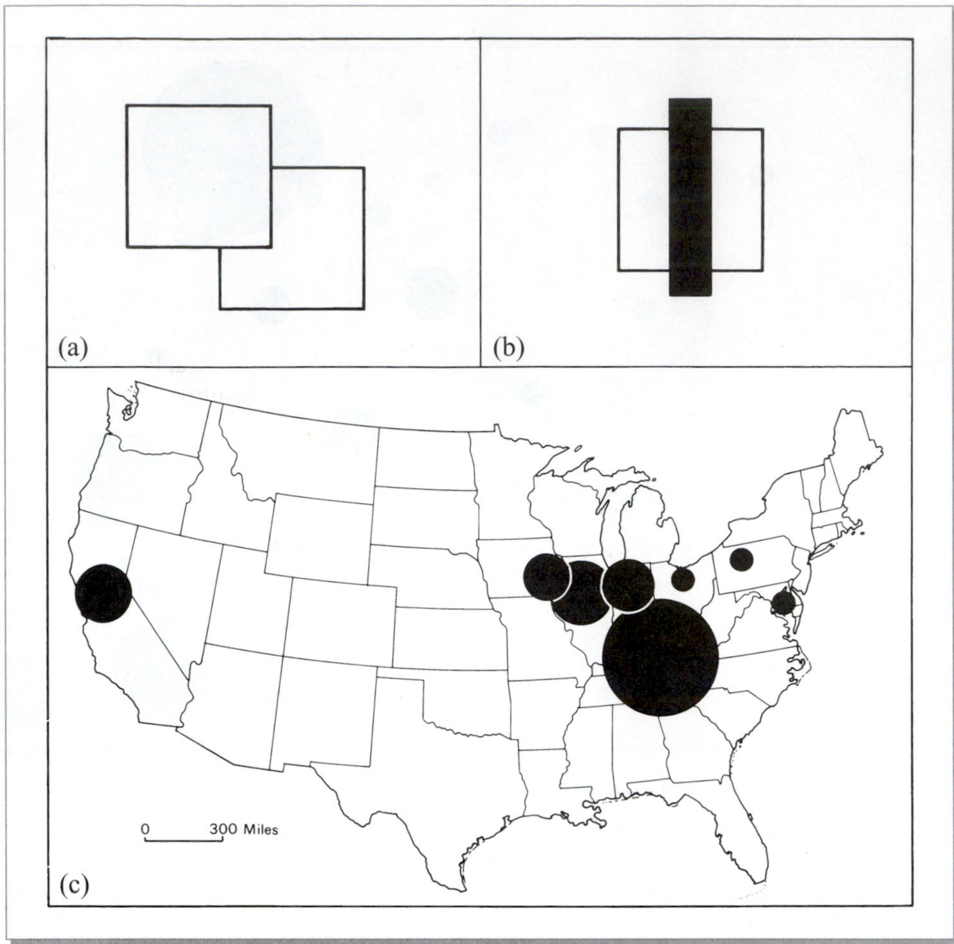

FIGURE 12.22 INTERPOSITION TO ASSIST IN FIGURE DEVELOPMENT.
In (a), we tend to see one square in front of the other one. In (b), the back rectangle is seen "over" the square. The objects whose edge contours continue unbroken are the ones seen as being on top. Interposition can be used on maps, as in (c). The back circles break coastlines or state boundaries, and appear "on top" of the land. This enhances their figure properties.

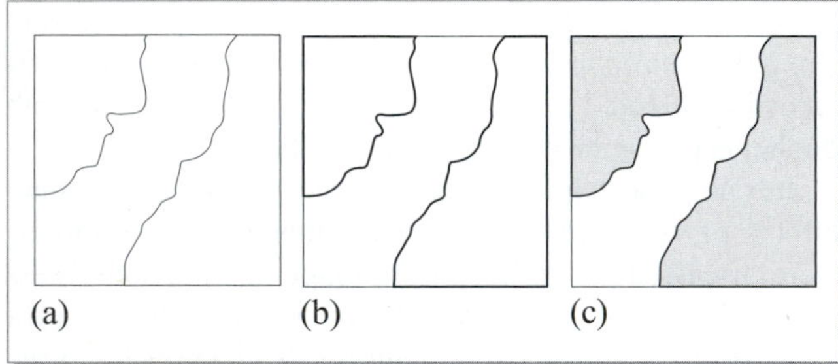

FIGURE 12.23 FIGURE DEVELOPMENT OF LAND AREAS.
In (a), the configuration of land and water areas and the graphic treatment of the map border hinder the land from becoming figure. In (b), the borders joining the land areas are treated as part of the land. This forms enclosing contours and strengthens the perception of the land areas as wholes. The addition of a solid fill (c) further enhances the figure development. Cartographic designers will need to apply different solutions to varying designs.

considerable choice in such elements as texture, **articulation,** edging, and interposition to accentuate objects as figures, but less freedom in the manipulation of figure size and map space. Constraints imposed by location, scale, and map size can preclude any adjustment to enhance figure areas of the map. In cases where these restrictions do not limit design choice, there are guidelines to assist in selecting the proper size ratios of figures and grounds (Crawford 1976). In one study, acceptable size ratios ranged from 1:2.18 to 1:3.56—non-figure areas may be from 2.18 to 3.56 times larger than figure areas without interfering with figure formation.

One approach to assure that the figure portion of the map is completely enclosed, in the case of land-water differentiation, is to provide an enclosing contour by continuing the shoreline as the map border (Head 1972; see Figure 12.23b). A combination of increasing the line thickness and adding a fill to the intended figure is recommended, as in Figure 12.23c.

THE SPECIAL CASE OF THE LAND-WATER CONTRAST

The principal concern of the thematic map designer is to communicate a spatial message effectively. To a great extent, the success of this effort depends on how the message is presented or arranged for the map reader. If the graphic material is presented in a clear and unambiguous manner, success in communication will probably result. On the other hand, an unclear and confusing graphic picture will make the reader frustrated and unreceptive to the message. The graphic elements of the map must therefore be arranged in such a way as to reduce any possible reader conflict. One way of eliminating possible confusion is to provide clues on the map to help the reader in determining geographical location.

A significant geographical clue is the differentiation between land and water, if the mapped area contains both. This distinction has been suggested as the first important process in thematic map reading (Head 1972). Maps that present confusing land-water forms deter the efficient and unambiguous communication of ideas (see Figure 12.24). Design solutions should never be visually distracting or lacking in clarity and stability in the intended order. Land-water differentiation usually aims to cause land areas to be perceived as figures and water areas as ground. In unusual cases, water areas are the focal point of the map and would therefore be given graphic treatment to cause them to appear as figures.

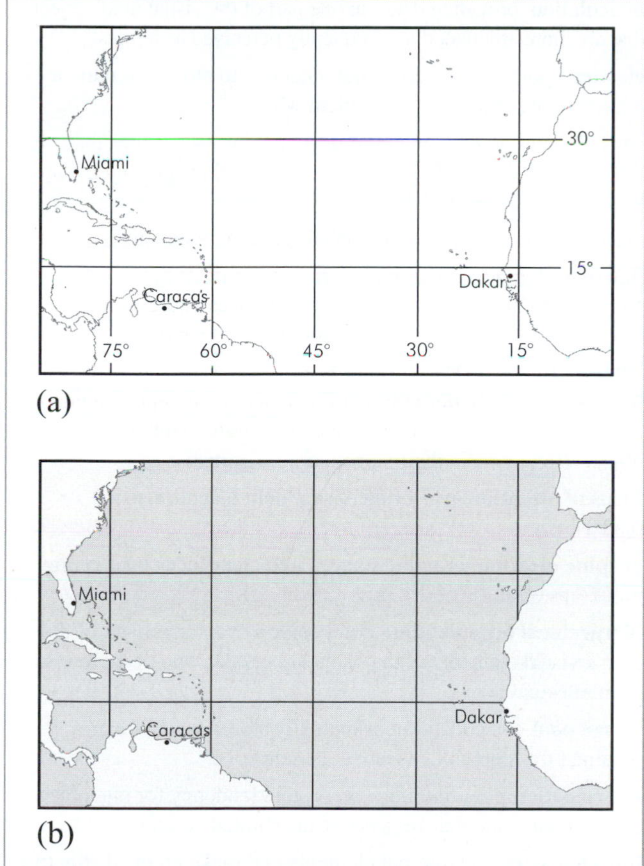

FIGURE 12.24 LAND AND WATER CONTRASTS.
In (a), the contrast between land and water is not clearly evident. The graticule creates a similar texture over the entire map, resisting figure formation. A very simple correction can overcome this. In (b), the graticule has been eliminated over land areas and a gray fill has been added to the ocean area, enhancing the figure and increasing the land-water contrast. In many cases, only simple corrections are necessary to solve problems of insufficient contrast between land and water.

On many thematic maps, water is often not depicted. Most thematic maps exclude water, and unless it supports the thematic variable, water (and other nonrelated general reference features) should be excluded.

Vignetting for Land-Water Differentiation

Historically, engravers, drafters, and cartographers solved the land-water differentiation problem by use of **vignetting** at the coastline. Vignetting (a term that was originally used for the manual photographic screening of a printing plate) is any graphic treatment emerging from an edge or border and resulting in a continuous gradient of brightness. Vignetting of the coast is common, although it has been used on maps at political or other borders for visual differentiation between land areas.

One way of creating coastal vignetting is to select a continuously changing color ramp in the color options of the GIS and mapping software. Otherwise, assign a light grayscale or light blue to the water area of the map. This provides for easy differentiation of land and water.

DESIGNING THE PAGE-SIZE MAP

An interesting exercise for students is to examine by example how the page-size, or smaller, map is designed, and especially how the various map elements are treated in this process. Similar procedures apply to larger maps, screen size, or posters for presentation. We elect to use the page-size map because of its popularity. This size is a matter of convenience and is often used by beginning cartographers or students. The following paragraphs discuss how the design of the map in Color Plate 6.4 was reached.

The primary purpose of this map is to illustrate the spatial distribution of relationship of Per Capita Income to that of Median Age. This bivariate map uses two sets of data that are classified using three classes each. A 3x3 legend matrix permits the reader to identify the nine possible combinations of the classed data. The counties of the contiguous United States are used as the enumeration unit. The placement of the legend in the lower left and the source statement along the lower right assist in achieving good map balance. The title is concise and yet contains all the information necessary for the reader to understand the topic presented. Generally, a map title should include the components of *where, what, and when,* to borrow from the traditions of journalism. This map represents effective use of the map elements producing efficient communication of the data.

Good map design is a result of an understanding of how the map elements interrelate and the balance achieved by their placement within the map space. This understanding is achieved through either years of experience as a cartographer or through a good introductory level cartography class.

REFERENCES

Antes, J., K. Chang, and C. Mullis, 1985. The Visual Effect of Map Design: An Eye-Movement Analysis, *American Cartographer* 12: 143–55.

Arnheim, R. 1965. *Art and Visual Perception*. Berkeley: University of California Press.

———. 1976. The Perception of Maps, *American Cartographer* 3: 5–10.

Brewer, C. 2005. *Designing Better Maps: A Guide for GIS Users*. Redlands, CA: ESRI Press.

Burdick Group, The. 1982. *Creativity: The Human Resource* (exhibit). San Francisco: Standard Oil Company of California.

Carlson, F. 1952. *Geography of Latin America*. New York: Prentice-Hall.

Crawford, P. 1976. Optimum Spatial Design for Thematic Maps, *Cartographic Journal* 13: 134–44.

Dember, W., and J. Warm. 1979. *Psychology of Perception*. 2d ed. New York: Holt, Rinehart and Winston.

Dent, B. 1972. Visual Organization and Thematic Map Communication, *Annals* of the Association of American Geographers 62: 79–93.

DiBiase, D. 1990. Visualization in the Earth Sciences, *Earth and Mineral Sciences* 59: 13–18.

Fisher, P., J. Dykes, and J. Wood. 1993. Map Design and Visualization, *Cartographic Journal* 3: 136–42.

Gilmartin, P. 1981. Influences of Map Context in Circle Perception *Annals* of the Association of American Geographers 71: 253–58.

Groop, R., and D. Cole. 1978. Overlapping Graduated Circles: Magnitude Estimation and Method of Portrayal, *Canadian Cartographer* 15: 114–22.

Haber, R., and M. Hershenson. 1973. *The Psychology of Visual Perception*. New York: Holt, Rinehart and Winston.

Hanks, K., L. Bellistan, and D. Edwards. 1978. *Design Yourself*. Los Altos, CA: William Kaufmann.

Head, C. 1972. Land-Water Differentiation in Black and White Cartography, *Canadian Cartographer* 9: 25–38.

Hochberg, L. 1964. *Perception*. Englewood Cliffs, NJ: Prentice Hall.

Karssen, A. 1980. The Artistic Elements in Design, *Cartographic Journal* 17: 124–27.

Keates, L. 1982. *Understanding Maps*. New York: Wiley.

Muehrcke, P. 1981. Maps in Geography, in *Maps in Modern Geography: Geographical Perspectives on the New Geography*, edited by Leonard Guelke. *Cartographica*, Monograph No. 27.

———. 1982. An Integrated Approach to Map Design and Production, *American Cartographer* 9: 109–22.

Potts, A. ed. 1972. *The Assessment of Visual Function*. St. Louis: Mosby.

Robinson, A. 1966. *The Look of Maps*. Madison: University of Wisconsin Press.

Samuels, M., and N. Samuels. 1975. *Seeing with the Mind's Eye*. New York: Random House.

Sharpe, D. 1974. *The Psychology of Color and Design*. Chicago: Nelson-Hall.

Southworth, M., and S. Southworth. 1982. *Maps: A Visual Survey and Design Guide*. A New York Graphic Society Book. Boston: Little, Brown.

Surrey, R. 1929. *Layout Techniques in Advertising*. New York: McGraw-Hill.

Wood, C. 1976. Brightness Gradients Operant in the Cartographic Context of Figure-Ground Relationship, *Proceedings of the American Congress on Surveying and Mapping*: 5–34.

———. 1992. *The Influence of Figure and Ground on Visual Scanning Behavior in Cartographic Context*. Unpublished Ph.D. dissertation, Department of Geography, University of Wisconsin—Madison.

Wright, J. 1942. Map Makers Are Human, *Geographical Review* 32: 527–44.

GLOSSARY

articulation providing detail in one part of the visual field; objects that are more articulated are frequently perceived as figures

closure spontaneous perceptual tendency to close contours or edges around objects to make them whole

contrast important element of design; contrasts of line, texture, detail, hue, value, and saturation are means through which maps become interesting and dynamic

creativity ability to see relationships among elements

figure and ground organization fundamental behavioral tendency to organize perception into figures and grounds; figures are dominant elements, and grounds serve as backgrounds for figures

figure-ground phenomenon a fundamental perceptual tendency in which visual fields are spontaneously divided into outstanding objects (figures) and their surroundings (ground)

focus of attention part of the visual field that attracts the reader's eye

graphic ideation bringing images into clear focus by creating mock-ups or drafts of designs

hierarchical organization composition or arrangement of the map's visual elements as they appear between two or more visual or intellectual levels

image pool the collection of mentally stored visual images obtained from previous visual experiences

interposition phenomenon perceptual tendency for one object to appear behind another because of interrupted contour

line character the internal elements that make up the distinctive qualities of a line: for example, dot, dot-dash, dash-dash-dot, and so on.

line weight the thickness of a line

map composition arrangement of the map's visual and intellectual components

map design process characterized by six stages: problem identification, preliminary ideas, design refinement, analysis, decision, and implementation

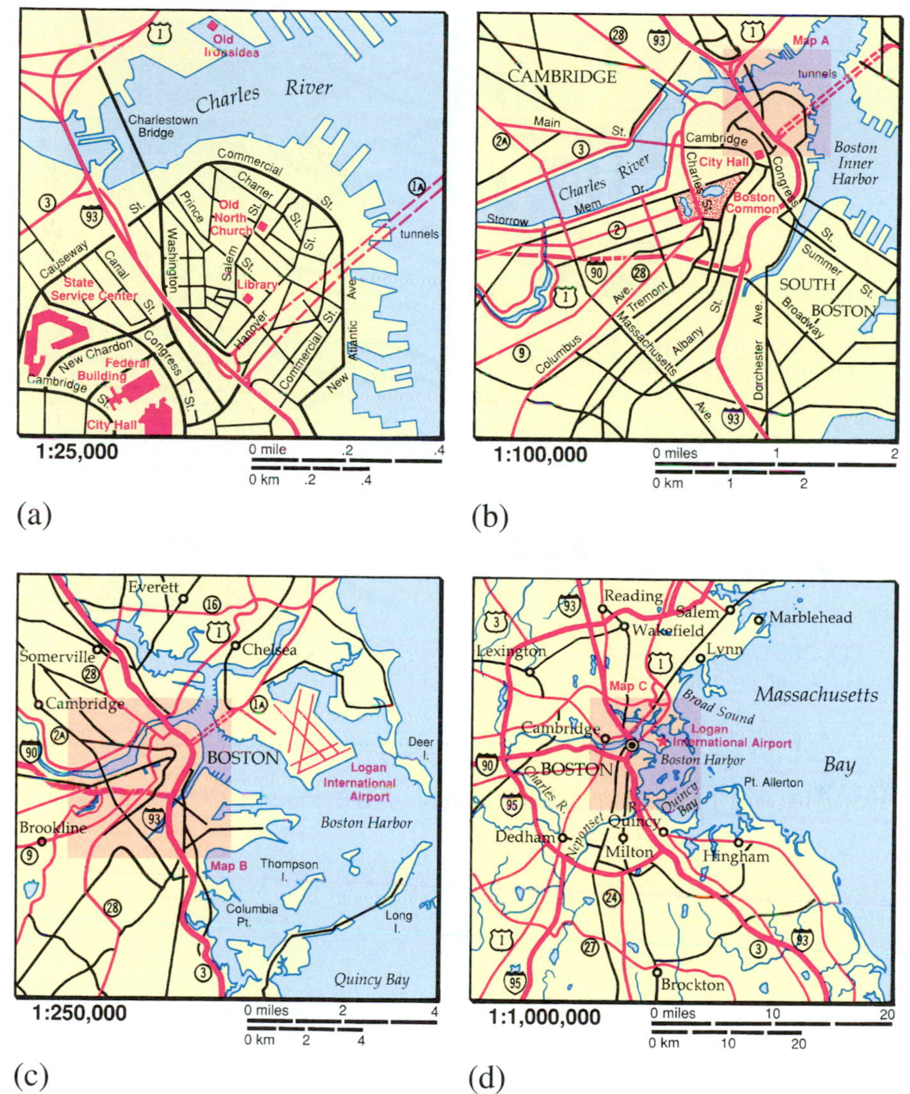

COLOR PLATE 1.1 SCALE AND MAP DETAIL.

Maps at relatively large scales such as 1:25,000 (a) show considerably more detail than those at small scales, for example, 1:1,000,000 (d). Notice that at the largest scale, individual buildings can be shown, but at smaller scales this is impossible. Maps at intermediate scales show more detail than those at smaller scales, but less than those at large scales. Scale sets limits on the types of questions that can be asked or answered from a map set at a definitive scale.

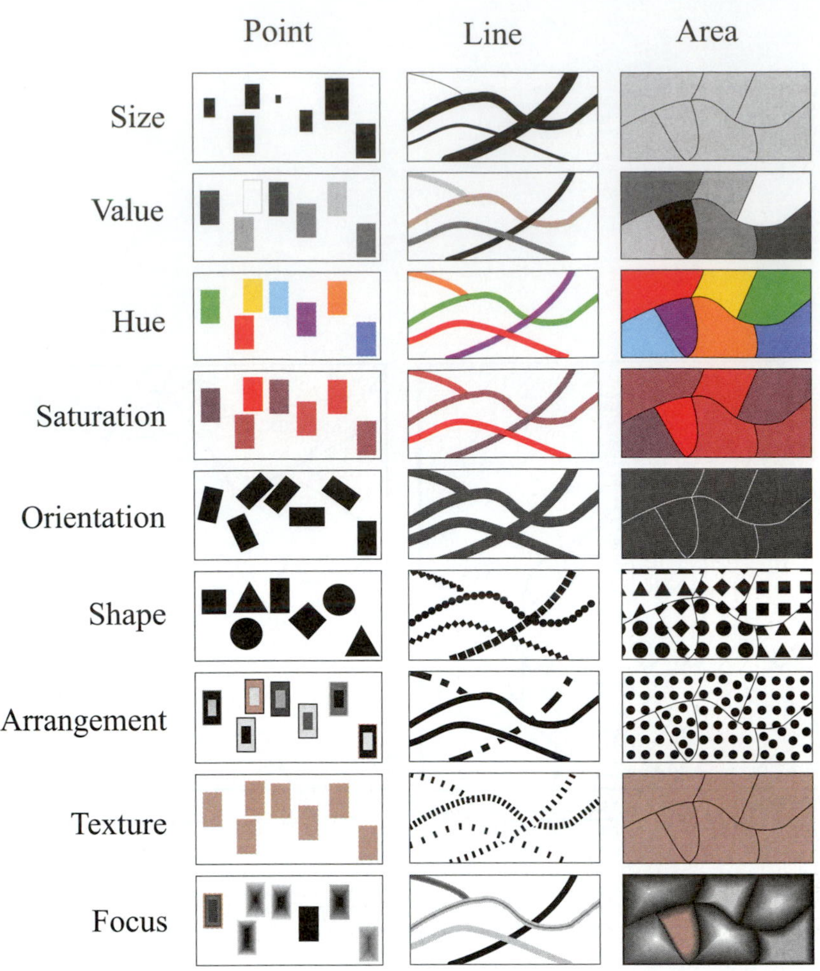

COLOR PLATE 4.1 VISUAL VARIABLES.
Source: MacEachren 1994, used by permission.

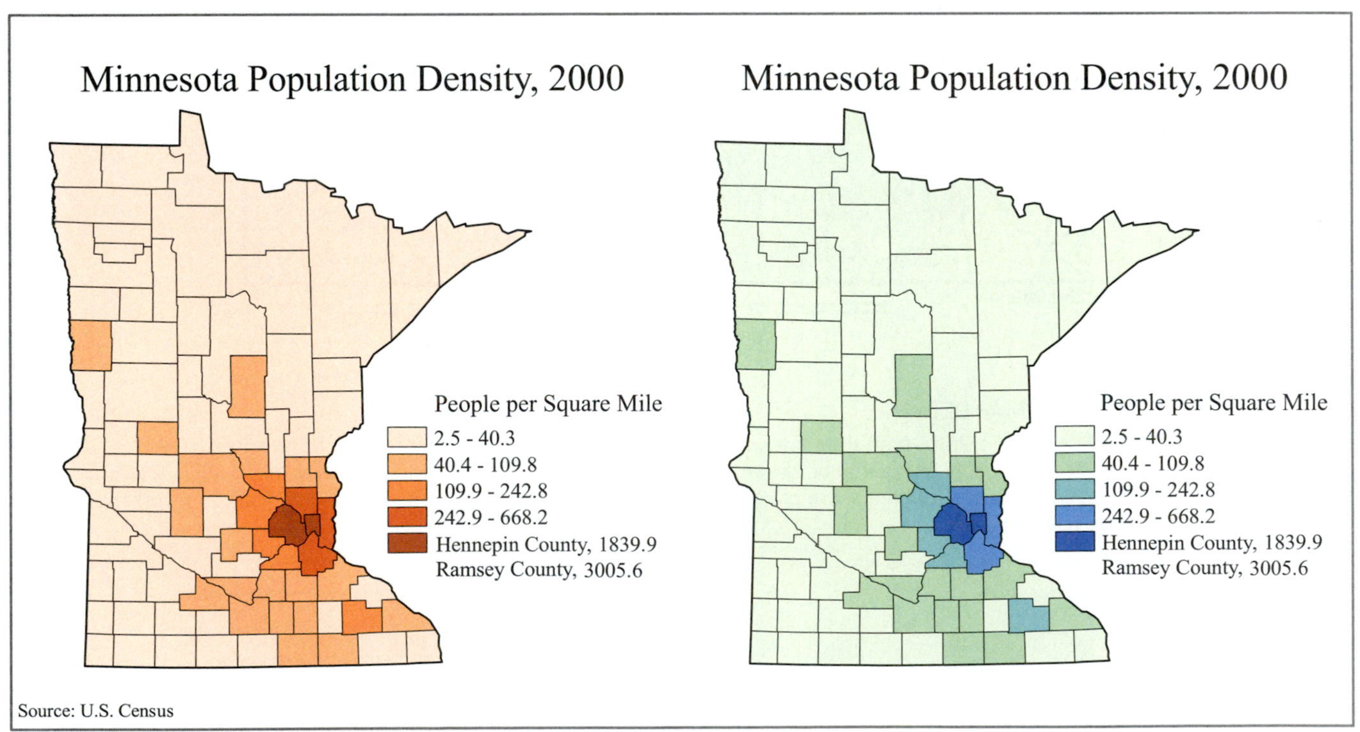

COLOR PLATE 6.1 USE OF HUE IN AREA SYMBOLS FOR CHOROPLETH MAPS.
An orange single hue scheme is illustrated in (a), and a green-blue dual hue scheme is illustrated in (b).

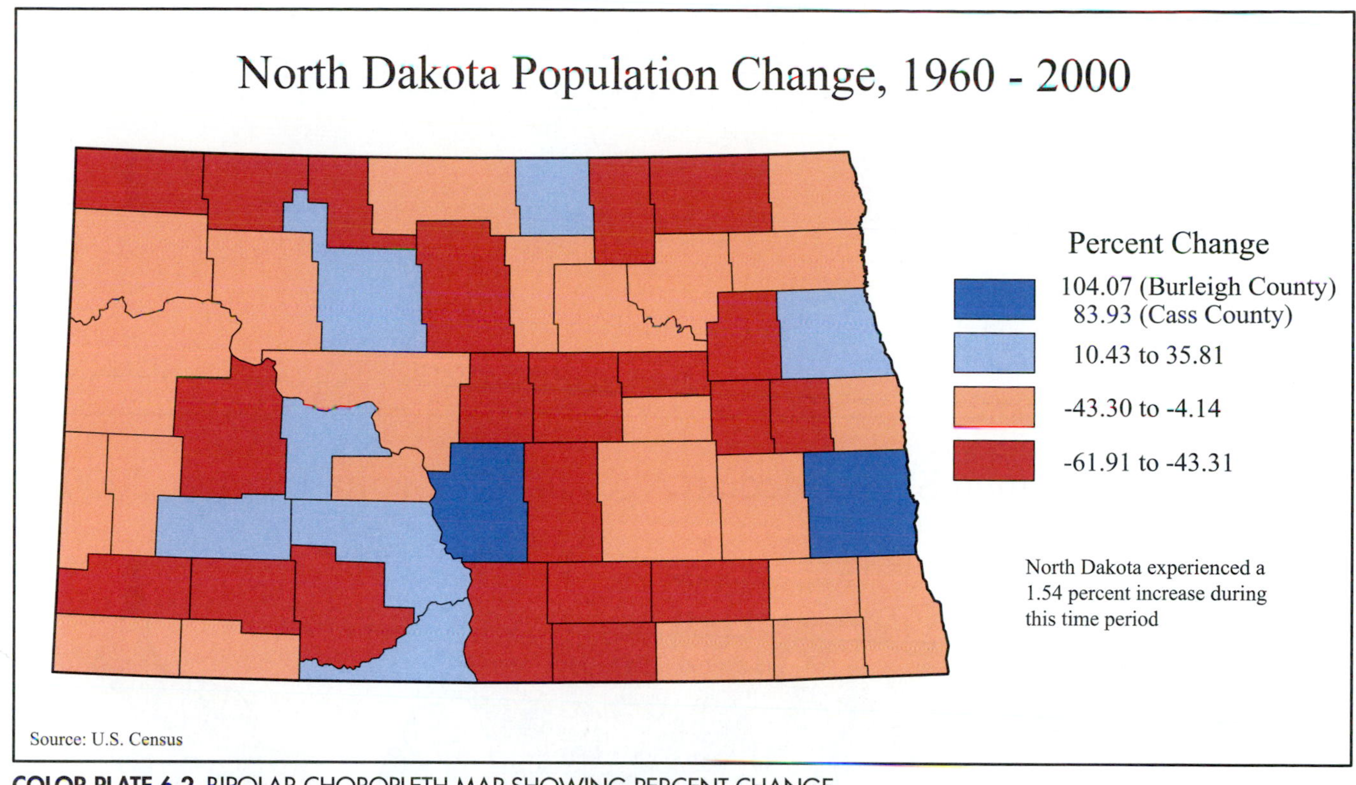

COLOR PLATE 6.2 BIPOLAR CHOROPLETH MAP SHOWING PERCENT CHANGE.
Illustrating increase/decrease as percent change is one of the most common applications of bipolar mapping.

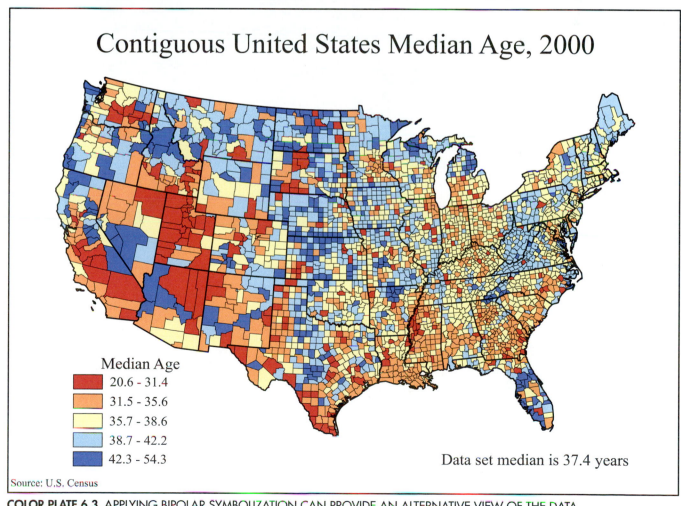

Contiguous United States Median Age, 2000

Median Age
20.6 - 31.4
31.5 - 35.6
35.7 - 38.6
38.7 - 42.2
42.3 - 54.3

Data set median is 37.4 years

Source: U.S. Census

COLOR PLATE 6.3 APPLYING BIPOLAR SYMBOLIZATION CAN PROVIDE AN ALTERNATIVE VIEW OF THE DATA.
This map is a bipolar representaion of Figure 6.6b. The median value is within the middle class when an odd number of classes is selected. The median value will form the class break when an even number of classes is used.

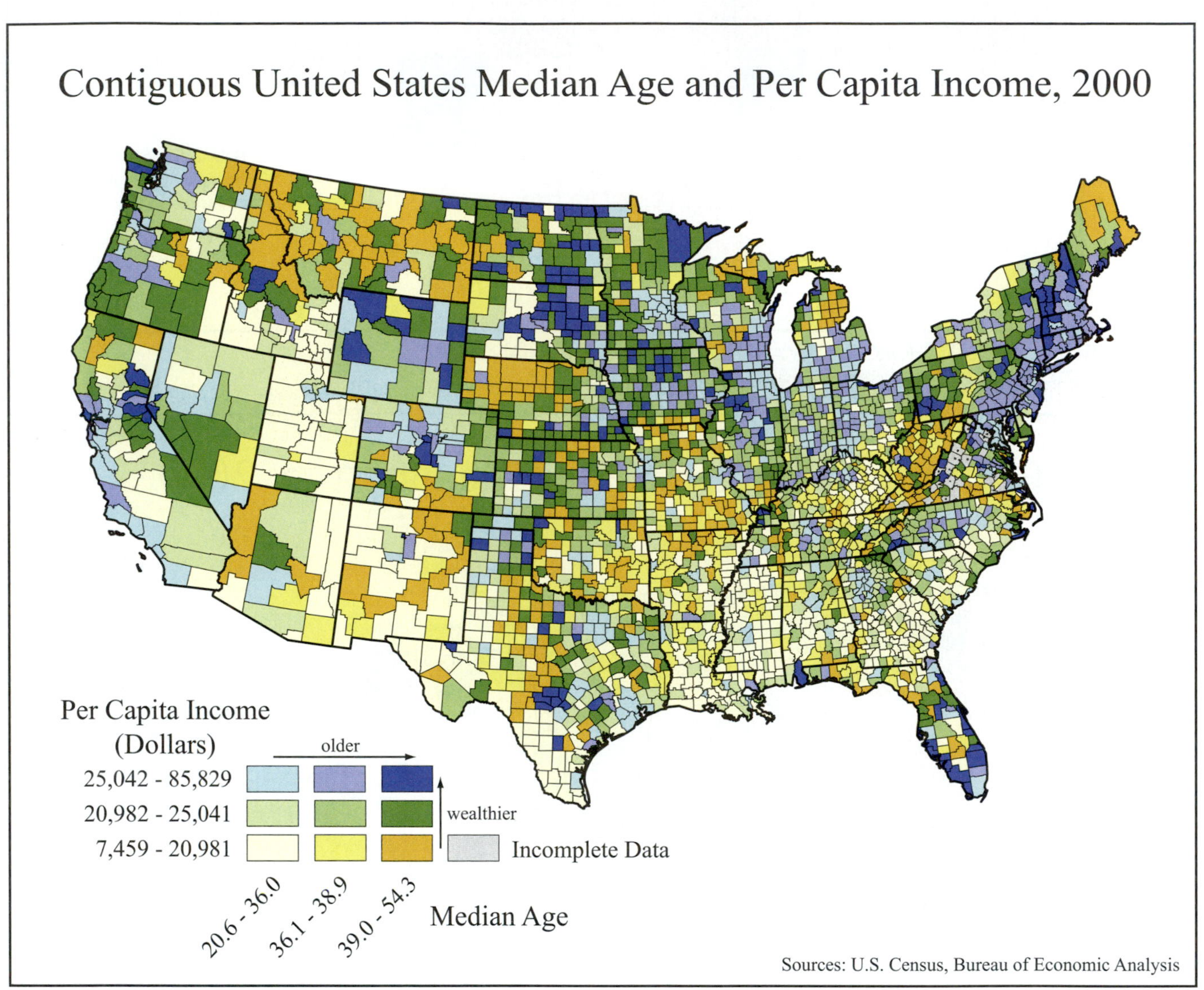

COLOR PLATE 6.4 BIVARIATE CHOROPLETH MAPS SHOW RELATIONSHIPS BETWEEN TWO VARIABLES.

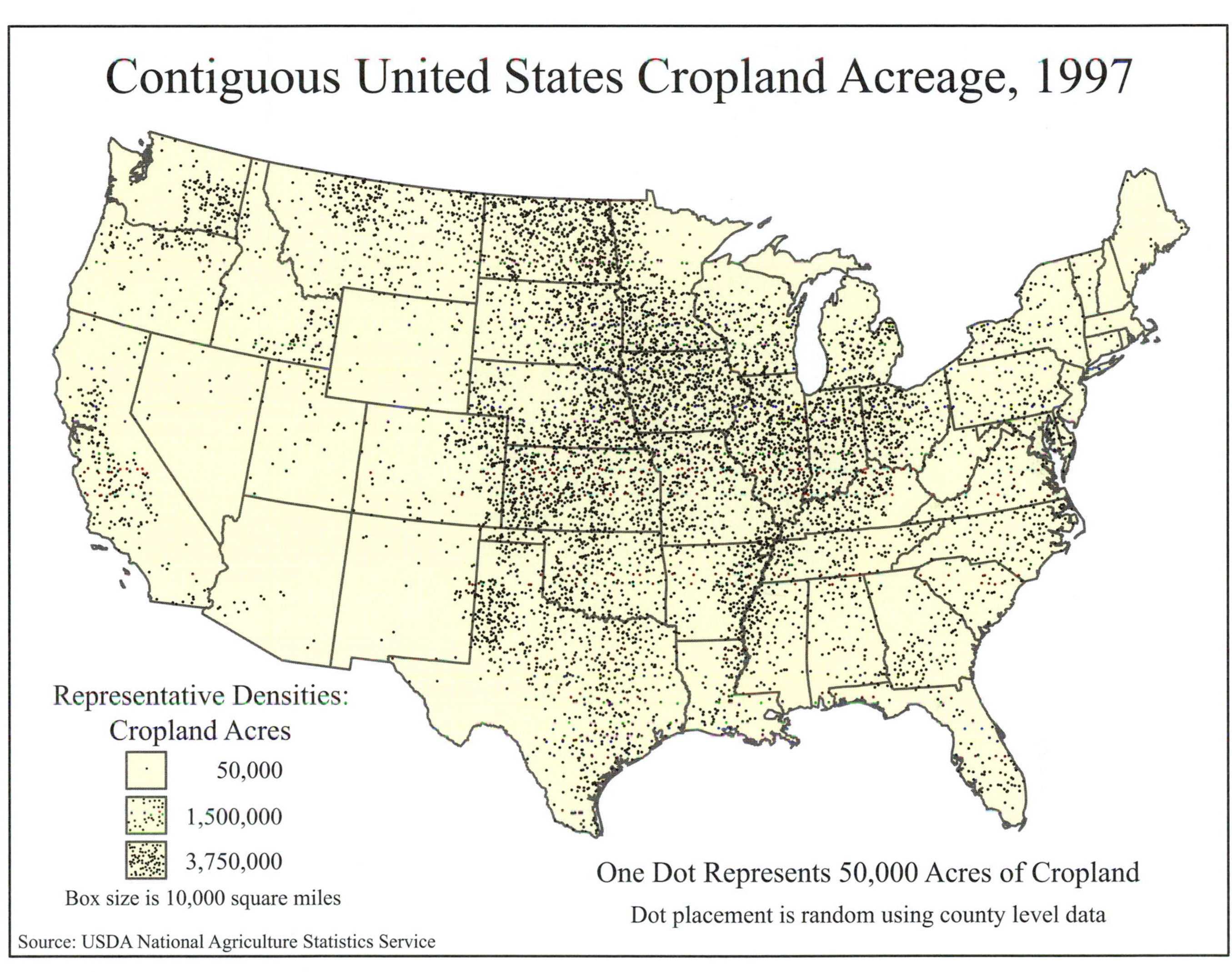

COLOR PLATE 7.1 A TYPICAL DOT DENSITY MAP.

Figure 7.1 with a light yellow fill to accentuate the dots. Other notable dot map elements include reference political boundaries taken from a higher level of enumeration (the state outlines), a legend that includes dot value and representative densities, and ancillary text stating the data enumeration units and the fact that the dots are randomized within those units.

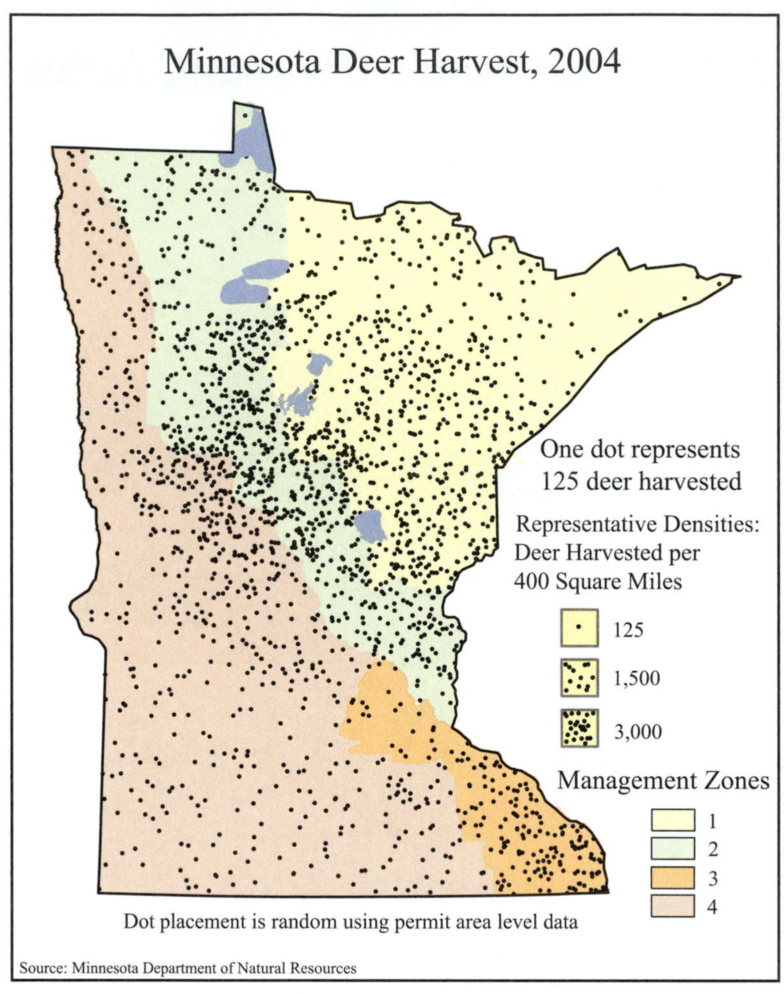

COLOR PLATE 7.2 DOTS PLACED ON SEVERAL HUES.

This map illustrates several notable facets, including good selection of dot size and value, use of locational filters to keep dots out of lakes, and enumeration by deer permit area (a region slightly smaller than a county). The inclusion of deer management zones gives the map an almost bivariate appearance. Dot map designed by Brian Lorenz and Alek Halverson, used by permission.

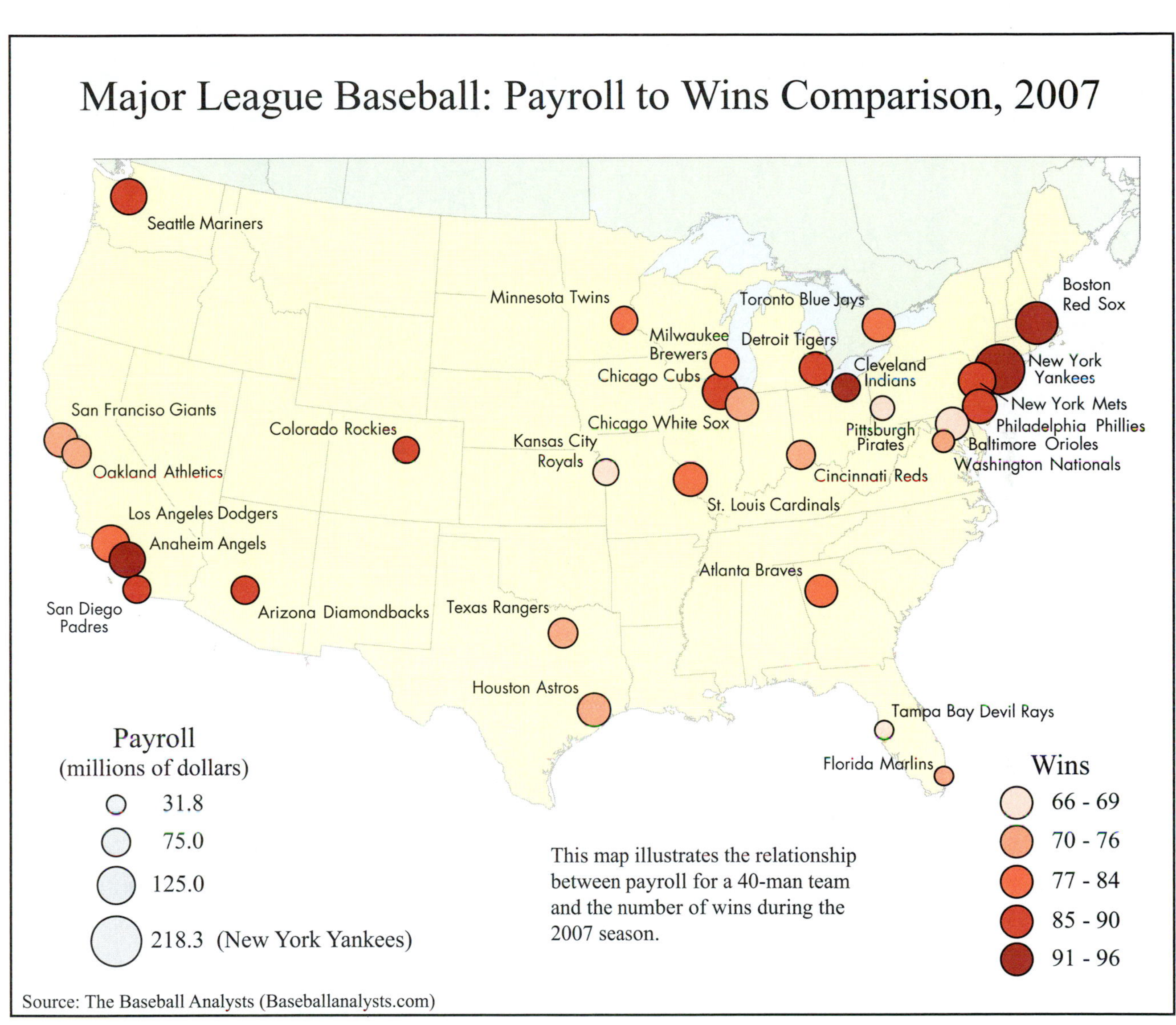

COLOR PLATE 8.1 BIVARIATE PROPORTIONAL SYMBOL MAP.

The symbol's hue values accommodate the second variable. It is often necessary to have a second legend in a bivariate map such as this.

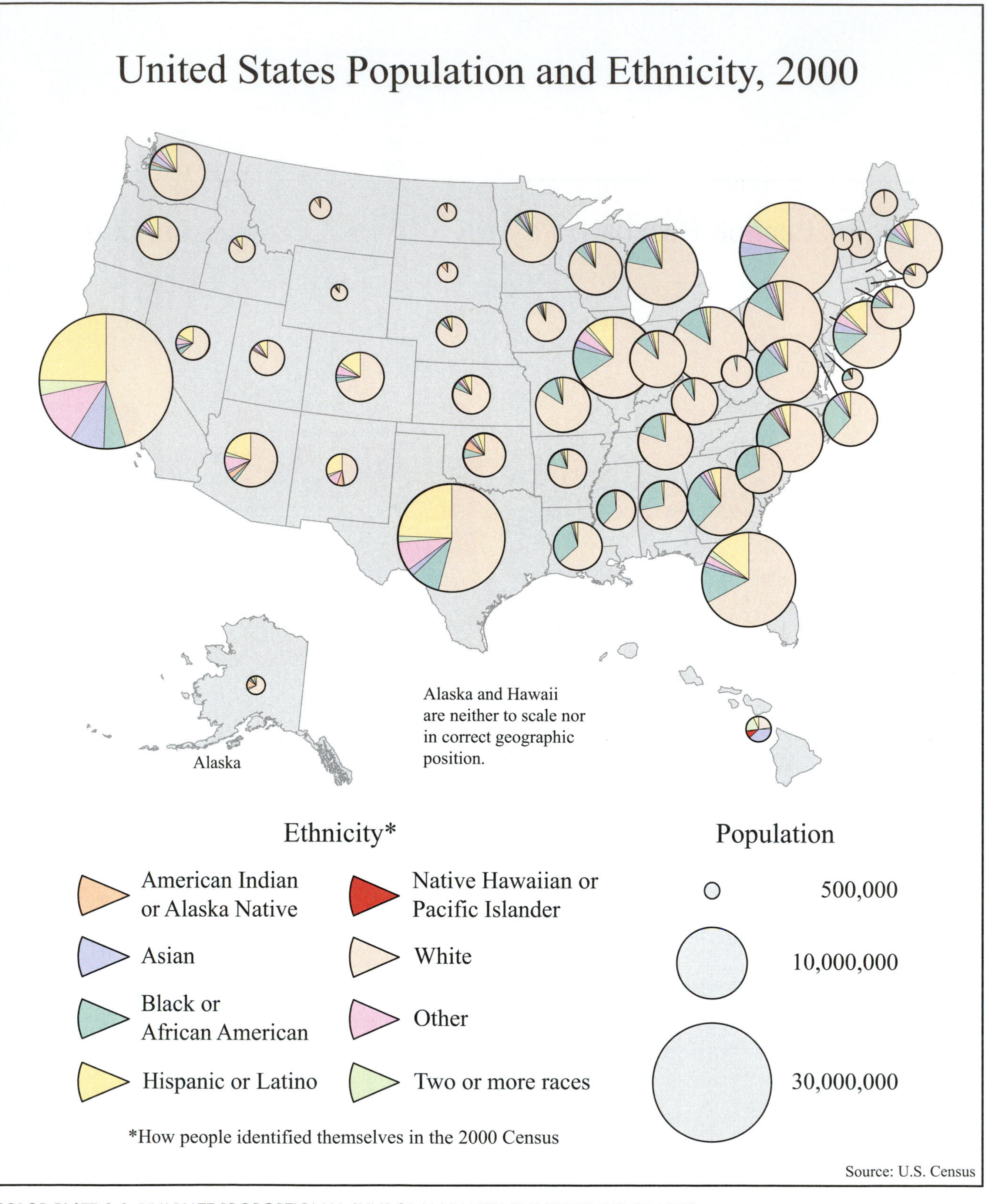

COLOR PLATE 8.2 BIVARIATE PROPORTIONAL SYMBOL MAP WITH EMBEDDED PIE GRAPHS.

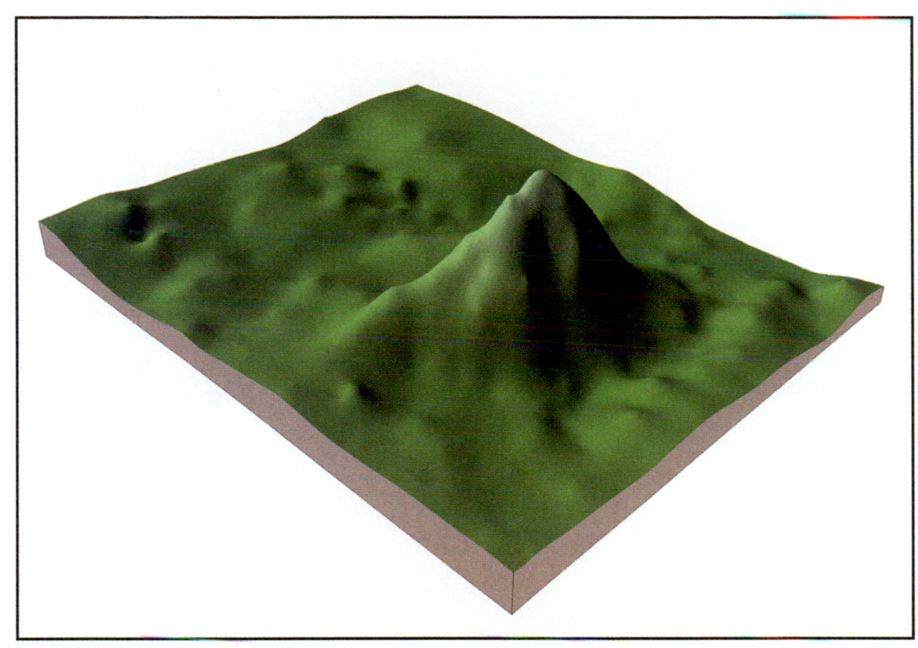

COLOR PLATE 9.1 SURFACE MAP OF STONE MOUNTAIN, GEORGIA.
(See Figure 9.13 for this figure using grayscale in place of color.)

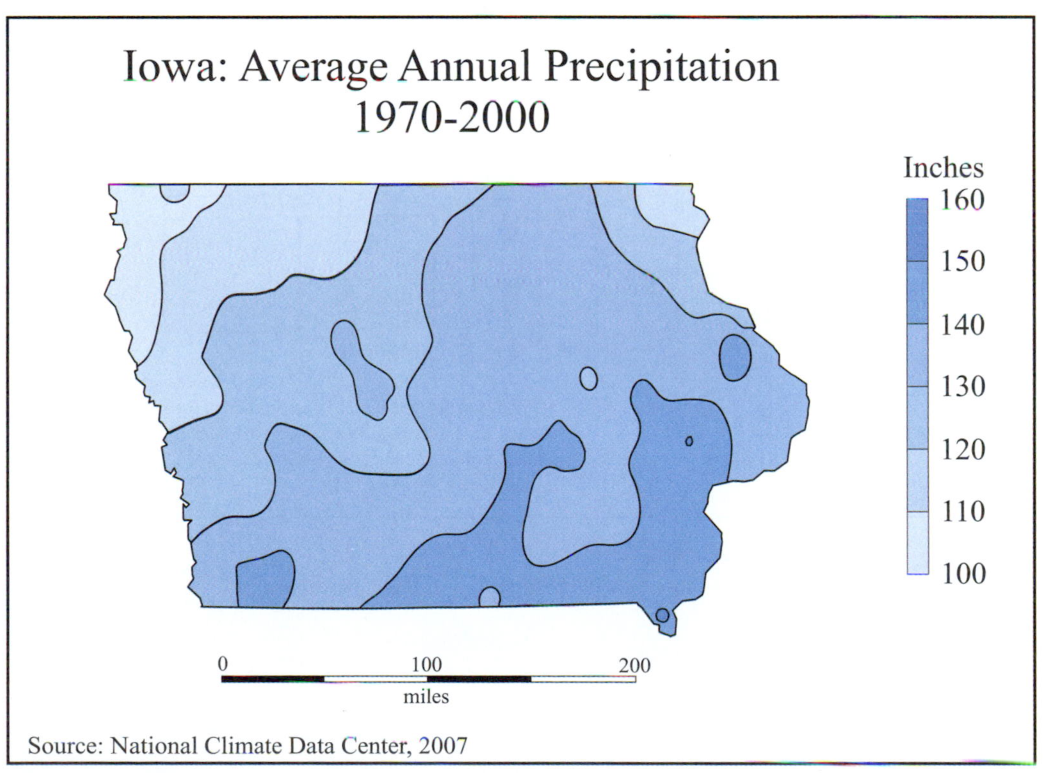

COLOR PLATE 9.2 SHADING BETWEEN ISARITHMS.
The addition of shades of blue between isarithms identifies areas with similar value range. The scale of the map is the vertical identification of color to data relationship. (See Figure 9.17 for this figure using grayscale in place of color.)

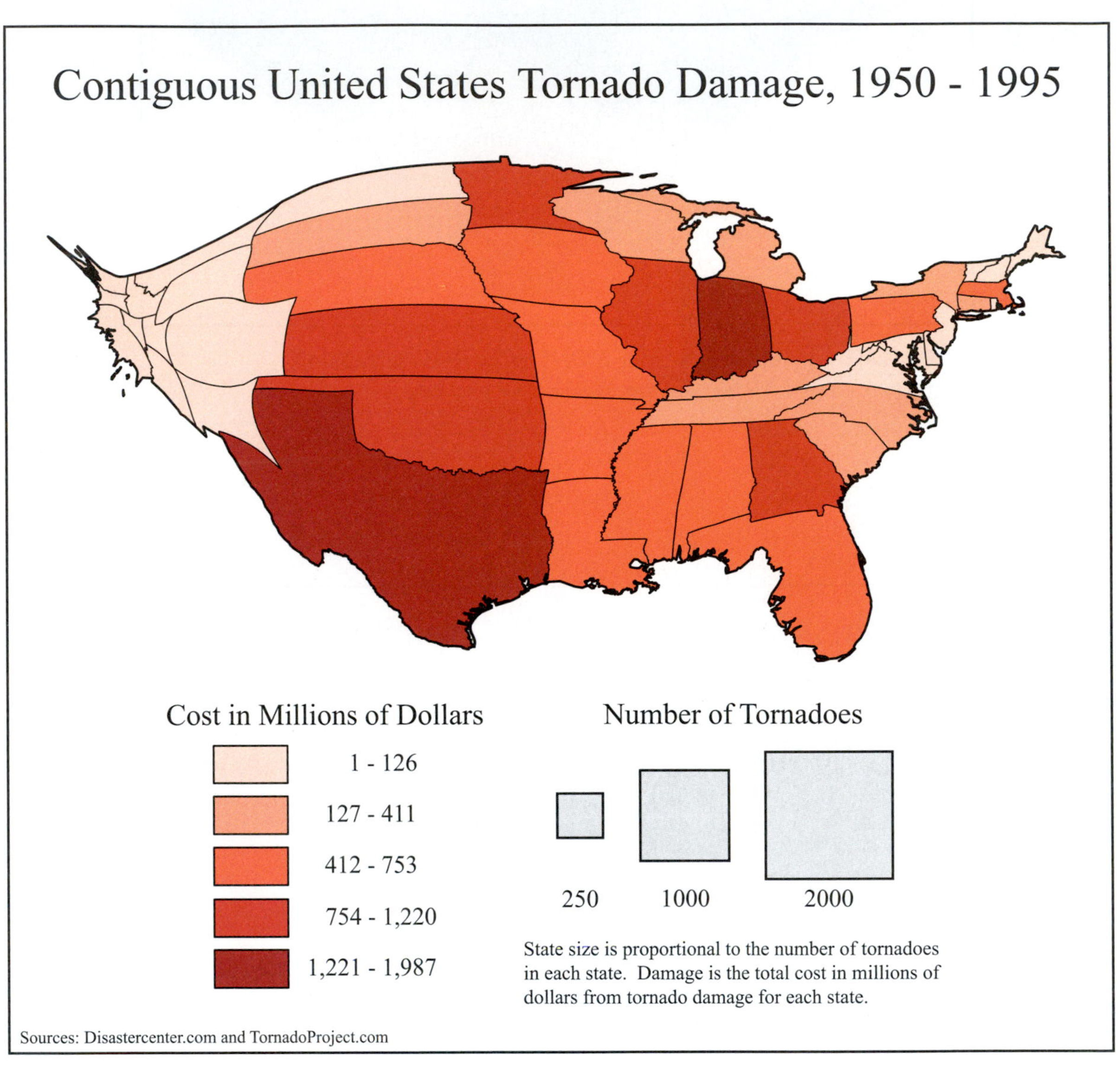

COLOR PLATE 10.1 BIVARIATE CARTOGRAM.

Source: Cartogram designed by Craig Wensmann and Nick Samuelson, used by permission.

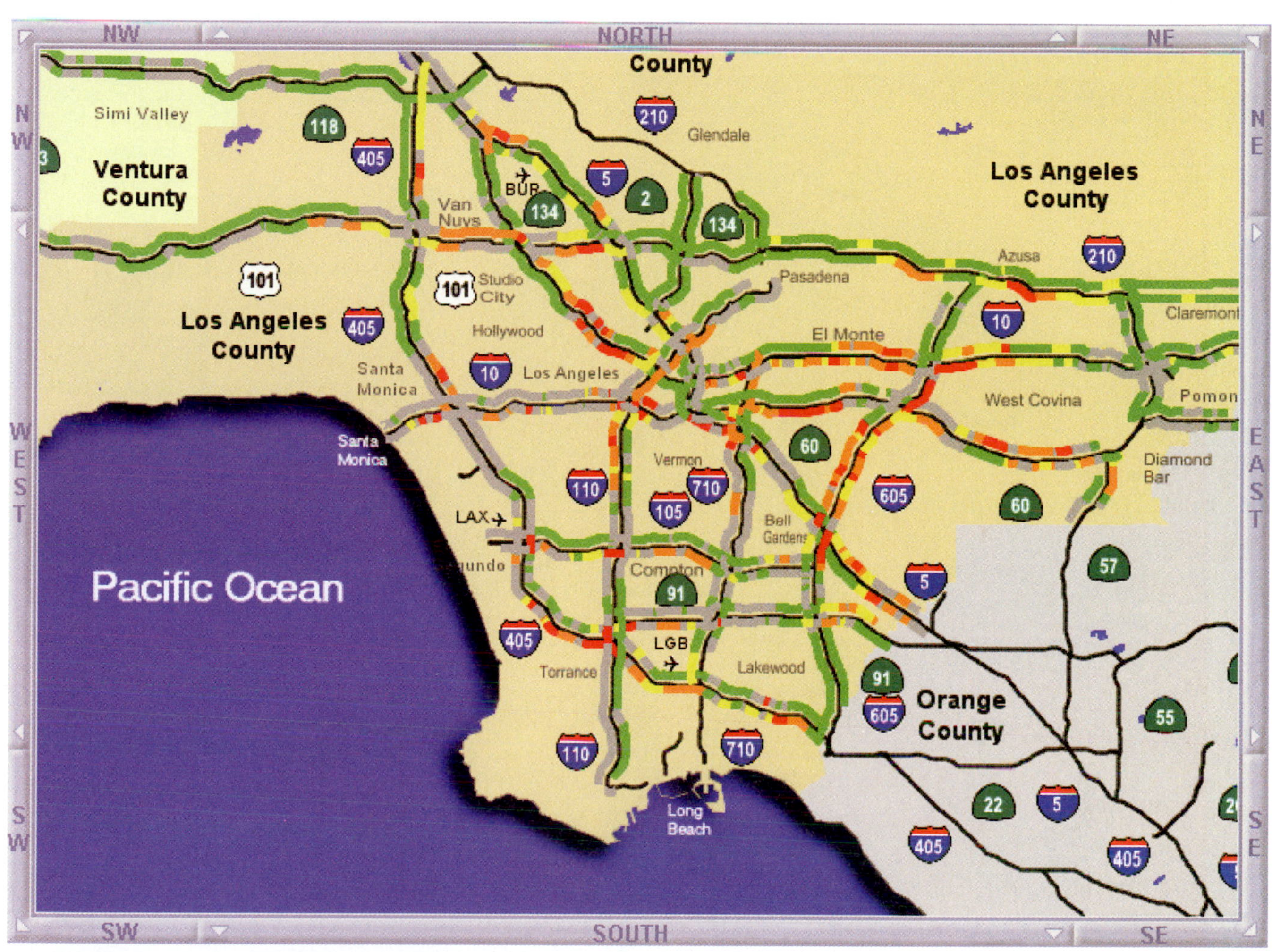

COLOR PLATE 11.1 TRAFFIC FLOW MAP USING COLOR AS THE DOMINANT VISUAL VARIABLE.

Online sites with regularly updated maps depicting traffic congestion for cities such as Los Angeles, California, are fairly common. They typically use signal light colors to indicate traffic flow—or lack thereof. *Source:* Used by permission from the Los Angeles County Metropolitan Transportation Authority.

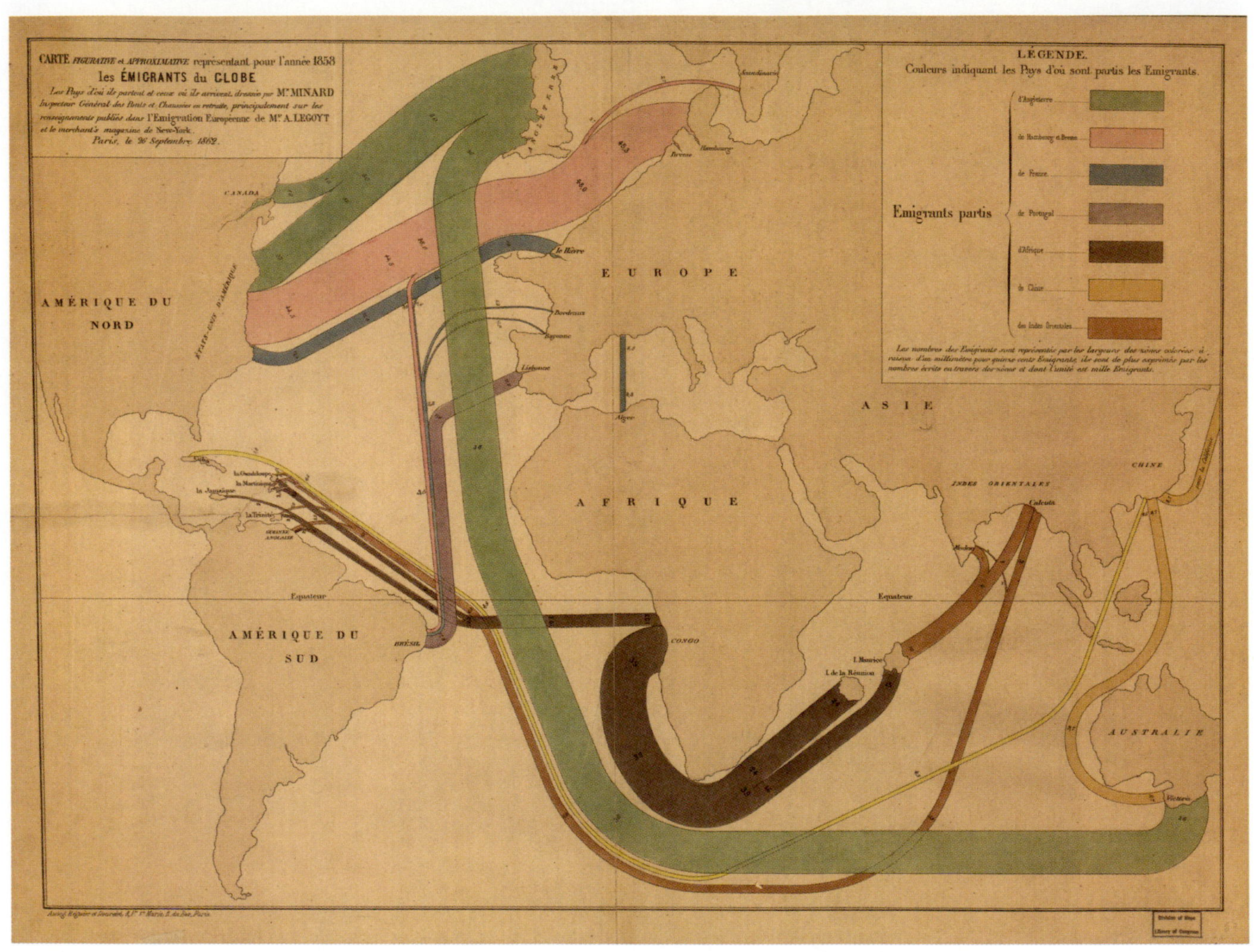

COLOR PLATE 11.2 MINARD FLOW MAP DEPICTING GLOBAL EMIGRATION FROM 1858.
Source: Minard 1862.

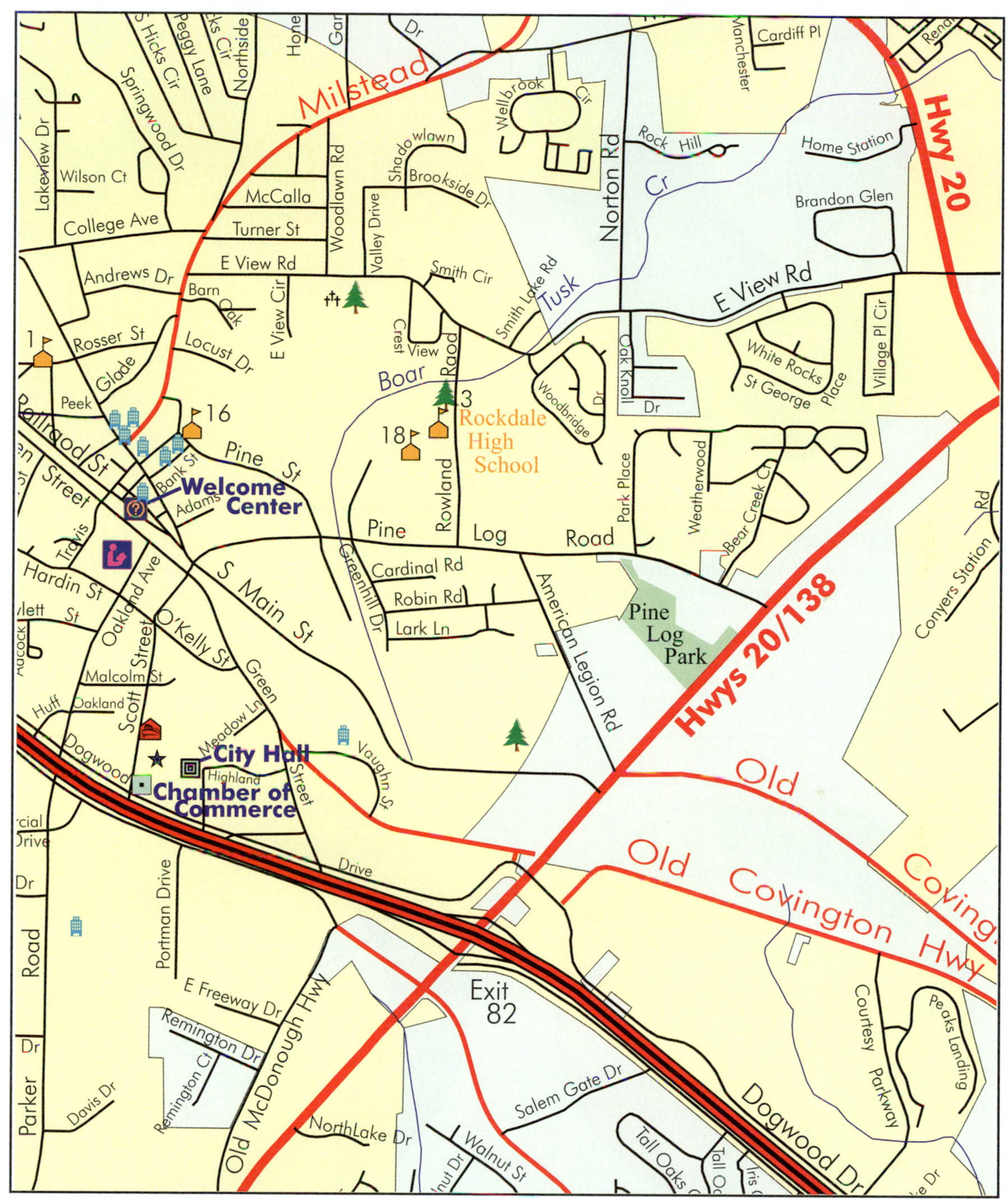

COLOR PLATE 13.1 PRACTICAL EXAMPLE OF TRADEOFFS ENCOUNTERED IN DESIGNING GENERAL REFERENCE MAPS.

See text for discussion.

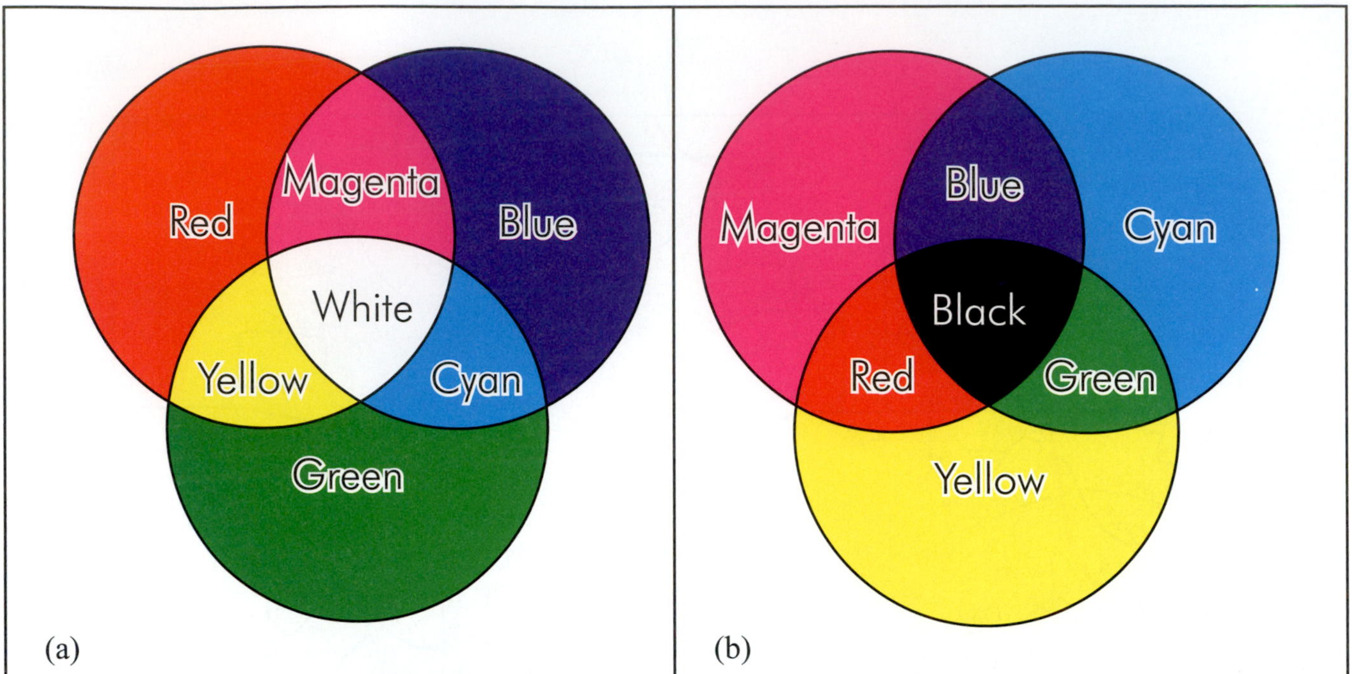

COLOR PLATE 14.1 ADDITIVE AND SUBTRACTIVE COLOR MODELS.

The additive model (a) depicts the colors generated by the mixing of the red, green, and blue color primaries. The subtractive model (b) depicts similar interactions between the process colors of magenta, cyan, and yellow.

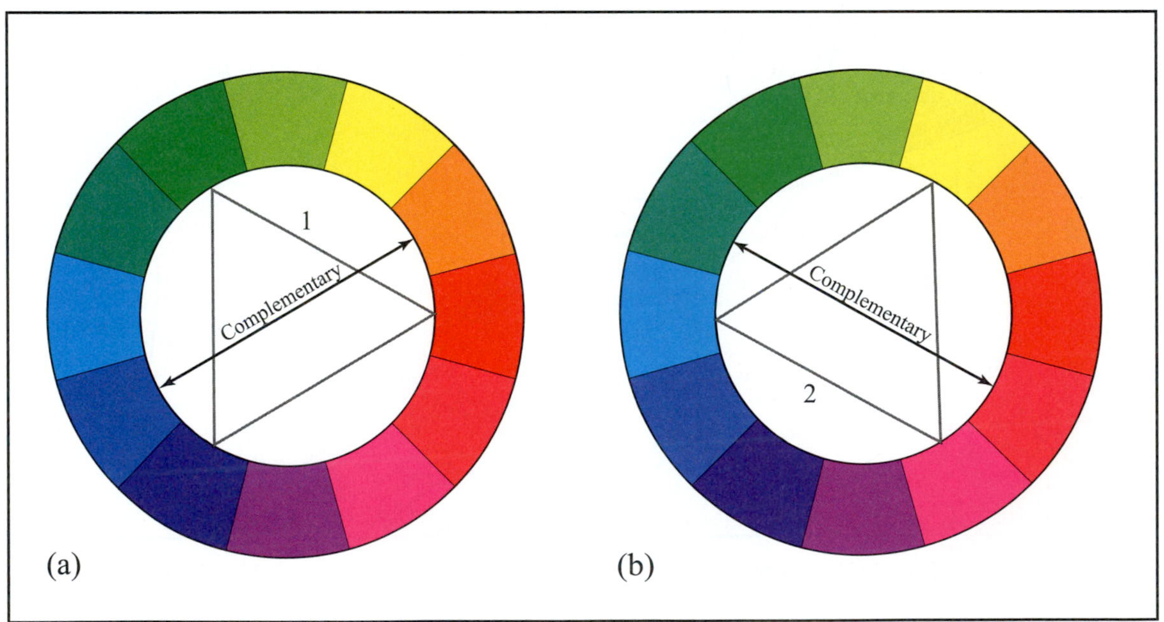

COLOR PLATE 14.2 THE COLOR WHEEL AND COMPLEMENTARY COLORS.

The color wheel is a technique for visualizing the continuum of color. The triangles are rotated to represent the additive primary colors (1) and the subtractive process colors (2). Any two colors that are directly across from each other on the wheel are known as complementary colors.

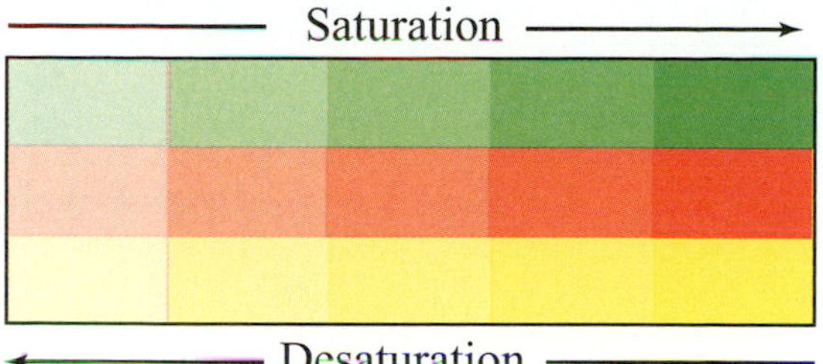

COLOR PLATE 14.3 SATURATION OF COLOR.

Saturation and desaturation of the colors are displayed. See text for the discussion.

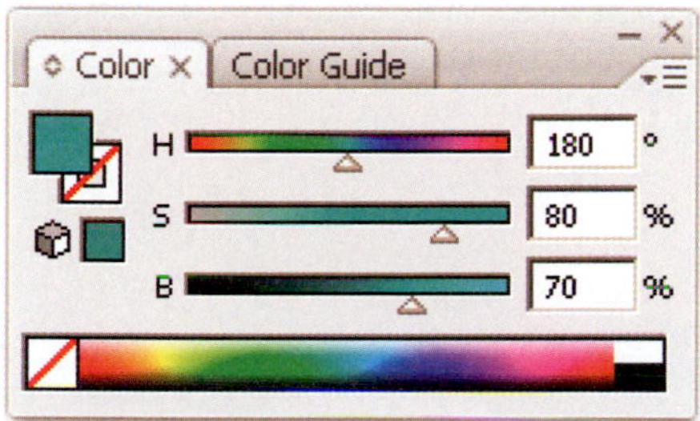

COLOR PLATE 14.4 COLOR SLIDERS FOR HSB COLOR MODEL.

The sliders are similar in most GIS and mapping software. Hue is selected by the angle on the color wheel and saturation and brightness are set as percentages with a range of zero to 180. These sliders are similar to HSV and other color models. (*Source:* Screen capture of the color model is reprinted with permission from Adobe Systems Incorporated.)

COLOR PLATE 14.5 THE COLOR CUBE.

All possible colors are displayed on and within the color cube. See text for discussion.

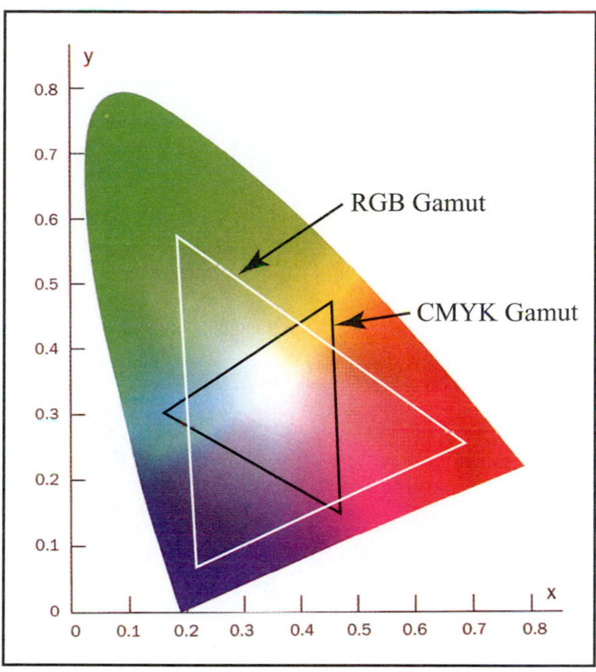

COLOR PLATE 14.6 THE CHROMATICITY DIAGRAM OF THE CIE COLOR SYSTEM.

See text for discussion. *Source:* Nyman, FOUR COLORS/ONE IMAGE: GETTING GREATER COLOR OUTPUT WITH PHOTOSHOP, QUARKXPRESS, AND CACHET, p. 25 figure 59, © 1993 by Mattlas Nyman and Software Plus. Reproduced by permission of Pearson Education, Inc. All rights reserved.

COLOR PLATE 15.1 FOUR-COLOR PROCESS SCREEN ANGLES.

Black is angled 45 degrees from the horizontal, yellow at 90 degrees, cyan at 15 degrees, and magenta at 75 degrees. These angles have been determined to provide the maximum inks of the four colors, thereby giving the purest colors on the printed page.

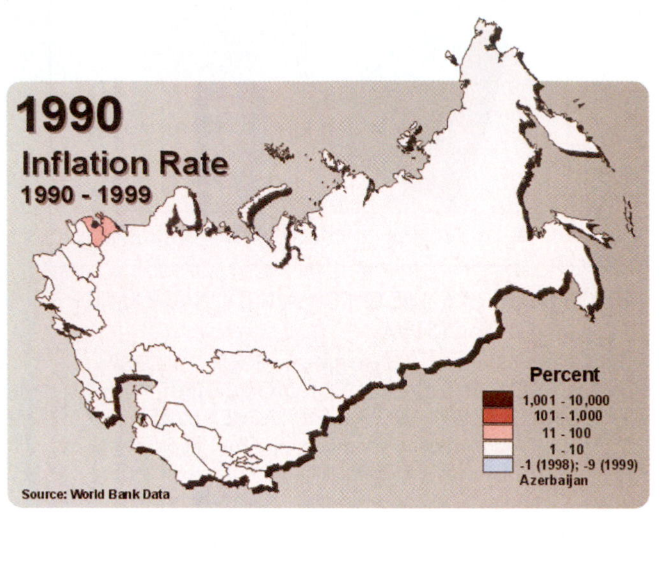

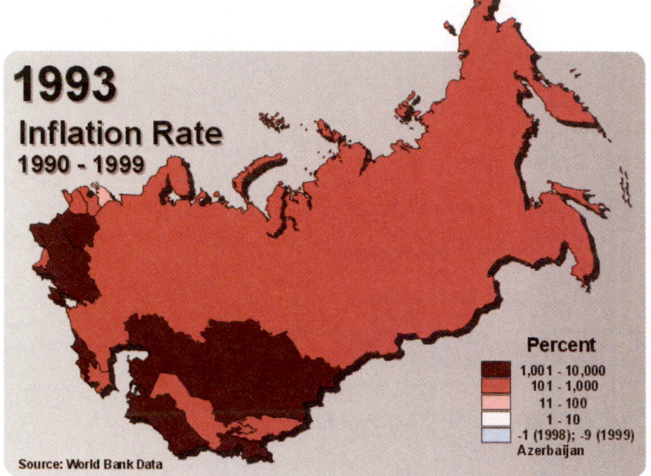

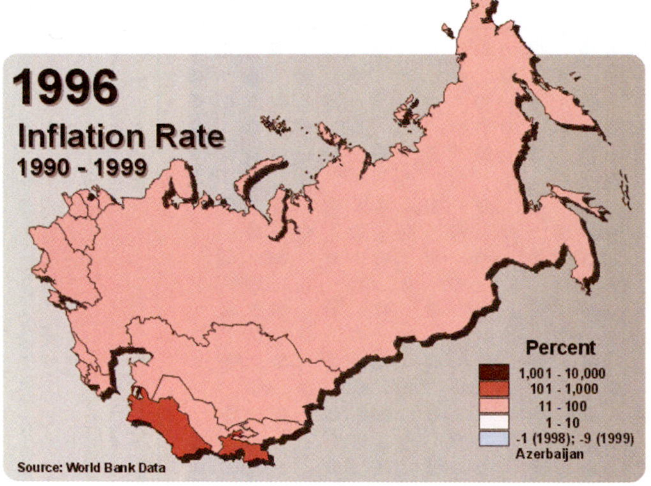

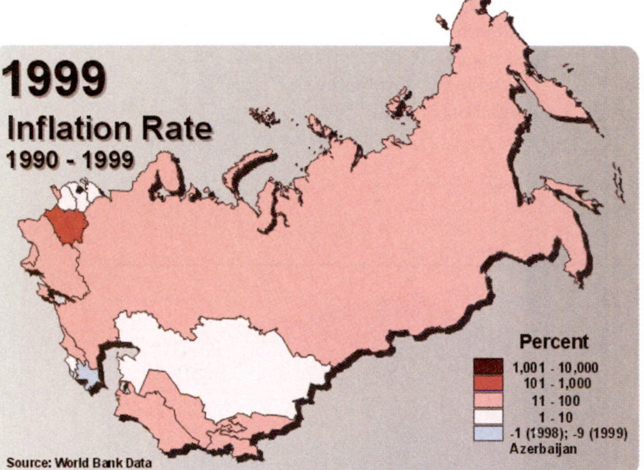

COLOR PLATE 16.1 SELECT FRAMES FROM A TEMPORAL 2-D MAP ANIMATION SEQUENCE.

Notice that the choropleth legend is unchanging and the years are clearly labeled. Reprinted by permission of the author.

map elements marks that make up the total visual image comprising a map, including the title, legend, scale, credits, mapped or unmapped areas, graticule, borders and neatlines, and symbols

mental imagining creating visual images in the mind's eye

optical center the place just above the geometric center in an image space; can be used to create visual balance

perceptual grouping mechanism identified by psychologists as leading to figure formation; includes grouping by shape, size, and proximity

planar organization composition or arrangement of the map's visual elements as they appear at one visual or intellectual level

resolution acuity ability to discern a separation between objects in the visual field; usually measured by linear distance in millimeters

scientific visualization transforming symbolic information into geometric forms by representing them on computer display devices, allowing the viewer to see structures not observable before

vignetting graphic treatment at an edge or border, resulting in a continuous gradient of value or brightness

visual acuity size threshold; the ability to discern an object in the visual field; usually measured by angular distance on the retinal image, as opposed to the physical size of the object

visual balance state in which all objects in a visual image appear in equilibrium

visual hierarchy the intellectual plan for the map and the subsequent graphic solution to satisfy the plan; may also be called the organization hierarchy

visualization process mental process in which the designer experiences whole new creations by rearranging previously stored visual images

13 MAKING THE MAP READABLE: THE INTELLIGENT USE OF TYPE

CHAPTER PREVIEW Map lettering is an integral part of the total design effort and should not be reduced to a minor role. Lettering on the map functions to bring the cartographer and map reader closer together and makes communication possible. To employ lettering properly, the cartographic designer should be familiar with letterform characteristics, sizes, spacing, type personalities, and legibility. Size is a most critical element in design, and lettering style and personality can affect the appearance of the map. Type classification is important for the cartographer in selecting a suitable typeface. Map typography includes lettering placement; overall lettering harmony can be achieved through adherence to established conventions of placement. Today, the computer provides us with a plethora of options relating to the type characteristics. A basic set of type families are included with most software and give the cartographer flexibility in lettering a map. ■

For many cartographic designers, the planning and application of map lettering remains the last task. This is unfortunate. In most instances, the appearance and mood of the entire map can be set by its lettering, so lettering should not be treated lightly in design. Although only a handful of cartographic researchers have examined map lettering in any detail, there is extensive general literature on type, its design, history, application, and production. This chapter deals with the fundamentals of lettering and attempts to illustrate the importance of lettering in thematic map design. Three major topics—the function of map lettering, the elements of type, and guidelines for type selection and placement—will be examined.

FUNCTIONS OF MAP LETTERING

All map lettering, whether on general-reference maps or thematic maps, serves to bring the cartographer and map user together, making communication possible. Only in the highly unusual situation where cartographer and map user discuss the map in person can written language be ignored.

On general-reference maps, lettering serves mainly to *name places* and to identify or *label things* (for example, scales, mountains, oceans, straits, and graticule elements). On thematic maps, lettering is also provided for titles, legends, and other explanatory marginal materials necessary to make the map content more comprehensible. Because all written language elements (letters and words) are symbols for meaning, they serve the same function on maps.

Map lettering should be viewed first as a *functional symbol* on the map, and only secondarily as an aesthetic object. Nevertheless, map lettering, if not done well, can hinder communication. Therefore, the cartographer should approach the employment of lettering with an appropriate regard for both function and form. Map lettering in this context refers to the *selection* of lettering type and its *placement* on the map. Questions of which words to use in titles, labels, or other marginalia are not addressed here.

Lettering can express the nature of a geographical feature by its *style,* the feature's importance by its *size,* the feature's location by its *placement,* and the feature's extent by its *spacing* (Raisz 1962). The variables of style and size are most important in the design of titles and legends; size, spacing, and placement are particularly important on the body of the map. These distinctions set apart map lettering from general text lettering, and the designer should keep them in mind.

TABLE 13.1 CLASSIFICATION OF MAP LETTERING ACCORDING TO USE

Use	Description and Use
Descriptive text	Reflects features that are symbolized on map face by point, line, area
Narrative	Names of objects
Descriptive	Additional property of feature ("scenic route")
Warning	Dangerous nature of feature ("sunken wreck")
Functional information	Locatable ground feature ("rescue post")
Regulatory	Legal information (area of land)
Analytical text	Links user with attribute of features
Confirmative	Spatial relations (distance between two towns or bearings on cadastre)
Determinative	Tables placed along map
Interpretive	Difficult to get information from map, so it is provided ("quickest route is . . .")
Reference	Text alongside map
Categorization	Categorization of a theme in codes (soil maps, geologic maps)
Positional Text	Text to describe or confirm location, in space or time
Geocoding	Grid reference notations
Measurement	Relative position (at edge of map, "Twenty miles to . . .")
Temporal Position	Text to give time of events (historic battles)
Metadata	Refer to nature of source data to map as a whole (reference ellipsoid)

(*Source:* Fairbairn 1993)

Cartographic writers have provided further convenient classifications to help in the understanding of the different uses of map type. Table 13.1 presents one such classification.

THE ELEMENTS OF TYPE

The selection of typeface and the placement of lettering are the two chief concerns of the designer. Proper selection can be made only if the cartographer fully understands the fundamentals of *type design.* As there are hundreds of individual typefaces from which to choose, the designer must recognize the elements and characteristics of their design, how they may be classified to make selection and use easier, and the fact that different typefaces can make different impressions on the reader. At first, these aspects of type can be bewildering, but familiarity comes easily with practice and experience.

Typeface Characteristics

Typeface design, type size, and letterforms are the principal characteristics of type with which the designer works. Letter and word spacing may be added, although strictly speaking these are not typeface elements. They are treated in this section because they can have such a tremendous impact on the perception of individual letters or words.

Letterform Components

All **typefaces** have elements in common regardless of the letter represented (see Figure 13.1a). Letters, both capital and lowercase, are begun on the **base line.** The height of the body of lowercase letters is referred to as the ***x*-height,** an important dimension in letter design because it often determines the readability of the type. Letter strokes that are higher than the *x*-height are **ascenders;** all ascenders in a lowercase alphabet will terminate at a common **ascender line. Descenders** are letter strokes that fall beneath the base line and terminate at the **descender line.** In most typefaces, capital letters are shorter than ascenders; there is therefore a **cap line** that defines the vertical dimension of capital letters.

A major element of some letterforms is the **serif.** Serifs are finishing strokes added to the end of the main strokes of the letter. Lettering styles that do not contain such finishing strokes are called **sans serif** styles. Serifs have different appearances, depending on the way they are joined to the main strokes. The serif may or may not be supported by a **bracket** (or *fillet*) (see Figure 13.1b). Not all designs have serifs, but it is important to note that in running text such as you are now reading, letterforms with serifs are easier to read (see Figure 13.2).

Counters, bowls, and loops of letters should be examined carefully when choosing type. **Counters** are the partially or completely enclosed areas of a letter, and **bowls** are the rounded portions of such letters as o, b, d, and the upper part of the lowercase g. The lower part of the lowercase g is referred to as a **loop.** These spaces can close up during the printing of the map, possibly because of reduction in size, printing resolution (see Chapter 15 for discussion on print resolution), or excessive ink in printing. Similar problems occur with static on-screen maps as the pixels used to generate the letters will fill in as well. The fewer pixels used to create the letter, the greater the chance that text quality will diminish. Normal screen resolution is 72 pixels (dots) per inch (see Chapter 16 for discussion of monitor resolution) while most quality printing occurs at 300 dots per inch or higher. This reduces the map reader's ability to differentiate between letters and can therefore impede communication. Map designers should choose typefaces with letterforms that are open and have few narrow bowls or loops that can lead to such problems.

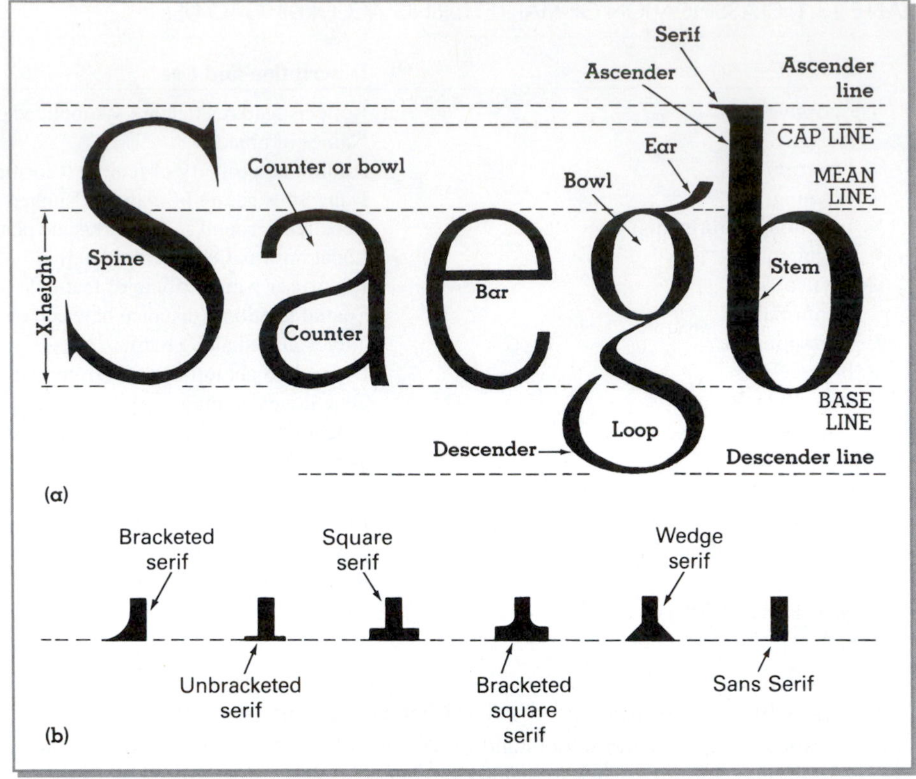

FIGURE 13.1 TYPE ELEMENTS.
The principal elements of letterforms are shown in (a), and (b) illustrates the different serif forms.

TIMES NEW ROMAN

Most geographical analyses involve point patterns (or centers of areas) that have weights attributed to them. Except at the simplest nominal scales, geographical point phenomena do not usually occur everywhere with equal value. An especially interesting study is to plot weighted means over time to discover a spatially dynamic pattern. Weights are most often socioeconomic data such as income, production, sales, or employment data.

FUTURA LIGHT

Most geographical analyses involve point patterns (or centers of areas) that have weights attributed to them. Except at the simplest nominal scales, geographical point phenomena do not usually occur everywhere with equal value. An especially interesting study is to plot weighted means over time to discover a spatially dynamic pattern. Weights are most often socioeconomic data such as income, production, sales, or employment data.

FIGURE 13.2 TWO LETTERING TEXT BLOCKS.
Times Roman, a roman, serified type, is easier to read in running text than lettering set in sans serif style, such as Futura Light.

Typeface Style and Classification

The cartographer who has a knowledge of the history of type design and sees how different typefaces relate to each other will be better prepared to select type for maps. Type classification is one way to begin. The system discussed here is one among many; it is selected for its simplicity and historical emphasis.

Prior to the invention of movable type and the printing of Gutenberg in Germany, writing in Europe was accomplished by hand, usually by a select group of clerical scribes. The alphabets had come to them from Latin alphabets, as modified through time. Manuscript letterforms were called **textura** by the Italians (Tschichold 1966). Early movable type designs simply attempted to replicate the manuscript forms. A system of type classification begins with the first movable type and places all subsequent typefaces into classes, based on when they were created. A modern typeface is placed in one of these classes according to the match of its style characteristics with those of faces designed earlier. A relatively simple classification, developed by Alexander Lawson (1971), contains eight major classes (see Table 13.2). For the most part, **Black Letter** is the term used for at least three or four different styles of early movable type that attempted to replicate manuscript forms. Another term used to describe Black Letter is *gothic,* but some confusion exists because many in the United States used this name for the early sans serif styles developed in the 1830s. Two styles currently used are Cloister Black (Old English) and Goudy Text, which closely resembles the type of Gutenberg's first Bible.

TABLE 13.2 TYPEFACE CLASSES

1. Black Letter
2. Oldstyle
   a. Venetian
   b. Aldine-French
   c. Dutch-English
3. Transitional
4. Modern
5. Sans serif
6. Script-cursive
7. Display-decorative

(*Source:* Lawson 1971, 27)

The initial differentiation between different fonts is referred to as the type's style. The cartographer normally selects type by the font name or face (see Type Font below); however, the style of the font is inherent in its characteristics. The four most common styles are **Oldstyle, transitional, modern,** and **sans serif.**

Oldstyle. This style is sometimes referred to as *Old face* or *Roman* type and is characterized by geometrically rectangular and circular letter characteristics. This style has serifs, small strokes at the end of letters,—for example, A, C, F, or T. The weight of the line components of this style is relatively heavy, with little contrast between thick and thin strokes (see Figure 13.3a).

There are three groups that comprise this style, *Venetian, Aldine-French, and Dutch-English. Venetian* is a style first developed in northern Italy some 20 years after Gutenberg, with Goudy Old Style as an example (Lawson 1971). The *Aldine-French* group is modeled after the Venetian but was originally characterized by a greater difference between thick and thin strokes and modifications of the serifs. This style influenced French types of the sixteenth century and dominated style efforts in Europe for nearly 150 years (Lawson 1971). Palatino is an example of this group. *Dutch-English* forms the third group in the Oldstyle class. The French type forms were popular in Europe but soon gave way to the enterprising Dutch, who in many cases redesigned type to facilitate printing. The contrast of thick and thin strokes was increased, and serifs were straightened and bracketed in the lowercase forms. Early Dutch designs were adopted in England and enjoyed enormous popularity because of William Caslon (Updike 1966; see Figure 13.3a). Caslon's foundry and others in England produced virtually all the type seen in America until the American Revolution. Caslon forms abound today; many manufacturers offer the style. The Dutch-English tradition is at the end of the development of the Oldstyle class of letterforms (see Figure 13.4).

Transitional. Such type forms make a historical bridge between the Oldstyle and Modern typefaces. John Baskerville, a printer and type founder in England around 1750, designed the type that bears his name (Updike 1966), often thought to

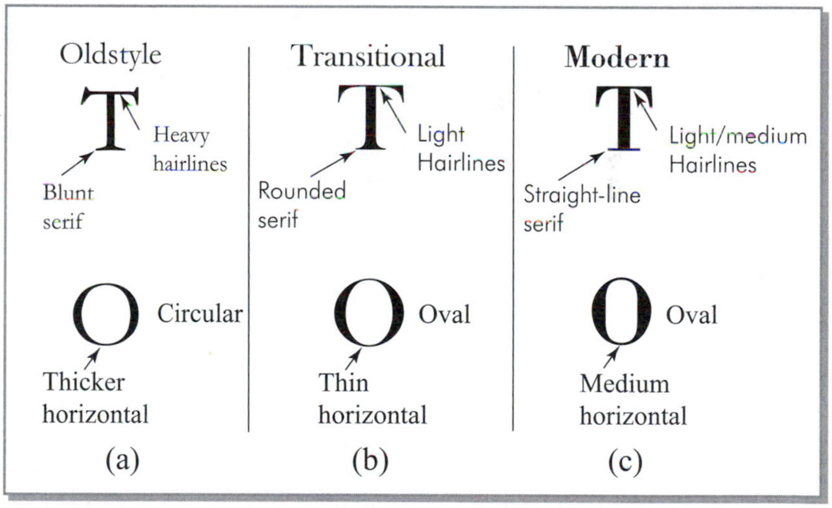

FIGURE 13.3 EXAMPLES OF (a) OLDSTYLE, (b) TRANSITIONAL, AND (c) MODERN TYPE STYLES.
The chief differences among the three main classes are their strokes and serifs.

Oldstyle	Transitional	Modern	Sans serif
Goudy Old Style	Baskerville	Bodoni	Arial
Adobe Caslon Pro	Garamond		Futura
Palatino			Helvetica
			Univers

FIGURE 13.4 EXAMPLES OF VARIOUS TYPEFACES FOR THE DIFFERENT CLASSES USED IN MANY APPLICATIONS.
See text for discussion.

be an excellent example of the Transitional style. The style is characterized by less serif bracketing than letterforms of the Oldstyle period (see Figure 13.3b).

Modern. The modern style of letterform began in the late eighteenth century and was considerably influenced by the Transitional forms. Baskerville's designs led Bodoni (see Figure 13.4) of Italy and Didot of France to design faces that have become identified with the modern movement. The development of typeface designs in the modern period was in response to several other parallel achievements, notably in printing, manufacturing, and advertising. The strokes of the copper engraver led to hairline serifs and thick cross strokes. There was an emphasis on vertical shading. The development of **display types** (those too large for book-text printing) was important because of the Industrial Revolution and the advertising that it brought. Type of the Modern period was still characteristically Roman in style, but brought to the extreme in geometric regularity. The strokes of the modern style produce vertically oval bowl components where the vertical stokes are significantly heavier than the light hairline thin horizontal strokes (see Figure 13.3c).

Sans Serif. The most notable type style to appear in the history of type design is sans serif, a letter having no serifs. Previous innovations in design had at least carried along the Latin alphabet's tradition of serifs, but the new sans serif styles marked a radical departure. They have remained in use since their introduction and have enjoyed widespread popularity. A Swiss type designer named Adrian Frutiger styled a complete family of sans serif alphabets called **Univers,** one of the most popular such forms ever designed. Unique to this family was Frutiger's use of a number series, rather than descriptive names, to designate weights. Those in the 40s are lightface, in the 50s medium, in the 60s bold, and in the 70s extra bold. Both roman and italic forms are included in each number series (see Figure 13.5).

Two other lettering styles round out this classification. **Script-cursive** and **display-decorative** fonts are similar in that they are not useful for ordinary text typography but find their place in decorative or advertising display. They are difficult to read in most circumstances and have had little use in ordinary cartography.

The fonts that accompany computers for word processing as well as GIS and mapping software permit the user to select from a variety of serif and sans serif type styles. Many of the popular fonts include Arial, Century Schoolbook, CG Times, Futura, Georgia, Helvetica, and Times New Roman. These fonts are appropriate and may be selected for either map labels, titles, or other text, such as legends and sources (see Figure 13.6).

Nearly any type class may be used effectively on the thematic map except Black Letter, script, or decorative forms. Certain designs may be more effective than others or may impart a more appropriate mood to the map. Certain principles regarding lettering class combinations and use of families are recommended and outlined later in the chapter. First, however, we will examine one important singular quality of letter designs—their personality.

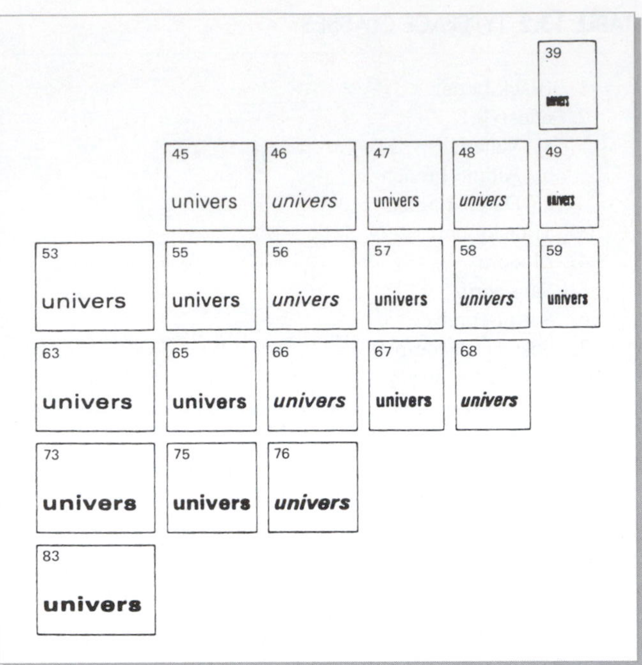

FIGURE 13.5 THE COMPLETE UNIVERS PALETTE.

A unique feature of this type family is the use of numbers instead of descriptive terms to designate weight. (The idea for representing the Univers family this way comes from Lawson 1971, 100. Redrawn with modifications from same source.)

FIGURE 13.6 POPULAR TYPE FACES.

These type faces are commonly available in GIS and mapping software. The example includes options of having the fonts displayed in their regular appearance.

Arial
Century Schoolbook
CG Times
Futura
Georgia
Helvetica
Times New Roman

The Personality of Type

Most graphic artists and designers seem to agree that typefaces are capable of creating moods, and that there is such a thing as **type personality.** There is no question that the selection of type can alter the appearance of a map, but it is exceedingly difficult to categorize each type design by the kind of response it is likely to produce. "Many experts suggest that typefaces can convey mood, attitude, and tone while having a distinct persona based on the font's unique features. Each document should be rendered in a font that connects mood, purpose, intended audience, and context of the document" (Shaikh *et al.* 2006, 1).

TABLE 13.3 PERSONALITY OF TYPEFACES

Dignity	Oldstyle typefaces
Power	Bold sans serif
Grace	Italics or scripts
Precision	Sans serif
Excitement	Mixtures of typefaces

(*Sources:* Nelson 1972; Shaikh *et al.* 2006)

In general, Oldstyle designs, with their emphasis on diagonal shading, create more mellow and restful designs (Biggs 1954). Bibles are more likely to be printed in Caslon or Bembo than Bodini. On the other hand, Modern faces, characterized by strong vertical and horizontal strokes, create abrupt, dynamic, and unsettling lettering. Lettering set in a sans serif face is dazzling, which makes it difficult for the eye to settle down. These observations are especially true in text typography but are also observable in display settings, such as in most cartographic applications. In cartographic applications, the selection of type is not as much to dazzle but to complement the thematic symbolization and give it context.

As a starting point, the student might wish to consider the contents of Table 13.3. These are only generalizations; the appearance of type in a cartographic environment may be significantly altered in the map's content and other style features. Nonetheless, an awareness of a type's influence on mood is important for the designer.

The Legibility of Type

Although the designer may be interested in the aesthetic qualities of letters and alphabets, the real concern is readability. If type on the map cannot be read easily, its employment has been a wasted effort. **Type legibility,** the term most often used in research dealing with readability, has been defined this way:

> *The characteristic of smooth and easy reading in regard to the design of a typeface and how groups of characters read* (Sinclair 1999, 480).

In cartography, **type discernibility**—the perception and comprehension of individual words not set in text lines—is more critical than legibility.

There is extensive literature dealing with text-type legibility, but little that deals specifically with type discernibility. Nonetheless, certain important facts are known to apply to cartographic lettering.

The individual lowercase s, q, c, and x are difficult to distinguish, and the letters f, i, j, l, and t are frequently mistaken for each other (Spencer 1968). The lowercase e, although a frequent letter in print, is often mistaken. In general, these legibility patterns have been recorded:

dmpw	high legibility
jrvxy	medium legibility
ceinl	low legibility

Current experience goes a step further. The resolution of the printer or monitor (see Chapters 15 and 16 respectively), font, type size (see "Type Size"), spacing and other decisions made by the cartographer can impact the legibility of type. The impact of these factors can often cause misinterpretation of other letter combinations. The letters m, n, and r may be misconstrued as a result of their roundness, whereas, letters such as b, c, d, and l have constraints created by both the bowl and ascender components of the letter. The size of the type, and the quality of display, can create a number of legibility problems. The cartographer must select map type that produces high readability. This may be done through experimentation and will become more intuitive with experience.

The legibility of individual letters becomes important in cartography because the map may be embedded with different backgrounds, textures, or colors that can make letters difficult to distinguish. In addition, legibility may be affected by the way the letters are shown; they can be screened or rendered in color.

Lettering in all capitals, more than any other factor, slows reading. Apparently, this is caused by a larger number of eye fixation pauses. Because capital letters occupy more space, the result is a reduction in the number of words perceived. Lowercase letterforms lead to quicker recognition. "Lowercase words impress the mind with their total silhouette while capitals are mentally spelled out by letter" (Tschichold 1966, 35; see Figure 13.7). Also, the upper halves of lowercase letters contribute more recognition cues than the lower halves. Text in all capital letters not only inhibits reading speed but is also perceived as "shouting" at the map reader. Unfortunately, many of the tops of the letters in the sans serif forms are identical or nearly so, creating problems of legibility.

In the cartographic context, placement of letters relative to their environments and the selection of letterforms can either add to or detract from overall legibility.

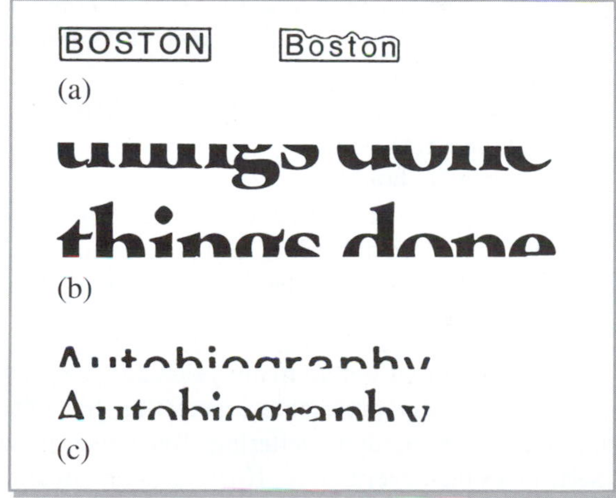

FIGURE 13.7 READING TYPE.

(a) Lowercase letters lead to easier word recognition because of the silhouette they produce. (b) It is easier to distinguish words by the tops of letters than by their bottoms. (c) Serif letterforms are frequently easier to read, because the letters are more distinct, especially along the upper halves of lowercase letters.

In the styles commonly used in text printing, legibility is not influenced by typeface style. This also applies to sans serif styles. Aesthetic preference might lead the designer away from sans serif, even though it might be equivalent in legibility (Tinker 1963). It is interesting to note that the United States Geological Survey selected a sans serif typeface (Univers) to appear on many of the 1:100,000 series maps, as well as others they produce (Gilman 1982). However, the current standards for USGS maps include a variety of type faces including Univers, Souvenir, Century, and others.

A significant research finding is that lettering in all italics slows reading, or at least is not preferred by most readers (Spencer 1968). Italic letterforms have historically been used for labeling hydrographic features, and this practice is not likely to change. Of course, the legibility studies of italic forms were in a text format. Similar findings might not apply in cartographic contexts.

Cartographic Requirements

Map lettering differs considerably from book and text lettering. Type set on lines with identical background is characteristic of book typesetting. Letters are **set solid,** and the **leading** (spacing) between lines of type is an important concern for the book designer. In cartography, however, letters are often spread out, placed on changing backgrounds (colors or tints), oriented in a variety of ways, and interrupted by lines or other symbols. The letters themselves may be rendered in different colors or tones, thereby causing considerable variation. As a result of these situations, the chief criterion of the cartographic designer in selecting a typeface is *that the individual letters be easily identifiable.*

One cartographer has identified the following considerations in selecting type for maps (Keates 1958).

1. The legibility of individual letters is of paramount importance, especially in smaller type sizes. Choose a typeface in which there is little chance of confusion between c and e or i and j.
2. Select a typeface with a relatively large *x*-height relative to lettering width.
3. Avoid extremely bold forms.
4. Choose a typeface that has softer shading; extreme vertical shading is more difficult to read than rounder forms.
5. Do not use decorative typefaces on the map; they are difficult to read.

It might be added that not so many years ago serif letterforms were normally preferable in meeting the unusual requirements of cartographic lettering. Today, we see more sans serif faces than ever before. However, individual letter recognition continues to be vital on maps. The serif can play an important role in this recognition, because it tends to complete letters optically and ties one letter visually to another (Gardiner 1964). Aesthetic considerations often affect the choice between serif and sans serif forms also.

The cartographer must make a variety of decisions in selecting the type to be included on both the body of the map and the title and legend. One progresses through the following topics in making these decisions: face, size, form, width, weight, and color.

Type Font and Type Families

Typographic nomenclature includes the word **font,** also known as the *type face,* which is a complete set of design characters of a typeface. A font of type normally includes numerals and special characters such as punctuation marks, in addition to the letters of the alphabet. The original set of digital fonts were bitmaps, comprised of individual pixels or dots that determine the clarity of the letter (see Chapter 16; and see Figure 13.8). With lower resolution printers or monitors, the letter may become pixilated, jagged edges, when enlarged (see Figure 13.9).

Most fonts are specified as either True Type or Open Type format. These fonts were originally designed for the Apple operating system but are now available on personal computers as well. These fonts are vector based and, like Adobe's PostScript, are scalable. True Type fonts provided more control and flexibility, which is especially valuable in sizes less than 15 point (Sinclair 1999). Open Type was jointly developed by Microsoft and Adobe and is cross-platform compatible (Doughty 2006). These fonts are designed to be easily used in a large variety of applications and computer platforms. Encapsulated PostScript fonts are saved within a graphics file ready for submission to an output device. Type designers often design variations of a basic font, making up the **family** of that font. A normal *type family* will include these letterform variants: size, form, width, weight, and color.

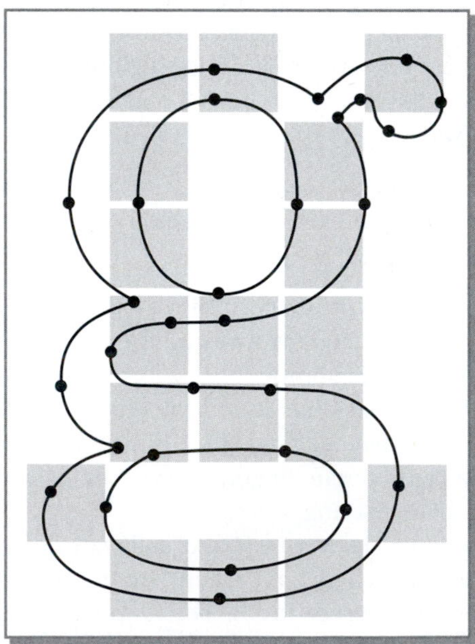

FIGURE 13.8 BIT-MAPPED VERSUS SCALABLE (VECTOR OUTLINE) FONT CHARACTERS.
Characters defined in bit-mapped format are formed by pixels, and are not as precise as those described by mathematical formulas, called scalable, or outline fonts.

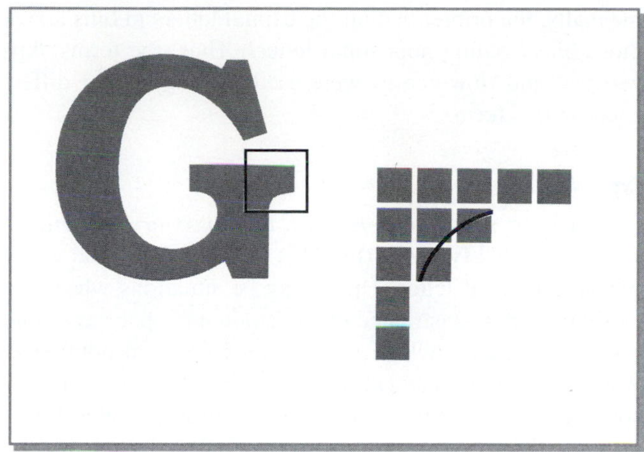

FIGURE 13.9 BIT-MAPPED FONTS.

If bit-mapped fonts are enlarged, the irregularity caused by the fixed pixel resolution causes the edges of the letterform to appear jagged.

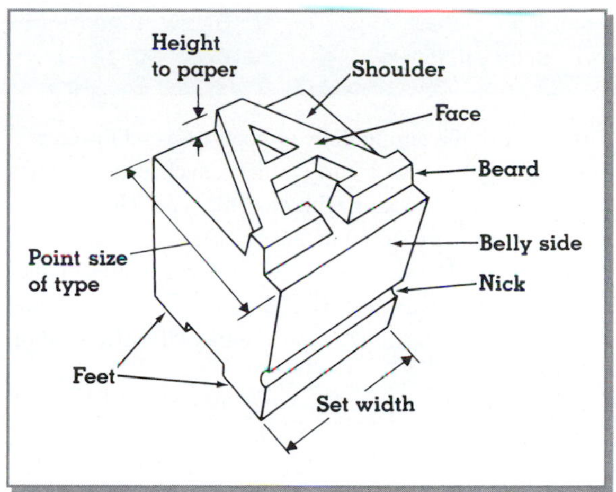

FIGURE 13.10 FOUNDRY TYPE CHARACTER AND COMMON TERMINOLOGY TO DESCRIBE ITS PARTS.

Notice that the point size refers to the size of the body height of the block, not the height of the face on the body; with the exception of point size, none of these terms apply to modern type specification. Computer versions of type size refer to the relative height of the type even though the terminology is referenced to this origin.

Type Size

Type size designation is related to the way type was originally produced on metal, or foundry, blocks (see Figure 13.10). The body or height of the block specifies the type size, although the actual letter on the block may not extend to the full height. The system of specifying type size in the United States and Britain is based on division of the inch into 72 parts called **points.** A point equals .0138 inch (.351 mm). Thus, 72 points essentially equal one inch. All type is measured in point sizes. Because the face of a particular type style may not occupy the whole height of the foundry body, the actual point size of the printed letter may be different from the nominal size given for a particular typeface.

Type size specifications based on foundry production of type are still used today, although we now make our selection of type size from software's toolbars or other menu options. The important point is that a given style and size of type is standard. The cartographer normally chooses type based on the actual size of the typeface and specifies the nominal type size as listed in the menu.

The designer must specify type that is large enough to be read easily. A number of studies have been conducted to determine the minimal size. Professional cartographers rarely use type as small as 4 or 5 points; safe practice is to set the lower limit at 6 points (.0828 in). This 6 point minimum guideline is based on the average map user's eyesight and the ability to discern the individual letters on printed maps. (Refer to the previous discussion on type discernibility.) Although the cartographer does not have control on how the map is used once it is distributed, the selection of type size should be made based upon the ultimate scale of the map. Although less critical on virtual maps where we can zoom in or out on a particular set of text, the printed map presents problems of the bowl letters filling in by the ink used in the printing process. It is up to the printing press operator to make sure that the proper amount of ink is applied, thus keeping the letters crisp and clean. Virtual maps present the same problem of the bowl filling in due to the low resolution of the monitor. Perhaps an 8 or even 10 point minimum would be better suited for virtual maps.

A guideline to follow is that no type on the map be smaller than 6 point when the map is produced at scale (see Figure 13.11).

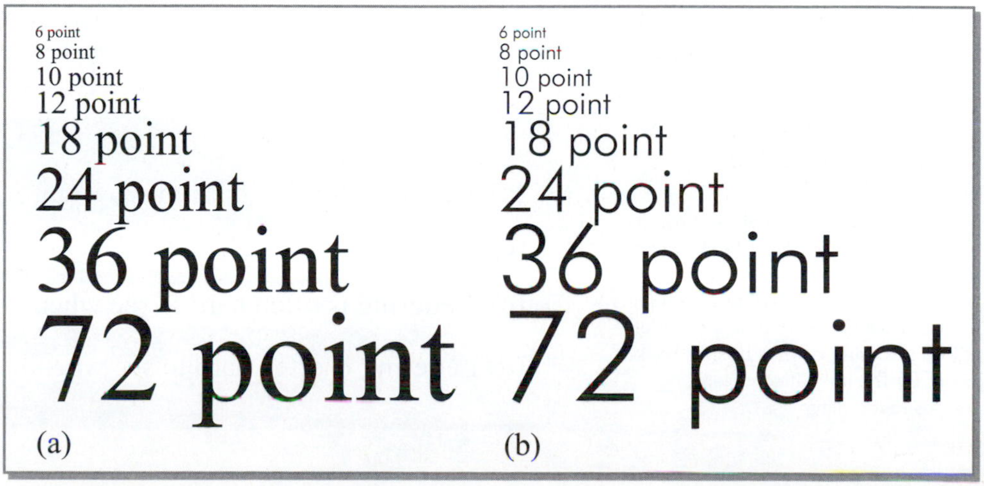

FIGURE 13.11 A SELECTION OF TYPE SIZES IN BOTH SERIF (a) AND SANS SERIF (b) FONTS.

The most common type size ranges between 6 and 18 point for page-size mapping.

TYPESETTING

The origin of the terminology of uppercase and lowercase lettering originates with early printers, such as newspapers of the Old West, working with movable type. The larger capital letters were stored in bins or cases on a second tier, or upper case. The small letters were stored in bins on the bottom tier, or lower case. This terminology caught on and has lead to our referring to type as either UPPERCASE or lowercase.

If the cartographer knows that the map is to be reduced in its final form by 50 percent, for example, the selection of 12 point type should be made as the smallest type on the map. Neither should the size be excessively large. The scale and complexity of the map, in either printed or virtual form, should dictate the range in type size on the map.

Type Form

The principal variants in a type family are **roman** versus **italic** and upper- and lowercase type forms. Roman is the basic, upright version of the design, and italic is a slanted version of its roman counterpart. The letters used to compose this sentence are of the roman form. Italicized lettering is created by either having the letters slope upward to the right, such as this *Century Schoolbook Italic*, or in a form that simulates handwriting, commonly referred to as cursive, as is found in this Script font. Modern graphics and word processing software provide the user with the option of type form via their formatting options. As will be discussed later in this chapter, the cartographer must determine the form of the type used on his or her map.

The second form of type is its case. The decision is whether to have the individual letters set in uppercase or lowercase (capital- or small-letter form) established by the rules of grammar. Such rules provide us with guidelines for using capital letters, such as in the first word of a sentence. The cartographer normally capitalizes the first letter of place names, or the entire word when trying to differentiate ordinal levels of importance. In earlier times when type was set manually, the printer had all the capital letters in bins above those bins holding noncapital letters. Thus, the terms "uppercase" and "lowercase" were used to designate the differences in this form.

Type Width

Type may be available in *regular, condensed,* or *expanded* versions, called **type widths.** This refers to the actual width of the individual letter. There may be situations where the cartographer is constrained by the amount of space available for placing a label while maintaining a constant point size. Another situation may call for stretching the letters or their spacing in order to create an expression of extended linear distance. The regular width is rarely identified since it is the normal width of a particular type face. However, many families of type include an indication of their width in the name of the type face. They will include terms such as *condensed, narrow, extended,* or *expanded* to provide you with a quick reference as to the form of the type.

Type Weight

Type weight refers to the relative blackness of a type created by the width of the lines that make up the lettering. It is customary to find three weights: normal, light, and bold. Just as with the width of type, font names often include the weight of the type in their name, such as Times Light, Standard, **ExtraBold.** Type weight can be created through software formatting options.

Type Color

Historically, type on a map has been set using black for all text color. Black text produces high contrast between the letters and the white background. This rarely produces legibility problems in itself. However, when black text is placed over an area containing a grayscale fill, the contrast can be lost. Figure 13.12 displays a gradient grayscale fill from light to dark. The black text becomes illegible when it is placed over the area with a level of blackness equal to that of the text (see Figure 13.12a). The reverse is also true if you select white text to increase the legibility in the darker area. In Figure 13.12b, the white text disappears when it is printed over the light background.

If the map contains a variety of base or background colors, including grayscale, it would be advisable to utilize a halo

(a) Lettering is often hard to see when using background fill
(b) Lettering is often hard to see when using background fill
(c) Lettering is often hard to see when using background fill
(d) Lettering is often hard to see when using background fill

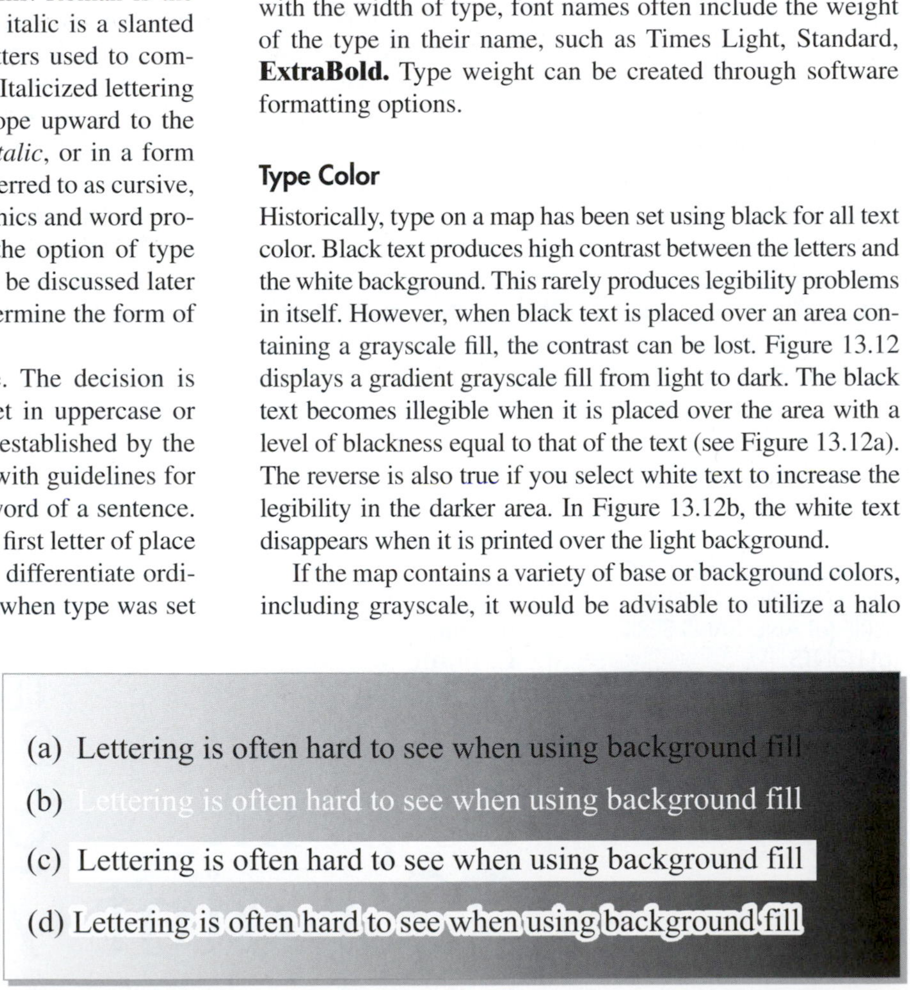

FIGURE 13.12 THE INTERACTION BETWEEN TYPE AND THE BACKGROUND FILL.

Black lettering creates a high contrast situation when printed on a light background but becomes difficult to read when the background darkens (a). (b) White lettering has the same result except in a reverse order. (c) Black text using a white box for background. (d) The halo effect of lettering increases readability.

effect where a silhouette of white is placed around the individual letters (see Figure 13.12d). A silhouette width can be selected that increases the blanking of the background to ensure that the letters are easily read no matter what the background density. The use of the halo effect is recommended over the creation of a rectangular box with the text placed inside the box (Hodler and Doyon 1984; see Figure 13.12c). The box may call undue attention to that individual text item. We advise the cartographer to consider all possible alternatives prior to its use.

If the background is very dark or all black, something that is not recommended—the use of white text—becomes an option. One example where this would be acceptable is in labeling the classic nighttime view of the United States or the world. The high contrast that the white letters have with the black background enhances the design of the map. Otherwise, GIS and mapping software allow you the option of selecting the color of the type associated with the layer of information being mapped. Certain map layers are associated with traditions in symbolization and lettering, such as red for transportation, blue for hydrologic features, and brown for contours. The cartographer can easily select the color of the type to be used on the map via the software's menu options. Care must be taken to select a color for type that has a high contrast from the background colors (see Color Plate 13.1). A dark color may be highly visible and legible in areas of lighter tones or grayscale but be lost where the background has darker tones or shades of gray. Here too the use of the halo effect is recommended to overcome these problems of contrast.

Care should be taken in selecting text colors when creating a noninteractive map. Certain bright colors, such as blue or red, are often associated with hyperlinks, and may create a false impression of interactivity to the map user. Yellow tends to be difficult to read when the font is quite thin or small in size. Care should be taken when using yellow for text.

Bright colors also have a tendency to distract from the overall design of all thematic maps. We recommend the use of black for most fonts except in such instances when the planned intent is to draw attention to the word or its position. The traditional black font helps to emphasize the thematic symbolization.

The cartographer is faced with a variety of decisions relating to the type being placed on the map. This is not a process to be taken lightly but should be done with forethought and based upon the overall design of the map. The graphic map provides its user with an understanding of the spatial variability of the information displayed. It is the type set on the map that conveys the specifics of historical, cultural, physical, or topographical information contained on the well-designed map.

Letter, Word, and Line Spacing

Strictly speaking, letter spacing and word spacing are not elements of typeface design. However, these features are extremely important, especially in the visual perception of letter pairs and words. Entire map projects can be weakened by poor spacing of letters and words; no other feature is so obviously incorrect at first glance. Practice and a critical eye are essential in providing correct spacing.

Letter spacing involves the appropriate distribution of spaces between letters (Tschichold 1966). Letter spacing is usually required for lines of type in all capitals but not in lines of lowercase letters. Letters are usually set solid (without letter spacing) in lines of lowercase. **Word spacing** is the proper distribution of spaces between words to achieve harmonious rhythm in the line. Words set in all capitals or all lowercase require word spacing.

Spacing of both letters and words is especially important in cartography because so much of the lettering is stretched across map spaces to occupy geographic areas. Some names cause little difficulty for the designer because they are set solid: town and city names and other labels applied to point phenomena.

The real culprits are capital letters that cause open spaces, such as A, J, L, P, T, U, W, and O. Notice that these contain few vertical strokes and have mostly oblique or rounded strokes. For some capital letter combinations, notably AV, AT, AW, AY, LT, LV, LW, and LY, **kerning** may be required. Kerning, also called **mortising,** is fitting letters closer together to achieve proper visual balance in letter spacing (see Figure 13.13a). According to the International Typeface Corporation, which copyrights and supplies many typefaces to third-party vendors, the 20 most frequently used kern pairs are: Yo We To Tr Ta Wo Tu Tw Ya Te P. Ty Wa yo we T. Y. TA PA and WA. This group suggests these generalizations (Haley 1991):

1. Commas, periods, and quotes almost always have to kern.
2. Cap and lowercase letters with outside diagonal strokes require kerning more often than not.
3. T, L, and P generally need to be kerned with non-ascending lowercase letters.

The use of all capital letters is employed in special instances, such as in country or state names or abbreviations, and in such instances where the design is to draw attention to a

(a) AVWJ (no kerning)
AVWJ (auto kerning)
AVWJ (optical kerning)

(b) Tracking at 0
Tracking at -25
Tracking at +100
Tracking at +200

FIGURE 13.13 LETTER SPACING AND WORD LENGTH CAN BE ADJUSTED BY (a) KERNING AND (b) TRACKING.
See text for discussion.

word or place. However, in such cases, kerning should be considered for the character pairs identified above.

Almost all GIS and graphics software provide the user with the option of modifying the spacing of letters. Such modification is referred to as **character spacing** or **tracking.** With character spacing, one can select the following examples in conjunction with a percentage value of adjustment: normal, expanded, or condensed. Tracking, which allows spacing between letters, is normally adjusted so that the letters are spread farther apart. The normal range options include settings between −100% and +500%, and 0 is the standard spacing. If you select a spacing value of 100%, an area equivalent to an entire letter width will be added between letters (see Figure 13.13b). Kerning is also available in these same software functions. One may select kerning, sometimes referred to as *optical spacing,* by simply checking the option box. Certain characters may be beyond the normal keyboard and are referred to as *glyphs*. A glyph is any font character and may represent letters of other languages, or symbols. Glyph spacing can also be accessed through the software text functions and can operate in the same manner as letter spacing.

Within these same software functions the spacing between lines of text can be adjusted. This spacing is referred to as **leading.** Paragraph text is normally set 12/14 or 12-point type on 14-point leading. This would provide two points of space between the top of one line and the top of the next line or more commonly 14 points of space between the bottom of consecutive lines (see Figure 13.14). If you are creating a narrative text box for your map, this is an important consideration. Care must be taken in selecting the leading as too much space—just as too little space—between lines can make the narrative difficult to read. Normally, leading that is two points larger than the type is adequate to provide balance in the design of the text. Leading can also be used to adjust the spacing between two line labels on a map. (The narrative that you are reading in this book is set 10 on 12 using the typeface of Times.)

> The first thing to realize is that the rhythm of a well-formed word can never be based on equal linear distances between letters. Only the visual space between letters matters. This unmeasurable space must be equal in size. But only the eye can measure it, not the ruler. The eye, not the brain, is the judge of all visual matter.
>
> *Source:* Tschichold, 1966, 29.

GUIDELINES FOR TYPE SELECTION AND PLACEMENT

The selection of map type and its proper placement have been developed mostly by tradition. There are few experimental studies that pertain solely to map design. This section presents the highlights of conventional practices and a set of recommended guidelines to assist the cartographer in making design decisions pertaining to the selection and placement of type.

The initial decision pertaining to the font to be used should be made taking into consideration the legibility of the type and typeface harmony. Legibility research traditionally deals with the characteristics of letter design and how they affect the speed and accuracy of reading. Because lettering on maps usually occurs in a much more complex environment (for example, the integration of text with the map background, colors, and symbolization), legibility is achieved not only by choosing the appropriate typeface but also by attending to good placement and adequate spacing of letters and words. Care and advanced planning help avoid problems.

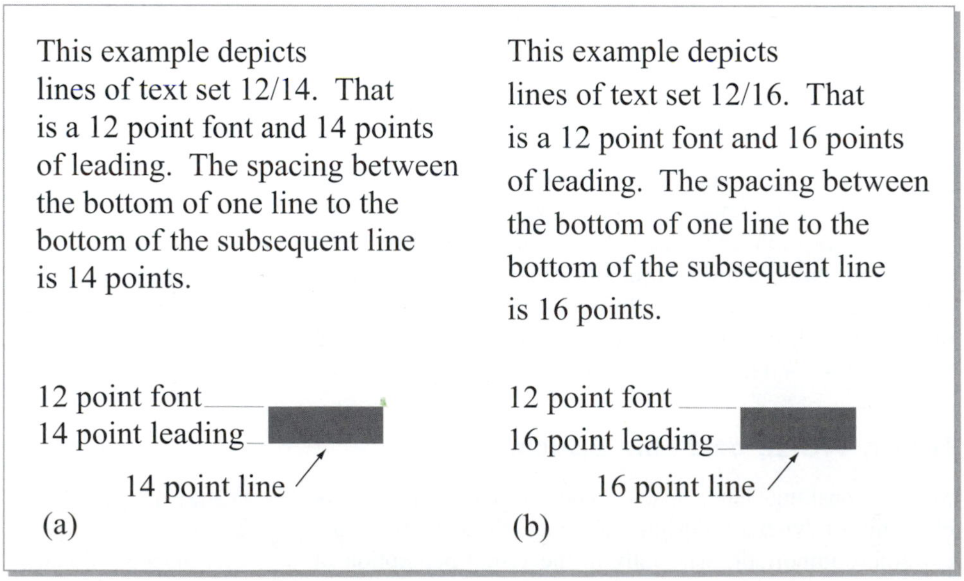

FIGURE 13.14 LEADING.
The spacing between the two examples depicts leading of 12 points and 14 points, respectively.

This example depicts
lines of text set 12/14. That
is a 12 point font and 14 points
of leading. The spacing between
the bottom of one line to the
bottom of the subsequent line
is 14 points.
12 point font
14 point leading
14 point line
(a)
This example depicts
lines of text set 12/16. That
is a 12 point font and 16 points
of leading. The spacing between
the bottom of one line to the
bottom of the subsequent line
is 16 points.
12 point font
16 point leading
16 point line
(b)

TypeBrewer, a Web-based program, allows for the experimentation of design using a variety of typographic alternatives (Sheesley 2006). Design options permit the manipulation of various elements for immediate evaluation by the cartographer. This program's interactivity provides the novice cartographer with a practical exercise founded on sound cartographic design principles.

The Use of Capital and Lowercase Letters

Good lettering design on the map can be achieved by contrast of capitals and lowercase. A map that contains only one form or the other is exceptionally dull and usually indicates a lack of planning. In general, capitals may be used to label larger features such as countries, oceans, and continents, and important items such as large cities, national capitals, and perhaps mountain ranges (Robinson 1966). Smaller towns and less important features may be labeled using mixed-case lettering.

One practice is to label features represented by point symbols with mixed-case lettering (see Figure 13.15a). Scale and hierarchy of importance are used in selecting the case of place names. At small scales, the city of Baltimore is represented by a dot and labeled in lowercase and smaller size. When the map scale increases and the location is represented as an area, the type remains mixed case while the size is increased. As Robinson indicated, state or national capitals may be set using all uppercase lettering. An exception is made for country names, which are normally set in uppercase. Country names, (continents and large oceans as well) are displayed using all uppercase lettering at any scale (see Figure 13.15b). This selection is maintained whether or not the place name fits within the country's border. One must consider the fact that many items being labeled are a part of a feature class, such as lakes or countries. A consistency in labeling the same feature class is important no matter where the text is placed in reference to the feature.

The Placement of Lettering

Careful lettering placement enhances the appearance of the map. There are several conventions, supported by a few experimental studies. The placement of type on a map should be carefully thought out and planned, using a series of guidelines presented here. Table 13.4 provides for a summary of these guidelines. Most decisions regarding lettering placement fall into one of the following categories: labeling point, line and areas, and the placement and design of titles and legend materials. Although guidelines for placing type on the body of a map have developed over time, Arthur Robinson (1960) and Erwin Raisz (1962) in their early landmark textbooks on cartography were among the first to draw attention to the need for guidelines for the positioning of map type. Some of the guidelines discussed in this section are attributable to their work.

Point-Symbol Labeling

Most professional cartographers agree that point symbols should be labeled with letters set solid (no letter spacing). On small-scale maps, these should follow the line of the parallels (if apparent). They should be set horizontally on large-scale maps (Imhof 1975; see Figure 13.16). Over thirty years ago Yoeli (1972) established a priority for the placement of labels adjacent to point symbols (see Figure 13.17). His system of rules was not based on empirical work but on his own preferred judgments. Imhof's rules, though less strict, agree with Yoeli's, especially regarding the most preferred location (up and to the right). Others have developed their own priority schemes, some similar

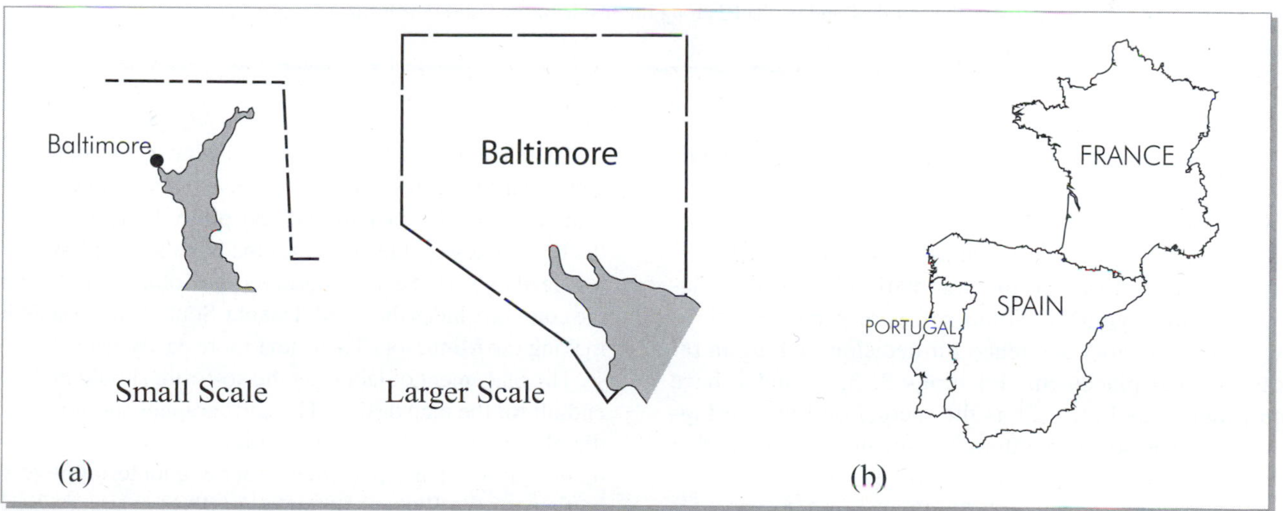

FIGURE 13.15 THE USE OF CAPITAL AND LOWERCASE LETTERFORMS.
Point phenomena (a) are customarily labeled with mixed case. At larger scales where inside lettering can be accommodated, mixed case is also suggested at a larger point size. Because of space problems, areal phenomena such as countries may have their labels either inside or outside, as in (b). In these cases, the convention of using capitals is retained.

TABLE 13.4 SUMMARY OF TYPE PLACEMENT GUIDELINES

The following guidelines for creating page-size maps will also apply to most map layouts.

General:
- No text smaller than 6 point
- Water features are labeled using italics
- Water features are labeled in blue when color is used
- Use of bold type should be carefully considered
- Type size indicates relative importance
- Use capital letters for names of countries, oceans, and continents
- Be consistent with lettering within same feature class
- Normally, a single font is used within the body of the map with a different font for titles, legends, and so forth

Point Symbols:
- Do not letter space
- Text, not used in conjunction with a grid, should be set horizontally
- Placement of label should be close to the symbol represented
- Place labels according to placement strategy of Imhof and Yoeli
- Labels should not overlap
- Text should be set parallel to the bottom or right-side neatline
- Lettering should be read from the lower quadrant viewing position
- Never have upside-down letters
- Place text all on land or all on water
- Place names on the side of the river where the feature is located
- Use call outs sparingly

Line Symbols:
- Water features set in italics
- Labels placed above the linear feature
- Name of the feature and its category (for example, Columbia River) may be separated in order to show extent
- Disassociated letters should be set along a constant arc

Area Symbols:
- Place the label inside the area where possible
- Labels within the same feature class may be inside larger areas but outside smaller areas; for example, lakes
- Labels should be placed with regard to the geometric extent of the area

Titles, Legends, Source, and Authorship:
- Titles are initially sized one-and-one-half to twice the size of the largest feature text on the body of the map and then modified for map balance
- Legend captions set at one to one-and-a-half the size of the largest feature
- Source and authorship text is the smallest and is placed along the inside of the bottom neatline

Source: After Robinson *et al.* 1995.

and some dissimilar to both Yoeli and Imhof but none universally adopted.

This placement priority has been established by most GIS and mapping software. Imhof's Position 1 has been given the first priority in the automatic placement of labels for point symbols. Some software group positions into a three-category sequence for decision making in regards to label placement. Positions 2, 3, 5, and 7 have been grouped and utilized as the second option (see Figure 13.18). The software attempt to locate the label in that numerical sequence until an available position is encountered. The third category includes Positions 4, 6, and 8 for the final option. The automatic placement of labels sequences through these position options until such time that the label can be placed without overwriting other labels or symbols. If such overlap occurs, the label may not be placed.

There are times when the label may have to be placed over a boundary line. In such a case, create a background box the same color as the background and place the text in front of the box. That way, the box hides the boundary and makes the text readable. This can be observed in Color Plate 8.1 where the color box hides the North Dakota-South Dakota state line, making the Minnesota Twins text more easily read.

The placement of labels by the software should not be the end-all for the map design. The cartographer should evaluate the placement of all type and utilize the user-defined placement options if the placement seems counterintuitive to a particular map. Wu and Buttenfield (1991) examined Yoeli's, Imhof's, and others' guidelines for automated type placement as applied to the map as a whole. They state:

> *The quality of name placement is not determined by the label's position alone, but by the integral pattern of map features and*

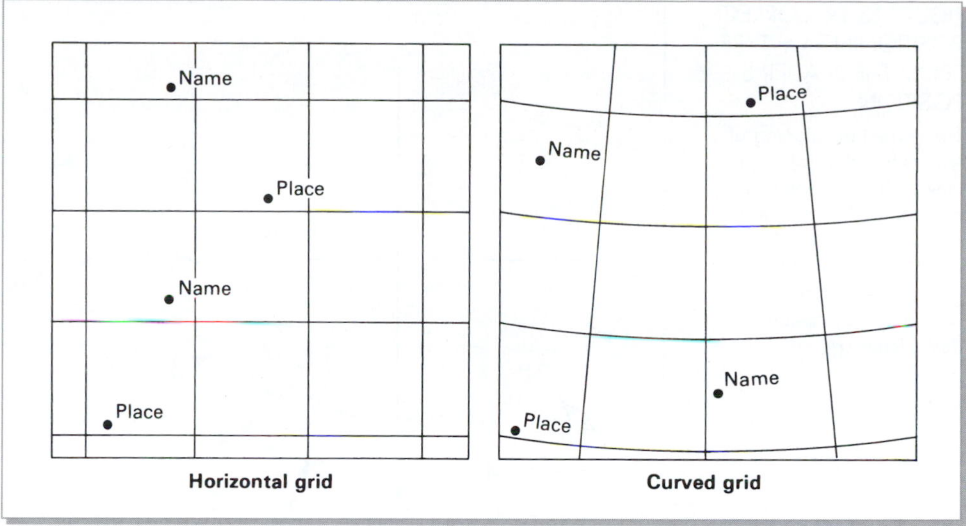

FIGURE 13.16 LETTERING PLACEMENT CONVENTIONS AND THE MAP GRATICULE.

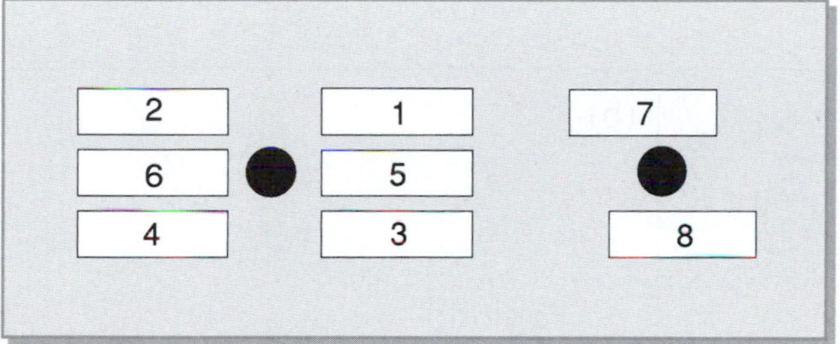

FIGURE 13.17 POINT FEATURE LABEL LOCATION PRIORITIES, AFTER YOELI.

Yoeli's priority for label placement has been used for automated labeling techniques in GIS software.

> *name placement components, e.g., the label's position, the distance from the symbol to the label, letter font and font size. That is, the optimization of a label's position must be incorporated within the graphic context of the entire map. (12)*

Automated label placement has come a long way in GIS and mapping software development. However, the cartographer's sense of design and balance should always outweigh the default placement options of the software. Some auto-labels must often be converted to graphics or be placed interactively in order to move them. In other words, the cartographer is encouraged to place the labels interactively and to evaluate the placement of all labels whether placed by the auto-label function or by a manual entry process.

Labels should be placed so that they parallel the bottom or right-side neatline (see Figure 13.19). If the map space is viewed as a series of quadrants centered on the body of the map, this positioning would have the map reader's viewing position at locations 1 or 2. If either of these viewing options is not viable, the second choice would be to have the label angled in a straight line outward from the point symbol so that the reading position is from within quadrant IV, or location 3. The last option for placing a label, if none of the first options work, is to place the text in a straight line outward from the point from a reading position in quadrant III. If the label is placed so that it is read from the left-side neatline, the label position should be altered so that it is read from location 2. The placement of any portion of a label that requires a viewing position from within the upper quadrants (I and II) or the left-side neatline is considered as *upside-down* type and must be avoided at all cost.

Names of ports and harbor towns should be placed seaward, if possible (Imhof 1975). Names located along coastlines or rivers should never be placed so they overlap the coastline. They should either be all on the land or all on the water (see Figure 13.20). Likewise, town names should be placed on the side of the river on which the town is located.

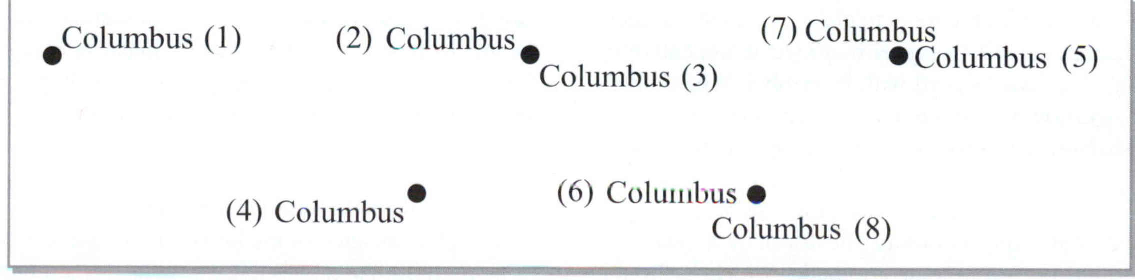

FIGURE 13.18 EXAMPLES OF YOELI'S PREFERENCE SEQUENCE OF POINT SYMBOL LABELING.

Placement of point symbol labels is shown in conjunction with their order of preference by Yoeli. The five positions displayed across the top are identified by him as the most preferred.

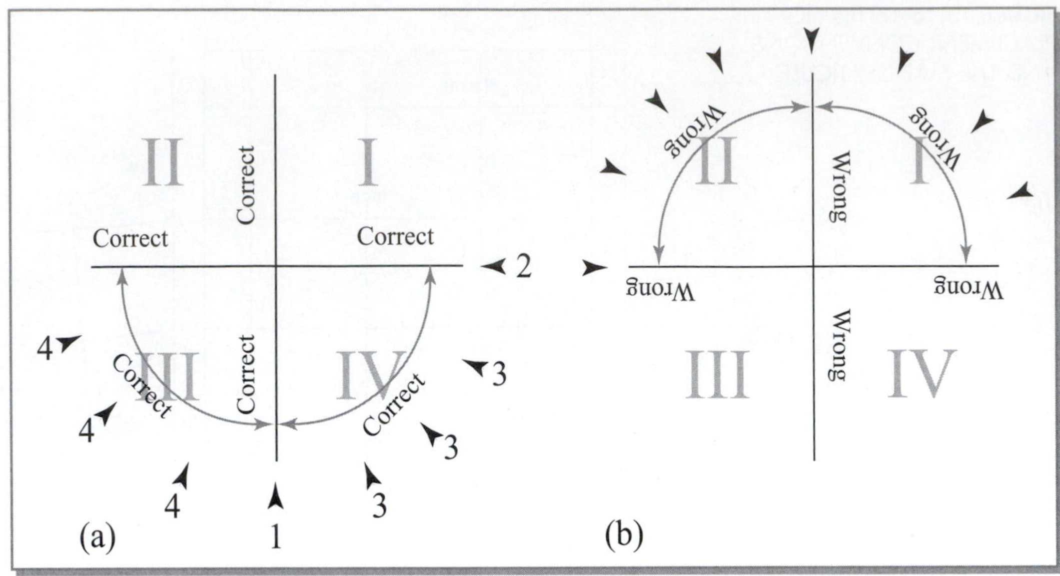

FIGURE 13.19 CORRECT POSITIONING OF TYPE FROM THE READER'S POSITION.
The correct positioning of type is found in (a). Type viewed from the bottom and right-side neatline is preferred, followed by quadrants IV and III. All type placement in (b) is considered as upside-down lettering.

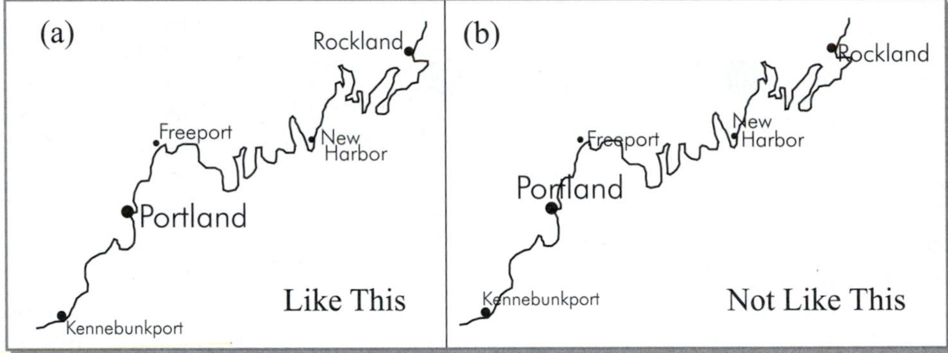

FIGURE 13.20 POSITIONING OF POINT LABELS IN ASSOCIATION WITH LINEAR FEATURES.
See text for discussion.

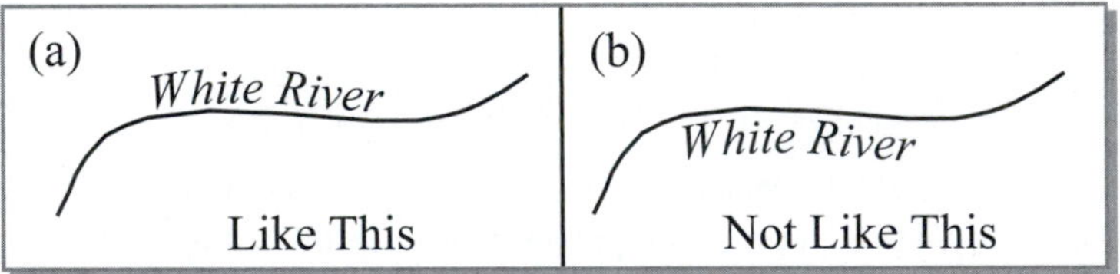

FIGURE 13.21 LABEL PLACEMENT FOR LINEAR FEATURES IS ON TOP OF THE FEATURE AS IN (a), NOT BELOW THE FEATURE (b).

If the river flows through the town, place the name on the side of the river with the larger population or political importance. This requires personal knowledge of the place being labeled.

Call Outs. On rare occasions the labeling of features may be very congested, leaving little room to place an additional label. When this happens the cartographer may elect to use **call outs** where the label is disassociated with its symbol. When this is done, the label should have a line connecting the word to the point symbol. Normally, this line is of a thickness equivalent to the downstroke of the font being used and may be broken if other labels must be crossed (refer to Color Plate 8.1).

Some call outs may occur in the form of a text box where informational comments are made. This box should also be connected with a line to the symbol or location to which it applies.

Linear Feature Labeling

Linear features on thematic maps include rivers, streams, roads, railroads, streets, paths, airlines, and many linear quantitative symbols. The general rule is that their labels should be set solid (no letter spacing) and repeated as many times along the feature as necessary to facilitate its identification. The type used for any hydrologic feature, whether linear or area symbolization, should be italicized. This is a long-standing tradition that has been used to differentiate water features from cultural features. Along with the use of italics, the color blue is also a tradition that should be used if the final map product is in color. Either way, the italic lettering identifies it as water.

The ideal location of the label for a linear feature is above it, along a horizontal stretch if possible (see Figure 13.21). Placement of the feature's label below the line symbol creates the impression that the name is hanging from that feature.

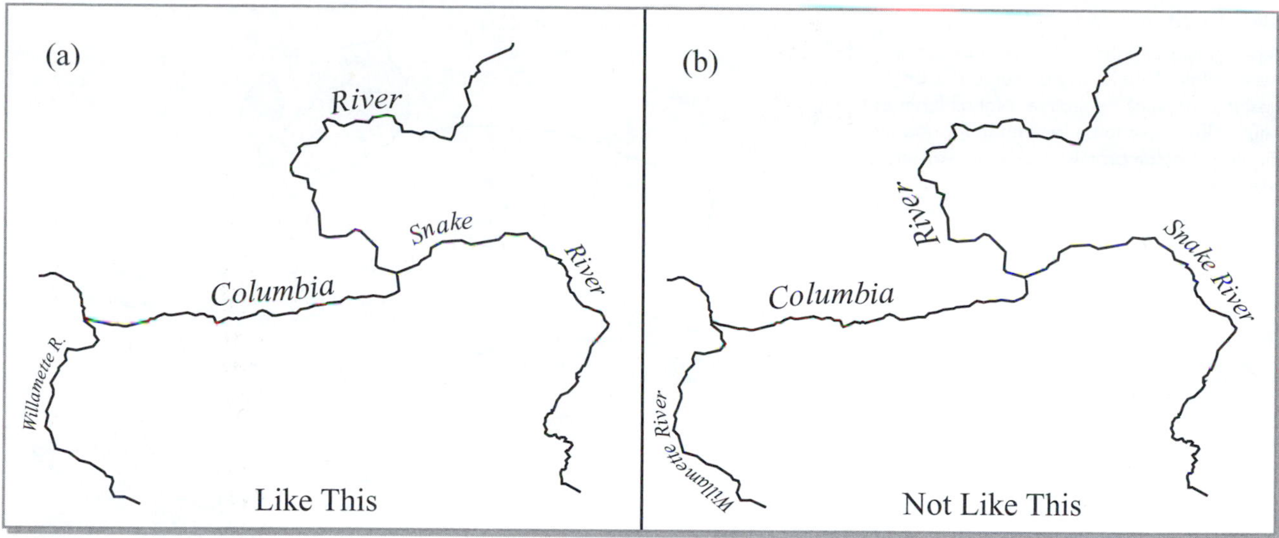

FIGURE 13.22 LABELING LINEAR FEATURES.

The separation of the name components allows for the proper interpretation of which lines are the main trunk and which are tributaries, as in (a). (b) lacks this separation, plus uses upside-down lettering.

This should occur only if other label positions on top of the line are not available. Do not crowd the label into the feature. Room must be reserved for lowercase descenders, if any. Most GIS and mapping software provide the user with the option of adding text to a curve, such as a river. Artistic drawing packages excel at providing even more text-curve options. Care should be taken not to have the individual letters bend with each meander, creating a mishmash of letters slanting in all directions. The software allows the user to create a spline curve of constant arc on which the river name is placed. This option allows for the gentle manipulation of the curvature of the label name.

The separating of the proper name of a river, for example, from its category (river) allows the cartographer to create the linear extent of the feature. The separation can be made so that the trunk of the feature can be followed even though tributaries may flow into it. The placement of the name of the feature may be separated by a modest distance but not so much that the relationship between the two is no longer made (see Figure 13.22). "Letter spacing should be 'normal' and visually uniform in each instance. Word spacing must be such that the component words of names (e.g., rivers, ridges, and valleys) are not so far apart that their relationship is not evident" (Wood 2000, 8). If the linear extent of the feature is excessive, the labeling of the feature may occur in more than one location. Care should be taken not to place a name too close to the neatline if that feature flows beyond the map dimensions.

Area Feature Labeling

If a space on the map is large enough to accommodate lettering completely inside it, it may be designated an areal feature. Examples include oceans and parts of oceans, large bays, lakes, continents, countries, states, forests, cities, and other geographic subdivisions. Label placement within large areas can parallel the neatline or be angled if the feature is elongated. Not all features within a category, such as lakes, require placement of labels within the area (see Figure 13.23). Not all lakes are big enough to put their names within their boundary. Some lake names may appear outside of the area and some areas may be too small to label but will still be mapped. Just as with linear features of large linear extent, letter-spaced or disassociated letters used to label large-area features should be placed on an imaginary curved line that maintains a smooth constant curve for the entire label length. The use of such curves shows that the letters were placed with a plan and not by mistake.

Labeling physical features with linear extent, such as a mountain range, should be done by using letter spacing or letters that are disassociated from each other, that is, with additional spacing between letters so that they appear to be standing alone on the feature (see Figure 13.24). This allows the spacing of the individual letters to draw the reader's eye to the elongated dimension of the feature. When *disassociated letters* are used they should never occur in a straight line but rather should be placed along a curve with a constant radius of arc.

Placement and Design of Titles and Legends

Map titles, legends, and other explanatory information, like any other graphic elements, need to fit into the whole plan for the map. Titles are generally the most important intellectually and should therefore be largest in type size. Any subordinate titles should have somewhat smaller point sizes. Legend materials are important elements on the thematic map but are subordinate intellectually and visually. Their lettering size should reflect their position in the hierarchy. Map sources, explanatory notes, and the like are the smallest in point size.

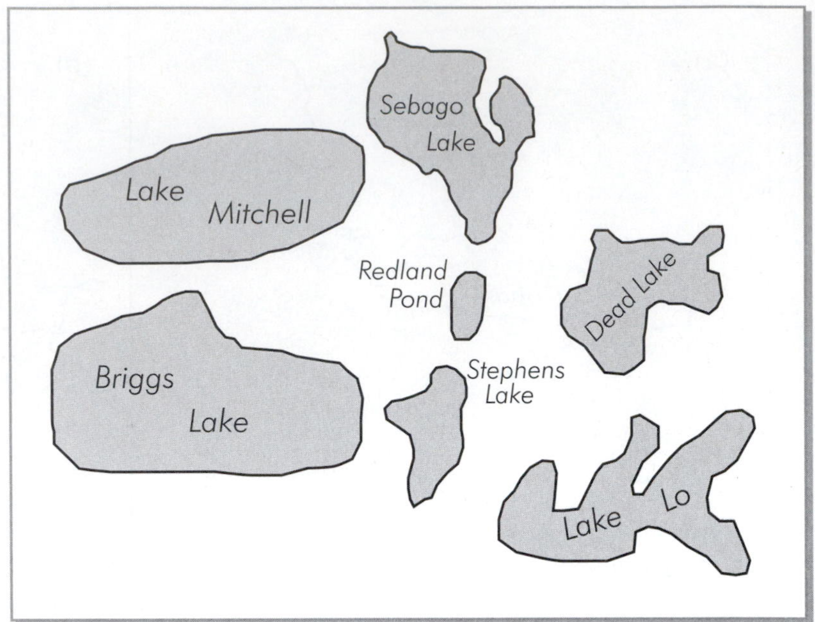

FIGURE 13.23 AREAL LABELS.
Features should be labeled inside of the area symbol where possible. Labels may be curved or angled to suggest the shape of the feature. Not all features within a feature class have to be labeled in this manner. Positioning is a compromise based on varying feature size.

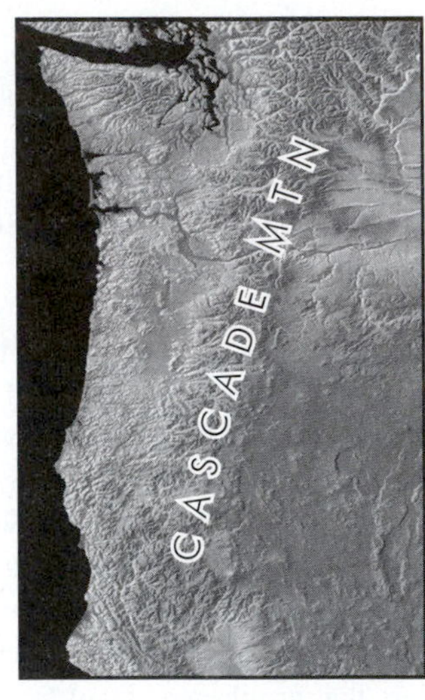

FIGURE 13.24 LABELING AREA FEATURES WITH LINEAR EXTENT.
Some features cover large areas and possess linear trends. When labeling features like mountain ranges, deserts, or even water bodies, extend the spacing of letters beyond letter spacing and always set the letters on a slight curve.
Source: National Geophysical Data Center, 2008.

The font used for titles and legends may be different from that used within the body of the map. While sans serif fonts work best within the map, serif fonts provide the reader with a more elegant presentation of the map's topic. As a general rule of thumb, the type size used in the title should be one-and-a-half to twice the largest point size used within the map body. This will be determined by the amount of space free for title placement and the overall size of the map space. This rule is appropriate for maps prepared for traditional page-size presentation. Similarly, legend titles and numerical information should be no greater than one-and-a-half the size of the text size within the map. For large format poster displays, this size may be expanded as needed. In a paper, report, or article, the title may appear as a figure caption.

The positioning of the title on the map is determined by the overall balance of the map. If ample space is available beyond the limits of the map body, the title can be placed in a position that is in harmony with the shape of the mapped area. Frequently, we elect to place the title at the top of the map. However, other positions are acceptable in order to maintain balance. When the body of the map includes all the physical space within the neatline of the page, the title should be placed in a location where it does not cover important features or spatial data characteristics. When such a location is identified, the title is placed overlying the map space so that it sits as a layer above the map space, thus establishing its place in the figure-ground association. We caution that this is a worst-case scenario and should not be attempted until considering all other options, such as a scale reduction in your map. The type color should allow for good contrast between the text and the background. If necessary, the type can be generated using the halo effect. White lettering is not as effective as black or darker colors.

Legends titles should work in combination with the map title. Do not be redundant with information already contained in the title. The legend caption should simply identify the features of data used in the thematic map. For example, if the data are numerical, the legend caption may simply state "Percent" or "Bushels Per Acre." If nominal land-use classification of the data is used, then the caption should simply read "Land-Use Categories." The legend caption should be placed so that it is centered above information, using both the symbol and the categorical information as the width in which to center the text. The positioning of the legend on the map should take a secondary position when compared to the title. It serves as an explanatory tool in assisting the map reader in interpreting the map.

Care should be taken when generating the title and legend caption when using GIS and mapping software. Frequently, the software uses the header in the data table as a default portion of the title or figure caption. Rarely is this an acceptable component for either of the features. Be sure to modify their content to communicate the topic being presented to the map user. The file name or header text does not serve this purpose. The use of the words "Legend" or "Map of" should be avoided. Such usage brings into question the user's ability to identify these elements and their purpose.

Scales and North Arrows

Text on north arrows and scale bars is coupled with the symbol as a graphic element of the map (see Chapters 1 and 13). The type used for the scale and north arrow may be either of the fonts used in labeling or titling the map. These type elements, like their symbols, are normally a smaller, less obtrusive point size. The numbers are normally placed above the scale bar, each centered above the mark indicating their position. The numbers units of measurement for the map may be placed either centered beneath the linear scale or to the right of the larger numerical value as if it were a part of the number set. The vertical thickness of the linear scale should be modest at best. The scales in Figure 13.25b are very thick and become distracting to the map reader. A scale should have at least one intermediate break whether it is labeled or not.

The purpose of the north arrow is to provide a sense of orientation when viewing the map. There are a number of north arrow options available in the GIS and mapping software. Care should be taken so that the north arrow is not bold or ostentatious. The selections in Figure 13.26b are bold and distracting whereas the options in Figure 13.26a present clean, crisp symbols that clearly identify the direction of north. The letter "N" should be placed at the top of the symbol in the direction of north.

These two elements are of lesser importance in the overall hierarchy of the map. In many thematic maps, they are not used at all. Their position on the map as well as the type size used should not draw attention to either element (see Figures 13.25 and 13.26).

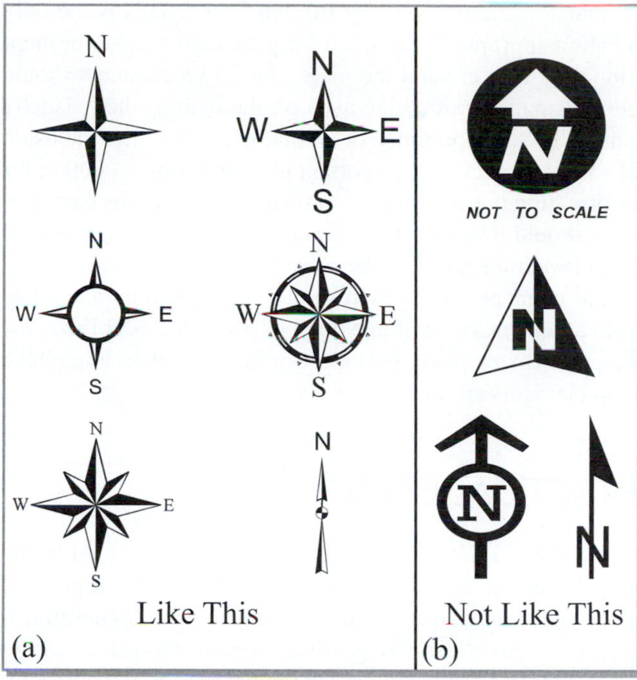

FIGURE 13.26 NORTH ARROWS.

North arrows can be somewhat decorative and still communicate the map's orientation. The letter "N" should always be at the outer extent of the symbol (a). North arrows are not intended to distract the map reader with by their bold design (b).

Source and Author Information

A brief indication of the source of the data used in the map and the authorship or copyright of the map should be placed at the very bottom of the map. The source statement is normally placed on the lower left portion (or corner) of the map, using a very brief indication such as "Source: U.S. Census of Population 1990." The authorship statement is then placed in the lower right portion of the map as symmetrically as possible on the same line as the source statement with a statement

FIGURE 13.25 LINEAR SCALE SELECTION.

(a) Linear scales that are relatively thin are less intrusive to the total map design than are bold, distracting ones (b).

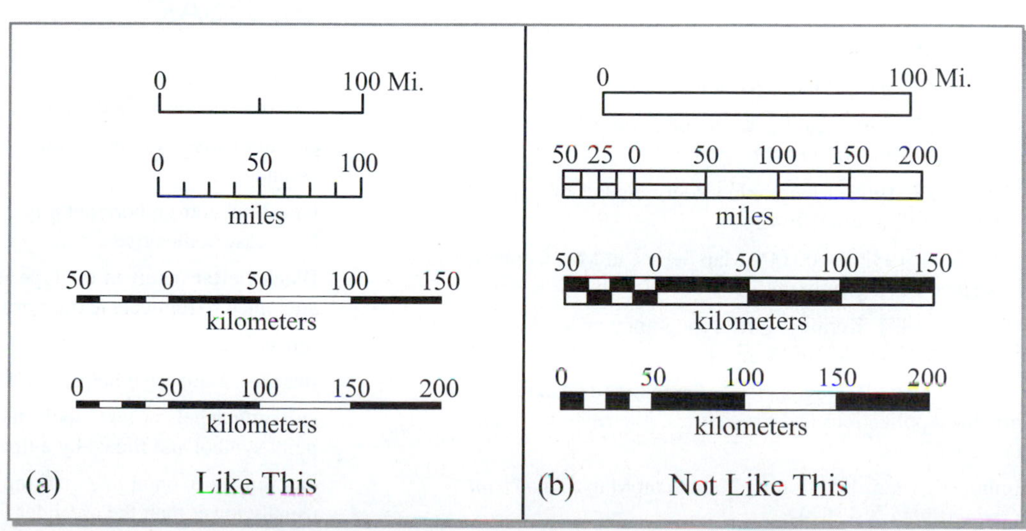

similar to "Cartography by Borden Dent." This is the only location appropriate for identifying the cartographer or mapping firm who created the map. These two elements should be the smallest point size used on the map as they visually carry a lesser importance of all map elements. The inclusion of source, however, is important to assure proper citation for information not collected by primary research. The cartographer should take credit for her or his graphic design but not for information supplied by others.

All elements of the map should be incased by a neatline with a line thickness of one point or less. This establishes the frame of the map and provides for the overall area in which map elements are in balance.

PRACTICAL EXAMPLE

Color Plate 13.1 is an example of a map produced in the private sector. It is shown because it displays a number of trade-offs that frequently occur when you are generating a map for a customer. Congestion is one of the primary concerns that the cartographer has to deal with as the customer wants as much "stuff" on the map as possible. They may want every street to be named, even if it means going well below the 6-point type minimum. Street length and name length also produce problems. Some streets may be only a block in length, but the name may contain multiple words that make it impossible to fit in the space provided. Symbols representing locations important to the customer, such as their library or city hall, are exaggerated in size beyond what would normally be considered acceptable. The interaction between symbols, labels, and color is often difficult to manage, and trade-offs result in the design process. This map is a general reference map, and very few thematic maps will have as many "rules" competing at the same time.

REFERENCES

Biggs, J. 1954. *The Use of Type*. London: Blandford Press.

Doughty, M. 2006. Mike's Sketch Pad. www.sketchpad.net

Fairbairn, D. 1993. On the Nature of Cartographic Text. *Cartographic Journal* 30: 104–11.

Gardiner, R. 1964. Typographic Requirements of Cartography. *Cartographic Journal* 1: 42–44.

Gilman, G. 1982. 1:100,000 Map Series, in Map Contemporary. *American Cartographer* 9.

Haley, H. 1991. Kerning: Fine Typography or Marketing Hype? *Upper and Lower Case* 18, no. 1.

Hodler, T., and R. Doyon. 1984. Spreads and Chokes: Cartographic Applications and Procedures. *American Cartographer* 11 (1): 63–67.

Imhof, E. 1975. Positioning Names on Maps. *American Cartographer* 2: 128–44.

Keates, J. 1958. The Use of Type in Cartography. *Surveying and Mapping* 18:75–76.

Lawson, A. 1971. *Printing Types: An Introduction.* Boston: Beacon Press.

Nelson, R. 1972. *Publication Design*. Dubuque, IA: William C. Brown.

Raisz, E. 1962. *Principles of Cartography.* New York: McGraw-Hill.

Robinson, A. 1960. *Elements of Cartography.* 2nd ed. New York: John Wiley & Sons.

———. 1966. *The Look of Maps: An Examination of Cartographic Design*. Madison: University of Wisconsin Press.

Robinson, A., J. Morrison, P. Muehrcke, A. Kimerling, and S. Guptell. 1995. *Elements of Cartography.* 6th ed. New York: John Wiley & Sons.

Shaikh, A., B. Chapparro, and D. Fox. 2006. *Perception of Fonts: Perceived Personality Traits and Uses.* http://psychology.wichita.edu/surl/usabilitynews/81/PersonalityofFonts.htm

Sheesley, B. 2006. http://www.typebrewer.org

Sinclair, J. 1999. *Typography on the Web*. San Diego: AP Professional.

Spencer, H. 1968. *The Visible Word.* New York: Hastings House.

Tinker, M. 1963. *Legibility of Print*. Ames: Iowa State University Press.

Tschichold, J. 1966. *Treasury of Alphabets and Lettering*. New York: Reinhold.

Updike, D. 1966. *Printing Types, Their History, Forms, and Use: A Study in Survivals.* 3d ed., 2 vols. Cambridge: Harvard University Press.

Wood, C. 2000. A Descriptive and Illustrated Guide for Type Placement on Small Scale Maps. *Cartographic Journal* 37 (1): 5–18.

Wu, C., and B. Buttenfield. 1991. Reconsidering Rules for Point-Feature Name Placement. *Cartographica* 28 (1): 10–27.

Yoeli, P. 1972. The Logic of Automated Map Lettering. *Cartographic Journal* 9: 99–108.

GLOSSARY

ascender letter stroke extending above the *x*-height or body of a lowercase letter, as in h, d, b

ascender line horizontal line marking the maximum extent of all ascenders

base line bottom horizontal line from which all capital letters and lowercase bodies rise

Black Letter name of the type style first used in movable type; attempted to replicate textura or manuscript forms

bowls rounded portions of such letters as o, b, and d

bracket a fillet that helps join serifs to the main stroke

call outs labels or text blocks that are disassociated from their point symbol and linked by a line

cap line horizontal line marking the height of all capital letters; usually lower than the ascender line

character spacing the selection of type so that the letters are spaced normal, expanded, or condensed, thus altering the length of a word

counters partially or completely enclosed areas of a letter

descender letter stroke extending below the *x*-height or body of a lowercase letter, as in y, g, q

descender line horizontal line marking the maximum extent of all descenders

display-decorative special typefaces used in art and advertising; not ordinarily useful in cartography

display types type sizes not ordinarily used in running text; usually greater than 12 points

family a series of lettering styles all related to the basic style; a family contains variations of the basic style in weights, widths, roman, and italic

font a complete set of all characters of one family and design of a typeface

italic slanted version of a type style; minor letter form differences may exist between an italic and its roman counterpart

kerning fitting letters closer than in the ordinary (geometric) arrangement in order to achieve a good visual result

leading spacing between horizontal rows of continuously running text

letter spacing additional horizontal spacing placed between letters; required when all capital letters are to be set but optional in lowercase

loop rounded portions of lowercase descenders such as j, g, and y

Modern type design developed in the early eighteenth century; these styles have a pronounced vertical shading that distinguishes them from earlier forms

mortising same as kerning

Oldstyle early type designs that did not try to replicate manuscript styles; Venetian (Italy), Aldine-French (mostly French), and Dutch-English are three variations of Oldstyle

point unit of measurement for type; 72 points to the inch (0.138 inches per point)

roman basic, upright version of a type style

sans serif a letterform containing no serifs

script-cursive type style that replicates handwriting; not ordinarily useful in cartography

serif finishing stroke added to the end of the main strokes of a letter

set solid no letter spacing

textura early manuscript alphabet form common at the time of Gutenberg's first printing of the Bible

tracking the process of modifying the spacing between characters in words or a block of words

transitional type design that bridged the gap between Oldstyle and Modern; Baskerville (English) is the best example of period

type discernibility perception and comprehension of individual words not set in running text; perhaps of greater theoretical use in cartographic design than type legibility

typeface particular style or design of letterforms for a whole alphabet

type legibility type design characteristics that can affect ease and speed of reading

type personality informal way of describing the impressions that type can elicit, for example, dignity, power, grace, precision, and excitement

type weight relative blackness of a type: normal, lightface, boldface

type width different width of a letter design, such as condensed or extended

Univers a popular sans serif typeface designed by Adrian Frutiger in the 1930s

word spacing space between words set horizontally

***x*-height** height of the body of lowercase letters

14 PRINCIPLES FOR COLOR THEMATIC MAPS

CHAPTER PREVIEW The application of color to the thematic map is one of the most exciting aspects of cartographic design, yet perhaps the least studied by cartographers. Color is produced by physical energy, but our reaction to it is psychological. Color has been found to have three dimensions: hue, saturation, and value. Our perception of color is influenced by its environment and by the subjective connotations we attach to colors. Map readers have preference for certain colors, but these change throughout life. Our understanding and use of color, especially in color research, are enhanced by learning how color can be specified. Color theorists may use either additive or subtractive color theories and the color models of HSV, RGB, CIE, CMYK, or the Munsell specifications. Cartographic designers are aware that colors can function in certain ways in design. Certain color conventions guide the cartographer, especially in quantitative mapping such as on choropleth maps. Design strategies to achieve figure and ground, the proper degree of contrast, and color harmony improve the use of color on the thematic map. ■

Introducing color into the design of a thematic map can be both exciting and troublesome. On the one hand, color provides so many design options that designers often quickly seize the opportunity to include it. Yet the inclusion of color invites many potential problems. Two individuals may view the same color but perceive it differently. Computer monitors attempt to produce the same color but generate colors with slight variations because they have various settings, such as video card resolution, color calibration, and the ambient light conditions that may impact the color displayed. The paper map printed in color will appear different when printed on a color laser jet printer than the way it appears on an ink jet printer. The quality of the paper on which the map is printed can also have an impact on the way the color is perceived by different users. Color is also affected by the ambient light conditions in which the map is viewed. In sum, the designer can never be totally certain as to how the reader(s) will respond to the color in a map. However, if allowed to choose, most map designers choose color mapping because of its inherent advantage of greater design freedom.

Color in thematic mapping is perhaps the most fascinating and least understood of the design elements. Color is subjective rather than objective:

> *The concept of color harmony by the color wheel is an objective conclusion arrived at by intellectual activity. The response to color, on the other hand, is emotional; thus there is no guarantee that what is produced in a purely intellectual manner will be pleasing to the emotions. Man responds to form with his intellect and to color with his emotions; he can be said to survive by form and to live by color* (Sharpe 1974, 123).

Another problem in dealing with color is that it is difficult, if not impossible, to set color rules. Certain standards for color use have been adopted for some forms of mapping, notably on USGS topographic maps and maps of other national mapping programs. No standards or rules for color use, except for a few conventions, exist for thematic maps. This situation is getting greater attention as the creation of color maps viewed on-screen can be designed from a palette of thousands of colors. On the other hand, rigid rules can also be restrictive in design.

Color is a complex subject and can be studied in many different ways. Physicists, chemists, physiologists, psychologists, philosophers, musicians, writers, architects, and artists approach color from different perspectives, with distinct purposes (Itten 1970). The physicist looks at the electromagnetic spectrum of energy and how it relates to color production. Chemists examine the physical and molecular structures of **colorants,** the elements in substances that cause color through reflection or absorption. The physiologist treats the mechanisms of color reception by the eye-brain pathway, and the psychologist deals with the meaning of color to human beings. Finally, the artist works with aesthetic qualities of color, using information gained from the physiologist and psychologist.

This chapter introduces the characteristics of color that are important in thematic mapping: color perception, color specification systems, cartographic conventions, and design strategies. The focus throughout is on the psychological and aesthetic aspects of color for both the maps viewed on-screen and in printed form.

LIGHT AND THE COLOR SPECTRUM

Light is that part of the electromagnetic energy spectrum (EMS) that is visible to the human eye (see Figure 14.1). This radiation spectrum is characterized by energy generated by the sun descending on us at different wavelengths. These wavelengths vary from very short (10^{-12} cm) to very long (10^5 cm, or 1 km). All visible light comprises a very small portion of this spectrum and varies from 400 nm (nanometers—a billionth of a meter, or 10^{-9} meters) to about 700 nm. The composite of all visible light wavelengths is referred to as *white light* and is colorless. **Color,** however, is simply light energy at different places along this visible light portion of the electromagnetic spectrum. When our eyes detect light energy at approximately 750 nm in wavelength, we see red; when we detect wavelengths at 350 nm, we see violet.

We see light that is either emitted by a source or reflected from an object. The sun is, of course, the source of the EMS and the source of visible light. Similarly, incandescent light fixtures, computer monitors, television screens, and other electronic devices also emit energy in the visible light portion of the spectrum. We see this light directly and we interpret the images and colors generated by that energy. However, we also see images based upon reflected energy. Reflected light bounces off objects and is somewhat modified by the process of reflecting. The colors that we see from printed maps result from light reflecting off of the paper and the printed ink forming the map image. Some of the ambient light is absorbed by the paper and ink and thus we see the wavelengths that remain after that absorption. The fact that we can see light either by direct emission from a computer monitor or indirectly from light reflected from a surface creates a problem for the cartographer in design and use. Although one might say that light is light no matter what its source, that is not altogether true. As we will see in this chapter, the origin of the light plays an important role in understanding the use of color in map design. Virtual maps displayed on-screen use color models that are different from color printed maps. As with all map design, the cartographer must consider the means of presentation when selecting map colors.

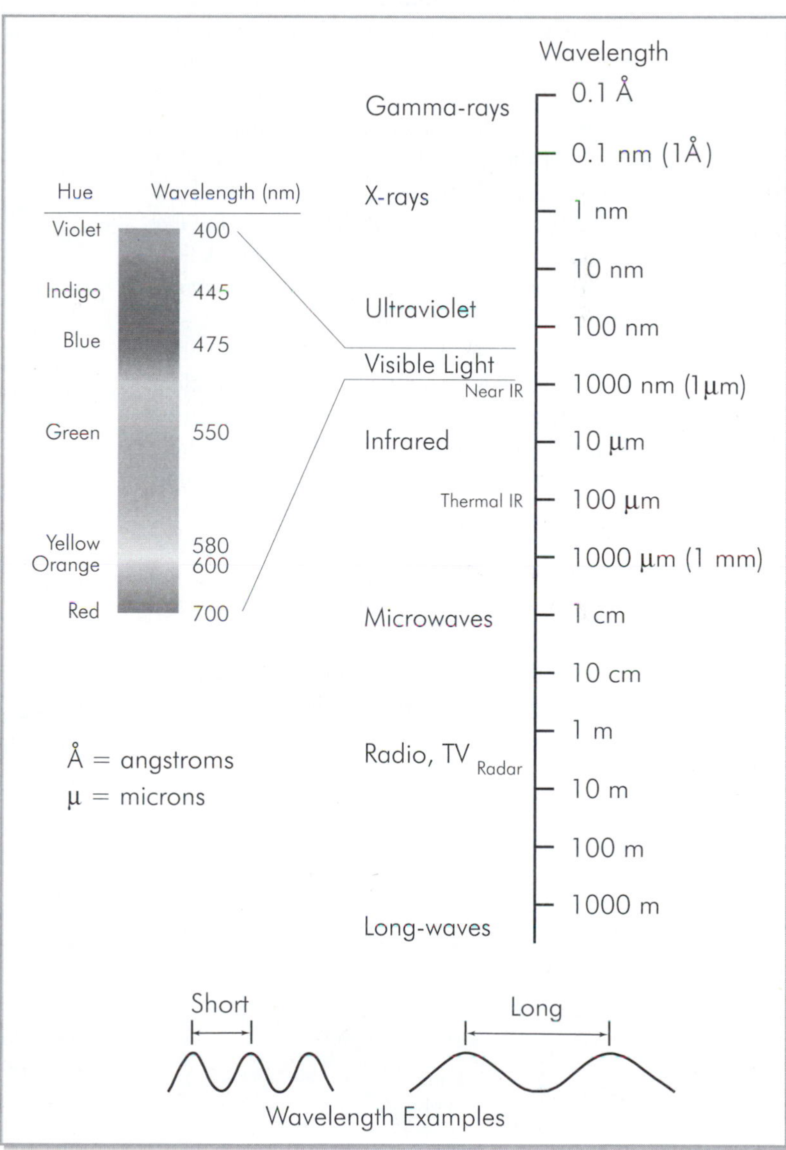

FIGURE 14.1 THE ELECTROMAGMETIC SPECTRUM (EMS) AND VISIBLE LIGHT.

The visible light portion of the electromagnetic spectrum lies between 400 nm and 700 nm.

COLOR PERCEPTION

Reading thematic maps is a process involving the eyes and brain of the map reader. Light emitted by the computer monitor or reflected off the map is sensed by the eyes, which report sensations to the brain, where cognitive processes begin. The sensing and cognitive processing of color is called **color perception.** Aspects of color perception treated in this section are the physiology of the human eye and the physical properties of color production.

The Human Eye

The human eyes are wondrous organs, often considered to be external linkages to the brain itself. The processing of light that falls on the human eye is like a vast, complex data-analyzing system. Each eye is a spherical body about 1 inch in diameter (see Figure 14.2). Light enters through the **cornea,** a transparent outer protective membrane. The amount of light entering the eye is controlled by the **iris,** a diaphragm-like muscle that alters the size of the pupil. Light is then focused by the transparent **lens** and passed onto the back wall of the eye chamber. The inside of the back of the eye chamber is covered by a thin tissue called the **retina.** A fluid (vitreous humor) fills the eye and serves to keep the distance between lens and retina nearly constant.

Light-sensitive cells make up the retina. **Rod cells,** numbering about 120 million, provide for only achromatic sensations and have no color discrimination. **Cone cells,** numbering about 6 million, are comprised of three types with different wavelength sensitivities. Some have peak sensitivity to blue, some to red, and others to green portions of the electromagnetic spectrum. As these rod and cone cells are triggered, electric impulses pass through a complex system of specialized cells to the **optic nerve,** which is a bundle of nerve cells that transmit the electrical impulses to the brain. Color perception takes place in the brain and involves cognitive processes that add meaning and substance to the light patterns sensed by the eye.

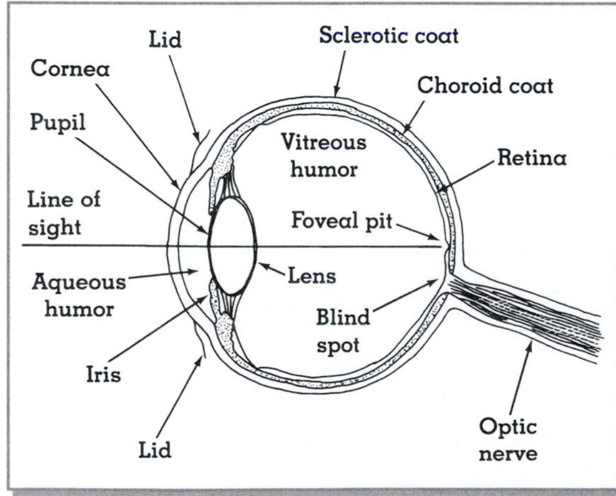

FIGURE 14.2 THE PARTS OF THE HUMAN EYE.

GENETICS AND COLOR

According to two recent studies in molecular biology, a difference in a single amino acid—the minimum genetic difference between two people—can cause a perceptible difference in color vision. Research teams at the University of Washington in Seattle and at Johns Hopkins University tracked the genetic basis of red photopigments, a type of protein, and showed that the newly discovered amino acid affects the part of the cone cells where color perception begins. The studies showed that there is nearly an infinite number of ways to see red alone. The variations are caused by subtle differences in genetic makeup, offering a biological explanation for the extreme subjectivity of chromatic responses. . . . With research pointing to complete individuality in perception, what theoretical models have a chance?

Source: Riley 1995, 1–2

The cartographic designer does not ordinarily need to pay particular attention to the details of the color-sensing apparatus of the map reader, except to make sure that graphic marks do not exceed the lower limits of visual acuity. For paper maps, illumination is very important. At low levels, only achromatic sensing is possible; color detection becomes possible as illumination increases. Virtual map colors may be impacted by the intensity of the energy emitted by the monitor. In both instances, the ambient light conditions may impact the perception of color by the user.

Color blindness is a facet of color perception that affects about eight percent of males and less than one percent of females. This is usually a hereditary condition, although not always (see boxed text above "Genetics and Color"). The manifestation is that the subject can see only blues and yellows and may have difficulty in perceiving distinctions between reds and greens and some yellows (Prust 1999). Although certainly not a major concern in map design, some mapping projects may call for the designer to accommodate color-blind readers, especially those maps that use hue differences to define symbol categories such as way-finding road maps, rapid-rail system maps, bus route maps, and others. There is more research being done on the design of color maps for the color blind, principally by using computer displays (Olson and Brewer 1997). This is not altogether unreasonable in this age of scientific and cartographic visualization, where much analysis takes place while viewing color maps, graphs, and satellite color images on-screen. ColorBrewer (Brewer 2006), an online guide to map color selection, includes color schemes that will not confuse people with red-green color blindness.

Physical Properties of Color

The generation of color will occur in either **illuminant mode** or **reflective mode.** The illuminant mode requires a light source and the eye-brain sensing system of the viewer while the reflective mode, also known as the *object mode,* requires three elements: a light source, an object, and the eye-brain system of the viewer. The illuminant mode applies to virtual maps generated for viewing on-screen with the computer monitor generating the energy for image display. The reflective mode occurs when a map is printed and the light striking the map reflects back to the eye of the map reader. These two modes create some difficulty in addressing the concept of color and map design as each has its own set of properties to consider.

Sir Isaac Newton showed that the visible portion of the energy spectrum is composed of the various colors, each having a different wavelength. Moreover, different light sources, such as the sun, a tungsten lightbulb, a fluorescent bulb, or the computer monitor, generate different spectral energy patterns. When we attempt to describe the physical properties (physical dimensions) of color, we need to be aware of which source is used. Computer monitors are calibrated according to their color temperature, and video card. Manual adjustments of the monitor's brightness, contrast, and gamma settings may alter the quality of the color image. Screen resolution impacts pixel size and the sharpness of the color. Therefore, identical hardware configurations have the potential of modifying the way the same color is displayed on-screen. In the illuminant mode, the monitor appears black when its power is turned off (no light is generated) and appears white when the maximum energy is produced.

For printed maps the physical characteristics of color are also affected by the quality of the object's surface. Some surfaces permit all light to pass through them, such as acetate film or laminate. These are called **transparent objects.** Most objects are reflective of portions of the visible spectrum; that part of the spectrum that is reflected defines the color of the surface. Surfaces that absorb all light are **opaque** and appear black. The amount of light that is reflected from surfaces can be plotted on a diagram that is called a **spectral reflectance curve.** Surfaces without ink appear white, assuming that the printed medium is white (see Figure 14.3). The perception of color in the reflective mode is influenced by the paper on which it is printed and the ambient light conditions in which the map is viewed. In a paint store, color chips are often

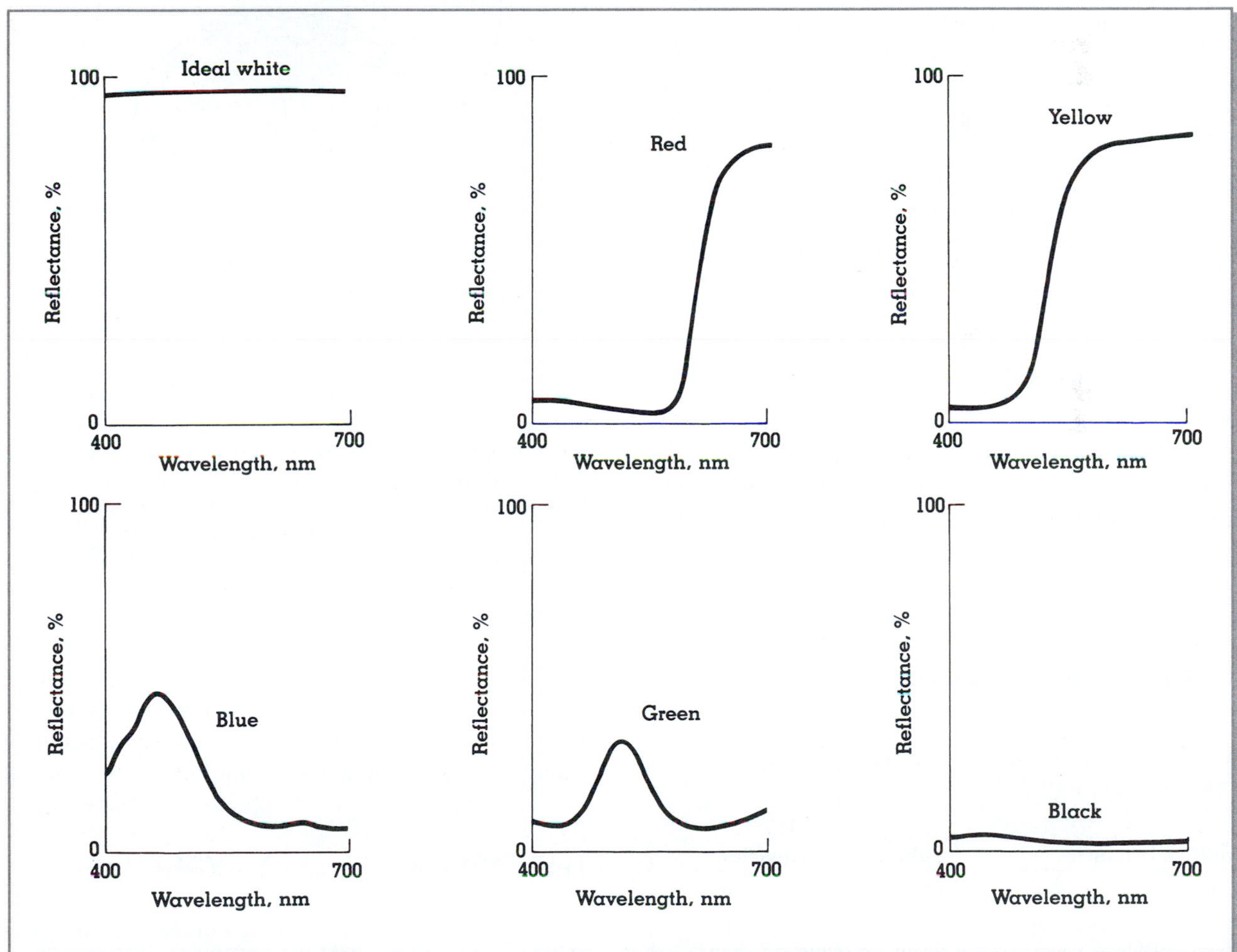

FIGURE 14.3 SPECTRAL REFLECTANCE (*R*) CURVES OF SEVERAL COLORED MATERIALS.

Notice that white reflects the most light and black the least. Also, the similarity in the curves of colors close to each other (red and yellow or blue and green) is quite remarkable.

displayed in such a way that the customer can switch on either incandescent or fluorescent lights to see how the colors behave in different light. It is no wonder that color specification and color choice are so difficult in cartography; color maps may be designed in one environment and viewed in another.

COLOR THEORIES

Depending on the whether the map is viewed on-screen or printed, two prominent color theories apply: **additive color theory** and **subtractive color theory.** The former specifically applies to light generated in the illuminant mode when color images are viewed on-screen, and the latter applies to the reflected mode.

Additive Color Theory

Although visible light is composed of a myriad of colors at various wavelengths, we consider white light to be made up of *three primary colors*—red, green, and blue (RGB)—because these cannot be made from combinations of other colors. Viewed individually we see the single color; however, when combined we can generate any number of other colors (see Table 14.1, Figure 14.4a, and Color Plate 14.1). In that area where all three colors overlap, we will see white; whereas in those areas where two colors overlap, we will see magenta, cyan, or yellow. Red, green, and blue are called the **additive primary colors** because they can, in various combinations, produce any other hue in the visible part of the energy spectrum. Colors produced by computer monitors, television screens, or movie film are the result of additive primaries of emitted light.

Subtractive Color Theory

Color produced by printing is not based on the additive primaries of emitted light but on inks or pigments laid down on paper. These inks reduce the wavelength of the energy being reflected, thus subtracting the energy being absorbed by the ink and reflecting the remaining energy. For example, red ink absorbs the blues and greens and reflects the red to the reader's eye. Thus the color of the paper or *other ink* is perceived by the reader (see Figure 14.4b). *Process color* printing uses the inks magenta, cyan, and yellow (see Table 14.1), which together can create any hue or recreate a continuous-tone color image. These pigment colors are called the **subtractive primary colors** (see Color Plate 14.1).

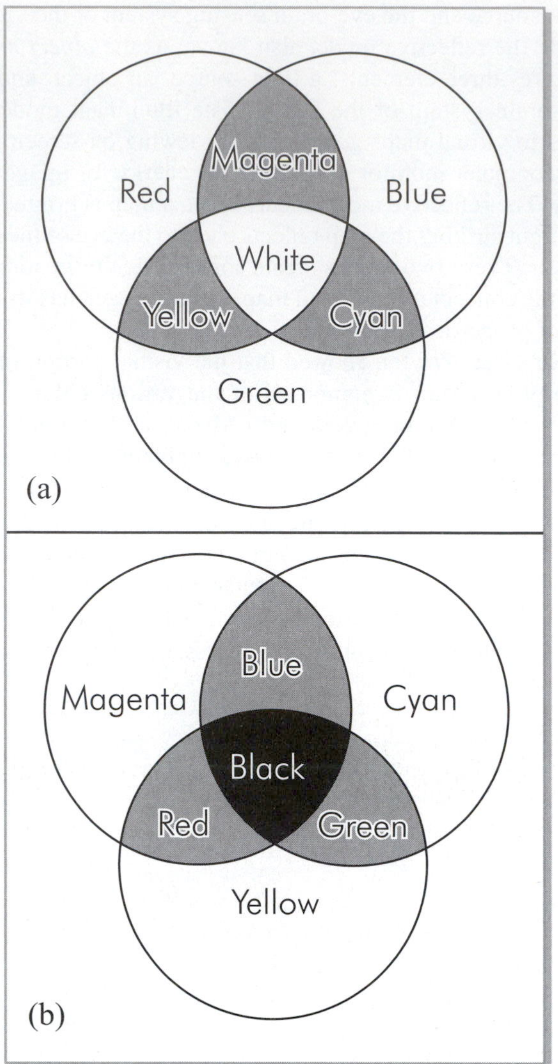

FIGURE 14.4 ADDITIVE AND SUBTRACTIVE COLOR MODELS. In the additive color model (a), red, green, and blue are the primary colors. Combining two additive primary colors produces magenta, cyan, or yellow. The combination of all three additive primaries produces white. In the subtractive color model (b), magenta, cyan, and yellow are the primary colors. Combining two subtractive primary colors (in printed form) produces red, green, or blue. The combination of all three subtractive primaries produces black. Refer to text for discussion of these models and to Color Plate 14.1.

TABLE 14.1 COLOR MIXING USING ADDITIVE AND SUBTRACTIVE PRIMARIES

Additive Primaries	Subtractive Primaries
Red + Green = Yellow	Magenta + Yellow = Red
Red + Blue = Magenta	Magenta + Cyan = Blue
Green + Blue = Cyan	Cyan + Yellow = Green
Red + Blue + Green = White	Magenta + Yellow + Cyan = Black
Absence of color = Black	Absence of color = White

THE DESERT ISLAND EXPERIMENT

The following experiment is retold in many books on color. A subject is asked to imagine that he or she is on a desert island that is covered by many large pebbles of different colors (Billmeyer and Saltzman 1966). The task is to arrange (classify) the pebbles by color, using any system. The subject first decides to divide the pebbles into two groups—those with color **(chromatic)** and those that are white, gray, or black, or without color **(achromatic).**

The next step in classification by this subject might be to arrange the achromatic pebbles (the whites, grays, and blacks) into an array from white to black, with the grays in between. This array, based on the single quality *lightness,* is called a value scale.

To deal with the sorting of the chromatic pebbles, the subject decides to make piles of red stones, blue stones, green stones, and so on. If greater precision is needed, there could be groups of blue-green pebbles, red-orange, and so forth. The chromatic pebbles are now divided into hue categories.

Within each pile, the subject now decides to array the colored pebbles by lightness or value. Thus, for example, the red stones could be arrayed from light red to dark red, from pinks to the deepest cherry reds.

In this experiment, a last judgment is made by the subject. After careful examination of, say, all the red pebbles, it may be noticed that even though two stones are similar in value (lightness), they still *look* different. A red radish looks somehow different from a tomato whose red is equally dark. After careful study, the subject decides that the two reds can be compared in terms of their neutral (or middle) gray content; all reds of the same value can be further divided into yet another spectrum. This is sometimes referred to as *saturation,* or chroma. A hue having less gray in it is more saturated.

This completes the desert island experiment. The subject has classified colors into groups based on appearance or **psychological dimensions.** Because these dimensions are based on perception, they have particular significance in thematic map design.

COMPONENTS OF COLOR

The components of color are essentially the same whether one is using additive or subtractive color theories. These components include terminologies that allow us to specify a name of a color, the intensity or vibrancy of a color, and the lightness of the color. The terminology used to describe these three components varies depending on presentation mode and the color model utilized. The presentation mode defines whether we view a map on-screen or in paper format. The color models are tied to these modes and are used by GIS, mapping, and artistic software in the design of maps. The basic components of color are hue, saturation, and value.

Hue

Hue is the name we give to various colors: the reds, greens, blues, browns, red-oranges, and the like. Each hue has its own wavelength in the visible spectrum. Now, although hues and their relationships to other hues may be illustrated in many ways, a customary way, especially by the artist, is on a **color wheel,** as in Figure 14.5 and Color Plate 14.2. The idea

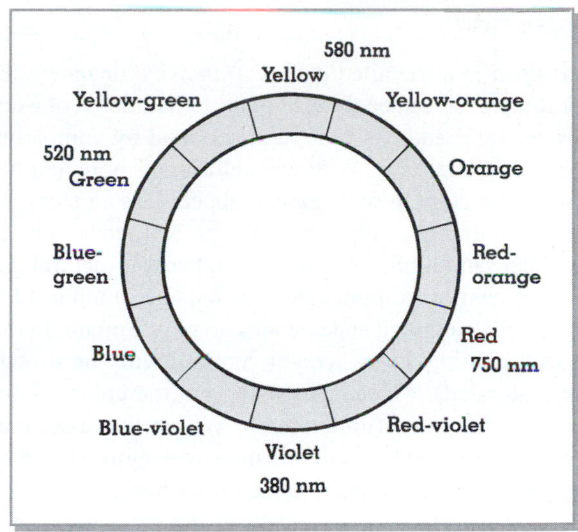

FIGURE 14.5 A 12-PART COLOR WHEEL, SHOWING APPROXIMATE WAVELENGTHS OF THE VARIOUS HUES.
Color wheels are typically used by artists but can be useful in describing certain color associations (such as complementary colors and opposites). Typical color wheels often contain 8 or 12 hues. The abbreviation nm stands for nanometers; one nanometer is one-billionth of a meter.

of using a wheel to conceptualize the sequence of colors was envisioned by Sir Isaac Newton as light was refracted through a prism. Thus, the color wheel has been used extensively in the discussion of color. This refraction is the same that we observe in a rainbow. Color on the wheel is sequential according to the wavelength in the visible spectrum. We normally think of the color sequence of red, orange, yellow, green, blue, indigo, and violet, which is conceptualized by the mnemonic device of a fictitious person, Roy G. Biv.

Theoretically, a color wheel can contain an almost infinite number of hues, but most include no more than 24. Eight or 12 are more common, 12 being especially useful for artists (Fisher 1974). The organization of hue on the color wheel is not based on the physical relationships of the hues but is simply a conventional way of showing visual hue relationships. It may be added here that some writers about color suggest that the color wheel is simply the human's "impulse" to simplify color into a tabular schema, much as the chemist places the elements into a periodic table (Riley 1995).

Although the color wheel normally contains 12 hues, the human eye can theoretically distinguish millions of different colors. In ordinary situations, however, we are not called on to do so; in cartographic design, our assortment of hues is certainly far less. In fact, it would be difficult to imagine any case where a map reader would have to identify more than a dozen (this is common on land-use or geologic maps). The United States census has done 16-color, two-variable choropleth maps (Olson 1981), but these have not found wide use because of the complexities inherent in reading them. However, multivariate mapping is common in cartography. Nevertheless, the selection of colors to insure that the reader can easily interpret the map content is often challenging.

Saturation

Saturation is also called *chroma, intensity,* or *purity.* This color dimension can be thought of as the vividness of a color (Brewer 2005) and can best be understood by comparing a color to a neutral gray. With the addition of more and more pigment of a color, it will begin to appear less and less gray, finally achieving a full saturation or brilliance. For any given hue, saturation varies from zero percent (neutral gray) to 100 percent (maximum color). At the maximum level, the color is fully saturated and contains no gray. You might think of it as "watering down" a color. Starting with the strongest color (saturated), we can add water and the color becomes weaker or less pure (unsaturated). Within the spectrum, a color is determined by its dominant wavelength. However, a color may appear saturated if the curve has a very narrow base, is steeply sloped, and peaks at the same point of the spectrum; whereas, a curve that has a wider base and is more gently sloped will produce a desaturated color. (Fraser *et al.* 2005; See Color Plate 14.3). Saturation levels vary with hues; the most intense yellow appears brighter than the most intense blue-green. Achromatic colors are said to have zero saturation. Fully saturated colors when viewed on a computer monitor will appear brighter than the same level of saturation of a printed map. The printer's ink rarely matches the purity level that can be displayed on-screen.

Value

Value is the quality of *lightness* or *darkness* of achromatic shades and chromatic colors. Conceptually easy to grasp, value can be thought of as a sequence of steps from lighter to darker displays of gray. The lowest value, specified as zero, will produce a light neutral gray, and 100 percent gray will produce black. Chromatically progressing up the value scale, the hue becomes successively darker. This provides us with *shades* of a particular color—for example, dark red compared to light red. It is this increase in value that cartographically presents increments of a hue in the design of choropleth maps.

Sensitivity to value is easily influenced by environment, and apparent lightness or darkness is not proportional to the reflected light of achromatic surfaces. A given gray, or hue of a specified value, looks one way when examined individually against a white background but differs when viewed in an array of colors on a computer monitor. This concept applies to both emitted and reflected light.

Value has been widely used as a component of color. Brewer (2005) points out, however, that "the term becomes confusing in quantitative work if you are also describing data values" (95). As a result, lightness is often used as a substitute for value in discussions of color. You are cautioned to be cognizant that the term may create some confusion when using it in multiple applications. In art, value is controlled by the addition of white or black pigment to a hue. If white is added to a hue, a **tint** results. When black is added, a **shade** is produced. A **tone** results from adding amounts of a hue, white, and black (Itten 1970). Tint, shade, and tone are discussed also in the section on harmony later in the chapter. These terms are mentioned in this chapter because they are used so frequently in informal discussions of color.

Munsell Approach

Albert Munsell, an artist and painter, recognized the three distinct qualities of color to be hue, value, and chroma (or saturation) (Munsell 1981). These qualities were incorporated into a system to help describe color using a series of notations that we now refer to as the **Munsell color solid** (Birren 1969). A color solid is a representation of color space**,** a way of integrating and showing the relationships of the three color dimensions (see Figure 14.6). The solid is only roughly a sphere, because of the unevenness of chroma.

The color solid of Munsell and the quantitative specifications derived from it are based on the whole array of colors available for human sensation. Steps or divisions between colors are based on psychological intervals, which were measured through extensive testing. As new color pigments are devised by science and industry, new color chips can be added to the model. This is a principal advantage of this color solid.

Ten hues make up the color band around the solid; each hue has a letter designation. Ten number divisions are associated with each hue—the "5" position designates the pure color of each hue. There are therefore 100 different hues on the color wheel; for example, 7R, 2Y, and 8B each specify a particular hue on the circle (see Figure 14.6a).

Psychophysical testing has determined that people do not perceive color value in an equal-step manner. Munsell's value scale, called an *equal-value scale,* records the reflectance percentages of what appear to be equal steps from white to black. Value on the solid is represented along a vertical scale (the axis) of the color solid, perpendicular to the plane of the color wheel. The top of this value scale is white (10) and the bottom position is black (0). Between these two "poles" are nine divisions, graduating from white to black, that yield different grays. The middle gray has a value of 5. Notation of value is by the position on this vertical scale, followed by a slash (7/, 2/, and so on; see Figure 14.6b).

With these two dimensions, dark red or middle red, for example, can be exactly specified. A designation of 10BG 2/ would indicate a blue-green closer to blue and dark (nearly black) in value.

Munsell also incorporated chroma designation in his color solid. Chroma or saturation is represented by steps going inward toward the value axis for each hue band. The chroma scale is also psychologically stepped. As a color gets closer to the axis, it becomes weaker (more gray); conversely, the farther out on the chroma scale a color is, the stronger and more saturated it becomes. Chroma designation is a number that follows the value slash, such as /7. A complete Munsell color designation would thus be, for example, 8P 7/5.

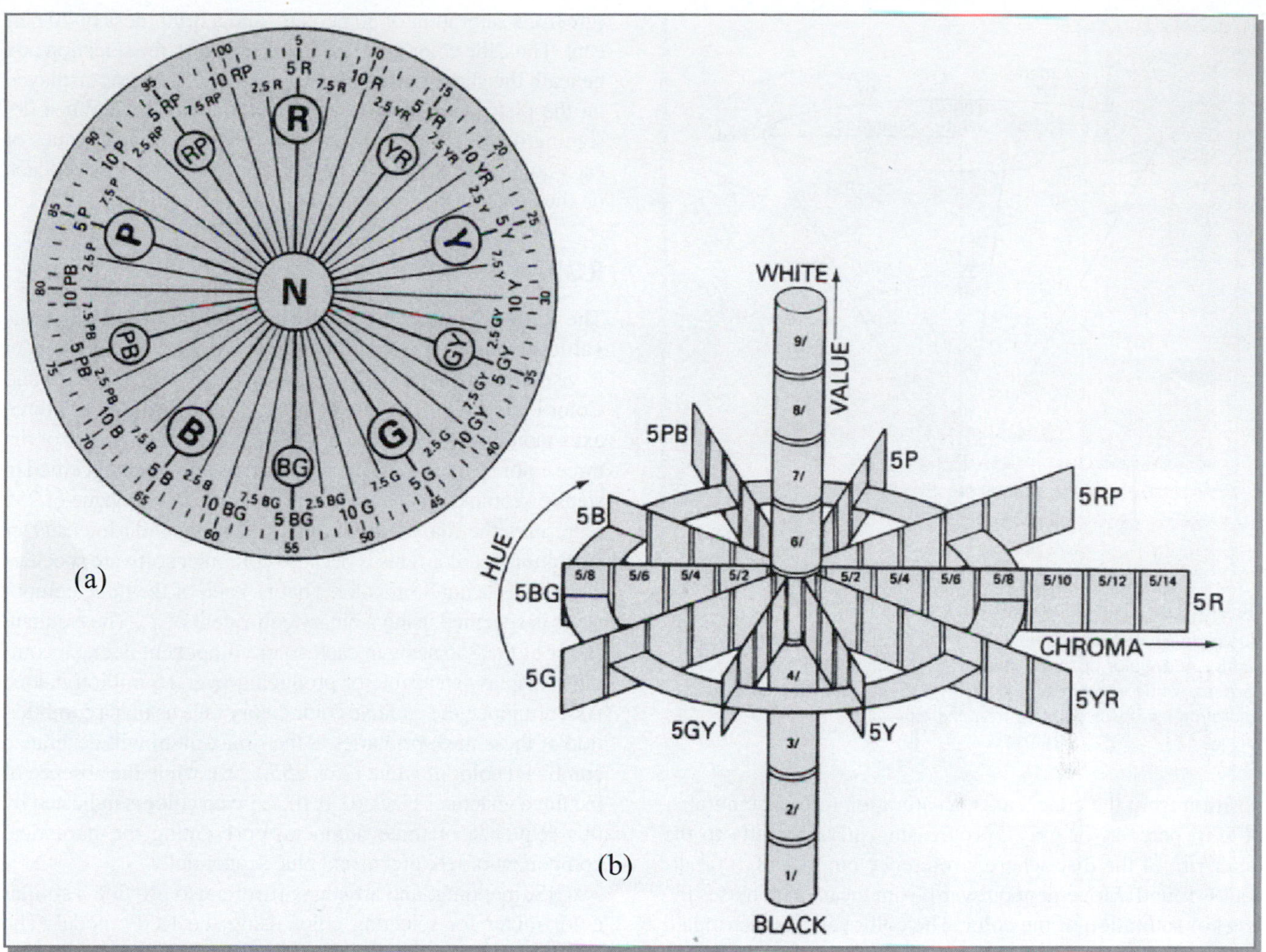

FIGURE 14.6 THE MUNSELL COLOR MODEL.

The hue symbols and their relation to one another on the color wheel are represented in (a). Hue, value, and chroma dimensions of the model are shown in (b).

Maximum chroma differs for the variety of hues and values. For example, for 5R, maximum chroma at a value of 8/ is 4, but for a value of 4/ it is 14. For 5PB, maximum chroma at value 8/ is 2, and for 2/ it is 6. These give the color solid an uneven and unbalanced appearance, causing it to deviate significantly from a perfect sphere.

There have been applications of the Munsell color system outside of the area of cartography. Practical uses of identifying color have been made in the textile industry and in soil science to describe the soil's color.

COLOR MODELS

The use of color in designing maps requires that we apply the appropriate color theory based upon the intended form of the final product, that is, paper copy or virtual screen image. In the various GIS, mapping, and artistic software, color is selected based upon a particular color model. The model is named using the key letters of the models components. The color models that are applicable for designing maps for computer displays include HSV, HSB, HSL, RGB, and CIE L*a*b. The color model that is selected within the software when the final product is a printed map is that of the CMY(K) model. The use of a grayscale model can be applied to both the printed and projected image maps.

HSV

This model refers to the components of color described above: hue, saturation, and value. This model may be visualized as an inverted cone (or hexacone) with the upper surface resembling the color wheel (see Figure 14.7). Around the circumference of the conical disc, color hues are arranged in wavelength sequence beginning with red, as the circle's origin, and progressing around the 360-degree space with the remaining hues. Thus, hues are distributed such that complementary colors are 180 degrees apart. Most observers are able to differentiate approximately 150 steps around the disc (Rossing and Chiaverina 1999). Saturation is indicated as a

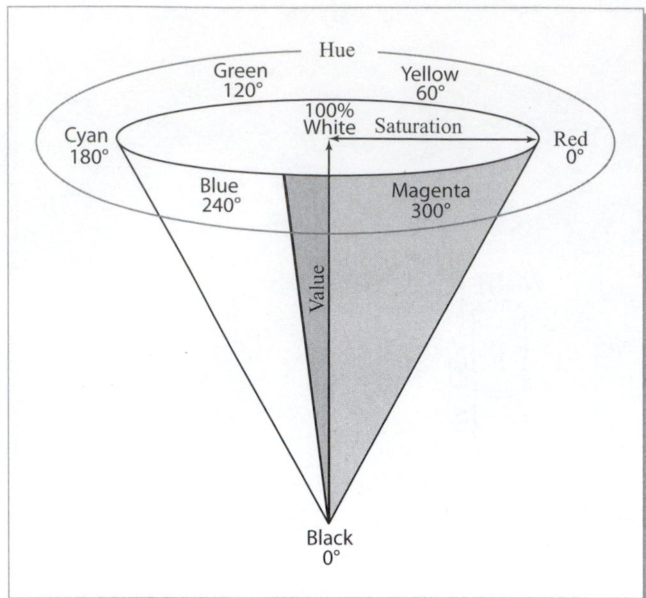

FIGURE 14.7 HSV COLOR MODEL CONE.

Hue is distributed along the 360° of the color wheel. Hue is identified by its angular position, in degrees, beginning with red at zero degrees. Value is achieved along the vertical axis of the cone, and saturation increases outward from the axis.

position from the cone's axis, a saturation reference number of zero percent (or gray), progressing incrementally to the outer rim of the disc where a reference number of 100 percent is found. Movement outward from the axis increases the level of saturation of the color. The color value is distributed vertically along the cone's axis. At the tip of the cone, a value of zero percent indicates black. You can increment up the axis, increasing the value until you reach the top of the conical disc or a value of 100 percent, indicating white.

HSB/HSL

Similar to the HSV model are those known as HSB and HSL. Included in these two models are hue and saturation. The remaining components refer to brightness and lightness. These words mean essentially the same thing. Lightness can be thought of as the amount of light reflecting off a white surface. In computer displays, this would be how much light is emitted from the screen, observed as a range of light to dark. Brightness, on the other hand, is referenced to black, indicating how much light is received. This is true of times when there isn't enough light to see color but we can still see variations in the amount of light. In such an instance, our eyes are utilizing their cones to differentiate between variations in darkness. These two color models, along with HSV, are very similar and for all practical purposes can be thought of as equal.

The computer software that we use for generating maps and graphics employs the use of color mixers that utilize sliders for each of the three components (see Color Plate 14.4). One can adjust the slider to specify a hue of 120 degrees (green), a saturation of 80 percent, and a brightness of 70 percent. Thus, the color generated will appear in the selection box beneath the slider portion of the window. The color displayed on the plate represents the combination specified. When designing maps, you can either specify the numerical values of each component or use the sliders for instantaneous feedback on the apparent changes in color made by sliding the bar.

RGB

The RGB color model specifically relates to the **tristimulus values** components of additive color theory. The color model is visualized using a RGB color cube (see Figure 14.8a and Color Plate 14.5). The cube has black as its origin with three axes radiating outward at 90°. Each axis represents one of the three color primaries. Numerically, each primary is specified in steps of 256 increments ranging from 0 to 255. A value of 255 represents the maximum amount of illumination intensity or saturation of color. This is because computer software specifies color steps using 8-bit color. That is, each of the three components is specified using a binary equivalent of 2^8. The resultant effect of the 256 steps in each color component is that a computer display is capable of producing over 16 million colors. As a reminder, the additive color theory tells us that a combination of these three primaries at their maximum will generate a combined color of white (255, 255, 255), while the absence of the three generates black (0, 0, 0). A given color is indicated by this sequence of three numbers representing the numerical component of red, green, and blue sequentially.

GIS, mapping, and artistic software also provide a similar color mixer for selecting color using the RGB model. The RGB combination necessary to represent the identical color created in the HSB model would be 0, 255, 0 (see Table 14.2). The color sliders and numerical specification are identical to the operation in the HSV/HSB/HSL models. The software converts the specifications internally, using mathematical transformation formulas. However, we recommend that you make all of your color selections within the appropriate model based upon the final format of your map, that is, printed or virtual image. Table 14.2 displays the relative setting values in RGB and other models for Color Plate 14.4.

CIE LAB

The models discussed above are implemented through specific GIS and computer mapping software. The CIE L*a*b model was developed by scientists at the International Commission on Illumination (*Commission Internationale de l'Eclairage*) or CIE (see Color Plate 14.6). The model is used to describe all color perceived by the human eye directly. The model uses three primaries to represent the lightness of color, relative position between magenta and green, and the relative position between yellow and blue. The lightness value (L) ranges between 0 (black) and 100 (white). The "a" represents how red or green something is with values ranging from −100 for green to +100 for magenta. The "b"

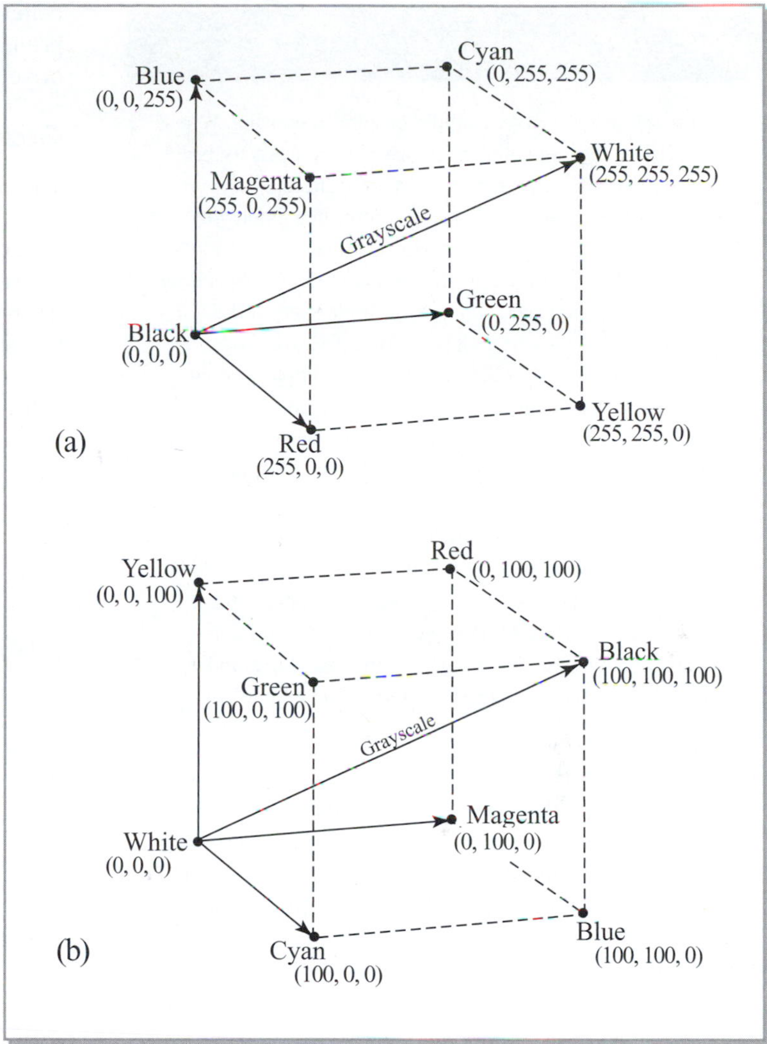

FIGURE 14.8 RGB AND CMY COLOR CUBES.
The cube represents a visualization of the way in which the color primaries are combined. The RGB model (a) has values ranging from 0 to 255 along each color axis. The CMY model (b) has values that increment by percent from 0 to 100. Refer to Color Plate 14.5.

component's values range between −100 for blue to +100 for yellow. The numbers can be depicted using very precise increments (see Table 14.2).

CMYK

The models described above all pertain to additive color theory or those colors viewed on your computer monitor.

TABLE 14.2 COMPARISON OF EQUIVALENT COLOR MIXER SETTINGS

Color Model	Settings for Model			
HSB	**120°H**	**100%S**	**100%B**	
HSV	120°H	100%S	100%V	
RGB	0 R	255 G	0 B	
CMYK	100% C	0% M	100% Y	0% K
CIE LAB	85.61 (L)	−91.49 (a)	73.96 (b)	

Depending on the color management system, the values achieved in color conversion may vary.

When color is printed on a paper medium, the subtractive color theory applies. The primary model for printed maps is that of the subtractive primary colors of CMY (cyan, magenta, yellow) (see Figure 14.8b). Each primary is specified according to percentage increments between 0 and 100. The combination of 100 percent of all three primaries produces the color black. The absence of these primaries produces white (although, in reality, the absence of the primaries results in the color of the paper). Potentially, the various combinations of one-percent increments can generate one million colors. However, in the printed medium too many other variables come into play which limit the number of colors you can reliably create. Such variables include the paper quality and color of the paper, purity of the printer ink, and ambient light conditions of the viewing mode.

The quality of the paper can determine the receptivity of ink and thus impact the density of the ink being applied. This will be discussed further in Chapter 15. The color of the paper has the potential of affecting the perceived color by the map user as it too provides absorption capabilities, thus reducing the wavelengths of light reflecting off its surface.

THE COLOR YELLOW

Yellow is one hue that you should be careful in using. When displayed on the computer monitor it may appear very brilliant and easily seen when used as a color fill in a polygon. However, if you use yellow as a line color in a graph with thin line weights, the yellow may be difficult to see. Yellow text is also difficult to see on a white background, depending upon the font, point size, and weight. In the case of the graph and text, yellow is a viable option if you are using a high contrast color as a background. For example, yellow will be quite visible on a dark blue background.

Solid yellow is one color that does not translate well from an RGB mode to a printed version in CMYK. The hue has a tendency to print much lighter than that seen on your monitor. Avoid using yellow for any line in a graph and for text. Yellow may be used as polygon fill. However, care should be taken to maintain proper figure-ground and color contrast. As a component of the CMYK color model, it is used in combination with the remaining components of the model to produce a large array of potential colors used in mapping.

The purity of the printer's ink can also impact the perceived color. In theory, the combination of the three primaries at maximum density will produce black. However, as a consequence of variable ink purity the resultant color when printing these colors over the top of each other may be a dark gray. We also find that the exact placement of the three layers over each other may be difficult if we were trying to print a thin black line, 0.25 point for example. As a result of the impurity of the ink and the possibility of misalignment of the primary separates, we add black ink (K) to the model. Thus, we frequently refer to this model as CMYK.

Because the printed map colors are viewed by the ink subtracting (absorbing) wavelengths of the ambient light, the intensity and type of the light reflecting off the map can produce a difference in the perceived color. Most colors are defined as if they were viewed under normal sunlight conditions. However, as the light intensity decreases because of low sun or cloudy conditions, the color viewed under normal conditions may appear different, normally darker. The wavelength of the ambient light also determines the perception of color. The wavelengths generated by tungsten and fluorescent lights are quite different from the wavelengths of normal sunlight. Therefore, the ambient light source can also impact the resultant color. As a cartographer, you will not have control over the viewing conditions in map use. Therefore, the designation of CMYK values should be made using increments of approximately ten percent. Research has shown that an increment of 10 percent is identifiable by most map users between 10 and 40 percent. Values above 40 percent should be made using increments of 15 to 20 percent to help ensure the user can identify the colors as different. We believe this to be true for combinations of the subtractive primaries or the single color component with the exception of yellow (see boxed text "The Color Yellow").

Grayscale

The grayscale uses an achromatic approach to presenting differences in shades of black. It is most applicably used for maps to be photocopied or printed in black and white on a laser printer or printing press. The incremental steps in selecting different levels of gray are similar to those discussed for printing in color. One should limit the number of grayscale levels to five or six on a given map. Beyond that, the ability of the map user to differentiate between gray levels is diminished. If you know that the map is to be produced in an achromatic manner, select the grayscale as your color model before beginning your map design. Conversion from a color model, especially an additive model, to grayscale is not always as successful in creating different gray levels as actually beginning your design with that model.

DEVICE LIMITATIONS

Most GIS, mapping, and artistic software provide the options for selecting among various color models. The shifting between these models is easily achieved by simply selecting the color mixer for specifying the color components. The color selection by the cartographer within the particular software is device dependent. Maps designed for viewing on-screen have the capacity for displaying a larger quantity of colors, with a greater probability that the map user will be able to distinguish between those colors. However, when the map is sent to a printer, the colors must be converted to the color management system used by that printer. Not all colors created on a computer can be printed in hard copy. Such unprintable colors are referred to as being out-of-gamut.

The **gamut** is determined by the hardware and software utilized and represents all the colors that the combination can display. For example, the digital camera may have the capability of capturing a wide range of color (its dynamic range). However, a scanner may be more restricted in its range of recognizing and producing colors. These input devices have different dynamic ranges, usually determined by their capability of brightness ranges. The color gamut of all perceivable colors is displayed in the form of a chromaticity diagram (see Color Plate 14.6). The triangular forms on the diagram depict the range of RGB color primaries visible on the monitor and CMYK color primaries that can be printed; that is, their gamut. Within the GIS or mapping software, some indicator is provided when the color selected is out-of-gamut. Frequently, an out-of-gamut color is identified by a small triangle or other symbol on the color mixer.

To avoid the use of colors that are out-of-gamut, start your design using the CMYK model for all maps to be produced in printed form. Check your printer documentation or your print service bureau for the starting color model and file format. The latter will be discussed in Chapter 15.

Web Safe Colors

Web safe colors are those colors that are displayed exactly the same on any computer monitor or web browser capable of displaying 8-bit color. Instead of 256 possible colors, only 216 colors are considered to be web safe. The colors contain RGB settings of 0, 51, 102, 153, 204, and 255 for *each* R, G, and B. These settings represent 20-percent increments between the 0 and 255 possible values. The six settings within each of the three color components produce a total of 216 colors. Although computer technology has made web safe colors somewhat less important, as a cartographer it is always better to feel a sense of security in one's design, knowing that it will appear on everyone's monitor in the same way. We suggest that you select web safe colors if your are preparing your maps for the Web.

COLOR MATCHING SYSTEMS

The selection of colors is often difficult and somewhat arbitrary. One can never be totally certain that what you see on-screen is what you will get (WYSIWYG) in a printed document or map. Computer software operates using color management systems that are device dependent. Traditionally, the approach to color selection has been done using a standardized color matching system. The one system used in color selection of printed maps that has served as an industry standard for many years is that of the PANTONE MATCHING SYSTEM (PMS) in which colors are selected from a swatch book, similar to individual paint chips one may examine at a paint or home supply store. The colors are referred to as *spot color* (a color printed in a single ink as opposed to process colors), with solid colors created by the mixing of ink to a specific hue. From the book, a specific PMS color could be specified, such as PANTONE red 032. This color is also cross-referenced to its CMYK process color designation. The book also contains examples of the color printed on coated and uncoated paper. The PANTONE colors are included in most GIS, mapping, and artistic software as color libraries.

Color Interaction

When we look around us, we see a world of many colors, patches of color surrounded by other colors, and colors adjacent to other colors. In many cases, maps are composed of more than one color. Reaction to a color patch is always modified by its environment. Seldom is the reader asked to look at a single color—even simple color maps usually contain variations of one hue, combined with black. There are at least three **color interactions** that must be considered in cartographic design—simultaneous contrast, successive contrast, and color constancy.

Simultaneous Contrast

When the eye spontaneously produces the complementary color (opposite on the color wheel) of a viewed color, the effect is called **simultaneous contrast** (Itten 1970). A color, if surrounded by another color, begins to appear tinged by the complementary color of the surrounding color. A simple experiment demonstrates this effect. Place a small gray patch inside a box of pure color. After you gaze at the surrounding color for a brief period, the gray square begins to look like the surrounding color's complementary color.

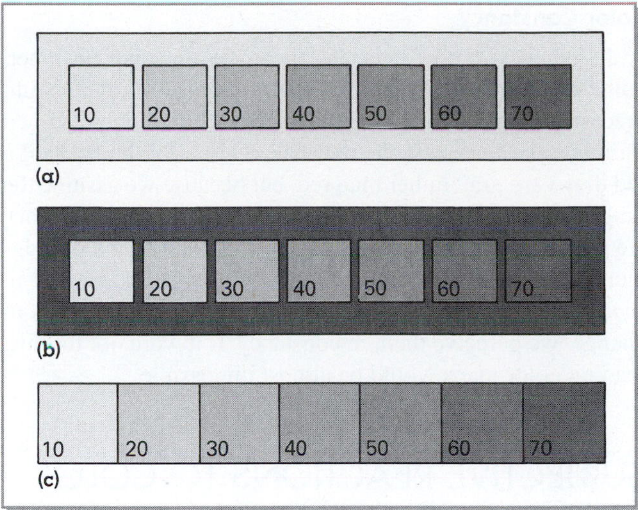

FIGURE 14.9 THE EFFECT OF THE ENVIRONMENT ON THE PERCEPTION OF VALUE.

The surrounding environments of (a) and (b) modify the appearance of the group (otherwise identical) in the strips they surround. Induction results when color patches of different values are closely juxtaposed, as in (c). Notice also that induction does not occur in (a) and (b) because the gray patches are separated by open spaces or black strips.

Simultaneous contrast (Figure 14.9) causes adjacent colors to be lighter in the direction of the darker adjacent colors, and darker in the direction of the lighter colors. This is sometimes referred to as **induction,** a special case of simultaneous contrast (Dember 1960). Induction is particularly bothersome in map design when several different values of the same color are juxtaposed on the map or in the legend. The effects can be reduced by separating color areas by white or black outlines (Robinson 1966). An interesting feature of induction, or any simultaneous contrast, is that it cannot be photographed; it is produced by the eye alone.

Successive Contrast

Successive contrast results when a color is viewed in one environment and then in another in quick succession (Beck 1972). The color will be modified relative to these new surroundings. It may appear darker or lighter, more or less brilliant, compared to each new environment. For example, a gray will look darker against a lighter background and lighter against a darker background. An orange will appear darker and more red on a yellow background and lighter and more yellow on a red background. New contrasts are therefore established in each new situation. Cartographers cannot control all color environments, but positive design attempts to reduce the number of successive contrasts on the map will help.

Color Constancy

Although it is not of central concern to the map designer, **color constancy** is a feature of color perception that should at least be noted. We tend to judge colors based on presumed illumination. For example, the shadow areas of the folds of a red drape are gray rather than red, but because we assume the drapes to be lit by a common source (for example, sunlight), we perceive these gray areas to be red also. This kind of judgment is made during map reading when color areas fall in shadows. Although the shadows cause actual color areas to change, we perceive them in constancy. If it were not for this, reading color maps would be almost impossible.

SUBJECTIVE REACTIONS TO COLOR

Human beings react to color for a number of reasons. Simultaneous and successive contrast are caused by our physiological system (eye-brain); most people have nearly identical response mechanisms to these. More variable, and much more difficult to control in design, are psychological reactions to color: color preferences, the meanings of color, and behavioral moods and color. Taken together, these are called the **subjective reactions to color.**

Color Preferences

An enormous amount of research has dealt with color preferences. Most of the formal research has come from the fields of psychology and advertising. Little research has been conducted by cartographers on color preference, so our understanding must be borrowed from the other disciplines.

There is abundant evidence that color preference is somewhat developmental. We must first define terms. **Warm colors** are those of the longer wavelengths (red, orange, and yellow), and **cool colors,** at the opposite end of the spectrum, have shorter wavelengths (violet, blue, and green). Warm and cool, of course, are psychological descriptions of these color ranges. Children about the age of four or five years prefer warm colors. Red is most popular, with blues and greens next (Sharpe 1974). Young children also prefer highly saturated colors, but this preference begins to drop off after about the sixth grade.

The findings regarding children's color choices have prompted one cartographic researcher to comment on color design for children's maps (Sorrell 1974):

1. As children are aware only of small ranges in hue, colors should be chosen from within the basic spectrum colors—blue, green, yellow, orange, and red.
2. Because school-age children begin to reject fully saturated colors, a step or two down the saturation range is more desirable.
3. Children appear to dislike dull unattractive colors, so color choice should avoid these. Stay close to the spectral hues.
4. Children generally reject achromatic color schemes—the gray scale. These reduce the attractiveness of the map.
5. Choosing colors that have greater compatibility with what is *expected* yields greater comprehension. Thus, for example, blue is better than red for water.

As we mature and leave childhood, our color preferences change. Generally, we tend to favor colors at the shorter wavelengths. Color preferences among North American adults are blue, red, green, violet, orange, and yellow, in that order. The greenish-yellow hues are the least liked by both men and women. Women show a slight preference for red over blue and yellow over orange, whereas men slightly prefer blue over red and orange over yellow. Both sexes choose saturated colors over unsaturated ones. Although color preference yields some insight into the broader realms of color psychology, a simple list of preferences may not be enough. Color preference tests performed by psychologists are conducted for the sake of color choice only and are not product related. Most color experts would agree that other variables, such as color environment, product name, packaging, context, and merchandising schemes, are also important in color choice (Dunn and Barbau 1978).

Parallels may be drawn for color use in cartographic design. Color conventions should overrule broader color preferences; cartographic design should otherwise seek to follow the preferences of the marketplace. Until detailed studies are provided, map designers need to follow the leads provided by merchandising, art, and related psychological literature.

Colors in Combination

In one particular study (Helson and Lansford 1970), 10 different hues, each with three different values and chromas, were selected as object colors. A random sample of 25 background colors was selected, and the subjects were asked color combination preferences. The general findings were these:

1. The most pleasant combinations result from large differences in lightness. Lightness (or value) contrast is necessary in pleasant object-background combinations.
2. A good background color must be either light or dark; being intermediate in lightness makes it poor.
3. Consistently pleasant object colors are hues in the green to blue range, or other hues containing little gray.
4. Consistently unpleasant object colors are in the yellow to yellowish-green range, or other hues containing considerable gray.
5. To be pleasant, an object color must stand out from its background color by being definitely lighter or darker. This is the single most important finding of the study for the cartographer.
6. Vivid colors combined with grayish colors tend to be judged as pleasant.
7. Good and poor combinations are found for all sizes of hue differences (that is, distances apart on the color wheel).

TABLE 14.3 CONNOTATIVE MEANINGS AND COLOR

Color	Connotations
Yellow	Cheerfulness, dishonesty, youth, light, hate, cowardice, joyousness, optimism, spring, brightness; strong yellow—warning
Red	Action, life, blood, fire, heat, passion, danger, power, loyalty, bravery, anger, excitement; strong red—warning
Blue	Coldness, serenity, depression, melancholy, truth, purity, formality, depth, restraint; deep blue—silence, loneliness
Orange	Harvest, fall, middle life, tastiness, abundance, fire, attention, action; strong orange—warning
Reddish browns, russets, and ochres (earth colors)	Warmth, cheer, deep worth, and elemental root qualities; can be friendly, cozy or dull, reassuring or depressing
Green	Immaturity, youth, spring, nature, envy, greed, jealousy, cheapness, ignorance, peace; midgreens—subdued
Purple (violet)	Dignity, royalty, sorrow, despair, richness; maroon—elegant, and painful
White	Cleanliness, faith, purity, sickness
Black	Mystery, strength, mourning, heaviness
Grays	Quiet and reserved, controlled emotions; can be used to create sophisticated atmosphere

Source: Littlefield and Kirkpatrick 1970, 204; Hackl 1981, 41–44; and Editions 1980, 138–41; see also Mahnke and Mahnke 1995, 11–16

These findings were not the result of a cartographic inquiry, so they can be used only as general guidelines. Nonetheless, they can serve the cartographer as a starting point for color selection.

Connotative Meaning and Color

Perhaps the most interesting design aspect of color is that people react differently to spectral energies. Two people look at the same red wavelength, but their responses may be entirely different. This presents the cartographic designer with a considerable challenge, so the subjective and connotative meanings of color should be studied.

Connotative responses to colors vary considerably. The literature, both in psychology and advertising, is sometimes vague and contradictory. Nonetheless, some generalization may be made (see Table 14.3). Psychologists suggest that reds, yellows, and oranges are usually associated with excitement, stimulation, and aggression; blues and greens with calm, security, and peace; black, browns, and grays with melancholy, sadness, and depression; yellow with cheer, gaiety, and fun; and purple with dignity, royalty, and sadness (Sharpe 1974). Advertising people are quick to point out that color context and copy also suggest the connotative meaning of color.

The implications of this in cartographic design are not altogether clear. There is a paucity of cartographic research into color meaning and map design. Until adequate research can lead the way, it appears that we must borrow from the psychologists and advertising people. In general, the map designer should attempt to select colors by carefully taking into account their connotative meanings in association with the map's message.

Advancing and Retreating Colors

The eye of the reader seems to perceive colors as **advancing** or **retreating.** The colors with the longer wavelengths, notably red, appear closer to the viewer when seen along with a color of shorter wavelength. There is some evidence of a physiological basis for this claim. The lens of the eye "bulges" when it refracts red rays. This same convex shaping of the lens takes place when we view objects close up. For this reason, red may have come to be associated with proximity. In general, warm hues (long wavelengths) advance, cools recede. For brightness or value, high values advance, and low values recede. In terms of saturation, deep or highly saturated colors advance, and less saturated colors recede.

There is at least one case in which this aspect of color has an effect on color selection. The development of figures and grounds on maps can be enhanced by choosing colors with their advancing and retreating characteristics in mind. Advancing colors should be applied to figural objects.

COLOR IN CARTOGRAPHIC DESIGN

Through the use of a variety of design strategies, several functional uses of color can be achieved on the map, as discussed in this final section of our introduction to the use of color in cartographic design.

The Functions of Color in Design

Arthur Robinson has written succinctly on the various functions of color in mapping (Robinson 1967), and is summarized as follows:

1. Color functions as a *simplifying* and *clarifying* agent. In this regard, color can be useful in the development of figure and ground organization on the map. Color can unify various map elements to serve the total organization of the planned communication.

2. Color affects the general perceptibility of the map. Legibility, visual acuity, and clarity (of distinctiveness and difference) are especially important functional results of the use of color.
3. Color elicits subjective reactions to the map. People respond to color, especially the hue dimension, with connotative and subjective overtones. Moods can be created with the use of color.

Thus, *structure, readability,* and the reader's *psychological reactions* can be affected by the use of color. It might be added that color functions to clarify thematic symbolization.

A summary of the functions of color in cartographic applications would include:

1. To establish figure-ground organization
2. To establish a visual hierarchy
3. To develop a balance in the map
4. To capture the user's attention
5. To identify and name locations
6. To identify categories
7. To provide emphasis
8. To show order and structure in layout
9. To enhance physical properties of the map
10. To reveal information for better communication

Design Strategies for the Use of Color

Map designers employ several strategies to use color to its fullest potential in map communication. Five will be treated here, with figure and ground development first.

Developing Figure and Ground

The figure and ground organization of the map can be enhanced by the use of color. Color provides contrast—a necessary component in figure formation. Perceptual grouping by similarity is also strengthened by the use of color. For example, similar hues are grouped in perception (although they may in fact be of different wavelengths). On a world map, for example, continents are more easily grouped as landmasses if they are rendered in similar hues. Colors of similar brightness or dullness are also grouped, as are warm colors, cool colors, or other like tints and shades (Sharpe 1974). Tints and shades are grouped with their primary colors. Perceptual grouping of colors is a strong tendency and should be a positive design element.

Generally, warm colors (reds, oranges, and yellows) tend to take on figural qualities better than cool colors (greens, blues, and purples), which tend to make good grounds. This may be partially explained by the tendency of the warm colors to advance and the cool colors to recede.

The selection of color must not be arbitrary and without forethought. Just as the old saying that "bigger is not necessarily better," brighter (bold or saturated color) is not necessarily better when using colors on a map. The color choices we make can degrade from the overall map design. A well-designed map maintains a balance in the placement of components as well as a balance in aesthetics of the colors selected. If a large area or polygon is filled with a highly saturated warm color, it may in fact degrade from the overall balance of the map. For the map user to observe the entirety of the map, stronger colors should be selected for smaller areas and less saturated colors selected for larger areas. The hue that has both high saturation and value when used in a larger polygon will actually draw too much attention away from the remaining categories. The selection of colors used in choropleth or other such quantitative maps is much less flexible, as the colors will be directly associated with the data which, in turn, is related to the polygon. We recommend that when selecting colors for quantitative maps you evaluate the distribution and impact of color and experiment with color ramps that are possibly less vibrant.

Color combinations also affect figure and ground development (refer to Color Plate 8.1 and Table 14.4). Yellow on black is noted to be the most visible color combination—yellow tends to be perceived as figure. The least visible, and therefore worst combination move in terms of figure and ground development, is red on green. The colors listed in Table 14.4 are to be used only as starting points for selection, because any modification of chroma and value will affect the results. The cartographic designer needs to balance other design elements with these color combinations to achieve an overall solution.

We frequently forget that areas that do not contain map components are comprised of a color, known as white space. This blank area also helps establish figure-ground organization. It allows the user to focus on the individual components, such as the title, legend, or body of the map without these components competing for visual attention. All colors, including white, are used to establish figure-ground relationships on two levels: symbols to the map body, and map body to the map background. White space is an active player in the organization of map space.

TABLE 14.4 COLOR COMBINATIONS USEFUL IN DEVELOPING FIGURE AND GROUND ORGANIZATION ON THE THEMATIC MAP

Figure Colors		Ground Colors
Yellow	Best	Black
White		Blue
Black		Orange
Black		Yellow
Orange		Black
Black		White
White		Red
Red		Yellow
Green		White
Orange		White
Red	Worst	Green

Source: Sharpe 1974, 107

The Use of Color Contrast

Contrast is the most important design element in thematic mapping. Contrast in the employment of color can lead to clarity, legibility, and better figure-ground development. A map rendered in color with little contrast is dull and lifeless and does not demand attention. Even in black-and-white mapping (or one color other than black), contrasts of line, pattern, value, and size are possible. With color, additional possibilities exist. One colorist lists six possible color contrasts (Itten 1970):

1. Contrast of hue
2. Contrast of value (light-dark)
3. Contrast of cold and warm colors
4. Complementary contrasts
5. Simultaneous contrast
6. Contrast of saturation (intensity or chroma)

Hue contrast can be used in cartographic design as a way to affect clarity and legibility and to generate different visual hierarchical levels in map structure. Some cartographers believe that hue is the most interesting dimension in color application in mapping, more so than value or chroma (Robinson 1967). Contrast of hue demands an overall color plan for the map and requires that some thought be given to color balance and harmony, topics not very well researched by cartographers. It is also important to note that identical hues appear differently, depending on their color environment.

Saturation and value are two contrasts that provide visual interest and, depending on the nature of the map, can carry quantitative information. Contrast of value is a fundamental necessity in structuring the color map's visual field into figures and grounds. Objects that are high in value (relatively light) tend to emerge as figures, provided other components in the field do not impede figure formation. It is difficult to talk of saturation and value in isolation; they are usually a part of any hue selection. Both are essential in the design of quantitative maps.

Contrast of cold and warm colors can also be used to enhance figure and ground formation on the map. Artists use this contrast to achieve the impression of distance; faraway objects are rendered in cold (blue/green) colors and nearby objects in warmer tones. Figures on maps should be rendered in colors of the warmer wavelengths, relative to the ground hues.

In most mapping cases the designer should avoid simultaneous contrast, because this effect is visually troublesome and distracting. For some, simultaneous contrast can be minimized, if not eliminated altogether, by separating color areas with black or white lines.

Many artists believe that providing complementary colors in a composition establishes stability. Complementary colors are opposite on the color wheel. The primary complementary colors are yellow-violet, blue-orange, and red-green. The eye will spontaneously produce the complementary of a color fixated on if it is not present. Because of this, it is said that harmony can be achieved in a composition by a balance of complementary colors. The eye does not then need to produce complementary colors on its own, so the image is more stable.

TABLE 14.5 LEGIBILITY OF COLORED LETTERING ON COLORED BACKGROUNDS

Color Combination (Object on Background)	
Black on yellow	Most legible
Green on white	
Blue on white	
White on blue	
Black on white	
Yellow on black	
White on red	
White on orange	
White on black	
Red on yellow	
Green on red	
Red on green	Least

Source: Hackl 1981, 41–44

Developing Legibility

The legibility of colored objects, especially lettering, is greatly influenced by their colored surroundings. Symbols in color must be placed on color backgrounds that do not affect their legibility. Black lettering on yellow (object on background) is a very high legibility combination, and green lettering on red is the least (see Table 14.5). The difficulty is exacerbated because the lettering is usually spread over several different background colors. Black lettering may become illegible as it crosses dull or gray colors. The designer must pay careful attention to lettering and all color environments in which it is placed. An otherwise good design can fail if caution is not exercised in this respect.

Color Conventions in Mapping

Conventional uses of color in mapping may be separated into qualitative and quantitative conventions. On most color thematic maps, these conventions must be observed because breaking a convention can be extremely disconcerting to the map reader.

Qualitative Conventions. Colors used on maps in a qualitative manner are those applied to lines, areas, or symbols that show nominal or ordinal information, not amount. Qualitative color conventions use color hue and saturation to show these nominal (and, in some cases, ordinal) classifications. Many conventions are quite old—such as showing water areas as blue—and the logic of their use is well established.

Robinson (1967) has itemized many of the conventional uses, as follows:

1. Blue for water
2. Red with warm and blue with cool temperature, as in climatic and ocean representations
3. Yellow and tans for dry and little vegetation
4. Brown for land surfaces (representation of uplands and contours)
5. Green for lush and thick vegetation

The color dimension most logically used on the qualitative map is hue. Hue shows nominal classification well and is especially appropriate because hue is difficult to associate psychologically with varying amounts or quantity of data. Caution must be exercised in using contrast of value to show nominal characteristics, because people tend to assign quantitative meanings to value differences.

If you are mapping polygons that are of different categorical data (such as land use or geological bedrock), then you should select colors that are of different hues. This is done by selecting colors from around the color wheel. They may be quite different, utilizing large jumps in hues or of more incremental steps. The latter allows you to indicate categories that are somewhat similar and yet different by selecting colors that are closely spaced spectrally. Spectrally ordered sequences provide you with eight to ten hue options that are significantly different. If a greater number of colors are required, then you can modify the saturation of the individual hues in order to increase the number used. It is recommended that you carefully select the saturation levels so as to maximize the color differences. It becomes quite challenging to select colors to map information that contains a large number of categories, such as land-use categories extracted from remotely sensed imagery. It may require combining color with pattern to achieve the largest diversity in the design of such maps.

Quantitative Conventions. Several International Statistical Congresses were held in Europe between 1853 and 1876. The question of statistical graphics (including maps) became an issue at several of these. In particular, at the Congress of 1857 in Vienna, the sixth section of one report contained several recommendations on the use of color for statistical maps. Among the most important to us were as follows (Funkhouser 1937, 314):

> *When one wishes to make use of the method [graphic], so often useful, of applying colors in various shades to geographic maps, he should have regard for the following points:*
>
> a. *Twelve classes or divisions permit the establishment of a gradation of tints easy to distinguish from the lightest to the darkest shades and present the further advantage of allowing the comparison to be limited to six, four, or three classes, when there is a desire to consider only the principal relations.*
>
> b. *The strongest proportions should be indicated by the darkest shades.*
>
> c. *When it is a question of showing opposite extremes, the intermediate shades should be omitted from the scale of colors used.*

It is indeed interesting to note that as long ago as 140 years, those in the emerging discipline of geographic mapping should have been so intuitive about how to use color value on the quantitative maps.

Conventions associated with color use on quantitative thematic maps today operate in terms of color choice or **color plan.** No conventions exist for color choice on quantitative maps (for example, population density maps are always blue, income maps are always green, and so on). A preconceived color plan is one way the designer chooses the color dimensions of hue, saturation, and value to symbolize varying amounts of data on the map (refer to Color Plates 6.1–6.3 for examples of single hue, bipolar, and bivariate color plans). It is especially important to use such a plan when creating atlases that involve many maps and/or graphics. Consistency in such products, whether they be printed or online, help produce a harmonious product with consistent design.

Quantitative data that are classified in order to generalize the cartographic presentation fall into two categories, either sequential or divergent. Sequential categories represent data that increase incrementally from a low value to its maximum value (refer to Color Plate 6.1). Examples of these data are the percentage of land in farms, with potential values ranging from 0 to 100 percent, or average farm value, with a possible range of 79 to 785 thousand dollars. The color plan selected for these data would be that of a single hue with incremental changes in value or saturation corresponding to the numerical gradations. This would produce colors that range from a light or pastel blue to a dark or fully saturated blue. The number of classes, and thus the number of colors necessary, will depend on the classification technique utilized. The accepted convention in color selection is that more fully saturated color represents higher data values and desaturated color represents lower values of the data.

Bipolar (also called *divergent*) data tend to increase in opposite directions from some neutral point, such as zero. An example would be the percentage change in population between two time periods. The classification of the data could create an equal number of classes with positive values as with negative values (refer to Color Plate 6.2). In this case, complementary hues from across the color wheel should be used to represent the data. These hues would increase in equal saturation increments from low to high data values.

Sometimes the divergent data are classified so that a midrange class represents an amount of modest change, for example, +5 to −5 percent change. In this case, the color selected for this middle category would be either a neutral gray or an equal low saturation amount of both hues. Normally, this classification technique generates an odd number of data classes.

In their research with the National Center of Health Statistics, Cynthia Brewer and her colleagues found spectral schemes "most pleasant and easy to read," and they go on to say "contrary to our expectations, spectral schemes are effective and preferred (the spectral scheme we tested included diverging lightness steps suited to the quintile-based classification of the mapped mortality data)" (Brewer *et al.* 1997, 87).

Brewer states, "The selection of thematic [color] schemes is assisted by the perceptual organization of the Munsell-based charts. Quantitative relationships between map categories are commonly represented by a progressive change in values accompanied by a systematic treatment of chroma" (Brewer 1989, 273). This is consistent with Mersey (1984, 1991) and many other cartographers on the use of color for

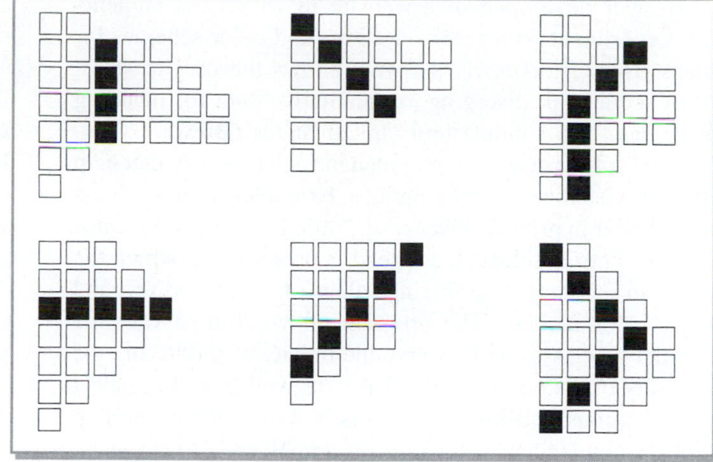

FIGURE 14.10 REPRESENTATIVE COLOR SERIES FOR QUANTITATIVE THEMATIC MAPS FROM MUNSELL COLOR CHARTS.

Either value or chroma alone, or together, may be selected from the chart. (*Source:* Redrawn from Brewer 1989, Figure 6)

such thematic maps. In her work, Brewer provides representative paths through the Munsell chips to achieve several appropriate color series; these are included in Figure 14.10.

Color Brewer. Reflecting considerable work of assimilation and amalgamation from many writers and color researchers by cartographers and others, and from her own expertise and experience, cartographer Cynthia Brewer provides a very useful organizational schemata for the use of color on thematic maps (Brewer 1994). This color schemata is provided by using hue (the name of a color, such as red, orange, and so forth) and "lightness" (a term she prefers instead of value, darkness, intensity, and so forth), and, only secondarily, saturation. These schemata are captured in Table 14.6.

The essential strength of Brewer's organization is that it is consistent with the overall approach that color use must align logically with the data being shown on the map. When color is used "appropriately" on a map, the organization of the perceptual dimensions of color corresponds to the logical organization in the mapped data.

As Brewer notes in her work, color application for quantitative mapping must take into consideration the data being displayed and the logical/perceptual organization of color. She notes that this is especially important for dynamic (animated) displays because events happen quickly and the logic between data and color must be clear and easily grasped. The coordination of these variables is essential in the use of color and map design.

In the more recent work of Brewer and her associates, they found that diverging schemes produce "better rate retrieval than spectral or sequential schemes. . . ." This is really a remarkable finding and bears close scrutiny in future quantitative color work and map application (Brewer *et al.* 1997).

TABLE 14.6 BREWER COLOR SCHEMATA FOR QUALITATIVE AND QUANTITATIVE MAPS

Schemata Type	Color Organization
One-variable map	
Qualitative	Hues of similar lightness
Binary	One hue and lightness step
Sequential—no hue	Lightness steps of neutral grays
Sequential—one hue	Lightness steps of a single hue
Sequential—hue transition	Lightness steps with a part-spectral transition in hue
Sequential—hue steps	Lightness steps with hue steps that progress through all spectral hues, but beginning in the middle of the spectrum, with low values represented by light yellow
Spectral	Not appropriate
Diverging	Two hues differentiate increase from decrease and lightness steps within each hue, colors lighter show minimal change
Diverging	Two hues with lightness steps diverging from the midpoint
Two-variable map	
Qualitative/binary	Different hues for the qualitative variable with a lightness step for the binary variable
Qualitative/sequential	Different hues for the qualitative variable crossed with lightness steps for the sequential variable
Sequential/sequential	Sequential/sequential scheme with cross of lightness steps of two complementary hues with mixtures producing a neutral diagonal
Combination of sequential schemes	Combinations of hue differences resulting in hue and lightness steps

Brewer developed of a website in which the students can experiment with color selections and color schemes for maps. Her ColorBrewer software allows the user to select from sequential, diverging, or qualitative data for mapping an area of the southeastern United States (Brewer 2006). Various **color ramps** are provided in each category that help the user visualize the color options. Examples of these charts also appear in print (Brewer *et al.* 2003; Brewer 2005). Once a color ramp is selected, a legend is presented in which the user can determine the mixing values of RGB, CMYK, and LAB color models. The program allows you to vary the number of classes and observe the impact of modifying the intensity of the polygon boundaries, as well as adding select city locations and highways. This is an excellent learning tool for any student studying cartography and developing a sense of color selection.

Color Harmony in Map Design

Color harmony traditionally is believed to belong to the realm of the artist and is therefore not considered a matter of concern to the cartographic designer. This is far from the truth, but this element of design has been overlooked in cartographic research. Some aspects of this important subject are introduced in this section.

Color harmony, as it applies to maps, is more than the suitable and pleasing relationship of hues. Color harmony relates to the overall color architecture for the entire map. Harmony includes these components:

1. Effectiveness of the *functional uses* of color on the map
2. Appropriateness of the *conventional uses* of color on the map
3. Overall appropriateness of *color selection* relative to map content
4. Effective use of the *quantitative color plan*
5. Effective employment of the *relationship of hues*

Effectiveness of functional uses relates to how well the designer has employed color as a simplifying and clarifying agent. A harmonious design can be judged also on its appropriate use of color convention. Has every care been taken to provide conventional color wherever possible? Do departures from convention restrict the ease of communication?

Another important component of overall color harmony is the proper selection of color relative to map content. Maps illustrating January temperatures should not be rendered in warm hues, because of convention and connotative aspects, and deserts having sparse vegetation should not be shown in green.

Achievement of an effective relationship of hue balance of all colors on the map requires great skill and careful planning. Any multicolor map has its different colors occupying areas of varying sizes. **Color balance** is the result of an artful blending of colors, their dimensions, and their areas so that dominant colors occupying large areas do not overpower the remainder of the map. Dominant colors are those that contrast greatly with their environment; any color can be dominant, depending on its surrounding colors.

Cartographic designers have their hands full when dealing with balance. In most instances, areas are fixed by geography, so only choice of color is possible when planning balance. Consideration must be given to the other elements of color harmony as well. The designer needs experience with color and knowledge of how colors behave in different environments to solve balance problems.

Color harmony, as artists and colorists use the term, refers to the pleasant combination of two or more colors in a composition. Harmony also refers to order—the arrangement of colors. According to Ostwald (1969), the simpler the order, the better. However, what is pleasant to one map reader may not be to another. Experimentation with the ideas of color harmony as developed by Munsell, Ostwald, Birren, Itten, and others would reveal differences.

The following thoughts regarding these relationships of harmony are derived from Birren (1956):

1. Most people like pure hues rather than modifications (reds, not purples).
2. Every color should be a good example in its category. If a red is chosen, it should not be possible to mistake it for a purple. Pure colors, if selected, must be brilliant and saturated.
3. Whites should be white, and blacks should be black.
4. Tints, shades, and tones should be easily seen as such, not confused with pure hues, blacks, or whites.

To these may be added the following general harmony rules:

5. Colors opposite on the color wheel tend to be harmonious.
6. Harmony can also be achieved by using colors adjacent on the color wheel.
7. Harmony can also be achieved by combining colors of the same hue but with different tints, shades, and tones.

The words *tint, shade,* and *tone* to treat harmony are not often used by cartographers but merit inclusion here because cartographers often use them in informal conversation. Professional cartographers must explore in considerable depth the relationships of colors. The task is made more difficult because cartography is not a purely expressive art. Color selection and employment are constrained by the map's other elements.

The next chapter is devoted to printing technology as it relates to the cartographer.

REFERENCES

Beck, J. 1972. *Surface Color Perception*. Ithaca, NY: Cornell University Press.

Billmeyer, F., and M. Saltzman. 1966. *Principles of Color Technology.* New York: Wiley.

Birren, F. 1956. *Selling Color to People*. New York: University Books.

———. 1969. *Munsell: A Grammar of Color*. New York: Van Nostrand Reinhold.

Brewer, C. 1989. The Development of Process-Printed Munsell Charts for Selecting Map Colors. *American Cartographer* 16 (4): 269–78.

———. 1994. Color Use Guidelines for Mapping and Visualization. In *Visualization in Modern Cartography,* eds. A. MacEachren and D.R. Fraser Taylor. New York: Elsevier: 123–47.

———. 2005. *Designing Better Maps: A Guide for GIS Users.* Redlands, CA: ESRI Press.

———. 2006. http://www.ColorBrewer.org

Brewer, C., G. Hatchard, and M. Harrower. 2003. ColorBrewer in Print: A Catalog of Color Schemes for Maps. *Cartography and Geographic Information Science* 30(1): 5–32.

Brewer, C., A. MacEachren, L. Pickle, and D. Herrmann. 1997. Mapping Mortality: Evaluating Color Schemes for Choropleth Maps. *Annals* Association of American Geographers 87: 411–38.

Dember, W. 1960. *Psychology of Perception.* New York: Holt, Rinehart and Winston.

Dunn, W., and A. Barbau. 1978. *Advertising: Its Role in Modern Marketing.* Hinsdale, IL: Dryden Press.

Editions, M. 1980. *Color.* New York: Viking Press.

Fisher, H. 1974. An Introduction to Color. In *Color in Art,* ed. James M. Carpenter. Cambridge: Fogg Art Museum, Harvard University.

Fraser, B., C. Murphy, and F. Bunting. 2005. *Real World Color Management, 2d ed.* Berkeley, CA: Peachpit Press.

Funkhouser, G. 1937. Historical Development of the Graphical Representations of Statistical Data. *Osiris* 3: 268–404.

Hackl, A. 1981. Hidden Meanings of Color, *Graphic Arts Buyer* 12: 41–44.

Helson, H., and J. Lansford. 1970. The Role of Spectral Energy of Source and Background Color in the Pleasantness of Object Colors. *Applied Optics* 9: 1513–62.

Itten, J. 1970. *The Elements of Color.* New York: Van Nostrand Reinhold.

Littlefield, J., and C. Kirkpatrick. 1970. *Advertising: Mass Communication in Marketing.* Boston: Houghton Mifflin.

Mahnke, F., and R. Mahnke. 1995. *Color and Light in Man-Made Environments*, New York: Van Nostrand Reinhold.

Mersey, J. 1984. *The Effects of Color Scheme and Number of Classes on Choropleth Map Communication.* Unpublished Ph.D. dissertation, Department of Geography, University of Wisconsin—Madison.

———. 1991. Color and Thematic Map Design. *Cartographica* 27 (Monograph, No. 41): 1–157.

Munsell, A. 1981. *A Color Notation.* 14th ed. Baltimore: Macbeth, A Division of Kollmorgen Corporation.

Nyman, M. 1991, 1993. *Four Colors/One Image*. Peachpit Press.

Olson, J. 1981. Spectrally Encoded Two-Variable Maps. *Annals* Association of American Geographers 71: 259–76.

Olson, J., and C. Brewer. 1997. An Examination of Color Selections to Accommodate Map Users with Color-Vision Impairment *Annals* Association of American Geographers 87: 103–34.

Ostwald, W. 1969. *The Color Primer.* New York: Van Nostrand Reinhold.

Prust, Z. 1999. Graphic Communications: The Printed Image. Tinley Park, Illinois: The Goodheart-Willcox Company, Inc.

Riley, C. 1995. *Color Codes.* Hanover, NH: University Press of New England.

Robinson, A. 1966. *The Look of Maps: An Examination of Cartographic Design.* Madison: University of Wisconsin Press.

———. 1967. Psychological Aspects of Color in Cartography. *International Yearbook of Cartography* 7: 50–59.

Rossing, T., and C. Chiaverina. 1999. *Light Science: Physics and the Visual Arts.* New York: Springer.

Sharpe, D. 1974. *The Psychology of Color and Design:* Chicago: Nelson-Hall.

Sorrell, P. 1974. Map Design—With the Young in Mind. *Cartographic Journal* 11: 82–91.

GLOSSARY

achromatic having no hue (such as red, green, blue), but having characteristics of levels of darkness or light (such as white, gray, or black)

additive color theory emitted light from the sun or a computer monitor generates colors based on the additive color primaries

additive primary colors red, green, and blue; additive colors are those that, when combined with other colors, yield new ones

advancing colors a hue of longer wavelength (notably red), a color of high value, or a highly saturated hue appears closer to the viewer; apparently caused by both physiological and learned mechanisms

chromatic having the quality of hue, such as red, green, and blue

color light energy at different wavelengths along the visible electromagnetic spectrum; red is about 750 nm in wavelength, and violet is about 350 nm

colorant the elements in substances that cause color, either through absorption or reflection

color balance the result of artful blending of colors and their areas so that dominant colors occupying large areas do not overpower the remainder of the composition; sometimes referred to as contrast of extension

color constancy the tendency to judge colors as being identical under different viewing conditions, such as different illumination

color harmony the pleasant combinations of two or more colors in a composition; can be developed by functional, conventional, and color selection plans

color interaction the way we perceive colors together; always modified by their environment

color perception cognitive process involving the brain, where meaning and substance are added to light sensation

color plan choosing the color dimensions of hue, value, and chroma to symbolize varying amounts of data

color ramps a sequence of colors used in qualitative and quantitative mapping; the ramp is frequently a component of GIS and mapping software color selection options

color wheel the organization of hues of the visible spectrum into a circle; many distinct hues can be shown, but 12 are customary, especially among artists

cone cells cells in the retina with peak sensitivity to blue, red, or green light energy

cool colors colors of the shorter wavelengths, such as violet, blue, and green

cornea transparent outer protective membrane over the lens of the eye

gamut display of all colors by a particular printer or other hardware device

hue the quality in light that gives it a color name such as red, green, or blue; a way of naming wavelength; one of the three color dimensions

illuminant mode when light is emitted from a computer monitor, movie projector, or other light source

induction a special case of simultaneous contrast; causes adjacent colors to be lighter in the direction of the darker adjacent colors, and darker in the direction of the lighter colors

iris diaphragm-like muscle that controls the amount of light coming into the eye

lens crystal-like tissue that focuses light onto the back of the inside of the eye

light that part of the electromagnetic spectrum that creates sensations in human eyes; light occupies wavelengths from 7.5×10^{-5} to 3.5×10^{-5} cm

Munsell color solid a three-dimensional geometric figure, roughly equivalent to a sphere, designed to show the interrelationships of hue, value, and chroma

opaque surfaces that absorb all light produce the color black

optic nerve bundle of nerve cells connecting the retina to the brain; conveys electrical impulses that carry light information

psychological dimensions (of color) hue, value, and chroma

reflective mode (also called object mode of viewing) the production of color in which electromagnetic energy is reflected off of an object reducing the wavelength of light

retina membrane that lines the back of the inside of the eye; contains light-sensitive cells

retreating colors a hue of shorter wavelength (blues), a color of low value, or a poorly saturated hue appears farther away to the viewer; apparently caused by both physiological and learned mechanisms

rod cells cells in the retina sensitive to achromatic light

saturation the chroma, intensity, or purity of a color; one of the three color dimensions

shade the result of mixing a hue with black

simultaneous contrast a color interaction; the eye produces the complementary color of the one being viewed

spectral reflectance curve graphic plot of light reflected from objects or surfaces

subjective reactions to color involve color preferences, meanings, and behavioral moods produced by colors

subtractive color theory light that strikes a paper map has some of the energy absorbed by the paper and ink, which reduces the energy that is reflected from the surface; color that is observed is created by the reduced wavelengths reflecting from the map surface

subtractive primary colors combinations of colors resulting from the absorption of a primary color or colors

successive contrast a color appears different to the eye on different backgrounds, especially when viewed successively

tint the result of mixing a hue with white

tone the result of mixing a hue, white, and black

transparent objects objects that transmit all light passing through them

tristimulus values amounts of red, green, and blue light necessary to match a test lamp color; used in RGB color specification

value an array of color based on lightness and darkness qualities; one of the three color dimensions

warm colors colors of the longer wavelengths such as red, orange, and yellow

PART IV

MAP PRODUCTION TECHNIQUES

Following the successful design of a map, one must prepare to distribute the product to others. This may be done in printed paper-copy format or in virtual form via the Web. This section deals specifically with these two topics. Chapter 15 explores the printing process, examining the familiar desktop laser and inkjet printers and the large-format plotters for printing a small number of copies. Following a brief examination of the history of commercial printing, the map production process is analyzed. A six-step sequence is examined for preparing documents for printing large-size copies and a large number of copies. The need for constant editing and proofing in every step of the process is stressed. In this chapter, we explore the differences in printing presses from web- to sheet-fed to digital presses. The production process of printing color maps uses the CMYK color model and, therefore, color separations in the use of process printing.

Chapter 16 examines virtual and web mapping techniques. These digital forms require an understanding of both raster and vector graphics and their related file types. This chapter considers static, interactive, and animated maps. A brief exploration of the mapping media includes the World Wide Web and graphics display monitors. The chapter concludes with a discussion of the constraints of the medium and the solutions and opportunities presented in virtual and web mapping.

15 PRINTING FUNDAMENTALS AND PREPRESS OPERATIONS FOR THE CARTOGRAPHER

CHAPTER PREVIEW Cartographic production and reproduction techniques are integral to the design planning of a thematic map. From the outset, the designer develops the graphic specifications based on a knowledge of how the map will be reproduced. A working knowledge of the technology of printing is essential for the well-trained cartographic designer. An understanding of color reproduction is especially important because this form of printing requires specific detail and is considerably more complex than black-and-white duplication. All phases of map production—from executing the design concept to final art form ready for printing—are referred to as "prepress operations."

Printing technology caused revolutionary changes in the printing industry in the late 1990s and the beginning of the twenty-first century. Desktop printing is the norm for most cartographers when relatively few copies of the maps are required. The quality of graphics produced by these printers continues to improve, and they provide color output that rivals documents created by the printing professional. Computer-to-plate technology has already revolutionized the printing process, whereas digital production presses are revolutionizing the type of presses of this century. Design work requires careful planning to accommodate such sophisticated printing technology, and cartographers need to work closely with service bureaus to successfully reach design goals. ■

This chapter covers several topics essential to the practice of thematic map production and reproduction. These topics play an important role in the overall design process. One cannot adequately approach a design task without at least an elementary knowledge of how maps are printed following their creation on-screen. Maps may be printed using desktop printers, stand-alone plotters, or professional photolithographic processes. This chapter includes a discussion on the preparation of the map for hard copy output, prepress planning, desktop printers, and free-standing plotters. The chapter also includes a brief discussion of the history of printing, a look at proofing of the final map product, and an examination of the map production process.

The cartographer has essentially two options when it comes to printing the final map or map layout. These options include personal printing via a printer or plotter connected to his or her computer or network—or through a commercial printing company. The choice between these options is based upon the number of copies to be printed and the size of the copy or copies. If the map is to be printed as a single copy or relatively few copies, the output will be created by laser and/or inkjet printers. Sometimes the graphic is of a size too large for these traditional "desktop" printers. In this situation, when the physical size of the map is large and the number of desired copies is small, a plotter may be used. A discussion of these desktop printers is found later in this chapter.

The commercial printing option results from the need for printing any size map or layout in large quantities, in the hundreds or thousands of copies. In this case, the cartographer creates the graphic in the GIS or mapping software and exports the document to the printer. At this point, the document is out of the hands of the cartographer and under the control of the printing professional. The printing of these maps is normally handled via a contractual arrangement between the cartographer and printer. Therefore, unlike the first option where the cartographer maintains personal control of the printing process, the need for good communication and interaction between the cartographer and printer is essential.

The distinction between these two options throughout the remaining part of this chapter is based on the quantity of printed documents desired. When the quantity is few, we will refer to desktop printing, and when the quantity is large, we will refer to commercial printing. The chapter concludes by examining the map production process for map design to post-press operations.

A caveat to keep in mind is that we design our maps using GIS, mapping, and artistic software in vector mode, lines comprised of a series of nodes that when linked together generate a line. Printers, by their very nature, use the raster mode in which the printer's resolution, measured in dots per linear inch, determines the quality of the printed line. The printing software converts the vector information input by the mapping software to a raster image prior to printing (refer to Chapter 16 for a discussion of raster and vector).

CARTOGRAPHY AND DIGITAL PRINTING

The production of a thematic map requires detailed planning so that the printed map will look in every way as the designer intended. The cartographer necessarily deals with content, generalization and symbolization, scale, and other mapping components, but the final map will not emerge as envisioned without careful attention to all phases of the production process. The purpose of this section is to present ways of organizing and managing this plan.

Getting Started

As we have stressed in previous chapters, planning is a crucial step in the generation of a well-designed map. Decisions have to be made from the very beginning. Depending upon the specific thematic map *type* being generated, one should determine the final map *form:* virtual on-screen, or printed hard copy.

For maps to appear on-screen, determine:

1. appropriate color model or grayscale
2. map projection
3. scale and physical size
4. map content
5. symbol selection
6. font and text size

For printed hard copy maps, determine:

1. CMYK color model or grayscale
2. map projection
3. scale and physical size
4. map content
5. which printing option will be used
6. symbol selection
7. font and text size
8. quantity to be printed

Color Model

Maps that are destined for printing will be designed using the CMYK color model. It is important that this color model be selected within the GIS and mapping software from the very beginning. This will help eliminate color selections that are out of range (gamut) for the output device. Printing of these maps will utilize each component of the color model as a separation used in the four-color process. The printer will access the individual C, M, Y, or K component and treat it as a layer in printing the hard copy. This color model is used whether you output your color maps using desktop printing or commercial printing.

DESKTOP PRINTING

For projects that require page-size color copies in small quantities, the normal mode of printing involves inexpensive printers that easily fit on the desktop. Such printers allow for color graphics to be printed in high resolutions of 600 to 1200 dots per inch (dpi) or higher. There are normally four classes of printers that meet the desktop category: laser, inkjet, plotter, and dye-sublimation printers.

Laser Printers

Laser (light amplification by simulated emission of radiation) was first produced by the Xerox Corporation in 1971. Color **laser printers** are modestly priced when additional memory is added as a result of the growing size of color graphic files. The printers are capable of relatively high-speed printing at 1200 dpi or higher (Adams 2002). The printer requires a warm-up operation in which the fuser element is heated. Laser diodes use the raster image to project an image onto a photoreceptor, creating an electrically charged latent image on the receptor surface. This image is exposed to the four-color toners which are electrostatically attached to the surface. The photoreceptor is rolled over the paper, transferring the image. Lastly, the paper passes through the fuser assembly which bonds the toner to the paper, using heat and pressure.

The cyan, magenta, yellow, and black toners are printed onto the paper surface, simulating the processes of a large printing press. Each ink is applied as dots printed in a linear fashion with a consistent number of dots per inch. In order to display an increase in the color value, the frequency of dots remains the same and the size of the dots is increased to represent the increased color density. As with lithographic printing technology, a different angle of printed dots is used for each CMYK component. This allows the dots to be printed without printing on top of one another. Lithographic printing technology will be discussed later in this chapter.

Inkjet Printers

The **inkjet printers** can be single-function printers or are currently combined as a printer/scanner/copier combination with high-quality output at very low cost. The system is comprised of a print head that passes across the paper, propelling small droplets of liquid ink onto the paper. The maintenance costs of these printers are comprised primarily in replacing or refilling the ink cartridges. Most inkjet printers have one cartridge of black and a second cartridge containing lesser volumes of cyan, magenta, and yellow ink. Ink is applied in a series of randomly noncontinuous patterns of CMYK dots of similar size. As the color value (tone) increases, the density of the dots is increased while the dot size remains constant. One disadvantage of this printer type is that the liquid ink can bleed laterally as a result of the capillary effect of low-quality paper. This effect can be minimized through the use of higher quality clay-treated printer paper. A second disadvantage is that the ink, once on the paper, may run (bleed) if it gets wet. Finally, the print head may clog if used infrequently, and will require cleaning. Usually, the printer software contains options for both checking color density and for unclogging the print head.

Plotters

The current use of the term **plotter** refers to a free-standing large-format inkjet printer. A common size of plotter ranges between 37 inches (94 cm) and 44 inches (112 cm) wide and uses a roll of print paper that allows for nearly unconstrained length to the plot. Raster graphic files are directed to the plotter, which then applies the CMYK inks to the paper in the same manner discussed under inkjet printer. In the past, a plotter was a vector graphics pen printing device that used a graphics language for such commands as pen up, pen down, move, and so on. In the truest sense of traditional cartography, this hardware device emulated the hand-drawn map of lines, points, and polygons, and was most associated with computer-aided design (CAD) drawing programs.

Dye-Sublimation Printers

Although these printers are less common on the personal desktop today, they are still used in high-quality printing of graphic images. The **dye-sublimation printer** uses vapors created by heat to transfer the dye, stored on a cellophane ribbon, to the paper. This sublimation process heats up the dye until it is turned into a gas which is diffused onto the paper and solidifies. These printers frequently use CMYO colors where black is replaced by a clear overcoat (the 'O' in CMYO) in order to protect the print from discoloration from exposure to UV light. The dye-sublimation printer is a continuous tone printer frequently used for photographic images.

COMMERCIAL PRINTING TECHNOLOGY

The desktop printers and free-standing plotters work well for generating copies both small in size and number of copies printed. For larger-sized graphics where maps are printed by the thousands and are laid out as pages of a book or components of a booklet, the commercial printing technology allows for high-quality printing of large press runs. The remainder of this chapter focuses on the history of commercial printing, printing techniques and presses, and the modern digital production presses.

Brief History of Commercial Printing

Map printing has paralleled the history of all printing. The development of **letterpress,** the process used in making the Gutenberg Bible in 1452, led to the first **woodcut map** in 1472 (Robinson 1975). From that time to the present, there has been a close working relationship between cartographers and printers. In fact, during the first 300 years of printing, the printer probably had as much to say about cartography as the cartographer did. This was no doubt a result of the strong influence of the printing craft guilds. It was not until the latter part of the twentieth century that cartographers began to control their own products. Designers now have access to a broad range of color-printing technology and can be confident that their maps will be printed using the highest quality control in both registration and color quality.

During the history of printing, three major printing methods have been used by cartographers: relief, intaglio, and planar. Each has its own merits and weaknesses for map reproduction. Most large sheet maps intended for mass production are reproduced today by planar methods, which will be examined in detail later in this chapter.

Relief—Letterpress

Letterpress, also called *relief* printing, is the oldest printing method. Although popularly thought to have been invented by Johann Gutenberg about 1450, the method was used by the Chinese in the eighth century (Adams and Faux 1977; Prust 2003). What Gutenberg actually invented was relief printing with movable type. In relief printing, ink is applied to a raised surface and pressed onto the paper (see Figure 15.1a). The relief blocks (letters) are reversed so that they will be right-reading after printing. For reproducing art and cartography products, this relief method is referred to as *woodcut*. End-grain wood was chiefly used. Portions of the image that would not be printed were chiseled out, leaving the print portions raised in relief.

Early maps reproduced from woodcuts suffered from a variety of problems inherent in the method. Lines were necessarily thick. Images were often smeared because each impression required a new inking. The paper had to be nearly smooth and free of imperfections. In addition, it was difficult to add lettering to maps—in some cases, holes were cut in the woodcut

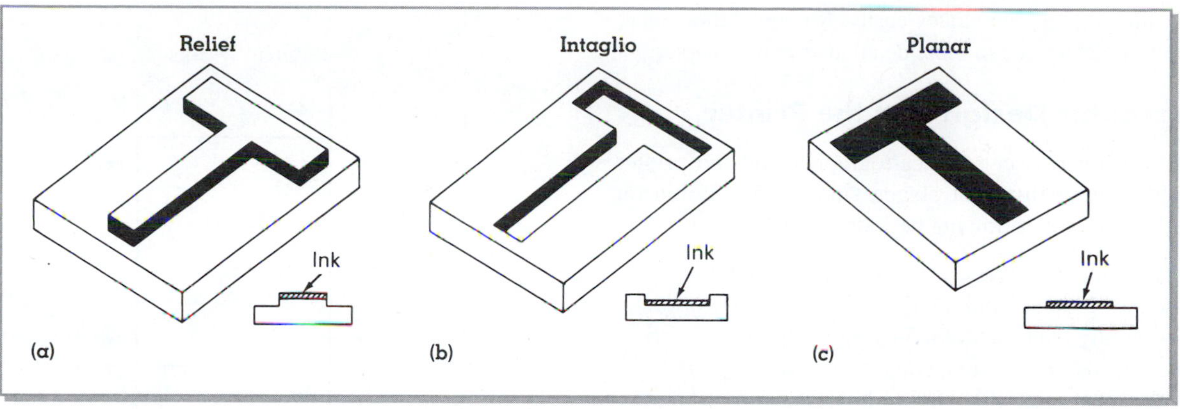

FIGURE 15.1 TRADITIONAL MANUAL APPROACH IN PRINTING.
The three principal ways of nonelectronic printing: (a) relief, (b) intaglio, and (c) planar.

and letter blocks dropped in. Also, large pieces of wood could not be used because of warping; large maps were often printed by piecing together the impressions made from smaller blocks. Gradation of tone was not really possible with the woodcut method, and the image had to be chiseled backward.

Current letterpress printing utilizes a metal plate coated with a light-sensitive chemical. The plate is exposed to a very bright light through a photographic negative of the image to be printed. Those areas of the coated plate that are exposed will harden, and the nonexposed areas can be washed away. This process leaves the image areas raised in relief; they receive and transfer the ink during printing.

Intaglio—Engraving

Intaglio (pronounced in-tal-yo) printing is also called *etching, engraving,* or *gravure*. The elements of the image are first cut out by hand or incised by acid (etching) (see Figure 15.1b). Ink is then applied and the plate cleaned. Some ink remains in the depressed portions; deeper depressions contain and transfer more ink than shallower depressions. When the plate is pressed onto the paper, the ink is transferred. Best results are achieved when the paper is dampened and considerable pressure applied during transfer. Actually, the paper is pressed into the depressions during printing, and afterward there is a slight raised relief on the paper where the ink adheres. This is one way of identifying materials prepared by the intaglio process.

This form of printing was first used in the early fifteenth century and had achieved prominence by 1700. The engraved copperplate map became standard in map printing for about 150 years. In copperplate engraving, much finer lines are possible than in woodcut, which was then its chief competitor. As intaglio became popular, such techniques as **mezzotint engraving** (working the surface to create tonal image), **stipple engraving** (creating tonal effects by specialized tools), and **aquatint engraving** (special etching to create tones) were introduced. These all had wide appeal to cartographers and were especially useful for vignetting (gradation of tone or texture) at coastlines.

Intaglio methods include various forms of gravure, and printing on high-speed web (continuous paper) presses is possible. Techniques of plate making for gravure include electromechanical or laser scanning of the image in order to break it into a pattern of dots used to control the mechanical engraving of small depressions (cups) in the plate. Similar techniques are used in conventional plate making, which involves the mechanical transfer, through a carbon tissue medium, of the image onto the plate. Original art is photographed through a gravure screen. The negative is placed in contact with a photographically sensitive gravure plate and exposed. After exposure, the image is chemically etched onto the plate. The image is created on the plate by a pattern of small depressions of varying size and depth. Plates are usually of polished copper, often chromium-plated for protection. Gravure is expensive and is generally economical only for large press runs. Most maps for reproduction are not directly prepared for printing.

Planar—Lithography

Planar printing was introduced in 1796 by Alois Senefelder in Germany. Planar printing is usually referred to as lithography. The Greek word *lithos* means stone, and limestone was first used to make the plates. Today it is often called **photolithography, offset lithography,** or photo-offset lithography. This printing method relies on the fact that water and oil (grease) do not mix well. On the printing stone or plate, the image area receives the greasy ink and the nonimage areas do not. When paper is pressed to the plate, only the inked areas will transfer to the paper. Unlike the relief or intaglio methods, planar printing creates no relief differences on the plate (see Figure 15.1c). Of course, lithography no longer uses stones for printing maps and graphics. This form of printing has come into wide use in practically all commercial applications, and most maps today are printed by this method.

As lithography was introduced to the printing industry, cartographers gradually began to see its advantages. Old copper engravings could easily be updated, transferred to stone, and then printed. Lithography also meant faster preparation of plates than copper engraving; most engraving craftspeople learned the new lithographic techniques easily. Modern lithographic techniques are examined in more detail later in the chapter.

Cartographic Design and the Printer

The relationship between the cartographer and the printer has gone through different stages since 1450. Robinson (1975) identifies the following periods:

1. During the period when woodcut maps predominated, most cartographers did not do their own woodworking. This usually led to better map designs, because woodcutters were better craftspeople than cartographers.
2. For the most part, this relationship remained unchanged during the time when intaglio was preeminent. A few cartographers—such as Mercator—were also engravers. When national map surveys first flourished, cartographers and in-house engravers worked more closely than ever. Map design was still mainly shaped by engravers' styles.
3. When the transfer process (making printing plates from right-reading material) and lithography were introduced, anything could be reproduced. Specialists such as engravers and cartographers could be bypassed entirely, replaced by others who did not possess any cartographic knowledge. Cartographic design suffered.
4. Since about 1950, printers and cartographers have once again been working closely together.

As the printing industry became more and more dominated by lithography (especially photolithography), the printer became primarily a duplicator. There were no craftspeople intervening to influence the look of maps. Cartographic education did not stress the aesthetic aspects of maps, except in rare cases. Today, the cartographer handles all aspects of the aesthetic aspect of map design. Knowledge of the printing industry and its varied techniques gives the cartographer greater freedom of choice in design.

COMMERCIAL PRINTING

Maps that are destined for printing using a commercial printing organization will also access the individual C, M, Y, or K component and treat each as a layer in printing the hard copy. **Registration** of these layers is essential in order to create the highest quality product. To assist the commercial printer in preparing the plates for proper registration, it is recommended that registration marks comprised of circled crosshairs, corner cut marks (crop marks), or other such techniques be applied to the graphic beyond the edges of the printed page (see Figure 15.2). This helps to assure proper registration of the final printed map.

Frequently, maps intended for printing in large quantities include large-format graphics that exceed the size limitations of desktop printers. The dimensions of the final paper size should also be set in the GIS, mapping, and artistic software prior to beginning the design process. If you are creating graphics that are large in size, we recommend that the graphic frequently be viewed on screen at 100 percent, that is, a 1:1 ratio to the final output. The line weights, text size, and symbols selected when viewing the graphic at a reduced size do not provide the cartographer with an adequate view of these items at the larger size. The relative sizes of these items may produce an undesirable design or balance when viewed at full size. It is important to design the map for the final printed scale.

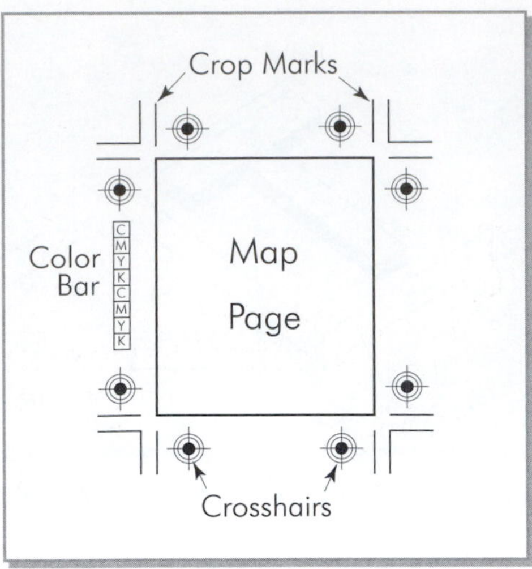

FIGURE 15.2 REGISTRATION MARKS USED IN PROCESS PRINTING.

Registration marks are added electronically to the four-color separations just off the map space. The most common techniques involve either crop marks or crosshair forms. Also shown are the color bars added for color checking during the press run.

Prepress Map Editing

Once the map design is complete and the cartographer is preparing to send the graphic to the printer, an initial edit of the map components must be made. Printing can be costly and the discovery of errors in the graphic once printed will either be retained and distributed to users or must be reprinted, increasing the cost substantially. The initial edit phase begins on-screen. All aspects of the map must be evaluated as if you are looking at the map for the first time. Overall layout of the map components should be the first item to be evaluated. Is the map balanced, has the figure-ground relationship been established, and have all intended map components been included? Line weights, symbols, colors, and font and type size should be reviewed. This is the crucial time for checking the placement and spelling of all map text. This editing step should be accomplished carefully.

The cartographer should create a test print of the graphic. Even at a smaller scale, items that are viewed in hard copy may reveal errors or discrepancies that were missed in the virtual view. If the graphic is to be printed in color, use a color printer to produce a composite image and print color separations for their quality and accuracy. This may be the last time the cartographer has physical control of her or his product.

The cartographer must constantly be on guard for errors. They may show themselves at any time and should be corrected immediately. Error corrections put off until a later phase are often forgotten and may potentially occur on the final map. Map editing is an on-going process and requires the cartographer's full attention.

File Preparation

Graphics to be printed are frequently sent to a printer or service provider for rendering and printing. The transfer of the digital file is normally done using either **Encapsulated PostScript (EPS)** or an **Editable Portable Document Format (PDF).** These file types are capable of embedding vector objects, raster images, text fonts, and symbols. When saving these files, be sure to imbed these fonts and symbol libraries used in the generation of the graphic. Without the inclusion of the fonts, the document used by the printing or **service bureau** may have substituted fonts used if they are specialty fonts and not of the common variety. The EPS and PDF formats are independent of the software in which they were created and allow the printer to prepare the document for the printing technique they use. Some service bureaus prefer **native format** files, meaning the file format that you would normally use for your particular software. All elements of a graphic file are converted to a raster format during the printing process, using a **Raster Image Processor (RIP)** as the printing process utilizes a raster system (Prust 2003).

Service Bureau and Prepress Proofing

After the production of the map has been thoroughly edited, it is then sent to a service bureau for negative preparation (if the designer's software does not produce its own copy negatives) or digital press proof. It is very important that a good relationship is maintained with this service provider. Normally, the workflow plan specifies the service bureau because some design decisions are made on the abilities of this vendor. Many vendors have their own preferences when it comes to file formats and will provide their customers with file specifications ahead of time. The formats described above are quite versatile and work across computer platforms. Most bureaus are equipped to handle documents created on any of the standard platforms. Cartographers must know which organization or service bureau will be printing their graphics before any production is begun. Communication between the cartographer and printing organization is essential and should begin prior to the production of the graphic.

THE MAP PRODUCTION PROCESS

Modern Offset Lithography

Lithography, as noted, had its beginnings around 1796. By the middle of the nineteenth century, the development of photography made it possible to put the image onto a thin metal plate that could be attached to a rotating cylinder. The impressions were made directly onto paper, requiring the image to be backward until printed. It was not until 1905, when Ira Rubel accidentally printed an impression on a blanket cylinder, that the *offset principle* was discovered (see Figure 15.3). Print, art, and map materials could be right-reading through the entire preparation process, making this phase much less risky.

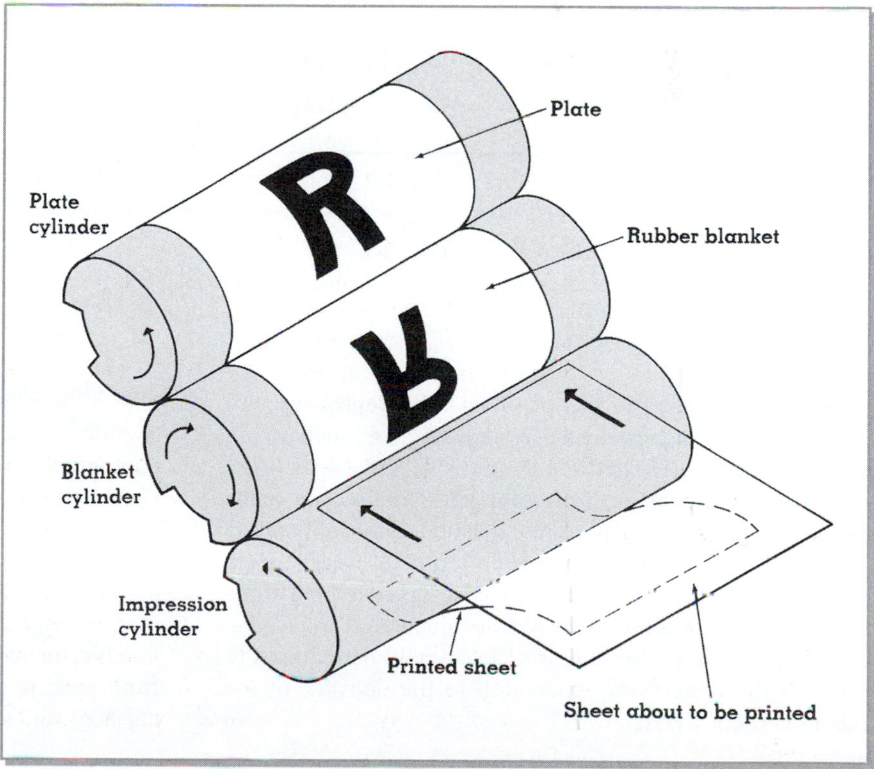

FIGURE 15.3 THE PRINCIPLE OF OFFSET PRINTING.
Notice that the image on the plate is right-reading, a feature that has made this form of printing very popular.

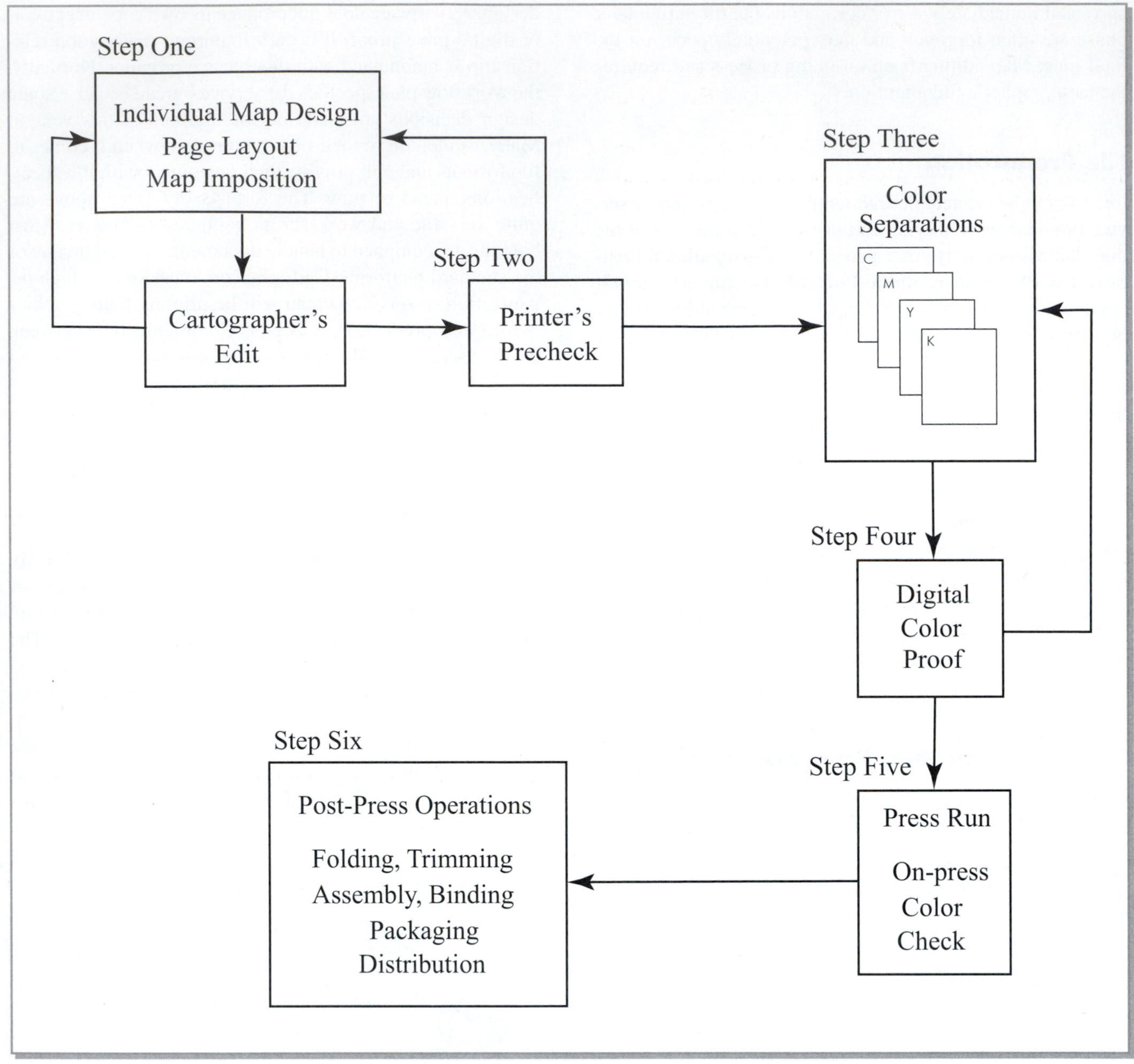

FIGURE 15.4 THE MAP PRODUCTION PROCESS.
The six-step production process from map design to distribution of the final product is presented. The editing stages are an extremely important part of printing in cartography.

To illustrate the process of offset lithography, we will use an example of printing a cartographic atlas, such as *The Atlas of Georgia* (Hodler and Schretter 1986). *The Atlas of Georgia* is a 288-page, full-color atlas focused on maps and supplemented by a narrative that adds additional information about the subject matter. In terms of average page space, the graphics-to-text ratio was designed to be five-to-one. If this or a similar atlas were generated today, the following six steps (see Figure 15.4) would be required to produce the atlas from conception to the delivery of the bound hardcopy atlas.

Step One: Design and Page Layout

A cartographic atlas is comprised of various maps, graphs, photographs, sketches, and narrative. These items are designed individually but intended to be placed on a page with other items. These items work together to form an organized whole that communicates a topic or theme. The layout of graphics on a page requires the use of a grid that is standard to every page in the atlas. Therefore, the positioning of these graphics or page elements will occur in the same location from page to page throughout the atlas. This brings both harmony and structure to the entire content of the atlas. The page grid is designed to incorporate as many different map scales, graphs, and narratives as desired. Page layout is accomplished through the GIS and mapping software as well as specialized layout software such as Adobe InDesign and Quark Xpress.

The individual graphic files are saved by the cartographer as either an Encapsulated PostScript (EPS) or PDF, or native format, depending on the requirements of the printer or service bureau. These files are arranged by the layout software that will be used by the printer in creating the actual document that is sent to the next step of the process.

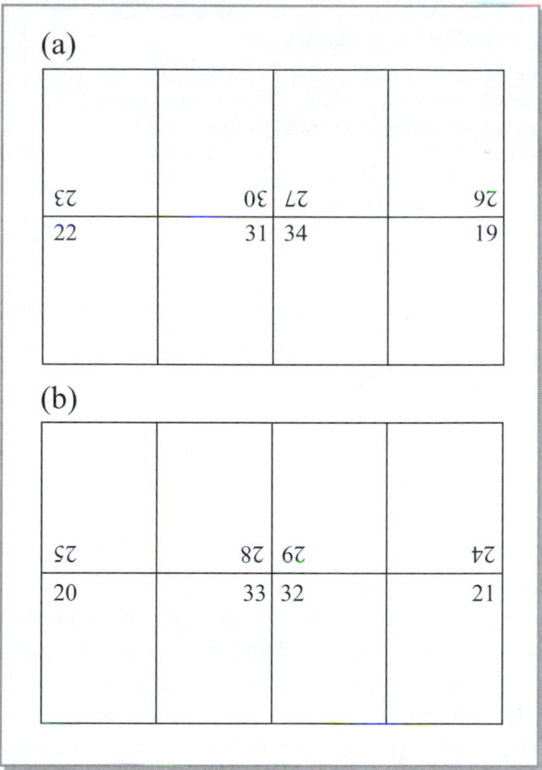

FIGURE 15.5 SIXTEEN-PAGE PRINTING SIGNATURE.
A sixteen-page signature consists of two sides (flats) on which eight pages are printed. Notice that the pages are assembled, not in numerical order, but are ordered so that once the signature is printed, folded, and trimmed the pages will occur in numerical sequence. (a) is one flat and (b) its opposite side.

Step Two: Printer's Precheck

The printer will process and check the page files by conducting a "pre-flight check." Task-specific software examines the files to verify that the specified graphics and fonts are included. This software also verifies the physical size of the page layout and examines color polygons to determine if they have proper overlap or if additional trapping is required. **Trapping** is the overlapping of colors to adjust for misaligned registration. The cartographer would have already checked for this in the test proof. The printer will perform this step to confirm the composition of the graphic is correct. At this point, the printer is looking for any "red flags" that may be problematic as the process continues. Anything that may require modification will return the process back to step one for editing. Lacking any problems, the printer will produce a color laser proof for the customer (cartographer) to approve.

Step Three: Color Separation and Plate Generation

During this stage, the printer will first convert the graphic files from vector to raster format using a raster image processor (RIP). Secondly, an imposition of the individual pages is created by grouping them together in correct placement on the **signature** or press sheet. This signature represents the pages that will be printed simultaneously on a single sheet of paper as it passes through the printing press. For example, in our atlas example, a signature contained 16 pages, eight printed to a side, or **flat.** A 288-page atlas will be comprised of eighteen 16-page signatures ganged together to form the book. Pages are organized on the signature in such a manner so as to have the pages occur in proper sequence once the signature is folded. Figure 15.5 depicts an example of the organization of pages on one side of a signature. Note that consecutively numbered pages rarely sit side-by-side on the signature and observe that orientation of a page will change from one half of the flat to the other.

The digital files of the signature are converted to color separations of the four CMYK layers. This process is referred to as **computer-to-plate (CTP)** or **direct-to-plate (DTP)** where the images of each of the separations are created on the plate used in the four-color process printing (see Step Four). The plates are commonly made of a thin sheet of aluminum that is coated with an emulsion layer to accept the latent image generated by the CTP process where a laser burns the image to the plate. Color is applied to the paper in the printing process through the use of dots arranged in a row at an angle in relation to the horizontal. The number of dots is controlled by the resolution established by the printer. Normal print resolution is 300 dots per inch (an equivalent to a 150-line screen). The value or tone of the color is controlled by the physical size of the dots. As the value percentage increases, the dot size increases. This, in turn, increases the amount of ink being applied in a given area. Darker changes in tone are created visually by increasing the amount of ink applied.

In order for the printed map to have a color that is comprised of different values of CMYK, the dots for each separation must be generated at a different angle (see Figure 15.6 and Color Plate 15.1). The angles from the horizontal for these separations are: 15 degrees for cyan; 75 degrees for magenta; 90 degrees for yellow; and, 45 degrees for black. The primary concern is for the screen angles for the cyan, magenta, and black to be 30 degrees apart. These angles, often referred to as screen angles, allow the inked dots to appear in the same area without totally overprinting. The eye views a composite color by combining the four separations such that we see one color without seeing the dots.

FIGURE 15.6 ANGLE OF DOTS FOR EACH OF THE CMYK COLOR SEPARATIONS.

Each angle has been selected so that when they are overprinted, most all of the dots will be used to generate the designed color. A thirty-degree separation is desired between the cyan, magenta, and black screens. Refer to Color Plate 15.1 for an example of the overprinting of the dots.

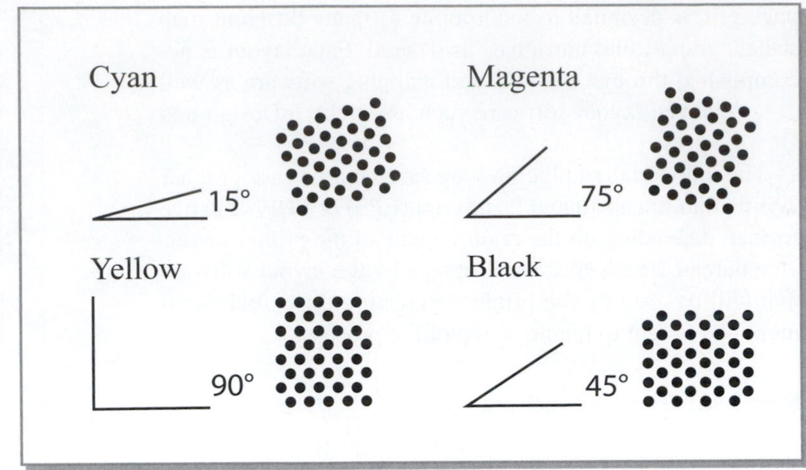

All color figures printed in this textbook are printed on a single signature and created using this four-color process printing technique. You may see the dot combinations by using an eight-power or higher hand lens or other magnification device and look carefully at an area of color in one of the figures. The colored dots represent one of the four separations generated in both the production and printing processes.

If it is determined that an area on one of the plates that is supposed to contain dots to accept color does not have those dots, the process must return to Step One and reconstruct the original file for separation in Step Two.

Step Four: Digital Color Proof

Following the creation of the color plates, a **Direct Digital Color Proof (DDCP)** is generated. This proof contains the color profile of the press on which the documents will be printed. As such, it depicts the signature (or page) in an exact match to the colors created in the printing process. This document serves as the final proofing stage for the cartographer. The proof is presented for approval and normally requires the cartographer to sign off on that proof. At this point, the proof becomes a contract proof between the cartographer and printer and serves as the legal agreement between the two parties. This stage protects both parties in the event of a later disagreement in the quality or shade of the color once it comes off the press.

Step Five: The Press Run

The primary variable that determines the type of press on which the atlas is to be printed is the number of copies to be printed. This is referred to as the **press run.** The quantity that separates a large press run from a smaller run is approximately 25,000 copies. Considering the cost factor involved in the press operator setting the plates in place, achieving proper registration of the plates in the make-ready process, and actually beginning the press run, the economy of scale comes into play. The larger the press run, the lower the per item cost in the printing of that signature. The primary cost of printing occurs in the initializing of the press. Once the press run begins, the cost is reduced based on the quantity printed.

Offset lithography occurs when ink is applied (rolled) onto the plate, which is attached to a rotating drum. The ink adheres to the latent image created when the plate was made. Ink that comes in contact with the part of the plate without a latent image will wash off using a water wash. The water-ink combination is recovered as a result of water and ink not mixing. The inked plate is then pressed onto a blanket of a second rotating drum. The blanket is comprised of a resilient material that is coated on a fabric. The use of the blanket aids in extending the life of the plate and in allowing the latent image on the plate to be oriented so that it reads correctly (right-reading). This sequence transfers the ink in a right-reading orientation to a reversed-reading orientation. It is the blanket that comes in contact with the paper, thus transferring the image to the paper again in a right-reading position. In order for the blanket to have a solid contact with the paper, an impression cylinder provides pressure to the opposite side of the paper so that ink is applied evenly. This sequence is the foundation on which offset lithography is based (Prust 2003; see Figure 15.7).

A printing press may print one side of the paper at a time, using one of the two signature flats. In many instances, a press may print both sides of the paper at the same time. This is referred to as a **perfecting press.** In this instance, the blanket cylinder on one side serves as the impression cylinder for the blanket cylinder on the opposite and vice versa (see Figure 15.7a–b).

Depending upon which printer is used, they may have a number of printing presses as options for your particular press run. Each plate-blanket set is used to apply one of the four process colors to the paper. A four-color press will have four sets of the plate-blanket combinations. The paper will come in contact with the four blankets where the inked dots are printed in order to achieve the final desired color. Some presses may have two sets of plate-blanket combinations and therefore are known as two-color presses. Larger presses may have six or more combinations in the event that a solid

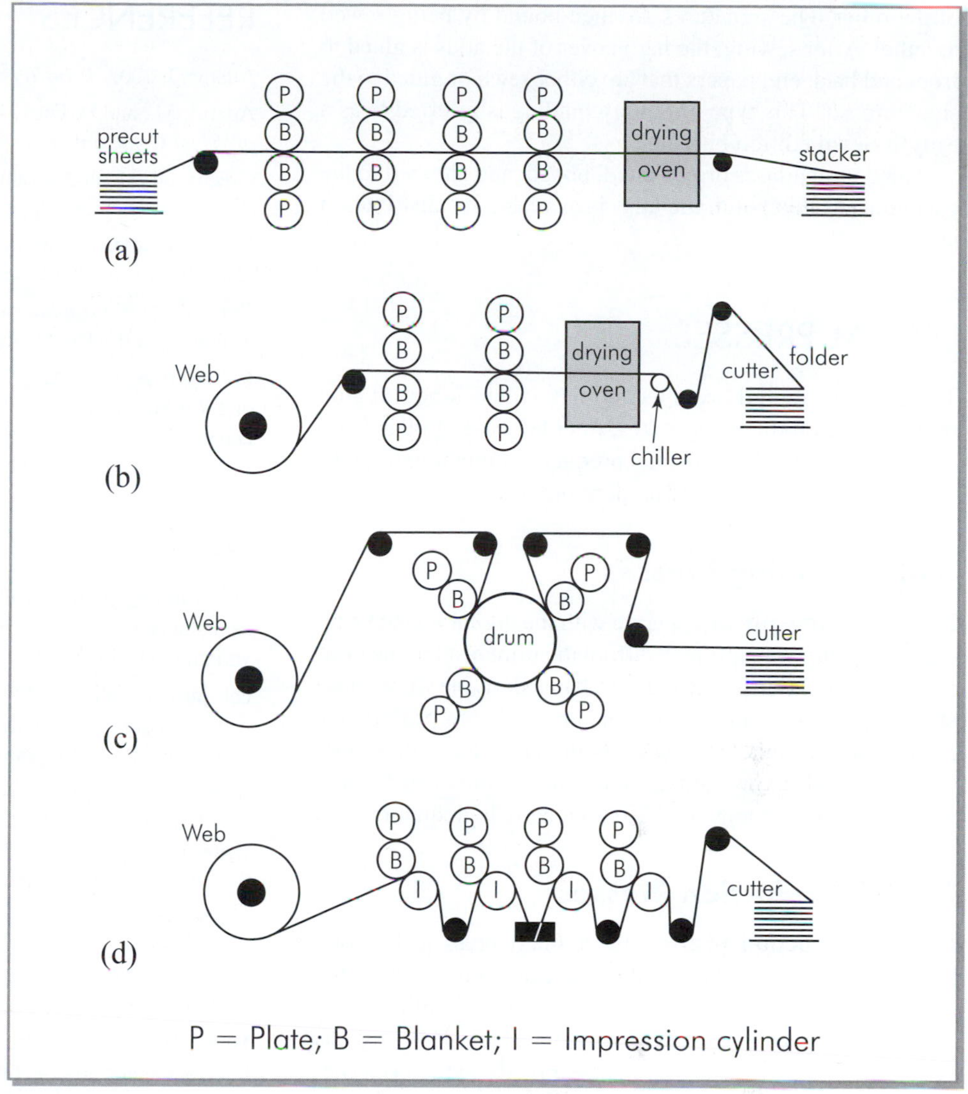

FIGURE 15.7 PRINTING PRESSES: SHEET-FED (a) AND WEB-FED (b-d) PRINTING.
Sheet-fed press (a) and web press (b) are also called perfecting presses, as they print on both sides of the paper at the same time. Web presses (b-d) utilize large rolls of paper stock as opposed to precut single sheets (a). The figures depicts three four-color presses and one two-color press (b).

Pantone match color is desired, or a shellac or other aqueous coating of the paper is planned.

The paper used in the printing of the atlas may come in either pre-sized individual sheets or a large roll of paper. This distinction allows for the designation of a press as either **sheet-fed** or **web-fed** (see Figure 15.7). Sheet-fed presses use both standard and nonstandard sizes of paper (Greenwald and Luttropp 2001) and are frequently used in press runs of 25,000 copies or less (see Figure 15.7a). Larger quantities are directed toward the use of the web-fed press where the paper stock comes on a large roll several feet in diameter (see Figure 15.7b–d). The registration capability and quality of final product is quite comparable in both systems.

The press operator will monitor the application of the ink to the paper during the printing run. The operator will be checking for dot gain, a process whereby dots become larger as they are printed, thus altering the value percentage of the color. Such dot gain is caused by the pressure exerted by the blanket on the paper during the printing sequence (Greenwald and Luttropp 2001). The operator can make adjustments at any time necessary, thus maintaining a quality product that matches the digital proof of Step Four.

Step Six: Post-Press Operations

In this step, the paper is trimmed, folded, assembled, and bound. Web-fed press output is transferred to a folder where it is trimmed and folded into signatures. Sheet-fed presses print a signature on each sheet. These copies are taken to a separate device for folding and an additional instrument to trim them to proper size. The folder equipment walks each sheet through a sequence where a single fold is made; for instance, the signature may be folded in half lengthwise. It then proceeds through the remaining folds until the signature has the sixteen pages in proper order and orientation. At this point, the signature is trimmed to proper size (Prust 2003). One would be able to thumb through these sixteen pages as if holding the complete atlas.

Once the eighteen signatures are printed, folded, and trimmed, they are assembled in proper sequence, one on top of the other. The signatures are then bound by being sewed together. After sewing, the hard cover of the atlas is glued to front and back end papers that are either sewn or glued to the signature set. This type of quality binding is referred to as a **Smyth Sewn** edition binding.

The bound atlases are counted, boxed, and shipped to the customer. At this point, the atlas is available for distribution and use.

DIGITAL PRESSES

The status of digital press technology includes both digital printing and production presses. In either case, a plate is no longer generated. And the final product is a function of high-end printing capabilities of modern presses.

Digital Printing Presses

Digital printing presses operate with the digital image being created by ink being projected from the print head to the print material. The laser image can be changed on the fly without stopping the press and creating additional setup procedures. An example of this technology is the Heidelberg press that uses a special inking unit for short press runs, usually less than 1,000 impressions, with exceptionally high quality.

Digital Production Presses

Digital production presses use a smart press technology where the final printed document is created directly from the digital file without the generation of plates and other traditional printing techniques. This system, developed early this century, creates an electrical charge directly to the print medium followed by dry ink sublimated into a vaporized toner and sprayed onto the paper as a single-point transfer. The electrically charged area is comprised of dots at the same angles and sizes as discussed in Step Three. These dots then attract the toner, holding the color in place. The final stage is to pass the printed medium through a heated fuser, securing the toner solidly to the paper.

This type of press currently handles smaller sheet sizes up to 12-inch by 18-inch or 14-inch by 19.5-inch. It is ideal for shorter press runs at a rate of approximately 6,000 full-process color impressions per hour. The print images are of high quality at a resolution of 600 dpi of 8-bit color. The advantages of this system are high quality in a very short time. Unlike a traditional printing press that requires the six steps discussed above, this system skips steps two through five. The color proof is generated by the first copy of the page. Once approved, the final product is generated in terms of hours instead of days. Larger-size documents, such as 16-page signatures for a high-quality thematic atlas will continue to require traditional offset lithographic presses. The digital production press, however, represents a positive technological advance for the realm of digital cartography.

REFERENCES

Adams, J. 2002. *Print Technology, 5th ed.* Albany, NY: Delmar.

Adams, M., and D. Faux. 1977. *Printing Technology: A Medium of Visual Communication.* North Scituate, MA: Duxbury Press.

Greenwald, M., and J. Luttropp. 2001. *Graphic Design and Production Technology.* Upper Saddle River, NJ: Prentice Hall.

Hodler, T., and H. Schretter. 1986. *The Atlas of Georgia.* Athens, GA: The University of Georgia.

Prust, Z. 2003. *Graphic Communications: The Printed Image.* Tinley Park, IL: The Goodheart-Wilcox Company, Inc.

Robinson, A. 1975. Mapmaking and Map Printing: The Evolution of a Working Relationship. In *Five Centuries of Map Printing*, ed. David Woodward. Chicago: University of Chicago Press.

GLOSSARY

aquatint engraving special method of engraving a copperplate using an etching process in order to create the impression of tonal variation when printed

computer-to-plate (CTP) the process whereby a printing plate is generated directly from a digital file bypassing the photographic process in offset lithography (see also direct-to-plate)

digital printing presses a digital printing technology whereby the digital file is fed directly into a press using a laser, which writes to the plate on the fly; the remaining printing technique is that of traditional lithographic process

digital production press a printing process whereby the digital file is fed directly into a press without the use of printing plates; the printing process utilizes electrically charged paper and sublimated dry ink sprayed onto the paper and fused by heat

Direct Digital Color Proof (DDCP) the generation of a hardcopy color proof generated directly from a digital file that will later be used to produce the printing plates

direct-to-plate (DTP) the process whereby a printing plate is generated directly from a digital file bypassing the photographic process in offset lithography (see also computer-to-plate)

dye-sublimation printer a printing process whereby a solid color dye is converted to a vapor without going through the liquid process; the vapors are produced from a wax-based color product

Editable Portable Document Format (PDF) a device-independent file format that will generate the same feature regardless of its origin or destination; it contains all the graphic instructions, fonts utilized, and color management necessary to print the graphic

Encapsulated PostScript (EPS) a file format that contains all the graphic instructions, fonts, and symbols utilized, raster images, and color management necessary to print the graphic in vector form; the file can be used in a variety of graphic programs

flat one side of a printing signature

inkjet printer a printer that utilizes liquid ink in the printing of the CMYK layers of a graphic

intaglio form of printing in which depressions are engraved in the block or plate; the depressions receive ink, which is transferred when pressed onto the paper; copperplate was a popular method of intaglio map printing during the seventeenth and eighteenth centuries

laser printer a printer that uses laser diodes to create an electrical charge in the paper prior to applying a powdered toner that is fused to the paper by heat

letterpress form of printing in which the printing surfaces are raised in relief from the printing block or plate; raised portions receive the ink

mezzotint engraving engraving of copper or steel by rubbing or scraping away material before printing

native format file format that is normally used for a particular software package; AI files are the native format for Adobe Illustrator

offset lithography an intermediate drum or roller on the press allows the plate image to be right-reading; the image on the blanket or offset roller is backward

perfecting press a printing press that prints on both sides of the paper at the same time

photolithography lithographic printing; so called because of the photographic preparation of the printing plate

planar printing relies on the fact that water and grease do not mix well; areas on the printing block or plate are at the same elevation, with ink either adhering or not, depending on the surface preparation between water and grease; a popular form is lithography

plotter a free-standing large-format printer that uses inkjet printing techniques and is capable of handling a wider format print medium

press run describes the sequence of printing many copies of a page or signature and is normally specified by the quantity of copies required

Raster Image Processor (RIP) a computer program that converts vector graphics to raster graphics; commonly used in preparation for printing the graphic

registration the proper alignment of CMYK printing layers in order to produce high-quality color graphics

service bureau a company that does digital cartographic and other computer services for a fee

sheet-fed a printing press that uses pre-sized individual sheets of paper on which a signature is printed

signature the arranging of several pages of a book or atlas in a format that allows for printing and folding of the pages so that they occur in proper page sequence

Smyth Sewn a group of signatures are sewn together to serve as the body of an atlas followed by gluing the signatures to the book cover using end papers

stipple engraving special method of engraving a copperplate using a specialized tool in order to create the impression of tonal variation when printed

trapping overlapping of color layers to ensure no gap or white space is left between the colors

web-fed a printing press that utilizes paper from a large roll that is continuously fed through the printing process

woodcut map method used to print maps during the fifteenth and sixteenth centuries; based on the relief method; similar to early letterpress

16 INTRODUCTION TO VIRTUAL AND WEB MAPPING

CHAPTER PREVIEW Much of the mapping done today is designed for the World Wide Web or other virtual environments. Common types include static, interactive, and animated maps. These maps are constructed in a variety of raster and vector formats, depending on the capabilities needed. The virtual maps are typically stored on a server, and are delivered to the map reader's monitor as a Web resource via the Internet—the "medium" for most virtual mapping today. The medium places some important constraints on virtual map design, perhaps the most important being screen resolution and the amount of screen "real estate." But perhaps even more powerful than the limitations are the opportunities that the medium affords. Interactive and animated cartography are popular today because of the medium, and can be a major part of the solution to the screen real-estate constraint. Future developments are likely, and the term cybercartography describes a research agenda that can embrace new technologies and cooperation among government, educational, and private agencies. ■

In the previous chapter, a variety of subjects dealing with modern printing techniques were presented. We now turn our attention to Web and virtual mapping, perhaps one of the most exciting and dynamic areas within the discipline of cartography. In Chapter One, we introduced the term **virtual map,** which in its modern use means any map that is presented on a monitor display or is projected onto a screen. Most virtual maps today are found on the World Wide Web. However, the television weather map, the map viewed from a CD or DVD, and a map displayed in a PowerPoint presentation are also examples of virtual maps. Even a map layout that appears on the monitor before the cartographer prints the map is also a virtual product. In fact, most virtual maps *can* be printed, if need be, but this chapter will place the most emphasis on virtual maps that are *designed* for virtual viewing on the map reader's monitor display.

VIRTUAL AND WEB MAPPING INTRODUCTION

Virtual maps are often created as **static, interactive,** or **animated** maps. The earliest digital atlases, such as Smith's (1988) *Electronic Atlas of Arkansas,* were produced before the popularity of the Web. These atlases were built primarily using static maps, which are the easiest of the three types to create. On the Web, it is still quite common to see static virtual maps. Many of these are essentially exported map layouts designed in a GIS or in artistic drawing software. Scanned printed maps, such as the images found at the Perry-Castañeda Library Map Collection (2007), are another example of static maps that are also quite common on the Web.

Interactive maps are those maps in which the map reader can click and interact with features or layers on a map, often in the same way that a GIS user might interact with map layers in a GIS. For example, some interactive maps allow the reader to click and identify features, turn on and off certain classes of features, or pan and zoom across the map. The specific type of interactivity is determined by the map author. With some interactive maps, the map reader might be not only interacting with the map on the Web browser but also with an underlying online GIS. For most interactive maps on the Web, however, the function and scope of the map is not as extensive as that of a geographic information system.

STRATEGIES FOR LEARNING TO BUILD WEB PAGES

One of the most common ways to put virtual maps online is to build a Web page. There are two aspects of building a Web page to consider: the design and layout of the page, and learning how to program the code behind the Web page. Both are incredibly deep topics, and mastering the skills to be able to wear the title of "Web programmer" or "Web designer" can take years (and is far beyond the scope of this text). However, building a *simple* Web page can be an extremely effective vehicle for display of online maps. This brief discussion is intended to provide a starting point for those beginning to develop basic Web skills.

Most of today's Web page layout software is very easy and intuitive. However, you don't need an expensive software package to make a Web page. A simple text editor (such as Notepad) can be used. *If* learning HTML coding is important to you, you should consider *not* using a professional design package that writes the code for you while you are first learning the language, because it can be too much of a crutch.

There are a number of great online tutorials that are free for Web design and HTML programming, such as the ones at w3schools.org. It can be really effective to view the instructions on the screen, where you can copy and paste examples of code from their page into yours, and see what that particular code does. If reading from the screen is not to your taste, there are hundreds of hardcopy books available at the library or your local bookstore. Both approaches work well for individuals who prefer learning at their own pace.

If classroom instruction is your learning style, you could consider taking a course at your local community college, vocational school, or university. If formal courses are too much, many communities offer low-cost introductory workshops on the basics of Web design and HTML. These types of classes can give a quick jump-start to getting your maps online.

Finally, remember that designing a simple Web site does not have to be complicated. In fact, most Web surfers tend to bypass complex and/or poorly organized pages. If you have a number of maps to present, consider making an index page with small-sized (both physically small and file-size small) thumbnail images of your maps that are hyperlinked to the main maps.

Animation puts an element of motion into the maps. In some cases, the entire map may seem to move, such as maps that change their projection. In other cases, just a select symbol or element may be animated, perhaps to draw extra attention to that feature. The most familiar animated maps are perhaps those found in news broadcasts where weather maps are "put in motion." During a select time period, the symbolization for precipitation will expand, contract, appear, and disappear in a few seconds, and weather front symbols will cross the map and high and low pressure cells will move accordingly. Some of these weather maps will even tilt, and the viewer is treated to a three-dimensional view of the area, which often includes a virtual fly-through of the terrain.

The most common venue for virtual maps is the Internet. People who "surf the Web" for maps do so on a Web **browser**—the program such as Internet Explorer or Mozilla Firefox that allows the reader to view and interact with information on the Internet (the distinction between the Internet and the World Wide Web will be discussed later). Most static maps that are generated in one of the more popular Web formats (also discussed later in the chapter) are viewable in the Web browser. However, some maps, particularly interactive maps, are not always directly supported by the browser. Browsers often require "plug-ins," (also called "add-ons" or "helper applications") in order to view some maps or run specialized Web applications. Plug-ins are programs that extend the capabilities of the browser to accommodate the various functions and display of specialized file types. For example, Flash is a popular plug-in that allows the playing of Adobe (formerly Macromedia) Flash (.swf) files, a popular format for interactive mapping. For file types that are not supported by the browser or its plug-ins, most browser software will prompt the user to either save the file or select a program from a list of programs currently installed on the computer that the Web surfer thinks will successfully open the file.

Creating virtual maps for the Internet requires an understanding of the basics of the technology and how that technology affects virtual map design. The next three sections provide a broad overview of three major areas with which the cartographer should have some degree of familiarity: how to select an appropriate format for the virtual map (Map Formats and Structures); the technology and terminology of the virtual map medium (Understanding the Medium); and some of the major design issues that are inherent in virtual mapping (Design Implications for Thematic Mapping). An introductory approach to these topics is taken here—the student is encouraged to see the linkages between these broad areas.

It should be noted that Web page design or programming techniques, although important, are beyond the scope of this text. Entire curricula are written on these topics. To those who wish to learn these skills, we point out that there are literally hundreds of texts available on these topics, as well as numerous online sources. Since many maps often have a Web page context, the boxed discussion, "Strategies for Learning to Build Web Pages" provides more information on how one can proceed if you wish to create your own Web site.

MAP FORMATS AND STRUCTURES

One of the largest challenges facing many map designers is to decide on the format for their map layout. There are scores of possible choices. One popular option is to save the map layout in the native format from which the cartographer is working. A native format is simply the preferred, designated, and often default format for the particular GIS, mapping, or artistic software being used by the cartographer. But when the layout is complete, what format choices are available so the map can be easily viewed on the Web? Does the format support interactivity or animation, or is the format best suited for static maps? Are there other functional and/or aesthetic differences between the major format types? In this section, we explore some of the more popular raster and vector formats used in Web and virtual mapping.

Raster Graphics

The raster format is one popular choice for Web and virtual-based cartography. The raster format is used in remotely sensed imagery and digital aerial photography, scanned maps and documents, picture images from digital cameras, and images of completed map layouts (see Figure 16.1). Many of today's online map collections, particularly those that contain volumes of scanned maps, such as those found at the Perry-Castañeda Library Map Collection (2007) mentioned earlier, contain hundreds of raster images.

In the early years of virtual cartography, most map layouts were presented in a raster format and were often referred to as bitmaps. Today, the raster image (or simply **image** among Web developers) remains a popular choice for map designers, even with the advent of vector Web formats that provide a potential for a higher degree of user-interactivity. There are four primary reasons that the raster image format is still commonly used among cartographers for their map layouts:

1. The most common image formats can be viewed in most browsers without the use of an plug-in.
2. Images are easily incorporated into Web pages as well as presentation software and other documents.

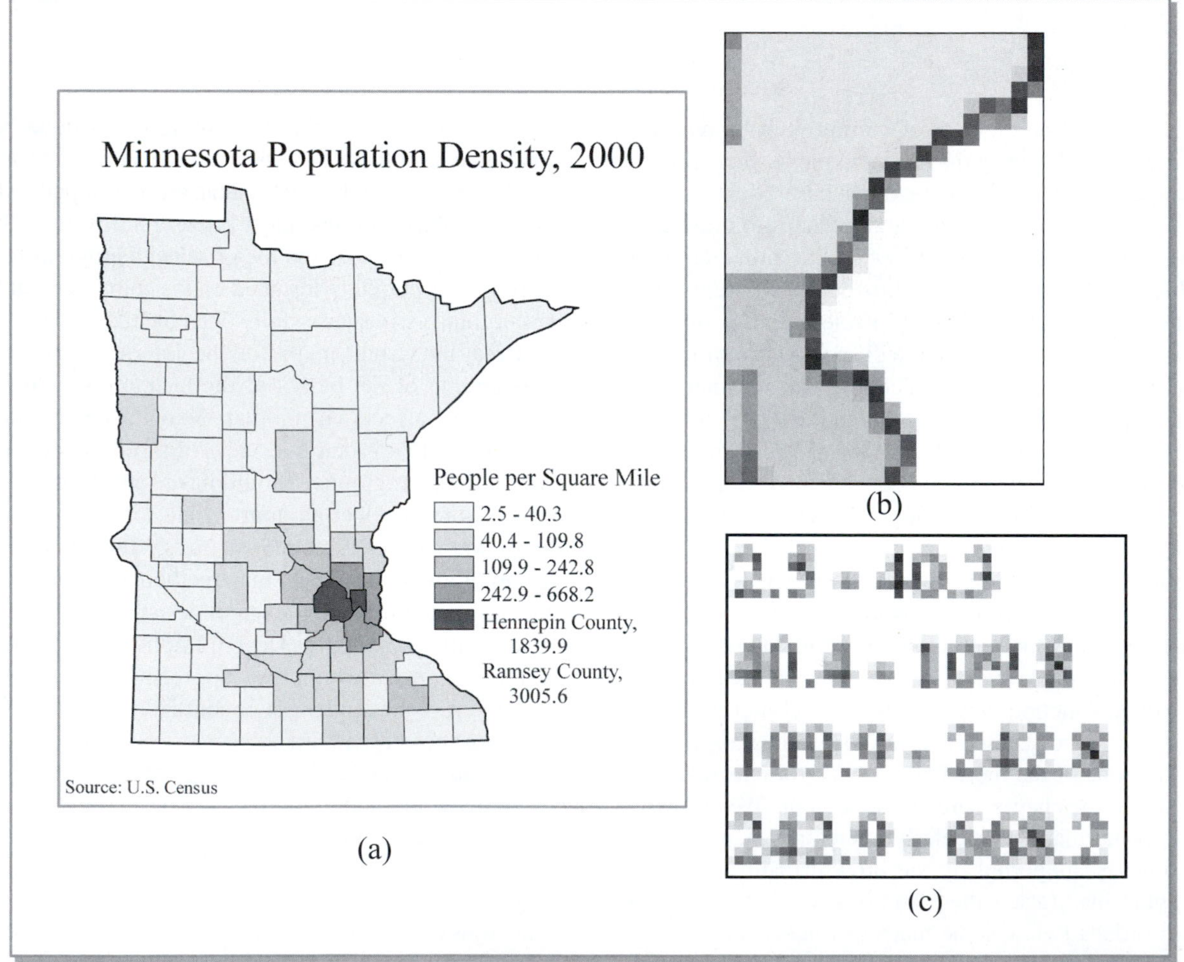

FIGURE 16.1 RASTER IMAGE OF FIGURE 1.8.
Figure 1.8 is converted to a raster image format in (a). When part of the image is zoomed into, the pixels are easily seen in (b) and (c).

3. GIS, mapping, and artistic drawing software can quickly and easily export a map layout to the most common raster formats.
4. Images can easily be extracted and saved from the Web page by the map reader.

If the cartographer is willing to accept the limitations of the raster image format, such as the fixed *resolution* and less potential for interactivity, the quickest and easiest solution for many cartographers is raster graphics.

Raster Concepts

The anatomy of most raster images is described using the same terminology as display monitors (see discussion), since monitors are also raster devices. Images have rows and columns of **pixels.** The number and size of the pixels in the image defines its **resolution.** Although images can be prepared at fairly high resolutions (usually for printing), images prepared for the Web are typically created at 72, 96, or 150 pixels per inch (ppi). Some Web designers will try to make the image's pixel dimensions less than a typical display monitor screen resolution, such as 800 * 600 pixels, in order to fit on a display screen.

The number of simultaneous colors that can be displayed is called the **bit depth.** The number of binary bits assigned to each pixel in an image determines the number of colors that can be displayed. Since there are only two possible numbers in a bit (each bit can have a value of 0 or 1), the number of potential colors is 2 raised to the power of the number of bits for each pixel. For example, if the bit depth is 8 bits, then 2^8 generates 256 possible combinations of zeroes and ones, usually expressed as color values of 0 – 255. Twenty-four-bit depth results in over 16 million colors, and is sometimes referred to as true color. In 24-bit color, 8 bits are allocated to each red, green, and blue color channel (256 intensity values for red, green, and blue, respectively). Table 16.1 illustrates the relationship between bit depth and number of colors.

Two hundred fifty-six colors are usually sufficient for most thematic maps. For example, a typical choropleth map may use only four or five hue values for the classes, along with black for text, white for a background, and perhaps gray for the enumeration unit borders. For maps or map layouts that incorporate imagery or photographs, a greater bit depth becomes necessary to accommodate a photograph's continuous tones, that is, the subtle but often numerous changes from pixel to pixel in an image. For these purposes, using a 24-bit depth image is common, not because all 16.7 million possible colors are required for photographic quality detail, but because it is the default bit depth on some of the most common true-color file formats (discussed later in "Selected Raster Image Formats").

There is some debate about whether bit depths greater than 24-bit are really necessary. As discussed in Chapter 14, the human eye can see only about 10 million colors. So why are values beyond 24-bit depths an option in so many graphic software programs? Two reasons are usually given. First, the higher bit depths do not always mean more colors. A 32-bit depth image, for example, refers to an image that allocates 24 bits for color but reserves the last 8 bits as an **alpha channel.** This channel is often used for creating masks in some software, or specifying transparency values. Second, for higher bit depths that do imply a greater number of colors (for example, a 48-bit depth that will produce trillions of colors), some graphic artists point out that the extra headroom allows for a greater editing flexibility (Fraser and Blatner 2005). However, most browsers and many monitors are not set up for the greater bit depths. Therefore, for most virtual mapping applications, 24-bit depth is usually the maximum.

An image's file size (for example, in terms of kilobytes [Kb], megabytes [Mb], and so on) is an important consideration for cartographers (see Table 16.2 for an explanation of storage sizes and terms). The final file size is determined primarily by the combination of the image resolution, the bit depth, and the compression (optional processing that reduces file size) used in a particular image file format type. A Web page with physically larger, higher resolution images with greater color depths and less compression will typically load more slowly and require more storage space than smaller, lower resolution images at lower bit depths and greater compression. Therefore, it is often desirable to minimize file size when preparing images for the Internet. Before surveying specific format types, it is important to briefly examine the impact of resolution, bit depth, and compression on an image's file size.

TABLE 16.1 BIT DEPTH AND NUMBER OF POSSIBLE COLORS

Bits per Pixel	Number of Colors	Comments
1	2	Black and white line work; simulated grayscale via dithering
2	4	
3	8	
4	16	
8	256	Color numbers 0–255 in a color palette
16	65,536	
24	16.7 million	Usually 8 bits per red, green, and blue channels
32	16.7 million	Additional 8 bits for transparency in the RGB model; 8 bits per channel if CMYK is used
48 (and more)	trillions (and more)	Useful if the image is going to be post-processed; otherwise is beyond human eye capabilities

TABLE 16.2 STORAGE TERMS AND SIZES

Term	Abbreviation	Actual Size in Bytes	Mathematical Power
Bit	b	0.125	2^0
Byte	B	1	2^3
Kilobyte	Kb or Kbyte	1,024	2^{10}
Megabyte	Mb	1,048,576	2^{20}
Gigabyte	Gb	1,073,741,824	2^{30}
Terabyte	Tb	1,099,511,627,776	2^{40}
Petabyte	Pb	1,125,899,906,842,624	2^{50}
Exabyte	Eb	1,152,921,504,606,846,976	2^{60}

The most significant component of a file's size is its resolution. A higher resolution results in a relatively large file size. Most artistic drawing software packages support a process called **resampling,** which is used to decrease the number of pixels in the image. Fewer pixels result in a decrease in the image's file size and reduce the quality and/or the physical size of the image.

A reduction in bit depth will also have an impact on file size, though usually not to the degree that resampling will. It is usually more common to reduce bit depths in images with more than 24 bits to make the image compatible with most browser/monitor combinations. Common bit depth transformations include a reduction to 24 bits when many colors are needed (such as when a photograph is in the layout or is included in the background, or if some sort of continuous shading is used) and a reduction to 8 bits when 256 colors will suffice (such as most thematic maps). When bit depth reductions are made, a **color palette** is often employed. A color palette is a collection of solid colors that are a subset of possible colors. The palette is usually not fixed, but can be adjusted to include the color options desired for a particular map. Palleted color is most common on lower-bit-depth images, especially 8 bits or less. Note that if *more* colors are needed than a lower-bit-depth palette can provide, it is often possible to "create" or emulate other colors by **dithering.** Dithering is the display of alternating pixels of two different hues or shades to create a visual impression of a color beyond what is on the color palette. For example, alternating black and white pixels will give a visual impression of gray. Unfortunately, patterns of dithering can often be seen if one looks closely enough at the image. Therefore, we usually recommend using sufficient bit depth (for example, 24) to include the needed colors for maps that incorporate photographic imagery.

Image compression will also reduce file size. Compression can fall into two categories, lossy and lossless compression. With **lossy compression,** image information is actually being lost in the compression process. The image's file size is reduced but at the expense of image quality. The designer will often have control over the amount of compression that is applied to the image. The greater the compression that is applied, the more information that is lost, resulting in a smaller file size and increasingly lower quality (see Figure 16.2). With **lossless compression,** the image's file size is reduced without losing image quality. Lossless compression algorithms will compress only information that is totally redundant, such as large homogeneous areas of a single color (for example, a large area of white space or a single color value for its pixels will allow for a great amount of compression in that part of the image). No information is actually lost in the compression process. The amount of compression that occurs is a function of the algorithm and is not under the direct control of the cartographer. Which type of compression (if any) that can be applied depends on the file format chosen for the image.

Selected Raster Image Formats

There are quite a number of raster image formats that a cartographic designer can use. These include formats that are proprietary to specific software, such as those for Adobe Photoshop (PSD), Corel PhotoPaint (CPT), Erdas Imagine (IMG), or those formats native to an operating system, such as the Windows Bitmap (BMP) format. Several types that are used extensively in Web and virtual cartography merit further discussion.

JPEG. The JPEG file, short for Joint Photographic Experts Group, is one of most common image formats on the Web today. It can be recognized from its file extension, usually .jpg, .jpeg, .jpe, or .jp2 (the latter is a newer modification of the jpeg format). At a 24-bit depth, the JPEG image can display over 16 million colors. This format also supports lossy compression. When used for satellite and aerial photographic imagery, the JPEG can be georeferenced with real-world coordinates to make it usable in GIS and mapping software.

This format is currently the most popular for distributing photographs, because it is effective for images with large hue and value ranges (particularly if little or moderate compression is applied). Thus, the JPEG format is popular for maps or map layouts that include imagery. It is important for map designers to note that lossy compression in JPEGs sometimes leaves some visual artifacts around letters and lines when a vector map in a GIS, mapping, or artistic drawing software is exported to a JPEG image.

GIF. The GIF image, along with the JPEG, is another important raster image format for the Web, since these formats are

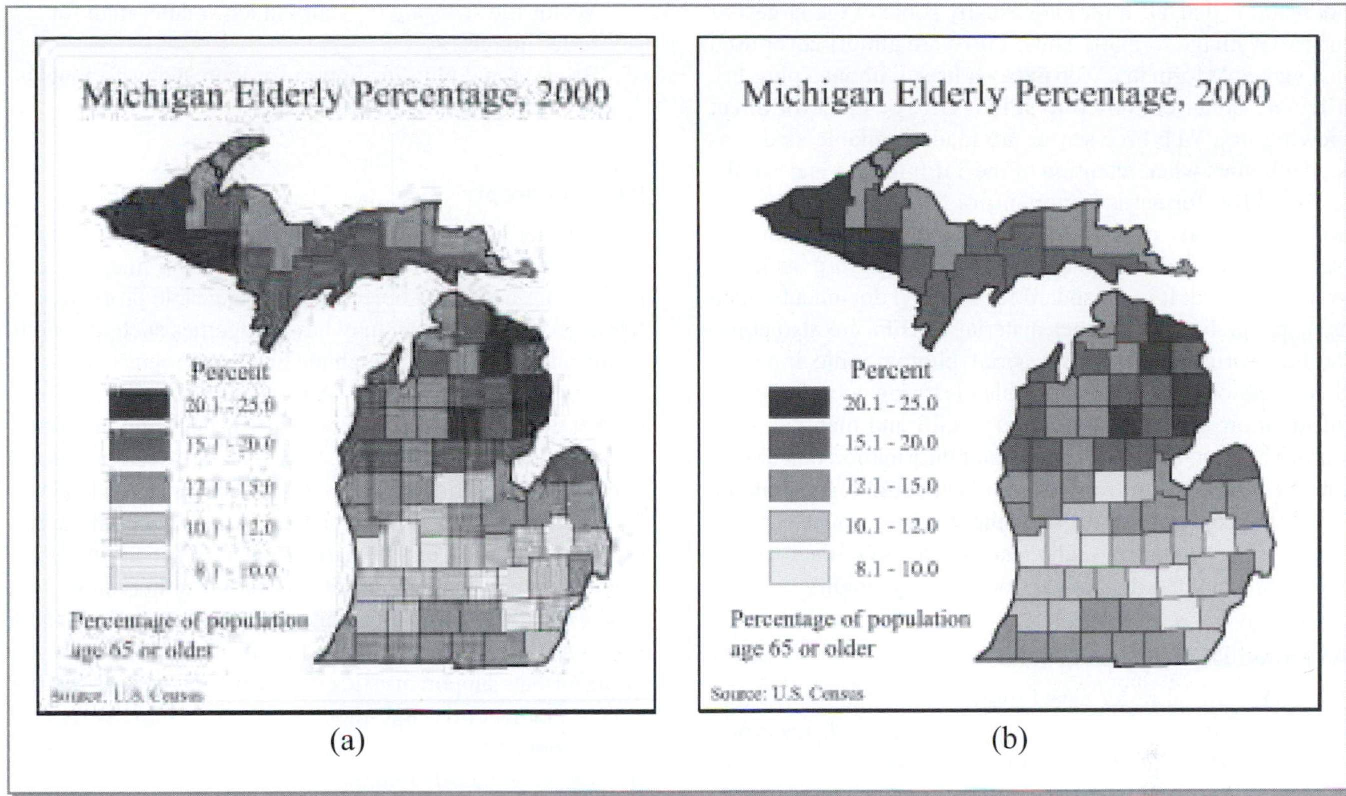

FIGURE 16.2 LOSSY COMPRESSION IN TWO JPEG IMAGES.

Figure 6.1 is converted to a compressed JPEG, resulting in image degradation in both (a) and (b). The amount of compression is greater in (a) than (b), but (b) will have a larger file size than (a). Caution is obviously urged if compression is to be used.

supported by all modern browser software. Short for Graphics Interchange Format, the GIF was developed in the late 1980s by CompuServe, and is now currently licensed by Unisys. The GIF employs an 8-bit depth, a 256-color palette, and a LZW (Lempel-Ziv-Welch) lossless compression algorithm. As a proprietary format, software manufacturers that employ GIFs or the LZW algorithm must pay a fee for their use.

Even at this relatively low-color-bit depth, the GIF is useful for cartographers primarily because many maps, such as a land-use map of 15 or 20 colors, do not require the use of more than 256 colors. The reduced file size from the lower bit depth and the universality of the GIF format make it a popular choice when exporting a map layout to an image format.

Two other capabilities add to the GIF's usefulness. First, the GIF format can also be used for short animation sequences. Although this capability is often associated with clip art that often clutters many Web pages, the format can be useful for making animated maps, usually of shorter duration (formats used for lengthier animations are discussed in the upcoming paragraphs). Second, the GIF file also allows for binary transparency, meaning that one color on the color palette can be set to a transparent setting. This feature allows backgrounds to show though every pixel that has a transparent value, although this can produce a jagged appearance between transparent and nontransparent areas.

PNG. The Portable Network Graphic, or PNG file, was designed as a nonproprietary Web standard that may someday replace the GIF (and possibly the TIFF file, discussed next). PNG supports both 24-bit true color and 8-bit paletted colors. Two hundred fifty-six levels of transparency are supported in the true color version of the format, providing a smoother transition between transparent and nontransparent pixels. Because lossless compression is used, PNG file sizes are larger than a comparable resolution and bit depth JPEG file.

Although PNG files do not support animation directly, because it is an open source (meaning that the code is nonproprietary and is widely available on the Internet), it is possible for software designers to create their own versions of the format, which could include animation. Most browsers support PNG files without the use of a plug-in, although a few browsers do not implement some of PNG's features (such as transparency) correctly.

TIFF. The Tagged Image File Format, or TIFF file, is another format that has been around since the 1980s but has seen many upgrades to its structure. These files can have extensions of .tif or .tiff. The TIFF can support a great number of color modes, bit depth levels, alpha channels, and other features, and can incorporate lossless or lossy compression (lossless is most common if compression is used at all). The net result is that TIFF files are usually some of the largest of the raster image formats. Thus, TIFFs are almost never used in a viewable form in a Web page (at least without a plug-in). They are either converted to JPEG, GIF, or PNG for direct viewing in a Web browser, or are made available as downloadable files when retention of the TIFF format is desired.

The TIFF format is important for cartographers for several reasons. Its robust structure and flexible capabilities have made it one of the most universal image formats in the graphics world. It is a standard for scanned documents, such as maps, pictures, and other materials. TIFFs are also popular for storing satellite and aerial photographic imagery. Some versions of TIFF are capable of storing georeferencing information, so they can be used in GIS and mapping software. They can also contain pyramid information that allows for display of the image at multiple resolutions. It should be noted that converting TIFFs to other image formats for easy display in a Web page usually results in a loss of the format's capacities and/or a degradation of its image quality.

Animated Raster Map Formats

There are several raster based formats available for creating animated maps. Some of the most popular include the cross-platform MPEG, Flash Video FLV, Microsoft Windows AVI or WMV, Apple's QuickTime format, and the Real Media format most often associated with digital video. Media so created are normally viewed in the browser (often with an appropriate plug-in) or in an external player. Animated maps stored in these formats are usually designed to be viewed in a linear fashion and for limited interactivity. That is, the map reader usually does not interact with the map beyond that which is provided in the standard controls for animated playback (for example, start, stop, pause, rewind, and so on).

Vector Graphics

Vector graphics are the format of choice for most cartographers. For most mapping endeavors, whether in a GIS or mapping package, in an artistic drawing program, in computer-aided design (CAD), or in some combination of these environments, vector graphics are involved for most symbolization and other basic map elements in a typical layout. There are some distinct advantages to the vector format for Web cartography:

1. Cartographic symbolization using vector point, line, and area symbols is a natural, more aesthetic expression for features and map elements, particularly if the map is resized or rescaled.
2. The interactive capabilities are much greater in vector formats, since map features are treated as discrete objects.
3. The format is extremely versatile, since most vector formats can also include text objects and raster imagery (in some cases, this includes animated raster graphics). Vector animation is also supported in some file formats.
4. Vector file sizes are typically much smaller than for raster images.
5. Browser and plug-in support for Web vector formats is increasing.

Vector Concepts

The familiar point, line, and area representations of features form the basis of the vector data model. Point, line, and area features are treated as objects with changeable properties. A line object, for example, may have properties such as length, width, and color (even though the line being represented only has one dimension–length). An area feature will not only have outline color and width but will also have fill characteristics such as color or pattern. Vector features can be resized without the pixilation that occurs in raster formats. However, the amount of zooming that can take place is not endless; it is usually incumbent on the cartographer to set limits to the amount of zooming that is practical for his or her map.

Beyond these basics, vector structures have a fairly high variation in capabilities, depending on the format selected. Some formats support artistic embellishments such as shadow effects, gradient fills, and smoothing of curves. Many formats also accommodate raster images and text objects. Still other structures can incorporate interaction and animation.

As with raster graphics, there are a number of formats that are native to certain software packages. The ESRI shapefile (SHP) or the AutoCAD Drawing file (DWG) are two common formats in the GIS and CAD worlds respectively; Adobe Illustrator (AI) and CorelDraw (CDR) files are common artistic drawing software formats. Native formats are often preferred while working on a project within a specific software package, but if the information needs to be transferred or formatted for Web viewing, the graphics are usually saved to a format that can be used in a Web browser, or exported to another finishing program.

There are a number of formats that are designed to be cross-platform (between software programs and operating systems), for transferring vector graphics from one program to another. Many of these transfer formats are more precisely termed **metafiles.** Metafiles are sets of drawing instructions that inform the software how to draw the objects, based on their properties, and can usually accommodate text and raster imagery.

It is important for cartographers to understand that once graphics are exported to a metafile format, the linkages to the data, including the map's coordinate system are usually lost during the transfer process. Also, the quality of the transferred graphics can vary tremendously depending on the source program, the destination program, and the format selected. If you are transferring files from one program to another, experimentation is usually required to find out which format works best with your software configuration.

A number of well-established metafile formats exist that are primarily designed for drawing software or for transferring graphics across a variety of platforms. One of the earliest cross-platform formats was the Computer Graphics Metafile (CGM). Windows Enhanced Metafiles (EMF) are a common structure for transferring objects between Windows software programs. Encapsulated Post Script (EPS) files, discussed in Chapter 15, are used to transport PostScript Language files and are often used in printing environments. Although not usually categorized as a metafile, EPS files are sometimes used like a metafile as a vehicle for transporting graphics from one software package to another.

Selected Vector Formats

Most of these formats discussed so far are usually not directly placed into a Web page. They are most commonly converted into a format supported by browsers or browser/ plug-in combination that can be used by many viewers. Four popular vector formats with a wide range of capabilities are discussed in more detail. For purposes of this discussion, they are treated as vector files but, as with metafiles, can often accommodate raster imagery and text as well.

Flash SWF. One of the first vector formats to be developed and arguably the most popular vector format for the Web is Flash SWF (sometimes pronounced "swiff"). SWF is a format for mixing vector, text, and raster objects (including controls for digital audio and video), and allows for a large degree of user interactivity and/or map animation. For example, maps stored as SWF files enable the map reader to pan and zoom on the map. Some other common options that map designers often include are point and click to obtain information about a specific feature, interactive legends to turn on and off features, and controls to start and stop a map object's movement. SWF files can be created in the Adobe Flash authoring system software, although some graphics programs can also export layouts to the SWF format.

For most of the features to work correctly, the map reader must have the most current Flash plug-in (Flash Player). Although SWF is a proprietary format, it is popular among cartographers for two primary reasons. First, Adobe asserts that over 90 percent of all browsers are Flash-enabled (Adobe Corporation 2007), so it is likely that the map reader will be able to see a Flash map. Second, basic map interactivity (panning and zooming) can be achieved without learning the program's scripting language (although one will have to learn the scripting language for more advanced functions).

SVG. Another vector format that cartographers should be aware of is the Scalable Vector Graphic, or SVG. SVG allows not only for combining vector, raster, and text objects but also for interactivity, animation, and other special effects (Neumann and Winter 2003). It is seen by many professionals as a vector Web alternative to SWF for those who do not wish to use (or perhaps are prohibited from using) a proprietary format.

SVG is based on XML, or Extensible Markup Language. Map layouts can be exported to SVG format directly from software such as ArcGIS, but to make the map interactive for the Web, the cartographer will have to write his or her own XML code. To view SVG files, a plug-in such as Adobe SVG Viewer is required.

PDF. Another popular format is the Adobe Portable Document (PDF) file. The PDF format is perhaps best known for cross platform distribution of text documents that can be viewed on the screen *or* printed at a fairly high quality. This duality of purpose is a hallmark trait of the PDF format (see discussion in Chapter 15 for its use in the printing and production aspects of cartography). Like SWF, SVG, and other metafile formats, the PDF file allows for the mixing of vector objects, text, and raster imagery. Zooming, panning, and insertion of hyperlinks are possible in a PDF file, but the format does not allow map animation or the degree of interactivity like that of the Flash format. Most browsers are enabled with Adobe Acrobat Reader, the plug-in necessary to view PDF files. Cartographers often choose to distribute maps via PDF when they want the map reader to be able to print out the map, or if the final use (virtual viewing versus printing) is unknown. A number of U.S. government Web sites distribute maps on diverse topics such as park trails, land use, and population density in the PDF format.

VRML. Finally, we mention Virtual Reality Markup Language (VRML). VRML is a format that allows for the display of three-dimensional vector graphics, although it can display two-dimensional graphics as well. These graphics can be rotated and scaled as desired by the map reader, and they can be used to interactively "explore" the surface, be it real or statistical. VRML graphics can be created from GIS and artistic drawing software packages that export to VRML, and can be viewed with an appropriate plug-in. VRML has held promise for the mapping community throughout the last decade, but has not yet realized its full potential in Web mapping.

UNDERSTANDING THE MEDIUM

In the previous sections, we have talked about the virtual map, particularly those maps that are meant to be viewed on a Web browser, and some of the most common formats in which they can be created. This section is written to give the cartographer a rudimentary understanding of the environment in which these virtual maps are used and displayed. The environment, or "medium," includes the Internet and the display monitor. The Internet is the medium in which most virtual maps travel, and the display monitor is the medium through which those same virtual maps are viewed. Both have an impact on virtual map design, which will be addressed in the last section of this chapter.

Key Internet Concepts

Perhaps no other development, save for the computer itself, has so dramatically affected the way maps are viewed, studied, read, responded to, constructed, and visualized as the introduction of the World Wide Web. It is important then, to have an understanding of the Web's basic framework and terminology, and understand how virtual maps can be delivered.

The origins of the Internet began in 1969 when experiments by the United States Department of Defense led to the development of Advanced Research Projects Administration (ARPANET) along with the first Internet protocol (IP). In the 1970s, industrial firms related to defense were the first to join the network, and then major universities came onboard. In the 1980s, the National Science Foundation became involved with the development of NSFNET, giving the system a temporary boost during a time of rapid Internet growth. By the late 1980s and into the 1990s, commercial establishments were allowed to join the growing "network of networks," and thus the infrastructure for the modern-day Internet was set in place.

The Internet's early success was due to four primary factors. The first was the ever increasing development of computer technology, notably increasing processing power and storage capability. The second was the underlying principle of **dynamic rerouting.** If one node on the Internet is shut down or becomes temporarily unavailable, then the information simply follows a different route. Third, by the late 1980s and early 1990s, most of the world had adopted transmission control protocol/Internet protocol (TCP/IP) for sending information across the Internet in "packets" (W3C 2007), and allowed disparate computer types to communicate with each other across a myriad of connection mechanisms. Fourth, by the 1990s, so many industries, commercial entities, independent organizations, government agencies, and educational institutions worldwide had joined the Internet that no single entity could claim "ownership" of it. Therefore, while a corporation or a country's government can restrict or shut down their particular node or nodes, takeovers, crashes, and the like of the entire Internet are not possible.

The World Wide Web

As successful as the concept of the Internet was, it would be in a form that most readers today would scarcely recognize—or perhaps even enjoy using. Prior to the early 1990s, the Internet was text-based and command-line driven, sometimes requiring the typing of many commands to accomplish simple things that we take for granted, such as emailing a friend, downloading a map (in the latter case, the map would have to be opened in an appropriate software package), or the easy searching for information via Google and other search engines that we enjoy today. It was not until the development of the World Wide Web and the modern graphical Web browser that "surfing the Web" would become the nearly universal concept that it is today.

The World Wide Web is a system that allows for linking of *resources* (text information, pictures, maps, videos, or other Web sites) on the Internet via **hypertext** (text that links to resources) and **hypermedia** (multimedia objects such as graphics that link to resources). The Web browser (as discussed in the first section above) allows for the Web surfer to view and interact with those resources. The Web and Web browser *combination* is what allows people to point and click on a certain highlighted text or graphic, and have new resources appear on the screen or become available for downloading.

Web resources are addressed and located according to a universal naming convention. This convention can be observed in examining a Web page's URL **(uniform resource locator),** which is usually listed at the top of the screen in the Web browser (W3C 2007). At a fictitious "My Site" Web site, for example, we can see the URL of some typical Web resources:

• My Site Web site	http://www.mysite.edu
• Web page at My Site	http://www.mysite.edu/population.html
• JPEG map of United States Population Density	http://www.mysite.edu/docs/USAPopDensity.jpg
• PDF version of the same map	http://www.mysite.edu/docs/USAPopDensity.pdf

The domain name, such as mysite.edu, census.gov, or ebay.com can provide a context for the resources being accessed. Specific types of Web sites can be identified by the suffix or extension, such as government sites (.gov), business sites (.com or .biz), educational sites (.edu), and independent or nonprofit organizations (.org). In some cases, there will be a zone designation for a particular county, state, or country.

The second URL above (population.html) is for an individual Web page at the My Site website. A Web site can be as simple as a single Web page, or the HTML extension indicates that the Web page is written in **hypertext markup language.** Although there are many Web languages, HTML is considered the *lingua franca* of the Web (W3C 2007). A typical Web page will combine text, graphics, and other multimedia elements depending on the page's purpose and readership. Figure 16.3 is a simple Web page with a map graphic. The figure illustrates both the image file embedded in HTML code (Figure 16.3a) and how it appears to the map reader in the Web browser (Figure 16.3b). As noted in the boxed text "Strategies for Learning to Build Web Pages," a Web page such as this can be created in dedicated Web page creation and editing software, or can be directly coded using a simple text editor.

Web sites are often structured in a manner similar to how files and folders (or directories) are set up on a personal computer. In the second and third examples, the choropleth map USAPopDensity is available both in

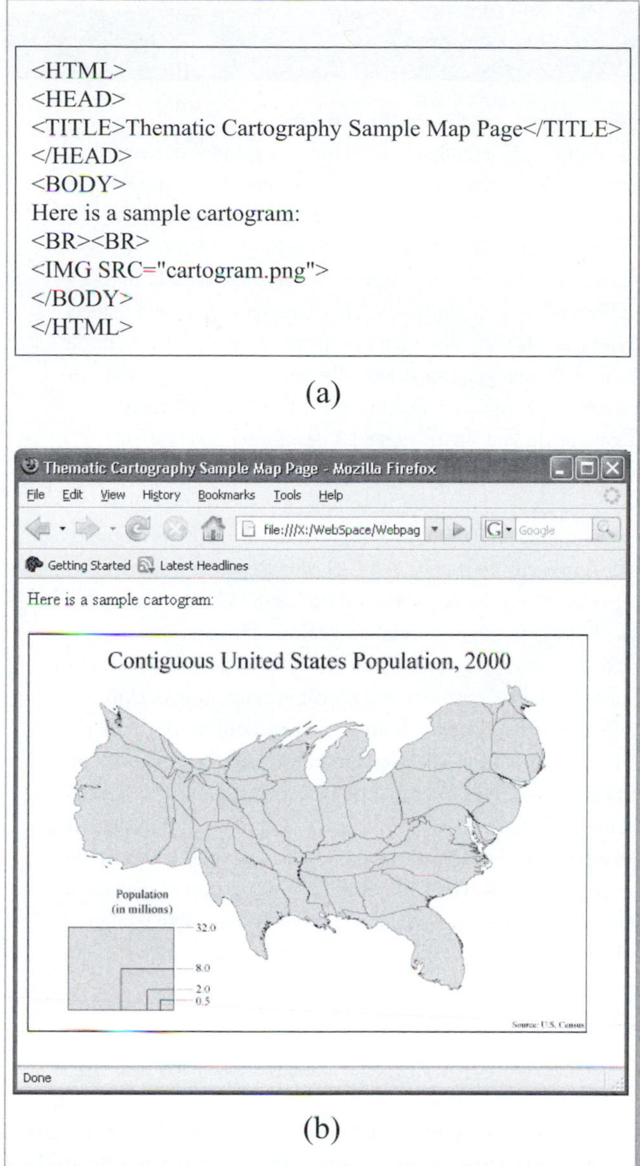

FIGURE 16.3 HTML EXAMPLE IN AND ITS WEB PAGE APPEARANCE.

In this simple Web page example, the HTML code and the reference to the map graphic (a) and how it appears in the Web browser (b).

JPEG and PDF formats. They are both stored in the "docs" directory. The directory (or directories) between the domain and the resource is called the path, similar to the path for files stored on a personal computer. When the resources are accessed, the JPEG will open directly in the browser window, as will the PDF file (although the window will be somewhat modified by the Adobe Acrobat plug-in). A Web site can be as simple as a single Web page or it can be as complex as hundreds of Web pages in an extensive directory structure. An example of the latter is the United States Geological Survey Web site (www.usgs.gov). This site is popular with geographers and cartographers because of the wealth of maps, data, and other information it contains.

When someone writes a Web page, the virtual maps and the HTML code (or other programming), and other resources have to be uploaded and stored on a *server* in order for the public to be able to view the Web page. A server is a computer (or group of computers acting in concert) that make the Web resources that are stored there available to potential users. For example, when a map reader clicks on a link to view a map, it the server's job to send the appropriate map resource back to the map reader's Web browser. This is the simplest model of how the process works. (See the boxed text "Server-Side Mapping" for a brief discussion on another model for online mapping and data distribution.)

The Web page and other resources travel over wireless networks, fiber optic cables, telephone lines, and other Internet transfer mechanisms. A network's **bandwidth** is its capacity for transferring data, including both the uploading of Web pages and other resources to a server and the downloading of resources to the Web browser or one's computer. Some networks have greater capacities than others for handling large volumes of data. It should be noted that video and raster imagery are generally the largest files in terms of storage, and therefore consume the greatest amount of bandwidth of the media types. Text, such as the HTML coding, consumes the least amount. It is important to recognize that in some areas of the world, the audience may be accessing the Internet with a slow modem or using an Internet service provider with limited bandwidth capabilities. In such cases, placing an extremely large map image on the opening (home) page of your Web site, for example, will probably dissuade the would-be reader from downloading your maps or even viewing your Web page.

The Graphics Display Monitor

Web pages and virtual maps are viewed on a display monitor. It is extremely important for the cartographer to realize that display monitors can vary greatly in physical size, resolution, quality, and display of color from monitor to monitor. This means the virtual map will not take on the same appearance in all display environments. This last part of "Understanding the Medium" is devoted to this essential topic.

Graphic monitor displays currently range from the largest format high definition displays, with a diagonal viewing size of over 60 inches (not including projection devices), down to the small personal digital assistant (PDA), with a diagonal view of just less than three inches (Wintges 2003; Peterson 2005) or even less with a cell phone display. For purposes of this chapter, we will be focusing on monitor displays that are typically found on laptops or desktop work environments. Most of these are between 14 and 28 inches diagonal viewing, and will either follow a 4:3 or 16:9 **aspect ratio**—the ratio of the horizontal dimension to the vertical dimension in

SERVER-SIDE MAPPING

In this chapter, we are focusing on the simplest case of designing virtual maps and making them available to the Web, by uploading the map(s), often in the context of a Web site, to a server. When the map is sent to the map reader's browser, all activity associated with the map (including interactivity and animation) happens inside the map reader's computer and is known as "client-side" processing.

Earlier in the chapter we alluded to the fact with some interactive maps, the map reader may be interacting with an underlying GIS or other data distribution software and not be aware of it. A primary example is the zoom-to city mapping Web sites such as MapQuest (www.mapquest.com). The map reader selects an address, zip code, region, or named place. This information is sent to the server and the server generates a new map based on the map reader's information. The server then returns a static gif image or other graphic to the browser. If the map reader wishes to zoom in further, or pan the map to a different location, that information is sent again to the server, and the new map result is presented.

This technology is referred to as server-side processing, and the software that makes it run is called a server-side application. In server-side processing, a majority of the work is done by the application on the server, and the result is then transferred to the browser. Maps so produced are usually not predesigned by the cartographer but are created "on-the-fly" based on the requirements of the map reader as in the MapQuest example. The "map design" at this point is all about developing the interface and the options for the map reader's viewing and interacting with the map. Note that most server-based map applications have a tendency to favor general purpose reference maps, such as those sites that serve zoom-to-city street level maps, or those that deliver satellite and air photo imagery.

This method of distribution is becoming increasingly popular among both private and government agencies eager to find new and better ways of distributing maps, images, data, and related documents. If your site is data or image intensive, (for example, the user of your site can potentially download gigabytes of information), or if you wish to allow the user to do GIS-based processing, then you should consider developing a server-side Web page. Note that developing server-side applications requires access to greater computer resources, increased costs, and programming skills. If GIS analytical functionality is required, access to software such as ArcGIS Map Server is also required.

What about the popular Google Earth? Google Earth, like MapQuest, distributes aerial and satellite imagery from a server as previously described. However, it is sometimes considered a *hybrid* of client and server architecture, because Google Earth allows anybody to create their own content (for example, from their desktop computer) and merge it with the imagery from the server in Google Earth (Plewe 2007). For example, Krygier (2007) developed an animated choropleth map designed for viewing in Google Earth. The map's data, in the form of KML (Keyhole Markup Language) code is stored on the user's computer. KML is a markup language that supports 3-D geographically referenced data. When activated, the map animation runs on top of the imagery.

a monitor display's viewing area. (see Figure 16.4). In many classrooms and university settings, the 4:3 format is more common. The 16:9 format monitors, while more expensive than their 4:3 counterparts, are popular for their ability to show movies in a theatrical (or high definition) ratio.

Display monitors are raster devices. Like the images discussed previously in this chapter's section on "Raster Graphics," display monitors have rows and columns of *pixels*. The number and size of the pixels in the display's viewable area defines its *resolution*. The display monitor's *bit depth* (the number of simultaneous colors that can be displayed) is often called **color depth** when applied to the display. Over 80 percent of all monitors on the Web today are set at 1024*768 or higher resolution, and almost all monitors are set up to display over 16 million colors (24-bit or higher color depth) (W3Schools 2007).

Most display monitors are either CRT (cathode ray tube) monitors or LCD (liquid crystal display) monitors. The CRT monitor uses a set of electron guns that charge sets of phosphor-coated dots that exist for each of the additive colors (red, green, and blue) on the inside of the monitor screen. The speed at which the screen is refreshed is called the **refresh rate.** Screens that have a refresh rate of less than 60 Hz (60 times a second) often have a noticeable flicker, and can produce eyestrain if viewed for an extended period of time. CRT monitor resolution (and color depth) can easily be changed.

The LCD monitor arranges pixels into red, green, and blue sub-pixels. Unlike the dots in a CRT, the pixels are hardwired into a specific resolution (for example, 1024 by 768), so there is a "best" resolution that is associated with each monitor. Setting the resolution to other than the manufacturer's recommendation can lead to results that are less than pleasing (if all of your maps look fuzzy on an LCD display, the screen resolution may be a place to start looking). Like CRTs, they are capable of displaying millions of colors, and the flicker often associated with CRTs is not an issue.

The LCD monitor is rapidly replacing the CRT in many computer environments, but the use of both types is still widespread. Some graphics enthusiasts point out that color and particularly motion is better on the CRT, and if you can obtain one, they are usually less expensive than the LCD

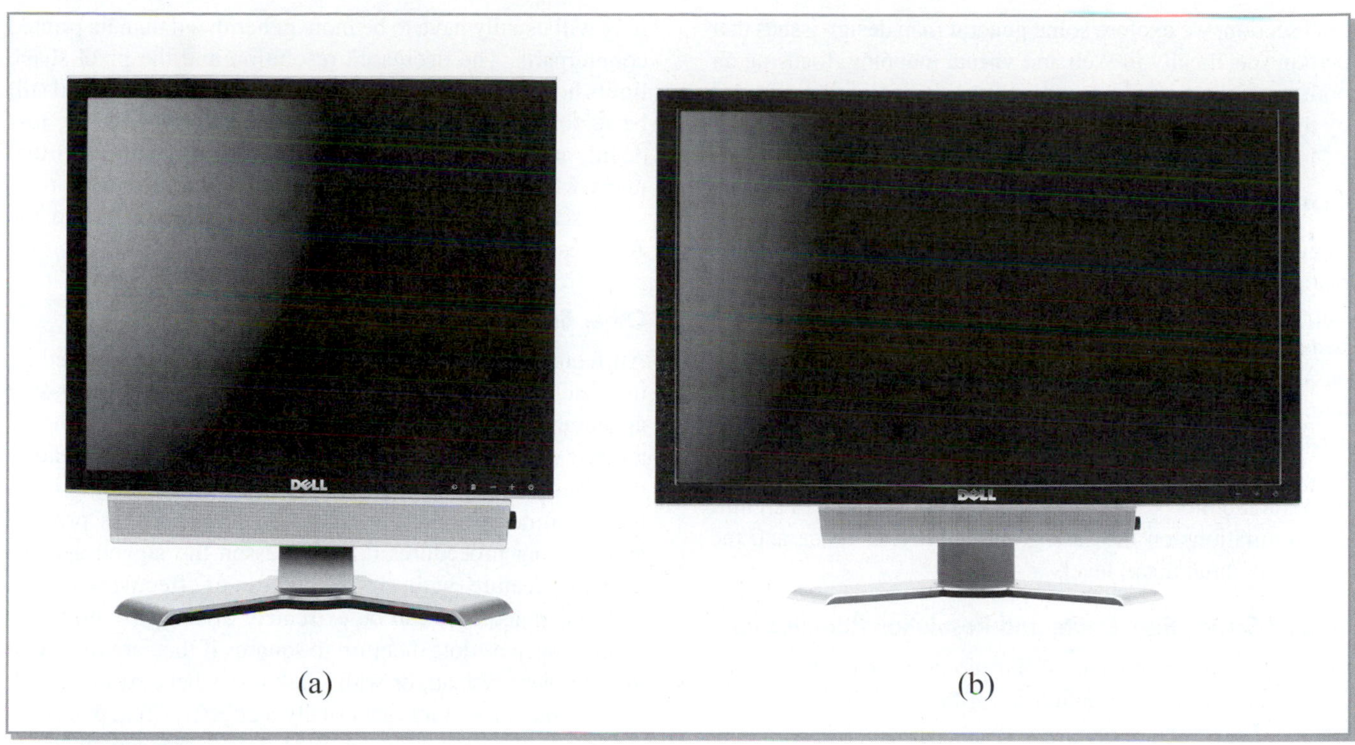

FIGURE 16.4 TYPICAL MONITORS AT BOTH 4:3 AND 16:9 ASPECT RATIOS.
The 4:3 monitor (a) is common in many academic settings, but the 16:9 monitor (b) is popular with those wanting the capability to view high definition (theatrical or wide-screen format) movies. Source: Courtesy of Dell Inc.

monitor. However, the CRT is bulkier and heavier than the LCD monitor, and uses more power as well. Even though the LCD monitor is hardwired for a "best" resolution, there is no flicker and therefore eye fatigue is noticeably reduced when using an LCD. It appears that the LCD will be the standard in the near future.

It should be noted that monitor resolutions are lower than printed graphics. Most display monitors operate at approximately 70 to 100 pixels per inch (depending on the size and resolution of the monitor), but printed graphics range from 600 dpi (dots per inch) in some inkjet printers, to 2400 dpi in many color laser printers, to over 4000 dpi in some high-end service bureau devices. This is one reason why the 72 dpi raster map (a common resolution for images on the Web) that looks so good on the screen often has such a poor printed appearance. Map designers should realize that limited screen resolution can be a liability, particularly with regard to text and in-map labels. Map design issues related to resolution will be discussed later in the chapter.

One other topic that is often overlooked in discussions about display monitors is the graphics card. The quality of the display, including the properties of resolution, color depth, and refresh rate (applied to a CRT) are a function of the *combination* of the monitor and the graphics card. The graphics card does the graphics processing and takes the load off the central processing unit (CPU). Advances in technology mean that the occasionally annoying trade-off between resolution and color depth is less of an issue now than a decade ago. Most graphics cards work reasonably well for two-dimensional graphics, but for three-dimensional animated graphics (such as a detailed terrain fly-through), a more expensive higher-end graphics card will produce noticeably better results.

The virtual map designer should also take notice that his or her map may take on a different appearance with different monitor brands and graphics card combinations, and with changes in settings to those hardware elements. Graphics on Apple monitors, for example, generally take on a darker appearance than with other brands. As noted in Chapter 14, display monitors have different internal settings that can be adjusted by the user, including contrast and brightness, and can radically change the intended look of a map. Again, understanding that the appearance of your virtual map will somewhat change from monitor display to monitor display is an important aspect of Web and virtual mapping.

DESIGN IMPLICATIONS FOR THEMATIC MAPPING

In the previous sections of this chapter, we have addressed some introductory concepts in Web and virtual cartography (including static, interactive, and animated maps); explored file formats and structures common for this type of mapping; and covered the basics of Internet, Web, and monitor technology through which these types of maps are delivered. In this final section, we explore some general map design issues that pertain specifically to Web and virtual mapping, focusing on both the limitations imposed and the opportunities afforded by the medium as discussed so far in the chapter.

Constraints of the Medium

Many issues are considered constraints in Web and virtual mapping. These can range from restrictions on map file sizes imposed by storage and/or bandwidth limitations to the expense and hassle of uploading the maps to a server (discussed earlier in "Understanding the Medium"). In virtual map *design,* however, the most important constraint is the resolution of the display monitor. The monitor's decreased resolution when compared with most printed maps, coupled with the monitor's size and fixed aspect ratio, places significant limitations on the cartographer's layout space and the map's potential detail level.

Limited Screen Real Estate and Resolution Ramifications

One of the major impacts of the monitor's resolution, size, and aspect ratio combination is on the overall screen "real estate." Decreased screen real estate lessens the area with which the thematic cartographer can create the map layout, including the map body, legend, title, and other basic map elements introduced in Chapter 1 (a complete discussion of basic map layout occurs in Chapter 12). Kraak (2001) notes that a number of Web maps lack many of these basic elements due to the lack of space, although we would also suggest that in many cases inadequate cartographic training and knowledge also plays a major role.

Some cartographers get around the space issue by placing all ancillary elements, such as the title, legend, and source statement outside the map graphic. For example, the title and source text may be placed in the HTML code of the actual Web page. In general, we recommend against this practice, particularly for raster image maps, because if the map reader extracts and saves the map from the Web site, important information will be lost when the map reader calls up the extracted map. Legends that are placed on a different Web page (requiring the reader to flip back and forth between pages), or below the map (requiring the map reader to scroll up and down the Web page) disrupt the legend's role as a visual anchor. Excessive flipping or scrolling between a map and legend page can be fairly annoying as well.

The decreased screen real estate also means that the rectangular aspect ratio of the display area makes it easier to create layouts in a landscape orientation. As such, political units with large north-south extents, such as California or Chile, can present quite a challenge to the cartographer. Units with modestly larger east-west extents, such as Colorado, Canada, or the continental United States, are easier to fit on display area, although units with more extreme east-west extents, such as Tennessee or Russia, can also be challenging.

If the cartographer is trying to limit or minimize the amount of scrolling, panning, or zooming that the map reader will have to use in order to take in the whole map, the virtual map will usually have to be more generalized than its printed counterparts. The decreased resolution and the pixel shape limits how small features and text can be on the map and still be legible. Ultimately, this means that fewer features, less detail, and less in-map text will be characteristic of virtual maps when comparing them to printed counterparts at the same scale. This resolution limitation has been called the Achilles heel of the medium (Monmonier 2005).

Other Screen Resolution Issues

All features—points, lines, areas, and text—are affected by the pixel's geometry, as evidenced in the **aliasing** (stair-step appearance) of lines on the map. Since the display monitor is a raster device, all maps are subject to aliasing. Any feature that does not have vertical or horizontal lines or edges is affected, although many artistic software packages provide anti-aliasing procedures that can lessen the jagged appearance of a feature, with varying degrees of effectiveness.

Smaller features can be extremely affected by pixel geometry, such as dots that turn to squares if they are too small in a dot-density map, or with small text labels. At this point, anti-aliasing procedures are usually ineffective. Text that is too small will lose its legibility, as the letters merge and the bowls on the individual letters begin to fill. If the text is rotated or otherwise placed on a curve, aliasing can be even more pronounced as the lines move away from a horizontal and vertical configuration, particularly with serif faces. To ensure legibility, we recommend a text size of at least 8 or 10 points for the final viewing scale of a map in virtual mapping applications.

If a map is originally created in vector format but is going to be exported to a raster image-format, it is wise to create the map at the final scale and size that you want it to be displayed. Images that will be embedded in a Web page can be resized via the HTML coding (or other Web languages). But rescaling or resizing rasterized linework usually results in excessive aliasing and pixilation. If the map ends up too large or too small for the Web page, then we recommend going back to the mapping software, changing the dimensions of the vector map, and re-exporting the map to a raster format. The result will be a more aesthetically pleasing map product.

Display Monitor Variations Limit Predictability

Because monitors vary in intensity, brightness, contrast, and so on, it is also important to revisit color (see Chapter 14 for discussion on color for monitor displays). Color depth can also vary, but since a majority of monitors are configured to accommodate over 16 million colors (24-bit depth or greater), the depth is not the issue that it was even a decade ago (W3Schools 2007). Ideally, it would be advantageous to view the maps on a number of display monitor and graphic card configurations. Since this is usually impractically, it is important that virtual maps be designed so that color levels used for features, area class fills (as in a choropleth map), and other map elements have sufficient difference between the color parameter hues, values, saturations so that individual monitor variations will not mask or otherwise adversely affect the intended map message.

ColorBrewer, the online color selection site discussed in Chapter 14 and in other chapters, can also assist with the selection of colors that work well with display monitors (Brewer 2007).

Solutions and Opportunities

With these limitations in mind, we turn our attention to some of the newer mapping techniques afforded by the medium. These include interactive mapping, map animation (virtual mapping concepts that were introduced in the first section), and cybercartography—a new term describing some other exciting possibilities for different kinds of Web mapping. Although any type of virtual map is subject to the medium's constraints, these areas provide at least partial solutions that even well-designed static maps cannot provide to the same extent.

Map Interactivity

Although static maps are currently the most common virtual form, the capability for the users to interact with maps is one of the hallmark benefits to virtual cartography, and can be part of the solution to the decreased screen resolution and real estate problem. Some of the most common functions for interactive maps tend to emulate basic GIS activities, such as panning and zooming, selecting layers of information to be turned on or off, and interaction with individual features. These functions can operate separately or in concert so that not all features need to be labeled or be visible at all times.

Panning and zooming techniques are especially popular for interactive vector reference maps that have not been generalized for smaller scales. Most map users are reasonably comfortable with the concept of panning and zooming (Harrower and Sheesley 2005). However, for most thematic maps, which are typically designed at medium or small scales, we suggest that maps that require *excessive* panning or zooming can interfere with the effort to communicate an overall distribution or pattern. Care in choosing the correct scale and generalization level is still one of the most important roles for the cartographer, and panning and zooming capability should not take away this responsibility.

Layer selection is another popular interactive map technique. With layer selection, layers of information, such as cities, population densities, rivers, and so forth can be turned on or off by the map reader. This allows the map reader to explore relationships between layers and to select a comfortable complexity level. Layer selection is a viable alternative to juxtaposing maps of related information (common in print cartography) in order to make map comparisons, since the limited screen real estate makes virtual multiple map layouts difficult.

Pointing and clicking on features to reveal information and attributes, such as clicking on a city or river to discover its name and population or average daily flow, is another common interactive function. This technique can be applied to most of the thematic map types covered in Part II of this text. For example, in an interactive choropleth map, the map reader could click on an enumeration unit to see its name and specific value.

For the most part, interactive mapping is done in the vector realm. However, there can be an element of interactivity that can be introduced into raster imagery. Web programmers refer to a clickable image as an *image map* (whether or not the image actually is a map). The pixels in an image have absolute X and Y locations. Using HTML, JavaScript, Java, or other programming applications, rectangular, circular, and even irregularly shaped areas can be made active within the image. The map reader can click on these active areas, and perform whatever function has been programmed by the Web page's designer. Thus, a map can be made to display another map, perhaps at a different scale, or have a feature be identified in a text box when clicked—even though it is not a feature-based structure.

Animation

In the first part of this chapter, we introduced the concept of map animation. Using motion as visual variable is an exciting prospect to many map designers. Temporal animations (animated maps that illustrate change over time) are currently the most common type of animated map (see Color Plate 16.1). However, animated maps can be nontemporal as well, such as an animation of one map projection of the world being changed into another to compare projection properties. Animated maps can also take a three-dimensional form. Since three-dimensional animations usually involve topography of an area, they are referred to as a terrain fly-through or fly-over, because the view is like flying though a virtual landscape in a simulated aircraft (see Figure 16.5). But note

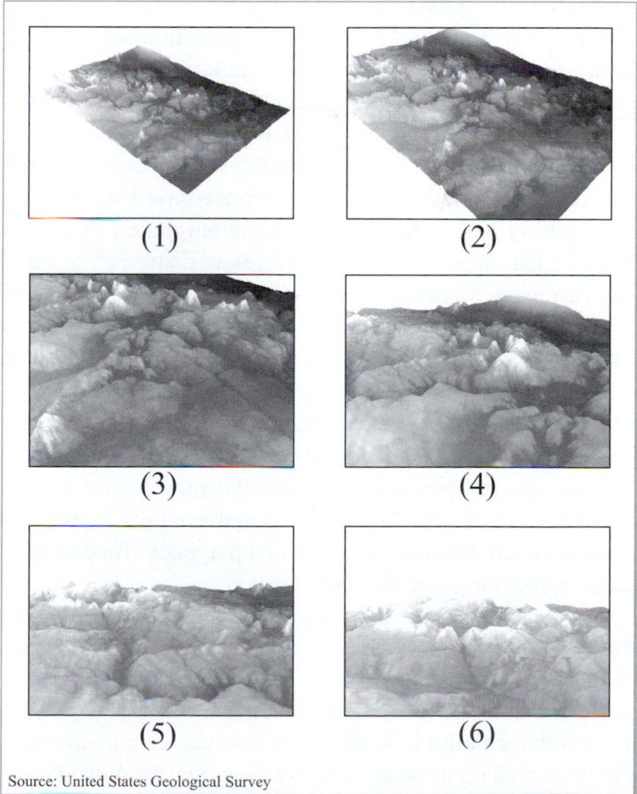

FIGURE 16.5 SELECT FRAMES FROM A 3-D TERRAIN FLY-THROUGH.

that any statistical surface that can be mapped in 3-D form can be made into a fly-over.

Like interactive maps, map animation can be part of the solution to the screen real estate problem, since animation is often used in place of displaying multiple maps side-by-side. For example, if the cartographer's thematic map topic is the change in an area's population density from 1900 to 2000, he or she could create four or five individual maps showing this change, or present the change in one animated map. The single animated map will fit much better on the display monitor than will multiple maps. From a communication standpoint, the current research is not definitive on whether multiple maps or an animation is "better." However, there is research suggesting that map animation can be a viable construct for communicating spatial information (Torguson 1993), even in the context of animation being an alternative to using multiple maps (Griffen *et al.* 2006).

Animated maps predate most current types of virtual maps, having been produced as early as the 1940s (Harrower 2004), albeit using film technology. The first academic paper on the subject by Thrower (1959) started a generation of animation research. This research often included the creation of map animations on a variety of subjects, but this research did not lead to the adoption of animation as a major cartographic form prior to the early 1990s (Campbell and Egbert 1990). Recent advances in technology have made it practical for cartographers to create this type of virtual map and distribute it on the Web, where animated maps of everything from tectonic plate movements to regional histories are now found in abundance. The combination of GIS, artistic drawing programs, and/or dedicated animation software can be used to generate animated maps, although creating them can sometimes be quite labor intensive.

There are two primary types of animated maps in terms of creation and storage (Peterson 1995; Harrower 2004). The first and conceptually simplest is **frame-based animation.** Gersmehl (1990) likened frame-based animation to a "flip-book." That is, a sequence of images is displayed in rapid succession, creating the illusion of motion. Individual frames are created in a GIS or drawing program, and are sequenced and stored as an animated GIF (for smaller animations) or in one of the other animated raster map formats discussed earlier. As noted in that section, animated maps stored in these formats are usually designed to be viewed in a linear fashion and for limited interactivity. Thus, the map reader usually does not interact with the map beyond that which is provided in the standard controls for animated playback (for example, start, stop, pause, rewind, and so on).

The second type of animated map is **cast-based animation,** which is accomplished in software that supports vector formats such as Flash or SVG (discussed earlier). In cast-based animations, cells are individual frames of animation that can have multiple layers. Objects—the cast members—can be placed on these layers, and can either be moved manually on a frame-by-frame basis or by a process called **tweening**—short for "in-betweening" (Peterson 1995). When tweening is used, the cartographer will set up a beginning and ending **key frame,** and the software will automatically generate intermediate frames, moving the object (and, in some cases, changing its shape or color) in the process (see Figure 16.6).

There are three distinct advantages in designing a map using the vector oriented cast-based animation. First, a vector structure can be easily rescaled, meaning that the animation can be played at multiple resolutions without additional pixilation or related problems. Second, the resulting files tend to be much smaller and therefore consume much less bandwidth than their raster counterparts (Harrower 2004). Third, and perhaps most important, the vector structure enables the map reader to be able to interact with the map. All of the benefits of interactive mapping can be combined with map animation. For example, if several objects are in motion, the map reader could turn each of them on or off to examine individual movement patterns or perhaps relationships between a selected subset of these objects.

A major design goal in map animation is usually to develop a reasonably fluid animation. If too much change occurs too quickly, then the eye may not perceive the detail intended by the cartographer, or the map may take on a jerky appearance. Conversely, there must be enough change to warrant using animation in the first place or the map may become uninteresting. In this case, the bored map reader may actually miss the few but perhaps important changes that are happening. Thus, it is important to provide controls that allow the map reader to start, stop, back up, or fast forward the animation, loop the playback, and control the frame rate, in order to absorb the information at his or her own pace (Torguson 1997; Harrower 2003).

It is also important to revisit some of the basic map elements (discussed in Chapters 1 and 12) when creating animated maps. As with most static thematic mapping, not all elements are necessarily going to be present on every animated map. One of the most important elements in animated *thematic* maps is the legend. The legend information needs to be in fairly close proximity to the map body, and cannot change position or form if it is to serve as a visual anchor for the symbolization. For example, in the animated choropleth map depicted in Color Plate 16.1, the same legend is used for every frame in the animation. The classification is based on the lowest and highest numbers in the data set, and the class ranges stay the same (see Chapter 6 for more discussion on legend construction for animation and multiple maps).

When the animated map is temporal in nature, it is important to have some form of indicator to represent passage of time. In the same way a graphic or representative fraction is used to depict spatial map scale, the indicator is used to depict temporal scale. Peterson (1995) identifies two expressions of temporal scale that are quite common today. The first is to display the date, year, or time of day in numeric form. As the animation progresses, the numbers will change

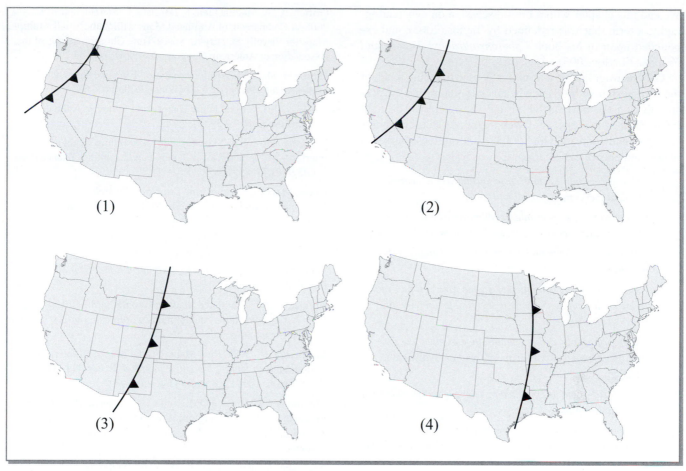

FIGURE 16.6 TWEENING BETWEEN KEY FRAMES IN A CAST-BASED ANIMATION.
Tweening (2 and 3) occurs between key frames (1 and 4).

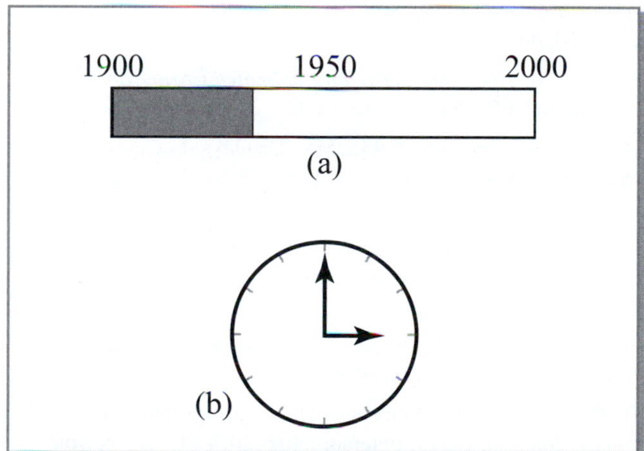

FIGURE 16.7 GRAPHIC TEMPORAL SCALES.
Using a timeline (a) and a clock (b). Source: After Kraak *et al*, 1996.

(see Color Plate 16.1). The second expression is analogous to the graphic or bar scale, and is illustrated in one of two ways (see Figure 16.7). In the timeline scale (Figure 16.7a), an indicator is included showing the location on the timeline that corresponds to the map at that moment in time. Temporal scale can also be illustrated as a clock (Figure 16.7b), which is common when the scale is measured in minutes or hours. For nontemporal maps, the animation player's built-in progress bar is usually sufficient to show how far the animation has progressed.

Finally, north arrows and inset maps are two basic elements that are commonly used in conventional cartography. Many small- and medium-scale thematic maps do not include these basic elements simply because, when the orientation is north and the map region is recognizable to the prospective map readers, they are often not necessary. However, 3-D terrain fly-throughs are often developed at relatively large scales. A small inset map can be included in the layout to show the mapped area's relative location. During the course of the fly-through, the orientation of the map may change numerous times. An animated north arrow indicating the current direction at that place in the animation should be considered.

Cybercartography

Throughout this text, we have pointed out numerous possibilities for virtual maps. And it is the exciting new and future possibilities that have given rise to a new term that encompasses both current and future prospects—**cybercartography.**

We end this chapter with a brief discussion on cybercartography, a term that was redefined by Taylor (2003), and expounded upon in his book *Cybercartography: Theory and Practice* (Taylor 2005).

Cybercartography is a term that is used to define much of the current state of Web and virtual cartography and the possibilities on the horizon for virtual cartography. The term includes all of the aspects of Web mapping discussed in this chapter—and more. Taylor (2003, 407) outlines the seven major elements of cybercartography as follows:

1. *Cybercartography is multisensory, using vision, hearing, touch, and eventually, smell and taste.*
2. *Cybercartography uses multimedia formats and new telecommunication technologies, such as the World Wide Web.*
3. *Cybercartography is highly interactive and engages the user in new ways.*
4. *Cybercartography is applied to a wide rage of topics of interest to the society, not only to location finding and the physical environment.*
5. *Cybercartography is not a stand-alone product like the traditional map, but part of an information/analytical package.*
6. *Cybercartography is compiled by teams of individuals from different disciplines.*
7. *Cybercartography involves new research partnerships among academia, government, civil society, and the private sector.*

Of course, many of the items on this list are things that geographers, cartographers, and other mapping scientists already do. But the *combination* of all of these elements can serve as a framework or roadmap for things to come in Web and virtual mapping. In other words, the term that encompasses these seven elements serves to "reassert and demonstrate the importance of maps and mapping and the centrality and utility of cartography in the information era" (Taylor 2005, 6)—something that most cartographers want for their profession.

REFERENCES

Adobe Corporation. 2007. Adobe—Flash Player Statistics. http://www.adobe.com/products/player_census/flashplayer/

Brewer, C. 2007. ColorBrewer—Selecting Good Color Schemes for Maps. http://www.ColorBrewer.org

Campbell, C., and S. Egbert. 1990. Animated Cartography: Thirty Years of Scratching the Surface. *Cartographica* 27(2): 24–46.

Fraser, B., and D. Blatner. 2005. *Real World Adobe Photoshop CS2*. Peachpit Berkeley, CA. Press.

Gersmehl, P. 1990. Choosing Tools: Nine Metaphors for Four-Dimensional Cartography. *Cartographic Perspectives* 5 (Spring 1990): 3–17.

Griffen, A., A. MacEachren, F. Hardisty, E. Steiner, and B. Li. 2006. A Comparison of Animated Maps with Static Small-Multiple Maps for Visually Identifying Space-Time Clusters. *Annals* of the Association of American Geographers 96(4): 740–53.

Harrower, M. 2003. Tips for Designing Effective Animated Maps. *Cartographic Perspectives* 44 (Winter 2003): 63–65.

———. 2004. A Look at the History and Future of Animated Maps. *Cartographica* 39 (3):33–42.

Harrower, M., and B. Sheesley. 2005. Designing Better Map Interfaces: A Framework for Panning and Zooming. *Transactions in GIS* 9 (2):77–89.

Kraak, M. 2001. Cartographic Principles. In *Web Cartography: Developments and Prospects*, eds. M. Kraak and A. Brown. New York: Taylor and Francis.

Kraak, M., R. Edsall, and A. MacEachren. 1996. Cartographic Animation and Legends for Temporal Maps: Exploration and or Interaction. *Proceedings of the 7th International Spatial Data Handling Conference* 1:17–28.

Krygier, J. 2007. Animated Maps in Google Earth. *Making Maps: DIY Cartography*. http://makingmaps.wordpress.com/2007/07/31/animated-maps-in-google-earth/

Moellering, H., 1984. Real Maps, Virtual Maps and Interactive Cartography. In *Spatial Statistics and Models*, eds. G.L. Gaile and C.J. Wilmott. 109–32. Boston: D. Riedel.

Monmonier, M. 2005. POMP and Circumstance: Plain Old Map Products in a Cybercartographic World. In *Cybercartography: Theory and Practice*, ed. D. Taylor. 15–34. Amsterdam: Elsevier.

Neumann, A., and A. Winter. 2003. Web Mapping with Scalable Vector Graphics (SVG): Delivering the Promise of High Quality and Interactive Web Maps. In *Maps and the Internet*, ed. M. Peterson. 197–220. Oxford: Elsevier Science.

Perry-Castañeda Library Map Collection (2007). http://www.lib.utexas.edu/maps/

Peterson, M. 1995. *Interactive and Animated Cartography.* Englewood Cliffs, NJ: Prentice-Hall.

———. 2005. Pervasive Public Map Displays. In *Cybercartography: Theory and Practice*, ed. D. Taylor. 349–71. Amsterdam: Elsevier.

Plewe, B. 2007. Web Cartography in the United States. *Cartography and Geographic Information Science* 34(2): 133–36.

Smith, R. 1988. *The Electronic Atlas of Arkansas*. Fayetteville, Ark: University of Arkansas Press.

Taylor, D. 2003. The Concept of Cybercartography. In *Maps and the Internet*, ed. M. Peterson. 405–20. Oxford: Elsevier Science.

———. 2005. The Theory and Practice of Cybercartography: An Introduction. In *Cybercartography: Theory and Practice*, ed. D. Taylor. 1–13. Amsterdam: Elsevier.

Thrower, N. 1959. Animated Cartography. *The Professional Geographer*. 11(6): 9–12.

Torguson, J. 1993. *Assessment of Cartographic Animation's Potential in an Electronic Atlas Environment*. Ph.D. diss., University of Georgia, Athens, GA.

———. 1997. User Interface Studies in the Virtual Map Environment. *Cartographic Perspectives* 28 (Fall 1997): 29–31.

W3C 2007. World Wide Web Consortium. www.w3.org

W3Schools 2007. Browser Display Statistics. http://www.w3schools.com/browsers/browsers_display.asp

Wintges, M. 2003. Geodata Communication on Personal Data Assistants (PDA). In *Maps and the Internet*, ed. M. Peterson. 397–402. Oxford: Elsevier Science.

GLOSSARY

aliasing the stair-step like appearance of nonhorizontal or nonvertical lines on a raster screen display or image

alpha channel bits in an image that are reserved for creating masks in some software, or for specifying transparency values; often the last 8 bits in a 32-bit image

animated map motion or appearance of motion in a map; in thematic cartography, often applied in cases where attributes change over time, called temporal animation

aspect ratio the ratio of the horizontal dimension to the vertical dimension in a monitor display's viewing area (typically 4:3 or 16:9) or in a raster image

bandwidth in any sort of network environment, the amount of information that can be processed at one moment in time; maps and other graphics consume more bandwidth than does text

bit depth the number of binary bits used to define color (or shade) in each pixel in the monitor or in a raster image

browser a program that allows the user to view and interact with information on the Internet; Internet Explorer, Mozilla Firefox, and Opera are examples of popular browsers

cast-based animation vector-based form of animation where objects are moved and manipulated in layers and/or with a background

color depth number of simultaneous colors that can be displayed on a display monitor at one time; see also bit depth

color palette any collection of solid colors that is a subset of a larger collection of colors; the term is usually associated with images of lower bit depths

cybercartography a view of cartography, technology, and Web in future terms; usually described as having seven elements (see text for the list)

dithering displaying alternating pixels of different hue or shade to create an impression of another color; usually associated with the use of a color palette in a raster image

dynamic rerouting Internet principle that allows packets of information to take on any one of a number of routes as information goes from the source to its destination

frame-based animation a sequence of images displayed in rapid succession, creating the illusion of motion; each image is called a frame

hypermedia like hypertext (below), links multimedia objects to other resources on the Web

hypertext text that points to other resources on the Web

hypertext markup language (HTML) language for the Web that allows for hypertext that is used to author a Web page; other languages are also used, but at present, HTML is considered the most basic and standard

image as a file format, synonymous with a raster image; in a cartographic layout, image data is sometimes associated with remotely sensed imagery or picture data

interactive map map in which the map reader can click and interact with features or layers, often in the same way that a GIS user might interact with map layers in a GIS

key frame a frame in animation in which the objects are fully designed by the cartographer; usually applies to animation in conjunction with tweening; See also tweening

lossless compression compression of a file in which no information is lost; this means no reduction in an image's quality on reduction

lossy compression compression of a file in which some information is lost; the greater the compression, the more degradation of the image's quality, but the image's file size will be smaller

metafiles files that are sets of "drawing instructions," which tell various software packages how to render and display, point, line, area, text, and raster objects; often used to transfer graphics from one program (or platform) to another

pixel smallest resolvable element on a CRT or LCD display, or in a raster image

refresh rate the rate or speed at which a CRT screen is refreshed; screens that have a refresh rate of less than 60 Hz (60 times a second) often have a noticeable flicker, and can produce eyestrain if viewed for an extended period of time

resampling changing the resolution and or number of pixels in an image (reduction is most common in Web mapping); this will alter its storage size, physical size, and clarity

resolution the number and size of the pixels on a display monitor or a raster image, often described in terms of the horizontal and vertical dimensions (in pixels) and/or pixels per inch (ppi)

static map a map that does not have animated or interactive qualities

tweening the process in animation that allows software to generate intermediate frames in between designated key frames

uniform resource locator (URL) the URL identifies the "Web address" of a particular resource; more specifically, it includes the naming scheme, the resource, and the host computer (W3C 2007)

virtual map a map that is viewable but impermanent, such as a map that is displayed on a computer monitor; virtual maps originally had several definitions, depending on the map's permanence and information storage (after Moellering 1984)

PART V

EFFECTIVE GRAPHING FOR CARTOGRAPHERS

The last part of this book contains only one chapter—a chapter that focuses on the development of effective graphs by the cartographer. Because many cartographers are frequently called upon to produce graphs, a section on the chief design principles is included. Interestingly, the figure-ground relationship plays an important part in the overall design, just as it did in map design. Graphs play an important role in the geographical sciences and therefore their proper design is of significance. Graphs that have two or three axes are examined in detail as are those graphs that do not have axes. Graphs play an important role in the understanding of data. They allow for communication of temporal or component information and serve to complement the map's spatial depiction of patterns. Finally, it is imperative that cartographers adopt ethical standards in graphing just as they do in making maps.

17 EFFECTIVE GRAPHING FOR CARTOGRAPHERS

CHAPTER PREVIEW The well-designed graph offers the cartographer an effective means of communicating statistical data. Such graphs can stand alone or exist as a part of books, atlases, and other creative works. There are certain guiding principles of design that can help. Some of these have come from a rather long history of graph design and some come from more recent studies of graph perception tasks. The design of the graph achieves graphical excellence when displaying multiple variables that are dynamic over time. Graphs can contain two or three axes or have no axes at all. Axis graphs have the common forms of simple, complex, or compound line or bar graphs, whereas the common pie graph contains no axis. The cartographer can create graphs about the data's characteristics using histograms and ogives for frequency distributions, scatter plots for understanding the range of the data, and box-whisker graphs for examining the data's hierarchical order in quartiles or percentiles. Trend data of multiple variables are most frequently displayed with line and bar graphs using arithmetic axes. Line graphs may also plot rates of change using a semi-logarithmic axis. The data in both line and bar graphs can be displayed in additive form so as to communicate the relative importance of each per time period. The pie graph allows for the display of the proportional importance of the components of the whole. The pie wedges permit the visualization of the percentage values of the individual component part. Graphical excellence is achieved by well-designed presentation of complex data with clarity and efficiency. The well-designed graph is an efficient tool of the cartographer for analyzing data. ■

The purpose of this chapter is to provide a brief look at graphing, paying particular attention to several common graphs and design suggestions for their construction. Whole books have been written about graphing, including presentations on the treatment of the data both in design and interpretation (Tufte 2001; Kosslyn 1994; Schmid 1983; Dickinson 1973). As cartographers are often called on to render graphs along with their maps, especially in publications such as atlases and other comprehensive works, some guiding principles are useful. The graph types presented here are ones that we consider most useful to many graphing tasks. At the end of the chapter, we provide some that are more unique (and innovative) and call for more experimentation in the exciting world of graphing.

At the very beginning, it may be useful to point out what is meant by the word *graph,* at least as it is used in this chapter. There appears to be no formal consensus in the application of the word *graph,* as both the words *chart* and *diagram* are used often in its place. Here we use the term ***graph*** when there is a mathematical basis for its construction. We prefer to use the terms *charts* and *diagrams* for those drawings, pictograms, and organization-based drawings for which no mathematical relations exist between the elements of the drawing. So, only graphs that have a mathematical structure are included here.

The underpinning philosophical basis for this chapter rests with two fundamental ideas. First is the idea of **graphicacy,** a term that refers to the intellectual skill necessary for effective communication of relationships that cannot be communicated well with words or mathematics (Balchin and Coleman 1965). It should be pointed out that these authors were discussing graphicacy in the context of communication with other modalities of *literacy, numeracy,* and *articulacy.* To them, graphicacy does not stand alone but works in conjunction with the other forms of communication. Graphicacy also is best suited to show spatial relationship ideas. This chapter rests on the idea that graphicacy can share as an equal partner in the communication of ideas.

Another philosophical foundation upon which we base this chapter is that graphic communication is a part of human communication and is composed of four components: (1) the communicator, that is, the one "sending", (2) the interpreter/reader, (3) the communication content, that is, the message, and (4) the communication situation or the perceived need to communicate (Dent 1972). As with maps, graphs must have a purpose, and that purpose is to affect change (learning) in some way in the reader. If the desired response is met, then the communication has been successful. Graphs also serve to assist the reader in visualizing the data, especially the display of temporal change of multivariable datasets. Graphs are designed to simplify the complexity of statistics and data. This simplification assists in the visualization process and the communication of information.

BRIEF HISTORY OF GRAPHING

Like maps, graphs have provided us with a visual display of information that varies with time and space. Tilling (1975) suggested that "although any map may be considered as a graph . . . we do not expect the shape of a coast-line to follow a mathematical law" (193). Graphs and statistical data are inherently intertwined. The history of graphing can be summarized by major events, which were identified by Friendly (2005) as *milestones* of data visualization. Figure 17.1 identifies time periods of these milestones from 1600 to the present, with the specific events displayed in Table 17.1.

The Industrial Revolution brought about new measuring devices and thus the proliferation of statistical data. The air and water thermometer, barometer, weather-clock, and mercury thermometer were all products of this period (Beniger and Robyn 1978). The weather-clock, devised by Wren, measured wind direction and rainfall, and automatically recorded temperature on a moving graph (Tilling 1975). Unfortunately, no record of this device remains, but it may well have been the first automated data correction and recording instrument. This early period of measurement and theory was also highlighted by van Langren's first visual representation of statistical data (Tufte 1997). In 1644, this linear graph displayed the longitudinal distance from Toledo, Spain, to Rome, Italy.

The 1700s brought about new graphic forms. Halley generated an isobar map that included lines of equal magnetic declination (isogons). William Playfair invented the line and bar graph and many other graph types used in analysis today. It was in his *Commercial and Political Atlas* (1785) that he displayed graphs depicting England's trade. Playfair is also credited with devising the pie and circle chart. The period of 1800 to 1850 represents the beginning of modern graphics. The use of multiple symbol types is credited to von Humboldt. He superimposed bar graphs, with each square representing

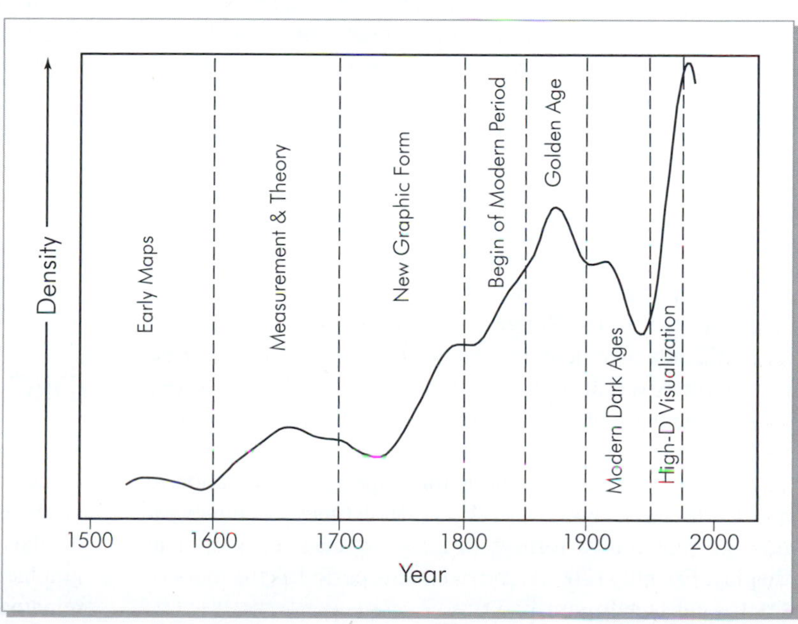

FIGURE 17.1 HISTORY OF GRAPHING.
Density is the estimated relative number of graphs that were produced in each era. (*Sources:* Friendly, 2005, 45; with kind permission of Springer Science & Business Media.)

TABLE 17.1 MAJOR EVENTS IN THE HISTORY OF GRAPHING

~3800 B.C.	Clay tablet of Mesopotamia (oldest known map)
~3200 B.C.	Coordinate systems by land surveyors in Egypt
A.D.	
1660	Weather-clock, automatic recording of temperature on moving graph (Wren)
1644	First visual representation of statistical data (van Langren)
1686	Bivariate plot of barometric readings versus altitude (Halley)
1701	Isobar map with lines of equal magnetic declination (Halley)
1760	Curve-fitting and interpolation (Lambert)
1765	Measurement error as deviation from a line (Lambert)
1785	Book with illustrated graphs, *The Commercial and Political Atlas* by Playfair
1786	Bar chart (Playfair)
1794	Printed coordinate graph paper
1796	Automated recording of bivariate data, pressure versus volume in steam engine (Watt)
1801	Pie chart, circle graph (Playfair)
1811	Subdivided bar graph, superimposed square (von Humboldt)
1819	Cartogram with shading of grayscale depicting distribution and intensity of literacy in France (Dupin)
1821	Ogive or cumulative frequency curve (Fourier)
1832	Curve-fitting to a scatter plot (Herschel)
1833	Histogram (Guerry)
1843	Contour map of data, temperature x hour x month (Lalanne)
1851	Map incorporating statistical diagrams (Minard)
1869	Three-dimensional graph called a stereogram (Zeuner)
1872	U.S. Congress appropriates funds for graphical treatment of statistics
1874	Age pyramid (Walker)
1884	Pictogram (Mulhall)
1910	Textbook in English devoted to statistical graphs (Peddle)
1913	College course in statistical graphical methods (Iowa State College)
1914	Published standards of graphical presentation for the United States (American Society of Mechanical Engineers)
	Pictogram of uniform size (concept of bar graph and pictogram of varying size [Brinton])
1969	Graphical innovation for exploratory data analysis, various graph types including box-whisker plots (Tukey)
1973	Chernoff faces
1980s	Development of the personal computer and software
1990s	Graphs displayed on the Web; map animation
2000s	Graphing software standard in education; interactive graphics

Source: After Friendly 2005; Beniger and Robyn 1978; Tilling 1975.

area information. The ogive or cumulative frequency curve, histogram, and curve-fitting to a scatter plot all had their origins during this time period. Lulanne is credited with the first three variable graphs in which temperature is plotted as contours against months on one axis and the hour of the day on the second axis. In the last half of the nineteenth century, the U.S. Congress appropriated funds for exploring the graphical treatment of statistics. Walker, Superintendent of the U.S. Census, produced the first age (population) pyramid that was included in the Statistical Atlas of the United States (1874) (Beniger and Robyn 1978).

Little advancement in graphical applications occurred during the first half of the twentieth century. The first course in statistical graphical methods was taught at Iowa State College in 1913, and the American Society of Mechanical Engineers developed standards for graphical presentation in 1914. Otherwise, World War II generated a need for numerical data in quantitative form with little emphasis on graphical display. Friendly (2005) referred to this period as the modern Dark Ages (see Figure 17.1).

The twentieth century concluded with an explosion of graphs on the Web, as well as developments in map and graph animation, interactive graphics, and the use of statistical and graphing software as standard tools in education. The time line brings us up to date with the "modern" era, a time when graphing can be done effortlessly on personal computers using a variety of graphing software. A fuller appreciation of graphing and its innovative history comes on the scene with books such as Edward Tufte's (2001) *The Visual Display of Quantitative Information.*

ORGANIZATION OF THIS CHAPTER

This chapter is organized to provide discussions around those graph types that you will most likely be called on to produce. At the very outset, a discussion of the data being displayed and the design principles that are used to achieve graphical excellence is given to form a foundation for the remainder of the chapter. Typical graphs include those that show features of statistical distributions, quantities proportions, and trends in data arrays; these also include the very typical logarithmic graph. Also, you will often be called on to show how two or more variables are related, with typical examples provided here. In many ways, graphing is an exciting part of graphic design, often calling for new solutions that require a great deal of creativity. Although certain standards do exist for standard data arrays, each graph is almost always a new and different challenge.

GRAPH DATA

Before talking about the graphs and the techniques for their creation, we should mention that all graphs present a visualization of the data. The data used to construct graphs are normally found in the worksheet, spreadsheet, or database of the GIS, mapping, or graphing software. These data are arranged in a row and column format with descriptive headers at the top of each column (refer to Figure 4.2). The first column is usually that of the independent variable, such as time, with the subsequent columns indicating the dependent variables to be graphed. The typical database that is suitable for graphing is numerical in format with each cell representing a specific value associated with a data year. Although it is preferred to have every cell populated with a value, it is possible to complete a graph with missing data. Caution should be taken to ensure that a zero in a cell is indeed a measured value and not a substitute for missing data. If the cell has a no data indicator, such as "nd," use a symbol or character that your software designates for no data (such as a blank) before attempting to graph such data.

The characteristics of the database will provide clues as to the type of graph to be created. For example, a database that has a sequence of years listed as a part of the first column is an indicator that the data can be displayed as trend or longitudinal information on either a line or bar graph. Here the changes in one variable are controlled by the changes in time. The years can be equally spaced along the time line, such as decadal census information, or unequally spaced as a result of data collection occurring along nonspecific intervals. If, however, the information of the independent variable is descriptive or topical, a bar or pie graph will be a more appropriate choice of graphic display. These graph categories are appropriate for data that display component parts as a proportion of a whole. These are achieved by the use of a group of bars or pie wedges.

GRAPHICAL EXCELLENCE

The concept of graphical excellence has been expertly presented by Tufte (2001) in *The Visual Display of Quantitative Information*. His vision of graphical excellence is for "statistical graphics to consist of complex ideas communicated with clarity, precision, and efficiency" (13). The data should be revealed in the graph, without distortion, in the clearest possible design. In order to achieve these goals, Tufte presents five principles that state that graphical excellence

1. is the well-designed presentation of interesting data—a matter of substance, of statistics, and of design.
2. consists of complex ideas communicated with clarity, precision, and efficiency.
3. is that which gives to the viewer the greatest number of ideas in the shortest time with the least ink in the smallest space.
4. is nearly always multivariate.
   and
5. requires telling the truth about the data (Tufte 2001, 51).

Two design concepts mentioned by Tufte (2001) include the idea of **data ink** and **chartjunk**. Data ink refers to that information that is critical to the display of the data graphically, which is the "non-erasable core of the graphic" (93). Such information includes the trend line of the data and the axis information necessary to interpret that data. Chartjunk includes such things as angled and cross-hatch patterns and other such symbolization used in place of color, or shades of gray for area fills beneath trend lines, inside of bars, pie wedges, or other locations where we wish to highlight the area for the graph user. These symbols are available to the cartographer in most GIS, mapping, and graphing software. Caution is advised in the use of such symbols as they tend to distract from the overall design of the graph and create moiré effects that lessen the quality of the design—and thus become chartjunk.

PRINCIPLES OF DESIGN APPLIED TO GRAPHS

This section examines the basic elements of a graph with axes, the types of scales used to define the axes, and the importance of planning in order to achieve the visual hierarchy. As we have expressed many times in this text, the planning of the graphical product is important. The level of communication and understanding is achieved through a well-conceived, well-planned design.

Elements of the Standard Graph

It is important in an introduction to graphing that some attention be paid to the typical elements of the standard graph. There are no set uniform labels attached to the elements, so you may encounter variations in the literature. Nevertheless, a brief discussion of these elements will be beneficial.

In the typical graph, there is a **data region,** that major part or area of the graph in which the graphic treatment of the data falls (see Figure 17.2). The data region is customarily bounded by the **scale lines,** those rulers along which we

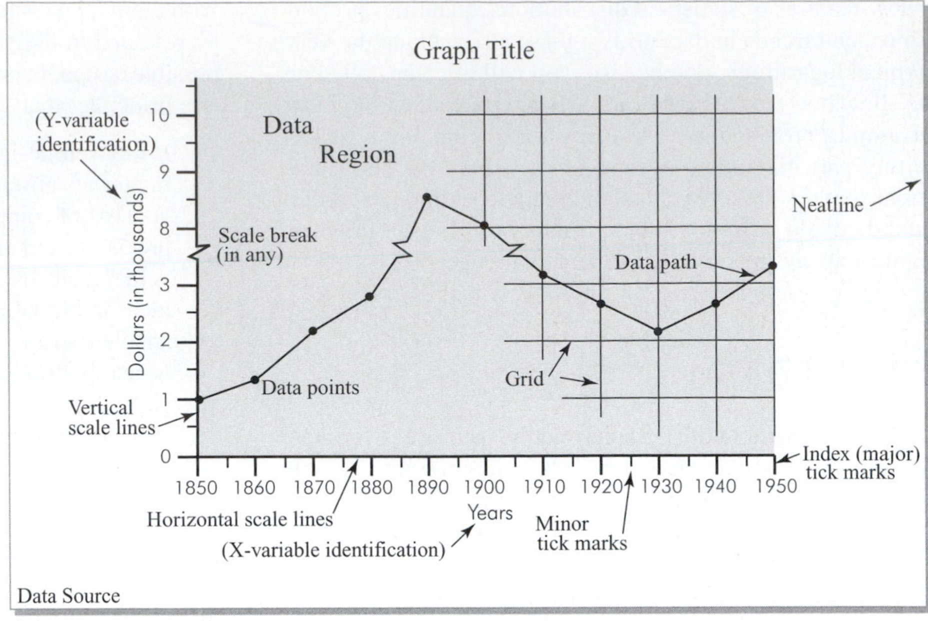

FIGURE 17.2 THE PARTS OF A TYPICAL GRAPH.
Not all graphs will have every part.

graph the data. Often the rulers are not included at the top or right of the data region if they are not needed for reader clarity. The scale to the left or right of the graph is called the **vertical scale** (y-axis), and the scale along the bottom or top is called the **horizontal scale** (x-axis). The x-axis displays the range of the independent variable (such as time) while the y-axis displays the range of the dependent variables (attributes). Each axis should display tick marks protruding out from the scale lines representing select values of the data to be labeled for easy understanding of the data. Along the x-axis, the tick marks should coincide with the time of data collection or other independent variable. Depending on the quantity of collection points, the designer may choose to use a combination of **index tick marks** in conjunction with minor tick marks. The index tick marks are those to be labeled, while minor marks are data collection dates but are not labeled. A tick mark frequently occurs on the x-axis for each time of collection. However, depending on the range of values and the number of points plotted, the minor tick marks may be left off, requiring the graph reader to interpolate between the labeled index marks. Index and minor tick marks are also used on the y-axis to delineate select data values in order for the graph user to interpret the numerical data values. This axis should include the entire range of the dataset with the labeled points being a logical sequence of numbers that assist in the interpretation of the data. Scale lines must have **scale labels,** clearly identifying the variable and its unit of measure, such as, crop yield (bushels per acre). These labels should be read from the bottom or the right side of the graph (refer to Chapter 13 concerning type placement) and centered on the axis.

The graph's grid may or may not be included within the data region, depending on the graph, but its *influence* is always present. The grid is intimately tied to the scale lines at the margins of the data region, and it is upon the grid that the data are plotted. Whether actually visible or merely implied by the tick marks, the grid should be easily detected by the reader.

Also in the data region, point symbols are used to indicate the location of each collected data value. These symbols may be connected by a line to form the **data path** that is used to guide the user through the trend of the data. In most graphs, the point symbol represents only the data location within the graph and has no numerical value assigned to it. This also applies to the line that is used to display the data path. The symbol type used for locating data points and the line for the data path should be unique for a single variable. When graphing multiple variables, each one should have a unique set of symbol-line characteristics, thus assisting the user in following the trend of each variable. For the data points, one may elect to use color (or grayscale) filled and open circles, squares, triangles or other appropriate point symbols. The trend lines may be differentiated by differences in line weights, color, and solid versus broken line patterns. Within the data region, there may be a **reference line,** a line identifying a certain place in the data region that is significant to the data presentation (for example, a mean line, or minimal-value line, and so forth).

Figure 17.3 displays four techniques of indicating the data path and data points. (Note that the graph is hypothetical and thus the vertical axis is left off.) Figure 17.3a depicts a graph constructed using the same symbolization for all lines. This makes for a difficult task in following the path of an individual line, which decreases the graph's utility. In Figure 17.3b, the symbolization of the lines is varied in thickness, value, and pattern, creating a much easier task of following the data paths. The symbols used to depict data locations along the path have been added to solid lines in graph (c). Finally, graph (d) combines the use of different point symbols and varying line symbols in order to increase the graph user's ability to follow individual data paths.

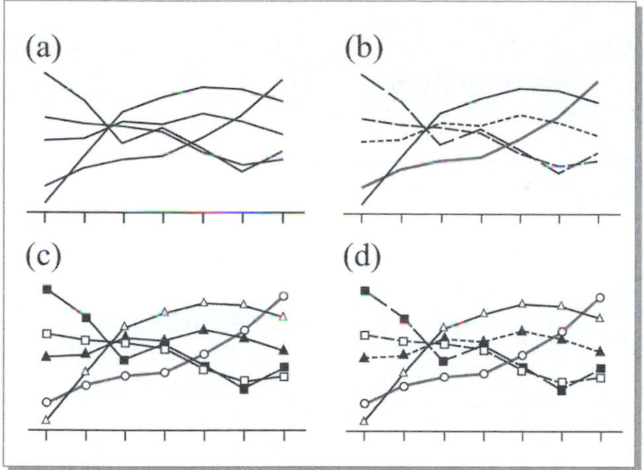

FIGURE 17.3 GRAPHIC MEANS OF DISPLAYING DATA PATHS.
Clustering of data paths are difficult to follow without varying symbolization (a). The data paths can be separated by varying line symbols (b), data point symbols (c), or both (d).

Many graphs require a **legend** (or key), which describes any part of the data presentation not immediately clear to the reader—for example, labels identifying the various data in the graph. Labels that mark the symbols on the graph are called **data labels.** These data labels may be placed just above and paralleling the line they represent or in a separate legend (see Figure 17.4).

Titles are a necessary part of every graph. For stand-alone graphs, the title is usually integral to the graph, and should be placed above (outside) the data region, centered over all components of the graph and within the frame or neatline. The title should be separate from the data region in order to provide the user with an unobstructed view of the data. For graphs that are part of narrative materials (such as in this textbook), the figure caption usually contains the graph title and appears beneath the graph. Titles need to be unambiguous, efficient, and economical. *Titles need as much thought devoted to their design as to the graph itself.* The graph title, like those of thematic maps, should communicate the where, what, and when of the data.

One last element of the typical graph is the **data source.** The source statement should be placed outside of the data region, along the lower frame or neatline. Data that does not originate with the author of the graph must be cited, giving credit to the individual or organization that supplied the data. Such citation may be brief, such as "Source: U.S. Census of Population, 2000."

Axis Types

There are three types of graph axes: arithmetic, semi-logarithmic, and log-log scaling. These scales also describe the characteristics of the grid underlying the placement of data information on the graph. The characteristics of the data will direct the cartographer in the selection of the *x*- and *y*-axis type.

Arithmetic Scaling

The *x*-axis frequently represents a time sequence and therefore should be graphed using an **arithmetic scale** or grid (see Figure 17.5a). Equal spacing of the grid lines represents equal steps in the continuum of time. Data years should be spaced in proportion to the length of time between successive

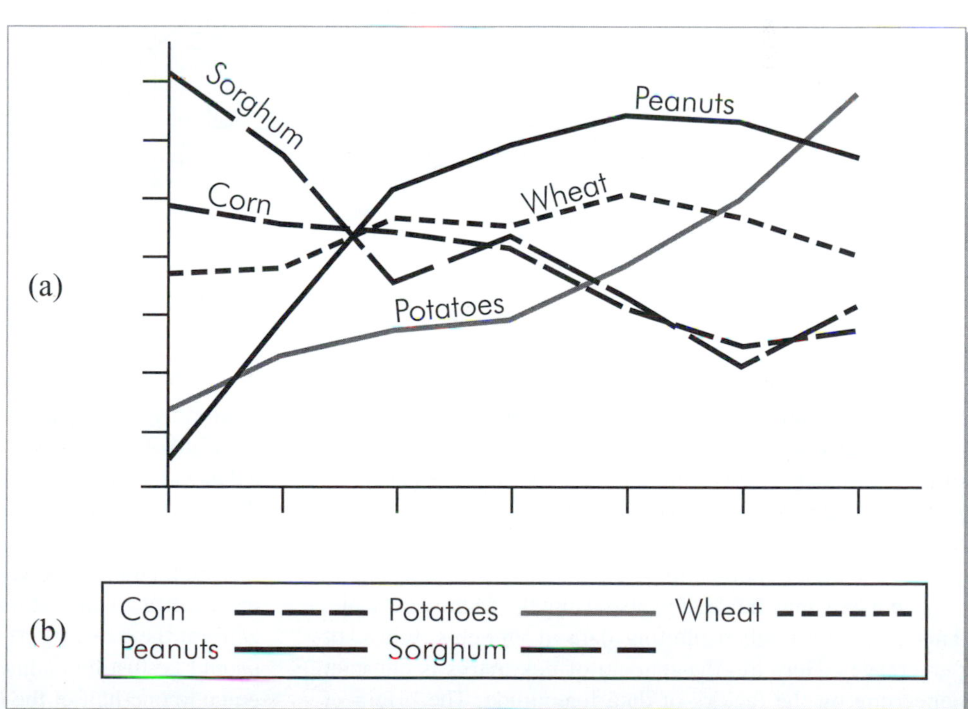

FIGURE 17.4 LABELING THE DATA PATH.
Labels can be added directly in the graph on the data paths (a) or in a legend (b). The *x*-axis is added for visual association only as the graph is focusing on how to label the data paths.

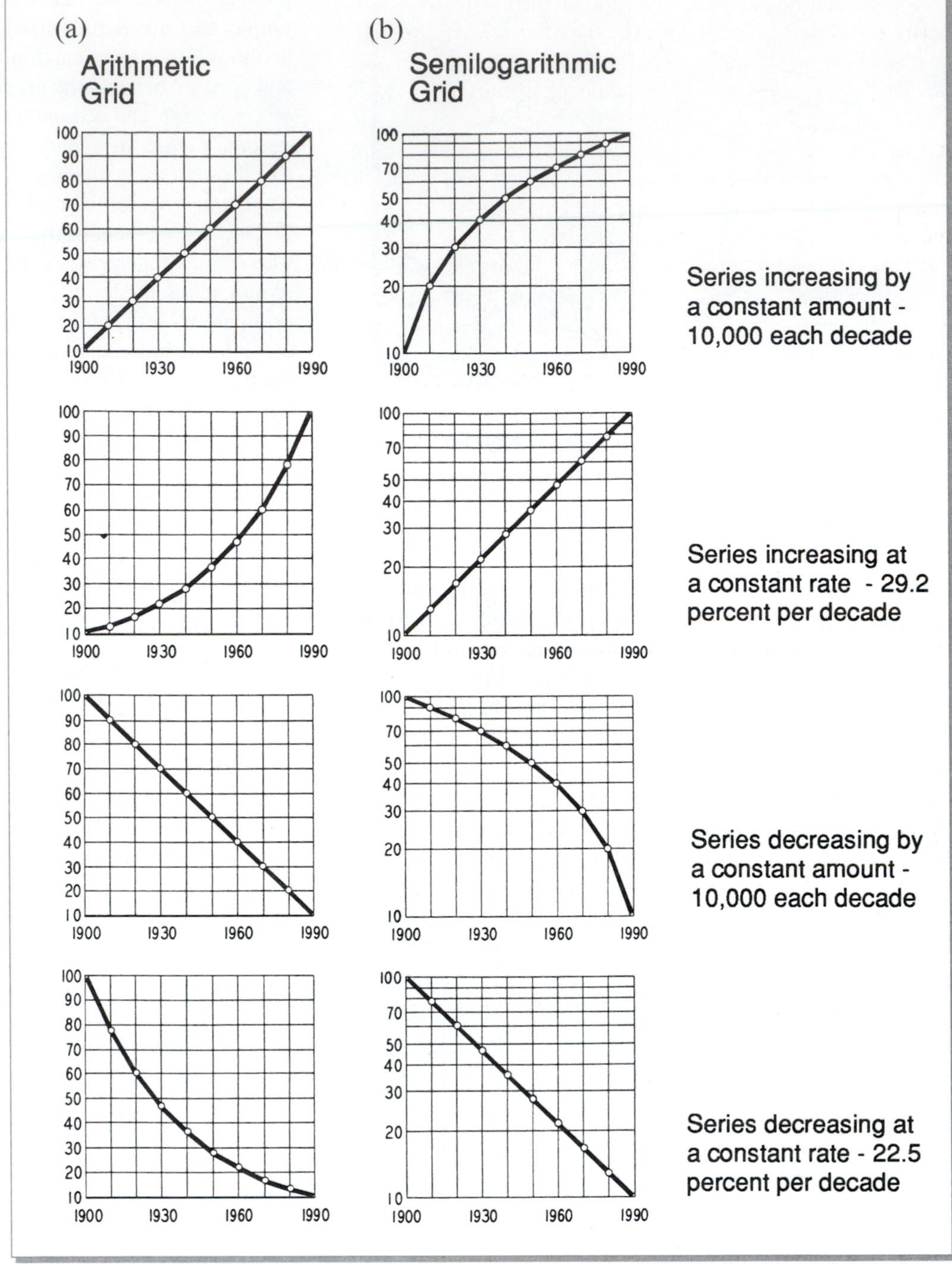

FIGURE 17.5 VARIOUS MATHEMATICAL PROGRESSIONS PLOTTED ON DIFFERENT GRIDS.
The arithmetic grid may be displayed on both axes (a) or only on one of the axes as with the semilogarithmic grid (b). *Source:* Reproduced with changes from Schmid and Schmid 1979, 158. Reprinted by permission of John Wiley and Sons, Inc.

reference years. Thus, the spacing between tick marks representing 1970 and 1977 should be seven-tenths of the length of tick marks representing 1960 and 1970. The positioning of tick marks along the arithmetic axis permits the cartographer to design the *x*-axis so that time maintains a constant rate. The origin point of the axis should represent the initial year of the data relationship.

The *y*-axis may be created also using the arithmetic scale. This is common when plotting data of amounts, totals, or percentages. Here too the spacing of tick marks is constant depending on the factors of data magnitude. The origin of the axis should represent the starting or zero point in the data (assuming all positive data). The extent of the axis must include all of the data, with the topmost tick mark representing a data value greater than the largest data value. The length of the axis line will be used to determine the number of major tick marks and their labeled values. If the data range is zero to 100, then the axis will be subdivided into ten equal parts, each worth a value of ten. The tick marks will be labeled as zero (at the axis origin), 10, 20 . . . 90, 100. These numbers should be justified adjacent to the tick marks leaving a gap equal to one-half of the number zero.

Large data ranges may cause the labeled sequence to be steps of 1,000 or 100,000 or even 1,000,000. It is advisable to avoid the use of multiple zero sets by dividing the label values by the step increments. Thus, the labels will be of fewer digits, making the values more easily understood by the user. If this is done, however, the title of the axis must indicate that the data displayed are in thousands, hundred thousands, or millions. This statement is normally placed in parentheses with the axis label. Data ranges comprised of values less than zero should include a zero followed by the decimal and the numerical value, such as 0.3 or 0.70. The number of places used to the right of the decimal must be consistent along the axis. If, for example, the labels include two decimal values, every label must include two decimal values, including the origin value of 0.00. Care should be taken in selecting the number of decimal places used on your axes. Decimals should not be substantively different from the attribute data being graphed. The use of too many decimals can imply a precision in the data that may not be there.

If the data set includes both positive and negative numbers, a zero line should be drawn parallel to the *x*-axis at some position along the *y*-axis. This line becomes the new secondary *x*-axis used to define positive numbers above the line and negative numbers below the line. The labeling of the tick marks will occur in the same manner as described above. Caution should be taken in the display of negative numbers. The negative sign (−) may be overlooked or misunderstood, especially when the labels are of a smaller point size. This can be overcome using one of two methods. The first is to always enclose negative numbers in parentheses, such as (−0.20). The second is to have the *y*-axis include two secondary titles indicating the positive and negative components of the axis.

Again, the arithmetic scaling of these axes allow for uniform sequencing of both the independent and dependent variables. Care should be taken to maintain the equal spacing and consistent axis labeling techniques in the design of the graph.

Semi-logarithmic Scaling

Semi-logarithmic graphs are created using an arithmetic scale for the *x*-axis and a logarithmic scale for the *y*-axis, or in some cases reversed, for example, flood stage recurrence intervals. The semi-logarithmic graph is used to depict either percentages or a rate of change of variables over time. The fundamental idea is that rate of change can be seen by inspecting the slope of the data path. The steeper the slope, the greater the rate of change, whereas the more gradual the slope, the less the rate of change (see Figure 17.5b).

A common logarithm is the power to which the number 10 must be raised to obtain a given number:

log of 10 = 1 (10^1 = 10*1 = 10)

log of 100 = 2 (10^2 = 10*10 = 100)

log of 1,000 = 3 (10^3 = 10*10*10 = 1,000)

In the event the number is not an exact power of 10, its log will be expressed as an approximation containing decimals:

log of 2 = 0.3010

log of 3 = 0.4771

There are two separate components of a logarithm: The *characteristic* component equals the integer of the logarithm, which is to the left of the decimal point. For example, the log of 20 is 1, and 200 = 2. The rule is that the characteristic of a number is always one less than the number of digits in the integer of the number. The *mantissa* equals the decimal part of the logarithm, which is to the right of the decimal point. The *antilogarithm* is the number corresponding to a given logarithm.

Assuming that you are using a semi-logarithmic graph whose vertical scale is logarithmic and whose horizontal scale is arithmetic (or linear), then

1. Equal vertical distances on the log scale always indicate equal percentages or rates of change.
2. Rates of change may be different along a data path, and these can be examined by simply looking at the slopes along the data path.
3. It is possible to compare two widely spaced data paths on the same logarithm scale because, regardless of the values, the rates of change (slopes) can be compared.
4. Amounts of change can be read from a log graph, but this is not its chief advantage.

Labeling the logarithmic axis of the graph is relatively easy. There is *no* zero baseline on the scale (the log of 0 is minus infinity). Instead, the bottom line can be labeled with any convenient value—usually slightly less than the smallest value in the data array. The scale should extend to that line that is slightly higher than the largest value in the array. For convenience, the bottom line is usually labeled 10 or some multiple of 10 (.0001, .001, .01, 1, 10, 100, 1,000, and so on). *The value of each successive bold line on the graph is 10 times that of the preceding value* (see Figure 17.6). Each series of multiples of 10 is called a *cycle,* and most preprinted log paper contains multiple cycles.

Log-Log Scaling

When both scales contain a logarithmic scale, it is called a **log-log graph.** Logarithmic graphs are also sometimes referred to as **ratio graphs.** The purpose of the log-log graph is to show the *rate of change* of one variable against the *rate of change* of another. The principles applied to the logarithmic side of the semi-logarithmic graph discussed above apply to both axes of the log-log graph. The title of each axis should carefully describe the concept that the data are displaying rate of change.

Graph Planning and the Visual Hierarchy

The same principles of design that were the focus of Chapters 12 and 13 are applied to graph design. Two chief design strategies are *contrast* and the *visual hierarchy.*

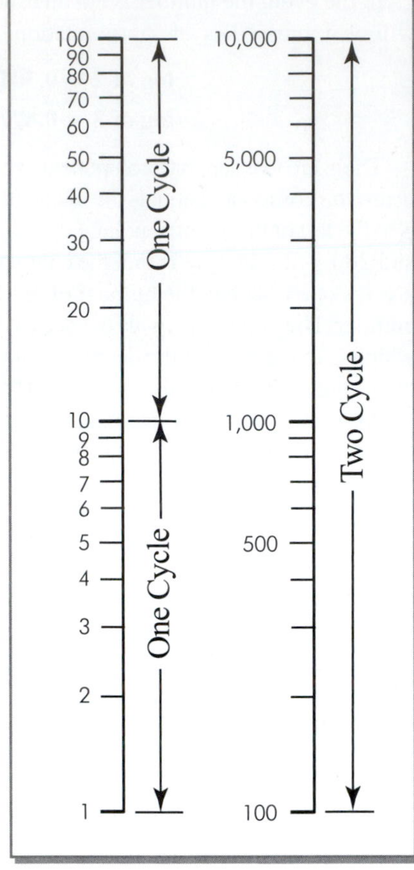

FIGURE 17.6
LABELING ON LOGARITHMIC AXES.

See text for discussion on when to use the logarithmic scale for one or both axes. If the scale is repeated in order to include the full range of the data, multiple cycles may be linked.

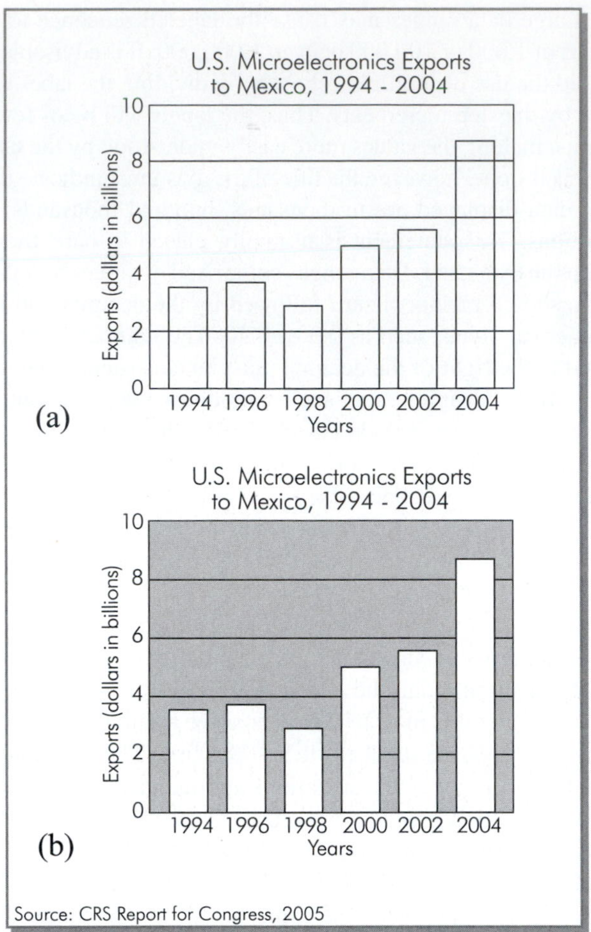

FIGURE 17.7 TYPICAL GRAPH DISPLAYING THE USE OF CONTRAST AND AREA FILL.

A graph with little contrast may be difficult to read (a). Using contrast and area fill improves the figure-ground association.

If these are properly integrated into the design, the graph will function successfully.

As with any thematic map, each graph should have a *purpose,* and the type of graph selected must be appropriate for the data it presents. Trend data should not be graphed with bar graphs, for example. Line graphs are not suited for showing component parts. Rate of change is best graphed on logarithmic graphs, not linear graphs. Graphs may be used to complement a data table by helping visualize complicated data. The placement of graphs directly on maps with a large number of enumeration units should be done with extreme caution. Even though most GIS and related software allow this functionality, the cartographer is advised against such inclusions. Examples include placing bar graphs or small trend graphs within enumeration units in the map. It is better to simply supply a table of this data in order to communicate the specific information. There are of course exceptions to every rule. If the map contains large areal units, such as regions of the United States, then graphs may be appropriate in some instances. However, when the enumeration units are large in number but small in size, the placing of graphs within the enumerations unit must be avoided. Each graphing project needs to have the designer ask basic questions about its purpose and appropriateness.

Contrast is perhaps the best design tool to use in any graphing design activity. As with maps, there can be contrast of line (weight and character), value (light and dark), detail, lettering, and color. The three that have the most significance are line, color, and value.

Figure 17.7a illustrates a typical graph with little line and texture contrast. This is a rather dull-looking graph, especially when compared to the one in Figure 17.7b, which is identical except for its contrast. With contrast, the image is not as bland, and appears more interesting. Also, by using contrast the designer is able to direct the reader's attention to certain parts of the graph, which serves to "scale" the importance of the different parts. Always use the element of contrast when designing graphs.

Data symbols, data paths, titles, axes and their labels are all very important to the graph and should be rendered as figures in the graph's visual hierarchy. Always important—but not needing the visual strength of the aforementioned—the graph's *grid* may be rendered in a subdued fashion. The data symbols and data path should appear to rest *on top* of the grid (see Figure 17.8). Coupled with contrast, developing a visual hierarchy for the graph calls for special attention to design detail, but will *always* yield a better-looking graph. Such a graph will have its data and design well integrated, as they should be.

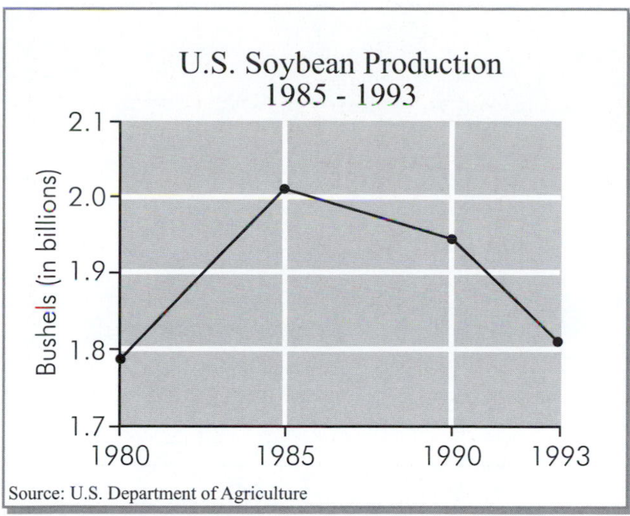

FIGURE 17.8 THE VISUAL HIERARCHY IN GRAPH DESIGN.
Data symbols, axis lines, labels, and other key components of the graph are portrayed as strong figures on the graph, with grids and other elements subdued.

GRAPH TYPES

For the purpose of discussion, we will describe the graph types as being a part of three categories: graphs about the data, graphs that use an *x*- and *y*-axis, and graphs that do not use an axis. The more familiar traditional graph types of these categories are histograms, line graphs, bar graphs, pie graphs, and other graph forms. While all three graph categories are generally recognized as a part of the GIS, mapping, or graphing software, not all specific types (such as trilinear graphs) are supported in all programs. Most graphs can be designed to be one of three levels of difficulty (clustered, complex, or compound) and/or to display the information in either a two- or three-dimensional design. Line and bar graphs employ the use of horizontal and vertical axes in order to delineate the ranges in time and numerical variables, whereas pie graphs and other graph forms are able to stand alone without the use of defining axes. A discussion of the difficulty levels and dimensional variations will be provided within the graph sections that follow.

Graphs of Numbers and Frequency

One of the first displays of numerical data is the simple array of the values in the data set. Often it is adequate to display the numbers in a table of some sort, but when this is not enough, a graphic method may be the solution. Statisticians often use this method and it is a subject frequently dealt with early on in statistics texts. Histograms, frequency distributions, ogives, and normal probability graphs are discussed here because they are fairly common and quite useful.

Histograms

Histograms are graphic records of the occurrence of data values in a statistical distribution. These graphs are developed after a researcher has already acquired the data—their purpose is to represent the statistical distribution better than

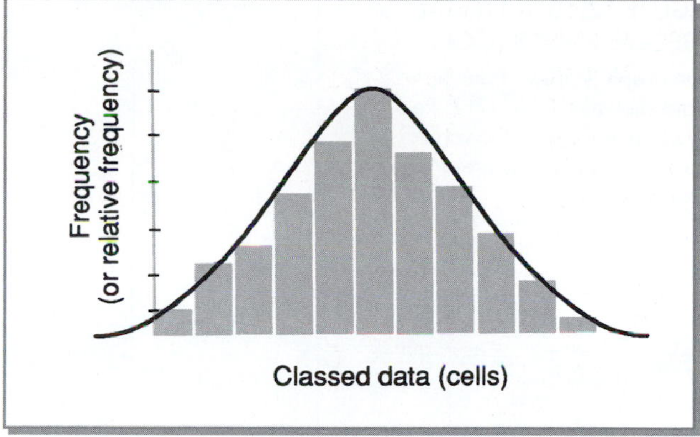

FIGURE 17.9 A TYPICAL HISTOGRAM WITH BOTH A FREQUENCY POLYGON AND A CURVE APPLIED.
Histograms are typically used for discrete data, and polygons are smoothed curves for continuous data. For a normal frequency distribution, areas under the curve are known, and standard deviations such as the ones shown are commonly displayed.

can be shown in a table of values. There are certain customs to follow when these graphs are constructed.

The horizontal scale of the histogram is divided into equal-width segments (classes), each of which represents a range of values and is repeated enough times to accommodate the entire range in the data distribution (see Figure 17.9). For reading the histogram, it is especially critical that the bars be of equal width and be drawn contiguous to each other. The vertical scale is marked off to represent frequencies of occurrence. Inside the data region of the graph, bars are placed over each segment on the horizontal scale and their heights scaled to the relative frequencies found in each bar's data range. It is important to understand that the data are represented by *equal units of area.*

It is a good idea to think of area when considering a histogram. If the width of each bar is considered to be one unit in width, and the heights of the bars are created proportional to relative frequency, then the total area under all the bars will equal "one." This is a good idea to keep in mind when considering probabilities associated with the normal frequency curve.

The distribution of values found in Table 17.2 was used to construct the graph of Figure 17.10.

TABLE 17.2 RANDOM NUMBERS FREQUENCY DISTRIBUTION

Mean = 2.0, *S* = 1.0

Frequency Class	Frequency	Cumulative Frequency
0–.403	6	.06
.403–.806	5	.11
.806–1.20	9	.20
1.20–1.612	15	.35
1.612–2.015	16	.61
2.015–2.418	9	.70
2.418–2.821	10	.80
2.821–3.224	9	.89
3.224–3.627	8	.97
3.627–4.030	3	1.00

FIGURE 17.10 A TYPICAL FREQUENCY POLYGON.

The graph is drawn from the data shown in Table 17.2. Relative frequency is on the vertical axis and data values are arrayed on the horizontal.

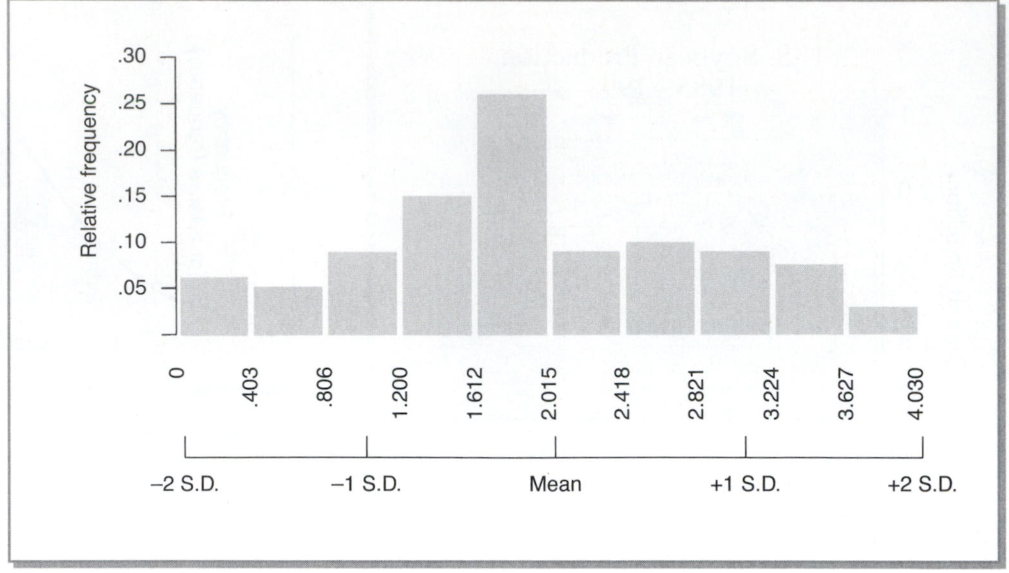

One reason for including a discussion here of the typical histogram is that cartographers should begin to display these graphs alongside any map made from the same data. This allows the map reader an opportunity to not only look at a spatial distribution of the data but also see a nonspatial array of the data. Histograms do this—they permit an easy inspection of how the data values are arrayed along a number line. They allow the map reader to gather a rather detailed picture of overall data characteristics.

Frequency Polygons and Frequency Distributions. If the centers of the tops of the bars in a histogram are joined by a line, a frequency polygon results (see Figure 17.9). If the number of observations in the array of values is very large and the class intervals very small, yielding very small bar widths, then the straight lines connecting the tops of the bars would appear smoothed, giving us a graph called a frequency curve. The frequency curve is of particular importance to the statistician because it too serves to enclose an area that accounts for all the values in the data distribution. The area under a frequency polygon and a frequency curve is equal to "one." It is possible to display such distribution characteristics as means and standard deviations directly on the frequency curve.

Again, it is important to remember that when thematic maps contain classed data, it is a good idea to illustrate these classes and the class boundaries on the frequency curve as well as in the map legend. This also provides more information to the map reader about the relationship between the statistical distribution and the map classes.

The data providing the frequency polygon may be reordered in ascending order and, if so, the polygon is called a **Lorenz curve.** These are used in some statistical applications.

Ogives and Probability Graphs. An extension of the relative frequency histogram is the cumulative frequency distribution graph, or cumulative percentage graph, more commonly called an **ogive** (pronounced ó-jive). Supposedly, this graph was called this by Sir Francis Galton (1822–1911) because of its resemblance to the curved rib of a Gothic vault (Kershner and DeMario 1980). The graph is constructed with the horizontal axis ruler laid out to show the upper limits of each class boundary arrayed ascending to the right, and the vertical axis ruler marked to show the cumulative percentage of observations in each class interval. A line graph is then constructed by placing data points at each respective place on the graph along with the cumulative percentage at each point, using straight lines to connect the points (see Figure 17.11).

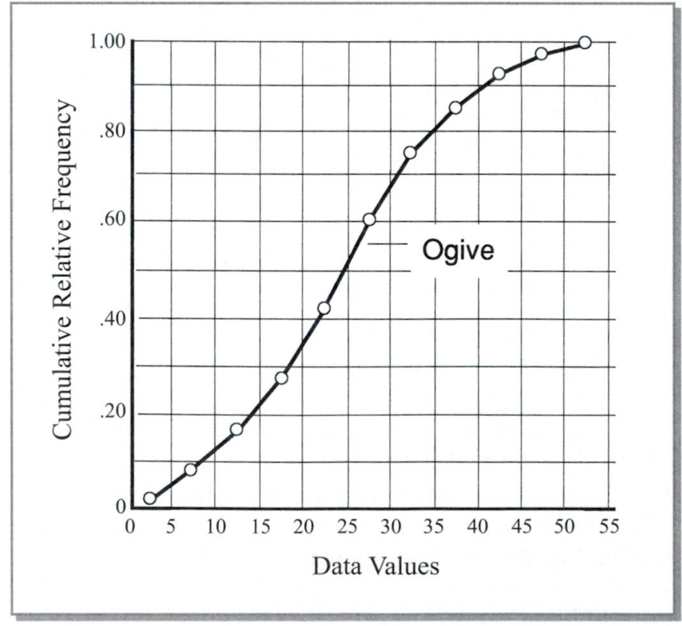

FIGURE 17.11 AN OGIVE.

Ogives are cumulative frequency graphs, and for normal frequency distributions a characteristic shape results, as in this graph.

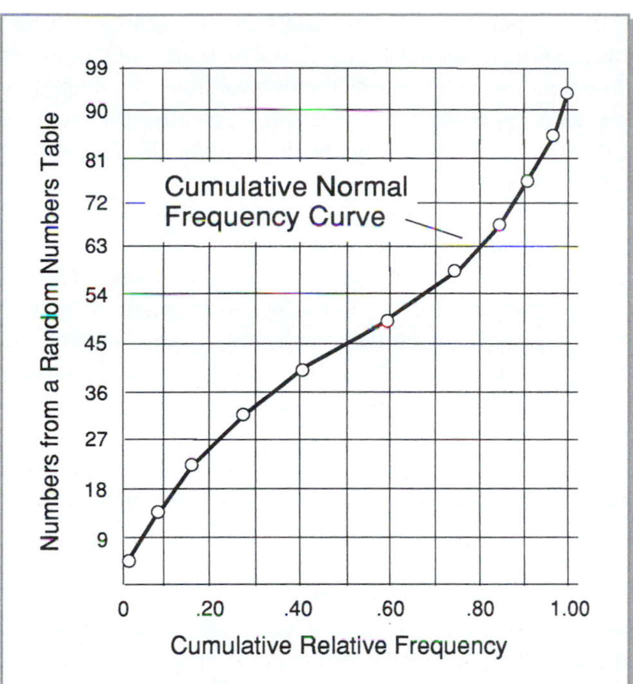

FIGURE 17.12 A REVERSED OGIVE.

When relative frequency is plotted on the horizontal axis, a reversed ogive results, and for normal distributions, a broad S-shape is formed.

At any point on the graph, the relative frequency (percent) for each upper limit of a class interval can show the percent of all observations falling *below* that interval boundary value. It is therefore a quick operation to determine reasonable approximations of the percent of cases falling below a particular observation. **Quantiles,** data arrays divided into groups with known proportions in each group, can be obtained from the ogive. Certainly, one of the uses of the ogive is to see quickly how the data are behaving—for example, is the cumulative percentage going up or down? Ogives are interesting and useful graphs for displaying frequency distributions.

A direct extension of the ogive is the *probability graph.* Here, the ogive is reversed so that the vertical axis becomes the ruler for the variable, and the horizontal axis the cumulative percentage ruler. This causes the curve to be reversed, as well (see Figure 17.12). If there were some way to scale the horizontal ruler so that the cumulative frequency curve of a *normal distribution* on the reversed ogive was straight, then the graph could be used to estimate whether other data plotted on the graph were normally distributed. Inasmuch as normality is the key to the use of many statistical tests, such a graph would be very useful.

Box-Whisker Graphs

A unique graph that shows the percentile summary of a data set is the box plot, sometimes called the **box-whisker graph** (see Figure 17.13). In this graph, a box is drawn so that certain horizontal lines within and at the edges of the box, aligned to the vertical axis of the graph, represent quartiles of the distribution. Customarily, the graph displays the position

FIGURE 17.13
A SIMPLE BOX-WHISKER GRAPH.

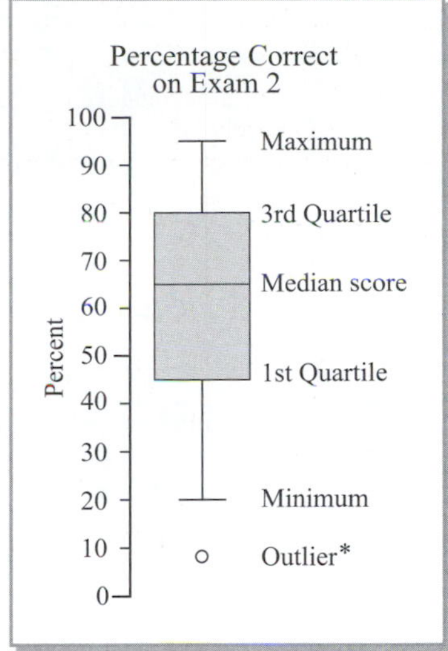

(in descending order) of the data maximum, third quartile, median, first quartile, and the data minimum. The construction of the box-whisker graph may vary depending upon the sophistication of your graphing software. Some software may graphically portray the location of the 90th and 10th percentile of the data. While other software plot the maximum value *or* the value of the third quartile plus one-and-a-half times the difference between values of the third and first quartiles. The same applies to the minimum value. In some cases, extreme values may lie beyond those points and are plotted on the graph as outliers.

A major use of the box graph is to show two or more data sets in the same graph in order to compare distributions. The compactness of the two distributions around their respective median values is immediately seen, as well as the values of the quartiles. The nature of the tails of the distributions is likewise easily seen.

A similar form of a box-whisker graph is used in graphically displaying the variability of stock prices. This is sometimes called a high-low-close graph. The top of the line represents the highest level that a particular stock achieved in a given period, while the bottom of the line depicts the lowest price value. The closing price is displayed by a horizontal line as a short tick mark. No box is associated with the high-low-close graph.

Scatter Plots

Many times geographers study the mathematical relationship between two variables. Most likely in their study they will want to develop a graph to show the relationship visually. If so, they construct a **scatter plot.** Scatter plots have a horizontal scale that is labeled to accommodate the values of the independent variable in the study, and a vertical scale that is labeled to account for the values in the dependent variable. Dots will be placed on the graph that corresponds to each pair of data values, which then represents the paired values. Scatter plots, then, show the *covariation* of two continuous variables (see Figure 17.14).

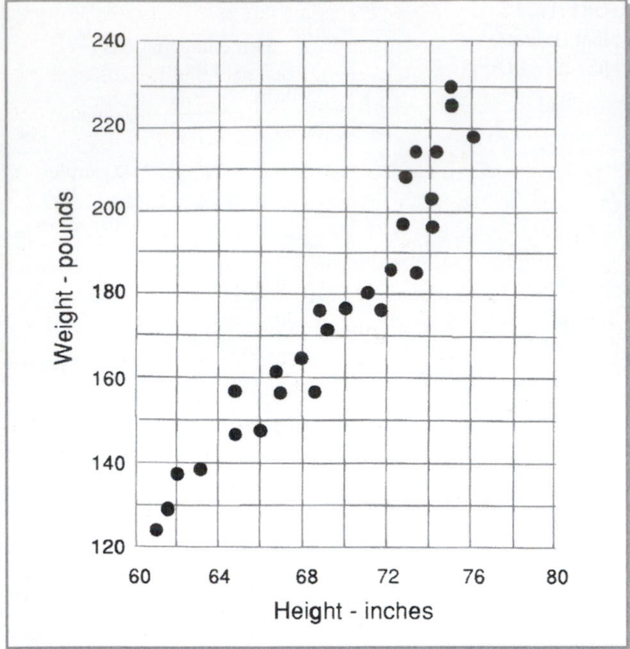

FIGURE 17.14 TYPICAL SCATTER PLOT.
It is important for these graphs to show the data symbols (dots) clearly by making them stand out from the grid.

It is important to understand that in a scatter plot the dots are *not* connected in any way. The scatter plot does not show trends—it shows how the two variables are related by their *location patterns* in the graph. Typically, a scatter plot is the first step in looking at the statistical relationships among variables.

Scatter plots can be a useful addition to a map layout when the map in some way depicts both variables in a map study. As with other graph designs, it is important in symbol selection for the data dots to be made to appear visually prominent on the graph and appear "above" the data grid. If the data values are classed in some way using the scatter of dots, these can actually be shown on the scatter plot.

The triple scatter plot (sometimes referred to as a **bubble graph**) is an interesting graph, and it probably is not used as often in geographical writing as it should be. The **triple scatter plot** is multivariate and depicts the relationship of *three* variables. It is very similar to the graduated symbol map. On this graph, one variable is displayed on the graph's horizontal axis and one along the vertical axis; the third is encoded in the sizes of the proportional circles placed at their correct positions within the *x–y* grid (Cleveland and McGill 1984) (see Figure 17.15).

FIGURE 17.15 A TRIPLE SCATTER PLOT.
These graphs are often called bubble graphs. Triple scatter plots graph three variables, as illustrated here.

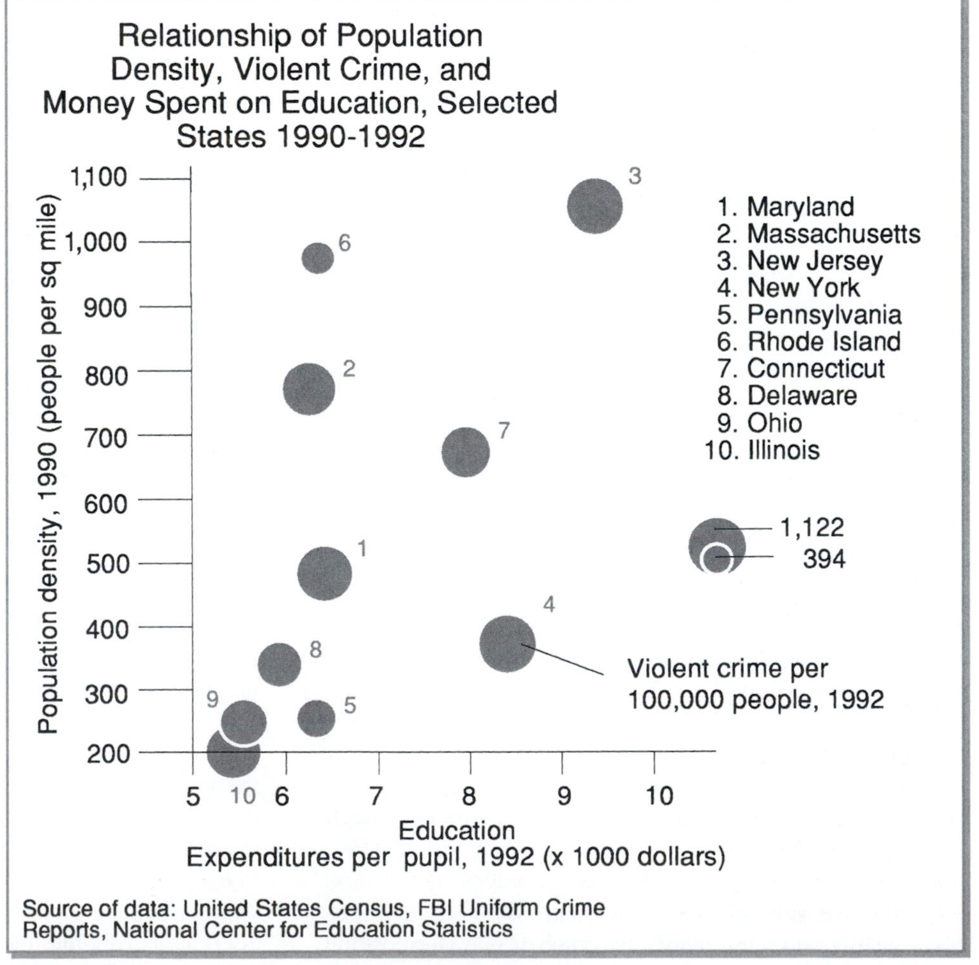

Line and Bar Graphs

The levels of difficulty in graph design are determined by the number of variables being graphed and the manner in which the data are displayed. **Simple** graphs tend to display the temporal tendency of only a single variable. Graphs that display multiple variables in the same array are of the **complex** variety. A **compound** graph displays data that are rearranged in order that the total magnitude of the combined data is represented by the topmost portion of the graph (see Figure 17.16).

Line Graphs

A quality **line graph** displays multiple variables and falls into the complex or compound varieties. The individual variables are identified by variations in line characteristics (such as thickness or style) and/or color. The use of the simple (see Figure 17.8) and traditional multivariate line graph (see Figure 17.16a) displays the trends of each variable over time. The lines for these variables may cross or, if of similar values, actually plot directly on top of each other, typical of a complex line graph. The thicker or darker line will dominate. The ordered compound graph and the complex percentage graph will not produce lines that cross but will be ordered sequentially from least to most.

The cartographer has the opportunity to communicate the total of the variables for each time period by organizing the variables such that they are ordered from least to most and summarily added together in sequence. Thus, the area beneath a line represents the summed values of all variables beneath that line. The top line in this graph will, therefore, represent the total of the data and may be labeled as such. The area between each line depicts the relative importance, or lack thereof, of each variable. Trends in the relative importance of the components can be identified as the thickness of the spacing between lines varies (see Figure 17.16b).

The graph can be modified to represent the percentage values of each time period as well. Each observation within a time period is divided by the total for that time period and multiplied by 100. This converts the data from observed numerical form to a derived percentage value. With the data in the least to most order, the lines will plot so that the area between them represents its percentage value of the total. The top line of this graphic design will be a straight line with a value of 100 percent (see Figure 17.16c).

Bar Graphs

Bar graphs utilize the height or length of the bar to represent the data. Even though this graph is displayed in two-dimensional space, it is considered a one-dimensional object. The bar widths have no value and are determined by the cartographer in the design of the graph. One can create several graphs of the same data using thin or very thick bar width. In doing so, the data being represented remains the same, the end of the bar. Unlike histogram bars, there is usually a spacing or gap between the bars.

As with line graphs, bar graphs are comprised of clustered, complex, and compound forms (see Figure 17.16d–f). Any number of variables can be represented in a complex bar graph. Normally, however, the number of time periods is somewhat limited as the combination of variables and observation periods must be distributed along the x-axis. Too many bars create congestion and decrease the interpretability of the graph. Each bar represents a single variable and the order of the bars is determined by the cartographer. The bars will be created by the software in the order they are presented in the database. The cartographer may want to sort the database either alphabetically or by numeric values to create the desired sequence. The order of the bars in any time period must match the order of the first time period. If an observation has a value of zero or has no data, a space representing the null value must be maintained within the bar set. We recommend using color fills or grayscale values in order to enhance the interpretability of the graph.

The labeling of the bars can be achieved in several effective techniques. The bars can be labeled directly across the independent variable axis aligned with the center of each bar. It is recommended that the labels be placed so that they parallel the orientation of the bars. Software provides the opportunity to position the labels at a user-defined angle. This creates a tendency for the label to be somewhat disassociated with the bar it is representing. The difference in the angle of the text from that of the bar causes the label to float free from its associated bar. The second technique is to add labels within the bar itself. This is appropriate only if all bars are long enough to accommodate the label. The third option is to add the labels directly above the bar within the graph space. Here too the labels should parallel the orientation of the bars. Lastly, the software allows the creation of a legend separate from the graph itself. Unfortunately, the legend is created using small squares to represent the data instead of elongated rectangles that simulate the bars. The symbol used in the legend should be of similar form to the symbol used in the graph. If the software does not permit this design option, the cartographer should utilize separate software to design his or her own legend that matches visually to the orientation, size, and order of the bars of the graph.

The compound bar graph is constructed similarly to that of the compound line graph. The dataset must be manipulated so that the data are added sequentially in a cumulative manner and the values stacked on top of each other within a single bar. The length of the bar will vary according to the total of each observation period. If the data are converted to percentages, the bar length will be equal, representing 100 percent per unit of time (see Figure 17.16d).

The axis orientation is at the discretion of the designer. Bars may increase in length either vertically or horizontally. Data being displayed may be both positive and negative. In such a case, the bars will extend beyond an intermediate line representing a value of zero. The data may also represent a

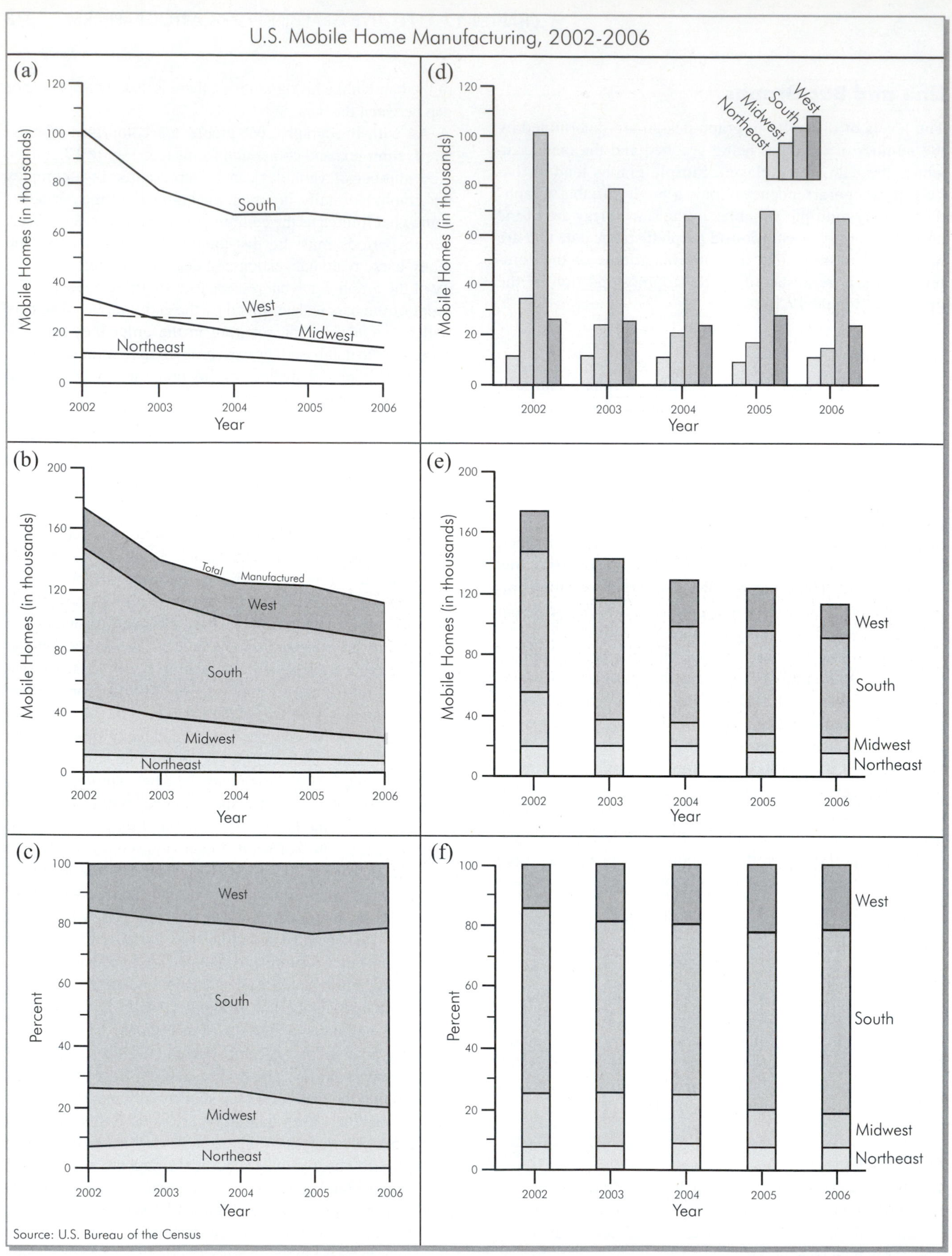

FIGURE 17.16 STYLES OF LINE AND BAR GRAPHS.

Line graphs may be clustered (a), complex by adding the values together so that the top line represents the total of the data (b), or compound by determining the percentage each value is per time period (c) with the top of the graph representing 100 percent. The same sequence of clustered (d), complex (e), and compound (f) applies to bar graphs as well.

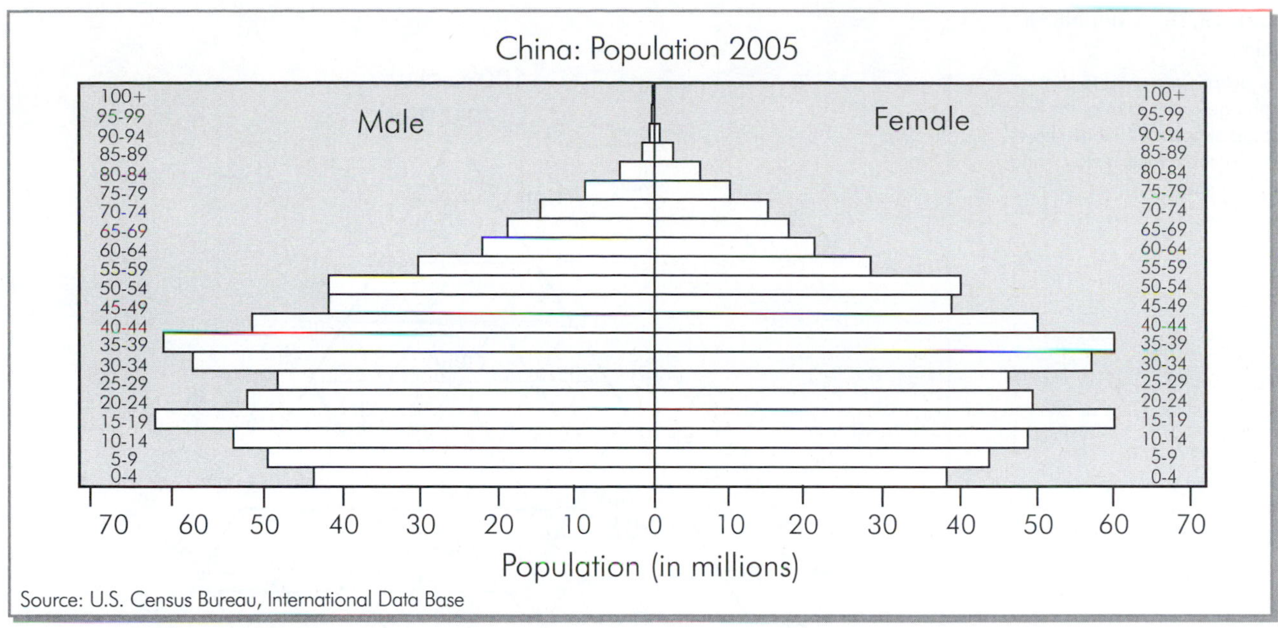

FIGURE 17.17 POPULATION PYRAMID OF CHINA 2005.
This graph type typically displays the population by gender and age group.

binary dataset, such as gender. The traditional population pyramid is a horizontal bar graph where the bars extend outward from a centerline in which the bar length represents the number of births, male or female, of a given age group. The shape of the pyramid is used by geographers and demographers to evaluate the age structure of a given population (see Figure 17.17).

Three Axes Graph

This type of graph, found in a number of applications in geography, is unique and perhaps could be used more often than it is. The **trilinear graph** is based on the geometric principle that the sum of the perpendiculars from any point within an equilateral triangle to the three sides is equal to the height of the triangle. The trilinear graph is used when there are three components (variables) that always add up to the whole—although each variable may account for a different part of the whole (and may change over time). Each side of the triangle serves as the base line for a variable. Customarily, the sides are marked off by a *percentage ruler.*

A common trilinear graph used in geography is the soil-texture graph (see Figure 17.18). This illustrates the component parts of a soil sample by noting the percentages of sand, silt, and clay in the soil. Another example is the percentage of cement, sand, and gravel in a concrete sample. Anything may be represented by the trilinear graph as long as there are three components that make up the sum of something.

Graphs Without Axes

Not all graphs are created to display linear trends over time. Graphs that do not apply an *x*- or *y*-axis to communicate total, proportion, or percentage values are used to represent data that occur at a given time period. These representations depict the relationship of the component parts of the whole. The geometric form most commonly associated with these graphs is that of the circle.

Pie Graphs

The components of any total may be displayed as wedges within a **pie graph,** also called a sector graph. The GIS, mapping, and graphic software utilize raw data, convert them into percentages and draw them as a proportion of a whole. The whole or 100 percent of the data is represented by the complete circle or 360 degrees. The wedges of the pie are calculated in terms of arc degrees around the circle. Each one percent value is represented on the circle as 3.6 degrees of arc (360/100). An initial component that represents 25 percent of the whole will have a wedge that is represented by a 90 degree arc or one-fourth of the circle's circumference. The cartographer must be aware that a circle divided into a large number of components is most difficult to design while maintaining a significant level of visualization. When the components are too numerous, many become an insignificant part of the whole. These components produce wedges that are slivers and most difficult to discern. It is often better to aggregate the small components into a category labeled on the graph as "other."

The design of a pie graph, like the circle itself, is quite simple. Each individual wedge should be labeled on the graph in order to identify the component it represents. We usually do not recommend indicating the percentage value of the component as the wedge size does that visually. Only in such instances where the percentage values are the focus

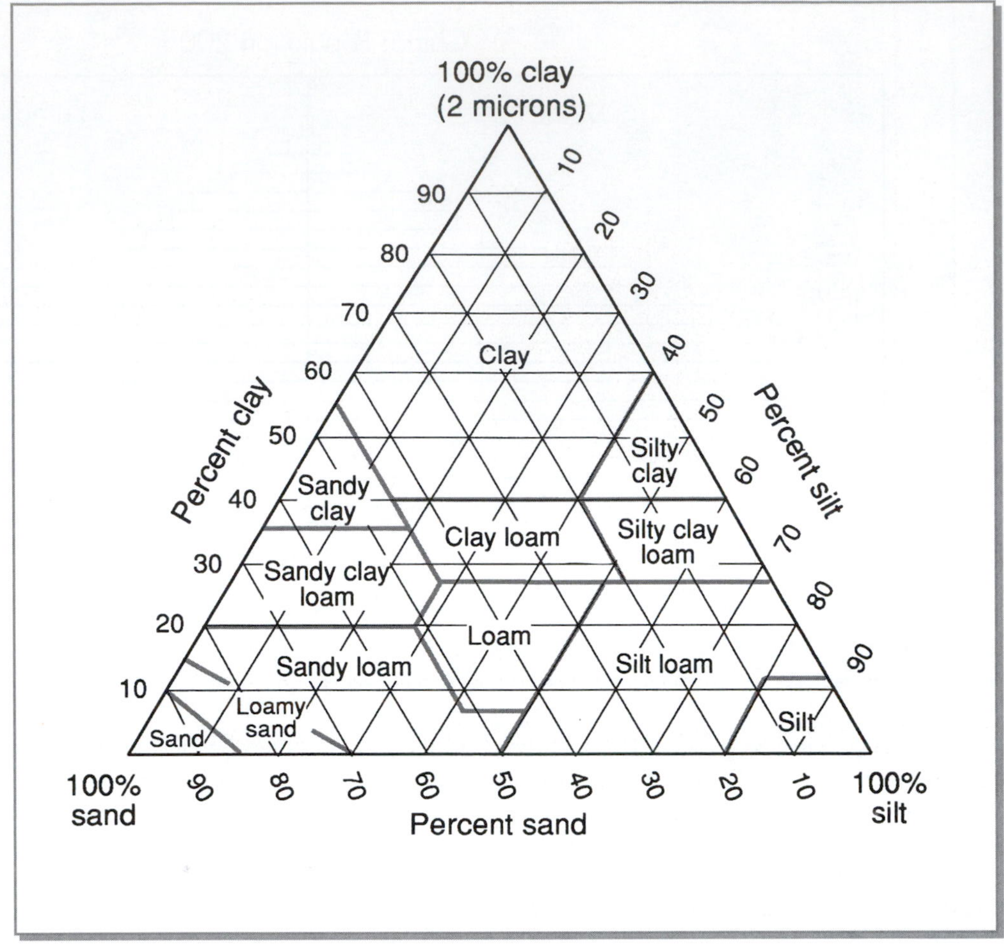

FIGURE 17.18 A TRILINEAR GRAPH.
These graphs are used to show percentage of the total of the three component parts. This one shows the constituent parts of clay, sand, and silt in soils.

of the graphic should they be included as a label. If you wish, you may indicate the raw data value along with the label of each wedge. The label should be placed outside of the circle in close proximity to the center of the wedge. Where the wedges decrease in size, it is frequently difficult to position all labels in this manner. If this occurs, the use of a line extension to the center edge of the wedge allows the label to be offset and not conflict with the other labels. The length of the line should be limited to the amount of offset the label requires for balance in positioning. Along with an appropriate title, it is recommended that the total numerical value of the whole be labeled adjacent to the circle. This allows the graph reader to estimate the value of each component (see Figure 17.19a).

The software will assign a color to each wedge of the pie. There is no special sequence used in assigning color in most GIS, mapping, or graphing software. The intent is to assign a different color to each wedge. Since the components are separate parts of the whole, separate hues are used to fill the individual wedges. The use of a single hue with varying saturation is not recommended in most cases. If the colors are simply used to differentiate the wedges within the pie, a sequence of four or five colors that may occur multiple times is permitted. Care must be taken, in this case, that the first and last adjacent wedges are not filled with the same color. In the case where the grayscale is used, this technique is recommended.

The size of the circle's diameter is relevant only when it comes to being able to see the individual components within the circle. *Bigger is not always better!* Too often the selected circle size fills an entire page. The scaling of the circle allows for a reduction in space used and still exhibits adequate communication of the proportional nature of the components. The only limit in reduction is determined by the point size and placement of the labels. Figure 17.19b displays the same information as in Figure 17.20a and uses less space. The graph communicates the information without being as large.

Attention to one or more of the components can be achieved in most software by "exploding" an individual wedge (see Figure 17.19c). This process offsets the wedge from the rest of the pie by a user-specified distance. It is not advisable to explode all the wedges. This creates an effect of disassociation and decreases the comparison of the individual parts. We recommend that you limit the number of exploded components to no more than two.

Clock Graph

Another graph that utilizes a circle for its major geometric form is the **clock graph.** This graph can be designed incorporating the use of bars that are arranged around a circular time diagram and scaled in length to a concentric grid (see Figure 17.20a).

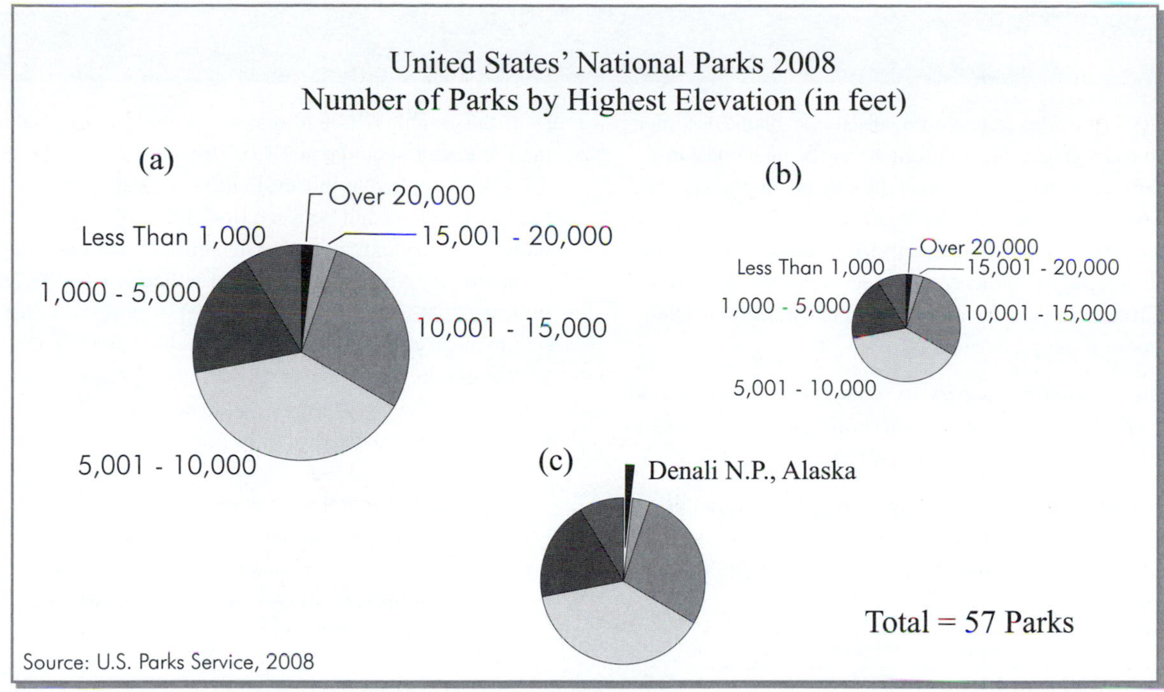

FIGURE 17.19 PIE GRAPH DESIGN.
The pie graph depicts the component parts of a total. The size may be larger (a) or smaller (b) and still communicate the relationships the same. To emphasize a particular segment of the data, one wedge may be "exploded" from the pie (c).

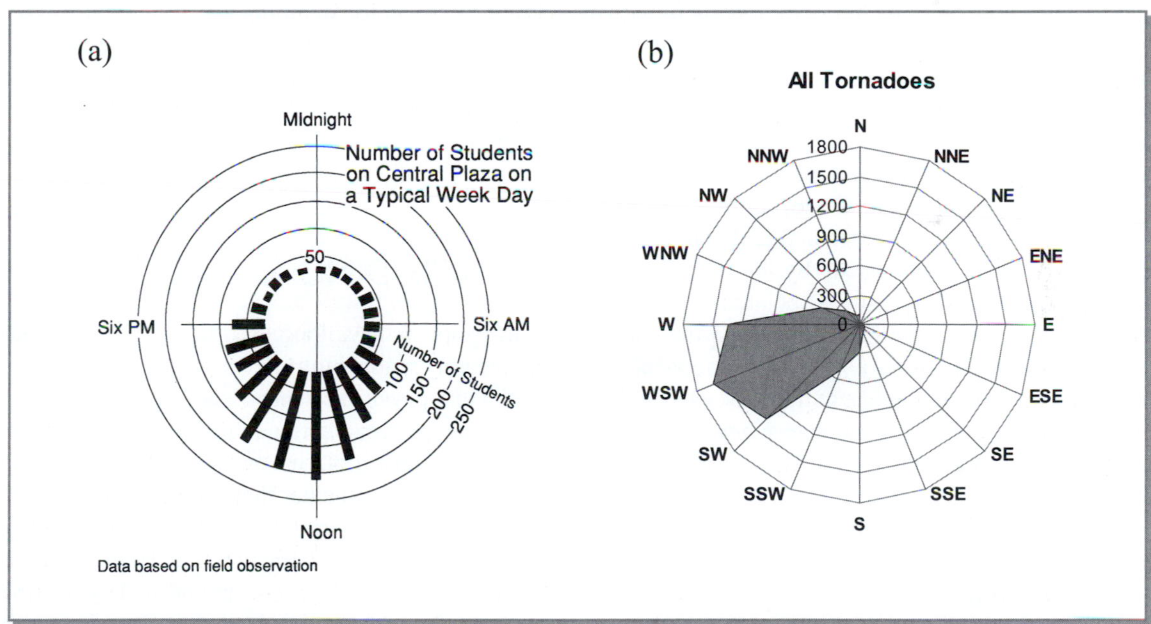

FIGURE 17.20 CLOCK-TYPE GRAPHS.
Clock-type graph using bars to display the data (a) or using a similar design to represent compass direction (b) as a radar diagram of tornado frequency and path. *Sources:* Suckling and Ashley, 2006, p. 24; (b) Copyright Association of American Geographers. Used by permission.

The clock graph may also be used for directional data. The wind rose graph used by meteorologists to show prevailing wind directions is a good example. Radar diagrams allow the meteorologist to plot frequency and path of weather phenomena, such as tornadoes (see Figure 17.20b). In fact, the clock graph should be explored for any number of unique uses.

Three-Dimensional Graphs

In almost every case, the GIS, mapping, and graphing software provide the user with the opportunity to create the same graph in three-dimensional space. The use of 3-D graphics is quite common in artistic renditions used in

THE LIE FACTOR

Tufte (2001) provides an interesting discussion on the distortion found in many graphs. Such distortion may be intentional in order to influence the perception of the user or simply accidental, a result of poor design. In his landmark text *The Visual Display of Quantitative Information,* he describes how the size of the effect shown in the graphic influences how individuals perceive the size of the effect in the data. He indicates that one of the principles that enhance graphical integrity is:

The representation of numbers, as physically measured on the surface of the graphic itself, should be directly proportional to the numerical quantities represented (56).

In order to evaluate whether a graphic conforms to this principle or not, the Lie Factor permits the measurement of the distortion found in a graph. The Lie Factor is determined by the formula:

$$\text{Lie Factor} = \frac{\text{Size of the effect shown in graphic}}{\text{Size of effect in data}}$$

If the graphic is free of distortion, the resulting value of the Lie Factor should equal 1.0. Tufte suggests that a +/− 5% can be applied in this evaluation. Therefore, acceptable Lie Factor values fall between 0.95 and 1.05. In the example included in his text, a graphic depicting "Fuel Economy Standards for Autos" generated a Lie Factor of 14.8. This indicates that the design visually exaggerates the information by nearly 15 times that found in the data (57). One can extract graphs found in newspapers and magazines and measure the Lie Factor. This is a good exercise for helping you to understand how the design of graphs can influence the perception of the reader.

It should be noted that most graphing software will generate 2-D graphics accurately, unless the author introduces intentional bias, or somehow accidentally generates the distortion through conversions between software.

reports, magazines, or other popular publications. Three-dimensional graphics are becoming more frequently used in the art of visualization.

Most personal computers have spreadsheet software such as Microsoft Excel included, and as a result, more people than ever are creating graphs. Excel's graphing options include eleven categories, each of which has both a 2-D and 3-D component. The categories that are appropriate for cartographic purposes have been addressed in this chapter. Most of the 3-D graphs are easily understood by the user. Three-dimensional graphs that are not easily associated with axes, which define the data values, are less effective in the communication of the data. Ribbon charts, the 3-D component of line graphs, are often difficult to relate to the axes, especially when presented in a clustered or stacked arrangement.

By having such software readily available, the creation of graphs can be done by any user. As with GIS and mapping software, the default settings allow for the creation of maps and, in this case, graphs, with a single keystroke. Such maps and graphs are not always cartographically sound in their design. Too frequently, the display of data may look aesthetically pleasing in 3D, but the communication and interpretation of the data is often lessened when the graphs are created. We suggest that 3-D graphs be used with caution. The third dimension creates unequal spacing of time and data within the graph. It is more of a visual rendition of the data than a scientific display and thus the perception of that data can frequently be distorted (Tufte 2001). See the boxed text "The Lie Factor" for a more complete description of this effect.

There are three-dimensional graphs that communicate effectively a more generalized purpose. The RGB color box diagram (refer to Color Plate 14.5) is an excellent example. Each of the color components is defined by an axis that ranges from 0 to 255. The use of three axes permits the visualization of the millions of colors produced by this system. The key to good three-dimensional graph design is to maintain directly the link between the data values and the symbols used.

CONCLUDING REMARKS

Cartographers have long used the map as a means of communicating spatial information. Today, practically all areas of social and physical sciences use maps and graphs to display locations, patterns or trends that are used for analysis. The map is also ubiquitous in our society. We are exposed to maps on a daily basis, whether they are in the print realm of newspapers and magazines or the virtual realm of maps of local weather or of some international location in the news. Maps are designed to be long lasting and widely distributed when we print them by the thousands, or when they are a part of a state or national atlas. They may also be designed to be in virtual form for display on computer monitors, the Web, or on television. Indeed, maps are a part of our daily lives.

Therefore, what we do as cartographers, from the types of data we use to the design choices for those data, really matters, and can have a great impact on the map reader. This text, which focuses on excellence in the design of thematic maps, strives to make sure that this impact is a positive one. Each of the text's five sections has its own area of focus in this endeavor.

In Part I of the text, we established a foundation between the data and the thematic map type, and the importance of the cartographer being knowledgeable of the data being mapped and having an understanding of the map's purpose. Maps are linked to the real world through the use of a coordinate system, scale, and a map projection. The nature of the data and steps taken to simplify complex data through classification and generalization were discussed.

In Part II, an overview of quantitative thematic mapping types was presented, including choropleth mapping, dot-density mapping, proportional symbol mapping, isarithmic and surface mapping, value-by-area or cartogram mapping, and flow mapping. The selection of form follows the kind of data being used. Attribute data that occur at points require the use of point symbolization, such as a proportional symbol map. Isarithmic mapping generates data contours based upon a distribution of sample points. When the attribute data are associated with lines, a flow map may be the result. Area data, such as those associated with enumeration units like counties, states, or countries, can become a choropleth map, dot-density map, cartogram, or even a proportional symbol map. In the latter case, the area data are associated with the unit's centroid. Since area data have several potential forms, other attribute characteristics are considered to assist in selection of map type. For example, densities are typically mapped using the choropleth form, while totals may be mapped with the dot-density, cartogram, or proportional symbols. Refer to Tables 4.3 and 11.1 for a quick synopsis of data type and map selection.

Part III of the book focused of the importance of layout and design. Understanding the role of typical thematic map elements can help the map designer decide what to include in his or her map (refer to Table 13.1). The map body is but one component of the overall map layout. Effective use of map text tells the reader what (the attribute), where (the location), and when (the time period of the data) is being mapped. Adding textural information is the cartographer's way of talking to the map reader. Therefore, the selection of a readable font, point size, and text placement is crucial to good map design. An understanding of the components of color and the color models is not only important for designing an effective thematic map, but is also essential in preparing the map for production.

Map production, the focus of Part IV, is the means of distributing our map product. Both print and virtual format were addressed. The means of production must be considered early in the map design process because it determines the color model selection, potential limitations on the level of detail on the map, and the file formats required for printing or for the Web.

This last chapter, Part V, has presented us with an understanding of graphing data. Graphing cannot only assist the cartographer in becoming familiar with his or her data, but can be a valuable addition to a map layout. The map provides us with the spatial patterns of the data while the graph provides us with the temporal trends or components of the data. The graph adds additional information that enhances the map layout.

The cartographer makes numerous decisions in the design of the map. Technology has given us the option to try multiple options in making those decisions. We have near instantaneous feedback in order to evaluate and/or modify the decisions made. A well-designed map includes all necessary components and communicates the data to the map reader. After all, that is what cartography is all about.

REFERENCES

Balchin, W. 1976. Graphicacy. *American Cartographer* 3: 33–38.

Balchin, W., and A. Coleman. 1965. Graphicacy Should Be the Fourth Ace in the Pack. *Times Educational Supplement,* November 5.

Beniger, J., and D. Robyn. 1978. Quantitative Graphics in Statistics: A Brief History. *The American Statistician,* 32 (1): 1–11.

Cleveland, W., and R. McGill. 1984. Graphical Perception: Theory, Experimentation, and Applications to the Development of Graphical Methods. *Journal of the American Statistical Association* 79: 531–54.

Dent, B. 1972. Visual Organization and Thematic Map Communication. *Annals* Association of American Geographers 62:79–93.

Dickinson, G. 1973. *Statistical Mapping and the Presentation of Statistics.* London: Edward Arnold Publishers.

Friendly, M. 2005. Milestones in the History of Data Visualization: A Case Study in Statistical Historiography. In *Classification: The Ubiquitous Challenge*. C. Weihs and W. Gaul, editors. New York: Springer-Verlag (34–52).

Kershner, H., and G. DeMario. 1980. *Understanding Basic Statistics*. San Francisco: Holden-Day.

Kosslyn, S. 1994. *Elements of Graph Design*. New York: Freeman.

Playfair, W. 1786. *The Commercial and Political Atlas*. London: T. Burton.

Schmid, C. 1983. *Statistical Graphics*. New York: Wiley.

Schmid, C., and S. Schmid. 1979. *Handbook of Graphic Presentation,* 2d ed. New York: Wiley.

Suckling, P., and W. Ashley. 2006. Spatial and Temporal Characteristics of Tornado Path Direction. *The Professional Geographer*, 58 (1): 20–38.

Tilling, L. 1975. Early Experimental Graphs. *The British Journal for the History of Science*, 8 (3): 193–213.

Tufte, E. 1997. *Visual Explanations: Images and Quantities, Evidence and Narrative*. Cheshire, Connecticut: Graphics Press.

———. 2001. *The Visual Display of Quantitative Information*. 2d ed. Cheshire, Connecticut: Graphics Press.

GLOSSARY

arithmetic scale a regularly spaced grid that is used for plotting a graph; value to distance relationship defined by the designer

bar graph a series of proportional bars drawn to represent different numerical values in a comparative study

box-whisker graph a unique graph that depicts the mean, several percentiles, and extreme values in a statistical distribution; sometimes referred to simply as a box graph

bubble graph another name for a triple scatter plot

chartjunk textured fills of bars such as angled or cross-hatched lines or patterns that distract from the graphic's aesthetic design

clock graph a circular bar graph arranged and scaled around a 24-hour clock

complex graph a multivariate graph in which data are displayed in a cumulative form; display can represent actual data values or be converted to percentages

compound graph a graph containing the data of two or more variables, or components of a compound variable

data ink information that is critical to the graphic display of the data, the core information of the graphic

data labels labels on a graph to identify different data symbols or data paths

data path a line of a graph that portrays the trend of the data

data region the area of a graph containing the grid and in which the graphic treatment of the data falls

data source the originator of primary or secondary data for which credit is given

graph a diagram, chart, or drawing that depicts one or more numerical data sets

graphicacy the intellectual skill necessary to effectively communicate numerical relationships

histogram a graphic record of the occurrence and frequency of data values in a statistical distribution

horizontal scale scale along the top or bottom of a graph

index tick marks the primary tick marks that are labeled on a graph's axis

line graph a time series graph that uses line symbols to represent the trend of variables

legend provides the user with a definition of the symbolization and other elements of the graph

log-log graph a graph in which both axes are logarithmic; used to compare rates of two variables

Lorenz curve a special frequency polygon graph

ogive a cumulative frequency curve where rank-ordered data are added together sequentially and the valued plotted

pie graph a circular graph that shows component parts; commonly called a pie chart

quantile value in data array that divides the array into groups with known proportions in each group; deciles, percentiles, quartiles, and quintiles are special quantiles

ratio graph another name for the logarithmic graph

reference line a line on a graph that identifies a certain place in the data region of significance to the data presentation

scale labels identifiers along the vertical or horizontal scales of a graph

scale lines the rulers that bound the data region on a graph

scatter plot a graph that presents by dots the relationship of two variables

semi-logarithmic graph a form of logarithmic graph with a log scale on one axis and an arithmetic scale on the other

simple graph a graph that shows time on the horizontal axis and a single variable on the vertical axis; sometimes called a time series graph

trilinear graph a special triangular graph that shows three component parts relative to the total

triple scatter plot a graph that portrays three variables by using two grids and proportional circles; also called a bubble graph

vertical scale scale along the left or right of a graph

APPENDIX A WORKED PROBLEMS

This appendix presents worked problems in the area of map scale. The intention is to give you additional experience in order to strengthen your understanding of scale.

MAP SCALE

Map scale problems occur routinely when you are dealing with maps. In today's world of cartography and thematic map design, we frequently find ourselves assembling maps in hardcopy and digital form that were previously generated by someone other than ourselves. These products serve as a foundation for compiling information and data to be used in our projects or research. The documents may be brought into our GIS and mapping software as a scanned raster image, such as an aerial photograph, which we use as a layer for on-screen digitizing. Even more common is using pre-existing vector planimetric maps, such as digital boundary files of enumeration units. We often use these products as our base map on which we compile additional attribute information. It is not uncommon to bring several documents together in compiling information for the generation of a new map. We may have to modify multiple documents of different scales by transforming them to a common scale. How much transformation is required will depend on the variation of the document scales of the original material.

In some software, measurement tools are provided that measure in linear distances between points. If we can measure along a road segment on our digital map, photograph, or image, for example, for which the Earth distance is known, we can calculate the scale. On existing hardcopy maps, similar measurements can be made manually using a ruler. It is this relationship between measurements on a map to the actual Earth distance that establishes the relative fraction (RF) scale.

The first thing to do in *any* map scale problem is to write down the fundamental relationship: map distance divided by Earth distance. Doing this first will usually solve most map scale problems.

Map Scale Problem 1:

A straight stretch of road is known to be 500 yards long. The same road segment measured on an aerial photograph is 3/4 of an inch long. What is the RF of the photograph?

1. First, put down the basic relationship:

$$\frac{\textbf{map distance}}{\textbf{Earth distance}}$$

2. Now substitute what you know to be true. Map distance in this case is .75 inch, and the Earth distance is 500 yards. You know that you must always deal in the same units, so convert the 500 yards into inches:

$$\textbf{500 yards} \times \textbf{3 feet/yard} \times \textbf{12 inches/foot} = \textbf{18,000 inches}$$

3. Simply place the number of units into the basic relationship:

$$\frac{\textbf{0.75 inch (map distance)}}{\textbf{18,000 inches (Earth distance)}}$$

Now, you know that the numerator in any RF scale is always expressed as one, so you must convert the fraction so that the number one is in the numerator.

4. To form a proportional relationship you must place two fractions equal to one another, but with one element unknown, and solve the equation for the unknown (X):

$$\frac{\mathbf{0.75}}{\mathbf{18{,}000}} = \frac{\mathbf{1}}{\boldsymbol{x}}$$

Solve for X:

$$(\mathbf{0.75})(\boldsymbol{X}) = (\mathbf{1})(\mathbf{18{,}000})$$
$$\mathbf{0.75}\boldsymbol{X} = \mathbf{18{,}000}$$
$$\boldsymbol{X} = \mathbf{24{,}000}$$

Thus, the scale of the aerial photograph is **1:24,000.**

Map Scale Problem 2:

In this problem, you have a map whose RF scale is 1:29,500. You are trying to find the distance between two cellular telephone towers that are 6.7 inches apart on the map. The first thing you do is to set down the fundamental relationship:

1. $$\frac{\textbf{map distance}}{\textbf{Earth distance}}$$
2. Now you substitute into the basic fraction those parts of the problem you already know:

$$\frac{\textbf{6.7 inches (map distance)}}{\textbf{Earth distance}}$$

Because you know their relationship to one another (for example, the map scale), you can write:

$$\frac{\textbf{6.7 inches}}{x} = \frac{\textbf{1}}{\textbf{29,500}}$$

Solve for X.

$$\textbf{(1)}(X) = \textbf{(6.7 inches)(29,500)}$$
$$X = \textbf{197,650}$$

3. It is unlikely that you will want the distance between the two towers in inches, so you can convert the inches into miles by dividing the total inches by 63,360 (the number of inches in 1 mile determined by 5,280 feet/mile × 12 inches/foot).

$$\frac{\textbf{197,650}}{\textbf{63,360}} = \textbf{3.12 miles}$$

Map Scale Problem 3:

Many times in mapping you are faced with the problem of expressing scale in terms that are different from what you have, and having to convert to different units. Imagine that a European friend of yours has a map of Europe, made in the United States, whose scale is expressed in inches to the mile, and she wishes you to convert it to centimeters to the kilometer. Her map has a verbal scale of 2 inches, representing 5 miles. There are several ways you could do this, but you decide to make this an educational lesson for your friend, so you first compute an RF.

1. First, you put down the basic relationship:

$$\frac{\textbf{map distance}}{\textbf{Earth distance}}$$

2. You already have these values, so you can substitute:

$$\frac{\textbf{2 inches}}{\textbf{5 miles}}$$

3. But you want to find the RF, so you set the two fractions equal, and solve for the denominator (X):

$$\frac{\textbf{2 inches}}{\textbf{5 miles}} = \frac{\textbf{1}}{x}$$

then

$$\textbf{(2 inches)}(X) = \textbf{(5 miles)(1)}$$

and converting miles to inches:

$$2X = \textbf{316,000}$$

then:

$$X = \textbf{158,000}$$

so the RF is **1:158,000.**

4. Because the RF shows only a relationship, it can be expressed in any units as long as they are the same on both numerator and denominator. In this case, whatever unit is applied to the 1 is also applied to the 158,000. You should not mix the units as you do in verbal scales.

   To get kilometers, simply divide 158,000 by the number of centimeters in 1 kilometer (100,000), and the answer becomes **1.58 kilometers.**

Map Scale Problem 4:

Conversions between fractional, verbal, and graphic scales are something every cartographer should know. The first three examples gave you practical examples that you frequently encounter when working with maps of differing scales or the use of aerial photography in conjunction with an existing map. This example is intended to give you experience in freely moving between the three scale type.

As was stated in Problem 2, there are 63,360 inches in a mile (12 inches × 5,280 feet). When we use this number in making our scale determination, we can use the word *equals* as in one inch represents 5 miles. At smaller scales, the amount of generalization makes the determination of such an exact statement difficult. The thickness of a line on the map may represent a distance of 50 feet of Earth distance. The quality of the ruler or how we estimate the position of a line in relation to that ruler in making a linear measurement may be quite varied. Therefore, at smaller scales we tend to think in terms of one inch *represents* approximately 5 miles. The level of generalization also allows us to use a different number of inches (62,500) to a mile in order to have our verbal scales result in whole numbers. In terms of the smaller scale maps, the difference of 860 inches makes little difference in our answer.

Converting RF to Verbal:

Convert the RF of 1:250,000 using both 63,360 inches and 62,500 inches per mile.

$$\frac{\textbf{250,000}}{\textbf{63,360}} = \textbf{3.95}$$

or

$$\frac{\textbf{250,000}}{\textbf{62,500}} = \textbf{4}$$

In the first instance, we would have a verbal scale that states one inch represents 3.95 miles. However, using the second instance, the resulting verbal scale is one inch represents *approximately* 4 miles. The first implies a greater accuracy that can actually be determined and therefore the approximated scale is acceptable.

What is the verbal scale of an RF of 1:750,000 using 62,500 inches in the conversion?

$$\frac{\mathbf{750{,}000}}{\mathbf{62{,}500}} = \mathbf{12}$$

or

One inch represents approximately 12 miles

Converting Verbal to RF:

What fractional scale is the equivalent of one inch represents approximately 7.25 miles?

$$7.25 \times 62{,}500 = 453{,}125 \text{ or } 1{:}453{,}425$$

Converting RF to Linear:

If we know that the fractional scale of a map is 1:200,000, how will we construct an equivalent linear scale if our software does not do it automatically?

The first step is to convert the RF to a verbal scale. In this case, one inch represents approximately 3.2 miles (200,000/62,500). It would be inappropriate for a linear scale to display 3.2 miles. A linear scale designed to depict a length of 5 miles is more appropriate and easier for the map reader to envision. To create the linear scale, we need to determine the line length for those distances.

$$\frac{\mathbf{5}}{\mathbf{3.2}} = \mathbf{1.56}$$

Therefore, we need to construct a line that is 1.56 inches in length that represents 5 miles. One half of this distance will, of course, represent 2.5 miles.

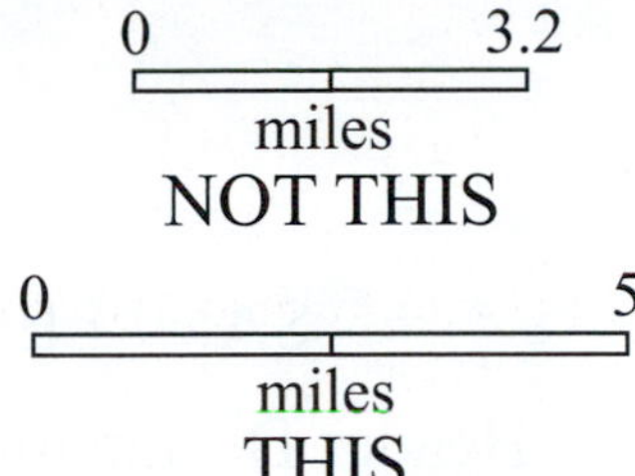

APPENDIX B GEORGIA DATABASE

SAMPLE DATABASE of GEORGIA DATA

Header Designations:

POP1990	Total county population for 1990
AREA	Area of each county in square miles
POP/SQ MILE	Population density (people per square mile) for 1990
BIRTHS	Total births in 1990
FEMALE_15_44	Total number of females per county ages 15 to 44
GFR	General Fertility Rate
BIRTHRATE	Number of live births per 1,000 women of childbearing age
HOUSING UNITS	Total number of housing units in a county
H_U_OCCUPIED	Number of total housing units that are occupied

COUNTY	FIPS	POP1990	POP2000	AREA	POP/SQ MILE	BIRTHS	FEMALE_15_44	GFR*	BIRTHRATE	HOUSING UNITS	H_U_OCCUPIED	COUNTY
Appling	13001	15744	17419	510.3680	34	278	3731	74.5	49.4	7854	6606	Appling
Atkinson	13003	6213	7609	345.7182	22	152	1706	89.1	62.3	3171	2717	Atkinson
Bacon	13005	9566	10103	280.8870	36	152	2136	71.2	47	4464	3833	Bacon
Baker	13007	3615	4074	338.0805	12	36	875	41.1	26.4	1740	1514	Baker
Baldwin	13009	39530	44700	276.7688	162	551	9557	57.7	40.1	17173	14758	Baldwin
Banks	13011	10308	14422	232.1243	62	207	3078	67.3	44.7	5808	5364	Banks
Barrow	13013	29721	46144	174.7151	264	768	10711	71.7	50.1	17304	16354	Barrow
Bartow	13015	55911	76019	461.6792	165	1380	17098	80.7	54.3	28751	27176	Bartow
Ben Hill	13017	16245	17484	241.0978	73	318	3658	86.9	56.3	7623	6673	Ben Hill
Berrien	13019	14153	16235	453.4271	36	239	3363	71.1	46.4	7100	6261	Berrien
Bibb	13021	149967	153887	259.5813	593	2520	35765	70.5	47.5	67194	59667	Bibb
Bleckley	13023	10430	11666	218.5517	53	157	2496	62.9	41.9	4866	4372	Bleckley
Brantley	13025	11077	14629	461.9328	32	135	3141	43.0	27.9	6490	5436	Brantley
Brooks	13027	15398	16450	494.6233	33	218	3371	64.7	42.4	7118	6155	Brooks
Bryan	13029	15438	23417	430.7970	54	386	5465	70.6	46.3	8675	8089	Bryan
Bulloch	13031	43125	55983	685.8501	82	684	15513	44.1	33.3	22742	20743	Bulloch
Burke	13033	20579	22243	859.4369	26	456	4931	92.5	60.7	8842	7934	Burke
Butts	13035	15326	19522	187.8932	104	277	3808	72.7	47.1	7380	6455	Butts
Calhoun	13037	5013	6320	284.7757	22	90	1087	82.8	54.2	2305	1962	Calhoun
Camden	13039	30167	43664	627.5547	70	673	10505	64.1	45.1	16958	14705	Camden
Candler	13043	7744	9577	258.7223	37	169	1829	92.4	59.2	3893	3375	Candler
Carroll	13045	71422	87268	507.1121	172	1353	20764	65.2	45.5	34067	31568	Carroll
Catoosa	13047	42464	53282	182.8218	291	708	11595	61.1	40.2	21794	20425	Catoosa
Charlton	13049	8496	10282	779.2063	13	129	2109	61.2	40.8	3859	3342	Charlton
Chatham	13051	216935	232048	424.9392	546	3492	52196	66.9	45.3	99683	89865	Chatham
Chattahoochee	13053	16934	14882	250.0710	60	270	3018	89.5	69.3	3316	2932	Chattahoochee
Chattooga	13055	22242	25470	309.8753	82	331	4794	69.0	45	10677	9577	Chattooga
Cherokee	13057	90204	141903	416.1042	341	2461	33155	74.2	49.9	51937	49495	Cherokee
Clarke	13059	87594	101489	128.0485	793	1301	31204	41.7	33.3	42126	39706	Clarke
Clay	13061	3364	3357	212.1015	16	45	616	73.1	44.1	1925	1347	Clay
Clayton	13063	182052	236517	145.5088	1625	4350	60643	71.7	50.3	86461	82243	Clayton
Clinch	13065	6160	6878	833.4575	8	124	1436	86.4	56	2837	2512	Clinch
Cobb	13067	447745	607751	352.5278	1724	10297	149708	68.8	47.1	237522	227487	Cobb
Coffee	13069	29592	37413	606.6497	62	675	8262	81.7	55.4	15610	13354	Coffee
Colquitt	13071	36645	42053	559.0215	75	716	8662	82.7	54.6	17554	15495	Colquitt
Columbia	13073	66031	89288	297.1142	301	1227	20081	61.1	38.9	33321	31120	Columbia
Cook	13075	13456	15771	233.1097	68	243	3295	73.7	48.4	6558	5882	Cook
Coweta	13077	53853	89215	454.4199	196	1510	20026	75.4	50.2	33182	31442	Coweta
Crawford	13079	8991	12495	325.9493	38	138	2735	50.5	32.9	4872	4461	Crawford
Crisp	13081	20011	21996	282.6688	78	344	4720	72.9	47.8	9559	8337	Crisp
Dade	13083	13147	15154	184.5125	82	170	3391	50.1	33.4	6224	5633	Dade
Dawson	13085	9429	15999	212.6014	75	231	3470	66.6	43.8	7163	6069	Dawson
Decatur	13087	25511	28240	624.7196	45	439	6105	71.9	47.9	11968	10380	Decatur
DeKalb	13089	545837	665865	257.3518	2587	11236	170917	65.7	46.4	261231	249339	DeKalb
Dodge	13091	17607	19171	497.8030	39	234	3683	63.5	41	8186	7062	Dodge
Dooly	13093	9901	11525	404.4778	28	191	2265	84.3	55.6	4499	3909	Dooly
Dougherty	13095	96311	96065	338.8529	284	1592	22494	70.8	48.1	39656	35552	Dougherty

COUNTY	FIPS	POP1990	POP2000	AREA	POP/SQ MILE	BIRTHS	FEMALE_15_44	GFR*	BIRTHRATE	HOUSING UNITS	H_U_OCCUPIED	COUNTY
Douglas	13097	71120	92174	200.1012	461	1398	21909	63.8	42.8	34825	32822	Douglas
Early	13099	11854	12354	515.0272	24	161	2552	63.1	40.9	5338	4695	Early
Echols	13101	2334	3754	424.4253	9	53	803	66.0	45.8	1482	1264	Echols
Effingham	13103	25687	37535	487.2826	77	567	8642	65.6	43.9	14169	13151	Effingham
Elbert	13105	18949	20511	377.1591	54	277	4184	66.2	43	9136	8004	Elbert
Emanuel	13107	20546	21837	690.2199	32	344	4613	74.6	48.2	9419	8045	Emanuel
Evans	13109	8724	10495	188.8901	56	205	2364	86.7	59.4	4381	3778	Evans
Fannin	13111	15992	19798	398.2520	50	228	3556	64.1	38.9	11134	8369	Fannin
Fayette	13113	62415	91263	194.6686	469	928	18736	49.5	28.9	32726	31524	Fayette
Floyd	13115	81251	90565	501.6981	181	1354	19463	69.6	46.8	36615	34028	Floyd
Forsyth	13117	44083	98407	238.8043	412	2083	22403	93.0	63.7	36505	34565	Forsyth
Franklin	13119	16650	20285	265.1060	77	227	4150	54.7	36.1	9303	7888	Franklin
Fulton	13121	648951	816006	551.6376	1479	13527	200511	67.5	47.1	348632	321242	Fulton
Gilmer	13123	13368	23456	429.1775	55	362	4577	79.1	51.1	11924	9071	Gilmer
Glascock	13125	2357	2556	148.4733	17	26	498	52.2	34.6	1192	1004	Glascock
Glynn	13127	62496	67568	387.8719	174	1003	13964	71.8	45.9	32636	27208	Glynn
Gordon	13129	35072	44104	352.6283	125	799	9623	83.0	55.7	17145	16173	Gordon
Grady	13131	20279	23659	453.7815	52	388	5035	77.1	50.2	9991	8797	Grady
Greene	13133	11793	14406	402.1014	36	197	2831	69.6	43.4	6653	5477	Greene
Gwinnett	13135	352910	588448	445.2067	1322	10556	144384	73.1	50.2	209682	202317	Gwinnett
Habersham	13137	27621	35902	276.3435	130	537	7065	76.0	49.8	14634	13259	Habersham
Hall	13139	95428	139277	415.0330	336	2882	30891	93.3	64	51046	47381	Hall
Hancock	13141	8908	10076	475.7145	21	116	1853	62.6	39.5	4287	3237	Hancock
Haralson	13143	21966	25690	286.6577	90	396	5333	74.3	48.8	10719	9826	Haralson
Harris	13145	17788	23695	477.1780	50	290	4862	59.6	37.3	10288	8822	Harris
Hart	13147	19712	22997	242.3430	95	262	4321	60.6	38.5	11111	9106	Hart
Heard	13149	8628	11012	308.0334	36	155	2375	65.3	43.6	4512	4043	Heard
Henry	13151	58741	119341	317.3110	376	1976	28257	69.9	47.4	43166	41373	Henry
Houston	13153	89208	110765	366.8960	302	1641	25755	63.7	43.2	44509	40911	Houston
Irwin	13155	8649	9931	363.0893	27	122	2007	60.8	39.6	4149	3644	Irwin
Jackson	13157	30005	41589	334.6115	124	662	8989	73.6	49.7	16226	15057	Jackson
Jasper	13159	8453	11426	388.0390	29	158	2342	67.5	42.6	4806	4175	Jasper
Jeff Davis	13161	12032	12684	335.3896	38	218	2637	82.7	53.7	5581	4828	Jeff Davis
Jefferson	13163	17408	17266	534.0236	32	284	3628	78.3	50.5	7221	6339	Jefferson
Jenkins	13165	8247	8575	351.5197	24	133	1793	74.2	47.9	3907	3214	Jenkins
Johnson	13167	8329	8560	309.5629	28	117	1674	69.9	46.2	3634	3130	Johnson
Jones	13169	20739	23639	395.3668	60	303	5207	58.2	38	9272	8659	Jones
Lamar	13171	13038	15912	191.5153	83	207	3566	58.0	38.5	6145	5712	Lamar
Lanier	13173	5531	7241	198.1203	37	103	1574	65.4	44.3	3011	2593	Lanier
Laurens	13175	39988	44874	817.8584	55	690	9564	72.1	47.1	19687	17083	Laurens
Lee	13177	16250	24757	346.3533	71	342	5804	58.9	39.3	8813	8229	Lee
Liberty	13179	52745	61610	521.6698	118	1421	15471	91.8	68.8	21977	19383	Liberty
Lincoln	13181	7442	8348	246.6001	34	95	1644	57.8	36.1	4514	3251	Lincoln
Long	13183	6202	10304	412.8838	25	195	2520	77.4	56.3	4232	3574	Long
Lowndes	13185	75981	92115	515.6110	179	1443	22526	64.1	45.6	36551	32654	Lowndes
Lumpkin	13187	14573	21016	282.2560	74	337	5254	64.1	45.7	8263	7537	Lumpkin
McDuffie	13189	20119	21231	261.2740	81	367	4613	79.6	51.3	8916	7970	McDuffie

COUNTY	FIPS	POP1990	POP2000	AREA	POP/SQ MILE	BIRTHS	FEMALE_15_44	GFR*	BIRTHRATE	HOUSING UNITS	H_U_OCCUPIED	COUNTY
McIntosh	13191	8634	10847	450.2233	24	174	2141	81.3	50.7	5735	4202	McIntosh
Macon	13193	13114	14074	392.0955	36	238	2796	85.1	54.7	5495	4834	Macon
Madison	13195	21050	25730	286.5073	90	347	5567	62.3	41	10520	9800	Madison
Marion	13197	5590	7144	366.0224	20	96	1492	64.3	41.2	3130	2668	Marion
Meriwether	13199	22411	22534	504.6656	45	301	4611	65.3	41.5	9211	8248	Meriwether
Miller	13201	6280	6383	289.7525	22	100	1252	79.9	50.8	2770	2487	Miller
Mitchell	13205	20275	23932	508.7475	47	359	4768	75.3	49.3	8880	8063	Mitchell
Monroe	13207	17113	21757	413.0969	53	296	4665	63.5	41.1	8425	7719	Monroe
Montgomery	13209	7163	8270	239.2082	35	106	1753	60.5	40.7	3492	2919	Montgomery
Morgan	13211	12883	15457	350.8100	44	224	3188	70.3	45.2	6128	5558	Morgan
Murray	13213	26147	36506	355.1524	103	571	8393	68.0	46.4	14320	13286	Murray
Muscogee	13215	179278	186291	224.0121	832	3178	41562	76.5	52	76182	69819	Muscogee
Newton	13217	41808	62001	273.1495	227	1105	14199	77.8	52.8	23033	21997	Newton
Oconee	13219	17618	26225	191.7150	137	311	5708	54.5	34.2	9528	9051	Oconee
Oglethorpe	13221	9763	12635	435.5708	29	166	2688	61.8	41	5368	4849	Oglethorpe
Paulding	13223	41611	81678	322.6718	253	1601	20393	78.5	56.4	29274	28089	Paulding
Peach	13225	21189	23668	151.6212	156	366	5635	65.0	44.6	9093	8436	Peach
Pickens	13227	14432	22983	224.1019	103	330	4719	69.9	45.3	10687	8960	Pickens
Pierce	13229	13328	15636	342.4013	46	241	3225	74.7	48.1	6719	5958	Pierce
Pike	13231	10224	13688	218.4159	63	155	2844	54.5	35.2	5068	4755	Pike
Polk	13233	33815	38127	324.9086	117	633	7800	81.2	53.7	15059	14012	Polk
Pulaski	13235	8108	9588	254.4721	38	117	2604	44.9	32	3944	3407	Pulaski
Putnam	13237	14137	18812	354.8207	53	279	3638	76.7	47.6	10319	7402	Putnam
Quitman	13239	2209	2598	160.3819	16	49	483	101.4	62.6	1773	1047	Quitman
Rabun	13241	11648	15050	385.4128	39	177	2611	67.8	41.2	10210	6279	Rabun
Randolph	13243	8023	7791	435.3282	18	90	1667	54.0	35.9	3402	2909	Randolph
Richmond	13245	189719	199775	337.3864	592	3269	46289	70.6	48	82312	73920	Richmond
Rockdale	13247	54091	70111	134.7088	520	995	15364	64.8	41.5	25082	24052	Rockdale
Schley	13249	3588	3766	170.9882	22	57	767	74.3	47.4	1612	1435	Schley
Screven	13251	13842	15374	646.5289	24	200	3135	63.8	40.4	6853	5797	Screven
Seminole	13253	9010	9369	260.8390	36	134	1917	69.9	46.7	4742	3573	Seminole
Spalding	13255	54457	58417	202.4376	289	914	12678	72.1	47.6	23001	21519	Spalding
Stephens	13257	23257	25435	179.3050	142	348	5268	66.1	43.6	11652	9951	Stephens
Stewart	13259	5654	5252	463.4725	11	57	970	58.8	36.8	2354	2007	Stewart
Sumter	13261	30228	33200	487.8176	68	525	7596	69.1	46.7	13700	12025	Sumter
Talbot	13263	6524	6498	382.3355	17	96	1328	72.3	44.9	2871	2538	Talbot
Taliaferro	13265	1915	2077	192.0350	11	21	384	54.7	33.5	1085	870	Taliaferro
Tattnall	13267	17722	22305	474.7020	47	324	3825	84.7	56	8578	7057	Tattnall
Taylor	13269	7642	8815	387.5258	23	125	1827	68.4	45.6	3978	3281	Taylor
Telfair	13271	11000	11794	445.3640	26	197	2118	93.0	60	5083	4140	Telfair
Terrell	13273	10653	10970	334.2052	33	190	2318	82.0	53.3	4460	4002	Terrell
Thomas	13275	38986	42737	543.5787	79	630	9074	69.4	45	18285	16309	Thomas
Tift	13277	34998	38407	268.2181	143	631	8445	74.7	50.4	15411	13919	Tift
Toombs	13279	24072	26067	373.8625	70	459	5461	84.1	54.6	11371	9877	Toombs
Towns	13281	6754	9319	185.4114	50	89	1558	57.1	35.3	6282	3998	Towns
Treutlen	13283	5994	6854	203.2830	34	84	1397	60.1	39.7	2865	2531	Treutlen
Troup	13285	55536	58779	451.6643	130	900	12668	71.0	46.4	23824	21920	Troup

COUNTY	FIPS	POP1990	POP2000	AREA	POP/SQ MILE	BIRTHS	FEMALE_15_44	GFR*	BIRTHRATE	HOUSING UNITS	H_U_OCCUPIED	COUNTY
Turner	13287	8703	9504	286.0231	33	158	2017	78.3	51	3916	3435	Turner
Twiggs	13289	9806	10590	355.7171	30	120	2325	51.6	33.7	4291	3832	Twiggs
Union	13291	11993	17289	335.2997	52	191	2831	67.5	40.4	10001	7159	Union
Upson	13293	26300	27597	331.8005	83	379	5652	67.1	43.7	11616	10722	Upson
Walker	13295	58340	61053	447.4386	136	728	12558	58.0	37.6	25577	23605	Walker
Walton	13297	38586	60687	328.9470	184	1012	13614	74.3	49.8	22500	21307	Walton
Ware	13299	35471	35483	908.8053	39	528	6935	76.1	49.3	15831	13475	Ware
Warren	13301	6078	6336	281.4669	23	88	1318	66.8	42.5	2767	2435	Warren
Washington	13303	19112	21176	693.3558	31	274	5202	52.7	36.1	8327	7435	Washington
Wayne	13305	22356	26565	637.7218	42	355	5339	66.5	43.6	10827	9324	Wayne
Webster	13307	2263	2390	211.6358	11	31	482	64.3	41.9	1115	911	Webster
Wheeler	13309	4903	6179	302.9115	20	67	1023	65.5	40.7	2447	2011	Wheeler
White	13311	13006	19944	231.2301	86	270	3978	67.9	43.7	9454	7731	White
Whitfield	13313	72462	83525	288.4606	290	1745	17950	97.2	65.6	30722	29385	Whitfield
Wilcox	13315	7008	8577	380.1750	23	109	1444	75.5	48.4	3320	2785	Wilcox
Wilkes	13317	10597	10687	474.3708	23	130	2115	61.5	39.9	5022	4314	Wilkes
Wilkinson	13319	10228	10220	446.7362	23	145	2209	65.6	43.3	4449	3827	Wilkinson
Worth	13321	19745	21967	576.8132	38	313	4638	67.5	43	9086	8106	Worth

*GFR = General Fertility Rate

INDEX

A

B

C

D

E

F

Q

R

S

T

U

V

W

X

Y

Z